## SI units and prefixes

### (a) SI units

| Quantity | Unit | SI symbol | Formula |
|---|---|---|---|
| **SI base units** | | | |
| Length | meter | m | — |
| Mass | kilogram | kg | — |
| Time | second | s | — |
| Temperature | kelvin | K | — |
| **SI supplementary unit** | | | |
| Plane angle | radian | rad | — |
| **SI derived units** | | | |
| Energy | joule | J | $N \cdot m$ |
| Force | newton | N | $kg \cdot m/s^2$ |
| Power | watt | W | J/s |
| Pressure | pascal | Pa | $N/m^2$ |
| Work | joule | J | $N \cdot m$ |

### (b) SI prefixes

| Multiplication factor | Prefix | SI symbol for prefix |
|---|---|---|
| $1,000,000,000,000 = 10^{12}$ | tera | T |
| $1,000,000,000 = 10^{9}$ | giga | G |
| $1,000,000 = 10^{6}$ | mega | M |
| $1,000 = 10^{3}$ | kilo | k |
| $100 = 10^{2}$ | hecto | h |
| $10 = 10^{1}$ | deka | da |
| $0.1 = 10^{-1}$ | deci | d |
| $0.01 = 10^{-2}$ | centi | c |
| $0.001 = 10^{-3}$ | milli | m |
| $0.000001 = 10^{-6}$ | micro | $\mu$ |
| $0.000000001 = 10^{-9}$ | nano | n |
| $0.000000000001 = 10^{-12}$ | pico | p |

## Conversion factors and definitions

### (a) Fundamental conversion factors

| | English unit | Exact SI value | Approximate SI value |
|---|---|---|---|
| Length | 1 in | 0.0254 m | — |
| Mass | 1 lbm | 0.45359237 kg | 0.4536 kg |
| Temperature | 1°R | $\frac{5}{9}$ K | — |

### (b) Definitions

| | |
|---|---|
| Acceleration of gravity | $Ig = 9.8066$ m/s$^2$ (32.174 ft/s$^2$) |
| Energy | Btu (British thermal unit) $\equiv$ amount of energy required to raise 1 lbm of water 1°F (1 Btu = 778.2 ft $\cdot$ lbf) |
| | kilocalorie $\equiv$ amount of energy required to raise 1 kg of water 1 K (1 kcal = 4187 J) |
| Length | 1 mil = 5280 ft; 1 nautical mile (nmi) = 6076.1 ft |
| Power | 1 hp $\equiv$ 550 ft $\cdot$ lbf/s |
| Pressure | 1 bar $\equiv 10^5$ Pa |
| Temperature | degree Fahrenheit $t_F = \frac{9}{5} t_C + 32$ (where $t_C$ is degrees Celsius) |
| | degree Rankine $t_R = t_F + 459.67$ |
| | Kelvin $t_K = t_C + 273.15$ (exact) |
| Kinematic viscosity | 1 P $\equiv$ 0.1 kg/(m $\cdot$ s) |
| | 1 St $\equiv$ 0.0001 m$^2$/s |
| Volume | 1 ft$^3$ = 7.48 gal |

### (c) Useful conversion factors

1 ft = 0.3048 m

1 lbf = 4.448 N

1 lbf = 386.1 lbm $\cdot$ in/s$^2$

1 kgf = 9.807 N

1 lbf/in$^2$ = 6895 Pa

1 ksi = 6.895 MPa

1 Btu = 1055 J

1 ft $\cdot$ lbf = 1.356 J

1 hp = 746 W = 2545 Btu/h

1 kW = 3413 Btu/h

1 qt = 0.000946 m$^3$ = 0.946 L

1 kcal = 3.968 Btu

# Fundamentals of Machine Elements

# McGraw-Hill Series in Mechanical Engineering

# Fundamentals of Machine Elements

## Second Edition

Bernard J. Hamrock
Ohio State University

Steven R. Schmid
The University of Notre Dame

Bo O. Jacobson
Lund University
Lund, Sweden

Boston    Burr Ridge, IL    Dubuque, IA    Madison, WI    New York    San Francisco    St. Louis
Bangkok    Bogotá    Caracas    Kuala Lumpur    Lisbon    London    Madrid    Mexico City
Milan    Montreal    New Delhi    Santiago    Seoul    Singapore    Sydney    Taipei    Toronto

**Higher Education**

FUNDAMENTALS OF MACHINE ELEMENTS, SECOND EDITION

Published by McGraw-Hill, a business unit of The McGraw-Hill Companies, Inc., 1221
Avenue of the Americas, New York, NY 10020. Copyright © 2005, 1999 by The
McGraw-Hill Companies, Inc. All rights reserved. No part of this publication may be reproduced
or distributed in any form or by any means, or stored in a database or retrieval system, without
the prior written consent of The McGraw-Hill Companies, Inc., including, but not limited to,
in any network or other electronic storage or transmission, or broadcast for distance learning.

Some ancillaries, including electronic and print components, may not be available to customers
outside the United States.

⊕ This book is printed on acid-free paper.

1 2 3 4 5 6 7 8 9 0 DOC/DOC 0 9 8 7 6 5 4

ISBN 0–07–246532–8

Publisher: *Elizabeth A. Jones*
Senior sponsoring editor: *Suzanne Jeans*
Developmental editor: *Amanda J. Green*
Marketing manager: *Dawn R. Bercier*
Senior project manager: *Sheila M. Frank*
Production supervisor: *Kara Kudronowicz*
Lead media project manager: *Audrey A. Reiter*
Media technology producer: *Eric A. Weber*
Cover designer: *Rick D. Noel*
Cover photo courtesy: *The SKF Group*
Lead photo research coordinator: *Carrie K. Burger*
Compositor: *The GTS Companies/York, PA Campus*
Typeface: *10/12 Times Roman*
Printer: *R. R. Donnelley Crawfordsville, IN*

**Library of Congress Cataloging-in-Publication Data**

Hamrock, Bernard J.
    Fundamentals of machine elements / Bernard J. Hamrock, Steven R. Schmid, Bo O. Jacobson.—2nd ed.
        p.    cm.
    Includes index.
    ISBN 0-07-246532-8 (hard copy : alk. paper)
    1. Machine design.    2. Mechanical movements.    I. Schmid, Steven R.    II. Jacobson, Bo O., 1942–.
III. Title.

TJ230.H245    2005                                                2004004560
621.8′15—dc22                                                     CIP

www.mhhe.com

# About the Authors

Bernard J. Hamrock joined the staff of Ohio State University as a professor of the Mechanical Engineering in 1985 and is now Professor Emeritus. Prior to joining Ohio State University he spent 18 years as a research consultant in the Tribology Branch of the NASA Lewis Research Center in Cleveland, Ohio. He received his Ph.D. and Doctor of Engineering degrees from the University of Leeds, Leeds, England. Professor Hamrock's research has resulted in a book with Duncan Dowson, *Ball Bearing Lubrication,* published in 1982 by Wiley Interscience, three separate chapters for handbooks, and over 150 archival publications. His second book, *Fundamentals of Fluid Film Lubrication* was published in 1993 by McGraw-Hill, a second edition, with co-authors Steven Schmid and Bo Jacobson, is scheduled for publication in early 2004. His awards include the 1976 Melville Medal from the American Society of Mechanical Engineers, the NASA Exceptional Achievement Medal in 1984, the 1998 Jacob Wallenberg Award given by The Royal Swedish Academy of Engineering Sciences, and the 2000 Mayo D. Hersey Award from the American Society of Mechanical Engineers.

Steven R. Schmid received his B.S. degree in Mechanical Engineering from the Illinois Institute of Technology in 1986. He then joined Triodyne, Inc., where his duties included investigation of machinery failures and consultation in machine design. He earned his Master's Degree from Northwestern University in 1989 and his Ph.D. in 1993, both in mechanical engineering. In 1993 he joined the faculty at The University of Notre Dame, where he teaches and conducts research in the fields of design and manufacturing. Dr. Schmid received the American Society of Mechanical Engineers Newkirk Award and the Society of Manufacturing Engineers Outstanding Young Manufacturing Engineer Award in 2000. He was also awarded the Kaneb Center Teaching Award in 2000 and 2003, and served as a Kaneb fellow in 2003. Dr. Schmid holds professional engineering (PE) and certified manufacturing engineer (C.Mfg.E) licenses. He is co-author (with S. Kalpakjian) of *Manufacturing Engineering and Technology* (2001) and *Manufacturing Processes for Engineering Materials* (2003), both published by Prentice Hall.

Bo O. Jacobson received his Ph.D. and D.Sc. degrees from Lund University in Sweden. From 1973 until 1987 he was Professor of Machine Elements at Luleå Technology University in Sweden. In 1987 he joined SKF Engineering & Research Centre in the Netherlands, while retaining a professorship at Chalmers University from 1987 to 1991 and at Luleå Technical University from 1992 to 1997. In 1997 he was appointed Professor of Machine Elements at Lund University, Sweden. Professor Jacobson was a NRC Research Fellow at NASA Lewis Research Center from 1981 to 1982. He has published four compendia used at Swedish universities. His text *Rheology and Elastohydrodynamic Lubrication* was published by Elsevier in 1991, and his book *Rolling Contact Phenomena* (with J. Kalker) was published in 2000 by Springer. Professor Jacobson has more than 100 archival publications. His awards include the prestigious Gold Medal given by the Institution of Mechanical Engineers, England, and the Wallenberg Award in 1984.

# Contents in Brief

# Contents

# Case Studies

# Preface

*Fundamentals of Machine Elements,* Second Edition provides undergraduate students with a clear, thorough understanding of both the theory and application of the fundamentals of machine elements. This text can also be used as a reference by practicing engineers. Familiarity with differential and integral calculus is needed to comprehend the material presented. Since the design of machine elements involves a great deal of geometry, the ability to sketch the various configurations that arise, as well as to draw a free-body diagram of the loads acting on a component, is also needed. The material covered in this text is appropriate as a third- or fourth-year engineering course for students who have studied basic engineering sciences, including physics, engineering mechanics, and materials.

## FEATURES

*Fundamentals of Machine Elements,* Second Edition has been carefully checked for accuracy and has been revised to include the following:

- Approximately 35% of the problems in the text are revised or are new
- NEW Chapter 15 on Helical, Bevel and Worm Gears
- NEW Chapter 20 on Elements of Microelectromechanical Systems (MEMS)
- NEW section 3.8, Effects of Manufacturing
- NEW section 7.10, Influence of Multiaxial Stress States
- NEW section 14.8, Gear Manufacture and Quality
- NEW web-based, Online Learning Center (see details below)

Pedagogical devices are used in each chapter to improve understanding and motivate the student:

- **Symbol List—** defines the symbols used within the chapter and gives their units for use in unit checks within equations.
- **Quotation and Photograph—** opens each chapter as an introduction to the topics in the chapter.
- **Introduction—** previews the material covered in the chapter.
- **Design Procedures—** shaded and boxed in order to emphasize important procedures used in design of machine elements.
- **Key Words—** presented in bold when first used and listed along with definitions at the end of the chapter.
- **Worked Examples—** presented when a new concept is developed to reinforce student understanding. There are over 200 worked examples, and each one uses a consistent problem-solving format.

- **Consistent Problem-Solving Methodology**— each example and problem is solved according to a consistent methodology. Students are encouraged to follow these four steps in solving examples and problems:
  1. *Sketch* - gives a graphical description of the problem.
  2. *Given* - presents the information from the problem statement in symbol form.
  3. *Find* - states what needs to be determined.
  4. *Solution* - indicates the method, procedure, and equations used to solve the problem.

- **Case Studies**— presented in select chapters. The Case Studies are design oriented and combine multiple concepts from the chapter. Many of the Case Studies involve situations encountered by practicing engineers. There are 17 cases included in the text. To find the location of the Case Studies, see the table of contents.

- **Summary**— recapitulates key information from the chapter.

- **Recommended Readings and References**— these lists are located at the end of each chapter as sources of further information. The author–date system is used for reference citations.

- **End-of-Chapter Problems**— over 600 homework problems are provided to solidify understanding of the chapter material and stimulate creativity. The problems range from simple to complex and many provide design-related opportunities for the student. The complex problems are identified with an asterisk (*). Also, answers to the majority of problems are given. Solutions to all of the homework problems are available to adopters of the text on the new, web-based Online Learning Center. The solutions are under password protection, for instructor use only.

## ORGANIZATION OF CONTENTS

The book is divided into two parts. Part 1 (Chapters 1 to 8) presents the fundamentals, and Part 2 (Chapters 9 to 20) uses the fundamentals in considering the design of various machine elements. The material in Part 1 is sequential; material presented in early chapters is needed in subsequent chapters. This building-block approach provides the foundation necessary to design the various machine elements considered in Part 2.

Chapter 1 introduces machine design and machine elements and covers a number of topics, such as safety factors, statistics, units, unit checks, and significant figures. In designing a machine element it is important to evaluate the kinematics, loads, and stresses at the critical section.

Chapter 2 describes the applied loads (normal, torsional, bending, and transverse shear) acting on a machine element with respect to time, the area over which the load is applied, and the location and method of application. The importance of support reaction, application of static force and moment equilibrium, and proper use of free-body diagrams is highlighted. Shear and moment diagrams applied to beams for various types of singularity function are also considered. Chapter 2 then describes stress and strain separately.

Chapter 3 focuses on the properties of solid engineering materials, such as the modulus of elasticity. (Appendix A gives properties of ferrous and nonferrous metals, ceramics, polymers, and natural rubbers. Appendix B explores the stress–strain relationships for uniaxial, biaxial, and triaxial stress states.)

Chapter 4 describes the stresses and strains that result from the types of load described in Chapter 2, while making use of the general Hooke's Law relationship developed in Appendix B. Chapter 4 also considers straight and curved members under four types of load. Ensuring that the design stress is less than the yield stress for ductile materials and less than the ultimate stress for brittle materials is important for a safe design. However, attention must also be paid to displacement (deformation), since a machine element can fail by excessive elastic deformation.

Chapter 5 attempts to quantify the deformation that might occur in a variety of machine elements. Some approaches investigated are the integral method, the singularity function, the method of superposition, and Castigliano's Theorem. These methods are applicable for distributed loads.

Stress raisers, stress concentrations, and stress concentration factors are investigated in Chapter 6. An important cause of machine element failure is cracks within the microstructure. Therefore, Chapter 6 covers stress levels, crack-producing flaws, and crack propagation mechanisms, and also presents failure prediction theories for both uniaxial and multiaxial stress states. The loading throughout Chapter 6 is assumed to be static (i.e., load is gradually applied and equilibrium is reached in a relatively short time).

Most machine element failures involve loading conditions that fluctuate with time. Fluctuating loads induce fluctuating stresses that often result in failure by means of cumulative damage. These topics, along with impact loading, are considered in Chapter 7.

Chapter 8 covers lubrication, friction, and wear. Not only must the design stress be less than the allowable stress and the deformation not exceed some maximum value, but lubrication, friction, and wear (tribological considerations) also must be properly understood for machine elements to be successfully designed. Stresses and deformations for concentrated loads, such as those that occur in rolling-element bearings and gears, are also determined in Chapter 8. Simple expressions are developed for the deformation at the center of the contact as well as for the maximum stress. Chapter 8 also describes the properties of fluid film lubricants used in a number of machine elements. Viscosity is an important parameter for establishing the load-carrying capacity and performance of fluid-film lubricated machine elements. Fluid viscosity is greatly affected by temperature, pressure, and shear rate. Chapter 8 considers not only lubricant viscosity but also pour point and oxidation stability, greases and gases, and oils.

Part 2 (Chapters 9 to 20) relates the fundamentals to various machine elements.

Chapter 9 deals with columns, which receive special consideration because yielding and excessive deformation do not accurately predict the failure of long columns. Because of their shape (length much larger than radius) columns tend to deform laterally upon loading; and if deflection becomes critical, they fail catastrophically. Chapter 9 establishes failure criteria for concentrically and eccentrically loaded columns.

Chapter 10 considers cylinders, which are used in many engineering applications. The chapter covers tolerancing of cylinders; stresses and deformations of thin-walled, thick-walled, internally pressurized, externally pressurized, and rotating cylinders; and press and shrink fits.

Chapter 11 considers shafting and associated parts, such as keys and flywheels. A shaft design procedure is applied to static and cyclic loading; thus, the material presented in Chapters 6 and 7 is directly applied to shafting. Chapter 11 also considers critical speeds of rotating shafts.

Chapter 12 presents the design of hydrodynamic bearings—both thrust and journal configurations—as well as design procedures for the two most commonly used slider bearings. The procedures provide an optimum pad configuration and describe performance parameters, such as normal applied load, coefficient of friction, power loss, and lubricant flow through the bearing. Similar design information is given for plain and nonplain journal bearings. The chapter also considers squeeze film and hydrostatic bearings, which use different pressure-generating mechanisms.

Rolling-element bearings are presented in Chapter 13. Statically loaded radial, thrust, and preloaded bearings are considered, as well as loaded and lubricated rolling-element bearings, fatigue life, and dynamic analysis. The use of the elastohydrodynamic lubrication film thickness is integrated with the rolling-element bearing ideas developed in this chapter.

Chapter 14 covers general gear theory and the design of spur gears. Stress failures are also considered. The transmitted load is used to establish the design bending stress in a gear tooth, which is then compared with an allowable stress to establish whether failure will occur. Chapter 14 also considers fatigue failures. The Hertzian contact stress with modification factors is used to establish the design stress, which is then compared with an allowable stress to determine whether fatigue failure will occur. If an adequate protective elastohydrodynamic lubrication film exists, gear life is greatly extended.

Chapter 15 extends the discussion of gears beyond spur gears as addressed in Chapter 14 to include helical, bevel, and worm gears. Advantages and disadvantages of the various types of gears are presented.

Chapter 16 covers threaded, riveted, welded, and adhesive joining of members, as well as power screws. Riveted and threaded fasteners in shear are treated alike in design and failure analysis. Four failure modes are presented: bending of member, shear of rivet, tensile failure of member, and compressive bearing failure. Fillet welds are highlighted, since they are the most frequently used type of weld. A brief stress analysis for lap and scarf adhesively bonded joints is also given.

Chapter 17 treats the design of springs, especially helical compression springs. Because spring loading is most often continuously fluctuating, Chapter 17 considers the design allowance that must be made for fatigue and stress concentration. Helical extension springs are also covered in Chapter 17. The chapter ends with a discussion of torsional and leaf springs.

Brakes and clutches are covered in Chapter 18. The brake analysis focuses on the actuating force, the torque transmitted, and the reaction forces in the hinge pin. Two theories relating to clutches are studied: the uniform pressure model and the uniform wear model.

Chapter 19 deals with flexible machine elements. Flat belts and V-belts, ropes, and chains are covered. Methods of effectively transferring power from one shaft to another while using belts, ropes, and chains are also presented. Failure modes of these flexible machine elements are considered.

Chapter 20 presents common elements of Micro-Electro-Mechanical Systems (MEMS). Design at macro and micro scales are very different, and the physics of small items requires different solutions to mechanical design problems. Chapter 20 is intended to introduce students to the design of MEMS and MEMS devices.

## ONLINE LEARNING CENTER

A new web-based, Online Learning Center containing book-related resources can be found at http://www.mhhe.com/hamrock2. The Online Learning Center provides reported errata, web links to related sites of interest, PowerPoint files of the text's images, FE review questions, lecture slides, and more!

Available to instructors only, the detailed solutions for all text problems will be delivered via CD-ROM, in our new, electronic, Complete Online Solution Manual Organization System. COSMOS is a database management tool geared toward assembling homework assignments, tests and quizzes.

## ACKNOWLEDGMENTS

Many people helped to produce this textbook. Professor Serope Kalpakjian (ret.) of the Illinois Institute of Technology and Professor Om Prakash Agrawal of Southern Illinois University both conducted extensive accuracy checks and reviews of the second edition manuscript. Special thanks are also due to Triodyne, Inc. personnel, especially Ralph L. Barnett and Brian D. King, whose insights were invaluable.

We would like to acknowledge the following reviewers who helped to prepare the second edition text through reviewing:

Om Prakash Agrawal, *Southern Illinois University, Carbondale*

Hamid Davoodi, *North Carolina State University*

Anoop K. Dhingra, *University of Wisconsin, Milwaukee*

Richard E. Dippery, Jr., *Kettering University*

Rollin Dix, *Illinois Institute of Technology*

William Dornfeld, *Fairfield University*

Thomas R. Grimm, *Michigan Technological University*

Karl-Heinrich Grote, *California State University, Long Beach*

Dennis W. Hong, *Purdue University*

Cecil O. Huey, Jr., *Clemson University*

Serope Kalpakjian, *Professor Emeritus*

Francis E. Kennedy, Jr., *Dartmouth College*

Kyu-Jung Kim, *University of Wisconsin, Milwaukee*

Peder Klit, *Technical University of Denmark*

Jesa Kreiner, *California State University, Fullerton*

Gregory G. Kremer, *Ohio University*

Stan Lukowski, *University of Wisconsin, Platteville*

Spencer P. Magleby, *Brigham Young University*

Noah D. Manring, *University of Missouri*

Roger Mayne, *State University of New York at Buffalo*

Lee Oberto, *Michigan Technological University*

John P. H. Steele, *Colorado School of Mines*

Lyndon S. Stephens, *University of Kentucky*

Brian S. Thompson, *Michigan State University*

Taher Saif, *University of Illinois at Urbana-Champaign*

Steven A. Velinsky, *University of California, Davis*

Robert C. Weber, *Marquette University*

Steven Yurgartis, *Clarkson University*

The dedication and continued help of our handler Amanda Green, Developmental Editor, whose support and encouragement greatly assisted us during the development process is acknowledged. Further thanks are due the staff at McGraw-Hill, including Jonathan Plant, Senior Sponsoring Editor, and Sheila Frank, Senior Project Manager.

The following publishers are recognized for the use of tables and illustrations: American Gear Manufacturers Association, American Society of Mechanical Engineers, Society of Tribologists and Lubrication Engineers, American Society for Testing and Materials, BHRA Fluid Engineering, Buttersworths, Elsevier Science Publishing Company, Engineering Sciences Data Unit, Ltd., Heinemann (London), Hemisphere Publishing Corporation, Macmillan Publishing Company, Inc., McGraw-Hill Higher Education, Mechanical Technology Incorporated, Non-Ferrous Founders Society, Oxford University Press, Inc., Penton Publishing Inc., Society of Automotive Engineers, Society of Tribologists and Lubrication Engineers, VCH Publishers, John Wiley & Sons, and Wykeham Publications (London), Ltd. The specific sources are identified in the text.

Bernard J. Hamrock
Ohio State University

Steven R. Schmid
The University of Notre Dame

Bo O. Jacobson
Lund University
Lund, Sweden

# FUNDAMENTALS

**Outline**

# INTRODUCTION

A rendition of the Mars Exploration Rover. (NASA)

*The invention all admir'd, and each, how he*
*To be th' inventor miss'd; so easy it seem'd,*
*Once found, which yet unfound most would have thought*
*Impossible*

John Milton

**SYMBOLS**

$n_s$     safety factor

$n_{sx}$     safety factor involving quality of materials, control over applied load, and accuracy of stress analysis

$n_{sy}$     safety factor involving danger to personnel and economic impact

$\sigma_{all}$     allowable normal stress, Pa

$\sigma_d$     design normal stress, Pa

## 1.1    WHAT IS DESIGN?

Design means different things to different people. A clothing manufacturer believes that incorporating different materials or colors into a new tuxedo style constitutes design. A potter paints designs onto china to complement its surroundings. An architect designs ornamental facades for residences. An engineer chooses a bearing from a catalog and incorporates it into a speed-reducer assembly. These design activities, although they appear to be fundamentally different, share a common thread: They all require significant creativity, practice, and vision to be done well.

"Engineering," wrote Thomas Tredgold [Florman (1987)], "is the art of directing the great sources of power in nature for the use and convenience of man." It is indeed significant that this definition of engineering is more than 60 years old—few people now use the words *engineering* and *art* in the same sentence, let alone in a definition. However, many products are successful for nontechnical reasons, reasons that cannot be proved mathematically. On the other hand, many problems are mathematically tractable, but usually because they have been inherently overconstrained. Design problems are, almost without exception, open-ended problems combining hard science and creativity. Engineering is indeed an art, even though parts of engineering problems lend themselves well to analysis.

For the purposes of this textbook, **design** is the transformation of concepts and ideas into useful machinery. A **machine** is a combination of mechanisms and other components that transforms, transmits, or uses energy, load, or motion for a specific purpose. If Tredgold's definition of engineering is accepted, design of machinery is the fundamental practice in engineering.

A machine comprises several different machine elements properly designed and arranged to work together as a whole. Fundamental decisions regarding loading, kinematics, and the choice of materials must be made during the design of a machine. Other factors, such as strength, reliability, deformation, tribology (friction, wear, and lubrication), cost, and space requirements, also need to be considered. The objective is to produce a machine that not only is sufficiently rugged to function properly for a reasonable time but also is economically feasible. Further, nonengineering decisions regarding marketability, product liability, ethics, politics, etc. must be integrated into the design process early. Since few people have the necessary tools to make all these decisions, machine design in practice is a discipline-blending human endeavor.

This textbook emphasizes one of the disciplines necessary in design—mechanical engineering. It therefore involves calculation and consideration of forces, energies, temperatures, etc.—concepts instilled into an engineer's psyche.

To "direct the great sources of power in nature" in machine design, the engineer must recognize the functions of the various machine elements and the types of load they transmit.

A **machine element function** may be as a normal load transmitter, a torque transmitter, an energy absorber, or a seal. Some normal load transmitters are rolling-element bearings, hydrodynamic bearings, and rubbing bearings. Some torque transmitters are gears, traction drives, chain drives, and belt drives. Brakes and dampers are energy absorbers. All the machine elements in Part 2 can be grouped into one of these classifications.

Engineers must produce safe, workable, good designs, as stated in the first fundamental canon in the *Code of Ethics for Engineers* [ASME (1997)]:

> Engineers shall hold paramount the safety, health, and welfare of the public in the performance of their professional duties.

Designing reasonably safe products involves many design challenges to ensure that components are large enough, strong enough, or tough enough to survive the loading environment. One subtle concept, but of huge importance, is that the engineer has a duty to protect the welfare of the general public. Welfare includes economic well-being, and it is well known that successful engineering innovations lead to wealth and job production. However, products that are too expensive are certain to fail in a competitive marketplace. Similarly, products that do not perform their function well will fail. Economics and functionality are always pressing concerns, and good design inherently means safe, economical, and functional design.

## 1.2    DESIGN OF MECHANICAL SYSTEMS

A **mechanical system** is a synergistic collection of machine elements. It is synergistic because as a design it represents an idea or concept greater than the sum of the individual parts. For example, a mechanical wristwatch, although merely a collection of gears, springs, and cams, also represents the physical realization of a time-measuring device. Mechanical system design requires considerable flexibility and creativity to obtain good solutions. Creativity seems to be aided by familiarity with known successful designs, and mechanical systems are often collections of well-designed components from a finite number of proven classes.

Designing a mechanical system is a different type of problem than selecting a component. Often, the demands of the system make evident the functional requirements of a component. However, designing a large mechanical system, potentially comprising thousands or even millions of machine elements, is a much more open, unconstrained problem.

To design superior mechanical systems, an engineer must have a certain sophistication and experience regarding machine elements. Studying the design and selection of machine elements affords an appreciation for the strengths and limitations of classes of components. They can then be more easily and appropriately incorporated into a system. For example, a mechanical system cannot incorporate a worm gear or a Belleville spring if the designer does not realize that these devices exist.

A toolbox analogy of problem solving can be succinctly stated as, "If your only tool is a hammer, then every problem is a nail." The purpose of studying machine element design is to fill the toolbox so that problem-solving and design synthesis activities can be flexible and unconstrained.

## 1.3   DESIGN AS A MULTIDISCIPLINARY ENDEAVOR

The quality revolution that hit the manufacturing sector in the early 1980s has forever changed the approach of companies and engineers toward product development. A typical design process of the recent past [Fig. 1.1(a)] shows that the skills involved in machine element design played an essential role in the process. This approach was commonly used in the United States in the postwar era. However, major problems with this approach became apparent in the 1970s and 1980s—some market-driven, others product-driven.

The term *over-the-wall (OTW) engineering* has been used to describe this design approach. Basically, someone would apply a particular skill and then send the product "over the wall" to the next step in development. A product design could sometimes flow smoothly from one step to the next and into the marketplace within weeks or months. This was rarely the case, however, as usually a problem would be discovered. For example, a manufacturing engineer might ask that workpieces be more easily clamped into a milling machine fixture. The design engineer would then alter the design and send the product back downstream. A materials scientist might then point out that the material chosen had drawbacks and suggest a different choice. The design engineer would make the alteration and resubmit the design. This process could continue *ad infinitum,* with the result that the product would take a long time to develop.

Figure 1.1(b) shows a more modern design approach. Here, there is still the recognized general flow of information from product conception through introduction into the marketplace, but there is immediate involvement of many disciplines in the design stage. Different disciplines are involved simultaneously instead of sequentially as with the over-the-wall approach. Some tasks are extremely technical, such as design analysis (the main focus of this book) or manufacturing. Others are nontechnical, such as market analysis. **Concurrent engineering** is the philosophy of involving many disciplines from the beginning of a design effort and keeping them involved throughout product development. Thus, redundant efforts are minimized, and higher-quality products are developed more quickly. Although design iterations still occur, the iteration loops are smaller and entail much less wasted time, effort, and expense. Also, design shortcomings can be corrected before they are incorporated. For example, service personnel can inform design engineers of excessive component failures during the conceptual design stage. No such mechanism for correcting design shortcomings ever existed in conventional design approaches or management structures.

Bringing high-quality products to market quickly was largely driven by the consumer electronics industry, where rapid change shortened useful market life to a few months. Further, new products introduced before their competitors' products enjoyed a larger share of the market and profits. Thus, manufacturers who could ship products weeks or even days faster than their competitors had a distinct sales advantage. Saving development time through concurrent engineering made companies much more competitive in the international marketplace.

Concurrent engineering has profoundly affected design engineers. They can no longer work alone and must participate in group discussions and design reviews. They need good communications skills. Designing machinery has become a cooperative endeavor.

Clearly, many disciplines now play a role in product development, but design engineers cannot merely focus on their discipline and rely on experts for the rest. They need to know other disciplines, at least from a linguistics standpoint, to integrate them into the design process. Thus, modern engineers may need to speak the language of materials science, law, marketing, etc., even if they are not experts in these fields.

## 1.4    DESIGN OF MACHINE ELEMENTS

Suggesting a mechanical system is only the beginning of the design synthesis process. Particular machine element classes need to be chosen, possibly leading to further design iterations. Designing a proper machine element usually involves the following steps:

1.  Select a suitable type of machine element from consideration of its function.

2.  Estimate the size of the machine element that is likely to be satisfactory.

3.  Evaluate the machine element's performance against the requirements.

4.  Modify the design and the dimensions until the performance is near to whichever optimum is considered most important.

Steps 3 and 4 in the process can be handled fairly easily by someone who is trained in analytical methods and understands the fundamental principles of the subject. Steps 1 and 2, however, require some creative decisions and, for many, represent the most difficult part of design.

After a suitable type of machine element has been selected for the required function, the specific machine element is designed by analyzing kinematics, load, and stress. These analyses, coupled with proper material selection, will enable a stress-strain-strength evaluation in terms of a safety factor (covered in Sec. 1.4.1). A primary question in designing any machine element is whether it will fail in service. Most people, including engineers, commonly associate **failure** with the actual breaking of a machine element. Although breaking is one type of failure, a design engineer must have a broader understanding of what really determines whether a part has failed.

A machine element is considered to have failed

1.  When it becomes completely inoperable

2.  When it is still operable but is unable to perform its intended function satisfactorily

3.  When serious deterioration has made it unreliable or unsafe for continued use, necessitating its immediate removal from service for repair or replacement

The role of the design engineer is to predict the circumstances under which failure is likely to occur. These circumstances are stress–strain–strength relationships involving the bulk of the solid members and such surface phenomena as friction, wear, lubrication, and environmental deterioration.

The principles of design are universal. An analysis is equally valid regardless of the size, material, and loading. However, an analysis by itself should not be looked on as an absolute and final truth. An analysis is limited by the assumptions imposed and by its range of applicability. Experimental verification of an analysis is always the preferred approach.

Design analysis attempts to predict the strength or deformation of a machine element so that the element can safely carry the imposed loads for as long as required. Assumptions have to be made about the material properties under different loading types (axial, bending, torsion, and transverse shear, as well as various combinations) and classes (static, sustained,

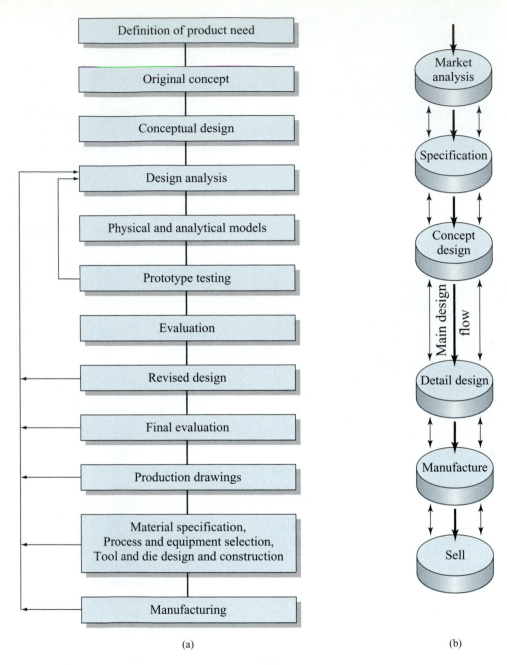

**Figure 1.1** Approaches to product development. (a) Classic approach, with large design iterations typical of the over-the-wall engineering approach. [Adapted from Kalpakjian and Schmid (2003)]. (b) A more modern approach, showing a main design flow with minor iterations representing concurrent engineering inputs. [Adapted from Pugh (1996).]

impact, or cyclic). These loading constraints may vary throughout the machine as they relate to different machine elements, an important factor for the design engineer to keep in mind.

## 1.4.1 SAFETY IN MECHANICAL DESIGN

The code of Hammurabi, a Babylonian doctrine over 3000 years old, had this requirement [Petroski (1992a)]:

> If a builder build a house for a man and do not make its construction firm, and the house which he has built collapse and cause the death of the owner of the house, that builder shall be put to death.

It could be argued that engineers are getting off a lot easier these days. Modern legal doctrines do not call for the death of manufacturers of unsafe products or of the engineers who designed them. Regardless of the penalty, however, engineers have a moral and legal obligation to produce reasonably safe products. A number of fundamental concepts and tools are available to assist them in meeting this challenge.

### Safety Factor

If 500 tension tests are performed on a specimen of one material, 500 different yield strengths will be obtained if the precision and accuracy of measurement are high enough. With some materials a wide range of strengths can be achieved; in others a reasonable guaranteed minimum strength can be found. However, this strength does not usually represent the stress that engineers apply in design.

Using results from small-scale tension tests, a design engineer prescribes a stress somewhat less than the semiempirical strength of a material. The **safety factor** can be expressed as

$$n_s = \frac{\sigma_{\text{all}}}{\sigma_d} \tag{1.1}$$

where $\sigma_{\text{all}}$ = allowable normal stress, Pa
 $\sigma_d$ = design normal stress, Pa

If $n_s > 1$, the design is adequate. The larger $n_s$ is, the safer the design is. If $n_s < 1$, the design may be inadequate and redesign may be necessary. In later chapters, especially Chapter 7, more will be said about $\sigma_{\text{all}}$ and $\sigma_d$. The rest of this section focuses on the left side of Equation (1.1).

It is difficult to accurately evaluate the various factors involved in engineering design problems. One factor is the shape of a part: For an irregularly shaped part there may be no design equations available for accurate stress computation. Another factor is the consequences of part failure: Life-threatening consequences require more consideration than non-life-threatening consequences.

Engineers use a safety factor to ensure against such uncertain or unknown conditions. The engineering student is often asked, What safety factor was used in the design, and which value should be used? Safety factors are sometimes prescribed by code, but usually they are rooted in design experience. That is, design engineers have established through a product's performance that a safety factor is sufficient. Future designs are often based on safety factors found adequate in previous products for similar applications.

Particular design experience for specific applications does not form a basis for the rational discussion of illustrative examples or for the guidance of engineering students. The

Pugsley method for determining the safety factor is a potential approach for obtaining safety factors in design, although the reader should again be warned that safety factor selection is somewhat nebulous in the real world and that the Pugsley method can be unconservative— that is, it predicts safety factors that are too low for reasonable design. Pugsley (1966) systematically determined the safety factor from

$$n_s = n_{sx}n_{sy} \tag{1.2}$$

where $n_{sx}$ = safety factor involving characteristics A, B, and C
A = quality of materials, workmanship, maintenance, and inspection
B = control over load applied to part
C = accuracy of stress analysis, experimental data, or experience with similar devices
$n_{sy}$ = safety factor involving characteristics D and E
D = danger to personnel
E = economic impact

Table 1.1 gives $n_{sx}$ values for various A, B, and C conditions. To use this table, estimate each characteristic for a particular application as being very good (vg), good (g), fair (f), or poor (p). Table 1.2 gives $n_{sy}$ values for various D and E conditions. To use this table, estimate each characteristic for a particular application as being very serious (vs), serious (s), or

**Table 1.1**    Safety factor characteristics A, B, and C

| Characteristic[a] | | B | | | |
|---|---|---|---|---|---|
| A | C | vg | g | f | p |
| vg | vg | 1.1 | 1.3 | 1.5 | 1.7 |
|  | g | 1.2 | 1.45 | 1.7 | 1.95 |
|  | f | 1.3 | 1.6 | 1.9 | 2.2 |
|  | p | 1.4 | 1.75 | 2.1 | 2.45 |
| g | vg | 1.3 | 1.55 | 1.8 | 2.05 |
|  | g | 1.45 | 1.75 | 2.05 | 2.35 |
|  | f | 1.6 | 1.95 | 2.3 | 2.65 |
|  | p | 1.75 | 2.15 | 2.55 | 2.95 |
| f | vg | 1.5 | 1.8 | 2.1 | 2.4 |
|  | g | 1.7 | 2.05 | 2.4 | 2.75 |
|  | f | 1.9 | 2.3 | 2.7 | 3.1 |
|  | p | 2.1 | 2.55 | 3.0 | 3.45 |
| p | vg | 1.7 | 2.15 | 2.4 | 2.75 |
|  | g | 1.95 | 2.35 | 2.75 | 3.15 |
|  | f | 2.2 | 2.65 | 3.1 | 3.55 |
|  | p | 2.45 | 2.95 | 3.45 | 3.95 |

[a]vg = very good, g = good, f = fair, and p = poor.
A = quality of materials, workmanship, maintenance and inspection.
B = control over load applied to part.
C = accuracy of stress analysis, experimental data, or experience with similar parts.

**Table 1.2**    Safety factor characteristics D and E

| Characteristic E[a] | D | | |
|---|---|---|---|
|  | ns | s | vs |
| ns | 1.0 | 1.2 | 1.4 |
| s | 1.0 | 1.3 | 1.5 |
| vs | 1.2 | 1.4 | 1.6 |

[a]vs = very serious, s = serious, and ns = not serious.
D = danger to personnel
E = economic impact

not serious (ns). Substituting the values of $n_{sx}$ and $n_{sy}$ into Equation (1.2) yields the safety factor.

Although it is a simple procedure to obtain safety factors, the Pugsley method illustrates the concerns present in safety factor selection. Many parameters, such as material strength and applied loads, may not be well known, and confidence in the engineering analysis may be suspect. For these reasons the safety factor has been called an "ignorance factor," as it compensates for ignorance of the total environment, a situation all design engineers encounter to some extent. Also, the Pugsley method is merely a guideline and is not especially conservative; most engineering safety factors are much higher than those resulting from Equation (1.2), as illustrated in Example 1.1.

**EXAMPLE 1.1**

**Given:**   A wire rope is used on an elevator transporting people to the 20th floor of a building. The design of the elevator can be 50% overloaded before the safety switch shuts off the motor.

**Find:**   What is the safety factor?

**Solution:**

A = vg, because life-threatening
B = f to p, since large overloads are possible
C = vg, due to being highly regulated
D = vs, people could die if the elevator fell from the 20th floor
E = vs, possible lawsuits

From Tables 1.1 and 1.2 the safety factor is

$$n_s = n_{sx}n_{sy} = (1.6)(1.6) = 2.56$$

Note that the value of 1.6 obtained from Table 1.1 was obtained by interpolation. By improving factors over which there is some control $n_{sx}$ can be reduced from 1.6 to 1.0 according to the Pugsley method, thus reducing the safety factor to 1.6. The Pugsley method could conceivably suggest that a passenger elevator wire rope be designed with a safety factor of 1.6. In actuality, the safety factor is prescribed by an industry standard (see Sec. 1.4.2) and cannot be lower than 7.6 and may be as high as 11.9 [ANSI (1995)].

### Product Liability

When a product is brought to the market, it is probable that safety will be a primary consideration. A design engineer must consider the hazards, or injury producers, and the risk, or likelihood of obtaining an injury from a hazard, when evaluating the safety of a system. Unfortunately, this is mostly a qualitative evaluation, and combinations of hazard and risk can be judged acceptable or unacceptable.

The ethical responsibilities of engineers to provide safe products are clear, but the legal system also enforces societal expectations through a number of legal theories that apply to designers and manufacturers of products. Some of the more common legal theories are the following:

• *Caveat emptor.*   Translated as "Let the buyer beware," this doctrine is founded on Roman laws. In the case of defective product or dangerous design, the purchaser or user of the product has no legal recourse to recover losses. In a modern society, such a philosophy is incompatible with global trade and high-quality products, and it is mentioned here only for historical significance.

- **Negligence.**   In negligence, a party is liable for damages if it failed to act as a reasonable and prudent party would have done under like or similar circumstances. For negligence theory to apply, the injured party, or plaintiff, must demonstrate that

    1. A standard of care was violated by the accused party, or defendant.

    2. This violation was the proximate cause of the accident.

    3. No contributory negligence of the plaintiff caused the misfortune.

- **Strict liability.**   Under the strict-liability doctrine, the actions of the plaintiff are not an issue; the main emphasis is placed on the machine. To recover damages under the strict-liability legal doctrine, the plaintiff must prove that

    1. The product contained a defect that rendered it unreasonably dangerous (e.g., an inadequately sized or cracked bolt fastening a brake stud to a machine frame).

    2. The defect existed at the time the machine left the control of the manufacturer. (The manufacturer used the cracked bolt.)

    3. The defect was a proximate cause of the accident. (The bolt broke, the brake stud fell off the machine, and the machine's brake didn't stop the machine, resulting in an accident.) Note that the plaintiff does not need to demonstrate that the defect was the proximate cause; the actions of the plaintiff that contribute to the accident are not considered under strict liability.

- **Comparative fault.**   Used increasingly in courts throughout the United States, juries are asked to assess the relative contributions that different parties had in relation to an accident. For example, a jury may decide that a plaintiff was 75% responsible for an accident and may reduce the monetary award by that amount.

- **Assumption of risk.**   Although rarely recognized, the assumption-of-risk doctrine states that a plaintiff has limited recourse for recovery of loss if he or she purposefully, knowingly, and intentionally conducted an unsafe act.

One important requirement on engineers is that their products must be reasonably safe for their intended uses as well as their reasonably foreseeable misuses. For example, a chair must be made structurally sound and stable enough for people to sit on (intended use). In addition, a chair should be stable enough that someone can stand on the chair to change a lightbulb, for example. This represents a reasonably foreseeable misuse of the chair and must be considered by designers. In the vast majority of states, misuses of a product that are not reasonably foreseeable do not have to be considered by the manufacturers.

The legal doctrines and ethical requirements that designers produce safe products are usually consistent. Sometimes, the legal system does create requirements that engineers cannot meet. For example, in the famous *Barker v. Lull* case in New Jersey, the court ruled that product manufacturers have a nondelegable duty to warn of the unknowable.

*Liability proofing* is the practice of incorporating design features with the intent of limiting product liability exposure without other benefits. This can reduce the safety of machinery. For example, one approach to liability proofing is to place a very large number of warnings onto a machine, with the unfortunate result that all the warnings are ignored by machine operators. The few hazards that are not obvious and can be effectively warned against are then "lost in the noise," and a compromise of machine safety can occur.

---

**Case Study 1.1** | MASON V. CATERPILLAR TRACTOR CO.

Wilma Mason brought action against Caterpillar Tractor Company and Patton Industries for damages after her husband received fatal injuries while trying to repair a track shoe on a Caterpillar tractor. Mr. Mason was repairing the track shoe with a large sledgehammer, when a small piece of metal from the track shoe shot out and resulted in fatal injuries. The plaintiff alleged that the tractor track was defective because the defendants failed to use reasonable methods of heat treatment, failed to use a sufficient amount of carbon in the steel, and failed to warn the decedent of "impending danger."

The trial and appellate courts both granted summary judgments in favor of the defendants. They ruled that the plaintiff failed to show evidence of product defect that existed when the machine left the control of the manufacturer. Mr. Mason used a large, 20-lb sledgehammer with a full swing, striking a raised portion of the track shoe. There was no evidence that the defendants were even aware that the track shoes were being repaired or reassembled by sledgehammers. It was also noted by the court that the decedent wore safety glasses, indicating his awareness of the risk of injury. (Illinois Appellate Court, 1985)

---

## Safety Hierarchy

A design rule that is widely accepted in general is the **safety hierarchy**, which describes the steps that a manufacturer or designer should take when addressing hazards. In order, a designer should attempt to

1. Eliminate hazards through design.
2. Reduce the risk or eliminate the hazard through safeguarding technology.
3. Provide warnings.
4. Train and instruct.
5. Provide personal protective equipment.

There is a general understanding that primary steps are more efficient in improving safety than later steps. That is, it is more effective to eliminate hazards through design than to use guards, which are more effective than warnings, etc. Clearly, the importance of effective design cannot be overstated.

Eliminating hazards through design can imply a number of different approaches. For example, a mechanical part that is designed so that its failure is not reasonably foreseeable is one method of eliminating a hazard or risk of injury. However, design of a system that eliminates injury producers or moves them away from people also represents a reasonable approach.

This book emphasizes mechanical analysis and design of parts to reduce or eliminate the likelihood of failure. As such, it should be recognized that this approach is one of the fundamental, necessary skills required by engineers to provide reasonably safe products.

## Failure Mode and Effects Analysis and Fault Trees

Some common tools available to design engineers are **failure mode and effects analysis (FMEA)** and fault tree analysis. FMEA addresses component failure effects on the entire system. It forces the design engineer to exhaustively consider reasonably foreseeable failure modes for every component and its alternatives.

FMEA is flexible, allowing spreadsheets to be tailored for particular applications. For example, an FMEA can also be performed on the steps taken in assembling components to identify critical needs for training and/or warning.

In **fault tree analysis**, statistical data are incorporated into the failure mode analysis to help identify the most likely (as opposed to possible) failure modes. Often, hard data are not available, and the engineer's judgment qualitatively identifies likely failure modes.

As discussed, machine designers are legally required to provide reasonably safe products and to consider the product's intended uses as well as foreseeable misuses. FMEA and fault tree analysis help identify unforeseeable misuses as well. For example, an aircraft designer may identify aircraft-meteorite collision as a possible loading of the structure. However, because no aircraft accidents have resulted from meteorite collisions and the probability of such occurrences is extremely low, the design engineer ignores such hypotheses, recognizing them as not reasonably foreseeable.

### Load Redistribution, Redundancy, Fail-safe, and the Doctrine of Manifest Danger

One potential benefit of failure mode and effects analysis and fault tree analysis is that they force the design engineer to think of minimizing the effects of individual component failures. A common goal is that the failure of a single component does not result in a catastrophic accident. The design engineer can ensure this by designing the system so that, upon a component failure, loads are redistributed to other components without exceeding their nominal strengths—a philosophy known as **redundancy** in design. For example, a goose or other large bird sucked into an aircraft engine may cause several components to fail and shut down the engine. This type of accident is not unheard of and is certainly reasonably foreseeable. Thus, modern aircraft are designed with sufficient redundancy to allow a plane to fly and land safely with one or more engines shut down.

Many designs are redundant. Redundant designs can be active (where two or more components are in use but only one is needed) or passive (where one component is inactive until the first component fails). An *active* redundant design is to use two deadbolt locks on a door: Both systems serve to keep the door locked. A *passive* redundant design is to add a chain lock on a door having a deadbolt lock: If the deadbolt lock fails, the chain will keep the door closed.

An often used philosophy is to design machinery with **fail-safe** features. For example, a brake system can be designed so that a pneumatic cylinder pushes the brake pads or shoes against a disk or drum, respectively. Alternatively, a spring could maintain pressure against the disk or drum, and a pneumatic system could work against the spring to release the brake. If the pressurized air supply were interrupted, such a system would force brake actuation and prevent machinery motion. This alternative design assumes that the spring is far more reliable than the pneumatic system.

The doctrine of **manifest danger** [Barnett (1992)] is a powerful tool used by machinery designers to prevent catastrophic losses. If danger becomes manifest, troubleshooting is straightforward and repairs can be quickly made. Thus, if a system can be designed so that imminent failure is detectable or so that single-component failure is detectable before other elements fail in turn, then a safer design results. A classic application of the doctrine of manifest danger is seen in the design of automotive braking systems, where the brake shoe consists of a friction material held onto a metal backing plate by rivets. By making the rivets long enough, an audible and tactile indication is given to the car driver when the brake system needs service (i.e., the friction material has worn to where the rivet contacts the disk or drum, indicating maintenance is required) long before braking performance is compromised.

### Reliability

Safety factors are a way of compensating for variations in loading and material properties. Another approach that can be extremely successful in certain circumstances is the application of reliability methods.

Manufacturing tension test coupons from the same extruded billet of aluminum would result in little difference in material properties from one test specimen to another. Thus, aluminum in general (as well as most metals) is a deterministic material, and deterministic methods can be used in designing aluminum structures if the load is known. For example, in a few hundred tensile tests a guaranteed minimum strength can be defined that is below the strength of any test specimen and that would not vary much from one test population to another. Deterministic methods are the approach used in most solid mechanics and mechanics courses. That is, a material has a single strength, and the loading is always well defined.

Most ceramics, however, would have a significant range of any given material property. Thus, ceramics are probabilistic, and an attempt to define a minimum strength for a population of ceramic test coupons would be an exercise in futility. There would not be a guaranteed minimum strength. One can only treat ceramics in terms of a likelihood or probability of strength exceeding a given value. There are many such probabilistic materials in engineering practice.

Some loadings, on the other hand, are well known and never vary much. Examples are the stresses inside intravenous bags during sterilization, the load supported by counterweight springs, and the load on bearings supporting centrifugal fans. Other loads can vary significantly, such as the force exerted on automotive shock absorbers (depends on the size of the pothole and the speed at impact) or on wooden pins holding a chair together (depends on the weight of the seated person or persons) or the impact force on the head of a golf club.

For situations where a reasonable worst-case scenario cannot be defined, reliability methods are sometimes a reasonable design approach. In reliability design methods the goal is to achieve a reasonable likelihood of survival under the loading conditions during the intended design life. This approach has its difficulties as well, including the following:

1. To use statistical methods, a reasonable approximation of an infinite test population must be defined. That is, mean values and standard deviations about the mean, and even the nature of the distribution about the mean, must be known. However, they are not usually known very well after only a few tests. After all, if only a few tests were needed to quantify a distribution, deterministic methods would be a reasonable, proper, and less mathematically intensive approach.

2. Even if strengths and loadings are known well enough to quantify their statistical distributions, defining a desired reliability is as nebulous a problem as defining a desired safety factor. A reliability of 99% might seem acceptable, unless that were the reliability of an elevator you happened to be occupying. A reliability of 100% is not achievable, or else deterministic methods would be used. A reliability of 99.9999 · · · % should be

recognized as an extremely expensive affair, as indicative of overdesign as a safety factor of 2000.

3. The mathematical description of the data has an effect on reliability calculations. A quantity may be best described by a Gaussian or normal distribution, a lognormal distribution, a binary distribution, or a Weibull distribution, etc. Often, one cannot know *a priori* which distribution is best. Some statisticians recommend using a normal distribution until it is proved ineffective.

The implications are obvious: Reliability design is a complicated matter and even when applied does not necessarily result in the desired reliability if calculated from insufficient or improperly reduced data.

This textbook will emphasize deterministic methods for the most part. The exceptions are the treatments of rolling-element bearings and reliability in fatigue design. For more information on reliability design refer to the excellent text by Lewis (1995), among others.

### 1.4.2 GOVERNMENT CODES AND INDUSTRY STANDARDS

In many cases engineers must rely on government codes and industry-promulgated standards for design criteria. Some of the most common sources for industry standards are

1. ANSI, the American National Standards Institute
2. ASME, the American Society of Mechanical Engineers
3. ASTM, the American Society for Testing and Materials
4. AGMA, the American Gear Manufacturers Association
5. AISI, the American Iron and Steel Institute
6. AISC, the American Institute of Steel Construction
7. ISO, the International Organization for Standardization
8. NFPA, the National Fire Protection Association
9. UL, Underwriters Laboratories

Government codes are published annually in the Code of Federal Regulations (CFR) and periodically in the Federal Register (FR) at the national level. States and local cities and towns have codes as well, although most relate to building standards and fire prevention.

Code compliance is important for many reasons, some of which have already been stated. However, one essential goal of industry standards is conformability. For example, bolt geometries are defined in ANSI standards so that bolts have fixed thread dimensions and bolt diameters. Therefore, bolts can be mass-produced, and manufacturing economy of scale results in inexpensive, high-quality threaded fasteners. Also, maintenance is simplified in that standard bolts can be purchased anywhere, making replacement parts readily available.

### 1.4.3 MANUFACTURING

Design and manufacturing are difficult to consider apart from each other. The tenet of "form follows function" suggests that shapes are derived only from applied loads in the design environment. As Petroski (1992b) eloquently discussed, this is not always the case, and the shapes of products are often natural progressions from arbitrary beginnings.

**Figure 1.2** Effect of manufacturing and assembly considerations on design of a reciprocating power saw. (a) Original design, with 41 parts and 6.37-min assembly time; (b) modified design, with 29 parts and 2.58-min assembly time. [Adapted from Boothroyd (1992).]

*Design for manufacturability (DFM)* is a well-established and needed tool for design engineers. Manufacturability plays a huge role in the success of commercial products. After all, a brilliant concept that cannot be manufactured cannot be a successful design (per our definition in Sec. 1.1). Also, because most manufacturing costs are determined by decisions made early in the design process, market success depends on minimizing these costs. Individual components should be designed to be easily fabricated, assembled, and constructed [*design for assembly (DFA)*]. Although manufacturing and assembly are outside the scope of this text, Figure 1.2 shows their effect on design.

Engineers must wear many hats. Some predominant concerns of a design engineer have been discussed, but many more exist, including these:

1. **Environmental design.** This issue addresses whether products can be produced that are less harmful to the environment. Biodegradable or easily recycled materials may need to be selected.

2. **Economics.** Deciding whether a product will lead to corporate profitability is of utmost concern.

3. **Legal considerations.** Violating patents and placing unreasonably dangerous products into the marketplace are not only ethically wrong but also legal suicide.

4. **Marketing.** The features of a product that attract consumers and how the product is presented to the marketplace play a significant role in a product's success.

5. **Serviceability.** If a part breaks, can repairs be done in the field, or must customers send the product back to the manufacturer at excessive expense? Unless such concerns are incorporated into design, long-term customer loyalty is compromised.

6. **Quality.** Approaches such as total quality engineering and Taguchi methods have been successfully applied to make certain that no defects are shipped.

These are merely a few of the concerns faced by design engineers.

The design process may appear so elaborate and involved that no one can master it. In actuality, one most important skill makes the design process flow smoothly. Communication between diverse disciplines involved in the product life cycle ensures that all voices are heard. Effective communication skills, written and oral, are today the most important trait of a good engineer. Although this text emphasizes the more analytical and technical sides of design, always remember that total design is not merely an analytical effort but one of human interaction.

## 1.5  COMPUTERS IN DESIGN

**Computer-aided design (CAD)** also means different things to different people, but in this text it is the application of computer technology to planning, performing, and implementing the design process. Computers allow us to virtually integrate all phases of the design process, whether technical or managerial activities. With sophisticated hardware and software, manufacturers can now minimize design costs, maximize efficiency, improve quality, reduce development time, and maintain an edge in domestic and international markets.

CAD allows the design engineer to visualize geometries without making costly models, iterations, or prototypes. These systems can now analyze designs of simple brackets to complex structures quickly and easily. Designs can be optimized and modified directly and easily at any time. Information stored by computer can be accessed and retrieved from anywhere within the organization.

Whereas some restrict the term *CAD* to drafting activities, others will generically group all computer-assisted functions arbitrarily as CAD. **Artificial intelligence (AI)** studies attempt to duplicate how the human mind works and apply it to processes on the computer. Sometimes, AI is used to describe the cases where computers are used as more than mere drafting tools and actually help in the intellectual design tasks. **Expert systems** are rule-based computer programs that solve specialized problems on an expert level and provide problem-solving skills to the design engineer. For example, an expert system could analyze a part drafted on a computer system for ease in manufacturing. If an excessively small tolerance is found, the expert system warns the engineer that manufacturing difficulties will ensue and suggests easing the tolerance. Similarly, an expert system could analyze a design to standardize parts (e.g., to make sure that an assembly uses only one bolt size instead of the optimum size for each location, thereby easing inventory and maintenance difficulties). Artificial intelligence is a more elaborate form of an expert system; AI really refers to computer systems that can learn new information.

The advent of *computer numerical control* (*CNC*) has made *group technology* (*GT*) and *cellular manufacturing* (*CM*) increasingly valuable. Group technology involves classifying and codifying parts so that those with similar design attributes can be quickly retrieved on a computer system. GT is extremely valuable when a desired design is merely an extension or modification of an existing component. The existing component's drawing can be quickly retrieved because of the logical classification, and the file can be modified with minimal redundant effort. Cellular manufacturing is the organization of machine tools so that a wide variety of designs can be fabricated under CNC with minimal lag time.

**Rapid prototyping** is another computer-driven technology that produces parts from computer geometry description files in hours or even minutes. Rapid prototyping has been especially helpful in design visualization and rapid detection of design errors. For example, a casting with an excessively thin wall is easily detected when a solid model is held in the hand, a subtlety difficult to discern when one is viewing a part drawing on a two-dimensional computer screen.

**Finite element analysis (FEA)** is the most prevalent computational method for solid and fluid mechanics analysis. The finite element computational method solves complex shapes, such as those found in machinery, and replaces the complex shape with a set of simple elements interconnected at a finite set of node points. In FEA a part geometry is sectioned into many subsections or elements. The stiffness of each element is known and is expressed in terms of a stiffness matrix for that element. By combining all the stiffness matrices, applying kinematic and stress boundary conditions, and solving for unknown stresses or displacements, complicated geometries and loading conditions can be easily analyzed.

## 1.6   CATALOGS AND VENDORS

Manufacturing concerns are latent in all design problems. Clearly, many machine elements are mass-produced because economic arguments favor hard automation for large production runs. Hard automation generally results in higher-quality, tighter-tolerance parts than does soft automation or hand manufacture, and can result in less expensive parts as well. In fact, the industry standards mentioned in Section 1.4.2 exist to prescribe geometries that can be mass-produced in order to advocate mass production. For example, a centerless grinder can produce many high-quality $\frac{1}{2}$-in-diameter bushings, whereas a single $\frac{1}{2}$-in bushing is difficult to manufacture. Therefore, machinery design often involves selecting mass-produced elements from suppliers, often as summarized in catalogs.

Mechanical designers know the importance of good vendor identification and readily available and up-to-date catalog information. Such catalogs are rarely available for the engineering student and are not needed if a mere familiarity with the design process is the main goal of study. However, for the design practitioner such material is invaluable.

Machine element vendors can be identified most easily through routine Internet searches or through the Thomas Register, a listing of corporations in the United States. Most, and probably all, engineering libraries maintain a copy of the Thomas Register, and this is an excellent resource. The internet makes product information readily available.

## 1.7   UNITS

The solutions to engineering problems must be given in specific units that correspond to the specific parameter being evaluated. Two systems of units are mainly used in this text:

1.  **Système International d'Unités (SI units).**   Force is measured in newtons, length in meters (sometimes millimeters are more convenient for certain problems), time in seconds, mass in kilograms, and temperature in degrees Celsius.

2.  **English units.**   Force is measured in pounds force, length in inches, time in seconds, mass in pounds mass, and temperature in degrees Fahrenheit.

In Chapter 8 an additional measure—viscosity—is given in the centimeter-gram-seconds (cgs) system.

---

**EXAMPLE 1.2**

**Given:**   A supercomputer has a calculation speed of 1 gigaflop $= 10^9$ floating-point operations per second. The calculation speed is determined only by the length of the wires. The electron speed for coaxial cables is 0.9 times the speed of light.

**Find:**   How long can the connecting wires be?

**Solution:**   If the speed is determined only by the cable length,

$$l = \frac{(0.9)(3)(10^8)}{10^9} = 0.27 \, \text{m} = 27 \, \text{cm}$$

The mean cable length must be less than 27 cm.

---

**EXAMPLE 1.3**

**Given:**   The distance from earth to $\alpha$-Centauri is 4 light-years.

**Find:**   How many terameters (Tm) is it?

**Solution:**

$$1 \, \text{yr} = (365)(24)(3600) \, \text{s} = (3.1536)(10^7) \, \text{s}$$
$$\text{Speed of light} = (3)(10^8) \, \text{m/s}$$

The distance from earth to $\alpha$-Centauri is

$$(4)(3.1536)(10^7 \, \text{s})(3)(10^8) \, \text{m/s} = 3.784 \times 10^{16} \, \text{m}$$

From Table 1.3(b), $T = 10^{12}$. Therefore,

$$(3.784)(10^{16}) \, \text{m} = 37,840 \, \text{Tm}$$

**Table 1.3** SI units and prefixes

### (a) SI Units

| Quantity | Unit | SI symbol | Formula |
|---|---|---|---|
| **SI base units** | | | |
| Length | meter | m | — |
| Mass | kilogram | kg | — |
| Time | second | s | — |
| Temperature | kelvin | K | — |
| **SI supplementary unit** | | | |
| Plane angle | radian | rad | — |
| **SI derived units** | | | |
| Energy | joule | J | $N \cdot m$ |
| Force | newton | N | $kg \cdot m/s^2$ |
| Power | watt | W | J/s |
| Pressure | pascal | Pa | $N/m^2$ |
| Work | joule | J | $N \cdot m$ |

### (b) SI Prefixes

| Multiplication factor | Prefix | SI symbol for prefix |
|---|---|---|
| $1,000,000,000,000 = 10^{12}$ | tera | T |
| $1,000,000,000 = 10^9$ | giga | G |
| $1,000,000 = 10^6$ | mega | M |
| $1000 = 10^3$ | kilo | k |
| $100 = 10^2$ | hecto | h |
| $10 = 10^1$ | deka | da |
| $0.1 = 10^{-1}$ | deci | d |
| $0.01 = 10^{-2}$ | centi | c |
| $0.001 = 10^{-3}$ | milli | m |
| $0.000\,001 = 10^{-6}$ | micro | $\mu$ |
| $0.000\,000\,001 = 10^{-9}$ | nano | n |
| $0.000\,000\,000\,001 = 10^{-12}$ | pico | p |

The data needed to solve problems are not always in the same system of units. It is therefore necessary to convert from one system to another. The SI units, prefixes, and symbols used throughout the text are shown in Table 1.3 as well as inside the front cover. The primary units of the text are SI, but problems are given in English units as well as in SI units to enable the student to handle either system.

Basic SI units, some definitions, and fundamental and other useful conversion factors are given in Table 1.4, which is also inside the front cover.

It is important to maintain the proper units throughout a problem to prevent numerical and conceptual errors. For example, it is important that units not be confused when one is converting between measurement systems. Also, it is important that proper measures be used. The most common example of the latter is the use of force units for mass measures or vice versa. It is highly recommended that students maintain the units throughout problem solutions to avoid these errors.

**Table 1.4**    Conversion factors and definitions

| (a) Fundamental Conversion Factors | | | |
|---|---|---|---|
| | **English unit** | **Exact SI value** | **Approximate SI value** |
| Length | 1 in | 0.0254 m | — |
| Mass | 1 lbm | 0.45359237 kg | 0.4536 kg |
| Temperature | 1°R | $\frac{5}{9}$ K | — |

**(b) Definitions**

Acceleration of gravity   1 g = 9.8066 m/s$^2$ (32.174 ft/s$^2$)

Energy   Btu (British thermal unit) = amount of energy required to raise 1 lbm of water 1°F (1 Btu = 778.2 ft · lb)
kilocalorie = amount of energy required to raise 1 kg of water 1 K (1 kcal = 4187 J)

Length   1 mi = 5280 ft; 1 nautical mile (nmi) = 6076.1 ft

Power   1 hp = 550 ft · lb/s

Pressure   1 bar = 10$^5$ Pa

Temperature   degrees Fahrenheit $t_F = \frac{9}{5}t_C + 32$
(where $t_C$ is degrees Celsius)
degrees Rankine $t_R = t_F + 459.67$
Kelvin $t_K = t_C + 273.15$ (exact)

Kinematic viscosity   1 P = 0.1 kg/(m · s)
1 St = 0.0001 m$^2$/s

Volume   1 ft$^3$ = 7.48 gal

**(c) Useful Conversion Factors**

1 ft = 0.3048 m
1 lb = 4.448 N
1 lb = 386.1 lbm · in/s$^2$
1 kgf = 9.807 N
1 lb/in$^2$ = 6895 Pa
1 ksi = 6.895 MPa
1 Btu = 1055 J
1 ft · lb = 1.356 J
1 hp = 746 W = 2545 Btu/h[a]
1 kW = 3413 Btu/h
1 qt = 0.000946 m$^3$ = 0.946 L
1 kcal = 3.968 Btu
SI Units   1 hp = 736 W

**Case Study 1.2** | LOSS OF THE MARS CLIMATE ORBITER

On December 11, 1998, the Mars Climate Orbiter was launched to start its nearly 10-month journey to Mars. The Mars Climate Orbiter was a $125 million satellite intended to orbit Mars and measure the atmospheric conditions on that planet over a planetary year. It was also intended to serve as a communications relay for the Mars Climate Lander which was due to reach Mars in December 1999.

*(continued)*

**Case Study**    (Concluded)

The Mars Climate Orbiter was destroyed on September 23, 1999, as it was maneuvering into orbit. The cause for the failure was quickly determined: The manufacturer programmed the entry software in English measurements. But the navigation team at NASA's Jet Propulsion Laboratory in Pasadena, California, assumed the readings were in metric units. As a result, trajectory errors were magnified instead of corrected by midcourse thruster firings. This painful lesson demonstrated the importance of maintaining and reporting units with all calculations.

## 1.8   UNIT CHECKS

Unit checks should always be performed during engineering calculations to make sure that each term of an equation is in the same system of units. The importance of knowing the units of the various parameters used in an equation cannot be overemphasized. In this text a symbol list giving the units of each parameter is provided at the beginning of each chapter. If no units are given for a particular phenomenon, it is dimensionless. This symbol list can be used as a partial check during algebraic manipulations of an equation.

### PROCEDURE FOR UNIT CHECKS

1.   Establish units of specific terms of an equation while making use of Table 1.3(a)
2.   Place units of terms into both sides of equation and reduce
3.   Unit check is complete if both sides of the equation have similar units.

**Given:**   The centrifugal force $P$ acting on a car going through a curve with a radius $r$ at a velocity $v$ is $m_a v^2/r$, where $m_a$ is the mass of the car. Assume a 1.3-ton car running at 100 km/h through a 100-m-radius bend.

**EXAMPLE 1.4**

**Find:**   Calculate the centrifugal force.

**Solution:**   Rewriting all units in metric units gives

$$m_a = 1.3\,\text{ton} = 1300\,\text{kg}$$

$$v = 100\,\text{km/h} = \frac{(100)(1000)\,\text{m}}{3600\,\text{s}} = 27.78\,\text{m/s}$$

$$r = 100\,\text{m}$$

The centrifugal force is

$$P = \frac{m_a v^2}{r} = \frac{(1300\,\text{kg})(27.78\,\text{m/s})^2}{100\,\text{m}}$$

$$= 10{,}030\,\text{kg} \cdot \text{m/s}^2 = 10{,}030\,\text{N}$$

The centrifugal force is 10,030 N.

## 1.9   SIGNIFICANT FIGURES

The accuracy of a number is specified by how many significant figures it contains. When accuracy has not been otherwise restricted, this text will use four significant figures. For example, 8201 and 30.51 each has four significant figures. When numbers begin or end with a zero, however, it is difficult to tell how many significant figures there are. To clarify this situation, the number should be reported by using powers of 10. Thus, the number 8200 can be expressed as $8.200 \times 10^3$ to represent four significant figures. Also, 0.005012 can be expressed as $5.012 \times 10^3$ to represent four significant figures.

---

**EXAMPLE 1.5**

**Given:**   A car with a mass of 1502 kg is accelerated by a force of 14.0 N.

**Find:**   Calculate the acceleration and answer with significant figures.

**Solution:**   Newton's equation gives that acceleration equals the force divided by the mass, or

$$a = \frac{P}{m_a} = \frac{14.0}{1502} = 0.00932091 \text{ m/s}^2$$

Since force is accurate to three figures, the acceleration can only be calculated with the accuracy of

$$\pm \frac{0.5}{140} = \pm 0.004 = \pm 0.4\%$$

The acceleration is $0.00932 \text{ m/s}^2$.

---

**Case Study 1.3**    DESIGN AND MANUFACTURE OF THE INVISALIGN ORTHODONTIC PRODUCT

Widespread health care and improved diet and living habits have greatly extended the expected lifetime of people within the last century. Modern expectations are not only that life will be extended, but also that the *quality* of life will be maintained late in life. One important area where this concern manifests itself is with teeth; straight teeth lead to a healthy bite with low tooth stresses, and they also lend themselves to easier cleaning and therefore are more resistant to decay. Thus, straight teeth, in general, last longer with less pain. Of course, there are aesthetic reasons that people wish to have straight teeth as well.

Orthodontic braces have been available to straighten teeth for over one hundred years. These involve metal, ceramic, or plastic brackets that are adhesively bonded to teeth, with fixtures for attachment to a wire that then forces compliance on the teeth and straightens them to the desired shape within a few years. Conventional orthodontic braces are a well-known and wholly successful approach to long-term dental health. However, there are many drawbacks to conventional braces, including these:

- They are aesthetically unappealing.

- The sharp wires and brackets can cause painful oral irritation to the teeth and gums.

- They trap food, leading to premature tooth decay.

- Brushing and flossing of teeth are far more difficult with braces in place, and therefore it is less effective for most individuals.

- Certain foods must be avoided because they will damage the braces.

## Case Study (Continued)

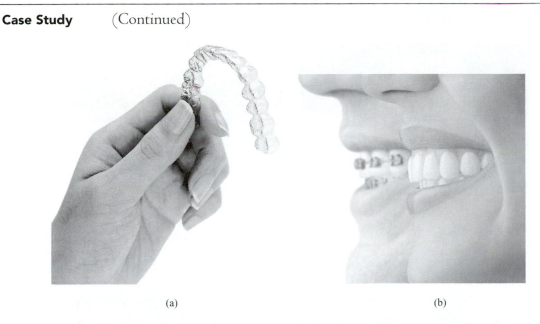

(a)

(b)

**Figure 1.3** The Invisalign® product. (a) An example of an Aligner; (b) a comparison of conventional orthodontic braces and a transparent Aligner. (From Align Technology, Inc.)

One innovative solution is the Invisalign product by Align Technologies. Invisalign consists of a series of Aligners, each of which the patient wears for approximately two weeks. Each Aligner (see Fig. 1.3) consists of a precise geometry which incrementally moves teeth to their desired positions. Because they are inserts that can be removed for eating, brushing, and flossing, most of the drawbacks of conventional braces are eliminated. Further, since they are produced from transparent plastic, they do not seriously affect the patient's appearance.

The Invisalign product uses an impressive combination of advanced technologies, and the production process is shown in Figure 1.4. The treatment begins with an orthodontist creating a polymer impression of the patient's teeth [Fig. 1.4(a)]. These impressions are then used to create a three-dimensional CAD representation of the patient's teeth, as shown in Figure 1.4(b). Proprietary computer-aided design software then assists in the development of a treatment strategy for moving the teeth in optimal fashion. Specially produced software called ClinCheck then produces a digital video of the incremental movements which can be reviewed by the treating orthodontist and modified, if necessary. This demonstrates the multidisciplinary blending of specialties; the CAD engineer and the orthodontist must work together to provide superior treatment to the patient.

Once a treatment plan has been designed, the computer-based information is used to produce the Aligners. This is done through a novel application of rapid prototyping technology. Stereolithography is a process that uses a focused laser to cure a liquid photopolymer. The laser cures only a small depth of the polymer, so a part can be built on a tray that is progressively lowered into a vat of photopolymer as layers or slices of the desired geometry are traced and rastered by the laser.

A number of materials are available for stereolithography, but these have a characteristic yellow-brown shade to them and are therefore unsuitable for direct application as an orthodontic product. Instead, the stereolithography machines produce patterns of the desired incremental positions of the teeth [Fig. 1.4(c)]. A sheet of clear polymer is then molded over these patterns to produce the Aligners.

*(continued)*

**Case Study**        (Continued)

(a)

(b)

(c)

(d)

**Figure 1.4**    The process used in application of Invisalign orthodontic treatment. (a) Impressions are made of the patient's teeth by the orthodontist and shipped to Align Technology, Inc. These are used to make plaster models of the patient's teeth. (b) High-resolution three-dimensional representations of the teeth are produced from the plaster models. The correction plan is then developed using computer tools. (c) Rapid-prototyped molds of the teeth at incremental positions are produced through stereolithography. (d) An Aligner is produced by molding a transparent plastic over the stereolithography part. Each Aligner is used for approximately two weeks. The patient is left with a healthy bite and a beautiful smile. [From Align Technology, Inc.]

*(continued)*

**Case Study**   (Concluded)

These are sent to the treating orthodontist, and new Aligners are given to the patient as needed, usually every two weeks or so.

The Invisalign product has proved very popular for patients who wish to promote dental health and preserve their teeth long into their lives. It depends on advanced engineering technologies, computer-aided design and manufacturing, and rapid prototyping and advanced polymer manufacturing processes.

## 1.10   SUMMARY

This chapter introduced the concept of design as it applies to machines and machine elements. The most important goal of the design process is to ensure that the design does not fail. The design engineer must predict the circumstances under which failure is first likely to occur. These circumstances are stress–strain relationships involving the material properties as well as surface phenomena, including friction, wear, lubrication, and environmental deterioration.

The concept of failure was quantified by using a safety factor, which is simply the allowable stress established for the material being used, divided by the maximum design stress that will occur. If the safety factor is less than 1, the design is inadequate and redesign is necessary. Besides the simple safety factor, other failure models, such as failure mode and effects analysis and fault tree analysis, were presented. Design was found to be a cooperative endeavor in which multidisciplinary approaches are invaluable. Failure prevention and engineering analysis of machine elements are two essential requirements of successful design teams. These topics are the focus of this text.

## KEY WORDS

**artificial intelligence (AI)**   an approach that attempts to duplicate how the human mind works in computer processes

**computer-aided design (CAD)**   application of computer technology to planning, performing, and implementing the design process

**concurrent engineering**   design approach wherein all disciplines involved with a product are in the development process from beginning to end

**design**   transformation of concepts and ideas into useful machinery

**English units**   system of units in which

        force is measured in pounds-force (lbf)
        length in inches (in)
        time in seconds (s)
        mass in pounds-mass (lbm)
        temperature in degrees Rankine (°R)

**expert systems**   computer programs that solve specialized problems on an expert level

**fail-safe**   design approach in which no catastrophic loss can occur as a result of a component failure

**failure**   the condition of a machine element when it is completely inoperable, cannot perform its intended function adequately, or is unreliable for continued safe use

**failure mode and effects analysis (FMEA)**   systematic consideration of component failure effects on the entire system

**fault tree analysis**   technique in which statistical data are used to identify the most likely failure modes

**finite element analysis (FEA)**   computational method used for solving for stress, strain, temperature, etc., in complex shapes, such as those found in machinery; replaces the complex shape with a set of simple elements interconnected at a finite set of node points

**machine**   combination of mechanisms and other components that transform, transmit, or use energy, load, or motion for a specific purpose

**machine element function**   normal load transmitter, torque transmitter, energy absorber, or seal

**manifest danger**   design approach where needed service is made apparent before catastrophic failure

**mechanical system**   synergistic collection of machine elements

**rapid prototyping**   parts are produced quickly from computer geometry description files

**redundancy**   additional capacity or incorporation of backup systems so that a component failure does not lead to catastrophic loss

**safety factor** ($n_s$)   ratio of allowable stress to design stress

**safety hierarchy**   the preferred approaches to treat a hazard in order to achieve reasonable safety

**SI units**   system of units where

> force is measured in newtons (N)
> length in meters (m)
> time in seconds (s)
> mass in kilograms (kg)
> temperature in kelvins (K)

## RECOMMENDED READINGS

### GENERAL ENGINEERING
Florman, S. C. (1976) *The Existential Pleasures of Engineering*, St. Martin's Press, New York.
Petroski, H. (1992) *To Engineer Is Human*, Vintage Books, New York.

### GENERAL DESIGN
Cagan, J., and Vogel, C. M. (2002) *Creating Breakthrough Products*, Prentice Hall, Upper Saddle River, NJ.
Haik, Y. (2003) *Engineering Design Process*, Brooks-Cole, Thomson, Australia.
Hyman, B. (2003) *Fundamentals of Engineering Design*, 2nd ed., Prentice Hall, Upper Saddle River, NJ.

Lindbeck, J. R. (1995) *Product Design and Manufacture*, Prentice Hall, Upper Saddle River, NJ.

Otto, K., and Wood, K. (2000) *Product Design: Techniques in Reverse Engineering and New Product Development*, Prentice Hall, Upper Saddle River, NJ.

Ullman, D. G. (2003) *The Mechanical Design Process*, McGraw-Hill, New York.

### MANUFACTURING/DESIGN FOR MANUFACTURE

Boothroyd, G. (1992) *Assembly Automation and Product Design*, Marcel Dekker, New York.

Boothroyd, G., Dewhurst, P., and Knight, W. (1994) *Product Design for Manufacture and Assembly*, Marcel Dekker, New York.

DeGarmo, E. P., Black, J. T., and Kohser, R. A. (1997) *Materials and Processes in Manufacturing*, 8th ed., Prentice Hall, Upper Saddle River, NJ.

*Design for Manufacturability (1992), Tool and Manufacturing Engineer's Handbook*, vol. 6, Society of Manufacturing Engineers, New York.

Dieter, G. E. (2000) *Engineering Design: A Materials and Processing Approach*, McGraw-Hill, New York.

Groover, M. K. (2002) *Fundamentals of Modern Manufacturing*, 2nd ed., Wiley, New York.

Kalpakjian, S., and Schmid, S. R. (2003) *Manufacturing Processes for Engineering Materials*, 4th ed., Prentice Hall, Upper Saddle River, NJ.

Kalpakjian, S., and Schmid, S. R. (2000) *Manufacturing Engineering and Technology*, 4th ed., Prentice Hall, Upper Saddle River, NJ.

Wright, P. K. (2001) *21st Century Manufacturing*, Prentice Hall, Upper Saddle River, NJ.

### CONCURRENT ENGINEERING

Clausing, D. (1994) *Total Quality Development*, Marcel Dekker, New York.

Nevins, J. L., and Whitney, D. E. (eds.) (1989) *Concurrent Design of Products and Processes*, McGraw-Hill, New York.

Prasad, B. (1996) *Concurrent Engineering Fundamentals*, Prentice Hall, Upper Saddle River, NJ.

Pugh, S. (1996) *Creating Innovative Products Using Total Design*, Addison-Wesley, Reading, MA.

Pugh, S. (1991) *Total Design*, Addison-Wesley, Reading, MA.

### COMPUTER-AIDED DESIGN

Groover, M., and Zimmer, W. (1984) *CAD/CAM Computer-Aided Design and Manufacture*, Prentice Hall, Upper Saddle River, NJ.

Wilson, C. E. (1997) *Computer Integrated Machine Design*, Prentice Hall, Upper Saddle River, NJ.

## REFERENCES

ANSI (1995) A17.1 Minimum Safety Requirements for Passenger Elevators, American National Standards Institute, New York.

ASME (1997) *Code of Ethics for Engineers*, Board on Professional Practice and Ethics, American Society of Mechanical Engineers, New York.

Barnett, R. L. (1992) *The Doctrine of Manifest Danger*, Triodyne Inc., Niles, IL.

Boothroyd, G. (1992) *Assembly Automation and Product Design*, Marcel Dekker, New York, pp. 239–240.

Burke, J. (1985) *The Day the Universe Changed*, Little, Brown and Company, Boston, pp. 303–305.

Florman, S. C. (1987) *The Civilized Engineer*, St. Martin's Press, New York, p. 44.

Kalpakjian, S., and Schmid, S. R. (2003) *Manufacturing Processes for Engineering Materials*, 4th ed., Prentice-Hall, Upper Saddle River, NJ. p. 7.

Lewis, E. E. (1987) *Introduction to Reliability Engineering*, Wiley, New York.

Petroski, H. (1992a) *To Engineer Is Human*, Vintage Books, New York, p. 3.

Petroski, H. (1992b) *The Evolution of Useful Things*, Alfred A. Knopf, New York.

Pugh, S. (1996) *Creating Innovative Products Using Total Design*, Addison-Wesley, Reading, MA, p. 11.

Pugsley, A. G. (1966) *The Safety of Structures*, Arnold, New York.

## PROBLEMS

### Section 1.1

**1.1**  Design transport containers for milk in 1-gal and 1-L sizes.

**1.2**  Design a kit of tools for campers so they can prepare and eat meals. The kit should have all the implements needed and be lightweight and compact.

### Section 1.3

**1.3**  Journal bearings on train boxcars in the early 19th century used a stink additive in their lubricant. If the bearing got too hot, it would have a noticeable odor, and an oiler would give the bearing a squirt of lubricant at the next train stop. What design philosophy does this illustrate? Explain.

**1.4**  Explain why engineers must work with other disciplines, using specific product examples.

### Section 1.4

**1.5**  A handheld drilling machine has a bearing to take up radial and thrust load from the drill. Depending on the number of hours the drill is expected to be used before it is scrapped, different bearing arrangements will be chosen. A rubbing bushing has a 50-h life. A small ball bearing has a 300-h life. A two-bearing combination of a ball bearing and a cylindrical roller bearing has a 10,000-h life. The cost ratios for the bearing arrangements are $1:5:20$. What is the optimum bearing type for a simple drill, a semiprofessional drill, and a professional drill?

**1.6**  For the handheld drill described in Problem 1.5, if the solution with the small ball bearing was chosen for a semiprofessional drill, the bearing life could be estimated to be 300 h until the first spall forms in the race. The time from first spall to when the whole rolling-contact surface is covered with spalls is 200 h, and the time from then until a ball cracks is 100 h. What is the bearing life

*a)*     If high precision is required?

*b)*     If vibrations are irrelevant?

*c)*     If an accident can happen when a ball breaks?

**1.7**  A car is being driven at 150 km/h on a mountain road where the posted speed limit is 100 km/h. At a tight turn, one of the tires fails (a blowout), causing the driver to lose control, and results in an accident involving property losses and injuries but no loss of life. Afterward, the driver decides to file a lawsuit against the tire manufacturer. Explain which legal theories give him a viable argument to make a claim.

**1.8** The dimensions of skis used for downhill competition need to be determined. The maximum force transmitted from one foot to the ski is 2500 N, but the snow conditions are not known in advance, so the bending moment acting on the skis is not known. Estimate the safety factor needed.

**1.9** A crane has a loading hook that is hanging in a steel wire. The allowable normal tensile stress in the wire gives an allowable force of 100,000 N. Estimate the safety factor that should be used.

*a)* The wire material is not controlled, the load can cause impact, and fastening the hook in the wire causes stress concentrations. (If the wire breaks, people can be seriously hurt and expensive equipment can be destroyed.)

*b)* The wire material is extremely well controlled, no impact loads are applied, and the hook is fastened in the wire without stress concentrations. (If the wire breaks, no people or expensive equipment can be damaged.)

**1.10** Give three examples of fail-safe products and three examples of fail-unsafe products.

**1.11** An acid container will damage the environment and people around it if it leaks. The cost of the container is proportional to the container wall thickness. The safety can be increased either by making the container wall thicker or by mounting a reserve tray under the container to collect the leaking acid. The reserve tray costs 10% of the thick-walled container cost. Which is less costly, to increase the wall thickness or to mount a reserve tray under the container?

## Section 1.7

**1.12** Calculate the following:

*a)* The velocity of hair growth in meters per second, assuming in 1 month hair grows 0.75 in.

*b)* The weight of a 1-in-diameter steel ball bearing in meganewtons.

*c)* The mass of a 1-kg object on the surface of the moon.

*d)* The equivalent rate of 4 hp of work in watts.

**1.13** The SI unit for dynamic viscosity is newton-seconds per square meter, or pascal-seconds $(N \cdot s/m^2 = Pa \cdot s)$. How can that unit be rewritten by using the basic relationships described by Newton's law for force and acceleration?

**1.14** The unit for dynamic viscosity in Problem 1.13 is newton-seconds per square meter $(N \cdot s/m^2)$, and the kinematic viscosity is defined as the dynamic viscosity divided by the fluid density. Find at least one unit for kinematic viscosity.

**1.15** A square surface has sides 1 m long. The sides can be split into decimeters, centimeters, or millimeters, where 1 m = 10 dm, 1 dm = 10 cm, and 1 cm = 10 mm. How many millimeters, centimeters, and decimeters equal 1 m? Also, how many square millimeters, square centimeters, and square decimeters equal 1 m$^2$?

**1.16** A volume is 1 tera (mm$^3$) large. Calculate how long the sides of a cube must be to contain that volume.

**1.17** A ray of light travels at a speed of 300,000 km/s $= 3 \times 10^8$ m/s. How far will it travel in 1 ps, 1 ns, and 1 $\mu$s?

## Section 1.8

**1.18** Two smooth flat surfaces are separated by a 10-$\mu$m-thick lubricant film. The viscosity of the lubricant is 0.100 Pa $\cdot$ s. One surface has an area of 1 dm$^2$ and slides over the plane surface with a velocity of 1 km/h. Determine the friction force due to shearing of the lubricant film. Assume the friction force is the viscosity times the surface area times the velocity of the moving surface, divided by the lubricant film thickness.

**1.19** A firefighter sprays water on a house. The nozzle diameter is small relative to the hose diameter, so the force on the nozzle from the water is

$$F = v \, \frac{dm_a}{dt}$$

where $v$ is the water velocity and $dm_a/dt$ is the water mass flow per unit time. Calculate the force that the firefighter needs to hold the nozzle if the water mass flow is 3 tons/h and the water velocity is 100 km/h.

**1.20** The mass of a car is 1346 kg. The four passengers in the car weigh 643, 738, 870, and 896 N. It is raining, and the additional mass due to the water on the car is 1.349 kg. Calculate the total weight and mass of the car, including the passengers and water, using four significant figures.

**1.21** During an acceleration test of a car, the acceleration was measured to be 1.4363 m/s². Because slush and mud adhered to the bottom of the car, the mass was estimated to be $1400 \pm 100$ kg. Calculate the force driving the car, and indicate the accuracy.

# LOAD, STRESS, AND STRAIN

Collapse of the Tacoma
Narrows Bridge in 1940.
© Bettmann/Corbis

*The careful text-books measure*
*(Let all who build beware!)*
*The load, the shock, the pressure*
*Material can bear.*
*So when the buckled girder*
*Lets down the grinding span,*
*The blame of loss, or murder,*
*Is laid upon the man.*
*Not on the stuff—The Man!*
Rudyard Kipling, "Hymn of Breaking Strain"

## SYMBOLS

| | |
|---|---|
| $A$ | area, $m^2$ |
| $d$ | diameter, m |
| $g$ | gravitational acceleration, $9.807 \text{ m/s}^2$ |
| $l$ | length, m |
| $M$ | moment, $N \cdot m$ |
| $m_a$ | mass, kg |
| $n$ | any integer |
| $P$ | force, N |
| $p$ | pressure, Pa |
| $q$ | load intensity function, N/m |
| $R$ | reaction force, N |
| $r$ | radius of Mohr's circle, m |
| **S** | stress tensor |
| **S'** | principal stress tensor |
| **T** | strain tensor |
| $T$ | torque, $N \cdot m$ |
| $V$ | transverse shear force, N |
| $W$ | normal applied load, N |
| $w_0$ | constant load per unit length, N/m |
| $x, y, z$ | Cartesian coordinate system, m |
| $x', y', z'$ | rotated Cartesian coordinate system, m |
| $\gamma$ | shear strain |
| $\delta$ | elongation, m |
| $\epsilon$ | normal strain |
| $\theta$ | angle representing deviation from initial right angle or angle at which force is applied, deg |
| $\mu$ | coefficient of friction |
| $\sigma$ | normal stress, Pa |
| $\tau$ | shear stress, Pa |
| $\tau_{1/2}, \tau_{2/3}, \tau_{1/3}$ | principal shear stresses in triaxial stress state, Pa |
| $\phi$ | angle of oblique plane, deg |

## SUBSCRIPTS

| | |
|---|---|
| $a$ | axial |
| avg | average |
| $b$ | biaxial stress |
| $c$ | center |
| cr | critical |
| $e$ | von Mises |
| $l$ | left |
| max | maximum |
| min | minimum |
| $n$ | normal surface |
| $r$ | roller |
| $t$ | triaxial stress; transverse |
| $x, y, z$ | Cartesian coordinates |
| $x', y', z'$ | rotated Cartesian coordinates |
| $\theta$ | angle representing deviation from initial right angle |
| $\sigma$ | normal stress |
| $\tau$ | shear stress |
| $\phi$ | angle of oblique plane |
| 1, 2, 3 | principal axes |

## 2.1   INTRODUCTION

The focus of this text is the design and analysis of machines and machine elements. Since machine elements carry **loads,** it follows that an analysis of loads is essential in machine element design. Proper selection of a machine element often is a simple matter of calculating the stresses and/or deformations expected in service and then choosing a proper size so that critical stresses or deformations are not exceeded. The first step in calculating the stress or deformation of a machine element is to accurately determine the load. Load, stress, and strain in all their forms are the focus of this chapter, and the information developed here is used throughout the text.

## 2.2   CRITICAL SECTION

To determine when a machine element will fail, the designer evaluates the stress, strain, and strength at the critical section. The **critical section,** or the location in the design where the largest internal stress is developed and failure is most likely, is often not intuitively known *a priori.*

To establish the critical section and the critical loading, the designer

1. Considers the external loads applied to a machine (e.g., an automobile)
2. Considers the external loads applied to an element within the machine (e.g., a cylindrical rolling-element bearing)
3. Locates the critical section within the machine element (e.g., the inner race)
4. Determines the loading at the critical section (e.g., contact stresses)

The first and second steps arise from system design. The third step is quite challenging and may require analysis of a number of locations or failure modes before the most critical mode is found. For example, a beam subjected to a distributed load might conceivably exceed the maximum deflection at a number of locations; thus, the deflection would need to be calculated at more than one position in the beam. The fourth step is a topic for future chapters.

In general, the critical section will often occur at locations of geometric nonuniformity, such as where a shaft changes its diameter along a fillet. Also, locations where load is applied or transferred are often critical locations. Finally, areas where the geometry is most critical are candidates for analysis. This topic will be expanded upon in Chapter 6.

---

**EXAMPLE 2.1**

**Given:** A simple crane shown in Figure 2.1(a) consists of a horizontal beam loaded vertically at one end with a load of 10 kN. At the other end the beam is pinned. The force at the pin and roller must not be larger than 30 kN to satisfy other design constraints.

**Find:** The location of the critical section and whether the load of 10 kN can be applied without damage to the crane.

**Solution:** Forces acting on the horizontal beam are shown in Figure 2.1(b). Summation of moments about the pin gives

$$1.0P = 0.25\,W_r$$
$$\therefore \quad W_r = 40\,\text{kN}$$

Summation of vertical forces gives

$$-W_p + W_r - 10\,\text{kN} = 0$$
$$\therefore \quad W_p = 30\,\text{kN}$$

(a)                                                  (b)

**Figure 2.1** A simple crane and forces acting on it. (a) Assembly drawing; (b) free-body diagram of forces acting on the beam.

The critical section is at the roller, since $W_r > W_p$. Also, since $W_r > W_{all}$, failure will occur. To avoid failure, the load at the end of the horizontal beam must be changed from 10 kN to

$$(10) \left( \frac{30}{40} \right) (10^3) N = 7.5(10^3) = 7.5 \text{ kN}$$

**Answer:**   The critical section occurs at the roller, where failure will occur unless the load applied to the end of the beam is 7.5 kN or less.

## 2.3   LOAD CLASSIFICATION AND SIGN CONVENTION

Any applied load can be classified with respect to time in the following ways:

1. **Static load.**   Load is gradually applied, and equilibrium is reached in a relatively short time. The structure experiences no dynamic effects.

2. **Sustained load.**   Load, such as the weight of a structure, is constant over a long time.

3. **Impact load.**   Load is rapidly applied. An impact load is usually attributed to an energy imparted to a system.

4. **Cyclic load.**   Load varies and even reverses itself in direction and has a characteristic period with respect to time.

The load can also be classified with respect to the area over which it is applied:

1. **Concentrated load.**   Load is applied to an area much smaller than the loaded member, as presented for nonconformal surfaces (Sec. 8.2). An example would be the contact between a caster and a support beam on a mechanical crane, where the contact area is 100 times smaller than the surface area of the caster. For these cases the applied force can be considered to act at a point on the surface.

2. **Distributed load.**   Load is spread over the entire area. An example would be the weight of a concrete bridge floor of uniform thickness.

Loads can be further classified with respect to location and method of application. Also, the coordinate direction must be determined before the sign of the loading can be established.

1. **Normal load.**   The load passes through the centroid of the resisting section. Normal loads may be tensile [Fig. 2.2(a)] or compressive [Fig. 2.2(b)]. The sign convention has tensile load being positive and compressive load negative.

2. **Shear load.**   The force $P$ is assumed to be collinear with the transverse shear force $V$. The separated bar in Figure 2.2(c) illustrates the action of positive shearing. A shear force is positive if the force direction and the normal direction are both positive or both negative. The shear force $V$ shown on the left surface of Figure 2.2(c) is in the positive $y$ direction, and the normal to the surface is in the positive $x$ direction. Thus, the shear force is positive. On the right surface of Figure 2.2(c) the shear force is also positive, since the direction of the shear force and the normal to the surface are both negative. A shear force is negative if the force direction and the normal direction have different

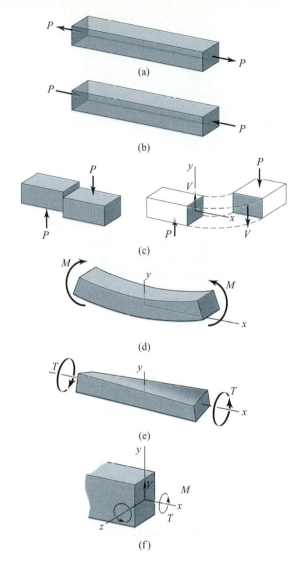

**Figure 2.2**  Load classified as to location and method of application. (a) Normal, tensile; (b) normal, compressive; (c) shear; (d) bending; (e) torsion; (f) combined.

signs. If the positive $y$ coordinate had been chosen to be downward (negative) rather than upward (positive) in Figure 2.2(c), the shear force would be negative rather than positive. Thus, to establish whether a shear force is positive or negative, the positive $x$ and $y$ coordinates must be designated.

3.  **Bending load.**   This commonly occurs when load is applied transversely to the longitudinal axis of the member. Figure 2.2(d) shows a member that is subject to equal

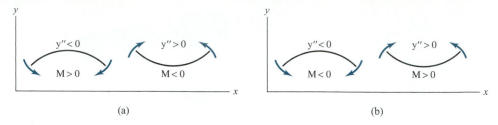

**Figure 2.3**     Sign conventions used in bending. (a) Positive moment leads to a tensile stress in the positive *y* direction; (b) positive moment acts in a positive direction on a positive face. The sign convention shown in (b) will be used in this book.

and opposite moments applied at its ends. The moment results in normal stresses in a cross section transverse to the normal axis of the member, as described in Section 4.5.

4.  **Torsion load.**     Load subjects a member to twisting motion [Fig. 2.2(e)]. The twist results in a distribution of shear stresses on the transverse cross section of the member. Positive torsion occurs in Figure 2.2(e). The right-hand rule is applicable here.

5.  **Combined load.**     Figure 2.2(f) shows a combination of two or more of the previously defined loads (e.g., shear, bending, and torsion acting on a member). Note that positive shear, bending, and torsion occur in this figure.

The sign convention for bending moments used in stress analysis should be briefly discussed. Two common sign conventions are used in engineering practice, as illustrated in Figure 2.3. The difference between these two sign conventions is in the sign of the moment applied, and each sign convention has its proponents and critics. The proponents of the sign convention shown in Figure 2.3(a) prefer that the stresses that arise in the beam follow the rule that a positive distance from the neutral axis results in a positive (tensile) stress. On the other hand, the sign convention shown in Figure 2.3(b) allows certain mnemonic methods for its memorization, such as a positive moment results in a deformed shape that holds water or has a positive second derivative. Perhaps the best reason for using the sign convention in Figure 2.3(b) is that the convention for bending moments is the same as that for applied shear forces—that a positive force or moment acting on a face with a positive, outward-pointing normal acts in a positive direction.

It should be recognized that the sign convention is arbitrary, and correct answers can be obtained for problems using either sign convention, as long as the sign convention is applied consistently within a problem. In this book, we will use the sign convention of Figure 2.3(b), but this should not be interpreted as mandatory for solution of problems.

---

**EXAMPLE 2.2**

**Given:**   The lever assembly shown in Figure 2.4(a).

**Find:**   The normal, shear, bending, and torsional loads acting at section B.

**Solution:**   Figure 2.4(b) shows the various loads acting on the lever, all in the positive direction. To the right of the figure, expressions are given for the loading at section B of the lever shown in Figure 2.4(a).

**Figure 2.4**    Lever assembly and results. (a) Lever assembly; (b) results showing (1) normal, tensile, (2) shear, (3) bending, and (4) torsion on section B of lever assembly.

---

**Given:**   A diver jumps on a trampoline.

**EXAMPLE 2.3**

**Find:**

(a)   The load type when the diver lands on the trampoline
(b)   The load type when the diver stands motionless, waiting for the signal to jump
(c)   The load type on the jumping tower just as the diver jumps
(d)   The load type on the jumping tower against the ground when no dynamic loads are acting

**Solution:**

(a)   Impact load—as the diver makes contact with the trampoline
(b)   Static load—when the diver is motionless
(c)   Cyclic load—when the trampoline swings up and down just after the dive
(d)   Sustained load—when gravity acts on the jumping tower, pressing it against the ground

---

## 2.4    SUPPORT REACTIONS

Reactions are forces developed at supports. For two-dimensional problems (i.e., bodies subjected to coplanar force systems), the types of support most commonly encountered, along with the corresponding reactions, are shown in Table 2.1. [Note that the direction of

**Table 2.1**    Four types of support with their corresponding reactions

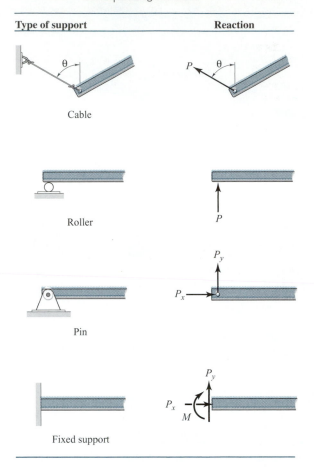

| Type of support | Reaction |
|---|---|

the forces and moments on each type of support and the reaction they exert on the attached member comply with the sign convention discussed in Sec. 2.3 and shown in Fig. 2.3(b).] One way to determine the support reaction is to imagine the attached member as being translated or rotated in a particular direction. If the support prevents translation in a given direction, a force is developed on the member in that direction. Likewise, if the support prevents rotation, a coupled moment is averted on the member. For example, a roller support prevents translation only in the contact direction, perpendicular (or normal) to the surface; thus, the roller cannot develop a moment on the member at the point of contact.

## 2.5    STATIC EQUILIBRIUM

**Static equilibrium** of a body requires both a balance of forces, to prevent the body from translating (moving) along a straight or curved path, and a balance of moments, to prevent the body from rotating. From statics it is customary to present these equations as

$$\sum P_x = 0 \qquad \sum P_y = 0 \qquad \sum P_z = 0 \tag{2.1}$$

$$\sum M_x = 0 \qquad \sum M_y = 0 \qquad \sum M_z = 0 \tag{2.2}$$

Often, in engineering practice the loading on a body can be represented as a system of coplanar forces. If this is the case, the coordinate system can be selected such that the forces lie in the $xy$ plane and the equilibrium conditions of the body can be specified by only three equations:

$$\sum P_x = 0 \qquad \sum P_y = 0 \qquad \sum M_z = 0 \tag{2.3}$$

Note that the moment $M_z$ is perpendicular to the plane that contains the forces. Successful application of the equilibrium equations requires complete specification of all the known and unknown forces acting on the body.

---

**Given:** A painter stands on a ladder that leans against the wall of a house. Assume the painter is at the midheight of the ladder. The ladder stands on a horizontal surface with a coefficient of friction of 0.3 and leans at an angle of 20° against the house, which also has a coefficient of friction of 0.3.

**EXAMPLE 2.4**

**Find:** Are the painter and ladder in static equilibrium, and what critical coefficient of friction $\mu_{cr}$ will not provide static equilibrium?

**Solution:** Figure 2.5 shows a diagram of the forces acting on the ladder due to the weight of the painter as well as the weight of the ladder. The mass of the ladder is $m_l$ and the mass of the painter is $m_p$.

If the ladder starts to slide down, the friction force will counteract this motion. Summation of horizontal and vertical forces gives

$$\sum P_x = \mu_{cr} P_1 - P_2 = 0 \tag{a}$$

$$\therefore \qquad P_2 = \mu_{cr} P_1$$

$$\sum P_y = P_1 - (m_l + m_p)g + \mu_{cr} P_2 = 0$$

Making use of Equation ($a$) gives

$$P_1 \left(1 + \mu_{cr}^2\right) = (m_l + m_p)g$$

$$\therefore \qquad P_1 = \frac{(m_l + m_p)g}{1 + \mu_{cr}^2} \tag{b}$$

Making use of Equation ($a$) gives

$$P_2 = \frac{\mu_{cr}(m_l + m_p)g}{1 + \mu_{cr}^2} \tag{c}$$

Taking the moment about point 0 and setting equal to zero give

$$P_1 l \sin 20° - P_2 l \cos 20° - (m_l + m_p)g \frac{l}{2} \sin 20° = 0 \tag{d}$$

**Figure 2.5**    Ladder in contact with the house and the ground while a painter is on the ladder.

where $l$ = ladder length, m. Substituting Equations (*b*) and (*c*) into Equation (*d*) gives

$$0 = \frac{(m_l + m_p)gl \sin 20°}{1 + \mu_{cr}^2} - \frac{\mu_{cr}(m_l + m_p)gl \cos 20°}{1 + \mu_{cr}^2} - (m_l + m_p)g\frac{l}{2}\sin 20°$$

$$0 = \frac{1}{1 + \mu_{cr}^2} - \frac{\mu_{cr}}{(\tan 20°)\left(1 + \mu_{cr}^2\right)} - \frac{1}{2}$$

$$0.5 = \frac{\tan 20° - \mu_{cr}}{(\tan 20°)\left(1 + \mu_{cr}^2\right)}$$

$$0.5 \tan 20° + 0.5\mu_{cr}^2 \tan 20° = \tan 20° - \mu_{cr}$$

$$\mu_{cr}^2 + \frac{\mu_{cr}}{0.5 \tan 20°} - 1 = 0$$

$$\mu_{cr} = 0.1763$$

Since $\mu$ is 0.3, the ladder will not move because $\mu_{cr} < \mu$, or $0.1763 < 0.3$.

**Answer:**    The painter and ladder are in static equilibrium. The critical coefficient where the ladder starts to slide is 0.1763.

## 2.6 FREE-BODY DIAGRAM

Any individual machine element, part of a machine element, or an entire machine can be represented as a free body. Static equilibrium is assumed at each level. The best way to account for the forces and moments in the equilibrium equations is to draw the free-body diagram. For the equilibrium equations to be correct, the effects of all the applied forces and moments must be represented in the free-body diagram.

A **free-body diagram** is a sketch of a machine, a machine element, or part of a machine element that shows all acting forces, such as applied loads and gravity forces, and all reactive forces. The reactive forces are supplied by the ground, walls, pins, rollers, cables, or other means. The sign of the reaction is initially guessed. If after the static equilibrium analysis the sign of the reactive force is positive, the direction initially guessed is correct; if it is negative, the direction is opposite to that initially guessed.

**Given:** The external rim brake shown in Figure 2.6(a).

**EXAMPLE 2.5**

**Find:** Draw a free-body diagram of each component.

**Solution:** Figure 2.6(b) shows each brake component and the forces acting on it. The static equilibrium of each component must be preserved, and the friction force acts opposite to the direction of motion on the drum and in the direction of motion on both shoes. The $4W$ value in Figure 2.6(b) was obtained from the moment equilibrium of the lever. Details of brakes are considered in Chapter 17, but in this chapter it is important to be able to draw the free-body diagram of each component.

**Figure 2.6**    External rim brake and forces acting on it. (a) External rim brake; (b) external rim brake with forces acting on each part. (Linear dimensions are in centimeters.)

**EXAMPLE 2.6**

**Given:**   A steel sphere [Fig. 2.7(a)] with a mass of 10 kg hangs from two wires. A spring attached to the bottom gives a downward force of 150 N.

**Find:**   The forces acting on the two wires. Also, draw a free-body diagram showing the forces acting on the sphere.

**Solution:**   Figure 2.7(b) shows the free-body diagram of the forces acting on the sphere. Summation of the vertical forces gives

$$2P \cos 60° - m_a g - 150 = 0$$

or

$$P = \frac{(10)(9.807) + 150}{2 \cos 60°} = 248.1 \text{ N}$$

**Figure 2.7**    Sphere and forces acting on it. (a) Sphere supported with wires from top and spring at bottom; (b) free-body diagram of forces acting on sphere.

## 2.7    SUPPORTED BEAMS

A **beam** is a structural member designed to support loading applied perpendicular to its longitudinal axis. In general, beams are long, straight bars having a constant cross-sectional area. Often, they are classified by how they are supported. Three major types of support are shown in Figure 2.8:

1.   A **simply supported beam** [Fig. 2.8(a)] is pinned at one end, and the roller is supported at the other.

2.   A **cantilevered beam** [Fig. 2.8(b)] is fixed at one end and free at the other.

3.   An **overhanging beam** [Fig. 2.8(c)] has one or both of its ends freely extending past the support.

Two major parameters used in evaluating beams are strength and deflection. Shear and bending are the two primary modes of beam loading. However, if the height of the beam

**Figure 2.8**    Three types of beam support.
(a) Simply supported; (b) cantilevered;
(c) overhanging.

is large relative to its width, elastic instability becomes important and the beam can twist under loading [see unstable equilibrium in Chapter 9 (Sec. 9.2.3)].

## 2.8    SHEAR AND MOMENT DIAGRAMS

Designing a beam on the basis of strength first requires finding the maximum shear and moment. One way to do this is to express the transverse shear force $V$ and the moment $M$ as functions of an arbitrary position $x$ along the beam's axis. These shear and moment functions can then be plotted as shear and moment diagrams from which the maximum values of $V$ and $M$ can be obtained.

PROCEDURE FOR DRAWING SHEAR AND MOMENT DIAGRAMS

1.  Draw a free-body diagram, and determine all the support reactions. Resolve the forces into components acting perpendicular and parallel to the beam's axis, which is assumed to be the $x$ axis.

2.  Choose a position $x$ between the origin and the length of the beam $l$, thus dividing the beam into two segments. The origin is chosen at the beam's left end to ensure that any $x$ chosen will be positive.

3.  Draw a free-body diagram of the two segments, and use the equilibrium equations to determine the transverse shear force $V$ and the moment $M$.

4.  Plot the shear and moment functions versus $x$. Note the location of the maximum moment. Generally, it is convenient to show the shear and moment diagrams directly below the free-body diagram of the beam.

Note that if $q(x)$ is the load intensity function in the $y$ direction, the transverse shear force is

$$V(x) = -\int_{-\infty}^{x} q(x)\,dx \qquad\qquad (2.4)$$

and the bending moment is

$$M(x) = -\int_{-\infty}^{x} V(x)\, dx = \int_{-\infty}^{x} \left[\int_{-\infty}^{x} q(x)\, dx\right] dx \qquad \text{(2.5)}$$

**EXAMPLE 2.7**

**Given:**   The bar shown in Figure 2.9(a).

**Find:**   Draw the shear and moment diagrams.

**Solution:**   For $0 \le x < l/2$ the free-body diagram of the bar section would be as shown in Figure 2.9(b). The unknowns $V$ and $M$ are drawn in their positive directions. Applying the equilibrium equations gives

$$\sum P_y = 0 \rightarrow V = -\frac{P}{2} \qquad \text{(a)}$$

$$\sum M_z = 0 \rightarrow M = \frac{P}{2}x \qquad \text{(b)}$$

**Figure 2.9**   Simply supported bar. (a) Midlength load and reactions; (b) free-body diagram for $0 < x < l/2$; (c) free-body diagram for $l/2 \le x < l$; (d) shear and moment diagrams.

For $l/2 \leq x < l$, the free-body diagram is shown in Figure 2.9(c). Again, $V$ and $M$ are shown in the positive direction.

$$\sum P_y = 0 \rightarrow \frac{P}{2} - P - V = 0 \qquad V = \frac{P}{2} \qquad \text{(c)}$$

$$\sum M_z = 0 \rightarrow M + P\left(x - \frac{l}{2}\right) - \frac{P}{2}x = 0$$

$$\therefore \qquad M = \frac{P}{2}(l - x) \qquad \text{(d)}$$

The shear and moment diagrams [Fig. 2.9(d)] can be obtained directly from Equations ($a$) to ($d$). As a check, $V = -dM/dx$, which is indeed valid for $0 \leq x < l$.

## 2.9   SINGULARITY FUNCTIONS

If the loading is simple, the method for obtaining shear and moment diagrams described in Section 2.8 can be used. Usually, however, this is not the situation. Differentiating or integrating across a discontinuity, such as a concentrated load or moment, presents difficulties. For more complex loading, methods such as **singularity functions** can be used. A singularity function of $x$ is written as

$$f_n(x) = \langle x - a \rangle^n \qquad \text{(2.6)}$$

where $n$ = any integer (positive or negative) including zero
  $a$ = constant distance on $x$ axis equal to value of $x$ where the discontinuity
      occurs, m

Some general rules relating to singularity functions and angular brackets are as follows:

1. If $n > 0$ and the expression inside the angular brackets is positive (that is, $x \geq a$), then $f_n(x) = (x - a)^n$. Note that the angular brackets to the right of the equals sign in Equation (2.6) are now parentheses.

2. If $n > 0$ and the expression inside the angular brackets is negative (that is, $x < a$), then $f_n(x) = 0$.

3. If $n < 0$, then

$$f_n(x) = \begin{cases} 1 & \text{for } x = a \\ 0 & \text{for } x \neq a \end{cases}$$

4. If $n = 0$, then $f_n(x) = 1$ when $x \geq a$ and $f_n(x) = 0$ when $x < a$.

5. If $n \geq 0$, the integration rule is $\int_{-\infty}^{x} \langle x - a \rangle^n \, dx = \langle x - a \rangle^{n+1}/(n + 1)$, the same as if there were parentheses instead of angular brackets.

6. If $n < 0$, the integration rule is $\int_{-\infty}^{x} \langle x - a \rangle^n \, dx = \langle x - a \rangle^{n+1}$.

7. When $n \geq 1$, then

$$\frac{d}{dx}\langle x - a \rangle^n = n\langle x - a \rangle^{n-1}$$

The advantage of using a singularity function is that it permits writing an analytical expression directly for the transverse shear and moment over a range of discontinuities.

Table 2.2 shows six singularity and load intensity functions along with corresponding graphs and expressions. Note in particular the inverse ramp example. A unit step is constructed beginning at $x = a$, and the ramp beginning at $x = a$ is subtracted. To have the negative ramp discontinued at $x = a + b$, a ramp beginning at this point is constructed.

**Table 2.2**     Singularity and load intensity functions with corresponding graphs and expressions

| Singularity | Graph of $q(x)$ | Expression for $q(x)$ |
|---|---|---|
| Concentrated moment | | $q(x) = M \langle x - a \rangle^{-2}$ |
| Concentrated force | | $q(x) = P \langle x - a \rangle^{-1}$ |
| Unit step | | $q(x) = w_0 \langle x - a \rangle^0$ |
| Ramp | | $q(x) = \dfrac{w_0}{b} \langle x - a \rangle^1$ |
| Inverse ramp | | $q(x) = w_0 \langle x - a \rangle^0 - \dfrac{w_0}{b} \langle x - a \rangle^1$ |
| Parabolic shape | | $q(x) = \langle x - a \rangle^2$ |

## PROCEDURE FOR DRAWING THE SHEAR AND MOMENT DIAGRAMS BY MAKING USE OF SINGULARITY FUNCTIONS

1. Draw a free-body diagram with all the singularities acting on the beam, and determine all support reactions. Resolve the forces into components acting perpendicular and parallel to the beam's axis.

2. Referring to Table 2.2, write an expression for the load intensity function $q(x)$ that describes all the singularities acting on the beam.

3. Integrate the negative load intensity function over the beam length to get the shear force. Integrate the negative shear force over the beam length to get the moment (see Sec. 5.2).

4. Draw shear and moment diagrams from the expressions developed.

---

**Given:** The same conditions as in Example 2.7.

**EXAMPLE 2.8**

**Find:** Draw the shear and moment diagrams by using a singularity function for a concentrated force located midway on the beam.

**Solution:** The load intensity function for a simply supported beam [Fig. 2.10(a)] is

$$q(x) = \frac{P}{2}\langle x \rangle^{-1} - P\left\langle x - \frac{l}{2} \right\rangle^{-1} + \frac{P}{2}\langle x - l \rangle^{-1}$$

The shear expression is

$$V(x) = -\int_{-\infty}^{x} \left( \frac{P}{2}\langle x \rangle^{-1} - P\left\langle x - \frac{l}{2} \right\rangle^{-1} + \frac{P}{2}\langle x - l \rangle^{-1} \right) dx$$

$$= -\frac{P}{2}\langle x \rangle^{0} + P\left\langle x - \frac{l}{2} \right\rangle^{0} - \frac{P}{2}\langle x - l \rangle^{0}$$

$$= V_1 + V_2 + V_3$$

Figure 2.10(a) shows the resulting shear diagrams. The diagram at the top shows individual shear, and the diagram below shows the composite of these shear components. The moment expression is, from Eq. (2.5)

$$M(x) = -\int_{-\infty}^{x} \left( -\frac{P}{2}\langle x \rangle^{0} + P\left\langle x - \frac{l}{2} \right\rangle^{0} - \frac{P}{2}\langle x - l \rangle^{0} \right) dx$$

$$= \frac{P}{2}\langle x \rangle^{1} - P\left\langle x - \frac{l}{2} \right\rangle^{1} + \frac{P}{2}\langle x - l \rangle^{1}$$

$$= M_1 + M_2 + M_3$$

Figure 2.10(b) shows the moment diagrams. The diagram at the top shows individual moments; the diagram at the bottom is the composite moment diagram. The slope of $M_2$ is twice that of $M_1$ and $M_3$, which are equal. The resulting shear and moment diagrams are the same as those found in Example 2.7.

**Figure 2.10**    (a) Shear and (b) moment diagrams for Example 2.8.

**EXAMPLE 2.9**

**Given:**  A simply supported beam shown in Figure 2.11(a) when

$$P_1 = 8 \text{ kN} \qquad P_2 = 5 \text{ kN} \qquad w_0 = 4 \text{ kN/m} \qquad l = 12 \text{ m}$$

**Find:**  The shear and moment expressions as well as their corresponding diagrams, using singularity functions.

**Solution:**  The first task is to solve for the reactions at $x = 0$ and $x = l$. The force representation is shown in Figure 2.11(b). Note that $w_0$ is defined as the load per unit length for the central part of the beam. From force equilibrium

$$0 = R_1 + P_1 + P_2 + R_2 - \frac{w_0 l}{2} - \frac{w_0 l}{8} \tag{a}$$

In Figure 2.11(b) and Equation (a) it can be shown that the unit step $w_0$ over a length of $l/2$ produces a resultant force of $w_0 l/2$ and that the positive ramp over the length of $l/4$ can be represented by a resultant vector of

$$w_0 \left(\frac{l}{4}\right)\left(\frac{1}{2}\right) \qquad \text{or} \qquad \frac{w_0 l}{8}$$

Also, note that the resultant vector acts at

$$x = \left(\frac{2}{3}\right)\left(\frac{l}{4}\right) = \frac{l}{6}$$

$$\therefore \qquad R_1 + R_2 = -P_1 - P_2 + \frac{5 w_0 l}{8} \tag{b}$$

**Figure 2.11** Simply supported beam. (a) Forces acting on beam when $P_1 = 8$ kN, $P_2 = 5$ kN; $w_0 = 4$ kN/m; $l = 12$ m; (b) free-body diagram showing resulting forces; (c) shear and (d) moment diagrams of Example 2.9.

Making use of moment equilibrium and the moment of a triangular section load gives

$$\frac{(P_1 + 2P_2)l}{4} - \frac{w_0 l^2}{4} - \frac{w_0 l}{8}\left(\frac{l}{6}\right) + R_2 l = 0$$

$$\therefore \qquad R_2 = \frac{13 w_0 l}{48} - \frac{P_1 + 2P_2}{4} \qquad\qquad (c)$$

Substituting Equation (c) into Equation (b) gives

$$R_1 = -\frac{3P_1}{4} - \frac{P_2}{2} + \frac{17 w_0 l}{48} \qquad\qquad (d)$$

Substituting the given values for $P_1$, $P_2$, $w_0$, and $l$ gives

$$R_1 = 8.5 \text{ kN} \qquad \text{and} \qquad R_2 = 8.5 \text{ kN} \qquad\qquad (e)$$

The load intensity function can be written as

$$q(x) = R_1 \langle x \rangle^{-1} - \frac{w_0}{l/4} \langle x \rangle^1 + \frac{w_0}{l/4} \left\langle x - \frac{l}{4} \right\rangle^1 + P_1 \left\langle x - \frac{l}{4} \right\rangle^{-1}$$

$$+ P_2 \left\langle x - \frac{l}{2} \right\rangle^{-1} + w_0 \left\langle x - \frac{3l}{4} \right\rangle^0 + R_2 \langle x - l \rangle^{-1} \tag{f}$$

Note that a unit step beginning at $l/4$ is created by initiating a ramp at $x = 0$ acting in the negative direction and summing it with another ramp starting at $x = l/4$ acting in the positive direction while assuming that the slopes of the ramps are the same. The second and third terms on the right side of Equation ($f$) produce this effect. The sixth term on the right side of the equation turns off the unit step. Integrating the load intensity function gives the shear force as

$$V(x) = -R_1 \langle x \rangle^0 + \frac{2w_0}{l} \langle x \rangle^2 - \frac{2w_0}{l} \left\langle x - \frac{l}{4} \right\rangle^2 - P_1 \left\langle x - \frac{l}{4} \right\rangle^0$$

$$- P_2 \left\langle x - \frac{l}{2} \right\rangle^0 - w_0 \left\langle x - \frac{3l}{4} \right\rangle^1 - R_2 \langle x - l \rangle^0$$

Integrating the negative shear force gives the moment, and substituting the values for $w_0$ and $l$ gives

$$M(x) = 8.5\langle x \rangle^1 - \frac{2}{9}\langle x \rangle^3 + \frac{2}{9}\langle x - 3 \rangle^3 + 8\langle x - 3 \rangle^1 + 5\langle x - 6 \rangle^1 + 2\langle x - 9 \rangle^2 + 8.5\langle x - 12 \rangle^1$$

The shear and moment diagrams are shown in Figure 2.11(c) and (d), respectively.

## 2.10   STRESS

One of the fundamental problems in engineering is the determination of the effect of a loading environment on a part. This determination is an essential part of the design process; one cannot choose a dimension or a material without first understanding the intensity of force inside the component being analyzed. **Stress** is the term used to define the intensity and direction of the internal forces acting at a given point on a particular plane. Strength, on the other hand, is a property of a material or a part and will be covered in later chapters.

For normal loading on a load-carrying member in which the external load is uniformly distributed over a cross-sectional area of a part, the magnitude of the average normal stress can be calculated from

$$\sigma_{\text{avg}} = \frac{\text{force}}{\text{cross-sectional area}} = \frac{P}{A} \tag{2.7}$$

Thus, the unit of stress is force per unit area, or newtons per square meter. Consider a small area $\Delta A$ on the cross section, and let $\Delta P$ represent the internal forces transmitted by this small area. The average intensity of the internal forces transmitted by the area $\Delta A$ is obtained by dividing $\Delta P$ by $\Delta A$. If the internal forces transmitted across the section are assumed to be continuously distributed, the area $\Delta A$ can be made increasingly smaller and will approach a point on the surface in the limit. The corresponding force $\Delta P$ will also become increasingly smaller. The stress at the point on the cross section to which $\Delta A$ converges can be defined as

$$\sigma = \lim_{\Delta A \to 0} \frac{\Delta P}{\Delta A} = \frac{dP}{dA} \qquad (2.8)$$

The stress at a point acting on a specific plane is a vector and thus has a magnitude and a direction. Its direction is the limiting direction $\Delta P$ as area $\Delta A$ approaches zero. Similarly, the shear stress can be defined in a specific plane. Thus, a stress must be defined with respect to a direction.

**Given:** As shown in Figure 2.12(a), a 3-m-long bar is supported at the left end (B) by a 6-mm-diameter steel wire and at the right end (C) by a 10-mm-diameter steel cylinder. The bar carries a mass $m_{a1} = 200$ kg, and the mass of the bar itself is $m_{a2} = 50$ kg.

**EXAMPLE 2.10**

**Find:** Determine the stresses in the wire and in the cylinder.

**Solution:**

$$A_B = \frac{\pi}{4}(6)^2 = 28.27 \text{ mm}^2$$

$$A_C = \frac{\pi}{4}(10)^2 = 78.54 \text{ mm}^2$$

$$m_{a1} = 200 \text{ kg}$$

$$m_{a2} = 50 \text{ kg}$$

Figure 2.12(b) shows a free-body diagram of the forces acting on the bar. From force equilibrium

$$R_B - m_{a1}g - m_{a2}g + R_C = 0$$

$$\therefore \quad R_B + R_C = g(m_{a1} + m_{a2}) = 9.81(200 + 50) = 2452.5 \text{ N}$$

$$\therefore \quad R_B = 2453 - R_C$$

(a)                    (b)

**Figure 2.12** Figures used in Example 2.10. (a) Load assembly drawing; (b) free-body diagram.

Moment equilibrium about point C gives

$$3R_B = 2(200)(9.81) + 1.5(50)(9.81) = 4660 \text{ Nm}$$

$$R_B = 1553 \text{ N}$$

$$R_C = 2453 - 1553 = 900 \text{ N}$$

The stresses at points B and C are

$$\sigma_B = \frac{R_B}{A_B} = \frac{1553}{28.27} = 54.93 \text{ N/mm}^2 = 54.93 \text{ MPa}$$

$$\sigma_C = -\frac{R_C}{A_C} = -\frac{900}{78.54} = -11.46 \text{ N/mm}^2 = -11.46 \text{ MPa}$$

## 2.11    STRESS ELEMENT

Figure 2.13 shows a stress element with the origin of stress placed inside the element. Across each of the mutually perpendicular surfaces there are three stresses, yielding a total of nine stress components. Of the three stresses acting on a given surface, the normal stress is denoted by $\sigma$ and the shear stress by $\tau$. A normal stress will receive a subscript indicating the direction in which the stress acts (for example, $\sigma_x$). A shear stress requires two subscripts, the first to indicate the plane of the stress and the second to indicate its direction (for example, $\tau_{xy}$). The **sign convention for normal stress** distinguishes positive for tension and negative for compression. A positive shear stress points in the positive direction of the coordinate axis denoted by the second subscript if it acts on a surface with an outward normal in the positive direction. The **sign convention for shear stress** is directly associated with the coordinate directions. If both the normal from the surface and the shear are in the positive

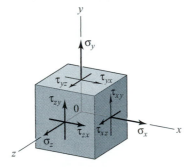

**Figure 2.13**    Stress element showing general state of three-dimensional stress with origin placed in center of element.

direction or if both are in the negative direction, then the shear stress has a positive sign. Any other combinations of the normal and the direction of shear will produce a negative shear stress. The surface stresses of an element have the following relationships:

1. The normal and shear stress components acting on opposite sides of an element must be equal in magnitude but opposite in direction.

2. Moment equilibrium requires that the shear stresses be symmetric, implying that the subscripts can be reversed in order, or

$$\tau_{xy} = \tau_{yx} \qquad \tau_{xz} = \tau_{zx} \qquad \tau_{yz} = \tau_{zy} \qquad (2.9)$$

thus reducing the nine different stresses acting on the element to six: three normal stresses $\sigma_x$, $\sigma_y$, $\sigma_z$ and three shear stresses $\tau_{xy}$, $\tau_{yz}$, $\tau_{xz}$.

The general laws of stress transformation, given in Appendix B (Sec. B.1), enable the determination of stresses acting on any new orthogonal coordinate system.

**EXAMPLE 2.11**

**Given:** The stress element shown in Figure 2.13 is put into a pressure vessel pressurized to 10 MPa. The shear stress acting on the bottom surface is directed in the positive $x$ direction.

**Find:** Are the stresses positive or negative?

**Solution:** The definition of a positive normal stress is a tensile stress, and a positive shear stress is directed in the positive coordinate direction when the normal to the surface is directed in the positive coordinate direction. The normal stress here is thus

$$\sigma = -10 \text{ MPa}$$

A shear stress acting on a surface with the normal in the negative coordinate direction is positive when the stress is directed in the negative coordinate direction. The shear stress acts on a surface with the normal in the negative $y$ direction, but the stress is directed in the positive $x$ direction. Thus, the shear is negative.

## 2.12 STRESS TENSOR

In engineering courses it is common to encounter scalar quantities, those that have numerical value. Vectors, such as force, have a magnitude as well as a direction. Stress requires six quantities for its definition; thus, stress is a tensor. From the stress element in Figure 2.13 and Equation (2.9), the stress tensor is

$$\mathbf{S} = \begin{pmatrix} \sigma_x & \tau_{xy} & \tau_{xz} \\ \tau_{xy} & \sigma_y & \tau_{yz} \\ \tau_{xz} & \tau_{yz} & \sigma_z \end{pmatrix} \qquad (2.10)$$

which is a symmetric tensor. A property of a **symmetric tensor** is that there exists an orthogonal set of axes 1, 2, and 3 (called *principal axes*) with respect to which the tensor elements are all zero except for those in the principal diagonal; thus,

$$\mathbf{S'} = \begin{pmatrix} \sigma_1 & 0 & 0 \\ 0 & \sigma_2 & 0 \\ 0 & 0 & \sigma_3 \end{pmatrix} \tag{2.11}$$

where $\sigma_1$, $\sigma_2$, and $\sigma_3$ = principal stresses ($\sigma_1 \geq \sigma_2 \geq \sigma_3$). Note that no principal shear stresses occur in Equation (2.11).

The combination of the applied normal and shear stresses that produces the maximum normal stress is called the *maximum principal stress* $\sigma_1$. The combination of the applied stresses that produces the minimum normal stress is called the *minimum principal stress* $\sigma_3$. Knowing the maximum and minimum **principal normal stresses** is important in engineering design. If the design satisfies these limits, it will satisfy any other stress.

## 2.13    PLANE STRESS

Many cases of stress analysis can be simplified to the case of plane stress, where one surface is comparatively free of stress. Another common practice is to instrument strain gages onto structural members. Since strain gages are easily placed onto free surfaces, they are placed in locations of plane stress. Although other locations in the geometry may have more complicated loadings, such measurements are often used to confirm loading environments or to ensure that stresses or strains are not excessive. Most stress analyses of machine elements involve one surface that is comparatively free of stress. The third direction can thus be neglected, and all stresses on the stress element act on two pairs of faces rather than three, as shown in Figure 2.14. This two-dimensional stress state is sometimes called **biaxial or plane stress.**

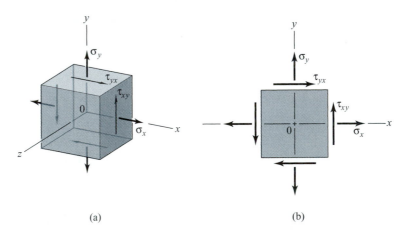

(a)                                                    (b)

**Figure 2.14**    Stress element showing two-dimensional state of stress.
(a) Three-dimensional view; (b) plane view.

**Figure 2.15**  Illustration of equivalent stress states. (a) Stress element oriented in the direction of applied stress; (b) stress element oriented in different (arbitrary) direction.

In comparing the two views of the plane stress element shown in Figure 2.14, note that all stresses shown in Figure 2.14(b) act on planes perpendicular to the paper, with the page being designated as either the $xy$ plane or the $z$ plane. The stresses shown in Figure 2.14 all have positive signs in compliance with the conventions presented in Section 2.11.

The magnitude of stress depends greatly on the orientation of the coordinate system. For example, consider the stress element shown in Figure 2.15(a). When a uniform stress is applied to the sheet, the stress state is clearly $\sigma_x = \sigma_0$, $\sigma_y = 0$, and $\tau_{xy} = 0$. However, if the original orientation of the element were as shown in Figure 2.15(b), this would no longer be the case, and all stress components in the plane would be nonzero. A profound question can be raised at this point: How does the material know the difference between these stress states? The answer is that there is no difference between the stresses of Figure 2.15, so that these stress states are equivalent. Obviously, it is of great importance to be able to transform stresses from one orientation to another, and the resultant stress transformation equations will be of great use throughout the remainder of the text.

If instead of the stresses acting as shown in Figure 2.14(b), they act in an oblique plane at angle $\phi$ as shown in Figure 2.16, then the stresses $\sigma_x$, $\sigma_y$, and $\tau_{xy}$ must be found in terms of the stresses on an inclined surface whose normal stress makes an angle $\phi$ with the $x$ axis, as shown in Figure 2.16. Angle $\phi$ is arbitrarily chosen, and the object of the analysis that follows is to establish what values $\sigma$ and $\tau$ have on the inclined surface.

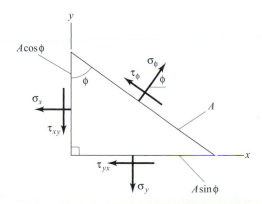

**Figure 2.16**  Stresses in an oblique plane at an angle $\phi$.

Note from Figure 2.16 that if the area of the inclined surface is $A$ (length of the surface times the thickness into the page), the area of the horizontal side of the triangular element will be $A \sin \phi$, and the area of the vertical side, $A \cos \phi$. From force equilibrium

$$\sigma_\phi A = \tau_{xy}(\sin \phi)(A \cos \phi) + \tau_{yx}(\cos \phi)(A \sin \phi) + \sigma_x(\cos \phi)(A \cos \phi)$$
$$+ \sigma_y(\sin \phi)(A \sin \phi)$$

This reduces to

$$\sigma_\phi = 2\tau_{xy} \sin \phi \cos \phi + \sigma_x \cos^2 \phi + \sigma_y \sin^2 \phi \tag{2.12}$$

By using trigonometric identities for the double angle, Equation (2.12) can be written as

$$\sigma_\phi = \frac{\sigma_x + \sigma_y}{2} + \frac{\sigma_x - \sigma_y}{2} \cos 2\phi + \tau_{xy} \sin 2\phi \tag{2.13}$$

Similarly, from force equilibrium the shear stress in the oblique plane can be expressed as

$$\tau_\phi = \tau_{xy} \cos 2\phi - \frac{\sigma_x - \sigma_y}{2} \sin 2\phi \tag{2.14}$$

Equations (2.13) and (2.14) have maximum and minimum values that are of particular interest in stress analysis. The angle $\phi_\sigma$, which gives the extreme value of $\sigma_\phi$, can be determined by differentiating $\sigma_\phi$ with respect to $\phi$ and setting the result equal to zero, giving

$$\frac{d\sigma_\phi}{d\phi} = -(\sigma_x - \sigma_y) \sin 2\phi_\sigma + 2\tau_{xy} \cos 2\phi_\sigma = 0$$

$$\tan 2\phi_\sigma = \frac{2\tau_{xy}}{\sigma_x - \sigma_y} \tag{2.15}$$

where $\phi_\sigma$ = angle $\phi$ where normal stress is extreme. Equation (2.15) has two roots, since the tangents of angles in diametrically opposite quadrants are the same. The roots are $180°$ apart; and for the double-angle nature of the left side of Equation (2.15), this suggests roots of $\phi_\sigma$ being $90°$ apart. One of these roots corresponds to the maximum value of normal stress; the other to the minimum value.

Substituting Equation (2.15) into Equations (2.13) and (2.14) gives the following, after some algebraic manipulation:

$$\sigma_1, \sigma_2 = \frac{\sigma_x + \sigma_y}{2} \pm \sqrt{\tau_{xy}^2 + \frac{(\sigma_x - \sigma_y)^2}{4}} \tag{2.16}$$

$$\tau_{\phi_\sigma} = 0 \tag{2.17}$$

At this stress element orientation, where the normal stresses are extreme, the shear stress is zero. The axes that define this orientation are called the principal axes, and the normal stresses from Equation (2.16) are called the **principal normal stresses.** Principal stresses are given numerical subscripts to differentiate them from stresses at any other orientation. A common convention is to order the principal stresses according to $\sigma_1 \geq \sigma_2 \geq \sigma_3$. In plane stress one of the principal stresses is always zero.

Another orientation of interest is the one where the shear stress takes an extreme value. Differentiating Equation (2.14) with respect to $\phi$ and solving for $\tau$ give the orientation $\phi_\tau$, where the shear reaches an extreme value. The stresses that result at this orientation are

$$\tau_{max}, \tau_{min} = \tau_1, \tau_2 = \pm\sqrt{\left(\frac{\sigma_x - \sigma_y}{2}\right)^2 + \tau_{xy}^2} \tag{2.18}$$

$$\sigma_{\phi\tau} = \frac{\sigma_x + \sigma_y}{2}$$

The shear stresses from Equation (2.18) are called **principal shear stresses.** Thus, on the stress element oriented to achieve a maximum shear stress, the normal stresses on the two faces are equal. Also, it can be shown that

$$|\phi_\tau - \phi_\sigma| = \frac{\pi}{4} \tag{2.19}$$

In summary, for a plane stress situation where $\sigma_x$, $\sigma_y$, and $\tau_{xy}$ are known,

1.  The normal and shear stresses $\sigma_\phi$ and $\tau_\phi$ for any oblique plane at angle $\phi$ can be determined from Equations (2.13) and (2.14).

2.  The principal normal and shear stresses $\sigma_1$, $\sigma_2$, $\tau_1$, and $\tau_2$ can be determined from Equations (2.16) and (2.18).

If the principal normal stresses $\sigma_1$ and $\sigma_2$ are known, the normal and shear stresses at any oblique plane at angle $\phi$ can be determined from the following equations:

$$\sigma_\phi = \frac{\sigma_1 + \sigma_2}{2} + \frac{\sigma_1 - \sigma_2}{2}\cos 2\phi \tag{2.20}$$

$$\tau_\phi = \frac{\sigma_1 - \sigma_2}{2}\sin 2\phi \tag{2.21}$$

In Equation (2.21) a second subscript is not present and is not needed because $\tau_\phi$ represents a shear stress acting on any oblique plane at angle $\phi$, as shown in Figure 2.16.

## 2.14 MOHR'S CIRCLE

**Mohr's circle** for a biaxial state of stress at a point was first constructed by a German engineer, Otto Mohr [Mohr (1914)], who noted that Equations (2.13) and (2.14) define a circle in a $\sigma\tau$ plane. This circle is used extensively as a convenient method of graphically visualizing the state of stress acting in different planes passing through a given point. The approach used in this text is first to apply Mohr's circle to a two-dimensional stress state and then in Section 2.15 to apply it to a three-dimensional stress state. Indeed, Mohr's circle is most useful for stress visualization in plane stress situations.

The main value of Mohr's circle is that it makes visual the realities of the stress state in a specific situation. Figure 2.17 shows a typical Mohr's circle diagram. Note that the circle is symmetric about the normal stress axis (the abscissa).

A number of points should be made regarding the Mohr's circle diagram:

1.  Normal stresses are plotted along the abscissa ($x$ axis), and shear stresses are plotted as the ordinate ($y$ axis).

2.  The circle defines all stress states that are equivalent.

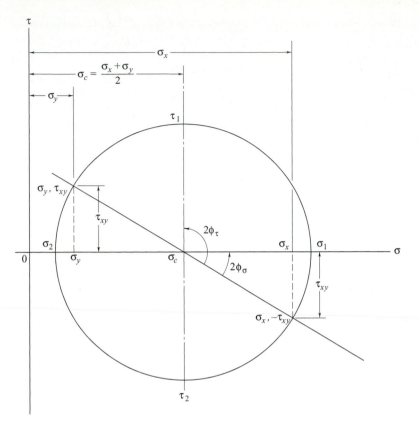

**Figure 2.17**    Mohr's circle diagram of Equations (2.13) and (2.14).

3. The biaxial stress state for any orientation can be scaled directly from the circle.

4. The principal normal stresses (i.e., the extreme values of normal stress) are at the locations where the circle intercepts the $x$ axis.

5. The maximum shear stress is equal to the radius of the circle.

6. A rotation from a reference stress state in the real plane of $\phi$ corresponds to a rotation of $2\phi$ from the reference points in Mohr's circle plane.

> **STEPS IN CONSTRUCTING AND USING MOHR'S CIRCLE IN TWO DIMENSIONS**
>
> 1. Calculate the plane stress state for any $xy$ coordinate system so that $\sigma_x$, $\sigma_y$, and $\tau_{xy}$ are known.
>
> 2. The center of Mohr's circle can be placed at
>
> $$\left(\frac{\sigma_x + \sigma_y}{2}, 0\right) \tag{2.22}$$

3. Two points diametrically opposite to each other on the circle correspond to the points $(\sigma_x, -\tau_{xy})$ and $(\sigma_y, \tau_{xy})$. Using the center and either point allows one to draw the circle.

4. The radius of the circle can be calculated from stress transformation equations or through trigonometry by using the center and one point on the circle. For example, the radius is the distance between point $(\sigma_x, -\tau_{xy})$ and the center, which directly leads to

$$r = \sqrt{\left(\frac{\sigma_x - \sigma_y}{2}\right)^2 + \tau_{xy}^2} \qquad (2.23)$$

5. The principal stresses have the values $\sigma_{1,2} = $ center $\pm$ radius.

6. The maximum shear stress is equal to the radius.

7. The principal axes can be found by calculating the angle between the $x$ axis in Mohr's circle plane and the point $(\sigma_x, -\tau_{xy})$. The principal axes in the real plane are rotated one-half this angle in the same direction relative to the $x$ axis in the real plane.

8. The stresses in an orientation rotated $\phi$ from the $x$ axis in the real plane can be read by traversing an arc of $2\phi$ in the same direction on Mohr's circle from the reference points $(\sigma_x, -\tau_{xy})$ and $(\sigma_y, \tau_{xy})$. The new points on the circle correspond to the new stresses $(\sigma_{x'}, -\tau_{x'y'})$ and $(\sigma_{y'}, \tau_{x'y'})$, respectively.

**EXAMPLE 2.12**

**Given:** The plane stresses $\sigma_x = 9$ ksi, $\sigma_y = 19$ ksi, and $\tau_{xy} = 8$ ksi.

**Find:** Draw Mohr's circle as well as the principal normal and shear stresses on the $xy$ axis. What is the stress state when the axes are rotated 15° counterclockwise?

**Solution:** We will demonstrate the eight-step approach given above, with the first step already done in the problem statement. Step 2 tells us to calculate the center of the circle and place it at $(\sigma_c, 0)$, where

$$\sigma_c = \frac{\sigma_x + \sigma_y}{2} = \frac{(9 + 19)10^3}{2}\,\text{psi} = 14\,\text{ksi}$$

Step 3 tells us to use either point $(\sigma_x, -\tau_{xy})$ or $(\sigma_y, \tau_{xy})$ to draw the circle. We use the point $(\sigma_x, -\tau_{xy}) = (9\,\text{ksi}, -8\,\text{ksi})$ to draw the circle as shown. From step 4 and from the triangle defined by the $x$ axis and the point $(\sigma_x, -\tau_{xy})$ we can calculate the radius as

$$r = \sqrt{(9 - 14)^2 + (-8)^2} = 9.43\,\text{ksi}$$

From step 5 the principal stresses have the values $\sigma_{1,2} = 14 \pm 9.43$, or $\sigma_1 = 23.43$ ksi and $\sigma_2 = 4.57$ ksi. Step 6 tells us that the maximum shear stress equals the radius, or $\tau_{\max} = 9.43$ ksi. We can calculate the principal stress orientation, if desired, from trigonometry. In Mohr's circle plane [Fig. 2.18(a)] the point $(\sigma_x, -\tau_{xy})$ makes an angle of $2\phi = \tan^{-1}(8/5) = 58°$ with the $x$ axis. To reach the point on the $x$ axis, we need to sweep an arc of this angle in the clockwise direction on Mohr's circle. Thus, the principal plane is $\phi = 29°$ clockwise from the $x$ axis.

Figure 2.18(b) shows an element of the principal normal stresses as well as the appropriate value of $\phi_\sigma$. Figure 2.18(c) shows an element of the principal shear stresses as well as the appropriate value of $\phi_\tau$. The stress at the center of Mohr's circle diagram is also represented in Figure 2.18(c) along with the principal shear stresses. Also, the stress element is always oriented from the $x$ axis to the principal shear stress on Mohr's circle diagram.

Finally, the stresses at an angle of $15°$ can be obtained by rotating $30°$ on Mohr's circle, leading to

$$\sigma_{x'} = 14\ \text{ksi} - (9.43\ \text{ksi})(\cos 28°) = 5.67\ \text{ksi}$$

$$\sigma_{y'} = 14\ \text{ksi} + (9.43\ \text{ksi})(\cos 28°) = 22.32\ \text{ksi}$$

$$\tau_{x'y'} = (9.43\ \text{ksi})(\sin 28°) = 4.43\ \text{ksi}$$

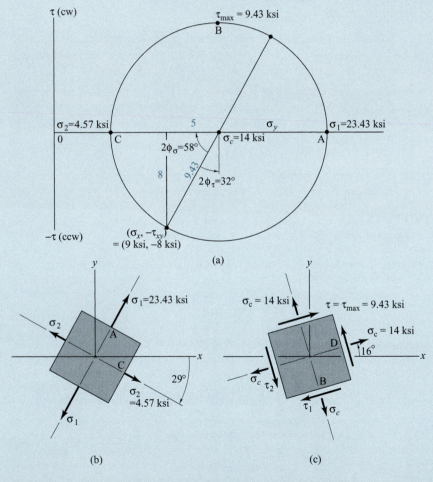

**Figure 2.18**    Results from Example 2.12. (a) Mohr's circle diagram; (b) stress element for principal normal stress shown in $xy$ coordinates; (c) stress element for principal shear stresses shown in $xy$ coordinates.

## 2.15    THREE-DIMENSIONAL STRESSES

Considering the general situation shown in Figure 2.13, the stress element has six faces, implying that there are three principal directions and three principal stresses $\sigma_1$, $\sigma_2$, and $\sigma_3$. Six stress components ($\sigma_x$, $\sigma_y$, $\sigma_z$, $\tau_{xy}$, $\tau_{xz}$, and $\tau_{yz}$) are required to specify a general state of stress in three dimensions, in contrast to the three stress components ($\sigma_x$, $\sigma_y$, and $\tau_{xy}$) that were used for two-dimensional (plane or biaxial) stress. Determining the principal stresses for a three-dimensional situation is more difficult. The process involves finding the three roots to the cubic equation

$$\sigma^3 - (\sigma_x + \sigma_y + \sigma_z)\sigma^2 + \left(\sigma_x\sigma_y + \sigma_x\sigma_z + \sigma_y\sigma_z - \tau_{xy}^2 - \tau_{yz}^2 - \tau_{zx}^2\right)\sigma$$
$$- \left(\sigma_x\sigma_y\sigma_z + 2\tau_{xy}\tau_{yz}\tau_{zx} - \sigma_x\tau_{yz}^2 - \sigma_y\tau_{zx}^2 - \sigma_z\tau_{xy}^2\right) = 0 \quad (2.24)$$

In most design situations many of the stress components are zero, making the full evaluation of this equation unnecessary.

A three-dimensional stress state is referred to as **triaxial stress.** Mohr's circle can be generated for triaxial stress states, but this is often unnecessary. In most circumstances it is not necessary to know the orientations of the principal stresses; it is sufficient to know their values. Thus, Equation (2.24) is usually all that is needed. Figure 2.19 shows Mohr's circle for a triaxial stress field. It consists of three circles, two externally tangent and inscribed within the third circle. The principal shear stresses shown in Figure 2.19 are determined from

$$\tau_{1/2} = \frac{\sigma_1 - \sigma_2}{2} \qquad \tau_{2/3} = \frac{\sigma_2 - \sigma_3}{2} \qquad \tau_{1/3} = \frac{\sigma_1 - \sigma_3}{2} \qquad (2.25)$$

The principal normal stresses must be ordered as mentioned earlier. Note that from Equation (2.25), the maximum principal shear stress is $\tau_{1/3}$.

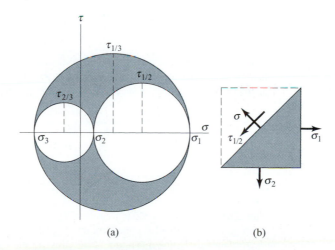

(a)                                        (b)

**Figure 2.19**    Mohr's circle for triaxial stress state. (a) Mohr's circle representation; (b) principal stresses on two planes.

**EXAMPLE 2.13**

**Given:** Assume that the principal normal stresses obtained in Example 2.12 are the same for triaxial consideration with $\sigma_3 = 0$.

$$\therefore \qquad \sigma_1 = 23.43 \text{ ksi} \qquad \sigma_2 = 4.57 \text{ ksi} \qquad \sigma_3 = 0$$

**Find:**

(a) Determine the principal shear stresses for a triaxial stress state and draw the appropriate Mohr's circle diagram.

(b) If the shear stress $\tau_{xy}$ is changed from 8 to 16 ksi, show how the Mohr's circles for the biaxial and triaxial stress states change.

**Solution:**

(a) From Equation (2.25) the principal shear stresses in a triaxial stress state are

$$\tau_{1/2} = \frac{\sigma_1 - \sigma_2}{2} = \frac{(23.43 - 4.57)10^3}{2} \text{psi} = 9.43 \text{ ksi}$$

$$\tau_{2/3} = \frac{\sigma_2 - \sigma_3}{2} = \frac{(4.57 - 0)10^3}{2} \text{psi} = 2.285 \text{ ksi}$$

$$\tau_{1/3} = \frac{\sigma_1 - \sigma_3}{2} = \frac{(23.43 - 0)10^3}{2} \text{psi} = 11.715 \text{ ksi}$$

Figure 2.20(a) shows the appropriate Mohr's circle diagram for the triaxial stress state.

(b) If the shear stress in Example 2.12 is doubled ($\tau_{xy} = 16$ ksi instead of 8 ksi), Equation (2.23) gives

$$\tau_1, \tau_2 = \pm \sqrt{\tau_{xy}^2 + \left(\frac{\sigma_x - \sigma_y}{2}\right)^2} = \pm 10^3 \sqrt{16^2 + \left(\frac{9 - 19}{2}\right)^2} \text{psi} = \pm 16.76 \text{ ksi}$$

The principal normal stresses for the biaxial stress state are

$$\sigma_1 = \sigma_c + \tau_1 = 30.76 \text{ ksi}$$

$$\sigma_2 = \sigma_c - \tau_2 = -2.76 \text{ ksi}$$

Figure 2.20(b) shows the resultant Mohr's circle diagram for the biaxial stress state. In a triaxial stress state that is ordered $\sigma_1 = 30.76$ ksi, $\sigma_2 = 0$, and $\sigma_3 = -2.76$ ksi, from Equation (2.25) the principal shear stresses can be written as

$$\tau_{1/2} = \frac{\sigma_1 - \sigma_2}{2} = \frac{(30.76 - 0)10^3}{2} \text{psi} = 15.38 \text{ ksi}$$

$$\tau_{2/3} = \frac{\sigma_2 - \sigma_3}{2} = \frac{(0 + 2.76)10^3}{2} \text{psi} = 1.38 \text{ ksi}$$

$$\tau_{1/3} = \frac{\sigma_1 - \sigma_3}{2} = \frac{(30.76 + 2.76)10^3}{2} \text{psi} = 16.76 \text{ ksi}$$

Figure 2.20(c) shows the Mohr's circle diagram for the second triaxial stress state. From Figure 2.20(b) and (c), the maximum shear stress in the biaxial stress state $\tau_1$ is equivalent to $\tau_{1/3}$, the maximum shear stress in the triaxial stress state. However, comparing Figures 2.18(a) and 2.20(a) shows that the maximum shear stress in the plane (or biaxial) stress state is not equal to that in the triaxial stress state. Furthermore, the maximum triaxial stress is larger than the maximum biaxial stress. Thus, if $\sigma_1$ and $\sigma_2$ have the same sign in the biaxial stress state, the triaxial maximum stress $\tau_{1/3}$ must be used for design considerations. However, if $\sigma_1$ and $\sigma_2$ have opposite signs in the biaxial stress state, the maximum biaxial and triaxial shear stresses will be the same and thus either one can be used in the analysis.

(a)

(b)

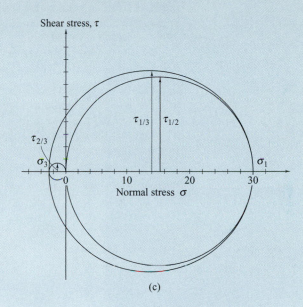

(c)

**Figure 2.20** Mohr's circle diagrams for Example 2.13. (a) Triaxial stress state when $\sigma_1 = 23.43$ ksi, $\sigma_2 = 4.57$ ksi, and $\sigma_3 = 0$; (b) biaxial stress state when $\sigma_1 = 30.76$ ksi and $\sigma_2 = -2.76$ ksi; (c) triaxial stress state when $\sigma_1 = 30.76$ ksi, $\sigma_2 = 0$, and $\sigma_3 = -2.76$ ksi.

## 2.16   OCTAHEDRAL STRESSES

Sometimes it is advantageous to represent the stresses on an octahedral stress element rather than on a conventional cubic element of principal stresses. Figure 2.21(b) and (c) shows the orientation of the eight octahedral planes that are associated with a given stress state. Each **octahedral plane** cuts across a corner of a principal element, so that the eight planes together form an octahedron [Fig. 2.21(b) and (c)]. The following characteristics of the stresses on an octahedral plane should be noted:

1. Identical normal stresses act on all eight planes. Thus, the normal stresses tend to compress or enlarge the octahedron but not to distort it.
2. Identical shear stresses act on all eight planes. Thus, the shear stresses tend to distort the octahedron without changing its volume.

The fact that the normal and shear stresses are the same for the eight planes is a powerful tool in failure analysis.

The octahedral normal and shear stress can be expressed in terms of the principal normal stresses, or the stresses in the $(x, y, z)$ coordinates, as

$$\sigma_{\text{oct}} = \frac{\sigma_1 + \sigma_2 + \sigma_3}{3} = \frac{\sigma_x + \sigma_y + \sigma_z}{3} \tag{2.26}$$

$$\tau_{\text{oct}} = \frac{1}{3}\left[(\sigma_1 - \sigma_2)^2 + (\sigma_2 - \sigma_3)^2 + (\sigma_3 - \sigma_1)^2\right]^{1/2}$$

$$= \frac{2}{3}\left(\tau_{1/2}^2 + \tau_{2/3}^2 + \tau_{1/3}^2\right)^{1/2} \tag{2.27}$$

$$\tau_{\text{oct}} = \frac{1}{3}\left[(\sigma_x - \sigma_y)^2 + (\sigma_y - \sigma_z)^2 + (\sigma_z - \sigma_x)^2 + 6\left(\tau_{xy}^2 + \tau_{yz}^2 + \tau_{xz}^2\right)\right]^{1/2} \tag{2.28}$$

Equations (2.26) to (2.28) are derived in Durelli et al. (1958).

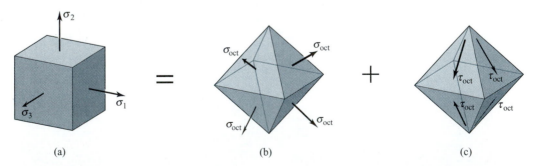

**Figure 2.21**   Stresses acting on octahedral planes. (a) General state of stress; (b) normal stress; (c) octahedral shear stress.

Note that the octahedral shear stress can be expressed in terms of the von Mises stress as

$$\tau_{oct} = \frac{\sqrt{2}}{3}\sigma_e \tag{2.29}$$

where $\sigma_e$ = von Mises stress, Pa. For a **uniaxial stress** state ($\sigma_2 = \sigma_3 = 0$)

$$\sigma_e = \sigma_1 \tag{2.30}$$

For a biaxial stress state ($\sigma_3 = 0$)

$$\sigma_e = \left(\sigma_1^2 + \sigma_2^2 - \sigma_1\sigma_2\right)^{1/2} \tag{2.31}$$

For a triaxial stress state

$$\sigma_e = \left[\frac{(\sigma_1 - \sigma_2)^2 + (\sigma_1 - \sigma_3)^2 + (\sigma_2 - \sigma_3)^2}{2}\right]^{1/2} \tag{2.32}$$

The triaxial maximum principal normal and shear stresses are greater than their octahedral counterparts.

---

**EXAMPLE 2.14**

**Given:** Both parts of Example 2.13.

**Find:** Determine the octahedral stresses.

**Solution:**

(a) The normal and octahedral stress can be written as

$$\sigma_{oct} = \frac{\sigma_1 + \sigma_2 + \sigma_3}{3} = \frac{(23.43 + 4.57 + 0)\,10^3}{3}\,\text{psi} = 9.33\,\text{ksi}$$

The octahedral shear stress from Equation (2.27) can be written as

$$\tau_{oct} = \frac{2}{3}\left[\tau_{1/2}^2 + \tau_{2/3}^2 + \tau_{1/3}^2\right]^{1/2} = \frac{(2)10^3}{3}(9.43^2 + 2.29^2 + 11.72^2)^{1/2}\text{psi} = 10.14\,\text{ksi}$$

Comparing these results with those from Examples 2.13 and 2.14 gives

$$\sigma_{1,b} = \sigma_{1,t} > \sigma_{oct} \qquad \text{and} \qquad \tau_{1,b} < \tau_{oct} < \tau_{1,t}$$

where subscript $b$ = biaxial stress
subscript $t$ = triaxial stress

(b) Only the shear stresses change in this part of the example.

$$\tau_{oct} = \frac{2}{3}\left(\tau_{1/2}^2 + \tau_{2/3}^2 + \tau_{1/3}^2\right)^{1/2} = \frac{(2)10^3}{3}(15.38^2 + 1.38^2 + 16.76^2)^{1/2}\text{psi} = 15.19\,\text{ksi}$$

Comparing this result with those from Examples 2.12 and 2.13 gives

$$\tau_{oct} < \tau_{1,b} = \tau_{1,t}$$

Thus, in comparing the principal triaxial stresses with the octahedral stresses,

$$\sigma_{1,t} > \sigma_{oct} \qquad \text{and} \qquad \tau_{1,t} = \tau_{1,3} > \tau_{oct}$$

## 2.17   STRAIN

**Strain** may be defined as the displacement per length produced in a solid as the result of stress. In designing a machine element, not only do engineers need to ensure that the design is adequate when considering the stress relative to the strength of the material, but they also need to ensure that the displacements, or deformations, are not excessive and are within limits from the standpoint of fracture or yielding. Depending on the application, these deformations may be either highly visible or practically unnoticeable. Note that for the same stress large deformations are clearly visible in components made of rubber but much less visible in the same components made of steel.

Just as the direction and intensity of the stress at any given point are important with respect to a specific plane passing through that point, the same is true for strain. Thus, as for stress, the state of strain is a tensor. Also, just as there are normal and shear stresses, so too there are normal and shear strains. **Normal strain,** designated by the symbol $\epsilon$, is used to describe a measure of the elongation or contraction of a linear segment of an element in which stress is applied. The average normal strain is

$$\epsilon_{avg} = \frac{\delta_{avg}}{l} = \frac{\text{average elongation}}{\text{original length}} \tag{2.33}$$

Note that strain is dimensionless. Furthermore, the strain at a point is

$$\epsilon = \lim_{\Delta l \to 0} \frac{\Delta \delta_{avg}}{\Delta l} = \frac{d\delta_{avg}}{dl} \tag{2.34}$$

Figure 2.22 shows the strain on a cubic element subjected to uniform tension in the $x$ direction. The element elongates in the $x$ direction while simultaneously contracting in the $y$ and $z$ directions, a phenomenon known as the *Poisson effect.* From Equation (2.34) the normal strain components can be written as

$$\epsilon_x = \lim_{x \to 0} \frac{\delta_x}{x} \qquad \epsilon_y = \lim_{y \to 0} \frac{\delta_y}{y} \qquad \text{and} \qquad \epsilon_z = \lim_{z \to 0} \frac{\delta_z}{z} \tag{2.35}$$

(a)                              (b)

**Figure 2.22**   Normal strain of a cubic element subjected to uniform tension in the x direction. (a) Three-dimensional view; (b) two-dimensional (or plane) view.

**Given:** A 12-in-long aluminum bar with 2.25-in diameter has an axial load of 32,000 lb applied to it. The axial elongation is 0.00938 in, and the diameter is decreased by 0.000585 in.

**EXAMPLE 2.15**

**Find:** The transverse and axial strains in the bar.

**Solution:** The axial strain is

$$\epsilon_a = \frac{\delta}{l} = \frac{0.00938}{12} = (0.7817)(10^{-3})$$

The transverse strain is

$$\epsilon_t = \frac{\delta_t}{d} = \frac{-(0.585)(10^{-3})}{2.25} = -(0.2600)(10^{-3})$$

The sign for the transverse strain is negative because the diameter decreased after the bar was loaded. The axial strain is positive because the axial length increased after loading.

Figure 2.23 shows the shear strain of a cubic element due to shear stress in both three-dimensional and two-dimensional (or plane) views. The **shear strain,** designated by $\gamma$, is used to measure angular distortion (the change in angle between two lines that are orthogonal in the undeformed state). The shear strain as shown in Figure 2.23 is defined as

$$\gamma_{yx} = \lim_{y \to 0} \frac{\delta_x}{y} = \tan \theta_{yx} \approx \theta_{yx} \tag{2.36}$$

where $\theta_{yx}$ = angle representing deviation from initial right angle.

The subscripts used to define the shear strains are like those used to define the shear stresses in Section 2.11. The first subscript designates the coordinate direction perpendicular to the plane in which the strain acts, and the second subscript designates the coordinate direction in which the strain acts. For example, $\gamma_{yx}$ is the strain resulting from taking adjacent planes perpendicular to the $y$ axis and displacing them relative to each other in the $x$ direction. The sign conventions for strain follow directly from those developed for stress.

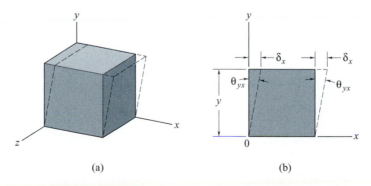

(a)                    (b)

**Figure 2.23** Shear strain of cubic element subjected to shear stress.
(a) Three-dimensional view; (b) two-dimensional (or plane) view.

A positive stress produces a positive strain, and a negative stress produces a negative strain. The shear strain shown in Figure 2.23 and described in Equation (2.36) is positive. The strain of the cubic element thus contains three normal strains and six shear strains, just as found for stresses. Symmetry reduces the shear strain elements from six to three, as for stress.

## 2.18   STRAIN TENSOR

For strains within the elastic range, the equations relating normal and shear strains with the orientation of the cutting plane are analogous to the corresponding equations for stress given in Equation (2.10). Thus, the state of strain can be written as a tensor as

$$\mathbf{T} = \begin{pmatrix} \epsilon_x & \frac{1}{2}\gamma_{xy} & \frac{1}{2}\gamma_{xz} \\ \frac{1}{2}\gamma_{xy} & \epsilon_y & \frac{1}{2}\gamma_{yz} \\ \frac{1}{2}\gamma_{xz} & \frac{1}{2}\gamma_{yz} & \epsilon_z \end{pmatrix} \tag{2.37}$$

In comparing Equation (2.37) with Equation (2.10), note that whereas $\epsilon_x$, $\epsilon_y$, and $\epsilon_z$ are analogous to $\sigma_x$, $\sigma_y$, and $\sigma_z$, respectively, it is one-half of the shear strain $\gamma_{xy}/2$, $\gamma_{yz}/2$, $\gamma_{zx}/2$ that is analogous to $\tau_{xy}$, $\tau_{yz}$, and $\tau_{zx}$, respectively. The general laws of strain transformation, given in Appendix B, enable the determination of strains acting on any orthogonal coordinate system.

## 2.19   PLANE STRAIN

Instead of the six strains for the complete strain tensor, here the effects of the components $\epsilon_z$, $\gamma_{xz}$, and $\gamma_{yz}$ are ignored, and thus only two normal strain components $\epsilon_x$ and $\epsilon_y$ and one shear strain component $\gamma_{xy}$ are considered. Figure 2.24 shows the deformation of an element caused by each of the three strains considered in plane strain. The normal strain components $\epsilon_x$ and $\epsilon_y$, shown in Figure 2.24(a) and (b), are produced by changes in element length in the $x$ and $y$ directions, respectively. The shear strain $\gamma_{xy}$, shown in Figure 2.24(c), is produced by the relative rotation of two adjacent sides of the element. Figure 2.24(c) also

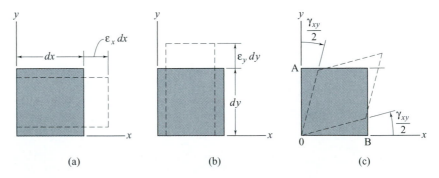

**Figure 2.24**   Graphical depiction of plane strain element. (a) Normal strain $\epsilon_x$; (b) normal strain $\epsilon_y$; and (c) shear strain $\gamma_{xy}$.

helps explain the physical significance of the fact that $\tau$ is analogous to $\gamma/2$ rather than to $\gamma$. Each side of an element changes in slope by an angle $\gamma/2$ when subjected to pure shear.

The following **sign convention** is to be used for strains:

1.  **Normal strains** $\epsilon_x$ and $\epsilon_y$ are positive if they cause elongation along the $x$ and $y$ axes, respectively. In Figure 2.24(a) and (b), $\epsilon_x$ and $\epsilon_y$ are positive.

2.  **Shear strain** $\gamma_{xy}$ is positive when the interior angle A0B in Figure 2.24(c) becomes smaller than 90°.

The principal strains, planes, and directions are directly analogous to those found earlier for principal stresses. The principal normal strains in the $xy$ plane, the maximum shear strain in the $xy$ plane, and the orientation of the principal axes relative to the $x$ and $y$ axes are

$$\epsilon_1, \epsilon_2 = \frac{\epsilon_x + \epsilon_y}{2} \pm \sqrt{\left(\frac{1}{2}\gamma_{xy}\right)^2 + \left(\frac{\epsilon_x - \epsilon_y}{2}\right)^2} \tag{2.38}$$

$$\gamma_{max} = \pm 2\sqrt{\left(\frac{1}{2}\gamma_{xy}\right)^2 + \left(\frac{\epsilon_x - \epsilon_y}{2}\right)^2} \tag{2.39}$$

$$2\phi = \tan^{-1}\frac{\gamma_{xy}}{\epsilon_x - \epsilon_y} \tag{2.40}$$

From here there are two important classes of problem. If the principal strains are known and it is desired to find the strains acting at a plane oriented at angle $\phi$ from the principal plane (numbered 1), the equations are

$$\epsilon_\phi = \frac{\epsilon_1 + \epsilon_2}{2} + \frac{\epsilon_1 - \epsilon_2}{2}\cos 2\phi \tag{2.41}$$

$$\gamma_\phi = (\epsilon_1 - \epsilon_2)\sin 2\phi \tag{2.42}$$

In Equation (2.42), $\gamma_\phi$ represents a shear strain acting on the $\phi$ plane and directed 90° from the $\phi$ axis. Just as for stress, the second subscript is omitted for convenience and no ambiguity results. Mohr's circle diagram can also be used to represent the state of strain.

The second problem of interest is the case where a normal strain component has been measured in three different but specified directions and it is desired to obtain the strains $\epsilon_x$, $\epsilon_y$, and $\gamma_{xy}$ from these readings. In this case the equation

$$\epsilon_\theta = \epsilon_x \cos^2\theta + \epsilon_y \sin^2\theta + \gamma_{xy}\sin\theta\cos\theta \tag{2.43}$$

is of great assistance. Here, $\epsilon_\theta$ is the measured strain in the direction rotated $\theta$ counterclockwise from the $x$ axis, and $\epsilon_x$, $\epsilon_y$, and $\gamma_{xy}$ are the desired strains. Thus, measuring a strain in three different directions gives three equations for the three unknown strains and is sufficient for their quantification. Strain gages are often provided in groups of three, called *rosettes,* for such purposes.

**EXAMPLE 2.16**

**Given:** A $0°$–$45°$–$90°$ rosette, as shown in Figure 2.25, is attached to a structure with the $0°$ gage placed along a reference $(x)$ axis. Upon loading, the strain gage in the $0°$ direction reads $+50\mu$m/m, the strain gage in the $45°$ direction reads $-27 \times 10^{-6}$, and the gage in the $90°$ direction reads 0.

**Find:** The strains $\epsilon_x$, $\epsilon_y$, and $\gamma_{xy}$.

**Solution:** Equation (2.43) can be applied three times to obtain three equations:

$$\epsilon_{0°} = \epsilon_x \cos^2 0° + \epsilon_y \sin^2 0° + \gamma_{xy} \sin 0° \cos 0° \qquad \text{or} \qquad \epsilon_x = \epsilon_{0°} = 50 \times 10^{-6}$$

$$\epsilon_{90°} = \epsilon_x \cos^2 90° + \epsilon_y \sin^2 90° + \gamma_{xy} \sin 90° \cos 90° \qquad \text{or} \qquad \epsilon_y = 0$$

$$\epsilon_{45°} = \epsilon_x \cos^2 45° + \epsilon_y \sin^2 45° + \gamma_{xy} \sin 45° \cos 45°$$

or

$$\gamma_{xy} = -54 \times 10^{-6} - 50 \times 10^{-6} - 0 = -104 \times 10^{-6}$$

**Figure 2.25**   Strain gage rosette used in Example 2.16.

---

**Case Study 2.1**  |  ANALYSIS OF A GLUE SPREADER SHAFT

**Given:** Figure 2.26 shows the expansion process used in making honeycomb materials. The process begins from the top left and continues to the right. Sheets are cut from the coil, and an adhesive is applied in parallel strips along the width of the sheet. These sheets are stacked to form a block and cured in an oven, enabling strong bonds to develop at the adhesive joints. The block is then cut into slices of desired dimension and stretched to produce a honeycomb structure. These stiff and lightweight construction materials are often used in aerospace applications.

This case study is particularly interested in the adhesive application to the sheet, or the glue spreader. This machine [Fig. 2.27(a)] spreads glue onto a sheet by exposing four polymer rollers to a polymer bath and then bringing them into contact with the sheet. By spacing the rollers properly, the desired glue pattern can be obtained. At the ends of the shaft are ball bearings, as shown in Figure 2.27(a).

Figure 2.27(b) shows a free-body diagram of the forces acting on the shaft along with the spacing between rollers. Here, the roller is represented by a concentrated force rather than a unit step. From moment and force equilibrium $R_A = R_B = 800$ N.

## Case Study (Concluded)

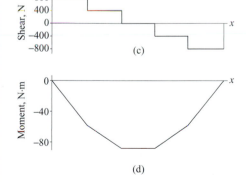

**Figure 2.26** Expansion process used in honeycomb materials. [Adapted from Kalpakjian and Schmid (2003).]

**Figure 2.27** Glue spreader case study.
(a) Machine; (b) free-body diagram (dimensions in millimeters); (c) shear diagram; (d) moment diagram.

### Find:

(a) Using singularity functions, determine the shear force and moment expressions and give the corresponding diagrams and the maximum moment.

(b) If the distance between the bearing and the first roller is $a$, as is also the distance between the last roller and the bearing, and the distance between rollers is $b$, find a general expression for the shear and moment for $n$ rollers.

### Solution:

(a) From Table 2.2 for a concentrated force, the load intensity function for the glue spreader shown in Figure 2.27(b) (when $x$ is expressed in meters) is

$$q(x) = -800\langle x\rangle^{-1} + 400\langle x - 0.075\rangle^{-1} \\ + 400\langle x - 0.145\rangle^{-1} + 400\langle x - 0.215\rangle^{-1} \\ + 400\langle x - 0.285\rangle^{-1} - 800\langle x - 0.360\rangle^{-1}$$

The shear force and moment expressions are

$$-V(x) = -800(\langle x\rangle^{0} + \langle x - 0.360\rangle^{0}) \\ + 400(\langle x - 0.075\rangle^{0} + \langle x - 0.145\rangle^{0} \\ + \langle x - 0.215\rangle^{0} + \langle x - 0.285\rangle^{0})$$

$$M(x) = -800(\langle x\rangle^{1} + \langle x - 0.360\rangle^{1}) \\ + 400(\langle x - 0.075\rangle^{1} + \langle x - 0.145\rangle^{1} \\ + \langle x - 0.215\rangle^{1} + \langle x - 0.285\rangle^{1})$$

Figure 2.27(c) and (d) shows the shear and bending moment diagrams. The maximum bending moment $M_{\max}$ is 88 N · m.

(b) The general expressions for the load intensity, shear force, and moment are

$$q(x) = -\frac{nP}{2}(\langle x\rangle^{-1} + \langle x - 2a - nb\rangle^{-1}) \\ + P\sum_{i=1}^{n}\langle x - a - b(i - 1)\rangle^{-1}$$

$$V(x) = \frac{nP}{2}(\langle x\rangle^{0} + \langle x - 2a - nb\rangle^{0}) \\ - P\sum_{i=1}^{n}\langle x - a - b(i - 1)\rangle^{0}$$

$$M(x) = -\frac{nP}{2}(\langle x\rangle^{1} + \langle x - 2a - nb\rangle^{1}) \\ + P\sum_{i=1}^{n}\langle x - a - b(i - 1)\rangle^{1}$$

where $n$ = number of rollers
$a$ = distance from bearing to roller
$b$ = distance between rollers

## 2.20   SUMMARY

This chapter described how load, stress, and strain affect the design of machine elements. If the proper type of machine element has been selected, after its function is considered, the primary cause of failure is the design stress exceeding the strength of the machine element. Therefore, it is important to evaluate the stress, strain, and strength of the machine element at the critical section. To do so requires first a determination of load in all its forms. The applied load on a machine element was described with respect to time, the area over which load is applied, and the location and method of application. Furthermore, the importance of support reactions, application of static force and moment equilibrium, and proper use of free-body diagrams was pointed out.

The chapter then focused on shear and moment diagrams applied to a beam. Singularity functions introduced by concentrated moment, concentrated force, unit step, ramp, inverse ramp, and parabolic shape were considered. Various combinations of these singularity functions can exist within a beam. Integrating the load intensity function for the various beam singularity functions over the beam length establishes the shear force. Integrating the shear force over the beam length determines the moment. From these analytical expressions the shear and moment diagrams can be readily constructed.

Stress defines the intensity and direction of the internal forces at a particular point and acting on a given plane. The stresses acting on an element have normal and shear components. Across each mutually perpendicular surface there are two shear stresses and one normal stress, yielding a total of nine stresses (three normal stresses and six shear stresses). The sign conventions for both normal and shear stresses were presented. It was found that the stress tensor is symmetric, implying that the tensor can be written with zero shear stress and the principal normal stresses.

In many engineering applications the stress analysis assumes that the surface is free of stress or that the stress in one plane is small relative to the stresses in the other two planes. The two-dimensional stress situation is called the *biaxial* (or *plane*) *stress state* and can be expressed in terms of two normal stresses and one shear stress ($\sigma_x$, $\sigma_y$, and $\tau_{xy}$). That the stresses can be expressed in any oblique plane is important in describing Mohr's circle diagram for a biaxial stress. It was found difficult to express analytically a general three-dimensional stress state. If the principal normal stresses are known, the triaxial stress state can be represented by Mohr's circle diagram when the shear stresses are expressed in terms of the principal normal stresses.

## KEY WORDS

**beam**   structural member designed to support loads perpendicular to its longitudinal axis

**bending load**   load applied transversely to longitudinal axis of member

**biaxial or plane stress**   condition where one surface is comparatively free of stress

**cantilevered beam**   support where one end is fixed and other end is free

**combined load**   combination of two or more previously defined loads

**concentrated load**    load applied to small nonconformal area

**critical section**    section where largest internal stress occurs

**cyclic load**    load varying throughout cycle

**distributed load**    load distributed over entire area

**free-body diagram**    sketch of part showing all acting forces

**impact load**    load rapidly applied

**load**    force, moment, or torque applied to mechanism or structure

**Mohr's circle**    method used to graphically visualize state of stress acting in different planes passing through given point

**normal load**    load passing through centroid of resisting section

**normal strain**    elongation or contraction of linear segment of element in which stress is applied

**overhanging beam**    support where one or both ends freely extend past support

**principal normal stresses**    A set of normal stresses on planes that have no shear stresses. These normal stresses are the maximum and minimum stresses for the stress state.

**principal shear stresses**    A set of shear stresses on planes that are 45° from the principal axes; these shear stresses are the extreme (maximum and minimum) shear stresses for the stress state.

**shear load**    load collinear with transverse shear force

**shear strain**    measure of angular distortion in which shear stress is applied

**sign convention for normal strain**    positive if elongation is in direction of positive axes

**sign convention for normal stress**    positive for tension and negative for compression

**sign convention for shear strain**    positive if interior angle becomes smaller after shear stress is applied

**sign convention for shear stress**    positive if both normal from surface and shear are in positive or negative directions; negative for any other combination

**simply supported beam**    support where one end is pinned and other is roller-supported

**singularity functions**    functions used to evaluate shear and moment diagrams, especially when discontinuities, such as concentrated load or moment, exist

**static load**    load gradually applied and equilibrium reached in short time

**static equilibrium**    is the state of a static body when the resultant of the forces acting on it is zero, also when the resultant of moments is zero.

**strain**    dimensionless displacement produced in solid as result of stress

**stress**    intensity and direction of internal force acting at given point on particular plane

**sustained load**    load constant over long time

**symmetric tensor**    condition where principal normal stresses exist while all other tensor elements are zero

**torsion load**     load subjected to twisting motion

**triaxial stress**     A three-dimensional stress state

**uniaxial stress**     condition where two perpendicular surfaces are comparatively free of stress

## RECOMMENDED READINGS

Bedford, A., and Liechtli, K. (2000) *Mechanics of Materials,* Prentice-Hall, Upper Saddle River, NJ.

Beer, F. P., Johnson, E. R., and Dewolf, J. T. (2001) *Mechanics of Materials,* McGraw-Hill, 3rd ed., New York.

Craig, R. R. (1996) *Mechanics of Materials,* Wiley, New York.

Crandall, S. H., and Dahl, H. C. (1954) *An Introduction to the Mechanics of Solids,* McGraw-Hill, New York.

Hibbeler, R. C. (2000) *Mechanics of Materials,* 4th ed., Prentice-Hall, Upper Saddle River, NJ.

Juvinall, R. C. (1967) *Stress, Strain, and Strength,* McGraw-Hill, New York.

Lardner, T. J., and Archer, R. R. (1994) *Mechanics of Solids: An Introduction,* McGraw-Hill, New York.

Popov, E. P. (1999) *Engineering Mechanics of Solids,* 2nd ed., Prentice-Hall, Upper Saddle River, NJ.

Popov, E. P. (1968) *Introduction to Mechanics of Solids,* Prentice-Hall, Englewood Cliffs, NJ.

Riley, W. F., Sturges, L. D., and Morris, D. H. (1999) *Mechanics of Materials,* 5th ed., New York, Wiley.

Shames, I. H., and Pitarresi, J. M. (2000) *Introduction to Solid Mechanics,* 3rd ed., Prentice-Hall, Upper Saddle River, NJ.

## REFERENCES

Durelli, A. J., Phillips, E. A., and Tsao, C. H. (1958) *Introduction to the Theoretical and Experimental Analysis of Stress and Strain,* McGraw-Hill, New York.

Kalpakjian, S., and Schmid, S. R. (2003) *Manufacturing Processes for Engineering Materials,* 4th ed., Prentice-Hall, Upper Saddle River, NJ.

Mohr, O. (1914) *Abhandlungen aus dem Gebiete der technischen Mechanik,* 2nd ed., Wilhelm Ernst und Sohn, Berlin, pp. 192–203.

## PROBLEMS

### Section 2.2

**2.1**   The stepped shaft A-B-C shown in sketch *a* is loaded with the forces $P_1$ and/or $P_2$. Note that $P_1$ gives a tensile stress $\sigma$ in B-C and $\sigma/3$ in A-B and that $P_2$ gives a bending stress $\sigma$ at B and $1.2\sigma$ at A. What is the critical section

    *a)*     If only $P_1$ is applied?

    *b)*     If only $P_2$ is applied?

    *c)*     If both $P_1$ and $P_2$ are applied?

Sketch a, used in Problems 2.1 and 2.2.

## Section 2.3

**2.2** The stepped shaft in Problem 2.1 (sketch $a$) has loads $P_1$ and $P_2$. Load $P_1$ is sinusoidal, and $P_2$ is the load from a weight. Find the load classification

- *a)* If only $P_1$ is applied.
- *b)* If only $P_2$ is applied.
- *c)* If both $P_1$ and $P_2$ are applied.

## Section 2.5

**2.3** A bar hangs freely from a frictionless hinge. A horizontal force $P$ is applied at the bottom of the bar until it inclines 30° from the vertical direction. Calculate the horizontal and vertical components of the force on the hinge if the acceleration due to gravity is $g$, the bar has a constant cross section along its length, and the total mass is $m_a$. *Ans.* $R_x = \frac{1}{2}gm_a \tan 30°$, $R_y = gm_a$.

**2.4** Sketch $b$ shows the forces acting on a rectangle. Is the rectangle in equilibrium? *Ans.* No.

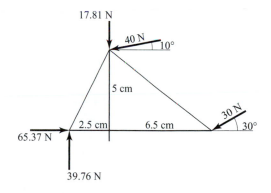

Sketch b, used in Problem 2.4

Sketch c, used in Problem 2.5

**2.5** Sketch $c$ shows the forces acting on a triangle. Is the triangle in equilibrium? *Ans.* Yes.

**2.6** Sketch $d$ shows a cube with side lengths $a$ and eight forces acting at the corners. Is the cube in equilibrium? *Ans.* Yes.

Sketch *d*, used in Problem 2.6

## Section 2.6

**2.7**  Given the components shown in sketches *e* and *f*, draw the free-body diagram of each component and calculate the forces.

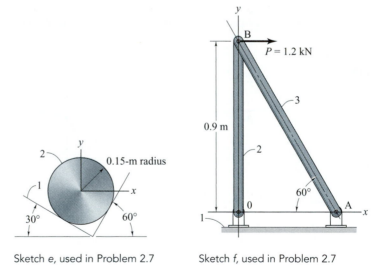

Sketch *e*, used in Problem 2.7          Sketch *f*, used in Problem 2.7

## Section 2.8

**2.8**  A 5-m-long bar is loaded as shown in sketch *g*. The bar cross section is constant along its length. Draw the shear and moment diagrams and locate the critical section. *Ans.* $|V|_{\max} = 8.8\,\text{kN}$, $|M|_{\max} = 6.4\,\text{kN} \cdot \text{m}$

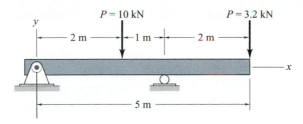

Sketch g, used in Problem 2.8

**2.9** Sketch *h* shows a 0.06-m-diameter steel shaft supported by self-aligning bearings (which can provide radial but not bending loads on the shaft). A gear causes each force to be applied as shown. All length dimensions are in meters. The shaft can be considered weightless. Determine the forces at A and B. Draw shear and moment diagrams. *Ans.* $M_{max} = 268$ N · m.

Sketch *h*, used in Problem 2.9

**2.10** A beam is loaded as shown in sketch *i*. Determine the reactions and draw the shear and moment diagrams for $P = 1000$ lb. *Ans.* Reactions $= 2000$ lb, $|M|_{max} = 4000$ ft · lb.

Sketch *i*, used in Problem 2.10                    Sketch *j*, used in Problem 2.11

**2.11** Sketch *j* shows a simply supported bar loaded with a force *P* at a position one-third of the length from one of the support points. Determine the shear force and bending moment in the bar. Also, draw the shear and moment diagrams. *Ans.* $M_{max} = \frac{2}{9}Pl$.

**2.12** Sketch *k* shows a simply supported bar loaded by two equally large forces *P* at a distance $l/4$ from its ends. Determine the shear force and bending moment in the bar, and find the critical

section with the largest bending moment. Also, draw the shear and moment diagrams. *Ans.*
$M_{max} = \frac{1}{4}Pl$.

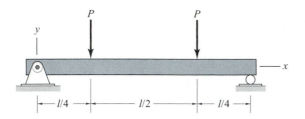

Sketch k, used in Problem 2.12                    Sketch l, used in Problem 2.13

**2.13**  The bar shown in sketch *l* is loaded by the force $P$. Determine the shear force and bending moment in the bar. Also, draw the shear and moment diagrams. *Ans.* $|M|_{max} = Pl$.

**2.14**  Sketch *m* shows a simply supported bar with a constant load per unit length $w_0$ imposed over its entire length. Determine the shear force and bending moment as functions of $x$. Draw a graph of these functions. Also, find the critical section with the largest bending moment. *Ans.* $M_{max} = \frac{1}{8}w_0 l^2$.

Sketch m, used in Problem 2.14                    Sketch n, used in Problem 2.15

**2.15**  Sketch *n* shows a simply supported beam loaded with a ramp function over its entire length, the largest value being $2P/l$. Determine the shear force and the bending moment and the critical section with the largest bending moment. Also, draw the shear and moment diagrams. *Ans.* $M_{max} = 2Pl/9\sqrt{3}$.

Sketch o, used in Problem 2.16

★  **2.16**  Sketch *o* shows a sinusoidal distributed force applied to a beam. Determine the shear force and bending moment for each section of the beam.

| *Indicates problems of greater difficulty

**2.17** Find the length $x$ that gives the smallest maximum bending moment for the load distribution shown in sketch $p$. *Ans.* $x = 0.207l$.

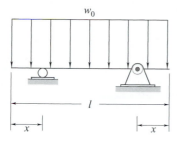

Sketch p, used in Problem 2.17          Sketch q, used in Problem 2.18

**2.18** Draw the shear and moment diagrams and give the reaction forces for the load distribution shown in sketch $q$. *Ans.* $R = w_0a$, $M_{max} = \frac{1}{2}w_0a^2$.

## Section 2.9

**2.19** The simply supported bar shown in sketch $r$ has $P_1 = 5$ kN, $P_2 = 10$ kN, $w_0 = 5$ kN/m, and $l = 12$ m. Use singularity functions to determine the shear force and bending moment as functions of $x$. Also, draw your results. *Ans.* $M_{max} = 52.5$ kN · m.

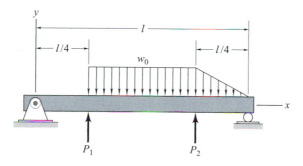

Sketch r, used in Problem 2.19

**2.20** Use singularity functions for the force system shown in sketch $s$ to determine the load intensity, the shear force, and the bending moment. From a force analysis determine the reaction forces $R_1$ and $R_2$. Also, draw the shear and moment diagrams. *Ans.* $R_1 = -7.14$ lb, $R_2 = 27.14$ lb, $|M|_{max} = 137$ in · lb.

Sketch s, used in Problem 2.20

**2.21** Use singularity functions for the force system shown in sketch *t* to determine the load intensity, the shear force, and the bending moment. Draw the shear and moment diagrams. Also, from a force analysis determine the reaction forces $R_1$ and $R_2$. *Ans.* $R_1 = 87$ lb, $R_2 = 48$ lb, $M_{max} = 480$ in · lb.

Sketch *t*, used in Problem 2.21

**2.22** Draw a free-body diagram of the forces acting on the simply supported bar shown in sketch *u*, with $w_0 = 5$ kN/m and $l = 15$ m. Use singularity functions to draw the shear force and bending moment diagrams. *Ans.* $M_{max} = 120$ kN · m.

Sketch *u*, used in Problem 2.22

★ **2.23** Sketch *v* shows a simply supported bar with $w_0 = 3$ kN/m and $l = 8$ m. Draw a free-body diagram of the forces acting along the bar as well as the coordinates used. Use singularity functions to determine the shear force and the bending moment. *Ans.* $M_{max} = 5.33$ kN · m.

Sketch *v*, used in Problem 2.23

★ **2.24** Redo the Case Study 2.1 problem if the 400-N force is evenly distributed along the width of the roller and a unit step is used to represent the loading. The width of the roller is 30 mm. Do both (a) and (b) portions of the case study while considering the unit step representation. *Ans.* $|M|_{max} = 88$ N · m.

**2.25** An extra concentrated force with an intensity of 20 kN is applied downward at the center of the simply supported bar shown in sketch $v$. Draw a free-body diagram of the forces acting on the bar. Assume $l = 8$ m and $w_0 = 4$ kN/m. Use singularity functions to determine the shear force and bending moments, and draw the diagrams. *Ans.* $M_{max} = 50.7$ kN $\cdot$ m.

## Section 2.11

**2.26** A steel bar is loaded by a tensile force $P = 25$ kN. The cross section of the bar is circular with a radius of 7 mm. What is the normal tensile stress in the bar? *Ans.* $\sigma = 162$ MPa.

**2.27** A stainless-steel bar of square cross section has a tensile force of $P = 15$ kN acting on it. Calculate how large the side $l$ of the cross-sectional area must be to give a tensile stress in the bar of 120 MPa. *Ans.* $l = 11.2$ mm.

**2.28** What is the maximum length $l_{max}$ that a copper wire can have if its weight should not give a higher stress than 75 MPa when it is hanging vertically? The density of copper is 8900 kg/m$^3$, and the density of air is so small relative to that of copper that it may be neglected. The acceleration of gravity is 9.81 m/s$^2$. *Ans.* $l_{max} = 859$ m.

**2.29** A machine weighing 5 tons will be lifted by a steel rod with an ultimate tensile strength of 860 MPa. A safety factor of 5 is to be used. Determine the diameter needed for the steel rod. *Ans.* $d = 19.1$ mm.

**2.30** A string on a guitar is made of nylon and has a cross-sectional diameter of 0.6 mm. It is tightened to a force $P = 15$ N. What is the stress in the string? *Ans.* $\sigma = 53.1$ MPa.

**2.31** Determine the normal and shear stresses in sections A and B of sketch $w$. The cross-sectional area of the rod is 0.050 m$^2$. Ignore bending and torsional effects. *Ans.* In AA, $\sigma = -173.2$ kPa, $\tau = -100$ kPa.

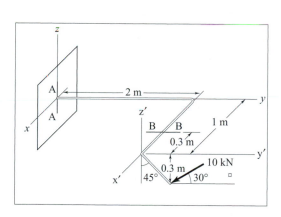

Sketch w, used in Problem 2.31

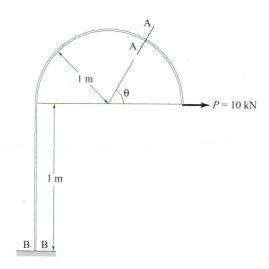

Sketch x, used in Problem 2.32

**2.32** Determine the normal and shear stresses due to axial and shear forces at sections A and B in sketch $x$. The cross-sectional area of the rod is 0.025 m$^2$ and $\theta = 30°$. *Ans.* At section AA, $\sigma = 200$ kPa, $\tau = 346$ kPa. At section BB, $\tau = 400$ kPa.

## Section 2.13

★ **2.33** Sketch $y$ shows a distributed load on a semi-infinite plane. The stress in polar coordinates based on the plane stress assumption is

$$\sigma_r = -\frac{2w_0 \cos \theta}{\pi r}$$

$$\sigma_\theta = \tau_{r\theta} = \tau_{\theta r} = 0$$

Determine the expressions $\sigma_x$, $\sigma_y$, and $\tau_{xy}$ in terms of $r$ and $\theta$. *Ans.* $\sigma_x = -\frac{2w_0 \cos^3 \theta}{\pi r}$.

Sketch y, used in Problem 2.33

**2.34** Sketch $z$ shows loading of an extremely thin and infinitely long plane. Determine the angle $\theta$ needed so that the stress element will have no shear stress. For plate thickness $t_w$, modulus of elasticity $E$, and Poisson's ratio $v$, find the reduction in thickness of the plate. *Ans.* $\theta = 0$, $\Delta t_w = -\frac{v t_w}{E}(\sigma_x + \sigma_y)$.

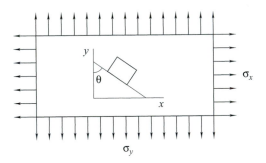

Sketch z, used in Problem 2.34

**2.35** A stress tensor is given by

$$\mathbf{S} = \begin{pmatrix} 178 & -83 & 0 \\ -83 & 12 & 0 \\ 0 & 0 & 0 \end{pmatrix}$$

where all values are in megapascals. The stress state is applied to a machine element made of AISI 1020 steel. Calculate the principal normal stresses and the principal shear stresses. *Ans.* $\sigma_1 = 212$ MPa, $\sigma_2 = -22.4$ MPa, $\tau = \pm 117.4$ MPa.

**2.36** A thin, square steel plate has normals to the sides in the $x$ and $y$ directions. A tensile stress $\sigma$ acts in the $x$ direction, and a compressive stress $-\sigma$ acts in the $y$ direction. Determine the normal and shear stresses on the diagonal of the square. *Ans.* $\sigma_{45°} = 0$, $\tau_{45°} = -\sigma$.

## Section 2.14

**2.37** A thin, rectangular brass plate has normals to the sides in the $x$ and $y$ directions. A tensile stress $\sigma$ acts on the four sides. Determine the principal normal and shear stresses. *Ans.* $\sigma_1 = \sigma_2 = \sigma$, $\tau = 0$.

**2.38** Given the thin, rectangular brass plate in Problem 2.37, but with the stress in the $y$ direction being $\sigma_y = -\sigma$ instead of $+\sigma$, determine the principal normal and shear stresses and their directions. *Ans.* $\sigma_1 = -\sigma_2 = \sigma$, $\tau = \pm\sigma$.

**2.39** For the following stress states draw the appropriate Mohr's circle, determine the principal stresses and their directions, and show the stress elements.

*a)* $\sigma_x = 20$, $\sigma_y = -20$, and $\tau_{xy} = 10$
*b)* $\sigma_x = 30$, $\sigma_y = -30$, and $\tau_{xy} = 10$
*c)* $\sigma_x = 50$, $\sigma_y = -50$, and $\tau_{xy} = 0$
*d)* $\sigma_x = \sigma_y = -10$

All stresses are in megapascals. *Ans.* (a) $\sigma_1, \sigma_2 = \pm 22.36$ MPa.

**2.40** Repeat Problem 2.39 for

*a)* $\sigma_x = 55$, $\sigma_y = -15$, and $\tau_{xy} = 40$
*b)* $\sigma_x = 0$, $\sigma_y = 30$, and $\tau_{xy} = 20$
*c)* $\sigma_x = -20$, $\sigma_y = 40$, and $\tau_{xy} = -40$
*d)* $\sigma_x = 30$, $\sigma_y = 0$, and $\tau_{xy} = -20$

All stresses are in megapascals. *Ans.* (a) $\sigma_1 = 73.1$ MPa, $\sigma_2 = -33.13$ MPa.

(a)　　　　　　　　　(b)

Sketch *aa*, used in Problem 2.41

**2.41** Given the state of stresses shown in the two parts of sketch *aa*, determine the principal stresses and their directions by using Mohr's circle and the stress equations. Show the stress elements. All stresses in sketch *aa* are in megapascals. *Ans.* (a) $\sigma_1 = 34$ MPa, $\sigma_2 = -38$ MPa.

**2.42** Given the normal and shear stresses $\sigma_x = 2$ ksi, $\sigma_y = 6$ ksi, and $\tau_{xy} = -4$ ksi, draw Mohr's circle diagram and the principal normal and shear stresses on the $xy$ axis. Determine the triaxial stresses and give the corresponding Mohr's circle diagram. *Ans.* $\sigma_1 = 8.47$ ksi, $\sigma_2 = 0$, $\sigma_3 = -0.47$ ksi.

**2.43** Given the normal and shear stresses $\sigma_x = 0$ ksi, $\sigma_y = 10$ ksi, and $\tau_{xy} = 12$ ksi, draw Mohr's circle diagram and the principal normal and shear stresses on the $xy$ axis. Determine the triaxial stresses and give the corresponding Mohr's circle diagram. *Ans.* $\sigma_1 = 18$ ksi, $\sigma_2 = 0$, $\sigma_3 = -8$ ksi.

**2.44** Given the normal and shear stresses $\sigma_x = 10$ ksi, $\sigma_y = 24$ ksi, and $\tau_{xy} = -6$ ksi, draw the Mohr's circle diagram and the principal normal and shear stresses on the $xy$ axis. Determine the triaxial stresses and give the corresponding Mohr's circle diagram. *Ans.* $\sigma_1 = 26.22$ ksi, $\sigma_2 = 7.78$ ksi, $\sigma_3 = 0$.

## Section 2.15

**2.45** In a three-dimensional stress field, the stresses are found to be $\sigma_x = 4$ ksi, $\sigma_y = 2$ ksi, $\sigma_z = 6$ ksi, $\tau_{xy} = -2$ ksi, $\tau_{zy} = -2$ ksi, and $\tau_{xz} = 2$ ksi. Draw the stress element for this case. Determine the principal stresses and sketch the corresponding Mohr's circle diagram. *Ans.* $\sigma_1 = 8.43$ ksi, $\sigma_2 = 2.92$ ksi, $\sigma_3 = 0.65$ ksi.

## Section 2.16

★ **2.46** Given the normal and shear stresses $\sigma_x = -10$ ksi, $\sigma_y = 15$ ksi, and $\tau_{xy} = 5$ ksi, determine or draw the following:

    *a)*       Two-dimensional Mohr circle diagram
    *b)*       Normal principal stress element on $xy$ axis
    *c)*       Shear principal stress on $xy$ axis
    *d)*       Three-dimensional Mohr circle diagram and corresponding principal normal and shear stresses

## Section 2.18

**2.47** The strain tensor in a machine element is

$$\mathbf{T} = \begin{pmatrix} 0.0012 & -0.0001 & 0.0007 \\ -0.0001 & 0.0003 & 0.0002 \\ 0.0007 & 0.0002 & -0.0008 \end{pmatrix}$$

Find the strain in the $x$, $y$, and $z$ directions, in the direction of the space diagonal $(1/\sqrt{3}; 1/\sqrt{3}; 1/\sqrt{3})$, and in the direction $\epsilon_x$, $\epsilon_y$, and $\epsilon_z$. *Ans.* In the direction of the diagonal, $\epsilon = 0.0005$.

**2.48** The strain tensor is

$$\mathbf{T} = \begin{pmatrix} 0.0023 & 0.0006 & 0 \\ 0.0006 & 0.0005 & 0 \\ 0 & 0 & 0 \end{pmatrix}$$

Calculate the maximum shear strain and the principal strains. *Ans.* $\epsilon_1 = 0.00248$, $\epsilon_2 = 0.00032$.

# INTRODUCTION TO MATERIALS AND MANUFACTURING

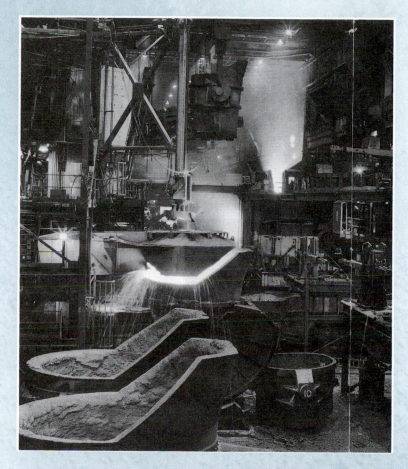

Molten iron is poured from a furnace to form a billet. (Courtesy of Weirton Steel Corporation)

*Give me matter, and I will construct a world out of it.*
Immanuel Kant

## SYMBOLS

| | |
|---|---|
| $A$ | area, $m^2$ |
| $\bar{a}$ | linear thermal expansion coefficient, $(°C)^{-1}$ |
| $C$ | reference line |
| $C_p$ | specific heat of material, $J/(kg \cdot °C)$ |
| $C_R$ | relative cost |
| $d$ | fiber diameter, m |
| $E$ | modulus of elasticity, Pa |
| %EL | elongation, percent |
| $G$ | shear modulus, Pa |
| $K$ | bulk modulus, Pa |
| $K_A$ | Archard wear constant, $(Pa)^{-1}$ |
| $K_t$ | thermal conductivity, $W/(m \cdot °C)$ |
| $k$ | spring rate, N/m |
| $l$ | length, m |
| $l_{cr}$ | critical length, m |
| $l_{fr}$ | length of specimen at fracture, m |
| $l_s$ | sliding distance, m |
| $l_0$ | length of specimen without load, m |
| $m_a$ | mass of body, kg |
| $P$ | force, N |
| $p$ | normal pressure, Pa |
| $p_l$ | limiting pressure, Pa |
| $\hat{Q}$ | quantity of heat, J |
| $r_i$ | inner radius, m |
| $r_m$ | mean radius, m |
| $r_o$ | atom size, m |
| $S$ | strength, Pa |
| $S_{fr}$ | fracture strength, Pa |
| $S_u$ | ultimate strength , Pa |
| $S_y$ | yield strength , Pa |
| $T$ | temperature, °C |
| $\Delta T$ | temperature change, °C |
| $t_h$ | thickness, m |
| $U_r$ | modulus of resilience, Pa |
| $v$ | volume fraction |
| $W$ | weight, N |
| $W_r$ | wear rate, $m^2$ |
| $x$ | volume of carbon fiber in plastic |
| $\alpha$ | rotational angle, deg |
| $\gamma$ | shear strain |
| $\delta$ | deformation; deflection, m |
| $\epsilon$ | strain |
| $v$ | Poisson's ratio |
| $\rho$ | density, $kg/m^3$ |
| $\sigma$ | normal stress, Pa |
| $\tau$ | shear stress, Pa |
| $\tau_f$ | fiber-matrix bond strength, Pa |

## SUBSCRIPTS

| | |
|---|---|
| $a$ | axial |
| all | allowable |
| avg | average |
| $c$ | composite; cross-sectional |
| cr | critical |
| $f$ | fiber |
| fr | at fracture |
| $g$ | glass transition |
| $i$ | inner |
| $m$ | matrix; magnesia |
| max | maximum |
| $s$ | steel |
| $t$ | transverse |
| 0 | without load |
| 1, 2, 3 | principal axes |

## SUPERSCRIPTS

| | |
|---|---|
| $c$ | compression |
| $t$ | tension |

## 3.1    INTRODUCTION

Choosing the solid material is an important step in designing machine elements. Being able to exploit a material's potential and characteristics is essential to ensuring that the best material is used for a particular machine element. Therefore, knowing the properties of solid materials is essential. This chapter will classify, give the properties of, and select solid materials in a general sense. A brief summary of the manufacturing processes that are available for material classes will be discussed, with an introduction to their relationship to mechanical design. Chapters 9 to 20, dealing with different machine elements, present more specific information regarding material selection for particular cases. The general knowledge obtained from this chapter will be of great use in later chapters.

## 3.2 DUCTILE AND BRITTLE MATERIALS

### 3.2.1 DUCTILE MATERIALS

**Ductility** measures the degree of plastic deformation sustained at fracture. A ductile material can be subjected to large strains before it ruptures. Designers often use ductile materials because they can absorb shock (or energy) and, if they become overloaded, will usually exhibit large deformations before failing. Also, stress concentrations (covered in Chap. 7) can be partially relieved through the deformations that can be achieved with ductile materials.

One way to specify a material as ductile is by the percent elongation (%EL)

$$\%EL = \frac{l_{\text{fr}} - l_0}{l_0} \times 100\% \tag{3.1}$$

where $l_{\text{fr}}$ = length of specimen at fracture, m
$l_0$ = length of specimen without load, m

A **ductile material** is one with a large %EL before failure, arbitrarily defined as 5% or higher for the purposes of this text. Table A.1 (App. A) shows that the %EL for low-carbon, medium-carbon, and high-carbon steels is 37%, 30%, and 25%, respectively. (Note that some of the material from tables in App. A also appears on the inside front cover.) Thus, steel is ductile because it far exceeds the 5% elongation described in Equation (3.1). Also note from Equation (3.1) that the original length of the specimen $l_0$ is an important value because a significant portion of the plastic deformation at fracture is confined to the neck region. Thus, the magnitude of %EL will depend on the specimen length. The shorter $l_0$, the greater the fraction of total elongation from the neck and, consequently, the higher the value of %EL. Therefore, $l_0$ should be specified when percent elongation values are cited. The length of a specimen without load is commonly defined as 50 mm (Table A.1).

Figure 3.1(a) shows a test specimen of a ductile material in which necking (decreasing cross-sectional area) is occurring. Figure 3.1(b) shows the same specimen just as fracture occurs. Note the considerable amount of plastic deformation at fracture.

**Given:** A flat plate is rolled into a cylinder with an inner radius of 100 mm and a wall thickness of 60 mm.

**Find:** Determine which of the three stainless steels in Table A.1 cannot be cold formed to the cylinder. Assume that the midplane of the plate does not experience either tension or compression and thus will not experience any elongation. Also, calculate the %EL for the three stainless steels.

**EXAMPLE 3.1**

**Solution:** The length without load is at the midplane of the plate, or

$$l_0 = 2\pi \left( r_i + \frac{t_h}{2} \right) = 2\pi \left( 100 + \frac{60}{2} \right) = 260\pi \text{ mm}$$

The length at fracture at the outer diameter of the cylinder is

$$l_{fr} = 2\pi r_o = 2\pi (r_i + t_h) = 2\pi (100 + 60)(10^{-3}) \text{ m} = 320\pi \text{ mm}$$

Thus, from Equation (3.1) the percent elongation is

$$\%\text{EL} = \frac{l_{fr} - l_0}{l_0}(100\%) = \frac{320\pi - 260\pi}{260\pi}(100\%) = 23.04\%$$

From Table A.1 the ferritic type of AISI 446 steel can experience only 20% EL. Therefore, this material would crack if it were cold-formed with a 23.1% deformation.

### 3.2.2    BRITTLE MATERIALS

A **brittle material** exhibits little (%EL < 5%) or no yielding before failure. Gray cast iron is an example of a brittle material whose %EL is so small that no listing is given in Table A.1. Figure 3.2 shows a brittle test specimen at failure. Little or no necking occurred prior to failure, in contrast to Figure 3.1.

(a)                     (b)

**Figure 3.1**   Ductile material from a standard tensile test apparatus. (a) Necking; (b) failure.

**Figure 3.2**   Failure of a brittle material from a standard tensile test apparatus.

**Given:** A malleable cast iron plate has bent due to an uneven cooling rate through the thickness of the plate during casting. The plate is 20 mm thick and is bent to a mean radius of 750 mm. Elastic springback is neglected.

**EXAMPLE 3.2**

**Find:** Is it possible to flatten the plate without cracking it?

**Solution:**

$$\%EL = \frac{t_h/2}{r_m - t_h/2}(100\%) = \frac{10}{750 - 10}(100\%) = 1.33\%$$

From Table A.1 the %EL = 10% for malleable cast iron. Since 1.33% < 10%, the plate can be flattened without risk of cracking.

## 3.3 CLASSIFICATION OF SOLID MATERIALS

Engineering materials fall into four major classes: metals, ceramics (including glasses), polymers (including elastomers), and composites. The members of each class generally have the following common features:

1. Similar properties, such as chemical makeup and atomic structure
2. Similar processing routes
3. Similar applications

### 3.3.1 METALS

**Metals** are combinations of metallic elements. They have large numbers of nonlocalized electrons (i.e., electrons not bound to particular atoms). Metals are extremely good conductors of electricity and heat and are not transparent to visible light; a polished metal surface has a lustrous appearance. Furthermore, metals are strong and usually deformable, making them extremely important materials in machine design.

Metals can be made stronger by alloying and by mechanical and heat treatment, and they are usually ductile. High-strength alloys, such as spring steel, can have percent elongation as low as 2%, but even this is enough to ensure that the material yields before it fractures. Some cast parts can have very low ductility, however. Because metals are usually ductile, they are often used in circumstances where cyclic loading is applied (more in Chap. 7), and they are generally resistant to corrosion. Ductile materials, such as steel, accommodate stress concentrations by deforming in a way that redistributes the load more evenly; therefore, they can be used under static loads within a small margin of their yield strength.

*Isotropy* refers to a material whose properties are the same in all directions; a material with directional properties is *anisotropic*. On a microscopic scale, metals form well-defined crystals with ordered packing of atoms. However, the crystals in a metal are vary small and are randomly oriented. Thus, while a crystal may be anisotropic, a metal should be considered as polycrystalline, and the averaged results of many crystals lead to a reasonable assumption of isotropy. However, the crystals, or grains, in a metal may be elongated or oriented, leading to anisotropic behavior.

Most metals are initially cast and then can be further processed to achieve the desired shape. Further processes include secondary casting (such as sand, shell, investment, or die casting), bulk forming (forging, extrusion, rolling, drawing), sheet forming (deep drawing, stretch forming, stamping), or machining (milling, turning, grinding, polishing). An additional option for metals is to produce metal powders and form desired shapes through powder metallurgy techniques.

### 3.3.2   Ceramics and Glasses

**Ceramics** are compounds of metallic and nonmetallic elements, most frequently oxides, nitrides, and carbides. For example, aluminum oxide (also known as alumina, carborundum, or, in single-crystal form, sapphire) is $Al_2O_3$. **Glasses** are made up of metallic and nonmetallic elements just as are ceramics, but glasses typically have no clear crystal structure. A typical soda-lime glass consists of approximately 70 wt % silicon dioxide ($SiO_2$), the balance being mainly soda ($Na_2O$) and lime ($CaO$). Both ceramics and glasses typically insulate against the passage of electricity and heat and are more resistant to high temperatures and harsh environments than are metals and polymers.

Ceramics and glasses, like metals, have high density. However, instead of being ductile as metals are, ceramics and glasses (at room temperature) are brittle. They are typically 15 times stronger in compression than in tension. Ceramics and glasses cannot deform and so are more challenging to incorporate into machine elements from, as are metals. Despite this, they have attractive features. They are stiff, hard, and abrasion-resistant (hence their use for bearings and cutting tools). Thus, they must be considered as an important class of engineering material for use in machine elements.

Ceramics have a well-defined crystalline structure, just as metals do, but the presence of ionic bonds between the metallic and nonmetallic parts of a ceramic limits its ability to deform plastically. Instead, when sufficient energy has been applied to a ceramic, it generally fractures.

Ceramics are usually obtained by forming a ceramic slurry with binders and then firing the ceramics to develop a bond between adjacent particles. Some machining operations are possible with ceramics, but ceramics are often too brittle to be successfully machined. Grinding and polishing are commonly performed successfully with ceramics.

**EXAMPLE 3.3**

**Given:**   A piece of ceramic is to be implanted into a stainless steel ring. The ceramic to be used is stabilized zirconia ($ZrO_2$), and the fit between the stainless steel and the zirconia is a medium press fit at room temperature. When the temperature fluctuates from room temperature to 500°C, the zirconia should not loosen.

**Find:**   The correct type of stainless steel to be used from those given in Appendix A (Sec. A.1).

**Solution:**   From Table A.3 zirconia has a coefficient of thermal expansion of $1 \times 10^{-5}/°C$. Thus, since the zirconia should not loosen from the steel when the temperature is increased, a slightly smaller coefficient of thermal expansion, but very close to the zirconia value, is desired for the stainless steel. From Table A.1 the martensitic type of stainless steel (AISI 410) has a coefficient of thermal expansion of $0.99 \times 10^{-5}/°C$ and is therefore the type of stainless steel preferred.

### 3.3.3 POLYMERS AND ELASTOMERS

**Polymers** and **elastomers** include plastic and rubber materials. Many polymers are organic compounds chemically based on carbon, hydrogen, and other nonmetallic elements. Furthermore, they have large molecular structures.

Polymers are of two basic types: thermoplastics and thermosets. In general, thermoplastics are more ductile than thermosets, and at elevated temperatures they soften significantly and melt. Thermosets are more brittle, do not soften as much as thermoplastics, and usually degrade chemically before melting. **Thermoplastics** are long-chain molecules, sometimes with branches, where the strength arises from interference between chains and branches. **Thermosets** have a higher degree of cross-linking, similar to the structure of a sponge.

Elastomers have a networked structure, but one that is not as elaborate as that for thermosets, so they see large deformations at relatively light loads. A common elastomer is a rubber band, which displays the typical characteristics of large elastic deformation but brittle fracture. Further, the elastic properties of rubber bands are highly nonlinear.

Polymers and elastomers can be extremely flexible with large elastic deformations. Polymers are roughly 5 times less dense than metals but have nearly equivalent strength/weight ratios. Because polymers creep (the time-dependent permanent deformation that occurs under stress) even at room temperature, a polymer machine element under load may, with time, acquire a permanent set. The properties of polymers and elastomers change greatly with variations in temperature. For example, a polymer that is tough and flexible at 20°C may be brittle at the 4°C of a household refrigerator and yet creep rapidly at the 100°C of boiling water.

The mechanical properties of polymers are specified with many of the same parameters used for metals (i.e., modulus of elasticity and tensile, impact, and fatigue strengths). However, polymers vary much more in strength, stiffness, etc., than do metals. The main reasons for this variation are that, even with the same chemical constituents, two polymers may have different chain lengths and a varying number of atoms may be in a crystalline versus amorphous state. In addition, the mechanical characteristics of polymers, for the most part, are highly sensitive to the rate of deformation, the temperature, and the chemical nature of the environment (the presence of water, oxygen, organic solvents, etc.). Therefore, the particular values given for polymer mechanical properties should be used with caution. For most polymeric materials the simple stress–strain test is used to characterize some of these parameters.

Polymers are as strong per unit weight as metals. They are easy to shape: Complicated parts performing several functions can be molded from a polymer in a single operation. However, injection molding operations are costly and can only be justified for large production runs. Large elastic deflections allow the design of polymer components that snap together, making assembly fast and cheap. Polymers are corrosion-resistant and have low coefficients of friction.

Polymers have the potential for the greatest variation in mechanical properties and for extremely high anisotropy. This can be explained by recognizing that a polymer is usually amorphous, but sometimes will orient itself into crystalline arrangements.

Thermoplastics and thermosets have very different manufacturing options and strategies. Thermoplastics are generally heated to a temperature above their melting points, formed into the desired shape, and then cooled. Examples of common manufacturing processes for thermoplastics include injection molding, blow molding, thermoforming, and

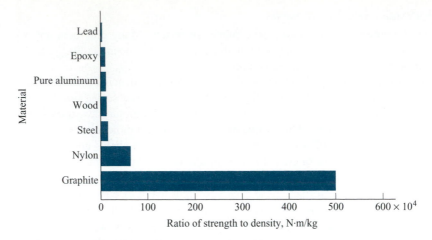

**Figure 3.3**    Strength/density ratio for various materials.

extrusion. Thermosets are blended from their constituents, molded to a desired shape, and then cured at elevated temperature and/or pressure to develop cross-links. Common manufacturing methods used for thermosets include reaction injection molding, compression molding, and potting (similar to casting).

### 3.3.4    COMPOSITES

Figure 3.3 compares a number of materials from a minimum-weight design standpoint (i.e., a larger strength/density ratio leads to a lighter design). The fibrous materials are far lighter than conventional extruded bars, molded plastics, and sintered ceramics. However, fibrous materials are notoriously susceptible to corrosion, even in air. For example, graphite fibers will oxidize readily in air and cannot provide their exceptional strength for long in an oxygen environment.

Many modern technologies require machine elements with unusual combinations of properties that cannot be met by conventional metal alloys, ceramics, and polymeric materials. Present-day technologies require solid materials that have low densities; are strong, stiff, and abrasion- and impact-resistant; and are not easily corroded. This combination of characteristics is rather formidable, considering that strong materials are usually relatively dense and that increasing stiffness generally decreases impact strength. Further, although fibers display some of these characteristics, they are easily corroded.

**Composite materials** combine the attractive properties of two or more material classes while avoiding drawbacks. A composite is designed to display a combination of the best characteristics of each component material. For example, graphite-reinforced epoxy acquires strength from the graphite fibers while the epoxy protects the graphite from oxidation. The epoxy also helps support shear stresses and provides toughness.

The three main types of composite material are

1. ***Particle-reinforced.***    Particles have approximately the same dimensions in all directions in a matrix, such as concrete.

**Figure 3.4** Cross section of fiber-reinforced composite material.

2. ***Discontinuous fiber-reinforced.*** Fibers have limited length/diameter ratio in a matrix, such as Fiberglas.

3. ***Continuous fiber-reinforced.*** Continuous fibers are constructed into a part by layers, such as graphite tennis rackets.

Figure 3.4 shows a cross section of a fiber-reinforced composite material where the fibers are assumed to be long relative to their diameter. Most such composites contain glass or carbon fibers and a polymer matrix. These composites cannot be used above 250°C because the polymer softens, but at room temperature their performance can be outstanding. Other disadvantages of using composites in machine elements are that they raise the price of the component considerably and are relatively difficult to form and join.

Composite materials have many characteristics that are different from those of the other three classes of material considered. Whereas metals, polymers, and ceramics are **homogeneous** (properties are not a function of position in the solid), **isotropic** (properties are the same in all directions at a point in the solid), or **anisotropic** (properties are different in all directions at a point in the solid), composites are *nonhomogeneous* and **orthotropic.** An orthotropic material has properties that are different in three mutually perpendicular directions at a point in the solid but has three mutually perpendicular planes of material symmetry. Consideration in this text is limited to simple, unidirectional, fiber-reinforced orthotropic composite materials, such as shown in Figure 3.4.

An important parameter in fiber-reinforced composites is the fiber length. Some critical fiber length is necessary for effective strengthening and stiffening of the composite material. The critical length $l_{cr}$ of the fiber depends on the fiber diameter $d$, its ultimate strength $S_u$, and the fiber-matrix bond strength $\tau_f$, according to

$$l_{cr} = \frac{S_u d}{2\tau_f} \tag{3.2}$$

The 2 occurs in Equation (3.2) because the fiber is embedded in the matrix and splits into two parts at failure. For a number of glass- and carbon-fiber-reinforced composites this critical length is about 1 mm, or 20 to 150 times the fiber diameter.

Composites share many of the manufacturing methods with thermosetting polymers and metals, depending on the matrix materials. Polymer matrix materials are commonly molded or placed onto a form (layup) and then cured in an oven, or formed through pultrusion, for example. Metal matrix composites use variants of casting or powder metallurgy techniques.

---

**EXAMPLE 3.4**

**Given:**   Fiber-reinforced plastic contains carbon fibers having an ultimate strength of 1 GPa and a modulus of elasticity of 150 GPa. The fibers are 3 mm long with a diameter of 30 $\mu$m.

**Find:**   How strong must the fiber-matrix bond be to ensure that the ultimate strength is safe?

**Solution:**   From Equation (3.2) the fiber-matrix bond strength can be expressed as

$$\tau_f = \frac{S_u d}{2 l_{cr}} = \frac{10^9 (30)(10^{-6})}{2(3)(10^{-3})} = 5,000,000 \text{ Pa} = 5 \text{ MPa}$$

The fiber-matrix bond strength must be greater than 5 MPa.

---

## 3.4    STRESS–STRAIN DIAGRAMS

The stress–strain diagram is important in designing machine elements because it yields data about a material's strength without regard to the material's physical size or shape. Because stress–strain diagrams differ considerably for the different classes of material covered in the previous sections, each class will be treated separately. The exception is that a stress–strain diagram for composites will not be presented because of the diverse natures of these materials.

### 3.4.1    METALS

Figure 3.5 shows the stress–strain diagram for a ductile metal. Although the stress shown in the figure is tensile, the stress–strain diagrams for metals are essentially the same for compression and tension. This feature is not true when one is dealing with polymers or ceramics.

Figure 3.6 better clarifies what is happening near the yield stress. A number of points presented in Figures 3.5 and 3.6 require some elaboration:

1. **Proportional limit** (point P) is the stress at which the stress–strain curve first deviates from a straight line (or the limit of linear-elastic theory, otherwise known as Hooke's law).

2. **Elastic limit** (point E) is the highest stress the material can withstand and still return exactly to its original length when unloaded.

3. **Yield strength** (point Y) is that **stress** at which significant deformation first occurs. If the load is removed after yielding to point Y, the specimen exhibits a 0.2% permanent elongation.

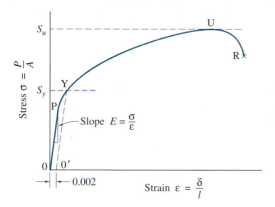

**Figure 3.5** Typical stress–strain curve for a ductile material.

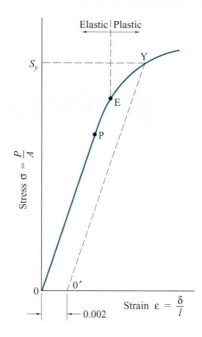

**Figure 3.6** Typical stress–strain behavior for ductile metal showing elastic and plastic deformations and yield strength $S_y$.

4. **Ultimate strength** (point U) is the maximum stress reached in the stress–strain diagram.
5. **Fracture stress** (point R) is the stress at the time of fracture or rupture. For some materials the ultimate stress is greater than the fracture stress.

Note that the elastic limit (point E) is not shown in Figure 3.5 but is shown in Figure 3.6 and clarifies what is happening near the elastic-plastic demarcation point. The true yield stress is difficult to determine experimentally. A solution is to use a small strain, usually 0.2%, and draw a straight line with its slope equal to the initial Young's modulus from the experiment. The point where this line intersects the stress–strain curve is known as the 0.2% offset yield point. In both figures the yield strength $S_y$ is determined by 0.002 strain. That is, the distance in Figure 3.6 between 0 and 0′ is 0.002 and is the permanent set. Loading occurs along 0PEY; unloading occurs along Y0′.

The strain portion of Figure 3.6 can be divided into elastic and plastic behavior. The demarcation point is point E, the elastic limit. Three different phenomena occur in plastic behavior:

1. **Yielding.** A slight increase in stress above the elastic limit (point E) will cause it to deform permanently (plastic deformation). For stress just slightly larger than the yield stress the specimen continues to elongate with little increase in load. This condition is

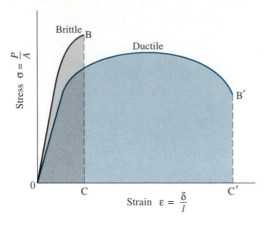

**Figure 3.7**   Typical tensile stress–strain diagrams for brittle and ductile metals loaded to fracture.

not accurately depicted in Figure 3.5 or 3.6 because the strain region while yielding is relatively small.

2. **Strain hardening.**   After yielding and before reaching the ultimate strength $S_u$, strain hardening occurs. From Figure 3.5 note that as the metal approaches $S_u$, the stress–strain curve flattens. A ductile metal usually becomes harder and stronger as it plastically deforms.

3. **Necking.**   Necking, or localized plastic deformation, results in a decreased cross-sectional area, and occurs after the ultimate stress is reached until fracture occurs, as shown in Figure 3.1.

Thus far, the discussion of stress–strain diagrams has focused on ductile metals. Figure 3.7 shows typical tensile stress–strain diagrams for a brittle and a ductile metal loaded to fracture. The brittle metal experiences little or no plastic deformation upon fracture, whereas the ductile metal is subjected to large strains before fracture occurs. Also, brittle materials have considerably higher (typically 10 times or greater) ultimate strength in compression than in tension. In contrast, ductile materials have essentially the same ultimate strength in compression and in tension.

---

**EXAMPLE 3.5**

**Given:**   A 1-in-long steel bar is stressed to its yield point and then released.

**Find:**   How much longer has it become by this loading?

**Solution:**   The definition of yield stress is that it leaves a permanent deformation of 0.2% of the original length.

$$\frac{2}{1000} \times 1 = 0.002 \text{ in}$$

This is the permanent deformation.

**Figure 3.8** Stress–strain diagram for ceramic in tension and in compression.

## 3.4.2 CERAMICS

The stress–strain behavior of ceramics is not determined by the tensile test used for metals. The reason for this is twofold. First, it is difficult to prepare and test specimens having the required geometry; second, there is a significant difference in results obtained from tests conducted in compression and in tension. Therefore, a more suitable transverse bending test is most frequently used, in which a specimen rod having either a circular or rectangular cross section is bent until fracture. Stress is computed from the specimen thickness, the bending moment, and the moment of inertia of the cross section. The maximum stress, or the stress at fracture, is known as the **modulus of rupture,** which is an important parameter used in determining the strength of ceramics.

The elastic stress–strain behavior for ceramic materials obtained in the transverse bending test is similar to the tensile test results for metals. A linear relationship exists for stress and strain. In Figure 3.8 the uniaxial stress–strain diagram for a ceramic shows that strength depends strongly on whether the loading is compressive or tensile. This is understandable because ceramics are notoriously brittle and, if they are loaded in tension, imperfections in the material become fracture initiation and propagation sites. In compression, on the other hand, defects such as microcracks are squeezed shut, so that they do not compromise the material's strength. This kind of behavior can also be seen with cast metals, which have large numbers of voids in their lattices. Strength for ceramics means fracture strength $S_{\text{fr}}^t$ in tension and crushing strength $S_{\text{fr}}^c$ in compression; typically, $S_{\text{fr}}^c \approx 15 S_{\text{fr}}^t$. Once the crushing strength is reached, the strain increases significantly but the stress decreases. The $x$ in the figure designates the elastic strain when the fracture strength in tension is reached.

---

**Given:** A bar, shown in Figure 3.9, consists of equal lengths and cross sections of magnesia and steel (AISI 1080). The magnesia and steel sections are glued together so that they bend together as one. The bar is simply supported, and bending is tested by applying a force at its center.

**EXAMPLE 3.6**

**Find:**   Is the bar strongest when the steel is at the bottom (as shown in Fig. 3.9) or at the top?

**Solution:**   From Table A.1 for AISI 1080 steel

$$E_s = 207 \text{ GPa} \qquad S_{ys} = 380 \text{ MPa} \qquad \text{and} \qquad S_{us} = 615 \text{ MPa}$$

From Table A.3 for magnesia

$$E_m = 207 \text{ GPa} \qquad \text{and} \qquad S_{um} = 105 \text{ MPa}$$

An important feature of ceramics is that they are 15 times stronger in compression than in tension, whereas steels have the same yield stress in compression or tension. In Figure 3.9 the top member is in compression while the bottom member is in tension. Thus, for the steel at the bottom and magnesia at the top, as shown in Figure 3.9, the magnesia is in compression with a compressive strength of $(15)(105)(10^6) = 1575$ MPa and is much stronger than the steel, which is in tension with a strength of 380 MPa. If the magnesia were at the bottom instead of the top, it would be in tension with a strength of 105 MPa and thus would be weaker than the steel, which would be in compression with a strength of 380 MPa. Taking the weakest members of the two bar combinations (since that is where it will fail), we find that the bar is about $380/105 = 3.642$ times stronger when the steel is at the bottom as shown in Figure 3.9.

**Figure 3.9**   Bending strength of bar used in Example 3.6.

### 3.4.3   POLYMERS

The testing techniques and specimen configurations used to develop stress–strain diagrams for metals must be modified for polymers, especially for highly elastic materials, such as rubbers. It also helps to distinguish between thermoplastics and thermosets. Thermosets typically have little plastic deformation, so that their stress–strain diagrams in tension are essentially the same as Figure 3.8.

Thermoplastics behave very differently. Figure 3.10 shows a stress–strain diagram for a thermoplastic polymer below, at, and above its glass transition temperature. The glass transition temperature $T_g$ is the temperature at which, upon cooling, a polymer transforms from a supercooled liquid to a solid. For $T \ll T_g$ the stress–strain diagram shows that the polymer fractures after relatively small strains. For $T \approx T_g$ the plastic can undergo enormous strains. Elastomers, like thermosetting polymers, are brittle but can survive large strains before fracture (as can be seen with a rubber band). Elastomers may be elastic, but they are highly nonlinear. At the yield strength $S_y$ the stress–strain diagram becomes markedly

**Figure 3.10**    Stress–strain diagram for polymer below, at, and above its glass transition temperature $T_g$.

nonlinear, having typically a strain of 0.01, as shown in Figure 3.10. This nonlinearity may be caused either by shear yielding, the irreversible slipping of molecular chains, or by crazing, the formation of low-density, cracklike volumes that scatter light, making the polymer look white. Polymers are slightly stronger (~20%) in compression than in tension. Finally, the deformation displayed for $T \ll T_g$ is totally elastic.

These stress–strain diagrams for metals, polymers, and ceramics reveal the following characteristics of the ultimate strength $S_u$ and the yield strength $S_y$ of these materials:

1. For brittle solids (ceramics, glasses, brittle polymers, and brittle metals) there is no easily defined yield strength, they are taken as the same value in design equations ($S_u = S_y$).

2. For metals, ductile polymers, and most composites the ultimate strength is larger than the yield strength by a factor of 1.1 to 4. The reason for this is mainly work hardening or, in the case of composites, load transfer to the reinforcement.

These are important observations to keep in mind when one is selecting solid materials.

---

**Given:**    A plastic cup is made of polymethylmethacrylate. The room-temperature elongation at breakage is 5%, and the glass transition temperature is 90°C. The cup is sterilized by being washed with 100°C superheated steam at a high pressure, stressing the plastic to 30 MPa.

**EXAMPLE 3.7**

**Find:**    Can the cup be expected to maintain its shape during the washing?

**Solution:**    Since the washing temperature is 10°C above the glass transition temperature, the plastic will deform by more than 5% (see Fig. 3.10). The stress is approximately one-half the ultimate strength at room temperature. It can therefore be concluded that the cup will deform during washing.

## 3.5    PROPERTIES OF SOLID MATERIALS

This section defines the various engineering properties of solid materials needed to select the proper materials for machine elements. Not all these properties may be important for each machine element considered in Chapters 9 to 20, but they are important for the wide range of applications of the various machine elements. For each property the relative behaviors of the classes of material are presented as well as the relative behaviors of the various materials within a specific class.

### 3.5.1    DENSITY

**Density** may be viewed as the mass per unit volume. The SI unit of density is kilograms per cubic meter. In the English system density is measured in pounds-mass per cubic inch. Typical densities of solid materials lie between $10^3$ and $10^4$ kg/m$^3$. Figure 3.11 illustrates the density ranking of various metals, polymers, and ceramics. Metals, such as lead, copper,

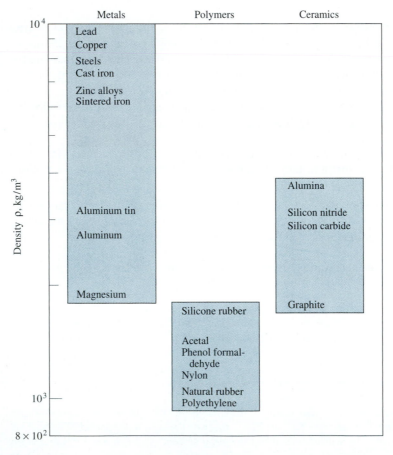

**Figure 3.11**    Density for various metals, polymers, and ceramics at room temperature (20°C; 68°F). [From *Properties of Common Engineering Materials* (1984).]

**Table 3.1**    Density for various metals, polymers, and ceramics at room temperature (20°C; 68°F).

| Material | Density $\rho$ | |
|---|---|---|
|  | kg/m³ | lbm/in³ |
| **Metals** | | |
| Aluminum and its alloys[a] | $2.7 \times 10^3$ | 0.097 |
| Aluminum tin | $3.1 \times 10^3$ | 0.11 |
| Babbitt, lead-based white metal | $10.1 \times 10^3$ | 0.36 |
| Babbitt, tin-based white metal | $7.4 \times 10^3$ | 0.27 |
| Brasses | $8.6 \times 10^3$ | 0.31 |
| Bronze, aluminum | $7.5 \times 10^3$ | 0.27 |
| Bronze, leaded | $8.9 \times 10^3$ | 0.32 |
| Bronze, phosphor (cast)[b] | $8.7 \times 10^3$ | 0.31 |
| Bronze, porous | $6.4 \times 10^3$ | 0.23 |
| Copper | $8.9 \times 10^3$ | 0.32 |
| Copper lead | $9.5 \times 10^3$ | 0.34 |
| Iron, cast | $7.4 \times 10^3$ | 0.27 |
| Iron, porous | $6.1 \times 10^3$ | 0.22 |
| Iron, wrought | $7.8 \times 10^3$ | 0.28 |
| Magnesium alloys | $1.8 \times 10^3$ | 0.065 |
| Steels[c] | $7.8 \times 10^3$ | 0.28 |
| Zinc alloys | $6.7 \times 10^3$ | 0.24 |
| **Polymers** | | |
| Acetal (polyformaldehyde) | $1.4 \times 10^3$ | 0.051 |
| Nylons (polyamides) | $1.14 \times 10^3$ | 0.041 |
| Polyethylene, high-density | $0.95 \times 10^3$ | 0.034 |
| Phenol formaldehyde | $1.3 \times 10^3$ | 0.047 |
| Rubber, natural[d] | $1.0 \times 10^3$ | 0.036 |
| Rubber, silicone | $1.8 \times 10^3$ | 0.065 |
| **Ceramics** | | |
| Alumina ($Al_2O_3$) | $3.9 \times 10^3$ | 0.14 |
| Graphite, high-strength | $1.7 \times 10^3$ | 0.061 |
| Silicon carbide (SiC) | $2.9 \times 10^3$ | 0.10 |
| Silicon nitride ($Si_2N_4$) | $3.2 \times 10^3$ | 0.12 |

[a]Structural alloys.

[b]Bar stock, typically $8.8 \times 10^3$ kg/m³ (0.03 lbm/in³).

[c]Excluding "refractory" steels.

[d]"Mechanical" rubber.

SOURCE: *Properties of Common Engineering Materials* (1984).

and steel, have the highest density. Polymers, such as nylon, natural rubber, and polyethylene, have the lowest density. Table 3.1 gives quantitative values of density at room temperature (20°C).

Metals are dense because they are made of heavy atoms in an efficient packing. Ceramics, for the most part, have lower densities than metals because ceramics contain oxygen, nitrogen, and carbon atoms, which are light, but they are also efficiently packed. Polymers have low densities because they are mainly made of carbon (atomic weight, 12) and hydrogen, and they are never totally crystalline. Thus, there is a lot of "wasted space" in a polymer.

Alloying changes the density of metals only slightly. To a first approximation the highest density of an alloy (metallic solid resulting from dissolving two or more molten metals in each other) is given by the **rule of mixtures** (i.e., by a linear interpolation between the densities of the alloy concentrations). The rule of mixtures may also be applied to composites.

### 3.5.2   MODULUS OF ELASTICITY, POISSON'S RATIO, AND SHEAR MODULUS

The **modulus of elasticity** (sometimes called the elastic constant or **Young's modulus**) is defined as the slope of the linear-elastic part of the stress–strain curve. In Figure 3.6 the linear portion of the stress–strain curve is between the origin and point P, or stress less than the proportional limit stress. Recall the description of average normal stress in a uniaxial stress state in Equation (2.7) and average normal strain in Equation (2.33). The modulus of elasticity can be written as

$$E = \frac{\sigma_{\text{avg}}}{\epsilon_{\text{avg}}} \tag{3.3}$$

Recall that the stress acts axially or normally. The SI unit of modulus of elasticity is that of stress, namely, pascals or newtons per square meter. Figure 3.12 illustrates the modulus of elasticity ranking for various metals, polymers, and ceramics at room temperature (20°C). The moduli of elasticity for metals and ceramics are high and quite similar, but those for polymers are considerably lower. Table 3.2 gives quantitative values of the moduli of elasticity, respectively, for various metals, polymers, and ceramics at room temperature. Note that in Table 3.2 the modulus of elasticity is in gigapascals and million pounds per square inch.

The moduli of elasticity of most materials depend on two factors: bond stiffness and bond density per unit area. A bond is like a spring: It has a spring rate $k$ (in newtons per meter). The modulus of elasticity $E$ is roughly

$$E = \frac{k}{r_o}$$

where $r_o$ = atom size (this can be obtained from the mean atomic volume $4\pi r_o^3/3$, which is generally known). The wide range of modulus of elasticity in Figure 3.12 and Table 3.2 is largely caused by the range of $k$ in materials. The covalent bond is stiff ($k = 20$ to 200 N/m); the metallic and ionic bonds are somewhat less stiff ($k = 15$ to 100 N/m). Diamond, although not shown in Figure 3.12 or Table 3.2, has a very high modulus of elasticity because the carbon atom is small (giving a high bond density) and its atoms are linked by extremely strong bonds (200 N/m). Metals have a high modulus of elasticity because close packing gives a high bond density and the bonds are strong, although not as strong as those of diamond. Polymers contain both strong diamondlike covalent bonds and weak hydrogen (or van der Waals) bonds ($k = 0.5$ to 2 N/m); these weak bonds stretch when a polymer is deformed, giving a lower modulus of elasticity. Elastomers have a low modulus of elasticity because the weak secondary bonds have melted, leaving only the extremely weak entropic restoring force, which is associated with tangled, long-chain molecules when the material is loaded.

Although no stress acts transversely to the axial direction, there will nevertheless be dimensional changes in the transverse direction, for as a bar experiences an axial loading and extends axially, it contracts transversely. The transverse average strain $\epsilon_{t,\text{avg}}$ is related

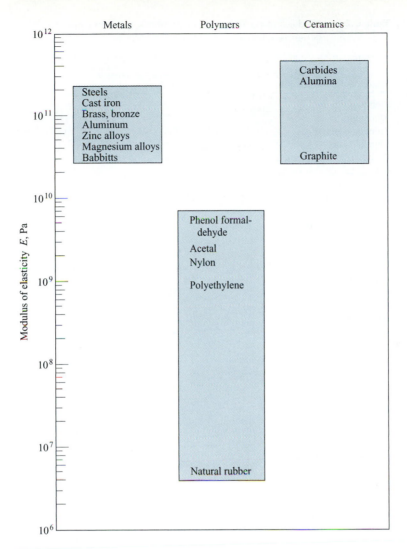

**Figure 3.12** Modulus of elasticity for various metals, polymers, and ceramics at room temperature (20°C; 68°F). [From *Properties of Common Engineering Materials* (1984).]

to the axial average strain $\epsilon_{a,\text{avg}}$ by **Poisson's ratio** $\nu$ such that

$$\epsilon_{t,\text{avg}} = -\nu\epsilon_{a,\text{avg}} \tag{3.4}$$

The minus sign simply means that the transverse deformation will be in the opposite sense to the axial deformation. Poisson's ratio is dimensionless. Table 3.3 gives quantitative values of Poisson's ratio for various metals, polymers, and ceramics at room temperature. The highest Poisson's ratio is 0.5 for rubber, and the lowest is 0.19 for silicon carbide and cemented carbides. In fact, Poisson ratio cannot be less than zero (or else the second law

**Table 3.2**    Modulus of elasticity for various metals, polymers, and
ceramics at room temperature (20°C; 68°F).

| Material | Modulus of Elasticity $E$ | |
| --- | --- | --- |
| | GPa | Mpsi |
| **Metals** | | |
| Aluminum | 62 | 9.0 |
| Aluminum alloys[a] | 70 | 10.2 |
| Aluminum tin | 63 | 9.1 |
| Babbitt, lead-based white metal | 29 | 4.2 |
| Babbitt, tin-based white metal | 52 | 7.5 |
| Brasses | 100 | 14.5 |
| Bronze, aluminum | 117 | 17.0 |
| Bronze, leaded | 97 | 14.1 |
| Bronze, phosphor (cast)[b] | 110 | 16.0 |
| Bronze, porous | 60 | 8.7 |
| Copper | 124 | 18.0 |
| Iron, gray cast | 109 | 15.8 |
| Iron, malleable cast | 170 | 24.7 |
| Iron, spheroidal graphite[b] | 159 | 23.1 |
| Iron, porous | 80 | 11.6 |
| Iron, wrought | 170 | 24.7 |
| Magnesium alloys | 41 | 5.9 |
| Steel, low alloys | 196 | 28.4 |
| Steel, medium and high alloys | 200 | 29.0 |
| Steel, stainless[c] | 193 | 28.0 |
| Steel, high-speed | 212 | 30.7 |
| Zinc alloys[d] | 50 | 7.3 |
| **Polymers** | | |
| Acetal (polyformaldehyde) | 2.7 | 0.39 |
| Nylons (polyamides) | 1.9 | 0.28 |
| Polyethylene, high-density | .9 | 0.13 |
| Phenol formaldehyde[e] | 7.0 | 1.02 |
| Rubber, natural[f] | .004 | 0.0006 |
| **Ceramics** | | |
| Alumina ($Al_2O_3$) | 390 | 56.6 |
| Graphite | 27 | 3.9 |
| Cemented carbides | 450 | 65.3 |
| Silicon carbide (SiC) | 450 | 65.3 |
| Silicon nitride ($Si_2N_4$) | 314 | 45.5 |

[a]Structural alloys.

[b]For bearings.

[c]Precipitation-hardened alloys up to 211 GPa (30 Mpsi).

[d]Some alloys up to 96 GPa (14 Mpsi).

[e]Filled.

[f]25% Carbon black "mechanical" rubber.

SOURCE: *Properties of Common Engineering Materials* (1984).

of thermodynamics would be violated), nor can it exceed 0.5 (or else a material's volume would increase when compressed).

Shear stress and strain are proportional to each other through the expression

$$\tau_{avg} = G\gamma_{avg} \tag{3.5}$$

where $G$ = shear modulus or modulus of rigidity, $N/m^2$. This relation is only true for the linear-elastic region of the shear stress–strain curve (from 0 to $P$ in Fig. 3.6). The unit of $G$ is the same as that for the modulus of elasticity $E$ because $\gamma$ is measured in radians, a dimensionless quantity.

The three material properties $E$, $G$, and $v$ are related by the equation

$$G = \frac{E}{2(1+v)} \tag{3.6}$$

Thus, when two parameters are known, the third can easily be determined from Equation (3.6). That is, if for a particular material the modulus of elasticity is obtained from Table 3.2 and Poisson's ratio from Table 3.3, the shear modulus can be obtained from Equation (3.6). The material presented thus far in this section is valid for metals, polymers, or ceramics. To establish the modulus of elasticity for a unidirectional fiber-reinforced composite in the direction of the fibers, it is assumed that the fiber-matrix interfacial bond is good, so that

**Table 3.3**      Poisson's ratio for various metals, polymers, and ceramics at room temperature (20°C; 68°F).

| Material | Poisson's ratio $v$ |
|---|---|
| **Metals** | |
| Aluminum and its alloys[a] | 0.33 |
| Brasses | 0.33 |
| Bronze | 0.33 |
| Bronze, porous | 0.22 |
| Copper | 0.33 |
| Iron, cast | 0.26 |
| Iron, porous | 0.20 |
| Iron, wrought | 0.30 |
| Magnesium alloys | 0.33 |
| Steels, | 0.30 |
| Zinc alloys | 0.27 |
| **Polymers** | |
| Nylons (polyamides) | 0.40 |
| Polyethylene, high-density | 0.35 |
| Rubber | 0.50 |
| **Ceramics** | |
| Alumina ($Al_2O_3$) | 0.28 |
| Cemented carbides | 0.19 |
| Silicon carbide (SiC) | 0.19 |
| Silicon nitride ($Si_2N_4$) | 0.26 |

[a]Structural alloys.

SOURCE: *Properties of Common Engineering Materials* (1984).

deformation of both matrix and fibers is the same. Under these conditions the total load sustained by the composite $P_c$ is equal to the loads carried by the matrix $P_m$ and the fiber $P_f$, as

$$P_c = P_m + P_f \qquad (3.7)$$

where subscripts $c$, $m$, and $f$ refer to composite, matrix, and fiber, respectively. Substituting Equation (3.7) into the definition of stress [Eq. (2.7)] gives

$$\sigma_{c,\text{avg}} = \sigma_{m,\text{avg}}\frac{A_m}{A_c} + \sigma_{f,\text{avg}}\frac{A_f}{A_c} \qquad (3.8)$$

If the composite, matrix, and fiber lengths are equal, Equation (3.8) becomes

$$\sigma_{c,\text{avg}} = \sigma_{m,\text{avg}}v_m + \sigma_{f,\text{avg}}v_f \qquad (3.9)$$

where $v_m$ = volume fraction of matrix
$v_f$ = volume fraction of fiber

Because the same deformation of matrix and fibers was assumed and the composite consists of only matrix and fibers (that is, $v_m + v_f = 1$), Equation (3.9) can be rewritten in terms of the modulus of elasticity, instead of the stress, as

$$E_c = E_m v_m + E_f v_f \qquad (3.10)$$

or $$E_c = E_m(1 - v_f) + E_f v_f \qquad (3.11)$$

Thus, Equation (3.11) enables the composite's modulus of elasticity to be determined when the moduli of elasticity of the matrix and the fiber and the volume fractions of each are known. It can also be shown that the ratio of the load carried by the fibers to that carried by the matrix is

$$\frac{P_f}{P_m} = \frac{E_f v_f}{E_m v_m} \qquad (3.12)$$

Recall that Equations (3.7) to (3.12) are only applicable for unidirectional, fiber-reinforced composites.

---

**EXAMPLE 3.8**

**Given:** A fiber-reinforced plastic contains 10 vol % glass fibers ($E = 70$ GPa, $S_u = 0.7$ GPa).

**Find:** Calculate how this fiber percentage has to be changed to give the same elastic properties if the glass fibers are changed to carbon fibers ($E = 150$ GPa, $S_u = 1$ GPa). For the matrix material $E_m = 2$ GPa.

**Solution:** According to Equation (3.10), the modulus of elasticity for a fiber composite is

$$E_c = E_m v_m + E_f v_f$$

where $E_m$ = modulus of elasticity for matrix material, which has volume fraction $v_m$
$E_f$ = modulus of elasticity for fiber material, which has volume fraction $v_f$

$$E_c = 2 \times 10^9 \times 0.9 + 70 \times 10^9 \times 0.1 = 8.8 \times 10^9 \text{ Pa} = 8.8 \text{ GPa}$$

This composite modulus of elasticity should also be maintained for the carbon-reinforced plastic.

$$8.8 \times 10^9 = 2 \times 10^9 \times x + 150 \times 10^9 (1 - x)$$
$$148x = 141.2$$
$$x = 0.954 \quad \text{or} \quad 1 - x = 0.046$$

The plastic should thus contain 4.6 vol % carbon fibers and 95.4 vol % matrix material to get the same elastic properties as those of the glass-fiber-reinforced plastic.

### 3.5.3  STRENGTH

The strength of a machine element depends on the class, treatment, and geometry of the specimen and on the type of loading that the machine element will experience. This section focuses on the strength of the various classes of material and then deals with the type of loading.

The various types of loading that a material experiences are important. Design deals with allowable stresses, or reduced value of strength. The allowable normal stress $\sigma_{all}$ and the allowable shear stress $\tau_{all}$ for ferrous and nonferrous metals for various types of loading may be represented by

Tension: $\qquad\qquad 0.45S_y \leq \sigma_{all} \leq 0.60S_y \qquad\qquad$ (3.13)

Shear: $\qquad\qquad\qquad \tau_{all} = 0.40S_y \qquad\qquad\qquad$ (3.14)

Bending: $\qquad\qquad 0.60S_y \leq \sigma_{all} \leq 0.75S_y \qquad\qquad$ (3.15)

Bearing: $\qquad\qquad\qquad \sigma_{all} = 0.9S_y \qquad\qquad\qquad$ (3.16)

These relationships can also be applied to polymers and ceramics if ultimate strength at the break and fracture strength, respectively, are substituted for yield strength in Equations (3.13) to (3.16). Recall from Sec. 1.4.1 that other factors besides type of loading are incorporated into the safety factor.

### Metals

Metals can be divided into ferrous and nonferrous alloys. Ferrous alloys are those in which iron is the primary component, but carbon as well as other alloying elements may be present. Nonferrous alloys are all those alloys that *are not* iron-based.

The strength of metals is directly associated with the yield strength of the material. The yield strength $S_y$ is determined by 0.002 of the strain after unloading (see Fig. 3.6). The strength of metals is essentially equal in compression or in tension. Tables A.1 and A.2 show the yield strengths for ferrous and nonferrous metals, respectively.

### Polymers

The strength of polymers is determined by a total strain of 0.010, as opposed to a nonrecoverable plastic strain of 0.002 for metals. Also, when dealing with polymers the strength of interest is the ultimate strength at the break rather than the yield strength, as for metals. The tensile strengths at the break for selected thermoplastic and thermosetting polymers are given in Table A.4. The other unique characteristic of polymers is that they are stronger (~20%) in compression than in tension.

*Ceramics*

The strength of interest for ceramics is the fracture strength. Ceramics, being brittle materials, are much stronger in compression (typically, 15 times) than in tension. Table A.3 gives fracture strength in tension for selected ceramic materials.

---

**EXAMPLE 3.9**

**Given:** For the contact pressure distribution between a sphere and a flat surface, H. Hertz calculated in 1881 that the maximum contact pressure in the middle of the contact area was approximately 10 times larger than the maximum tensile stress at the edge of the contact area. When a steel ball is dropped on a thick glass plate, a circular crack is formed, the size of which approximately equals the maximum Hertzian contact size.

**Find:** How much stronger is the glass in compression than in tension?

**Solution:** Since the glass is not crushed by the high contact pressure but by the tensile stress, it is concluded that the compressive strength of the glass is at least 10 times the tensile strength.

---

### 3.5.4   RESILIENCE AND TOUGHNESS

**Resilience** is the capacity of a material to absorb energy when it is deformed elastically and then, upon unloading, to release this energy. The modulus of resilience $U_r$ is the strain energy per unit volume required to stress a material from an unloaded state to the point of yielding. Mathematically, this is expressed as

$$U_r = \int_0^{\epsilon_y} \sigma \, d\epsilon \tag{3.17}$$

where $\epsilon_y$ = strain corresponding to yield strength $S_y$. For the linear-elastic region the area below the stress–strain diagram is the modulus of resilience or

$$U_r = \frac{S_y \epsilon_y}{2} \tag{3.18}$$

The SI unit of modulus of resilience is pascals (Pa).

By making use of Equation (3.3) while dropping the "average" subscript we get Equation (3.18) as

$$U_r = \frac{S_y^2}{2E} \tag{3.19}$$

Thus, resilient materials are those having high yield strengths and low moduli of elasticity. An example of a material with high modulus of resilience is high-carbon steel. Table A.1 shows that the yield strength for high-carbon steel is highest when the modulus of elasticity is high, but the difference between the highest and lowest values given is small. Thus, high-carbon steels have a high modulus of resilience. This property is extremely useful in selecting a material for springs (see Chap. 17), making high-carbon steel alloys primary candidate materials for springs.

**Toughness** is a material's ability to absorb energy up to fracture. Specimen geometry and manner of load application are important in determining the toughness of a material. Fracture toughness indicates a material's resistance to fracture when a crack is present.

For the static (low-strain-rate) situation, toughness can be obtained from the stress–strain curve (e.g., see Fig. 3.5) up to the point of fracture or rupture. Resilience is the strain energy per unit volume up to the yield strength of the material (point Y in Fig. 3.5); whereas toughness is energy per unit volume to rupture (point R in Fig. 3.5). The unit of toughness is the same as that for resilience (pascals).

For a material to be tough, it must display both strength and ductility; and often, ductile materials are tougher than brittle materials, as demonstrated in Figure 3.7. Even though the brittle material has higher yield and ultimate strengths, by virtue of lack of ductility, it has lower toughness than the ductile material. That is, the area of a ductile material (given as 0B′C′) is considerably larger than the area of a brittle material (given as 0BC).

---

**EXAMPLE 3.10**

**Given:** In a mining operation iron ore is dumped into a funnel that guides the ore into a train boxcar. The inside of the funnel wears rapidly because the impact of the ore causes plastic deformations in the funnel surface. A change of surface material is considered.

**Find:** Which is the most efficient choice of funnel surface material, hard steel or rubber?

**Solution:** A key parameter to be used in this evaluation is the resilience of the two materials. For high-carbon steel (AISI 1080) from Table A.1

$$S_y = 380 \text{ MPa} \quad \text{and} \quad E = 207 \text{ GPa}$$

From Equation (3.19) the modulus of resilience for steel is

$$(U_r)_{\text{steel}} = \frac{S_y^2}{2E} = \frac{(380)^2(10^{12})}{2(207)(10^9)} = 348,800 \text{ Pa} = 0.3488 \text{ MPa}$$

For natural rubber from Table A.5 and Equation (3.4)

$$S_u = 30 \text{ MPa} \quad \text{and} \quad E = 0.004 \text{ GPa}$$

$$\therefore \quad (U_r)_{\text{rubber}} = \frac{(30)^2(10^{12})}{2(4)(10^6)} = 112.5(10^6) \text{ Pa} = 112.5 \text{ MPa}$$

Rubber is over 2 orders of magnitude more resilient than steel. The rubber is able to elastically deform on impact, whereas the steel plastically deforms and thus has a higher wear rate. The funnel should have a rubber lining.

---

### 3.5.5 THERMAL CONDUCTIVITY

The rate at which heat is conducted through a solid at steady state (meaning that temperature does not vary with time) is a measure of the **thermal conductivity** $K_t$. When two bodies at different temperatures are brought together, the faster-moving molecules of the warmer body collide with the slower-moving molecules of the cooler body and transfer some of their motion to the latter. The warmer body loses energy (drops in temperature) while the cooler one gains energy (rises in temperature). The transfer process stops when the two bodies reach the same temperature. This transfer of molecular motion through a material is called *heat conduction*. Materials differ in how fast they let this transfer go on. The SI unit of thermal conductivity $K_t$ is watts per meter per degree Celsius; and the English unit is British thermal units per foot per hour per degree Fahrenheit.

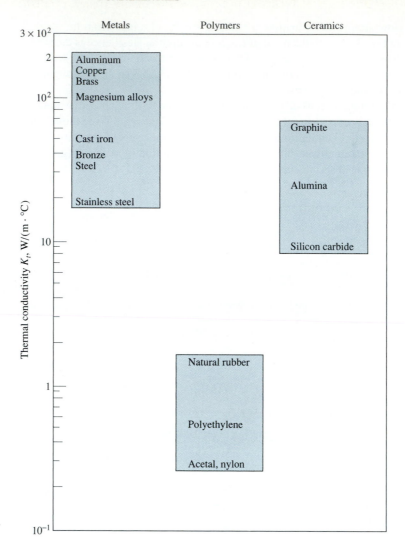

**Figure 3.13**    Thermal conductivity for various metals, polymers, and ceramics at room temperature (20°C; 68°F). [From *Properties of Common Engineering Materials* (1984).]

Figure 3.13 illustrates the thermal conductivity ordering for various metals, polymers, and ceramics. The metals and ceramics in general are good conductors (high $K_t$), and the polymers are good insulators (low $K_t$). Table 3.4 quantifies the thermal conductivity results given in Figure 3.13. In Figure 3.13 and Table 3.4, unless otherwise stated, the temperature is assumed to be room temperature (20°C; 68°F).

### 3.5.6    LINEAR THERMAL EXPANSION COEFFICIENT

Different materials expand at different rates when heated. A solid object elongates by a certain fraction for each degree rise in temperature. This rate is constant over a fairly large

**Table 3.4**    Thermal conductivity for various metals, polymers, and ceramics at room temperature (20°C; 68°F).

| Material | Thermal conductivity, $K_t$ | |
| --- | --- | --- |
|  | W/(m · °C) | Btu/(ft · h · °F) |
| **Metals** | | |
| Aluminum | 209 | 120 |
| Aluminum alloys, cast[a] | 146 | 84 |
| Aluminum alloys, silicon[b] | 170 | 98 |
| Aluminum alloys, wrought[c] | 151 | 87 |
| Aluminum tin | 180 | 100 |
| Babbitt, lead-based white metal | 24 | 14 |
| Babbitt, tin-based white metal | 56 | 32 |
| Brasses[a] | 120 | 69 |
| Bronze, aluminum[a] | 50 | 29 |
| Bronze, leaded | 47 | 27 |
| Bronze, phosphor (cast)[d] | 50 | 29 |
| Bronze, porous | 30 | 17 |
| Copper[a] | 170 | 98 |
| Copper lead | 30 | 17 |
| Iron, gray cast | 50 | 29 |
| Iron, spheroidal graphite | 30 | 17 |
| Iron, porous | 28 | 16 |
| Iron, wrought | 70 | 40 |
| Magnesium alloys | 110 | 64 |
| Steel, low alloys[e] | 35 | 20 |
| Steel, medium alloys | 30 | 17 |
| Steel, stainless[f] | 15 | 8.7 |
| Zinc alloys | 110 | 64 |
| **Polymers** | | |
| Acetal (polyformaldehyde) | 0.24 | 0.14 |
| Nylons (polyamides) | 0.25 | 0.14 |
| Polyethylene, high-density | 0.5 | 0.29 |
| Rubber, natural | 1.6 | 0.92 |
| **Ceramics** | | |
| Alumina ($Al_2O_3$)[g] | 25 | 14 |
| Graphite, high-strength | 125 | 72 |
| Silicon carbide (SiC) | 15 | 8.6 |

[a] At 100°C.

[b] At 100°C [~ 150 W/(m · °C) at 25°C].

[c] 20 to 100°C.

[d] Bar stock, typically 69 W/(m · °C).

[e] 20 to 200°C.

[f] Typically 22 W/(m · °C) at 200°C.

[g] Typically 12 W/(m · °C) at 400°C.

SOURCE: *Properties of Common Engineering Materials* (1984).

temperature range and, once measured, can be used to determine how much an object will expand for a given change in temperature. This rate is given for each material by a number called the linear expansivity or linear **thermal expansion coefficient** $\bar{a}$. The SI unit of $\bar{a}$ is $(°C)^{-1}$; the English unit is $(°F)^{-1}$. If the material is isotropic, the volume expansion per degree is $3\bar{a}$. Figure 3.14 illustrates the linear thermal expansion coefficient ordering for various metals, polymers, and ceramics applied over the temperature range 20 to 200°C (68 to 392°F). The polymers have the highest $\bar{a}$, followed by the metals and then the ceramics. Table 3.5 gives quantitative values of $\bar{a}$ for various metals, polymers, and ceramics from 20 to 200°C (68 to 392°F).

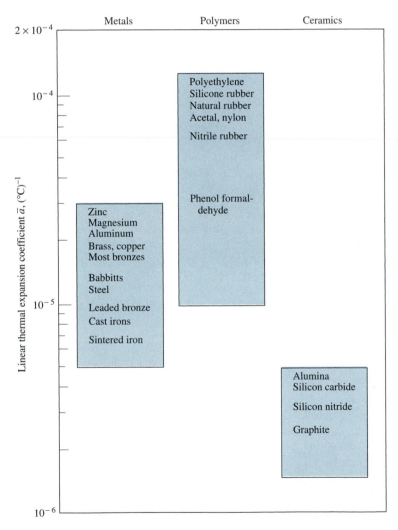

**Figure 3.14**    Linear thermal expansion coefficient for various metals, polymers, and ceramics at room temperature (20°C; 68°F). [From *Properties of Common Engineering Materials* (1984).]

**Table 3.5**  Linear thermal expansion coefficient for various metals, polymers, and ceramics at room temperature (20°C; 68°F).

| Material | Linear thermal expansion coefficient $\bar{a}$ | |
|---|---|---|
| | $(°C)^{-1}$ | $(°F)^{-1}$ |
| **Metals** | | |
| Aluminum | $23 \times 10^{-6}$ | $12.8 \times 10^{-6}$ |
| Aluminum alloys[a] | $24 \times 10^{-6}$ | $13.3 \times 10^{-6}$ |
| Aluminum tin | $24 \times 10^{-6}$ | $13.3 \times 10^{-6}$ |
| Babbitt, lead-based white metal | $20 \times 10^{-6}$ | $11 \times 10^{-6}$ |
| Babbitt, tin-based white metal | $23 \times 10^{-6}$ | $13 \times 10^{-6}$ |
| Brasses | $19 \times 10^{-6}$ | $10.6 \times 10^{-6}$ |
| Bronzes | $18 \times 10^{-6}$ | $10.0 \times 10^{-6}$ |
| Copper | $18 \times 10^{-6}$ | $10.0 \times 10^{-6}$ |
| Copper lead | $18 \times 10^{-6}$ | $10.0 \times 10^{-6}$ |
| Iron, cast | $11 \times 10^{-6}$ | $6.1 \times 10^{-6}$ |
| Iron, porous | $12 \times 10^{-6}$ | $6.7 \times 10^{-6}$ |
| Iron, wrought | $12 \times 10^{-6}$ | $6.7 \times 10^{-6}$ |
| Magnesium alloys | $27 \times 10^{-6}$ | $15 \times 10^{-6}$ |
| Steel, alloys[b] | $11 \times 10^{-6}$ | $6.1 \times 10^{-6}$ |
| Steel, stainless | $17 \times 10^{-6}$ | $9.5 \times 10^{-6}$ |
| Steel, high-speed | $11 \times 10^{-6}$ | $6.1 \times 10^{-6}$ |
| Zinc alloys | $27 \times 10^{-6}$ | $15 \times 10^{-6}$ |
| **Polymers** | | |
| Thermoplastics[c] | $(60\text{–}100) \times 10^{-6}$ | $(33\text{–}56) \times 10^{-6}$ |
| Thermosets[d] | $(10\text{–}80) \times 10^{-6}$ | $(6\text{–}44) \times 10^{-6}$ |
| Acetal (polyformaldehyde) | $90 \times 10^{-6}$ | $50 \times 10^{-6}$ |
| Nylons (polyamides) | $100 \times 10^{-6}$ | $56 \times 10^{-6}$ |
| Polyethylene, high-density | $126 \times 10^{-6}$ | $70 \times 10^{-6}$ |
| Phenol formaldehyde[e] | $(25\text{–}40) \times 10^{-6}$ | $(14\text{–}22) \times 10^{-6}$ |
| Rubber, natural[f] | $(80\text{–}120) \times 10^{-6}$ | $(44\text{–}67) \times 10^{-6}$ |
| Rubber, nitrile[g] | $34 \times 10^{-6}$ | $62 \times 10^{-6}$ |
| Rubber, silicone | $57 \times 10^{-6}$ | $103 \times 10^{-6}$ |
| **Ceramics** | | |
| Alumina ($Al_2O_3$)[h] | $5.0 \times 10^{-6}$ | $2.8 \times 10^{-6}$ |
| Graphite, high-strength | $1.4\text{–}4.0 \times 10^{-6}$ | $0.8\text{–}2.2 \times 10^{-6}$ |
| Silicon carbide (SiC) | $4.3 \times 10^{-6}$ | $2.4 \times 10^{-6}$ |
| Silicon nitride ($Si_3N_4$) | $3.2 \times 10^{-6}$ | $1.8 \times 10^{-6}$ |

[a] Structural alloys.

[b] Cast alloys can be up to $15 \times 10^{-6}(°C)^{-1}$.

[c] Typical bearing materials.

[d] $25 \times 10^{-6}(°C)^{-1}$ to $80 \times 10^{-6}(°C)^{-1}$ when reinforced.

[e] Mineral-filled.

[f] Fillers can reduce coefficients.

[g] Varies with composition.

[h] 0 to 200°C.

SOURCE: *Properties of Common Engineering Materials* (1984).

### 3.5.7    SPECIFIC HEAT CAPACITY

The nature of a material determines the amount of heat transferred to or from a body when its temperature changes by a given amount. Imagine an experiment in which a cast iron ball and a babbitt (lead-based white metal) ball of the same size are heated to the temperature of boiling water and then laid on a block of wax. The cast iron ball would melt a considerable amount of wax, but the babbitt ball, in spite of its greater mass, would melt hardly any. It therefore would seem that different materials, in cooling through the same temperature range, give up different amounts of heat.

The quantity of heat energy given up or taken on when a body changes temperature is proportional to the mass of the object, the amount that its temperature changes, and a characteristic number called the **specific heat capacity** of the material the body is made

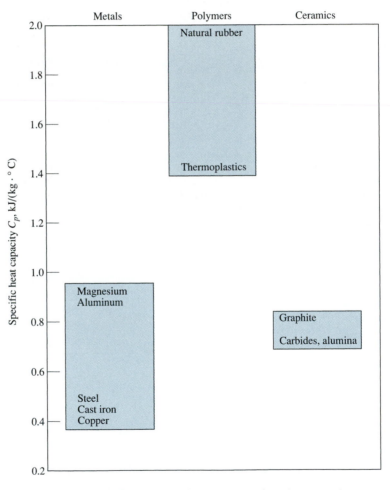

**Figure 3.15**    Specific heat capacity for various metals, polymers, and ceramics at room temperature (20°C; 68°F). [From *Properties of Common Engineering Materials* (1984).]

from:

$$\hat{Q} = C_p m_a (\Delta T) \qquad \text{(3.20)}$$

where $\hat{Q}$ = quantity of heat, J
$C_p$ = specific heat of material, J/(kg · °C)
$m_a$ = mass of body, kg
$\Delta T$ = temperature change, °C

Figure 3.15 illustrates the specific heat capacity ordering of various metals, polymers, and ceramics at room temperature (20°C; 68°F). Polymers have considerably higher specific heat than metals or ceramics. Table 3.6 quantifies the information presented in Figure 3.15. From the simple experiment given at the beginning of this section and the values in Table 3.6, the specific heat capacity $C_p$ of cast iron is 2.8 times that of lead babbitt. Table 3.1 shows lead babbitt to have 1.4 times the mass density of cast iron. Because cast iron's specific heat ratio is considerably higher relative to its mass than the babbitt's, it is clear why the cast iron ball melts a considerable amount of wax and the babbitt ball melts hardly any.

**Table 3.6**  Specific heat capacity for various metals, polymers, and ceramics at room temperature (20°C; 68°F).

| Material | Specific heat capacity $C_p$ | |
|---|---|---|
| | kJ/(kg · °C) | Btu/(lbm · °F) |
| **Metals** | | |
| Aluminum and its alloys | 0.9 | 0.22 |
| Aluminum tin | 0.96 | 0.23 |
| Babbitt, lead-based white metal | 0.15 | 0.036 |
| Babbitt, tin-based white metal | 0.21 | 0.05 |
| Brasses | 0.39 | 0.093 |
| Bronzes | 0.38 | 0.091 |
| Copper[a] | 0.38 | 0.091 |
| Copper lead | 0.32 | 0.076 |
| Iron, cast | 0.42 | 0.10 |
| Iron, porous | 0.46 | 0.11 |
| Iron, wrought | 0.46 | 0.11 |
| Magnesium alloys | 1.0 | 0.24 |
| Steels[b] | 0.45 | 0.11 |
| Zinc alloys | 0.4 | 0.096 |
| **Polymers** | | |
| Thermoplastics | 1.4 | 0.33 |
| Rubber, natural | 2.0 | 0.48 |
| **Ceramics** | | |
| Graphite | 0.8 | 0.2 |
| Cemented carbides | 0.7 | 0.17 |

[a]Aluminum bronze up to 0.48 kJ/(kg · °C) [0.12 Btu/(lbm · °F)].
[b]Rising up to 0.55 kJ/(kg · °C) [0.13 Btu/(lbm · °F)] at 200°C (392°F).
SOURCE: *Properties of Common Engineering Materials* (1984).

**EXAMPLE 3.11**

**Given:** A thermos made of steel where the inner bottle weighs 200 g and is filled with 500 g of boiling water. The initial temperature of the thermos is 20°C. The specific heat capacity for water $C_p = 4180$ J/(kg · °C).

**Find:**

(a) The maximum temperature of the water in the thermos when the heat has spread to the thermos walls

(b) The maximum temperature if the first hot water is emptied from the bottle and replaced by 500 g of new boiling water

**Solution:**

(a) No heat is dissipated to the surrounding area.

$$(m_a C_p)_{\text{steel}}(T - 20°C) = (m_a C_p)_{\text{water}}(100°C - T)$$

or

$$T = \frac{100(m_a C_p)_{\text{water}} + 20(m_a C_p)_{\text{steel}}}{(m_a C_p)_{\text{water}} + (m_a C_p)_{\text{steel}}}$$

$$= \frac{100(0.5)(4.180) + 20(0.2)(0.45)}{0.5(4.180) + 0.2(0.45)} = 96.7°C$$

(b) The only change in the above is that 20°C is replaced with 96.7°C.

$$T = \frac{100(0.5)(4.180) + 96.7(0.2)(0.45)}{0.5(4.180) + 0.2(0.45)} = 99.9°C$$

There is no need to empty the bottle and replace with new boiling water.

### 3.5.8   ARCHARD WEAR CONSTANT

Wear is harder than the other seven properties of solid materials to quantify, partly because it is a surface phenomenon and not a bulk phenomenon and partly because wear involves interactions between two materials, not just the property of one material. When solids slide, the volume of material lost from one surface, per unit distance slid, is called the *wear rate* $W_r$. The wear resistance of the surface is characterized by the **Archard wear constant** $K_A$ (SI unit of square meters per newton, or inverse pascals, and English unit of square inches per pound-force) defined by

$$\frac{W_r}{A} = K_A p \tag{3.21}$$

where $A$ = area of surface, m²

$p$ = normal pressure pressing surfaces together, Pa

The equations of wear are covered in greater detail in Chapter 8.

## 3.6 STRESS–STRAIN RELATIONSHIPS

As discovered earlier in this chapter, the stress–strain diagrams for most engineering materials exhibit a linear relationship between stress and strain within the elastic range. Therefore, an increase in stress causes a proportionate increase in strain. First discovered by Robert Hooke in 1678, this linear relationship between stress and strain in the elastic range is known as *Hooke's law*. Thomas Young, in 1807, suggested that the ratio of stress to strain is a measure of a material's stiffness. This ratio, called Young's modulus or the modulus of elasticity, is the slope of the straight-line portion of the stress–strain diagram and is given in Equation (3.3). Thus, for a uniaxial stress state, Hooke's law can be expressed as

$$\sigma = E\epsilon \tag{3.22}$$

This equation establishes the proportionality between stress and strain in simple tension or compression, discussed in Section 3.5.2.

Chapter 2 discussed stress and strain separately, and this chapter discusses material properties such as the modulus of elasticity. Appendix B explores the stress–strain relationships for biaxial and triaxial states. Here, the discussion is limited to solids loaded in the elastic range. Furthermore, only isotropic materials are considered (i.e., materials having identical physical properties, in particular, elastic properties, in any direction). Many engineering materials are isotropic, with the most commonly cited exceptions of wood and reinforced concrete.

---

**EXAMPLE 3.12**

**Given:** A 1-m-long rigid beam shown in Figure 3.16 is pinned at its left end, carries a 3-kN vertical load at its right end, and is kept horizontal by a vertical pillar located 0.3 m from the left end. The pillar is a 0.5-m-long steel tube with an outer diameter of 0.1 m and a wall thickness of 5 mm. The modulus of elasticity of the steel used throughout is 205 GPa.

**Find:** How much does the right end of the beam deflect due to the force?

**Solution:** Moment equilibrium about the hinge pin gives

$$P_1 l_1 - P_2 l_2 = 0$$

$$\therefore \quad P_2 = \frac{P_1 l_1}{l_2} = \frac{(3000)(1)}{0.3} = 10{,}000 \text{ N}$$

Because the beam can rotate only about the hinge pin, the deflection $\delta_1$ at $P_1$ can be described by the rotational angle $\alpha$, and this angle is the same when one is describing the deflection $\delta_2$ at $P_2$.

$$\therefore \quad \delta_1 = l_1 \alpha \quad \text{and} \quad \delta_2 = l_2 \alpha$$

This implies that

$$\delta_1 = \frac{l_1 \delta_2}{l_2} \tag{a}$$

**Figure 3.16**     Rigid beam assembly use in Example 3.12.

The compression of the vertical pillar $\delta_2$ can be obtained by using Hooke's law or

$$\delta_2 = \frac{l_3 P_2}{E A} = \frac{l_3 P_2}{\pi E \left(r_o^2 - r_i^2\right)} \tag{b}$$

Substituting Equation $(b)$ into Equation $(a)$ gives

$$\delta_1 = \frac{l_1 l_3 P_2}{\pi l_2 E \left(r_o^2 - r_i^2\right)} = \frac{1(0.5)(10{,}000)}{\pi (0.3)(205)(10^9)(0.05^2 - 0.045^2)} = 5.45 \times 10^{-5} \text{ m} = 54.5 \,\mu\text{m}$$

## 3.7   TWO-PARAMETER MATERIALS CHARTS

Material properties limit the performance and life of machine elements. Performance and life seldom depend on just one property of the solid materials. Thus, the information presented in Section 3.5 on the individual properties of solid materials is not adequate for selecting a material for a particular application. Instead, one or several combinations of properties are needed. Some important property combinations are

- Stiffness versus density ($E$ versus $\rho$)
- Strength versus density ($S$ versus $\rho$)
- Stiffness versus strength ($E$ versus $S$)
- Wear rate versus limiting pressure ($K_A$ versus $p_l$)

A number of other combinations might be useful in material selection, but these are the primary considerations in designing machine elements. The material presented in this section comes from Ashby (1999).

### 3.7.1   STIFFNESS VERSUS DENSITY

Modulus of elasticity and density (corresponding to stiffness and weight) are familiar properties in selecting a solid material. Lead is heavy; steel is stiff; rubber is compliant: These are the effects of modulus of elasticity and density. Figure 3.17 shows the full range of modulus of elasticity $E$ and density $\rho$ for engineering materials. Data for members of a particular class of material cluster together and are enclosed by a heavy line. The same class cluster appears on all the diagrams. Table 3.7 shows the various classes and their members with the short name for each member.

Figure 3.17 shows that the moduli of elasticity for engineering materials span five decades from 0.01 to 1000 GPa (1.47 to $147 \times 10^3$ ksi); the density spans a factor of 200, from less than 100 to 20,000 kg/m$^3$. The shear modulus $G \approx 3E/8$, and the bulk modulus $K \approx E$ for all materials except elastomers, for which they can be approximated as $G = E/3$ and $K \gg E$.

The chart helps in common problems of material selection for applications in which weight must be minimized. For example, consider a simple tension member where the weight is to be minimized under a constraint that the strain cannot exceed a given value $\epsilon_{cr}$. If the material is loaded up to its yield point, the stress is given by $\sigma = P/A$. Also, from Hooke's law

$$\sigma = E\epsilon_{cr}$$

Equating the stresses gives

$$A = \frac{P}{E\epsilon_{cr}} \tag{3.23}$$

The weight of the member is

$$W = Agl\rho$$

Substituting Equation (3.23) into the above equation gives

$$\frac{W}{g} = \frac{Pl}{\epsilon_{cr}} \frac{1}{E/\rho} \tag{3.24}$$

Note that the first fraction contains design constraints and the second fraction contains material properties. Thus, the optimum material for minimizing Equation (3.24) is one that maximizes the quantity $E/\rho$. In Figure 3.17 one would draw lines parallel to the $E/\rho = C$ reference. Those materials with the greatest value in the direction normal to these lines are the optimum materials (i.e., those farthest toward the top and left of the charts). In Figure 3.17 the reference lines refer to the minimum-weight design subjected to strain requirements under the following conditions:

| | |
|---|---|
| $\dfrac{E}{\rho} = C$ | Minimum-weight design of stiff tension members |
| $\dfrac{E^{1/2}}{\rho} = C$ | Minimum-weight design of stiff beams and columns |
| $\dfrac{E^{1/3}}{\rho} = C$ | Minimum-weight design of stiff plates |
| $\left(\dfrac{E}{\rho}\right)^{1/2} = C$ | Wave speed in material |

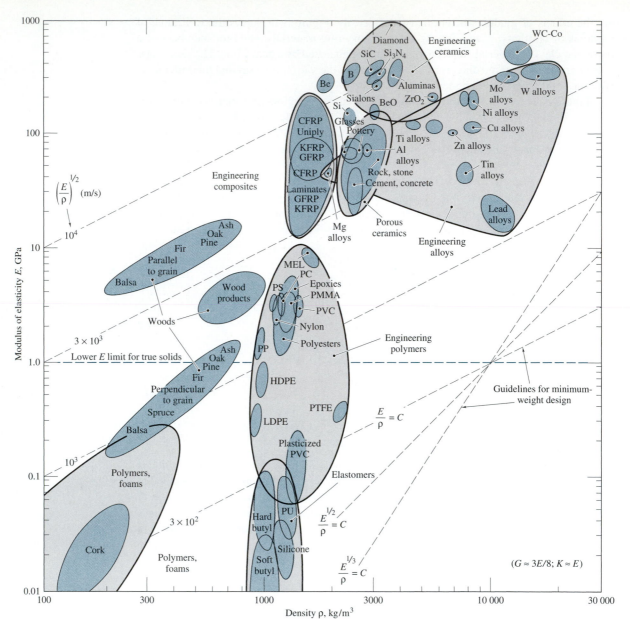

**Figure 3.17**     Modulus of elasticity plotted against density. The heavy envelopes enclose data for a given class of material. The diagonal contours show the longitudinal wave velocity. The guidelines of constant $E/\rho$, $E^{1/2}/\rho$, and $E^{1/3}/\rho$ allow selection of materials for minimum-weight, deflection-limited design. [Ashby, 1999.]

**Table 3.7** Material classes and members and short names of each member.

| Class | Members | Short name |
|---|---|---|
| **Engineering alloys** (the metals and alloys of engineering) | Aluminum alloys | Al alloys |
| | Copper alloys | Cu alloys |
| | Lead alloys | Lead alloys |
| | Magnesium alloys | Mg alloys |
| | Molybdenum alloys | Mo alloys |
| | Nickel alloys | Ni alloys |
| | Steels | Steels |
| | Tin alloys | Tin alloys |
| | Titanium alloys | Ti alloys |
| | Tungsten alloys | W alloys |
| | Zinc alloys | Zn alloys |
| **Engineering polymers** (the thermoplastics and thermosets of engineering) | Epoxies | EP |
| | Melamines | MEL |
| | Polycarbonate | PC |
| | Polyesters | PEST |
| | Polyethylene, high-density | HDPE |
| | Polyethylene, low-density | LDPE |
| | Polyformaldehyde | PF |
| | Polymethylmethacrylate | PMMA |
| | Polypropylene | PP |
| | Polytetrafluoroethylene | PTFE |
| | Polyvinyl chloride | PVC |
| | Alumina | $Al_2O_3$ |
| | Diamond | C |
| | Sialons | Sialons |
| | Silicon carbide | SiC |
| | Silicon nitride | $Si_3N_4$ |
| | Zirconia | $ZrO_2$ |
| **Engineering composites** (the composites of engineering practice) A distinction is drawn between the properties of a ply (uniply) and of a laminate (laminates). | Carbon-fiber-reinforced polymer | CFRP |
| | Glass-fiber-reinforced polymer | GFRP |
| | Kevlar-fiber-reinforced polymer | KFRP |
| **Porous ceramics** (traditional ceramics cements, rocks, and minerals) | Brick | Brick |
| | Cement | Cement |
| | Common rocks | Rocks |
| | Concrete | Concrete |
| | Porcelain | Pcln |
| | Pottery | Pot |
| **Glasses** (ordinary silicate glasses) | Borosilicate glass | B-glass |
| | Soda glass | Na-glass |
| | Silica | $SiO_2$ |

*continued*

**Table 3.7**      Continued

| Class | Members | Short name |
|---|---|---|
| **Woods** | Ash | Ash |
| Separate clusters | Balsa | Balsa |
| describe properties | Fir | Fir |
| parallel to the grain | Oak | Oak |
| and normal to it and | Pine | Pine |
| wood products. | Wood products (ply, etc.) | Wood products |
| **Elastomers** | Natural rubber | Rubber |
| (natural and | Hard butyl rubber | Hard butyl |
| artificial rubbers) | Polyurethanes | PU |
| | Silicone rubber | Silicone |
| | Soft butyl rubber | Soft butyl |
| **Polymer foams** | Cork | Cork |
| (foamed polymers of | Polyester | PEST |
| engineering) | Polystyrene | PS |
| | Polyurethane | PU |

I  SOURCE: Ashby, 1999.

---

**EXAMPLE 3.13**

**Given:**   A fishing rod is to be made of a material that gives low weight and high stiffness.

**Find:**   From Figure 3.17 determine which is better, a rod made of plastic (without fiber reinforcement) or a split-cane rod (bamboo fibers glued together).

**Solution:**   Figure 3.17 shows that only very special polymers have moduli of elasticity as high as those of the best wooden fibers. The polymers are also 2 to 3 times more dense than wood. A split-cane rod will therefore give a lower weight for a given stiffness than any plastic.

---

### 3.7.2   STRENGTH VERSUS DENSITY

Weight is represented by density. *Strength,* on the other hand, means different things for different classes of solid material. For metals and polymers it is the *yield strength,* which is the same in tension and compression. For brittle ceramics it is the *crushing strength in compression,* not that in tension, which is about 15 times smaller. For elastomers strength means the *tear strength.* For composites it is the *tensile failure strength* (compressive strength can be lower because of fiber buckling).

Figure 3.18 shows these strengths, for which the symbol $S$ is used (despite the different failure mechanisms involved), plotted against density $\rho$. The class clusters for brittle materials are enclosed by broken lines as a reminder of this. The considerable vertical extension of the strength bubble for an individual material reflects its wide range, caused by the degree of alloying, work hardening, grain size, porosity, etc.

Figure 3.18 is useful for determining optimum materials based on strength where deformation under loading is not an issue. Just as before, one chooses a reference line, and

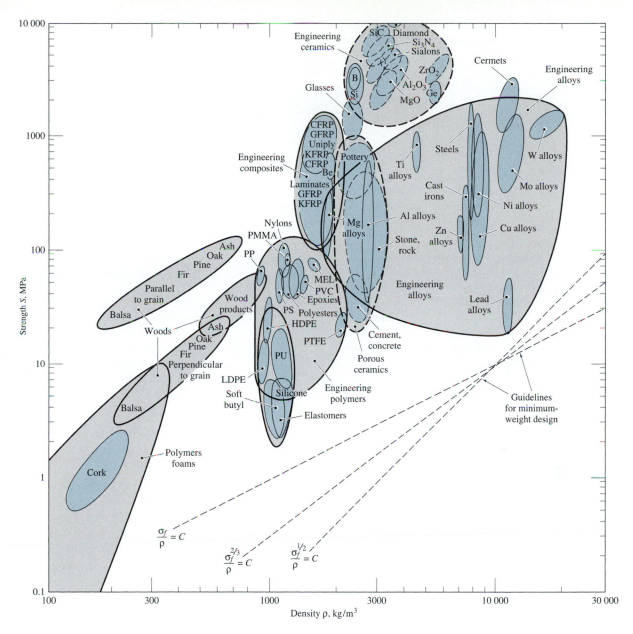

**Figure 3.18** Strength plotted against density (yield strength for metals and polymers, compressive strength for ceramics, tear strength for elastomers, and tensile strength for composites). The guidelines of constant $S/\rho$, $S^{2/3}/\rho$, and $S^{1/2}/\rho$ allow selection of materials for minimum-weight, yield-limited design. [Ashby, 1999.]

materials located at the greatest distance from this line (up and to the left) are superior. The following circumstances correspond to the reference lines in Figure 3.18:

$$\frac{\sigma}{\rho} = C \qquad \text{Tension members}$$

$$\frac{\sigma^{2/3}}{\rho} = C \qquad \text{Beams and shafts}$$

$$\frac{\sigma^{1/2}}{\rho} = C \qquad \text{Plates}$$

The range of strength for engineering materials spans five decades, from 0.1 MPa (foams used in packaging and energy-absorbing systems) to $10^4$ MPa (diamond). The range of density is the same as in Figure 3.17.

---

**EXAMPLE 3.14**

**Given:** The fishing rod given in Example 3.13 is manufactured in the form of a tapered tube with a given wall thickness distributed along its length.

**Find:** The material that makes the rod as strong as possible for a given weight.

**Solution:** Figure 3.18 shows that the strongest materials for a given density are diamond and silicon carbide and other ceramics. It is difficult and expensive to use these as fishing rod materials. The best choice is carbon-fiber-reinforced plastic or glass-fiber-reinforced plastic that has 800- to 1000-MPa strength for a density of 1500 kg/m$^3$.

---

### 3.7.3    STIFFNESS VERSUS STRENGTH

Figure 3.19 plots stiffness, or modulus of elasticity, versus strength. The qualifications on strength are the same as those in Figure 3.18. The ranges of the variables, too, are the same. Contours of normalized strength $S/E$ appear as a family of straight parallel lines.

Engineering polymers have normalized strengths between $10^{-2}$ and 0.1 and remarkably high elasticity relative to metals, for which the values are at least a factor of 10 smaller. Even ceramics in compression are not as deformable, and in tension they are far weaker (by a further factor of 15 or so). Composites and woods lie on the $10^{-2}$ contour, as good as the best metals. Elastomers, because of their exceptionally low moduli, have larger values of $S/E$ than any other class of material at 0.1 to 10.

The reference lines in Figure 3.19 are useful for the following circumstances:

$$\frac{S}{E} = C \qquad \text{Design of seals and hinges}$$

$$\frac{S^{3/2}}{E} = C \qquad \text{Elastic components such as knife-edges and diaphragms}$$

$$\frac{S^2}{E} = C \qquad \text{Elastic energy storage per volume (for compact energy adsorption)}$$

**Figure 3.19** Modulus of elasticity plotted against strength. The design guidelines help with the selection of materials for such machine elements as springs, knife-edges, diaphragms, and hinges. [Ashby, 1999.]

**EXAMPLE 3.15**

**Given:**  The springs in a car suspension can be made of rubber, steel, or carbon-fiber-reinforced plastic. The geometries of the different suspension springs are quite different, depending on allowable elastic deformations.

**Find:**  The maximum elastic strains in the three types of spring if the rubber is polyurethane (PU), the steel has a strength of 1 GPa, and the carbon-fiber-reinforced plastic is uniply.

**Solution:**  From Figure 3.19 PU rubber has a strength of 30 MPa and a modulus of elasticity of 0.05 GPa. The maximum elastic strain is

$$\left(\frac{S}{E}\right)_{rubber} = \frac{30}{50} = 0.60$$

Likewise, for steel and carbon-fiber-reinforced plastic

$$\left(\frac{S}{E}\right)_{steel} = \frac{1}{205} = 0.005$$

$$\left(\frac{S}{E}\right)_{plastic} = \frac{1}{200} = 0.005$$

The rubber has a maximum elastic strain of 60%, whereas the steel and carbon-fiber-reinforced plastic springs have a maximum elastic strain of 0.5%. Also, from Figure 3.18 the steel spring will be 5 times heavier than the carbon-fiber-reinforced plastic spring.

### 3.7.4    WEAR RATE VERSUS LIMITING PRESSURE

Wear presents a new set of problems in attempting to choose a solid material. If the materials are unlubricated, sliding motion is occurring, and if one of the surfaces is steel, the *wear rate* is defined as

$$W_r = \frac{\text{volume of material removed}}{\text{sliding distance}} \tag{3.25}$$

The wear rate $W_r$ thus has the SI unit of square meters. At low limiting pressure $p_l$ (the force pressing the two surfaces together, divided by the area of contact)

$$W_r = K_A A p_l \tag{3.26}$$

where $K_A$ = Archard wear constant, $Pa^{-1}$
       $A$ = area of contact, $m^2$
       $p_l$ = limiting pressure, Pa

Figure 3.20 shows the constant $K_A$ plotted against limiting pressure $p_l$. Each class cluster shows the constant value of $K_A$ at low $p_l$ and the steep rise as $p_l$ is approached. Materials cannot be used above $p_l$.

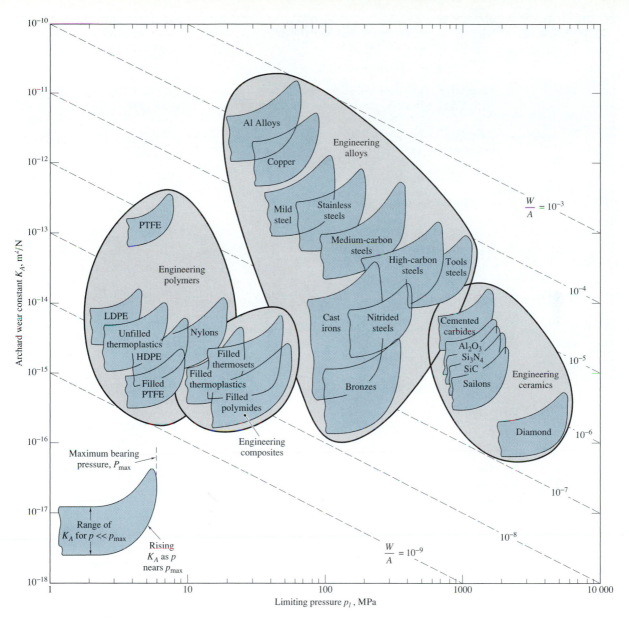

**Figure 3.20** Archard wear constant plotted against limiting pressure. [Ashby, 1999.]

**EXAMPLE 3.16**

**Given:** A polytetrafluoroethylene (PTFE, or Teflon) slider is in contact with high-carbon steel. The sliding distance is 300 m, and the thickness of the Teflon layer allowed to be worn away is 3 mm.

**Find:** How large the PTFE slider surface has to be so that it will not have excessive wear and so that the limiting pressure will not be exceeded if the load carried is 10 MN.

**Solution:** From Figure 3.20 the limiting pressure for PTFE on steel is $p_l = 8$ MPa, and the Archard wear constant is $K_A = 2 \times 10^{-13}$ m$^2$/N. From Equation (3.26)

$$\frac{W_r}{A} = K_A p_l = 2(10^{-13})(8)(10^6) = 1.6 \times 10^{-6}$$

The worn volume of the material is

$$At_h = W_r l_s$$

where $l_s$ = sliding distance, m
$\quad\quad t_h$ = wear depth, m

$$\therefore \quad \frac{W_r}{A} = \frac{t_h}{l_s} = \frac{0.003}{300} = 10^{-5}$$

The pressure can be written as

$$p = \frac{W_r}{A}\frac{1}{K_A} = \frac{10^{-5}}{2 \times 10^{-13}} = 0.5 \times 10^8 \text{ Pa} = 50 \text{ MPa}$$

Since $p \gg p_l$, the limiting pressure is needed to determine the size of the slider.

$$\therefore \quad p_l A = (10)(10^6) \text{ N} = 10^7 \text{ N}$$

$$\therefore \quad A = \frac{10^7}{p_l} = \frac{10^7}{8(10^6)} = 1.25 \text{ m}^2$$

The surface area has to be 1.25 m$^2$ to avoid too high a compressive stress. For these conditions the wear depth will be only 0.48 mm.

### 3.7.5    Young's Modulus versus Relative Cost

In practice, design engineers consider cost much more than it has been considered thus far in this text. Figure 3.21 shows the stiffness of a material versus the relative cost (i.e., the cost per weight of the material, divided by the cost per weight of mild steel). The reference lines are useful for the following:

$$\frac{E}{C_R \rho} = C \quad \text{Minimum-cost design of stiff tension members}$$

$$\frac{E^{1/2}}{C_R \rho} = C \quad \text{Minimum-cost design of stiff beams and columns}$$

$$\frac{E^{1/3}}{C_R \rho} = C \quad \text{Minimum-cost design of stiff plates}$$

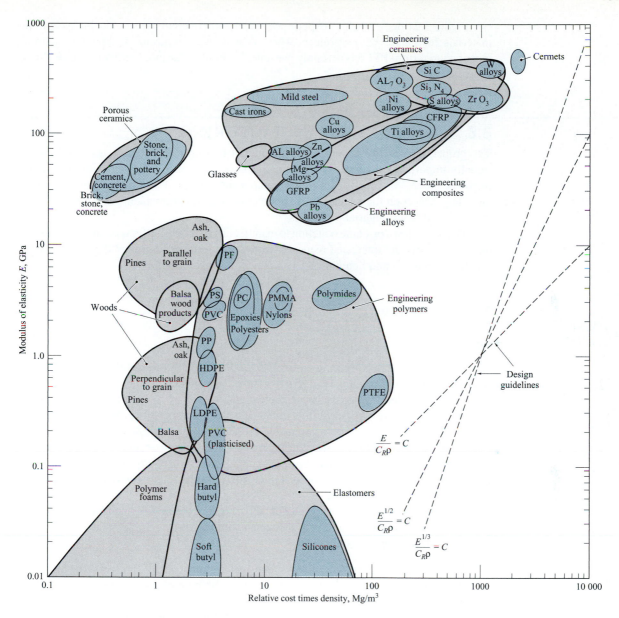

**Figure 3.21** Modulus of elasticity plotted against cost times density. The reference lines help with selection of materials for machine elements. [Ashby, 1999.]

Figure 3.21 does much to explain why steel and concrete are so valuable as building materials for public works projects where cost is to be minimized. Although a bridge manufactured from PTFE is certainly possible, it would be far more costly than a steel-and-concrete bridge.

## 3.8    EFFECTS OF MANUFACTURING

The proper selection of an engineering material is a critical task for successful design. Equally important for its performance and economic impact is the selection of a manufacturing process or processes for each component. Selection of a manufacturing process has a large effect on the material's microstructure and can dramatically affect the strength, ductility, and other material properties.

This section is a brief introduction to manufacturing process effects on machine element design. As such, the interested reader is strongly recommended to read the texts by Schey (2000), Kalpakjian and Schmid (2001, 2003), or Groover (2002).

### 3.8.1    MANUFACTURE OF METALS

*Casting*

A wide variety of casting processes are available, all of which require molten metal to solidify within a mold. Upon solidification, grains can be large or small depending on the presence or absence of nucleating agents during cooling and the rate of cooling. Regardless, castings in general have a microstructure that contains a large number of micropores. In tension, these pores act as stress risers, while in compression, the pores close upon themselves; the main result is a much higher strength in compression than in tension. Castings can have limited ductility because of this microstructure, but these properties can be improved by annealing. The main advantages to castings are low cost, especially for moderate production runs, and highly intricate shapes.

Casting processes are usually classified as expendable mold-expendable pattern; expendable mold-permanent pattern; or permanent mold processes. Some of the most common of these processes are as follows:

- **Expendable mold, expendable pattern.**    Investment casting and evaporative pattern casting are common examples of this class of casting process. In investment casting, a pattern is created from a low-melting-point solid such as a thermoplastic or wax. This pattern is then coated by successive dips into a slurry of ceramic material with binders. Once a desired coating thickness has been developed, the coated pattern is placed in an oven, melting the pattern and leaving a cavity. Molten metal is then poured into the cavity, which solidifies in the shape of the original pattern.

    In evaporative pattern casting, a polystyrene foam pattern is produced with the desired shape of the metal part; this foam is then buried in sand. Molten metal is poured onto the foam, evaporating the polystyrene and displacing it in the sand cavity. After solidification, the part is removed.

(a) Mechanical drawing of part

(b) Cope pattern plate

Core prints

(c) Drag pattern plate

(d) Core boxes

(e) Core halves pasted together

(f) Cope ready for sand

(g) Cope after ramming with sand and removing pattern, sprue, and risers

(h) Drag ready for sand

(i) Drag after removing pattern

(j) Drag with core set in place

(k) Cope and drag assembled ready for pouring

(l) Casting as removed from mold; heat-treated

(m) Casting ready for shipment

**Figure 3.22** Schematic illustration of the sand casting process. [From Steel Founders' Society of America.]

- **Expendable mold, permanent pattern.** This class of operations includes sand casting (shown in Fig. 3.22), shell casting, and plaster-mold and ceramic-mold casting. Sand casting is the most prevalent form of casting, with typical applications including machine tool bases, engine blocks, and machine housings. In sand casting and other such processes, a mold is created from a pattern, but the pattern can be reused for many parts.

- **Permanent mold.** Including processes such as die casting, pressure casting and centrifugal casting, permanent mold processes have high tooling costs and are therefore limited to large production runs. In these operations, the desired part shape is produced into a mold of a metal with a higher melting temperature than that of the workpiece, or graphite in some applications. Molten metal is injected under high pressure into the mold cavity, so that it fills the mold completely before solidification. Examples of parts produced in permanent mold operations are transmission housings, valve bodies, hand tools, computer housings and camera frames, and toys.

**Figure 3.23**     An example of the steps in forging a connecting rod for an internal combustion engine, and the die used. [Courtesy Forgin Industry Association, Cleveland, Ohio.]

### Bulk Metal Forming

In bulk deformation processes, materials are subjected to large strains in order to achieve the desired shape. Thus, the material must be quite ductile to be formed without fracturing. Some of the common bulk forming operations are

- **Forging.**    Forging is controlled deformation of metal through the application of compressive stresses. In *open die forging*, simple die shapes such as flats and rounds are used to obtain a rough shape in the workpiece. In *closed die forging*, as shown in Figure 3.23, a cavity is carefully prepared in a die, so that the metal will conform to and acquire the cavity shape during forging. Note that the workpiece in forging will develop *flash* which must be trimmed off; the flash is necessary to ensure that the metal completely fills the die.

- **Extrusion.**    In extrusion, shown in Figure 3.24(a), a *billet* is pushed through a die to produce a product with a constant cross section. Structural shapes are commonly produced through extrusion, and smaller parts can be produced by cutting extrusions to desired lengths.

- **Rolling.**    Arguably the most common bulk forming operation, rolling is performed on approximately 90% of metals. In *flat rolling,* a billet is reduced in thickness through the compressive stress applied by two rollers. Rolling can also be used to produce structural shapes, tubes, rings (such as bearing races), spheres (such as balls in rolling-element bearings), and screw threads.

Bulk deformation processes include forging (involving plastic deformation through compressive stresses), extrusion (where a workpiece is forced through compressive stresses through a die), rolling, and drawing (pulling a workpiece through a die). Bulk deformation can take place at elevated temperatures (hot working) to exploit increases in material ductility and decreases in strength at elevated temperatures. Cold working has associated with it superior surface finish, improved mechanical properties due to strain hardening, and a textured microstructure.

(a)

(b)

**Figure 3.24**    The extrusion process. (a) Schematic illustration of the forward or direct extrusion process [from Kalpakjian and Schmid (2003)]. (b) Examples of cross sections commonly extruded [Courtesy Almay Aluminium, Inc., Brampton, Ontario.]

### Sheet Forming

One of the main advantages of metals is that they display sufficient ductility to be formed or rolled into thin sheets. These sheets are then used for further processing. This allows the inexpensive production of high-quality parts with large aspect ratios. Since sheet metals are cold-rolled, they will usually have small, elongated grain size, some anisotropy, and higher strength than bulk forms of the same material.

The most common sheet forming operations are bending, stretch forming, and deep drawing. In bending, a sheet or tube is forced around a mandrel to a desired shape, or sheet metal is plastically deformed in a die with the desired bend shape. In stretch forming, a sheet is forced between two dies with the desired profile; the sheet metal plastically deforms to match the die profile (or it fractures, a common defect in stretch forming). In deep drawing, a cup-shaped part is produced by forcing a steel sheet or blank into a die cavity; the punch and die have a clearance that is slightly larger than the sheet thickness to avoid shearing the blank. Typical deep drawn parts include cookware, oil pans, beverage containers, and artillery shells.

### Machining

Material removal processes such as metal cutting, grinding, or electrical discharge machining are used to remove material from a bulk form (or near net-shaped form) to achieve desired surface finishes, tolerances, or shapes that are difficult to obtain otherwise. The machining process affects part design in a number of ways, including these:

- All machining operations result in feed marks on the workpiece, which can limit the fatigue life. This topic is discussed in detail in Chapter 7.

- Very smooth surfaces or very low tolerances can be produced only through expensive manufacturing operations. Thus, a goal of designers is to specify rough surfaces whenever possible in order to have maximum economy in their designs.

- Machining does not affect the material microstructure. Thus, a brittle material will not become ductile because its surface was machined.

- Machining requires parts to be held in fixtures during machining. Therefore, designers need to incorporate clamping and fixturing locations in their parts to allow machining.

- In machining, chamfers are preferable over radii, while designers often specify radii because they are easier to analyze. Internal cavities with sharp corners should be avoided.

- It is difficult to produce sharp external corners in casting or bulk forming operations, but it is easier to produce sharp external corners than shoulders in machining.

### Powder Metallurgy (P/M)

Metals can be obtained in powder form through a number of methods. These powders can be processed by:

- **Pressing and sintering.**   The powder is placed inside a die cavity and compressed under high pressure; the part that is ejected is called a *green compact* and has a strength comparable to that of chalk. These parts are then sintered, or heated in a controlled atmosphere furnace at up to 90% of their absolute melting temperature for up to 4 hours. In the sintering furnace, the powder particles fuse and develop a strong bond.

- **Metal injection molding.**   The powder is mixed with a polymer binder and processed by injection molding, as described below.

- **Cold and hot isostatic pressing.**   The powder is placed in a compliant mold and then placed into a pressurized chamber. The high pressures that result yield a P/M part that is strong and has tight tolerances.

  P/M parts are very porous; they can have up to 20% porosity. As opposed to castings, the pores in P/M parts are not isolated; they are all interconnected. As such, P/M parts can be thought of as a sponge, in that once they are infiltrated by a lubricant, the lubricant is always present. This is one of the main reasons that P/M parts are very popular for tribological applications such as gears, cams, bearings, and sleeves.

### 3.8.2   MANUFACTURE OF POLYMERS

Polymers are produced by a wide variety of manufacturing operations; only a few very popular approaches are described here. There are two basic forms of polymers, and these determine the manufacturing strategy that will be used. Thermoplastics are polymers that have a defined melting point; heating them greatly reduces their strength and allows them to flow into desired shapes. Once they are cooled, they regain their strength and hardness. Thermosets chemically degrade when heated long before they flow. The basic strategy used with thermosets is to form the polymer constituents into the desired shape, then through the application of heat and/or pressure cause cross-links to grow. After a cure cycle, the thermoset part is ready for use.

### Thermoplastic Manufacture

Thermoplastics are very common, versatile materials, available in a wide variety of shapes and colors. Thermoplastics are produced from their chemical constituents and are available

**Figure 3.25**    Schematic illustration of a typical extruder. [Source: *Encyclopedia of Polymer Science and Engineering*, 2nd ed. Copyright 1985. Reprinted by permission of John Wiley & Sons, Inc.]

in bulk, pellet, or powder form in order to process them into their final desired shape. Note that a "melted" polymer is still quite viscous; its consistency is closer to that of bread dough or soft taffy than a liquid. This should be understood when one is trying to visualize the processes described here.

Among the common methods of processing thermoplastics are:

- **Extrusion.**    In extrusion, a polymer in pellet or powder form is heated in an extruder, shown in Figure 3.25. The polymer is melted in the extruder barrel by the heat input from heating elements as well as the friction between pellets and screw. As a result, a liquid polymer is forced out through a die with a desired cavity, resulting in a part with a constant cross section. The polymer is then cooled, usually by forced convection of cool air. Polymer tubes, structural members, and rods are produced through extrusion.

- **Injection molding.**    Instead of forcing the polymer through a die opening, the polymer is injected periodically into a die cavity. Usually, metal dies are used that incorporate channels for coolant, so that the heat from the polymer is quickly removed. Once solid, the polymer part is ejected, and another cycle can begin. Injection molding is extremely popular; a wide variety of automotive, consumer electronic, and home-use products are produced by injection molding.

- **Thermoforming.**    This process involves the use of an extruded film that is heated and draped over a (usually) metal die with an intended shape. The soft thermoplastic complies with the die, cools because of contact with the die, and hardens. Plastic packaging, advertising signs, refrigerator liners, and the Invisalign product in Case Study 1.3 are examples of thermoformed products.

- **Blow molding.**    In extrusion blow molding, an extruded tube is clamped and expanded by internal pressure against a die-defined cavity. In injection blow molding, a short tubular piece called a parison is first injection-molded and then transferred to a blow molding die. Blow molding produces hollow containers such as plastic beverage bottles.

### *Manufacture of Thermosets*

Thermosets are also very commonly applied materials. Compared to thermoplastics, thermosets in general have advantages in strength (especially at elevated temperatures) and strength/weight ratios. As discussed above, thermosets are manufactured by blending polymer constituents, usually in powder or clay consistency, producing the desired shape, and then curing the plastic. Some of the important thermoset manufacturing operations are

1. **Reaction injection molding.**   Similar to injection molding described above, reaction injection molding involves injecting the thermoset into a heated die; the elevated temperature then allows curing of the thermoset.

2. **Compression molding.**   Compression molding shares many similarities with forging. A claylike-consistency polymer is placed into a heated mold and is compressed to fill the cavity. The polymer then cures into the shape defined by the mold. Compression-molded parts have good tolerances and surface finish, and yield a polymer with high molecular weight and crystallinity (and therefore strength) compared to other processes.

### 3.8.3   MANUFACTURE OF CERAMICS

Ceramics share many processing similarities to P/M techniques. Ceramics are processed from powder form, usually by mixing with water and binders. In all processing techniques, the ceramics will be somewhat porous, and the mechanical properties will depend on this porosity. The most common processes for producing ceramic parts are

- **Slip casting.**   In slip casting, a ceramic slurry is poured into a permeable mold. Some of the water in the slurry diffuses into the mold, leaving a locally high slurry concentration next to the mold wall. After a few minutes, the slurry is poured from the mold, leaving a coating adhering to the mold. After a few hours, more of the water has been extracted from the slurry, and it is removed from the mold. It is very fragile at this step and is fired in a furnace or kiln to fuse the ceramic particles.

- **Dry or wet pressing.**   This approach is similar to pressing of metal powders; the resultant compact needs to be fired in a furnace or kiln to develop strength.

- **Doctor-blade process.**   In this operation, a claylike consistency of ceramic is spread by a blade. The ceramic can be spread into a sheet or can be formed into a desired shape such as plates or bowls (jiggering).

- **Pressing and injection molding.**   These are similar to the P/M and polymer processes of the same name, respectively. The part from these operations needs to be fired to develop strength.

Ceramics and P/M parts share many design considerations. Since there is significant shrinkage during sintering or firing, there is a possibility for warpage in large parts, and thin cross sections are likely to fracture. Sharp corners will crumble while the part is in the compact stage, and chamfers are preferable to radii because of tooling design issues.

### 3.8.4 SELECTION OF MANUFACTURING PROCESSES

Manufacturing process selection is a difficult task, and it must incorporate design parameters as well as economic considerations. A full discussion of manufacturing process selection is beyond the scope of this text, and the interested reader is again directed to the text by Kalpakjian and Schmid (2003). However, it should be recognized in the design of machine elements that there are certain trends in processes used. For example:

- Processes that are advantageous for small production runs usually have low capital equipment costs, but high labor costs, whereas hard automated machinery is expensive but has low labor costs. Therefore, the cost per part can be greatly reduced if the parts are produced in quantities large enough to justify purchase of expensive tooling. In practice, this means that bolts, gears, bearings, etc., are mass-produced in standard sizes at far greater economy than if produced to order. From a design standpoint, it is therefore important to specify sizes to standard sizes, such as 1 inch instead of 0.9456 inch, or 50 mm instead of 43.6 mm.

- It should be recognized that product quality and robustness also increase with large production runs.

- Certain materials are commercially available in some forms (see Table 3.8). The desire to use materials in other forms is not impossible to satisfy, but this is probably not economically advisable.

- Manufacturing processes are determined, to a great extent, by the design requirements of size, strength, tolerance, and surface finish. Figure 3.26 can be used as a guide to help select manufacturing processes based on tolerance and roughness.

**Table 3.8**    Commercially available forms of materials

| Material | Available forms[a] |
|---|---|
| Aluminum | B, F, I, P, S, T, W |
| Ceramics | B, p, s, T |
| Copper and brass | B, f, I, P, s, T, W |
| Elastomers | b, P, T |
| Glass | B, P, s, T, W |
| Graphite | B, P, s, T, W |
| Magnesium | B, I, P, S, T, w |
| Plastics | B, f, P, T, w |
| Precious metals | B, F, I, P, t, W |
| Steels and stainless steels | B, I, P, S, T, W |
| Zinc | F, I, P, W |

[a] B = bar and rod; F = foil; I = ingot; P = plate and sheet; S = structural shapes; T = tubing; W = wire. Lowercase letters indicate limited availability. Most of the metals are also available in powder form, including prealloyed powders.

SOURCE: From Kalpakjian and Schmid (2003).

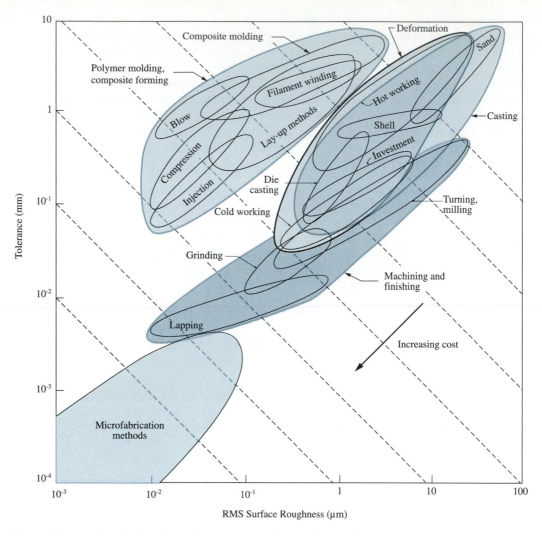

**Figure 3.26**    A plot of achievable tolerance versus surface roughness for assorted manufacturing operations. The dashed lines indicate cost factors; an increase in precision corresponding to the separation of two neighboring lines gives an increase in cost for a given process, or a factor of 2. [Ashby, 1992.]

## 3.9    SUMMARY

Eight important mechanical properties of solid materials were discussed in this chapter. The differences between ductile and brittle materials were presented. It was found that, at fracture, ductile materials exhibit considerable plastic deformation whereas brittle materials exhibit little or no yielding before failure.

Four major classes of solid material were described: metals; ceramics and glasses; polymers and elastomers; and composites. The members of each class have common features, such as similar chemical makeup and atomic structure, similar processing routes, and similar applications.

A stress–strain diagram was presented for each class of solid material because the classes differ significantly. The stress–strain diagrams for metals are essentially the same for compression and tension. This feature is not true for polymers or ceramics. Results of transverse bending tests for ceramics were found to be similar to tensile test results for metals. Strength for ceramics means fracture strength in tension and crushing strength in compression. It was also found that the stress–strain diagram for a polymer becomes markedly nonlinear at a strain of 0.01, whereas for metals this point occurs at 0.002.

A number of solid material properties used in choosing the correct material for a particular application were presented: density, modulus of elasticity, Poisson's ratio, shear modulus, strength, resilience, toughness, thermal conductivity, linear thermal expansion coefficient, specific heat capacity, and Archard wear constant. These parameters were presented for the three major classes of solid material: metals, polymers, and ceramics. The results were represented in a one-dimensional manner.

Two-parameter materials charts were also presented. Stiffness versus weight, strength versus weight, stiffness versus strength, and wear constant versus limiting pressure for the various classes of material can give a better idea of the best material for a particular machine element.

# Key Words

**anisotropic**   material having different properties in all directions at a point in the solid

**Archard wear constant**   wear property of material

**brittle material**   material that fractures at strain below 5%

**ceramics**   compounds of metallic and nonmetallic elements

**composite materials**   combinations of two or more materials, usually consisting of fiber and thermosetting polymer

**density**   mass per unit volume

**ductile material**   material that can sustain elongation greater than 5% before fracture

**ductility**   degree of plastic deformation sustained at fracture

**elastic limit**   stress above which material acquires permanent deformation

**elastomers**   polymers with intermediate amount of cross-linking

**fracture stress**   stress at time of fracture or rupture

**glasses**   compounds of metallic and nonmetallic elements with no crystal structure

**homogeneous**   material having properties that are not a function of position in solid

**isotropic**   material having same properties in all directions at a point in solid

**metals**   combinations of metallic elements

**modulus of elasticity**   proportionality constant between stress and strain

**modulus of rupture**   stress at rupture from bending test, used to determine strength of ceramics

**necking**   decreasing cross-sectional area that occurs after ultimate stress is reached and before fracture

**orthotropic**   material having different properties in three mutually perpendicular directions at a point in the solid and having three mutually perpendicular planes of material symmetry

**Poisson's ratio**   absolute value of ratio of transverse to axial strain

**polymers**   compounds of carbon and other elements forming long-chain molecules

**proportional limit**   stress above which stress is no longer linearly proportional to strain

**resilience**   capacity of material to release absorbed energy

**rule of mixtures**   linear interpolation between densities of alloy concentration

**specific heat capacity**   ratio of heat stored per mass to change in temperature of material

**strain hardening**   increase in hardness and strength of ductile material as it is plastically deformed

**thermal conductivity**   ability of material to transmit heat

**thermal expansion coefficient**   ratio of elongation in material to temperature rise

**thermoplastics**   polymers without cross-links

**thermosets**   polymers with highly cross-linked structure

**toughness**   ability to absorb energy up to fracture

**ultimate strength**   maximum stress achieved in stress–strain diagram

**yielding**   onset of plastic deformation

**yield strength**   stress level defined by intersection of reference line (with slope equal to initial material elastic modulus and $x$ intercept of 0.2%) and material stress-strain curve

**Young's modulus**   (see modulus of elasticity)

---

# RECOMMENDED READINGS

Ashby, M. J. (1999) *Materials Selection in Mechanical Design,* Elsevier Science, Oxford.

*ASM Metals Handbook,* 8th ed. (1973) American Society for Metals, Metals Park, OH.

Budinski, K. (1979) *Engineering Materials, Properties and Selection,* Prentice-Hall, Englewood Cliffs, NJ.

Cottrell, A. H. (1964) *Mechanical Properties of Matter,* Wiley, New York.

Crane, F. A. A., and Charles, J. A. (1984) *Selection and Use of Engineering Materials,* Butterworths, London.

Dieter, G. E. (1976) *Mechanical Metallurgy,* McGraw-Hill, New York.

Farag, M. M. (1990) *Selection of Materials and Manufacturing Processes for Engineering Materials,* Prentice-Hall, Englewood Cliffs, NJ.

Flinn, R. A., and Trojan, P. K. (1986) *Engineering Materials and Their Applications,* Houghton Mifflin, Boston.

Kalpakjian, S., and Schmid, S. R. (2003) *Manufacturing Processes for Engineering Materials,* 4th ed., Prentice-Hall, Upper Saddle River, NJ.

Kalpakjian, S., and Schmid, S. R. (2001) *Manufacturing Engineering and Technology,* 4th ed., Prentice-Hall, Upper Saddle River, NJ.

Lewis, G. (1990) *Selection of Engineering Materials,* Prentice-Hall, Englewood Cliffs, NJ.

Schey, J. A. (2000) *Introduction to Manufacturing Processes,* 3rd ed., McGraw-Hill, New York.

# REFERENCES

Ashby, M. F. (1999) *Materials Selection in Mechanical Design,* Elsevien Science, Oxford.

Groover, M. P. (2002) *Fundamentals of Modern Manufacturing,* John Wiley & Sons, New York.

Kalpakjian, S., and Schmid, S. R. (2003) *Manufacturing Processes for Engineering Materials,* 4th ed., Prentice-Hall, Upper Saddle River, NJ.

Kalpakjian, S., and Schmid, S. R. (2001) *Manufacturing Engineering and Technology,* 4th ed., Prentice-Hall, Upper Saddle River, NJ.

*Properties of Common Engineering Materials,* Data Item 84041, ESDU International, London, 1984.

Schey, J. A. (2000) *Introduction to Manufacturing Processes,* 3rd ed., McGraw-Hill, New York.

# PROBLEMS

## Section 3.2

**3.1** A 2-m-long polycarbonate tensile rod has a diameter of 130 mm. It is used to lift a tank weighing 65 tons from a 1.8-m-deep ditch onto a road. The vertical motion of the crane's arc is limited to 4.2 m. Will it be possible to lift the tank onto the road?

## Section 3.3

**3.2** Materials are normally classified according to their properties, processing routes, and applications. Give examples of common metal alloys that do not show some of the typical metal features in their applications.

★ **3.3** Equation(B.56) gives the relationship between stresses and strains in isotropic materials. For a polyurethane rubber the elastic modulus at 100% elongation is 7 MPa. When the rubber is exposed to a hydrostatic pressure of 10 MPa, the volume shrinks 0.5%. Calculate Poisson's ratio for the rubber. *Ans.* $\nu = 0.499$.

**3.4** A fiber-reinforced plastic has fiber-matrix bond strength $\tau_f = 10$ MPa and fiber ultimate strength $S_u = 1$ GPa. The fiber length is constant for all fibers at $l = 1.5$ mm. The fiber diameter $d = 30\,\mu$m. Find whether the fiber strength or the fiber-matrix bond will determine the strength of the composite.

**3.5** Given the same material as in Problem 3.4 but with fiber length $l = 1$ mm, calculate if it is possible to increase the fiber stress to $S_u = 1$ GPa by making the fiber rectangular instead of circular, maintaining the same cross-sectional area for each fiber.

*Indicates problems of greater difficulty.

## Section 3.4

**3.6** A copper bar is stressed to its ultimate strength $S_u = 200$ MPa. The cross-sectional area of the bar before stressing is 120 mm$^2$, and the area at the deformed cross section where the bar starts to break at the ultimate strength is 70 mm$^2$. How large a force is needed to reach the ultimate strength? *Ans.* 24 kN.

**3.7** AISI 440C stainless steel has ultimate strength $S_u = 807$ MPa and fracture strength $S_{fr} = 750$ MPa. At the ultimate strength the cross-sectional area of a tension bar made of AISI 440C is 80% of its undeformed value. At the fracture point the minimum cross-sectional area has shrunk to 70%. Calculate the real stresses at the point of ultimate strength and at fracture. *Ans.* At fracture, $\sigma = 1071$ MPa; at ultimate strength, $\sigma = 1009$ MPa.

★ **3.8** According to sketch $a$, a beam is supported at point A and at either B or C. At C the silicon nitride tensile rod is lifting the beam end with force $P = S_{fr}^t A_c$, where $A_c$ is the cross-sectional area of the rod. Find the distance AB such that the silicon nitride rod would be crushed if it took up a compressive force at B instead of a tensile force at C. Note that $S_{fr}^c = 15 S_{fr}^t$ for silicon nitride. Also, find the reaction forces at A for the two load cases. *Ans.* $l' = l/15$; $A_y = \frac{120}{7} P$.

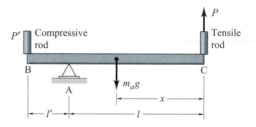

Sketch a, for Problem 3.8

**3.9** Polymers have different properties depending on the relationship between the local temperature and the polymer's glass transition temperature $T_g$. The rubber in a bicycle tire has $T_g = -12°C$. Could this rubber be used in tires for an Antarctic expedition at temperatures down to $-70°C$?

## Section 3.5

**3.10** Given an aluminum bronze with 30 wt % aluminum and 70 wt % copper, find the density of the aluminum bronze. *Ans.* $\rho_{bronze} = 7071$ kg/m$^3$.

**3.11** The glass-fiber-reinforced plastic in Example 3.8 (Sec. 3.5.2) is used in an application where the bending deformations, caused by the applied static load, will crack the plastic by overstressing the fibers. Will a carbon-fiber-reinforced plastic also crack if it has the same elastic properties as the glass-fiber-reinforced plastic?

**3.12** In Problem 3.11 carbon fibers were used to reinforce a polymer matrix. The concentration of fibers was decreased in Example 3.8 (Sec. 3.5.2) to give the same elastic properties for the carbon-fiber-reinforced polymer as for the glass-fiber-reinforced polymer. If instead the fiber concentration were kept constant at 10% when the glass fibers were changed to carbon fibers, how much smaller would the deformation be for the same load, and would the fibers be overstressed or not? The material properties are the same as those in Example 3.8.

**3.13** A bent beam, shown in sketch $b$, is loaded with force $P = 148,000$ N. The beam has a square cross section with length of 30 mm. Lengths $l_1 = 50$ mm and $l_2 = 100$ mm. The yield strength

$S_y = 350$ MPa (medium-carbon steel). Find whether the stresses in tension and shear are below the allowable stresses. Neglect bending.

Sketch b, for Problem 3.13            Sketch c, for Problem 3.14

**3.14** A tough material, such as soft stainless steel (AISI 316), has yield strength $S_y = 207$ MPa, ultimate strength $S_u = 552$ MPa, and 60% elongation. Find the ratio of the material toughness to the resilience, assuming that the stress–strain curve consists of two straight lines according to sketch $c$. *Ans.* Toughness/resilience = 1062.

## Section 3.6

**3.15** A steel cube has sides with length $l = 0.1$ m, modulus of elasticity $E = 206$ GPa, and Poisson's ratio $v = 0.3$. Find the compressive stresses needed on four of the cube faces to give the same elongation perpendicular thereto as a stress $\sigma$ in that perpendicular direction. *Ans.* $\sigma_c = -1.67\sigma$

**3.16** For the stressed steel cube in Problem 3.15 calculate the volume ratio $(v_{tension}/v_{compression})$ when $\sigma = 500$ MPa. *Ans.* $v_t/v_c = 1.0042$

**3.17** Hooke's law describes the relationship between uniaxial stress and uniaxial strain. How large is the ratio of deformation for a given load within the group of materials considered in this chapter?

## Section 3.7

**3.18** According to Archard's wear equation, the wear depth is proportional to the sliding distance and the contact pressure. How will the contact pressure be distributed radially for a disk brake if the wear rate is the same for all radii? *Ans.* Pressure is inversely proportional to radius.

**3.19** Given a brake pad for a disk brake on a car, and using Archard's wear constant, determine how the wear is distributed over the brake pad if the brake pressure is constant over the pad. *Ans.* Wear rate is proportional to radius.

# CHAPTER

# 4

# STRESSES AND STRAINS

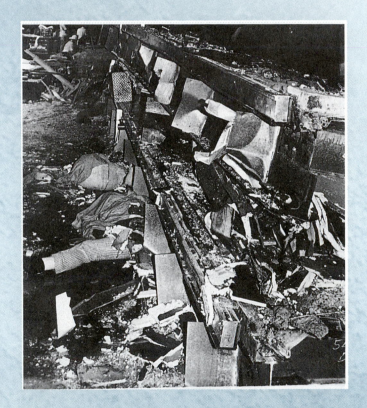

The failed Hyatt Regency Hotel walkway [Kansas City, 1981] that was directly attributable to changes in design that over-stressed structural members of the walkway. (AP/Wide World Photos)

*I am never content until I have constructed a mechanical model of the subject I am studying. If I succeed in making one, I understand; otherwise I do not.*

William Thomson (Lord Kelvin)

## SYMBOLS

| | |
|---|---|
| $A$ | cross-sectional area, $m^2$ |
| $A'$ | partial cross-section, $m^2$ |
| $a$ | width of cross-section, m |
| $b$ | height of cross-section, m |
| $c$ | distance from neutral axis to outer fiber of solid, m |
| $d_x, d_y$ | distance between two parallel axes one of which contains centroid of area, m |
| $E$ | modulus of elasticity, Pa |
| $e$ | eccentricity, distance separating centroidal and neutral radii of curved member, m |
| $G$ | shear modulus of elasticity, Pa |
| $h$ | height of triangular cross-section, m |
| $h_p$ | power, W |
| $I$ | area moment of inertia, $m^4$ |
| $I_m$ | mass moment of inertia, $kg \cdot m^2$ |
| $J$ | polar area moment of inertia, $m^4$ |
| $\bar{J}$ | polar area moment of inertia about centroidal coordinates, $m^4$ |
| $k$ | spring rate, N/m |
| $k_a$ | angular spring rate, $N \cdot m/rad$ |
| $l$ | length, m |
| $M$ | bending moment, $N \cdot m$ |
| $m_a$ | mass, kg |
| $N_a$ | rotational speed, rpm |
| $P$ | force, N |
| $Q$ | first moment about neutral axis, $m^3$ |
| $R$ | reaction force, N |
| $r$ | radius, m |
| $\bar{r}$ | centroidal radius, m |
| $r_c$ | radius of cross-section, m |
| $r_g$ | radius of gyration, m |
| $r_n$ | radius of neutral axis, m |
| $T$ | torque, $N \cdot m$ |
| $u$ | velocity, m/s |
| $u_f$ | velocity, ft/min |
| $V$ | transverse shear force, N |
| $w_t$ | width, m |
| $x, y, z$ | Cartesian coordinate system, m |
| $\bar{x}, \bar{y}, \bar{z}$ | centroidal coordinate system, m |
| $x', y'$ | coordinates parallel to $x$ and $y$ axes, respectively, m |
| $Z_m$ | section modulus, $I/c$, $m^3$ |
| $\alpha$ | angle of sector, rad |
| $\gamma$ | shear strain |
| $\epsilon$ | normal strain |
| $\theta$ | angle of twist, rad |
| $\sigma$ | normal stress, Pa |
| $\tau$ | shear stress, Pa |
| $\omega$ | rotational speed, rad/s |

## SUBSCRIPTS

| | |
|---|---|
| avg | average |
| $i$ | inner |
| max | maximum |
| $o$ | outer |
| $x, y, z$ | Cartesian coordinates |
| $\bar{x}, \bar{y}, \bar{z}$ | centroidal coordinates |
| $x', y'$ | coordinates parallel to $x$ and $y$ axes |

## 4.1    INTRODUCTION

In Section 2.3 of this book, the different types of loading were described: normal, torsional, bending, and transverse shear. This chapter describes the stresses and strains resulting from these types of loading while making use of the general Hooke's law relation developed in Chapter 3. The general theory developed in this chapter is applicable to any machine element. For the purposes of this chapter, however, the member is assumed to be straight, to have a symmetrical cross section, and to be made of a homogeneous, linear-elastic material. Later in the chapter a curved member is also considered for bending. More complicated geometries are considered in Chapter 6.

First, six concepts important to this chapter as well as to the overall text need to be defined.

## 4.2   DEFINITIONS

The centroid of an area, the moment of inertia of an area, the parallel-axis theorem, the radius of gyration, the section modulus, and the mass moment of inertia are important concepts that need to be defined before the main topics of this chapter can be studied. These concepts are used throughout the text.

### 4.2.1   CENTROID OF AREA

The **centroid of an area** (Fig. 4.1), or the center of gravity of an area, refers to the point that defines the geometric center of the area. Mathematically, it is that point at which the sum of the first moments of an area about the axis through it is zero, or

$$\int_A (y - \bar{y})\, dA = 0 \tag{4.1}$$

$$\int_A (x - \bar{x})\, dA = 0 \tag{4.2}$$

Solving for $\bar{x}$ and $\bar{y}$ gives

$$\bar{y} = \frac{\int_A y\, dA}{\int_A dA} = \frac{\int_A y\, dA}{A} \tag{4.3}$$

$$\bar{x} = \frac{\int_A x\, dA}{\int_A dA} = \frac{\int_A x\, dA}{A} \tag{4.4}$$

A complicated area can usually be divided into simple subareas, and Equations (4.3) and (4.4) can be applied by making the numerator equal to the sum of the first-moment integrals of the separate pairs. The denominator is the total area.

$$\therefore \quad \bar{y} = \frac{A_1\bar{y}_1 + A_2\bar{y}_2 + \cdots}{A_1 + A_2 + \cdots} \tag{4.5}$$

$$\bar{x} = \frac{A_1\bar{x}_1 + A_2\bar{x}_2 + \cdots}{A_1 + A_2 + \cdots} \tag{4.6}$$

These equations describe the centroid of the composite area.

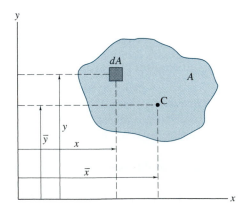

**Figure 4.1**   Centroid of area.

EXAMPLE 4.1

**Given:** Figure 4.2 shows a rectangular hole having dimensions $c \times d$ within a rectangular cross section $a \times b$. It also shows the coordinates and location of this hole within the rectangular cross section. The dimensions are $a = 10\,\text{cm}$, $b = 5\,\text{cm}$, $c = 3\,\text{cm}$, $d = 1\,\text{cm}$, $e = 9\,\text{cm}$, and $f = 3\,\text{cm}$.

**Find:** Centroid of section.

**Solution:** Making use of Equations (4.5) and (4.6) gives

$$\bar{y} = \frac{ab(b/2) - cd(f + d/2)}{ab - cd} = \frac{(10)(5)(5/2) - (3)(1)(3 + 1/2)}{(10)(5) - (3)(1)} = 2.436\,\text{cm}$$

$$\bar{x} = \frac{ab(a/2) - cd(e - c/2)}{ab - cd} = \frac{(10)(5)(10/2) - (3)(1)(9 - 3/2)}{(10)(5) - (3)(1)} = 4.840\,\text{cm}$$

**Figure 4.2**   Rectangular hole within a rectangular section used in Example 4.1.

## 4.2.2   AREA MOMENT OF INERTIA

The terms **area moment of inertia** and *second moment of an area* are used interchangeably. In Section 4.2.1, the first moment of an area $\int_A y \, dA$ was associated with the centroid of an area, and in this section the second moment of an area $\int_A y^2 \, dA$ is associated with the moment of inertia of an area.

Figure 4.3 shows the coordinates that describe the area moments of inertia, which are designated by the symbol $I$. Thus, the moments of inertia with respect to the $x$ and $y$ axes, respectively, can be expressed as

$$I_x = \int_A y^2 \, dA \qquad \text{and} \qquad I_y = \int_A x^2 \, dA \tag{4.7}$$

When the reference axis is normal to the plane of the area, through 0 in Figure 4.3, the integral is called the *polar moment of inertia, J* and can be written as

$$J = \int_A r^2 \, dA = \int_A (x^2 + y^2) \, dA = \int_A x^2 \, dA + \int_A y^2 \, dA \tag{4.8}$$

$$\therefore \qquad J = I_x + I_y \tag{4.9}$$

Note that the $x$ and $y$ axes can be any two mutually perpendicular axes intersecting at zero. The unit of moment of inertia and polar moment of inertia is length (commonly in meters or inches) raised to the fourth power.

**Figure 4.3**    Area with coordinates used in describing area moment of inertia.

**EXAMPLE 4.2**

**Given:**   Figure 4.4 shows a circular cross section with radius $r$ and $x$, $y$ coordinates.

**Find:**   The area moment of inertia of the circular area about the $x$ and $y$ axes and the polar moment of inertia about the centroid.

**Solution:**   From Equation (4.7) the area moment of inertia of the circular area about the $x$ axis is

$$I_x = \int y^2 \, dA = \int y^2 2r \cos \phi \, dy$$

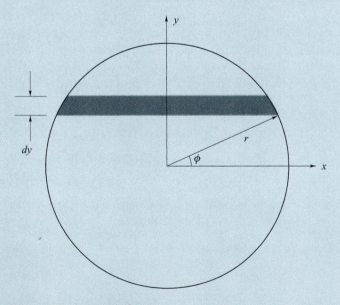

**Figure 4.4**    Centroid of area.

But because $y = r \sin \phi$ and $dy = r \cos \phi \, d\phi$,

$$I_x = 2r^4 \int_{-\pi/2}^{\pi/2} \sin^2 \phi \cos^2 \phi \, d\phi = \frac{r^4}{2} \int_{-\pi/2}^{\pi/2} \frac{1 - \cos 4\phi}{2} \, d\phi = \frac{\pi r^4}{4}$$

Because of symmetry the moment of inertia with respect to the $y$ axis is

$$I_y = \frac{\pi r^4}{4}$$

Thus, the area moments of inertia about the $x$ and $y$ axes are identical and equivalent to $\pi r^4/4$. From Equation (4.9) the polar moment of inertia about the centroid is

$$J_z = I_x + I_y = 2I_x = \frac{\pi r^4}{2}$$

## 4.2.3   PARALLEL-AXIS THEOREM

When an area's moment of inertia has been determined with respect to a given axis, the moment of inertia with respect to a parallel axis can be obtained by means of the parallel-axis theorem, provided that one of the axes passes through the centroid of the area. Figure 4.5 shows the coordinates and distances to be used in deriving the parallel-axis theorem.

The moment of inertia of the area $A$ about the $x'$ axis is

$$I_{x'} = \int_A (y + d_y)^2 \, dA = \int_A y^2 \, dA + 2d_y \int_A y \, dA + d_y^2 \int_A dA$$

$$I_{x'} = I_x + 2d_y \int_A y \, dA + Ad_y^2 \qquad (4.10)$$

But the integral $\int_A y \, dA$ is the first moment of the area with respect to the $x$ axis, or the centroid. Thus, if the $x$ axis passes through the centroid of the area, the first moment is zero and Equation (4.10) reduces to

$$I_{x'} = I_x + Ad_y^2 \qquad (4.11)$$

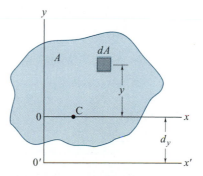

**Figure 4.5**   Coordinates and distance used in describing parallel-axis theorem.

where $I_x$ = moment of inertia of area with respect to axis parallel to $x'$ axis and passing through centroid, $m^4$

$d_y$ = distance between two axes $x$ and $x'$, of which $x$ contains centroid of area, m

$A$ = cross-sectional area, $m^2$

Similarly, for an axis $y'$ parallel to the $y$ axis that goes through the centroid and is separated by a distance $d_x$,

$$I_{y'} = I_y + Ad_x^2 \qquad\qquad (4.12)$$

Thus, the **parallel-axis theorem** states that an area's moment of inertia with respect to any axis is equal to the second moment of the area with respect to a parallel axis through the centroid of the area, added to the product of the area and the square of the distance between the two axes.

**EXAMPLE 4.3**

**Given:** Figure 4.6 shows a triangular cross section having a hole within it.

**Find:** The area moment of inertia and the centroid about the $x$ axis.

**Solution:** Assume that the $y$ axis starts at the midwidth of the base and is positive in the upward direction. The height of the triangle is

$$h = \frac{6}{2}\tan 60° = 5.196 \text{ cm}$$

The triangle is defined as $a$ and the circle as $b$. The centroids and areas of the triangle and the circle can be expressed separately as

$$\bar{y}_a = \frac{h}{3} = \frac{5.196}{3} = 1.732 \text{ cm}$$

$$A_a = \frac{1}{2}bh = \frac{1}{2}(6)(5.196) = 15.59 \text{ cm}^2$$

$$A_b = \frac{\pi d^2}{4} = \frac{\pi(2)^2}{4} = \pi = 3.142 \text{ cm}^2$$

**Figure 4.6**   Triangular cross section with circular hole in it, used in Example 4.3.

The centroid of the composite figure is

$$\bar{y} = \frac{\bar{y}_a A_a - \bar{y}_b A_b}{A_a - A_b} = \frac{(1.732)(15.59) - (2)(3.142)}{15.59 - 3.142} = 1.664 \text{ cm}$$

The moments of inertia of the areas of the triangle and circle are, respectively,

$$I_a = \frac{bh^3}{36} = \frac{6(5.196)^3}{36} = 23.38 \text{ cm}^4$$

$$I_b = \frac{\pi d^4}{64} = \frac{\pi (2)^4}{64} = 0.7854 \text{ cm}^4$$

From the parallel-axis theorem, the moment of inertia of the composite area about the centroidal axis is

$$I_x = I_a + (\bar{y} - \bar{y}_a)^2 A_a - I_b - (\bar{y} - \bar{y}_b)^2 A_b$$

$$= 23.38 + (1.664 - 1.732)^2(15.59) - 0.7854 - (1.664 - 2)^2(3.142)$$

$$= 22.31 \text{ cm}^4 = 2.231 \times 10^{-7} \text{ m}^4$$

**Given:**  Figure 4.7 shows a circular cross section of radius $r$ relative to $x'$, $y'$ coordinates.

**EXAMPLE 4.4**

**Find:**  The area moments of inertia $I_{x'}$ and $I_{y'}$ and the polar moment of inertia $J'_z$, relative to the $x'$, $y'$, and $z'$ axes.

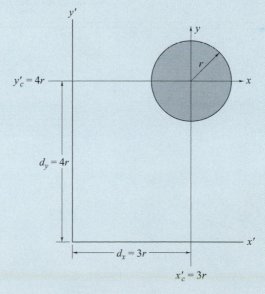

**Figure 4.7**  Circular cross-sectional area relative to $x'$, $y'$ coordinates, used in Example 4.4.

**Solution:**   Using Equations (4.11) and (4.12) and the results from Example 4.2 gives

$$I_{x'} = I_x + A d_y^2 = \frac{\pi r^4}{4} + \pi r^2 (4r)^2 = 16.25 \pi r^4$$

$$I_{y'} = I_y + A d_x^2 = \frac{\pi r^4}{4} + \pi r^2 (3r)^2 = 9.25 \pi r^4$$

$$J_{z'} = I_{x'} + I_{y'} = 25.5 \pi r^4$$

### 4.2.4   RADIUS OF GYRATION

An area's **radius of gyration** with respect to a specific axis is the length that, when squared and multiplied by the area, will give the area's moment of inertia with respect to the specific axis:

$$I = r_g^2 A \tag{4.13}$$

This, then, is another way of expressing the area's moment of inertia. The radius of gyration can be written as

$$r_g = \sqrt{\frac{I}{A}} \tag{4.14}$$

The radius of gyration is not the distance from the reference axis to a fixed point in the area (such as the centroid), but it is a useful property of the area and the specified axis.

---

**EXAMPLE 4.5**

**Given:**   The same situation as in Example 4.4.

**Find:**   The radius of gyration with respect to the $x'$, $y'$, and $z'$ axes.

**Solution:**   Using Equation (4.14) and the results from Example 4.4 gives

$$r_{gx'} = \sqrt{\frac{I_{x'}}{A}} = \sqrt{\frac{16.25 \pi r^4}{\pi r^2}} = 4.031r$$

$$r_{gy'} = \sqrt{\frac{I_{y'}}{A}} = \sqrt{\frac{9.25 \pi r^4}{\pi r^2}} = 3.041r$$

$$r_{gz'} = \sqrt{\frac{J_{z'}}{A}} = \sqrt{\frac{25.5 \pi r^4}{\pi r^2}} = 5.05r$$

When the circle is small relative to the distances $d_x$ and $d_y$, the radii of gyration are just a little larger than the centroidal distances $3r$, $4r$, and $5r$.

### 4.2.5 SECTION MODULUS

The **section modulus** is simply equivalent to an area's moment of inertia, divided by the farthest distance from the centroidal axis to the outer fiber of the solid $c$, or

$$Z_m = \frac{I}{c} \tag{4.15}$$

The area moment of inertia $I$ in Equations (4.13) to (4.15) can be applicable to $I_x$ and $I_y$ with appropriate changes in $r_g$ and $Z_m$. Thus, $r_{gx}$ and $Z_{mx}$ would correspond to the use of $I_x$, and $r_{gy}$ and $Z_{my}$ would correspond to the use of $I_y$.

Table 4.1 gives the centroid, area, and moment of inertia for seven different cross sections. Using the subscripts $\bar{x}$ and $\bar{y}$ for the area moment of inertia means that the area moment of inertia was taken with respect to the centroidal axis so designated. Also, $\bar{J}$ implies that the area polar moment of inertia was taken with the centroidal coordinates; $J$ without an overbar indicates that the area polar moment of inertia was taken with respect to coordinates $x$ and $y$. The formulas expressed in this table are dependent on the coordinate location relevant to the specific cross section.

### 4.2.6 MASS MOMENT OF INERTIA

The **mass moment of inertia** of an element of mass is the product of the element's mass and the square of the element's distance from the axis. Figure 4.8 shows a mass element in three-dimensional coordinates and the distance from the three axes. From this figure the mass moments of inertia can be expressed with respect to the $x$, $y$, and $z$ axes as

$$I_{mx} = \int (y^2 + z^2)\, dm_a \tag{4.16}$$

$$I_{my} = \int (x^2 + z^2)\, dm_a \tag{4.17}$$

$$I_{mz} = \int (x^2 + y^2)\, dm_a \tag{4.18}$$

The SI unit of mass moment of inertia is the kilogram-meter squared and the English unit is pound-mass–inch squared. If instead of the three-dimensional object given in Figure 4.8 we have a thin plate (see Fig. 4.9) so that only the $xy$ plane needs to be considered, then the mass moments of inertia become

$$I_{mx} = \int y^2\, dm_a \tag{4.19}$$

$$I_{my} = \int x^2\, dm_a \tag{4.20}$$

$$J_0 = \int r^2\, dm_a = \int (x^2 + y^2)\, dm_a = I_{mx} + I_{my} \tag{4.21}$$

The polar mass moment of inertia is given in Equation (4.21) as $J_0$. Table 4.2 gives the mass and mass moment of inertia of six commonly used shapes. Note in this table that the origin and coordinates are important for the equations relative to a specific shape.

**Table 4.1**     Centroid, area moment of inertia, and area for seven cross sections

| Cross section | Centroid | Area moment of inertia | Area |
|---|---|---|---|
| **Circular area** <br> | $\bar{x} = 0$ <br> $\bar{y} = 0$ | $I_x = I_{\bar{x}} = \dfrac{\pi}{4}r^4$ <br> $I_y = I_{\bar{y}} = \dfrac{\pi}{4}r^4$ <br> $J = \dfrac{\pi}{2}r^4$ | $A = \pi r^2$ |
| **Hollow circular area** <br> | $\bar{x} = 0$ <br> $\bar{y} = 0$ | $I_x = I_{\bar{x}} = \dfrac{\pi}{4}\left(r^4 - r_i^4\right)$ <br> $I_y = I_{\bar{y}} = \dfrac{\pi}{4}\left(r^4 - r_i^4\right)$ <br> $J = \dfrac{\pi}{2}\left(r^4 - r_i^4\right)$ | $A = \pi\left(r^2 - r_i^2\right)$ |
| **Triangular area** <br> | $\bar{x} = \dfrac{a+b}{3}$ <br> $\bar{y} = \dfrac{h}{3}$ | $I_x = \dfrac{bh^3}{12},\ I_{\bar{x}} = \dfrac{bh^3}{36}$ <br> $I_y = \dfrac{bh(b^2 + ab + a^2)}{12}$ <br> $I_{\bar{y}} = \dfrac{bh(b^2 - ab + a^2)}{36}$ <br> $\bar{J} = \dfrac{bh}{36}(b^2 + h^2 + a^2 - ab)$ | $A = \dfrac{bh}{2}$ |
| **Rectangular area** <br> | $\bar{x} = \dfrac{b}{2}$ <br> $\bar{y} = \dfrac{h}{2}$ | $I_x = \dfrac{bh^3}{3},\ I_{\bar{x}} = \dfrac{bh^3}{12}$ <br> $I_y = \dfrac{hb^3}{3},\ I_{\bar{y}} = \dfrac{hb^3}{12},$ <br> $\bar{J} = \dfrac{bh}{12}(b^2 + h^2)$ | $A = bh$ |
| **Area of circular sector** <br> | $\bar{x} = \dfrac{2}{3}\dfrac{r\sin\alpha}{\alpha}$ | $I_x = \dfrac{r^4}{4}\left(\alpha - \dfrac{1}{2}\sin 2\alpha\right)$ <br> $I_y = \dfrac{r^4}{4}\left(\alpha + \dfrac{1}{2}\sin 2\alpha\right)$ <br> $J = \dfrac{1}{2}r^4\alpha$ | $A = r^2\alpha$ |
| **Quarter-circular area** <br> | $\bar{x} = \bar{y} = \dfrac{4r}{3\pi}$ | $I_x = I_y = \dfrac{\pi r^4}{16}$ <br> $I_{\bar{x}} = I_{\bar{y}} = \left(\dfrac{\pi}{16} - \dfrac{4}{9\pi}\right)r^4$ <br> $J = \dfrac{\pi r^4}{8}$ | $A = \dfrac{\pi r^2}{4}$ |
| **Area of elliptical quadrant** <br> | $\bar{x} = \dfrac{4a}{3\pi}$ <br> $\bar{y} = \dfrac{4b}{3\pi}$ | $I_x = \dfrac{\pi ab^3}{16},\ I_{\bar{x}} = \left(\dfrac{\pi}{16} - \dfrac{4}{9\pi}\right)ab^3$ <br> $I_y = \dfrac{\pi a^3 b}{16},\ I_{\bar{y}} = \left(\dfrac{\pi}{16} - \dfrac{4}{9\pi}\right)a^3 b$ <br> $J = \dfrac{\pi ab}{16}(a^2 + b^2)$ | $A = \dfrac{\pi ab}{4}$ |

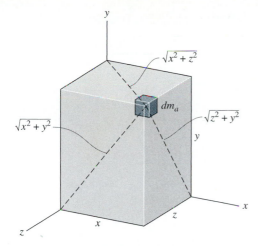

**Figure 4.8** Mass element in three-dimensional coordinates and distance from the three axes.

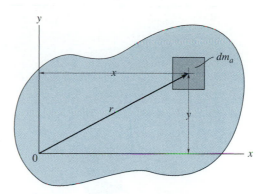

**Figure 4.9** Mass element in two-dimensional coordinates and distance from the two axes.

## 4.3 NORMAL STRESS AND STRAIN

The bar shown in Figure 4.10 supports forces $P$ and $N$ and is in tension. Such a force increases the length of the bar. For a section away from the ends, an average intensity of the normal force, which is also known as the **average normal stress** $\sigma_{avg}$ on the cross section, can be written as

$$\sigma_{avg} = \frac{P}{A} \qquad (4.22)$$

where $A =$ cross-sectional area, m$^2$. A subtlety is associated with the real application of loads. Had the section been cut near the ends, where the shape is no longer a prism, the situation would be more complicated and the stress system would no longer be simple tension uniformly distributed over the cross section. Fortunately, the stresses farther from the load application point are fairly uniform [i.e., stress concentrations (covered in detail in Chap. 6) are relieved as the distance from them increases]. This tendency was first noted by St.-Venant, so it is known as St.-Venant's principle. Its importance should not be underestimated; it makes the application of this chapter's equations possible. This chapter deals with circumstances far from the stress concentrations where the stresses are uniform.[1] The stress concentration itself is dealt with in Chapter 6.

Consistent with the sign convention presented in Chapter 2, a positive sign is used to designate a tensile normal stress; and a negative sign, a compressive normal stress. A compressive stress decreases the length of the bar in the direction of the bar. The total length change in a uniform bar caused by an axial load is called the *elastic deformation* $\delta$. The

---

[1] The exact distance that one must travel from a stress concentration before stresses can be considered uniform varies with the material and can range from 1 characteristic length (diameter, grip length, etc.) to over 10 characteristic lengths for some composite materials.

**Table 4.2**    Mass and mass moment of inertia of six solids

| Shape | Equations |
|---|---|
| Rod | $m_a = \dfrac{\pi d^2 l \rho}{4}$<br><br>$I_{my} = I_{mz} = \dfrac{m_a l^2}{12}$ |
| Disk | $m_a = \dfrac{\pi d^2 t_h \rho}{4}$<br><br>$I_{mx} = \dfrac{m_a d^2}{8}$<br><br>$I_{my} = I_{mz} = \dfrac{m_a d^2}{16}$ |
| Rectangular prism | $m_a = abc\rho$<br><br>$I_{mx} = \dfrac{m_a(a^2 + b^2)}{12}$<br><br>$I_{my} = \dfrac{m_a(a^2 + c^2)}{12}$<br><br>$I_{mz} = \dfrac{m_a(b^2 + c^2)}{12}$ |
| Cylinder | $m_a = \dfrac{\pi d^2 l \rho}{4}$<br><br>$I_{mx} = \dfrac{m_a d^2}{8}$<br><br>$I_{my} = I_{mz} = \dfrac{m_a(3d^2 + 4l^2)}{48}$ |
| Hollow cylinder | $m_a = \dfrac{\pi l \rho (d_o^2 - d_i^2)}{4}$<br><br>$I_{mx} = \dfrac{m_a(d_o^2 - d_i^2)}{8}$<br><br>$I_{my} = I_{mz} = \dfrac{m_a(3d_o^2 + 3d_i^2 + 4l^2)}{48}$ |
| Sphere | $m_a = \dfrac{\pi d^3 \rho}{6}$<br><br>$I_{mx} = I_{my} = I_{mz} = \dfrac{m_a d^2}{10}$ |

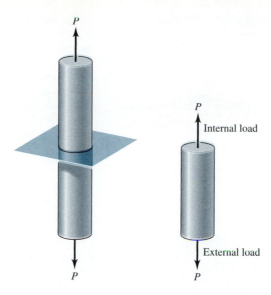

**Figure 4.10** Circular bar with tensile load applied.

**normal strain** is

$$\epsilon = \frac{\delta}{l} = \frac{\text{elastic deformation}}{\text{length without any load}} \tag{4.23}$$

Although the strain is dimensionless, it is given in terms of inch per inch or meter per meter.

From Hooke's law for a uniaxial normal loading,

$$\sigma = \epsilon E \qquad \text{or} \qquad \epsilon = \frac{\sigma}{E} \tag{4.24}$$

where $E$ = modulus of elasticity, Pa. The modulus of elasticity is a constant for a given material.

Substituting Equations (4.22) and (4.24) into Equation (4.23) gives

$$\delta = \epsilon l = \frac{\sigma}{E}l = \frac{Pl}{AE} \tag{4.25}$$

Equations (4.24) and (4.25) are valid for either tension or compression. Tensile stress, corresponding to a length increase, is considered positive; compressive stress, corresponding to length decrease, is considered negative.

The **spring rate** is the normal load, divided by the elastic deflection, or for normal axial loading

$$k = \frac{P}{\delta} = \frac{AE}{l} \tag{4.26}$$

The unit of spring rate for axial loading is newtons per meter or pounds per inch. Equations (4.22) to (4.26) are valid for any cross-sectional area as long as the area remains constant over length $l$.

**EXAMPLE 4.6**

**Given:** A hollow carbon steel shaft 50 mm long must carry a normal force of 5000 N at a normal stress of 100 MPa. The inside diameter is 0.65 of the outside diameter.

**Find:** The outside diameter, the axial deformation, and the spring rate.

**Solution:** The cross-sectional area can be expressed as

$$A = \frac{\pi}{4}\left(d_o^2 - d_i^2\right) = \frac{\pi d_o^2}{4}\left[1 - \left(\frac{d_i}{d_o}\right)^2\right] = 0.4536 d_o^2$$

From Equation (4.22)

$$A = \frac{P}{\sigma} = \frac{5000}{10^8} = 5 \times 10^{-5} \text{ m}^2$$

Therefore,

$$d_o^2 = 1.102 \times 10^{-4} \text{ m}^2$$

$$d_o = 1.05 \times 10^{-2} \text{ m} = 10.5 \text{ mm}$$

For carbon steel the modulus of elasticity is $2.07 \times 10^{11}$ Pa. From Equation (4.25) the elastic deformation is

$$\delta = \frac{Pl}{AE} = \frac{(5000)(0.05)}{(5)(10^{-5})(2.07)(10^{11})} = 24.15 \times 10^{-6} \text{ m} = 24.15 \,\mu m = 0.02415 \text{ mm}$$

From Equation (4.26) the spring rate is

$$k = \frac{P}{\delta} = \frac{5000}{0.02415} = 2.07 \times 10^5 \text{ N/mm} = 2.07 \times 10^8 \text{ N/m}$$

**EXAMPLE 4.7**

**Given:** A fisher catches a salmon with a lure fastened to a 0.45-mm-diameter nylon line. When the fish bites, the line is 46 m long from the fish to the reel. The modulus of elasticity of the line is 4 GPa, and its ultimate strength is 70 N. The salmon pulls with a force of 50 N.

**Find:** The elastic elongation of the line, the spring rate, and the tensile stress in the line.

**Solution:** The cross-sectional area of the line is

$$A = \frac{\pi d^2}{4} = \frac{\pi (0.45)^2}{4} = 0.1590 \text{ mm}^2$$

The stress being exerted on the line by the salmon is

$$\sigma = \frac{P}{A} = \frac{(50)(10^6)}{0.159} \text{ Pa} = 0.3144 \text{ GPa}$$

The elongation of the line is

$$\delta = \epsilon l = \frac{\sigma l}{E} = \frac{0.3144(46)}{4} = 3.615 \text{ m}$$

The spring rate of the line is

$$k = \frac{P}{\delta} = \frac{50}{3.615} = 13.83 \text{ N/m}$$

## 4.4 TORSION

A shaft is a slender element that is mainly loaded by an axial moment or twist (torque), which causes torsional deformation and shear stresses. Thus, **torsion** is loading resulting in twisting of the shaft. Shafting in general is dealt with in Chapter 11; the focus here is on the twisting, or torsion, that shafts are subjected to and the stresses that result. A major use of the shaft is to transfer, or transmit, mechanical power from one point to another. Engineers are interested primarily in the twisting moment that can be transmitted by the shaft without damaging the material or exceeding maximum deformations. Hence, they wish to know the stresses in the shaft and the angle of twist. Solid circular members are of primary concern because most torque-transmitting shafts are of this shape. For cross sections that are not circular, or for hollow cross sections, the interested reader is referred to the classic text by Timoshenko and Goodier (1970).

### 4.4.1 STRESS AND STRAIN

Torque is a moment that tends to twist a member about its longitudinal axis. Figure 4.11 shows the twisting of a member due to an applied torque. The circular shaft deforms such that each plane cross section originally normal to the axis remains plane and normal and does not distort within its own plane. The shaft is fixed at the top, and a torque is applied to the bottom end. The angle of twist is defined as $\theta$. The extension strains are assumed to be zero. Thus, the five strains are zero ($\epsilon_r = \epsilon_\theta = \epsilon_z = \gamma_{r\theta} = \gamma_{rz} = 0$), and the only nonzero strain is

$$\gamma_{\theta z} = r\frac{d\theta}{dz} \approx \frac{r\theta}{l} \tag{4.27}$$

**Figure 4.11** Twisting of member due to applied torque.

Equation (4.27) can also be obtained from Figure 4.11 by observing the common surface as it relates $l\gamma_{\theta z}$ and $r\theta$.

From Hooke's law the stress is related to the strain by

$$\tau_{\theta z} = G\gamma_{\theta z} = Gr\frac{d\theta}{dz} \approx \frac{Gr\theta}{l} \tag{4.28}$$

where $G$ = shear modulus of elasticity, Pa. In Equation (4.28) the shear strain $\gamma_{\theta z}$ and the shear stress $\tau_{\theta z}$ vary linearly with respect to the rate of twist $d\theta/dz \approx \theta/l$. Also, the shear stress does not change in the $\theta$ direction (because of symmetry) or in the $z$ direction (because the deformation and the stress pattern are uniform along the shaft's length).

The rate of twist $d\theta/dz$, or $\theta/l$ in Equation (4.28), is still unknown. To solve for it, the stresses must meet the conditions of equilibrium. The applied twisting moment, or torque, is

$$T = \int_A r\,(\tau_{\theta z}\,dA) = \frac{G\theta}{l}\int_A r^2\,dA \tag{4.29}$$

But the area polar moment of inertia is

$$J = \int_A r^2\,dA \tag{4.30}$$

Substituting Equation (4.30) into Equation (4.29) gives the torque, or the angle of twist, as

$$T = \frac{G\theta J}{l} \qquad \text{or} \qquad \theta = \frac{Tl}{GJ} \tag{4.31}$$

From Table 4.1 for a solid circular shaft of outer radius $r_o$ and diameter $d$,

$$J = \frac{\pi r_o^4}{2} = \frac{\pi d^4}{32} \tag{4.32}$$

Substituting Equations (4.29) and (4.30) into Equation (4.28) gives

$$\tau_{\theta z} = \frac{rT}{J} \tag{4.33}$$

The maximum stress is

$$\tau_{\max} = \frac{cT}{J} \tag{4.34}$$

where $c$ = distance from neutral axis to outer fiber, m. The angular spring rate can be expressed as

$$k_a = \frac{T}{\theta} = \frac{JG}{l} \tag{4.35}$$

EXAMPLE 4.8

**Given:** A 50-mm-long, circular, hollow shaft made of carbon steel must carry a torque of 5000 N · m at a maximum shear stress of 70 MPa. The inside diameter is 0.5 times the outside diameter.

**Find:** The outside diameter, the angle of twist, and the angular spring rate.

**Solution:** From Table 4.1 for a hollow cylinder

$$J = \frac{\pi}{2}\left(r^4 - r_i^4\right) = \frac{\pi r^4}{2}\left[1 - \left(\frac{r_i}{r}\right)^4\right] \tag{a}$$

From Equation (4.15) the section modulus is

$$Z_m = \frac{J}{c} = \frac{T}{\tau_{max}} = \frac{5000}{70(10^6)} = 71.43 \times 10^{-6}\ \text{m}^3 \tag{b}$$

Note that

$$\frac{r_i}{r} = \frac{r_i}{r_o} = 0.5 \quad \text{and} \quad c = r = \frac{d_o}{2} \tag{c}$$

Therefore, by making use of Equations (a) and (c), Equation (b) becomes

$$\frac{\pi r_o^3}{2}\left[1 - (0.5)^4\right] = 71.43 \times 10^{-6}\ \text{m}^3$$

$$r_o^3 = 48.51 \times 10^{-6}\ \text{m}^3$$

$$r_o = 0.03647\ \text{m} = 36.47\ \text{mm}$$

$$d_o = 72.94\ \text{mm}$$

The shear modulus of rigidity for carbon steel is 80 GPa. Making use of Equation (4.31) gives the angular twist due to torsion as

$$\theta = \frac{Tl}{GJ} = \frac{(5000)(2)(0.050)}{8(10^{10})(71.43)(10^{-6})(0.07294)} = 0.0012\ \text{rad}$$

Recall that $1° = \pi/180$ rad and 1 rad $= 57.296°$.

$$\therefore \quad \theta = 0.0687°$$

From Equation (4.35) the angular spring rate is

$$k_a = \frac{T}{\theta} = \frac{5000}{0.0012} = 4.167 \times 10^6\ \text{N} \cdot \text{m/rad}$$

## 4.4.2 Power Transfer

It is convenient to present power transfer directly after considering torsional stresses and deformation. One of the most common uses of a circular shaft is in power transfer; therefore, no discussion of torsion would be adequate without including this topic.

**Power** is the rate of doing work, or

$$h_p = \text{force} \times \text{velocity} = Pu \tag{4.36}$$

### English Units

The English unit of power is horsepower (hp), which is defined as 33,000 ft · lb/min.

$$h_p = \frac{Pu_f}{33,000} \tag{4.37}$$

where  $P$ = force, lb
       $u_f$ = velocity, ft/min

Equation (4.37) is applicable only for the unit specified.
     For rotating systems the force can be replaced by torque if

$$P = \frac{T}{r} \tag{4.38}$$

Also, the linear velocity $u_f$ at radius $r$ can be expressed as

$$u_f = \frac{2\pi r N_a}{12} \qquad \text{ft/min} \tag{4.39}$$

where $N_a$ = rotational speed, rpm
       $r$ = radius, in

Substituting Equations (4.38) and (4.39) into Equation (4.37) gives

$$h_p = \frac{(T/r)(2\pi r N_a/12)}{33,000} = \frac{TN_a}{63,025} \qquad \text{hp} \tag{4.40}$$

or
$$T = \frac{63,025 h_p}{N_a} \qquad \text{lb} \cdot \text{in} $$

### SI Units

The SI unit of work is a joule, or a newton-meter. The SI unit of power is a watt, or a joule per second.

$$\therefore \qquad h_p = Pu = T\omega \tag{4.41}$$

where $P$ = force, N
       $u$ = velocity, m/s
       $T$ = torque, N · m
       $\omega$ = rotational speed, rad/s

or
$$T = \frac{h_p}{\omega} \tag{4.42}$$

The torque obtained from Equation (4.40) is dependent on units of power and rotational speed, but the torque given in Equation (4.42) is in newton-meters, which is the SI unit of torque.

**Given:** A shaft carries a torque of 10,000 in · lb and turns at 900 rpm.

EXAMPLE 4.9

**Find:** The power transmitted in both horsepower and kilowatts.

**Solution:** From Equation (4.40)

$$h_p = \frac{TN_a}{63,025} = \frac{(10,000)(900)}{63,025} = 142.8 \text{ hp} \tag{a}$$

The torque can be expressed in newton-meters if

$$T = 10,000 \text{ in} \cdot \text{lb} \left(\frac{4.448 \text{ N}}{1 \text{ lb}}\right)\left(\frac{0.0254 \text{ m}}{1 \text{ in}}\right) = 1130 \text{ N} \cdot \text{m}$$

From Equation (4.42)

$$h_p = T\omega = 1130\frac{900(2\pi)}{60} = 1.065 \times 10^5 \text{ W} = 106.5 \text{ kW} \tag{b}$$

A conversion from English to SI units is provided by

$$\text{Horsepower} = \frac{\text{kilowatts}}{0.7457} \tag{c}$$

For example, 10 hp is equal to 7.457 kW. Thus, a check on the solution provided in Equations (*a*) and (*b*) is

$$142.8 \text{ hp} = \frac{106.5 \text{ kW}}{0.7457} = 142.8 \text{ hp}$$

Therefore, it checks out.

## 4.5 BENDING STRESS AND STRAIN

The bending of a long, thin member, commonly called a beam, is often a concern to engineers. Bending occurs in horizontal members of buildings (joists) subjected to vertical floor loading, in leaf springs on a truck, and in wings of an airplane in supporting the weight of the fuselage. In each of these applications, stress and deformation are important design considerations.

Throughout this section the following assumptions are made:

1. The cross section is symmetric in the plane of loading (about the *y* axis).

2. The solid material that the beam is made of is homogeneous and linear-elastic.

A straight member is considered initially and then a curved member.

### 4.5.1 STRAIGHT MEMBER

Figure 4.12 shows bending occurring in a highly deformable material, such as rubber, which is well suited for demonstration purposes. Figure 4.12(a) shows an undeformed bar with square cross sections marked by longitudinal and transverse grid lines. In Figure 4.12(b) a moment is applied. The longitudinal lines become curved while the transverse lines remain

**Figure 4.12**    Bar made of elastomeric material to illustrate effect of bending. (a) Undeformed bar; (b) deformed bar.

straight and yet undergo a rotation. The longitudinal lines have a radius when the bar is deformed, even though initially they were straight. Figure 4.13 further demonstrates the effect of bending. The bending moment causes the material in the bottom portion of the member to stretch, or be in tension, and that in the top to be in compression. Consequently, between these two regions there must be a surface, called the *neutral surface,* in which the longitudinal fibers of the material will not undergo a change in length. On this neutral surface no bending stress is occurring, and it is neither in tension nor in compression.

Figure 4.14 shows undeformed and deformed elements when bending occurs. The normal strain along line segment $\Delta s$ is

$$\epsilon = \lim_{\Delta s \to 0} \frac{\Delta s' - \Delta s}{\Delta s} \qquad\qquad \textbf{(4.43)}$$

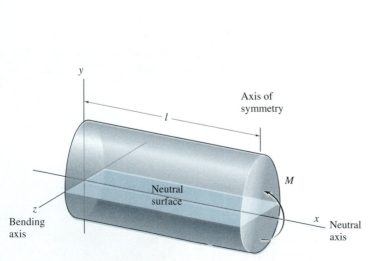

**Figure 4.13**    Bending occurring in cantilevered bar, showing neutral surface.

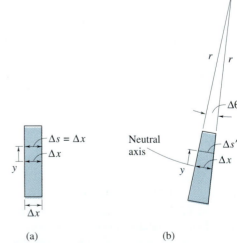

**Figure 4.14**    (a) Undeformed and (b) deformed elements in bending.

From Figure 4.14, where $r$ is the radius of curvature of the element's longitudinal axis,

$$\Delta x = \Delta s = r\,\Delta\theta$$
$$\Delta s' = (r - y)\,\Delta\theta$$

Substituting these equations into Equation (4.43) gives

$$\epsilon = \lim_{\Delta s \to 0} \frac{(r - y)\,\Delta\theta - r\,\Delta\theta}{r\,\Delta\theta} = -\frac{y}{r} \tag{4.44}$$

The longitudinal normal strain will vary linearly with $y$ from the neutral axis. The farthest distance from the neutral axis to the outer fiber is defined as $c$. The maximum strain occurs at the outermost fiber, located at $c$ from the neutral axis.

$$\therefore \qquad \frac{\epsilon}{\epsilon_{max}} = -\frac{y/r}{c/r}$$

or

$$\epsilon = -\frac{y}{c}\epsilon_{max} \tag{4.45}$$

Similarly, a linear variation of normal stress over the cross-sectional area occurs, or

$$\sigma = -\frac{y}{c}\sigma_{max} \tag{4.46}$$

Figure 4.15 shows a profile view of the normal stress. For positive $y$ the normal stress is compressive, and for negative $y$ the normal stress is tensile. The normal stress is zero at the neutral axis.

From force equilibrium

$$0 = \int_A dP = \int_A \sigma\,dA = \int_A -\frac{y}{c}\sigma_{max}\,dA = -\frac{\sigma_{max}}{c}\int_A y\,dA$$

Because $\sigma_{max}$ is not equal to zero,

$$\int_A y\,dA = 0 \tag{4.47}$$

This equation implies that the first moment of the member's cross-sectional area about the neutral axis must be zero.

The moment may be expressed as

$$M = \int_A y\,dP = \int_A y\sigma\,dA = -\frac{\sigma_{max}}{c}\int_A y^2\,dA$$

**Figure 4.15**  Profile view of bending stress variation.

But the area moment of inertia is

$$I = \int_A y^2 \, dA$$

$$\therefore \qquad \sigma_{max} = -\frac{Mc}{I} \qquad\qquad (4.48)$$

The stress at any intermediate distance $y$ is

$$\sigma = -\frac{My}{I} \qquad\qquad (4.49)$$

Making use of Equations (3.22), (4.44), and (4.49) gives

$$\frac{1}{r} = \frac{M}{EI} \qquad\qquad (4.50)$$

From Equation (4.50) when the bending moment is positive, the curvature is positive [i.e., concave in the $y$ direction (see Fig. 2.2)].

If the distribution of the normal stress in a given cross-section is not affected by the deformation caused by the shear stresses, the moment can be rewritten as $Vx$. Thus, Equation (4.49) becomes

$$\sigma = -\frac{Vxy}{I} \qquad\qquad (4.51)$$

where $V$ = transverse shear force, N.

The normal stress is thus proportional to the distance $x$ from the load to the section considered. Thus, the maximum compressive stress in the bar occurs at $x = l$ at the top of the bar and at the outer fiber. At $x = l$ the bar is mounted to a wall, thus producing a cantilever arrangement.

---

**EXAMPLE 4.10**

**Given:** An aluminum alloy beam with the cross-section shown in Figure 4.16 experiences positive bending by an applied moment $M$. The allowable stress is 150 MPa.

**Find:**

(a)  The maximum moment that can be applied to the beam.
(b)  Stresses at points A, B, and C when the maximum moment is applied.

**Solution:**

(a)  The cross-sectional area, if subscript 1 refers to a horizontal rectangle and subscript 2 refers to a vertical rectangle, is

$$A = A_1 + 2A_2 = 8(80 - 16 + 240) = 2432 \text{ mm}^2$$

The centroids of the cross section are

$$\bar{y} = \frac{\bar{y}_1 A_1 + 2\bar{y}_2 A_2}{A} = \frac{4(64)(8) + 2(60)(8)(120)}{2432} = 48.21 \text{ mm}$$

**Figure 4.16**  U-shaped cross section experiencing bending moment, used in Example 4.10.

The distances from the neutral axis to the centroids of the horizontal and vertical bars are

$$d_{n1} = 48.21 - 4 = 44.21 \text{ mm}$$

$$d_{n2} = 60 - 48.21 = 11.79 \text{ mm}$$

The area moment of inertia of the composite structure is

$$I = I_1 + A_1 d_{n1}^2 + 2\left(I_2 + A_2 d_{n2}^2\right)$$

$$= \frac{64(8)^3}{12} + (64)(8)(44.2)^2 + 2\left[\frac{8(120)^3}{12} + 8(120)(11.79)^2\right]$$

$$= 3,574,335 \text{ mm}^4 = 357.4 \text{ cm}^4$$

The distances from the neutral axis to the points where the stress is to be evaluated are

$$d_{nA} = 48.21 \text{ mm} \quad d_{nB} = 48.21 - 8 = 40.21 \text{ mm} \quad d_{nC} = 120 - 48.21 = 71.79 \text{ mm}$$

Point C is the farthest from the neutral axis and is thus the location where the stress is the largest. Furthermore, from Section 2.3, positive bending implies that the portion of the composite structure above the neutral axis is in compression while the portion below it is in tension. From Equation (4.48) the maximum moment is

$$M_{\max} = \frac{\sigma_{\text{all}} I}{d_{nC}} = \frac{(150)(10^6)(357.4)(10^{-8})}{(71.79)(10^{-3})} = 7468 \text{ N} \cdot \text{m}$$

(b)  The stresses at the various points of interest are

$$\sigma_A = \frac{M_{\max} d_{nA}}{I} = \frac{(7468)(48.21)(10^{-3})}{(357.4)(10^{-8})} = 100.7 \text{ MPa}$$

$$\sigma_B = \frac{M_{\max} d_{nB}}{I} = \frac{(7468)(40.21)(10^{-3})}{(357.4)(10^{-8})} = 84.0 \text{ MPa}$$

$$\sigma_C = \sigma_{\text{all}} = -150 \text{ MPa}$$

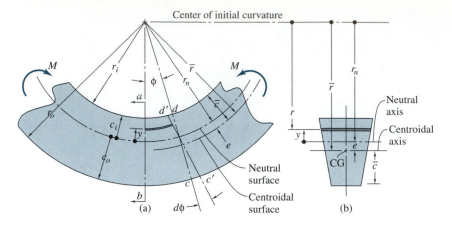

**Figure 4.17**   Curved member in bending. (a) Circumferential view; (b) cross-sectional view.

### 4.5.2    CURVED MEMBER

In a straight member, the normal stress and strain vary linearly from the neutral axis, as shown in Equations (4.45) and (4.46), respectively. In a curved member, the stress and strain are not linearly related. Examples of curved members are hooks and chain links, which are not slender but have a high degree of geometric curvature.

Figure 4.17 shows a curved member in bending. Positive bending causes surface $dc$ to rotate through $d\phi$ to $d'c'$. Note that $y$ is negative from the neutral axis to the outer radius $r_o$, the region where the member is experiencing tension, and positive from the neutral axis to the inner radius $r_i$, the region where the member is experiencing compression. At the neutral axis no bending stress is occurring; thus, the member is neither in compression nor in tension. The neutral axis occurs at a radius of $r_n$. From Figure 4.17 note that the neutral radius $r_n$ and the centroidal radius $\bar{r}$ are not the same, although they are the same for a straight member. The difference between $r_n$ and $\bar{r}$ is the eccentricity $e$. Also, the radius $r$ locates an arbitrary area element $dA$, as shown in the cross-sectional view of Figure 4.17.

The strain for an arbitrary radius $r$ can be expressed as

$$\epsilon = \frac{(r - r_n)\, d\phi}{r\phi} \tag{4.52}$$

The strain is zero when $r$ is equivalent to the neutral radius $r_n$ and is largest at the outer fiber, or $r = r_o$. The normal stress can be simply written as

$$\sigma = \epsilon E = \frac{E(r - r_n)\, d\phi}{r\phi} \tag{4.53}$$

For $r$ less than $r_n$ the stress is compressive, and for $r$ greater than $r_n$ the stress is tensile.

Solving for the strain and stress in Equations (4.52) and (4.53) requires the location of the neutral axis. This location is obtained by taking the sum of the normal stresses acting on the section and setting it to zero. Thus, making use of Equation (4.53) gives

$$\int_A \sigma \, dA = \frac{E \, d\phi}{\phi} \int_A \frac{r - r_n}{r} \, dA = 0 \qquad (4.54)$$

This equation reduces to

$$A - r_n \int_A \frac{dA}{r} = 0$$

or

$$r_n = \frac{A}{\int_A (dA/r)} \qquad (4.55)$$

Equation (4.55) clearly indicates that the neutral radius is a function of the sectional area.

### Rectangular Cross-Sections

Figure 4.18 shows a rectangular cross-section of a curved member, with its centroidal and neutral axes. From Equation (4.55) the neutral radius for a rectangular cross-section is

$$r_n = \frac{b(r_o - r_i)}{\int_{r_i}^{r_o} (b \, dr/r)} = \frac{r_o - r_i}{\ln(r_o/r_i)} \qquad (4.56)$$

The centroidal radius is

$$\bar{r} = \frac{r_i + r_o}{2} \qquad (4.57)$$

The eccentricity is

$$e = \bar{r} - r_n = \frac{r_i + r_o}{2} - \frac{r_o - r_i}{\ln(r_o/r_i)} \qquad (4.58)$$

**Figure 4.18** Rectangular cross-section of curved member.

### Circular Cross-Sections

The neutral radius for a circular cross-section is

$$r_n = \frac{\bar{r} + \sqrt{\bar{r}^2 - r_c^2}}{2} \tag{4.59}$$

where $r_c$ = radius of cross-section, m

$\bar{r}$ = distance from centroid of cross-section to center of curvature, m

Having established the location of the neutral radius for two different cross-sections of a curved member, we are ready to return to the stress equation given in Equation (4.53). The bending moment is just the moment arm $r - r_n$, multiplied by the force $\sigma \, dA$ while integrating over the cross-sectional area, or

$$M = \int_A (r - r_n)(\sigma \, dA)$$

Making use of Equation (4.53) gives

$$M = \frac{E \, d\phi}{\phi} \int_A \frac{(r - r_n)^2 \, dA}{r} = \frac{E \, d\phi}{\phi} \int_A \frac{\left(r^2 - 2rr_n + r_n^2\right) \, dA}{r}$$

$$= \frac{E \, d\phi}{\phi} \left( \int_A r \, dA - r_n A - r_n A + r_n^2 \int_A \frac{dA}{r} \right)$$

From Equation (4.55) this equation reduces to

$$M = \frac{E \, d\phi}{\phi} \left( \int_A r \, dA - r_n A \right) \tag{4.60}$$

From the definition of a centroid

$$\bar{r} = \frac{1}{A} \int_A r \, dA$$

Equation (4.60) then becomes

$$M = E \frac{d\phi}{\phi} Ae \tag{4.61}$$

where

$$e = \bar{r} - r_n \tag{4.62}$$

From Equation (4.53), Equation (4.61) becomes

$$M = \frac{r\sigma Ae}{r - r_n}$$

or

$$\sigma = \frac{M(r - r_n)}{Aer} = \frac{-My}{Ae(r_n - y)} \tag{4.63}$$

where

$$y = r_n - r \tag{4.64}$$

The stress distribution is hyperbolic. The maximum stress occurs at either the inner or the outer surface:

$$\sigma_i = -\frac{Mc_i}{Aer_i} \qquad (4.65)$$

$$\sigma_o = \frac{Mc_o}{Aer_o} \qquad (4.66)$$

---

**EXAMPLE 4.11**

**Given:** A rectangular cross-section of a curved member, as shown in Figure 4.18, has the dimensions $b = 1$ in and $h = r_o - r_i = 3$ in and is subjected to a pure bending moment of 20,000 in · lb. No other type of loading is acting on the member. Positive bending occurs.

**Find:** The maximum stress for the following geometries:

  (a)  A straight member
  (b)  A member whose centroidal axis has a radius of 15 in
  (c)  A member whose centroidal axis has a radius of 3 in

**Solution:**

  (a)  For a straight member

$$I = \frac{bh^3}{12} \qquad c = \frac{h}{2} \qquad \therefore \qquad \frac{I}{c} = \frac{bh^2}{6}$$

$$\sigma = \frac{Mc}{I} = \frac{(20,000)6}{(1)(3)^2} = 13,333 \text{ psi}$$

$$\therefore \qquad \sigma_i = -13,333 \text{ psi}$$

$$\sigma_o = 13,333 \text{ psi}$$

  (b)  The outer and inner radii relative to the centroidal radius of 15 in are

$$r_o = \bar{r} + \frac{h}{2} = 15 + \frac{3}{2} = 16.5 \text{ in}$$

$$r_i = \bar{r} - \frac{h}{2} = 15 - \frac{3}{2} = 13.5 \text{ in}$$

From Equation (4.56) the neutral radius is

$$r_n = \frac{r_o - r_i}{\ln (r_o/r_i)} = \frac{3}{\ln (16.5/13.5)} = 14.95 \text{ in}$$

$$e = \bar{r} - r_n = 15 - 14.95 = 0.05 \text{ in}$$

The distances from the neutral axis to the inner and outer fibers are

$$c_o = \frac{h}{2} + e = 1.5 + 0.05 = 1.55 \text{ in}$$

$$c_i = \frac{h}{2} - e = 1.5 - 0.05 = 1.45 \text{ in}$$

The corresponding normal stresses are

$$\sigma_i = -\frac{Mc_i}{Aer_i} = -\frac{(20,000)(1.45)}{3(1)(0.05)(13.5)} = -14,321 \text{ psi}$$

$$\sigma_o = \frac{Mc_o}{Aer_o} = \frac{(20,000)(1.55)}{3(1)(0.05)(16.5)} = 12,525 \text{ psi}$$

Because the radius of curvature is so large, there is little difference from the stresses found for a straight beam in part (a).

(c)   The outer and inner radii relative to the centroidal radius of 3 in are

$$r_o = \bar{r} + \frac{h}{2} = 3 + \frac{3}{2} = 4.5 \text{ in}$$

$$r_i = \bar{r} - \frac{h}{2} = 3 - \frac{3}{2} = 1.5 \text{ in}$$

$$r_n = \frac{r_o - r_i}{\ln(r_o/r_i)} = \frac{4.5 - 1.5}{\ln(4.5/1.5)} = 2.73 \text{ in}$$

$$e = \bar{r} - r_n = 3 - 2.73 = 0.27 \text{ in}$$

$$\therefore \quad c_o = \frac{h}{2} + e = 1.5 + 0.27 = 1.77 \text{ in}$$

$$c_i = \frac{h}{2} - e = 1.5 - 0.27 = 1.23 \text{ in}$$

$$\sigma_i = -\frac{Mc_i}{Aer_i} = -\frac{(20,000)(1.23)}{(1)(3)(0.27)(1.5)} = -20,247 \text{ psi}$$

$$\sigma_o = \frac{Mc_o}{Aer_o} = \frac{(20,000)(1.77)}{(1)(3)(0.27)(4.5)} = 9712 \text{ psi}$$

Thus, the more curved the surface is, the more different the stresses are from the straight-beam results.

## 4.6    TRANSVERSE SHEAR STRESS AND STRAIN

In addition to the bending stresses considered in Section 4.5, the loads on a member can cause shear stresses within the member. Figure 4.19 attempts to illustrate how transverse shear is developed. Figure 4.19(a) shows three boards that are not bonded together. The application of a force $P$ will cause the boards to slide relative to one another, and the beam will deflect as shown, with the ends no longer flush as they were when no load was applied. On the other hand, if the boards are bonded together, the longitudinal shear stress between the boards will prevent them from sliding, and consequently the beam will act as a single unit, as shown in Figure 4.19(b). In a solid beam the elements do not slide on each other, but the shear stress tending to make them do so is present. As a result of the internal shear stress distribution, shear strains develop, and they will tend to distort the cross section in a rather complex manner. An undeformed bar, shown in Figure 4.20(a), made of a highly deformable material and marked with horizontal and vertical grid lines tends to deform

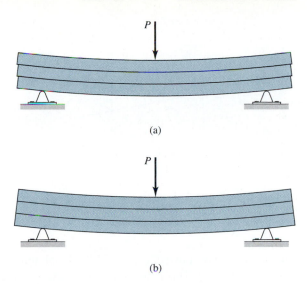

(a)

(b)

**Figure 4.19**  Development of transverse shear. (a) Boards not bonded together; (b) boards bonded together.

when a shear force is applied, changing these lines into the pattern shown in Figure 4.20(b). The squares near the top and bottom of the bar retain their undeformed shapes. The strain on the center square of the bar will cause it to have the greatest deformation. The transverse shear stress causes the cross-section to warp, or to not remain plane. Figure 4.20(b) shows that the deformation caused by transverse shear is much more complex than that caused by another type of loading (axial, torsion, or bending).

The transverse shear formulas also apply to Figure 4.21. The shaded top segment of the element has been sectioned at $y'$ from the neutral axis. This section has a width $w_t$ at the section and has cross-sectional sides each having an area $A'$. Because the resultant moments on each side of the element differ by $dM$,

$$\sum P_x = 0$$

(a)                                    (b)

**Figure 4.20**  Cantilevered bar made of highly deformable material and marked with horizontal and vertical grid lines to show deformation due to transverse shear. (a) Undeformed; (b) deformed.

**Figure 4.21**    Three-dimensional and profile views of moments and stresses associated with shaded top segment of element that has been sectioned at $y'$ about neutral axis. Shear stresses have been omitted for clarity. (a) Three-dimensional view; (b) profile view.

will only be satisfied from Figure 4.21 if a longitudinal shear stress $\tau$ acts over the bottom face of the segment. This longitudinal shear stress is also responsible for the results shown in Figure 4.19(a). The expression for the shear stress in the member at the point located a distance $y'$ from the neutral axis is

$$\tau = \frac{VQ}{Iw_t} \tag{4.67}$$

where  $V$ = transverse shear force, N

$I$ = moment of inertia of entire cross-section computed about neutral axis, m$^4$

$w_t$ = width at point where $\tau$ is determined, m

$Q$ = first moment about neutral axis of shaded portion of Fig. 4.21

$= \int_{A'} y\, dA = \bar{y}' A'$, m$^3$

$A'$ = cross-sectional area of top portion shown in Figure 4.21, m$^2$

$\bar{y}'$ = distance to centroid of $A'$, measured from neutral axis, m

Applying the shear formula given in Equation (4.67) for a rectangular cross-section gives

$$\tau = \frac{6V}{bh^3}\left(\frac{h^2}{4} - y^2\right) \tag{4.68}$$

where $b$ = base of rectangular section, m

$h$ = height of rectangular section, m

From Equation (4.68) the shear stress intensity for a rectangular cross-section varies from zero at the top and bottom ($y = \pm h/2$) to a maximum at the neutral axis ($y = 0$). At other intermediate values it is parabolic. For $A = bh$ and $y = 0$, Equation (4.68) reduces to

$$\tau_{\text{max}} = 1.5\frac{V}{A} \tag{4.69}$$

The maximum shear stress depends on the shape of the cross-section. Table 4.3 summarizes maximum values of $\tau$ for common cross-sectional shapes. In all cases, shear stress is zero

**Table 4.3** Maximum shear stress for different beam cross-sections.

| Cross section | Maximum shear stress |
|---|---|
| Rectangular | $\tau_{max} = \dfrac{3V}{2A}$ |
| Circular | $\tau_{max} = \dfrac{4V}{3A}$ |
| Round tube | $\tau_{max} = \dfrac{2V}{A}$ |
| I-beam | $\tau_{max} = \dfrac{V}{A_{web}}$ |

on extreme fibers, is maximum on the neutral axis, and obtains a parabolic distribution through the thickness.

**Given:** A cantilevered beam having a square cross-section is loaded by a shear force perpendicular to the beam centerline at the free end. The sides of the square cross-section are 50 mm, and the shear force is 10,000 N.

**EXAMPLE 4.12**

**Find:** The bending shear stress at the beam centerline

(a) If the shear force is parallel with two sides of the square cross-section
(b) If the shear force is parallel with one of the diagonals of the square cross-section

**Solution:** The area moment of inertia is

$$I = \frac{bh^3}{12} = \frac{(50)(50)^3}{12} = 520,833 \text{ mm}^4$$

This moment is valid in all directions due to symmetry. Thus, the shear force and the area moment of inertia are the same for parts (a) and (b).

(a)   The width of the point where $\tau$ is applied is 50 mm. The evaluation of $Q$ is

$$Q = \int_A y \, dA = \int_0^{25} y(50) \, dy = \frac{50(25)^2}{2} = 15{,}625 \text{ mm}^3$$

From Equation (4.67) the bending shear stress is

$$\tau = \frac{VQ}{Iw_t} = \frac{(10{,}000)(15{,}625)}{(520{,}833)(50)} = 6.0 \text{ N/mm}^2 = 6 \text{ MPa}$$

(b)   The width at the point where $\tau$ is applied is

$$w_t = \sqrt{2(50)^2} = 50\sqrt{2} = 70.71 \text{ mm}$$

$$Q = \int_0^{25\sqrt{2}} 2y(25\sqrt{2} - y) \, dy = \left[ 25\sqrt{2}y^2 - \frac{2y^3}{3} \right]_{y=0}^{y=25\sqrt{2}} = 14{,}731 \text{ mm}^3$$

From Equation (4.67) the bending shear stress is

$$\tau = \frac{VQ}{Iw_t} = \frac{(10{,}000)(14{,}731)}{(520{,}833)(70.71)} = 4.0 \text{ N/mm}^2 = 4 \text{ MPa}$$

---

**EXAMPLE 4.13**

**Given:**   A shaft is loaded by the forces and torques shown in Figure 4.22, which result from the actions of helical gears and the shaft's rolling-element bearing supports. The shaft has a diameter of 1.00 in.

**Find:**   Determine the location in the shaft where the stresses are highest. What are the principal stresses at this location?

**Solution:**   First, note that with a diameter of 1.00 in, the following can be calculated:

$$A = \frac{\pi d^2}{4} = \frac{\pi (1.00)^2}{4} = 0.7854 \text{ in}^2$$

$$I = \frac{\pi d^4}{64} = \frac{\pi (1.00)^4}{64} = 0.04909 \text{ in}^4$$

$$J = \frac{\pi d^4}{32} = \frac{\pi (1.00)^4}{32} = 0.09817 \text{ in}^4$$

**Figure 4.22**   Shaft with loading considered in Example 4.13.

**Figure 4.23** (a) Shear force, (b) normal force, and (c) bending moment diagrams for the shaft in Figure 4.22.

The shear and moment diagrams are shown in Figure 4.23. At first, it is not clear where the critical location is; the shear force is highest between 0 in $< x <$ 5 in, but the moment is highest at $x = 5$ in. Also, the torque is highest in the range 5 in $< x <$ 10 in, and the axial load is tensile between 5 in $< x <$ 10 in but compressive for 0 $< x <$ 5 in.

In practice, a designer must analyze *all* potential critical locations to determine the most critical. We will analyze the location just to the right of the gear that acted at $x = 5$ in, where $V = 50$ lb, $P = 75$ lb, $T = 1000$ in·lb, and $M = 1250$ in · lb. These result in the following stresses:

| Loading | Resultant maximum stress | Reference | Value |
|---|---|---|---|
| Axial | $\sigma_x = \dfrac{P}{A} = \dfrac{75}{0.7854}$ | Equation (2.7) | 95.49 psi |
| Bending | $\sigma_x = \dfrac{Mc}{I} = \dfrac{(1250)(0.50)}{0.04909}$ | Equation (4.48) | 12,732 psi |
| Vertical shear | $\tau_{xy} = \dfrac{4V}{3A} = \dfrac{4(50)}{3(0.7854)}$ | Table 4.3 | 84.88 psi |
| Torsion | $\tau = \dfrac{rT}{J} = \dfrac{(0.5)(1000)}{0.09817}$ | Equation (4.33) | 5093 psi |

Consider the cross section of the shaft shown in Figure 4.24, with the stress element locations shown. At location A, the axial stress is tensile, but the bending stress is compressive; the resultant normal stress is $\sigma_x = 95.49 - 12,732 = -12,637$ psi. The shear stress distributions for torsion and shear are shown in Figure 4.25; clearly at the top of the shaft, the shear stress due to the vertical shear force can be ignored, while the shear stress due to torsion is at its maximum value. Thus, $\tau_{xz} = 5093$ psi.

At location B, the bending stress is zero (see Fig. 4.15). Thus, the normal stress is due to axial force only, or $\sigma_x = 95.49$ psi. The shear stresses due to torsion and vertical shear are additive; thus $\tau_{xy} = 84.88 + 5093 = 5178$ psi.

At location C, the normal stresses are both tensile; thus $\sigma_x = 95.49 + 12,732 = 12,828$ psi. The shear stress is the same as at location A, but it is now negative, so that $\tau_{xz} = -5093$ psi.

At location D, the bending stress is zero, so the normal stress is the same as at location B, or $\sigma_x = 95.49$ psi. The shear stress due to torsion is zero, but the shear stress due to the vertical shear force has its maximum value of 84.88 psi.

From consideration of these stress elements, we conclude that the element at location C is critical; it has the largest normal and shear stresses. It has the stress state $\sigma_x = 12,828$ psi, $\sigma_y = \sigma_z = 0$,

**Figure 4.24**   Cross-section of shaft at $x = 5$ in, with identification of stress elements considered in Example 4.13.

$\tau_{xz} = -5093$ psi, and $\tau_{xy} = \tau_{yz} = 0$. This is a two-dimensional stress state; therefore, one of the principal stresses is zero. The other two principal stresses can be obtained from Mohr's circle or from Equation (2.16), using proper subscripts, as

$$\sigma = \frac{\sigma_x + \sigma_z}{2} \pm \sqrt{\tau_{xz}^2 + \frac{(\sigma_x - \sigma_z)^2}{4}} = \frac{12{,}828}{2} \pm \sqrt{5093^2 + \frac{12{,}828^2}{4}} = 64184 \pm 8190 \text{ psi}$$

Therefore, the principal stresses are, in their proper order, $\sigma_1 = 14{,}604$ psi, $\sigma_2 = 0$, and $\sigma_3 = -1776$ psi.

It is left for the student to verify that the stresses at $x = 10$ in are not as serious as these in Problem 4.36.

(a)     (b)

**Figure 4.25**   Shear stress distributions. (a) Shear stress due to a vertical shear force; (b) shear stress due to torsion.

## Case Study 4.1 | DESIGN OF SHAFT FOR COIL SLITTER

**Given:** Flat rolled sheets are produced in wide rolling mills, but many products are manufactured from strip stock. Figure 4.26(a) depicts a coil slitting line, where large sheets are cut into ribbons or strips. Figure 4.26(b) shows a shaft supporting the cutting blades. The rubber rollers ensure that the sheet does not wrinkle. For such slitting lines the shafts that support the slitting knives are a highly stressed and critical component. Figure 4.26(c) is a

(a)

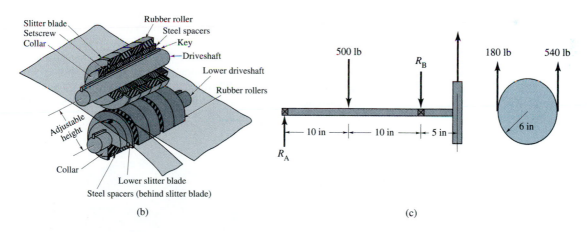

(b)                                             (c)

**Figure 4.26**   Design of shaft for coil slitting line. (a) Illustration of coil slitting line; (b) knife and shaft detail; (c) free-body diagram of simplified shaft for case study. [Illustrations (a) and (b) are adapted from *Tool and Manufacturing Engineers Handbook,* 4th ed., vol. 2, *Forming,* 1984. Reprinted with permission of the Society of Manufacturing Engineers, copyright 1984.]

*(continued)*

## Case Study      (Continued)

free-body diagram of a shaft for a short slitting line where a single blade is placed in the center of the shaft and a motor drives the shaft through a pulley at the far right end.

**Find:** If the maximum shear stress is 6000 psi and the largest gage sheet causes a blade force of 500 lb, what shaft diameter is needed? *Hint:* Use Mohr's circle to determine the maximum shear stress as a function of diameter, and equate this to the shear stress limit.

**Solution:** Figure 4.27 shows the shear and bending moment diagrams. The reactions are found through statics to be $R_A = 430$ lb and $R_B = 650$ lb. The reaction forces are shown in Figure 4.26(c). The maximum shear occurs just to the left of the pulley and equals 720 lb. The maximum bending moment is 4300 in·lb. In addition, there is a torque of 2160 in·lb between the pulley and the knife blade. Two locations must be analyzed: the location in the shaft where the moment is largest and the location where the shear is largest.

(a) *Moment.* The normal stress in the $x$ direction at the location of maximum moment is given by Equation (4.48) as

$$\sigma_x = \frac{Mc}{I} = \frac{4300\,(d/2)}{\pi d^4/64} = \frac{43{,}800}{d^3}$$

The shear stress due to the torque exerted on the pulley is, from Equation (4.34),

$$\tau_{xy} = \frac{Tc}{J} = \frac{11{,}000}{d^3}$$

Mohr's circle can be constructed as discussed in Chapter 2. Mohr's circle for this case is shown in

Figure 4.28 and has a radius of 24,510 in · lb/$d^3$. Setting this equal to the maximum allowable shear stress of 6000 psi yields

$$\frac{24{,}510}{d^3} = 6000 \rightarrow d = 1.60 \text{ in}$$

(b) *Shear.* The maximum shear stress at the location of maximum shear is, from Table 4.1,

$$\tau_{max} = \frac{4V}{3A} = \frac{4(720)}{3(\pi d^2/4)} = \frac{1222}{d^2}$$

At one end of the shaft, the torsion-induced shear stress is subtracted from this shear stress; at the other end, the effects are cumulative. Thus, the total shear is

$$\tau_{tot} = \frac{1222}{d^2} + \frac{11{,}000}{d^3} = 6000 \text{ psi}$$

Solving numerically gives $d = 1.279$ in.

(c) *Discussion.* As in most shaft applications the normal stresses due to bending determine the shaft diameter, so that a shaft with a diameter not less than 1.60 in should be used. A number of points should be made regarding this analysis:

- When the normal stress due to bending was calculated, the shear stress due to shear was neglected, even though there was shear in the shaft at that location. However, the distribution of normal stress is such that it is extreme at the top and the bottom, where the shear stress is zero.
- When the maximum shear stress due to vertical shear was calculated, the effects of bending

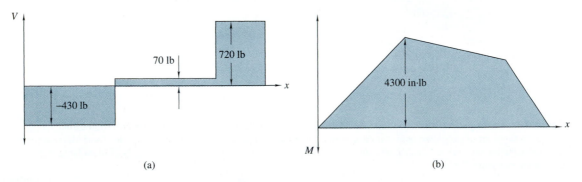

|     | (a) |     |     |     | (b) |     |

**Figure 4.27**    Shear diagram (a) and moment diagram (b) for idealized coil slitter shaft.

**Case Study** (Concluded)

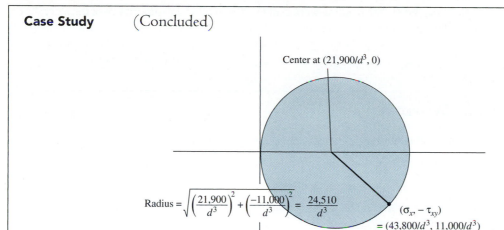

Center at $(21{,}900/d^3, 0)$

$\text{Radius} = \sqrt{\left(\dfrac{21{,}900}{d^3}\right)^2 + \left(\dfrac{-11{,}000}{d^3}\right)^2} = \dfrac{24{,}510}{d^3}$

$(\sigma_x, -\tau_{xy})$
$= (43{,}800/d^3, 11{,}000/d^3)$

**Figure 4.28** Mohr's circle for location of maximum bending stress.

were ignored. The bending stress is zero at the neutral axis, the location of the maximum shear stress.

• There are two general shaft applications. Some shafts are extremely long, as in this problem, whereas others are made much shorter to obtain compact designs. A coil slitter can have shafting more than 20 ft long, but more supporting bearings would be needed for stiffness. This shaft

was used only as an illustrative example; in actuality the supporting bearings would be placed much closer to the load application, and more than two bearing packs would probably be appropriate.

• The 10-in clearance between the leftmost bearing and the slitting knives is totally unnecessary and would lead to larger shaft deflections, as discussed in Chapter 5.

## 4.7 SUMMARY

The first four chapters have provided the essentials needed to describe the stress and strain for the four types of loading (normal, torsional, bending, and transverse shear) that might occur. These stresses and strains will be used in designing machine elements later in the text. Each chapter attempted to build from the knowledge learned in previous chapters. It was assumed throughout the first four chapters that the member experiencing one of the four types of loading had a symmetric cross section and was made of a homogeneous linear-elastic material. Chapter 4 started by defining centroid, moment of inertia, parallel-axis theorem, radius of gyration, and section modulus. These concepts needed to be understood before stresses and strains resulting from normal, torsional, bending, or transverse shear could be explored. These definitions, which are used throughout the text, were followed by the stresses and strains found in the four types of loading. While we were evaluating the stress and strain due to normal loading, it was convenient to include the axial displacement and the spring rate. For torsional loading the stresses and strains as well as the angular twist and the angular spring rate were presented. Also, in discussing torsion the relevant equations associated with power transfer were given in both English and SI units. The stresses and strains associated with bending were explained, as well as the importance of the neutral axis. Both a straight and a curved member were analyzed. Deformations associated with bending and combined transverse loading were not covered in this chapter but are considered more fully in Chapter 5. Section 4.6 defined the stress associated with transverse shear.

## KEY WORDS

**area moment of inertia**    represented by integral $\int r^2 \, dA$, m$^4$

**average normal stress**    average normal load, divided by cross-sectional area, Pa

**centroid of area**    geometric center of area, m

**mass moment of inertia**    product of element's mass and square of element's distance from the axis, kg · m$^2$

**normal strain**    elastic deformation divided by undeformed length

**parallel-axis theorem**    theorem that area's moment of inertia with respect to any axis is equal to second moment of area with respect to parallel axis through centroid added to product of area and square of distance between two axes

**power**    rate of doing work, or force times velocity, N · m/s

**radius of gyration**    radius that, when squared and multiplied by area, gives area moment of inertia, m

**section modulus**
area's moment of inertia, divided by farthest distance from centroidal axis to outer fiber of solid, m$^3$

**spring rate**    normal load, divided by elastic deformation, N/m

**torsion**    loading resulting in twisting of shaft

## RECOMMENDED READINGS

Beer, F. P., Johnson, E. R., and Dewolf, J. T. (2001) *Mechanics of Materials,* McGraw-Hill, 3rd ed., New York.
Craig, R. R. (1996) *Mechanics of Materials,* Wiley, New York.
Juvinall, R. C. (1967) *Stress, Strain, and Strength,* McGraw-Hill, New York.
Lardner, T. J., and Archer, R. R. (1994) *Mechanics of Solids: An Introduction,* McGraw-Hill, New York.
Popov, E. (1968) *Introduction to Mechanics of Solids,* Prentice-Hall, Englewood Cliffs, NJ.
Timoshenko, S. P., and Goodier, J. N. (1970) *Theory of Elasticity,* 3rd ed., McGraw-Hill, New York.
Ugural, A. C., and Fenster, S. K. (2003) *Advanced Strength and Applied Elasticity,* 4th ed., Prentice-Hall, Englewood Cliffs, NJ.

## REFERENCES

Marshall, R. D., et al. (1982) *Investigation of the Kansas City Hyatt Regency Walkways Collapse,* U.S. Department of Commerce, National Bureau of Standards, Report NBSIR 82–2465.

Timoshenko, S. P. and Goodier J. N. (1970) *Theory of Elasticity,* 3rd ed., McGraw-Hill, New York.
*Tool and Manufacturing Engineers Handbook* (1984), Society of Manufacturing Engineers, Dearborn, MI.

## PROBLEMS

### Section 4.2

**4.1** An area in the $xy$ coordinate system, as shown in sketch $a$, consists of a large circle having radius $r$ out of which are cut three smaller circles having radii $r/3$. Find the $x$ and $y$ coordinates for the centroid. The radius $r$ is 20 cm. *Ans.* $\bar{x} = 20$ cm, $\bar{y} = 17.78$ cm.

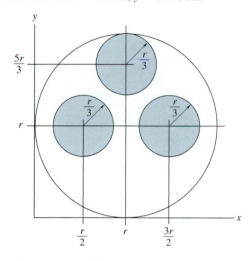

Sketch a, for Problems 4.1 and 4.2.

**4.2** The circular surface in sketch $a$ has circular cutouts glued onto it diagonally below the top cutout such that the centroids of the three cutouts are at $\bar{x} = r$ and $\bar{y} = r/3$. Find the $x$ and $y$ coordinates for the centroid ($r = 20$ cm). *Ans.* $\bar{x} = 20$ cm, $\bar{y} = 14.08$ cm.

**4.3** The rectangular area shown in sketch $b$ is situated symmetrically in the $xy$ coordinate system. The lengths of the rectangle's sides are $a$ in the $x$ direction and $b$ in the $y$ direction. Find the moments of inertia $I_x$ and $I_y$ and the polar moment of inertia $J$ for the rectangular surface.

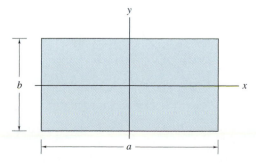

Sketch b, for Problem 4.3.

**4.4** Derive the area moment of inertia for the hollow circular area shown in Table 4.1.

★ **4.5** Derive the area moment of inertia for the elliptical quadrant shown in Table 4.1. *Ans. $I_x = ab^3\pi/16$, $I_y = a^3b\pi/16$.*

**4.6** Derive the area moment of inertia for the triangular section shown in Table 4.1. *Ans. $I_x = bh^3/12$, $I_y = hb(a^2 + ab + b^2)/12$.*

Sketch c, for Problem 4.7.

**4.7** Derive the area moment of inertia for a rectangular section with a cutout as shown in sketch *c*. *Ans. $I_y = 3.960 \times 10^{-6}$ m$^4$, $I_x = 1.929 \times 10^{-5}$ m$^4$.*

**4.8** Derive the area moment of inertia and the polar moment of inertia for the two relationships shown in sketch *d*. *Ans. $I_x = 6.646$ m$^4$, $I_y = 15.39$ m$^4$, $J = 22.04$ m$^4$.*

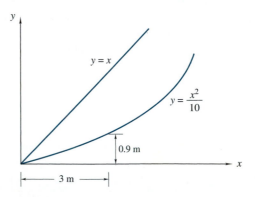

Sketch d, for Problem 4.8.

| * Indicates problem of greater difficulty.

## Section 4.3

**4.9** An elevator is hung by a steel rope. The rope has a cross-sectional area of 200 mm$^2$ and a modulus of elasticity of 70 GPa. The upward acceleration when the elevator starts is 3 m/s$^2$. The rope is 100 m long, and the elevator weighs 1000 kg. Determine the stress in the rope, the elongation of the rope due to the elevator's weight, and the extra elongation due to the acceleration. *Ans.* $\sigma = 64.05$ MPa, $\delta_{wt} = 70.1$ mm, $\delta_{acc} = 21.4$ mm.

**4.10** The 800-m-long cables in a suspension bridge are stressed to 220-MPa tensile stress. The total force in each cable is 11 MN. Calculate the cross-sectional area, the total elongation of each cable, and the spring rate when the modulus of elasticity is 75 GPa. *Ans.* $A = 0.05$ m$^2$, $\delta = 2.35$ m, $k = 4.69$ MN/m.

**4.11** A 1.08-m-long steel piston in a hydraulic cylinder exerts a force of 4000 kgf (39.24 MN). The piston is made of AISI 1080 steel and has a diameter of 50 mm. Calculate the stress in the piston, the elongation, and the spring rate. *Ans.* $\sigma = -20$ MPa, $\delta = -0.104$ mm, $k = 376$ MN/m.

**4.12** A steel pillar supporting a highway bridge is 12 m high and made of AISI 1040 steel tubing having an outer diameter of 1.5 m and a wall thickness of 30 mm. The weight carried by the pillar is 15 MN. Calculate the deformation of the pillar, the spring rate, and the stress in the pillar. *Ans.* $\sigma = -108.3$ MPa, $\delta = -6.28$ mm, $k = 2.389$ GN/m.

★ **4.13** The foundation of a bronze statue is made of a 3-m-high conical tube of constant wall thickness (8 mm). The tube's outer diameter is 200 mm at the top (just under the statue) and 400 mm at the ground. The tube material is AISI 316 stainless steel. The statue weighs 16,000 N. Calculate the deformation of the tube, the spring rate, and the maximum and minimum compressive stresses in the tube. *Ans.* $\delta = -35.3$ $\mu$m, $k = 453$ MN/m, $\sigma_{max} = -3.32$ MPa, $\sigma_{min} = -1.63$ MPa.

**4.14** Calculate the deflection at point A of the hanging cone shown in sketch *e*. The cone has a density of 3000 kg/m$^3$.

Sketch *e*, for
Problem 4.14.

Sketch *f*, for Problem 4.15.

**4.15** An aluminum core having a diameter $d_i$ of 30 mm is placed within a tubular steel shaft having a diameter $d_o$ of 50 mm. See sketch *f*. A flange is welded to the end of the shaft, and a pulling force of 100 kN is applied. The shaft is 100 mm long. Find the deflection at the end of the shaft and the stresses induced in the aluminum and steel sections of the shaft. Assume that the moduli of elasticity are $2 \times 10^{11}$ Pa for steel and $0.7 \times 10^{11}$ Pa for aluminum. *Ans.* $\delta = 0.3324$ mm.

**4.16** A bar of weight $W$ is supported horizontally by three weightless rods, as shown in sketch $g$. Assume that the cross-sectional areas, the moduli of elasticity, and the yield stresses are the same for the three rods. What is the maximum weight that can be supported? *Ans.* $W = 2.5 S_y A$.

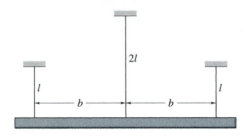

Sketch $g$ for Problem 4.16.

## Section 4.4

**4.17** An electric motor transmits 100 kW to a gearbox through a 50-mm-diameter solid steel shaft that rotates at 1000 rpm. Find the torque transmitted through the shaft and the angular torsion of the 0.5-m-long shaft. *Ans.* $T$=955 N · m, $\theta = 0.56°$

**4.18** The torque-transmitting shaft in Problem 4.17 is too heavy for the application, so it is exchanged for a circular tube having a 50-mm outside diameter and a 40-mm inside diameter. Find the angular torsion of the tube-formed shaft, which is 0.5 m long, when 100 kW is transmitted at 1000 rpm. The shear modulus is 80,000 N/mm². Also find the maximum shear stress in the tube and the percentage of weight decrease from the solid shaft. *Ans.* $\theta = 0.944°$, 64% weight savings.

**4.19** A torque-transmitting, hollow steel shaft with a circular cross section has an outer diameter of 50 mm and an inner diameter of 40 mm. Find the maximum length possible for the shaft if the torsion should be below 10° at a torque of 2000 N·m. *Ans.* $l = 2.52$ m.

**4.20** The bronze statue described in Problem 4.13 is asymmetric, so that when a gale-force wind blows against it, a twisting torque of 1000 N·m is applied to the tube. Calculate how much the statue twists. The tube's wall thickness, 6 mm at the top and 12 mm at the ground, is assumed to be proportional to its diameter. *Ans.* $\theta = 0.0196°$.

**4.21** Determine the minimum diameter of a solid shaft used to transmit 500 kW of power from a 2000-rpm motor so that the shear stress does not exceed 50 MPa. *Ans.* $d = 62.4$ mm.

**4.22** A steel coupling is used to transmit a torque of 40,000 N · m. The coupling is connected to the shaft by a number of 6-mm-diameter bolts placed equidistant on a pitch circle of 0.4-m diameter. The inner diameter of the coupling is 0.1 m. The allowable shear stress on the bolts is 500 MPa. Find the minimum number of bolts needed. *Ans.* 15.

**4.23** A shaft and a coupling are to transmit 50 kW of power at an angular speed of 1000 rpm. The coupling is connected to the shaft by 10 bolts, 10 mm in diameter, placed on a pitch circle of 200 mm. For an allowable stress on the bolts of 100 MPa, are the bolts able to transfer this power?

## Section 4.5

★ **4.24** A beam transmitting a bending moment $M$ of 5000 N · m has a square cross section 100 by 100 mm. The weight is decreased by making either one or two axial circular holes along the beam. Determine whether one hole or two holes give the lowest weight for the beam at a given bending stress. Neglect the transverse shear stress. *Ans.* Two holes give lower weight.

**4.25** A straight beam is loaded at the ends by moments $M$. The area moment of inertia for the beam is $I = a^3b/12$. Find the bending stress distribution in the beam when $M = 2000$ N · m, $a = 4$ cm, and $b = 6$ cm. Also find the radius of curvature to which the beam is bent. The beam's modulus of elasticity is $2.05 \times 10^{11}$ Pa. *Ans.* $\sigma = (6.25 \text{ GN/m}^3)y$, $r = 32.8$ m.

**4.26** The beam in Problem 4.25 is bent in a perpendicular direction so that $I = ab^3/12$. Find the bending stress distribution and the radius of curvature to which the beam is bent. *Ans.* $r = 73.8$ m.

**4.27** A curved bar has a rectangular cross section with height $h = r_o - r_i = 50$ mm and width $b = 100$ mm. Its inner radius is 100 mm. Find the distance between the neutral axis and the centroid. *Ans.* $e = 1.68$ mm.

**4.28** The curved bar in Problem 4.27 is loaded with a bending moment of 3000 N · m. Find the stress at the innermost and outermost radii. *Ans.* $\sigma_i = -83.29$ MPa, $\sigma_o = 63.52$ MPa.

**4.29** Two beams with rectangular cross sections $a \times b$ are placed on top of each other to form a beam having height $2a$ and width $b$. Find the area moments of inertia $I_x$ and $I_y$ for the two beams

a)    When they are welded together along the length
b)    When they are not welded together

★ **4.30** Two thin steel plates having width $b/20$ and height $5a$ are placed one at each side of the two beams in Problem 4.29. See sketch $h$. Find the moments of inertia around the $x$ and $y$ axes

a)    When the plates are not welded together
b)    When they are welded together as in sketch $h$
c)    When they are welded together to form a closed tube $5a \times 1.1b$

Sketch *h*, for Problem 4.30

## Section 4.6

**4.31** A cantilevered beam with a rectangular cross section is loaded by a force perpendicular to the beam centerline at the free end. The cross section is 75 mm high and 25 mm wide. The vertical load at the beam end is 40,000 N. Calculate how long the beam should be to give tensile and compressive stresses 10 times higher than the maximum shear stress. Also, calculate these stresses. *Ans.* $l = 187.5$ mm.

**4.32** A cantilevered beam with a circular-tube cross section has an outer diameter of 130 mm and a wall thickness of 10 mm. The load perpendicular to the beam is 20,000 N, and the beam is 1.3 m long from the point of force to the wall where the beam is fastened. Calculate the maximum bending and shear stresses. *Ans.* $\sigma_{max} = 247.3$ MPa, $\tau_{max} = 10.61$ MPa.

**4.33** A fishing rod is made of glass-fiber-reinforced plastic in the form of a tube having an outer diameter of 10 mm and a wall thickness of 1.5 mm. The glass fibers are parallel to the tube axis, so that the bending shear stress is carried by the plastic and the bending stresses are carried by the fibers. The fishing rod is 2 m long. Determine whether the rod fails from tensile stresses in the fibers or from shear overstressing in the plastic. The bending strength of the fiber-reinforced plastic is 800 MPa, and its shear strength is 3.2 MPa.

**4.34** Two aluminum beams, like the one shown in Figure 4.16, are welded together to form a closed cross section with dimensions of 240 mm by 80 mm. The weld was badly done, so that the wall is only 2 mm thick instead of 8 mm. The allowable shear stress in the weld is 50 MPa, whereas the maximum allowable bending stress is 150 MPa. Find out whether the 2-m-long beam first fails at the outermost fibers or at the welds.

**4.35** To decrease as much as possible the weight of beams subjected to bending, the center of gravity of the cross section is placed as far away from the beam center of gravity as possible. A beam supporting a floor has a cross section as shown in sketch $i$. The bending moment acting on the beam is 3000 N · m, and the shear force is 1000 N. Calculate the maximum bending stress and maximum shear stress. *Ans.* $\sigma_{max} = 19.7$ MPa, $\tau_{max} = 1.74$ MPa.

Sketch $i$, for Problem 4.35

**4.36** Obtain the largest principal stresses at $x = 10$ in for the shaft considered in Example 4.13.

# DEFORMATION

High-speed photographs
of a golf ball just before
and after impact.
(Courtesy of Cordin
Camera Company)

*Knowing is not enough; we must apply.*
*Willing is not enough; we must do.*
Johann Wolfgang von Goethe

**SYMBOLS**

| | |
|---|---|
| $A$ | area, m$^2$ |
| $a$ | length dimension, m |
| $b$ | length dimension, m |
| $C_1, C_2$ | constants |
| $c$ | distance from neutral axis to outer fiber of solid, m |
| $E$ | modulus of elasticity, Pa |
| $G$ | shear modulus of elasticity, Pa |
| $h$ | height, m |
| $I$ | area moment of inertia, m$^4$ |
| $J$ | polar area moment of inertia, m$^4$ |
| $l$ | length, m |
| $M$ | moment, N $\cdot$ m |
| $P$ | force, N |
| $Q$ | load applied at point of deformation, N |
| $q$ | load intensity, N/m |
| $R$ | reaction force, N |
| $T$ | torque, N $\cdot$ m |
| $U$ | strain energy, N $\cdot$ m |
| $V$ | shear force, N |
| $v$ | volume, m$^3$ |
| $w_0$ | unit step load distribution, N/m |
| $x, y, z$ | Cartesian coordinates, m |
| $\gamma$ | shear strain |
| $\delta$ | deformation (deflection), m |
| $\delta_{\max}$ | maximum deformation (deflection), m |
| $\delta_P$ | deflection at location of applied load, m |
| $\epsilon$ | strain ratio |
| $\theta$ | slope |
| $\nu$ | Poisson's ratio |
| $\sigma$ | normal stress, Pa |
| $\tau$ | shear stress, Pa |

**SUBSCRIPTS**

| | |
|---|---|
| $A, B, C$ | points at which reactions and moments occur |
| $a, b$ | solids $a$ and $b$ |
| $H$ | horizontal |
| max | maximum |
| $V$ | vertical |
| $x, y, z$ | coordinates |

## 5.1   INTRODUCTION

The focus of Chapters 2 through 4 has been on describing load, stress, and strain for various conditions that may occur in machine elements. Determining the design stress and making sure it is less than the yield stress for ductile materials and less than the ultimate stress for brittle materials are important for a safe design. However, attention also needs to be paid to strain limitation and displacement, since a machine element can fail if a part elastically deforms excessively. For example, in high-speed machinery with close tolerances, excessive deflections can cause interference between moving parts. This chapter attempts to quantify the deformation that may occur in a great variety of machine elements. Note that in this paragraph *displacement, deflection,* and *deformation* are used to mean essentially the same thing.

Chapter 4 described the deformation for normal stresses [Eq. (4.25)] and defined the angle of twist for torsional stress [Eq. (4.31)] as well as the spring rate and the angular spring rate for normal and torsional stresses [Eqs. (4.26) and (4.35), respectively]. These derivations were described in Chapter 4 because they are simple and could easily be included with the stress development. The derivations are not repeated in this chapter.

Chapter 5 focuses on describing the deformation for distributed loading such as occurs in a beam. Some approaches investigated are the integral method, the singularity function, the method of superposition, and Castigliano's theorem.

## 5.2 MOMENT-CURVATURE RELATION

Figure 4.14(b) shows a deformed element of a straight beam in pure bending. The **radius of curvature** $r$ can be expressed in Cartesian coordinates as

$$\frac{1}{r} = \frac{d^2y/dx^2}{[1 + (dy/dx)^2]^{3/2}} \tag{5.1}$$

Since $dy/dx$ is much less than 1, we can write

$$\frac{1}{r} = \frac{d^2y}{dx^2} \tag{5.2}$$

Substituting Equation (5.2) into Equation (4.50) gives

$$\frac{d^2y}{dx^2} = \frac{M}{EI} \tag{5.3}$$

This equation relates the transverse displacement to a bending moment. Even though an approximation to the curvature was used in reducing Equation (5.1) to Equation (5.2) that is valid only for small bending angles, this is a reasonable approximation for most beams in machinery element applications. Equation (5.3) is the **moment-curvature relation** and is sometimes referred to as the *equation of the elastic line.*

It is convenient to display the load intensity, shear force, moment, slope, and deformation in the following group of ordered derivatives:

$$\frac{q}{EI} = \frac{d^4y}{dx^4} \tag{5.4}$$

$$-\frac{V}{EI} = \frac{d^3y}{dx^3} \tag{5.5}$$

$$\frac{M}{EI} = \frac{d^2y}{dx^2} \tag{5.3}$$

$$\theta = \frac{dy}{dx} \tag{5.6}$$

$$y = f(x) \tag{5.7}$$

If SI units are used in these equations, the appropriate units are newtons per meter for load intensity, newtons for shear force, newton-meters for moment, and meters for deformation. Slope is dimensionless, hence it is given in radians.

Integrating Equation (5.3) gives the dimensionless slope at any point $x$ as

$$EI\frac{dy}{dx} = EI\theta = Mx + C_1 \tag{5.8}$$

Integrating Equation (5.8) gives the deflection at any point $x$ as

$$EIy(x) = M\frac{x^2}{2} + C_1x + C_2 \tag{5.9}$$

Note that integrating the load intensity function will produce the negative of the shear force distribution, or

$$-V(x) = \int_0^x q(x)\,dx \qquad\qquad\text{(5.10)}$$

Furthermore, integrating the shear force gives the moment

$$M(x) = -\int_0^x V(x)\,dx \qquad\qquad\text{(5.11)}$$

Equations (5.3) to (5.11) will henceforth be used in directly obtaining the deflection due to any type of loading.

**EXAMPLE 5.1**

**Given:**   A perpendicular force $P$ acts at the end of a cantilevered beam with length $l$, as shown in Figure 5.1. Assume that the cross section is constant along the beam and that the material is the same throughout, thus implying that the area moment of inertia $I$ and the modulus of elasticity $E$ are constant.

**Find:**   The deformation for any $x$, the slope of the beam, and the location and value of the maximum slope.

**Solution:**   The moment is $M = -Px$. From Equation (5.3)

$$\frac{d^2 y}{dx^2} = \frac{M}{EI} = -\frac{Px}{EI}$$

Integrating once gives

$$\frac{dy}{dx} = -\frac{Px^2}{2EI} + C_1$$

Integrating again gives

$$y = -\frac{Px^3}{6EI} + C_1 x + C_2$$

**Figure 5.1**   Cantilevered beam with concentrated force applied at free end; used in Example 5.1.

The boundary conditions are

1. $x = l$     $y = 0$

2. $x = l$     $\dfrac{dy}{dx} = 0$

From boundary condition 2

$$C_1 = \frac{Pl^2}{2EI}$$

$$\therefore \quad y = -\frac{Px^3}{6EI} + \frac{Pl^2 x}{2EI} + C_2$$

Making use of boundary condition 1 gives

$$C_2 = \frac{Pl^3}{6EI} - \frac{Pl^3}{2EI} = -\frac{Pl^3}{3EI}$$

$$\therefore \quad y = \frac{P}{6EI}\left(-x^3 + 3l^2 x - 2l^3\right)$$

and

$$\frac{dy}{dx} = \frac{P\left(l^2 - x^2\right)}{2EI}$$

The maximum slope occurs at $x = 0$ and is given by

$$\left(\frac{dy}{dx}\right)_{x=0} = \frac{Pl^2}{2EI}$$

## 5.3   SINGULARITY FUNCTIONS

A **singularity function** permits expressing in one equation what would normally be expressed in several separate equations with boundary conditions.

### DEFLECTION BY SINGULARITY FUNCTIONS

1. Draw a free-body diagram showing the forces acting on the system.

2. Use force and moment equilibrium to establish reaction forces acting on the system.

3. Write an expression for the load intensity function for all the loads acting on the system while making use of Table 2.2.

4. Integrate the negative load intensity function to give the shear force, and then integrate the negative shear force to obtain the moment.

5. Make use of Equation (5.9) to describe the deflection at any value.

6. Plot the following as a function of $x$:

   (a) Shear

   (b) Moment

   (c) Slope

   (d) Deflection

Examples 5.2 through 5.4 show how to determine the deflection by using singularity functions. These three examples are for general situations that will be used throughout the text.

---

**EXAMPLE 5.2**

**Given:**  A bar with a unit weight between simply supported ends.

**Find:**  The deflection for any $x$ by using singularity functions.

**Solution:**  Figure 5.2 shows a free-body diagram of the simply supported ends for a complete bar and a portion of the bar. From Table 2.2 for concentrated forces, the load intensity equation for the forces shown in Figure 5.2(b) can be written as

$$q(x) = \frac{Pb}{l} \langle x \rangle^{-1} - P \langle x - a \rangle^{-1} \tag{a}$$

Integrating twice gives the moment as

$$M(x) = \frac{Pb}{l} \langle x \rangle^{1} - P \langle x - a \rangle^{1} \tag{b}$$

Making use of Equation (5.3) gives

$$EI \frac{d^2 y}{dx^2} = M(x) = \frac{Pb}{l} \langle x \rangle^{1} - P \langle x - a \rangle^{1} \tag{c}$$

(a)

(b)

**Figure 5.2**    Free-body diagram of force anywhere between simply-supported ends. (a) Complete bar; (b) portion of bar.

Because $EI$ is constant along the bar, integrating Equation ($c$) gives

$$EI \frac{dy}{dx} = \frac{Pb}{2l} \langle x \rangle^2 - \frac{P}{2} \langle x - a \rangle^2 + C_1 \qquad \text{(d)}$$

Integrating again gives

$$EIy = \frac{Pb}{6l} \langle x \rangle^3 - \frac{P}{6} \langle x - a \rangle^3 + C_1 x + C_2 \qquad \text{(e)}$$

The boundary conditions are

1.  $y = 0$ at $x = 0 \rightarrow C_2 = 0$
2.  $y = 0$ at $x = l$ gives

$$C_1 = -\frac{Pbl}{6} + \frac{Pb^3}{6l} = \frac{Pb}{6l}(b^2 - l^2) = -\frac{Pb}{6l}(l^2 - b^2) \qquad \text{(f)}$$

$$\therefore \qquad y(x) = -\frac{P}{6EI}\left[ \frac{xb}{l}(l^2 - x^2 - b^2) + \langle x - a \rangle^3 \right] \qquad \text{(g)}$$

Note from Equation ($g$) that when $x \le a$, the last term on the right side of the equation is zero, and when $x > a$, the angular brackets become round. For more information about angular brackets see Chapter 2 (Sec. 2.9). Also note that from Figure 5.2 positive $y$ is upward but that in Equation ($g$) $y$ is negative, meaning the deflection is downward.

---

**Given:**  A cantilevered bar has a unit step distribution over a part of the bar, as shown in Figure 5.3. The bar is clamped or built into the structure at A and is free at C. The unit step distribution begins at the free end.

**EXAMPLE 5.3**

**Find:**  Derive a general expression for the deflection at any $x$ and at the free end by using singularity functions. Also, describe the deflection for the following two special cases:

(a)  When no unit step load is applied
(b)  When the unit step load is applied over the entire length of the beam

**Solution:**  The load intensity equation for the forces and moments shown in Figure 5.3(c) can be expressed by making use of Table 2.2 as

$$q(x) = w_0 b \langle x \rangle^{-1} - w_0 b \left( a + \frac{b}{2} \right) \langle x \rangle^{-2} - w_0 \langle x - a \rangle^0 \qquad \text{(a)}$$

Integrating twice gives

$$M(x) = w_0 b \langle x \rangle^1 - w_0 b \left( a + \frac{b}{2} \right) \langle x \rangle^0 - \frac{w_0}{2} \langle x - a \rangle^2 \qquad \text{(b)}$$

Making use of Equation (5.3) gives

$$EI \frac{d^2 y}{dx^2} = M(x) = w_0 b \langle x \rangle^1 - w_0 b \left( a + \frac{b}{2} \right) \langle x \rangle^0 - \frac{w_0}{2} \langle x - a \rangle^2$$

**Figure 5.3**    Cantilevered bar with unit step distribution over part of bar.
(a) Loads and deflection acting on cantilevered bar; (b) free-body diagram of
forces and moments acting on entire bar; (c) free-body diagram of forces and
moments acting on portion of bar.

Integrating once gives

$$EI \frac{dy}{dx} = \frac{w_0 b}{2} \langle x \rangle^2 - w_0 b \left( a + \frac{b}{2} \right) \langle x \rangle^1 - \frac{w_0}{6} \langle x - a \rangle^3 + C_1 \qquad \text{(c)}$$

Integrating again gives

$$EI\, y(x) = \frac{w_0 b}{6} \langle x \rangle^3 - \frac{w_0 b}{2} \left( a + \frac{b}{2} \right) \langle x \rangle^2 - \frac{w_0}{24} \langle x - a \rangle^4 + C_1 x + C_2 \qquad \text{(d)}$$

The boundary conditions are

1. $\dfrac{dy}{dx} = 0$ at $x = 0 \rightarrow C_1 = 0$

2. $y = 0$ at $x = 0 \rightarrow C_2 = 0$

$$y = \frac{w_0}{EI}\left[\frac{bx^3}{6} - \frac{bx^2}{2}\left(a + \frac{b}{2}\right) - \frac{1}{24}\langle x - a \rangle^4\right] \qquad \text{(e)}$$

The deflection at the free end, or where $x = a + b$, is

$$y_l = -\frac{w_0}{EI}\left[\frac{b^4}{24} + \frac{b(a+b)^2}{2}\left(a + \frac{b}{2}\right) - \frac{b(a+b)^3}{6}\right]$$

$$= -\frac{w_0 b}{EI}\left(\frac{a^3}{3} + \frac{3}{4}a^2 b + \frac{ab^2}{2} + \frac{b^3}{8}\right) \qquad \text{(f)}$$

(a) For the special case of $b = 0$, $y_{max} = 0$ (no deflection occurs).

(b) For the special case of $a = 0$,

$$y_l = -\frac{w_0 b^4}{8EI} \qquad \text{(g)}$$

In this situation the unit step extends completely across the length $l$.

---

**Given:** A cantilevered bar has its other end simply supported, and a concentrated force acts at any point along the bar, as shown in Figure 5.4(a).

**EXAMPLE 5.4**

**Find:** Determine a general expression for the deflection of the bar by using singularity functions.

**Solution:** Note from Figure 5.4(b) that there are three unknowns: $R_A$, $R_C$, and $M_C$. The force and moment equilibrium conditions produce two equations. A solution to this dilemma is to take one of the reaction forces as an unknown and express the other two in terms of that unknown.

$$\therefore \qquad R_C = P - R_A \qquad \text{(a)}$$
$$M_C = -Pb + R_A l \qquad \text{(b)}$$

The load intensity equation for the forces shown in Figure 5.4(a) can be expressed as

$$q(x) = R_A \langle x \rangle^{-1} - P \langle x - a \rangle^{-1} \qquad \text{(c)}$$

Integrating twice gives the moment as

$$M_B(x) = R_A \langle x \rangle^1 - P \langle x - a \rangle^1 \qquad \text{(d)}$$

Making use of Equation (5.3) gives

$$\therefore \qquad EI\frac{d^2 y}{dx^2} = M(x) = R_A x - P \langle x - a \rangle^1 \qquad \text{(e)}$$

**Figure 5.4**    Cantilevered bar with other end simply supported and with concentrated force acting anywhere along bar. (a) Sketch of assembly; (b) free-body diagram of entire bar; (c) free-body diagram of part of bar.

Integrating once gives

$$EI\,\frac{dy}{dx} = R_A\frac{x^2}{2} - \frac{P}{2}\,\langle x - a\rangle^2 + C_1 \tag{f}$$

Integrating again gives the deflection as

$$EI y = R_A\frac{x^3}{6} - \frac{P}{6}\,\langle x - a\rangle^3 + C_1 x + C_2 \tag{g}$$

The boundary conditions are

1.   $y = 0$ at $x = 0 \to C_2 = 0$

2. $\dfrac{dy}{dx} = 0$ at $x = l$ gives

$$C_1 = -R_A \frac{l^2}{2} + \frac{P}{2}(l-a)^2 \tag{h}$$

$$\therefore \quad EI\,y = R_A \frac{x^3}{6} - \frac{P}{6}\langle x-a\rangle^3 - \frac{R_A l^2 x}{2} + \frac{Px}{2}(l-a)^2 \tag{i}$$

3. $y = 0$ at $x = l$ gives

$$R_A = \frac{Pb^2}{2l^3}(3l-b) \tag{j}$$

The general expression for the deformation is

$$y = \frac{P}{6EI}\left[\frac{xb^2}{2l^3}(-3l^3 + 3lx^2 + 3bl^2 - bx^2) - \langle x-a\rangle^3\right] \tag{k}$$

Substituting Equation ($j$) into Equation ($d$) gives

$$M_B(x) = \frac{Pxb^2}{2l^3}(3l-b) - P\langle x-a\rangle^1 \tag{l}$$

Note that

$$M_B(x=0) = 0 \tag{m}$$

$$M_B(x=a) = \frac{Pb^2 a}{2l^3}(3l-b) \tag{n}$$

$$M_B(x=l) = \frac{Pb^2 l}{2l^3}(3l-b) - Pb = -\frac{Pab}{2l^2}(2l-b) \tag{o}$$

From Equations ($n$) and ($o$) the moment is positive at $x = a$ and negative at $x = l$. It is not clear whether the maximum magnitude exists at $x = a$ or at $x = l$. When $a = (\sqrt{2}-1)l = 0.414l$, the magnitude of the bending moment at B equals that at C. When $a < 0.414l$, the greater moment occurs at B; and when $a > 0.414l$, the greater moment occurs at C.

## 5.4   METHOD OF SUPERPOSITION

The **method of superposition** uses the principle that the deflection at any point in a bar is equal to the sum of the deflections caused by each load acting separately. Thus, if a bar is bent by $n$ separate forces, the deflection at a particular point is the sum of the $n$ deflections, one for each force. This method depends on the linearity of the governing relations between the load and the deflection, and it involves reducing complex conditions of load and support to a combination of simple loading conditions for which solutions are available. The solution of the original problem then takes the form of a superposition of these solutions. The solution assumes that the deflection of the bar is linearly proportional to the applied load. Thus, for $n$ different loads Equation (5.3) can be written as

$$EI\,\frac{d^2 y}{dx^2} = EI\,\frac{d^2}{dx^2}(y_1 + y_2 + \cdots + y_n) = M_1 + M_2 + \cdots + M_n \tag{5.12}$$

Table 5.1 gives solutions for some simple bar deflection situations that may be combined to produce the deflection for a more complex situation.

**Table 5.1**    Deflection for three different situations when one end is fixed and one end is free and two different situations of simply supported ends.

| Type of loading | Deflection for any $x$ |
|---|---|
| Concentrated load at any $x$  | $$y = -\frac{P}{6EI}\left((x-a)^3 - x^3 + 3x^2a\right)$$ |
| Unit step distribution over part or all of bar  | $$y = \frac{w_0}{EI}\left[\frac{bx^3}{6} - \frac{bx^2}{2}\left(a + \frac{b}{2}\right) - \frac{1}{24}(x-a)^4\right]$$ |
| Moment applied to free end  | $$y = -\frac{Mx^2}{2EI}$$ |
|  | $$y = \frac{P}{6EI}\left(\frac{b}{l}(x^3) - (x-a)^3 + 3a^2x - 2alx - \frac{a^3x}{l}\right)$$ |
|  | $$y = \frac{w_0 b}{24lEI}\left\{4\left(c + \frac{b}{2}\right)(x)^3 - \frac{l}{b}\left[(x-a)^4 - (x-a-b)^4\right]\right.$$ $$\left. + x\left[b^3 + 6bc^2 + 4b^2c + 4c^3 - 4l^2\left(c + \frac{b}{2}\right)\right]\right\}$$ |

**Given:** Figure 5.5 shows a cantilevered bar fixed at one end and free at the other end. A moment is applied at the free end, and a concentrated force is applied at any distance from the fixed end.

**Find:** Use the method of superposition to determine the deflection at the free end.

**Solution:** Figure 5.5(b) depicts the deflection with one end fixed and one end free for a concentrated force at any point within the length of the bar. Figure 5.5(c) shows a moment applied to the free end and the deformation. From Table 5.1 the individual deflections can be obtained directly as

$$y_{l,1} = -\frac{Pa^2}{6EI}(3l - a) \qquad \text{(a)}$$

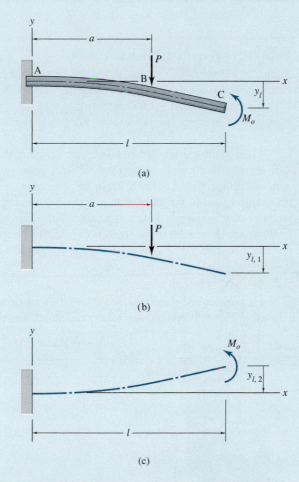

(a)

(b)

(c)

**Figure 5.5** Bar fixed at one end and free at other with moment applied to free end and concentrated force at any distance from free end. (a) Complete assembly; (b) free-body diagram showing effect of concentrated force; (c) free-body diagram showing effect of moment.

**EXAMPLE 5.5**

and
$$y_{l,2} = \frac{Ml^2}{2EI}$$
(b)

The resultant deflection from applying the method of superposition is

$$y_l = y_{l,1} + y_{l,2} = \frac{-Pa^2(3l-a) + 3Ml^2}{6EI}$$
(c)

The deflection at any point on the bar is

$$y = -\frac{P}{6EI}[(x-a)^3 - x^3 + 3x^2a] + \frac{Mx^2}{2EI}$$
(d)

## 5.5    STRAIN ENERGY

Statically indeterminate beams and beams of varying material properties or cross-sections cannot be successfully analyzed by using the methods discussed so far. Also, when a loading is energy-related, such as an object striking a beam with a given initial velocity, the exact forces in the loadings are not known. For this reason energy methods are often extremely useful.

When loads are applied to a machine element, the material of the machine element will deform. In the process the external work done by the loads will be converted by the action of either normal or shear stress into internal work called **strain energy,** provided that no energy is lost in the form of heat. This strain energy is stored in the body. The unit of strain energy is newton-meters in SI units and pounds-inch in English units. Strain energy is always positive even if the stress is compressive because stress and strain are always in the same direction. The symbol $U$ is used to designate strain energy.

### 5.5.1    NORMAL STRESS

When a tension test specimen is subjected to an axial load, a volume material element (shown in Fig. 5.6) is subjected to an axial stress, and the stress develops a force

$$dP = \sigma_z\, dA = \sigma_z\, dx\, dy$$
(5.13)

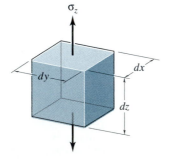

**Figure 5.6**    Element subjected to normal stress.

on the top and bottom faces of the element after the element undergoes a vertical elongation $\epsilon_z \, dz$.

Work is determined by the product of the force and the displacement in the direction of the force. Because the force $\Delta P$ is increased uniformly from 0 to its final magnitude $dP$ when the displacement $\epsilon_z \, dz$ is attained, the average force magnitude $dP/2$ times the displacement $\epsilon_z \, dz$ is the strain energy or

$$dU = \left( \frac{1}{2} \, dP \right) \epsilon_z \, dz$$

Making use of Equation (5.13) and the fact that $dv = dx \, dy \, dz$ gives

$$dU = \frac{1}{2} \sigma_z \epsilon_z \, dv \tag{5.14}$$

In general, then, if the member is subjected only to a uniaxial normal stress, the strain energy is

$$U = \int_v \frac{\sigma \epsilon}{2} \, dv \tag{5.15}$$

Also, if the material acts in a linear elastic manner, Hooke's law applies and Equation (3.3) can be substituted into Equation (5.15) to give

$$U = \int_v \frac{\sigma^2}{2E} \, dv \tag{5.16}$$

For axial loading of a bar of length $l$, making use of Equation (2.7) gives

$$U = \int_v \frac{P^2}{2EA^2} \, dv \tag{5.17}$$

However, since $dv = A \, dx$,

$$U = \int_0^l \frac{P^2}{2AE} \, dx = \frac{P^2 l}{2AE} \tag{5.18}$$

For a bending moment subjected to a given loading, making use of Equation (4.49) gives

$$U = \int_v \frac{\sigma^2}{2E} \, dv = \int \frac{M^2 y^2}{2EI^2} \, dv$$

But $dv = dA \, dx$, where $dA$ represents an element of the cross-sectional area. Also, recall that $M^2/(2EI^2)$ is a function of $x$ alone; then

$$U = \int_0^l \frac{M^2}{2EI^2} \left( \int y^2 \, dA \right) dx$$

Making use of Equation (4.7) gives

$$U = \int_0^l \frac{M^2}{2EI} \, dx \tag{5.19}$$

**Figure 5.7**   Element subjected to shear stress.

## 5.5.2   SHEAR STRESS

Consider the volume element shown in Figure 5.7. The shear stress causes the element to deform such that only the shear force $dV = \tau\, dx\, dy$, acting on the top of the element, is displaced $\gamma\, dz$ relative to the bottom surface. Only the vertical surfaces rotate, and therefore the shear forces on these surfaces do not contribute to the strain energy. Thus, the strain energy stored in the element due to a shear stress is

$$dU = \frac{1}{2}(\tau\, dx\, dy)\,\gamma\, dz$$

or

$$dU = \frac{1}{2}\tau\gamma\, dv \qquad\qquad (5.20)$$

Integrating over the entire volume gives the strain energy stored in the member due to shear stress as

$$U = \int_v \frac{\tau\gamma}{2}\, dv \qquad\qquad (5.21)$$

Using Hooke's law for shear stress, expressed in Equation (4.28), gives

$$U = \int_v \frac{\tau^2}{2G}\, dv \qquad\qquad (5.22)$$

For strain energy in torsion of circular shafts, Equations (4.27) to (4.33) give

$$U = \int_v \frac{T^2 r^2}{2GJ^2}\, dv$$

Since $dv = dA\, dx$, and $T^2/(2GJ^2)$ is a function alone of $x$,

$$U = \int_0^l \frac{T^2}{2GJ^2}\left(\int r^2\, dA\right) dx$$

Making use of Equation (4.30) gives

$$U = \int_0^l \frac{T^2}{2GJ}\, dx \qquad\qquad (5.23)$$

If the shaft has a uniform cross section,

$$U = \frac{T^2 l}{2GJ} \qquad\qquad (5.24)$$

Recall that Table 4.1 contains values of $J$ for a number of different cross-sections, but Equation (5.24) can be used only for cross-sections with circular symmetry.

---

EXAMPLE 5.6

**Given:** A 1-m-long solid shaft with circular cross-section has a diameter of 40 mm over 0.5 m of the length and a diameter of 30 mm over the rest of the length. The shaft is loaded with a torque of 1100 N · m. The shaft material is AISI 1080 high-carbon steel.

**Find:** Calculate the strain energy in the shaft, and calculate the ratio of the strain energies stored in the thinner and thicker parts of the shaft.

**Solution:** From Equation (5.23)

$$U = \int_0^l \frac{T^2}{2GJ}\, dx = \frac{T^2}{2G}\left(\frac{1}{J_1}\int_0^{0.5} dx + \frac{1}{J_2}\int_{0.5}^{1.0} dx\right)$$

From Table A.1 for AISI 1080 high-carbon steel

$$E = 207\ \text{GPa} \quad \text{and} \quad \nu = 0.3$$

From Equation (3.6) the shear modulus is

$$G = \frac{E}{2(1+\nu)} = \frac{207(10^9)}{2(1+0.3)}\ \text{Pa} = 79.62\ \text{GPa}$$

For a solid circular cross section

$$J_1 = \frac{\pi r^4}{2} = \frac{\pi}{2}(0.02)^4 = 25.13\left(10^{-8}\right)\text{m}^4$$

$$J_2 = \frac{\pi}{2}(0.015)^4 = 7.952\left(10^{-8}\right)\text{m}^4$$

The strain energy becomes

$$U = \frac{1100^2}{2(79.62)\left(10^9\right)}\left[\frac{10^8}{25.13} + \frac{10^8}{7.952}\right]\left(\frac{1}{2}\right) = 62.90\ \text{N} \cdot \text{m}$$

The ratio of the energies stored in the two parts can be expressed in terms of the two terms in brackets as

$$\frac{U_1}{U_2} = \frac{J_2}{J_1} = \frac{7.952\left(10^{-8}\right)}{25.13\left(10^{-8}\right)} = 0.3164$$

---

## 5.5.3 TRANSVERSE SHEAR STRESS

The strain energy due to the shear stress can be obtained from Equation (4.68). For a rectangular cross section with width $b$ and height $h$,

$$\tau = \frac{3V}{2A}\left(1 - \frac{y^2}{c^2}\right) = \frac{3V}{2bh}\left(1 - \frac{y^2}{c^2}\right) \tag{5.25}$$

Substituting this equation into Equation (5.22) and integrating give

$$U = \frac{1}{2G}\left(\frac{3V}{2bh}\right)^2 \int \left(1 - \frac{y^2}{c^2}\right)^2 dv$$

**Table 5.2**    Strain energy for four types of loading.

| Loading type | Factors involved | Strain energy for special case where all three factors are constant with $x$ | General expression for strain energy |
|---|---|---|---|
| Axial | $P, E, A$ | $U = \dfrac{P^2 l}{2EA}$ | $U = \displaystyle\int_0^l \dfrac{P^2}{2EA}\,dx$ |
| Bending | $M, E, I$ | $U = \dfrac{M^2 l}{2EI}$ | $U = \displaystyle\int_0^l \dfrac{M^2}{2EI}\,dx$ |
| Torson | $T, G, J$ | $U = \dfrac{T^2 l}{2GJ}$ | $U = \displaystyle\int_0^l \dfrac{T^2}{2GJ}\,dx$ |
| Transverse shear (rectangular section) | $V, G, A$ | $U = \dfrac{3V^2 l}{5GA}$ | $U = \displaystyle\int_0^l \dfrac{3V^2}{5GA}\,dx$ |

Setting $dv = b\,dx\,dy$ results in

$$U = \frac{9V^2}{8Gbh^2} \int_{-c}^{c} \left( 1 - \frac{2y^2}{c^2} + \frac{y^4}{c^4} \right) dy \int_0^l dx$$

Integrating yields

$$U = \frac{9V^2 l}{8Gbh^2} \left( y - \frac{2y^3}{3c^2} + \frac{y^5}{5c^4} \right)_{y=-c}^{y=c}$$

Evaluating and simplifying gives

$$U = \frac{6V^2 lc}{5Gbh^2}$$

Recall that $c = h/2$,

$$\therefore \qquad U = \frac{3V^2 l}{5Gbh} \tag{5.26}$$

Equation (5.26) gives the strain energy due to transverse shear stress for a rectangular cross section.

Table 5.2 summarizes the strain energy for the four types of loading. Recall that the transverse shear is valid for only a rectangular cross-section. For torsion Table 4.1 should be used for $J$, the polar area moment of inertia for a circular cross-section. For bending $I$ corresponds to $I_x$ in Table 4.1.

### 5.5.4    GENERAL STATE OF STRESS

The total strain energy due to a general state of stress can be expressed as

$$U = \int_v \left( \frac{\sigma_x \epsilon_x}{2} + \frac{\sigma_y \epsilon_y}{2} + \frac{\sigma_z \epsilon_z}{2} + \frac{\tau_{xy} \gamma_{xy}}{2} + \frac{\tau_{yz} \gamma_{yz}}{2} + \frac{\tau_{xz} \gamma_{xz}}{2} \right) dv \tag{5.27}$$

Making use of Equation (B.44) gives

$$U = \int_v \left[ \frac{1}{2E} \left( \sigma_x^2 + \sigma_y^2 + \sigma_z^2 \right) - \frac{\nu}{E} \left( \sigma_x \sigma_y + \sigma_y \sigma_z + \sigma_z \sigma_x \right) + \frac{1}{G} \left( \tau_{xy}^2 + \tau_{yz}^2 + \tau_{zx}^2 \right) \right] dv$$

(5.28)

If only the principal stresses $\sigma_1$, $\sigma_2$, and $\sigma_3$ act on the elements, Equation (5.28) reduces to

$$U = \int_v \left[ \frac{1}{2E} \left( \sigma_1^2 + \sigma_2^2 + \sigma_3^2 \right) - \frac{\nu}{E} \left( \sigma_1 \sigma_2 + \sigma_2 \sigma_3 + \sigma_3 \sigma_1 \right) \right] dv \qquad (5.29)$$

## 5.6 CASTIGLIANO'S THEOREM

It is often necessary to calculate the elastic deformation of distributed loads that are not as simple as those presented thus far. Castigliano's theorem can solve a wide range of deflection problems. Extensive use will be made here of the strain energy material presented in Section 5.5.

**Castigliano's theorem** states that when a body is elastically deformed by a system of loads, the deflection at any point $p$ in any direction $a$ is equal to the partial derivative of the strain energy (with the system of loads acting) with respect to a load at $p$ in the direction $a$, or

$$y_i = \frac{\partial U}{\partial Q_i} \qquad (5.30)$$

The load $Q_i$ is applied to a particular point of deformation and therefore is not a function of $x$. Thus, it is permissible to take the derivative with respect to $Q_i$ before integrating for the general expressions for the strain energy. Also, the load may be any of the loads presented first in Chapter 2 (Sec. 2.3) and throughout the text: normal, shear, bending, and transverse shear. Table 5.2 shows the strain energy for the various types of loading.

---

**PROCEDURE FOR USING CASTIGLIANO'S THEOREM:**

1. Obtain an expression for the total strain energy including the following:
   (a) Loads $(P, M, T, V)$ acting on the object (see Table 5.2)
   (b) A fictitious force $Q$ acting at the point and in the direction of the desired deflection
2. Obtain deflection from $y = \partial U / \partial Q$.
3. If $Q$ is fictitious, set $Q = 0$ and solve the resulting equation.

---

The best way to understand how to apply Castigliano's theorem is to observe how it is used in a number of different examples. Examples 5.7 through 5.10 demonstrate various features of Castigliano's approach.

**EXAMPLE 5.7**

**Given:**   The simply supported bar shown in Figure 5.2 with the force $P$ applied at $x = l/2$.

**Find:**   Determine the deflection at the location of the applied force $\delta_P$ by using Castigliano's theorem. Consider both bending and transverse shear.

**Solution:**   Because of symmetry the deflection at the point of applied force can be obtained by doubling the solution from 0 to $l/2$. The two types of loading being applied to the bar are

(a)   Bending with

$$M = \frac{Px}{2} \quad \text{and} \quad \frac{dM}{dP} = \frac{x}{2} \tag{a}$$

(b)   Transverse shear with

$$V = \frac{P}{2} \quad \text{and} \quad \frac{dV}{dP} = \frac{1}{2} \tag{b}$$

By making use of Table 5.2 the total strain energy can be expressed as

$$U = 2 \int_0^{l/2} \frac{M^2}{2EI} \, dx + 2 \int_0^{l/2} \frac{3V^2}{5GA} \, dx \tag{c}$$

By making use of Equations ($a$) and ($b$) the above equation becomes

$$U = \int_0^{l/2} \frac{P^2 x^2}{4EI} \, dx + \int_0^{l/2} \frac{3P^2}{10GA} \, dx$$

From Castigliano's theorem

$$-\delta_P = \frac{\partial U}{\partial P} = \int_0^{l/2} \frac{Px^2}{2EI} \, dx + \int_0^{l/2} \frac{3P}{5GA} \, dx$$

where the minus sign has been taken to reflect that the load causes a deflection in the $-y$ direction. Because $P, E, I, G,$ and $A$ are not functions of $x$, this equation becomes

$$\delta_P = -\frac{Pl^3}{48EI} - \frac{3Pl}{10GA} \tag{d}$$

The first term on the right side is due to bending, and the second is due to transverse shear.

To develop a better understanding of the bending and transverse shear contributions to the total deflection at the applied force, assume that the bar has a rectangular cross section. From Table 4.1 for a rectangular shape $I = bh^3/12$ and $A = bh$. Substituting these expressions into Equation ($d$) gives

$$\delta_P = -\frac{Pl^3}{4Ebh^3} - \frac{3Pl}{10Gbh} \tag{e}$$

which can be rewritten as

$$\delta_P = -\frac{3Pl}{10Gbh} \left[ \left(\frac{5}{6}\right) \left(\frac{G}{E}\right) \left(\frac{l}{h}\right)^2 + 1 \right] \tag{f}$$

Recall that the first term within the brackets is due to bending and the second is due to transverse shear.

For carbon steel $G/E = 0.383$. The length-to-height ratio $l/h$ of a bar is typically at least 10. Assume the smallest value, $l/h = 10$. After the values for $G/E$ and $l/h$ are substituted into Equation $(f)$, the first term within the brackets becomes 32 times the second term. Thus, in most applications the transverse shear term will be considerably smaller than the bending moment term (typically, less than 3%).

EXAMPLE 5.8

**Given:** A cantilevered bar with a concentrated force acting at a distance $b$ from the free end, as shown in Figure 5.8.

**Find:** Determine the deflection at the free end by using Castigliano's theorem. Assume that the transverse shear can be neglected.

**Solution:** Note from Figure 5.8 that since no force acts at the free end, a fictitious force $Q$ is created. The moment for any $x$ can be expressed as

$$M = -Qx - P\langle x - b\rangle$$

The only force contributing to the total strain is the bending moment, and from Table 5.2

$$U = \int_0^l \frac{M^2\,dx}{2EI} = \frac{1}{2EI}\int_0^l \left[Q^2x^2 + 2QxP\langle x-b\rangle + (P\langle x-b\rangle)^2\right]dx$$

(a)

(b)

**Figure 5.8** Cantilevered bar with concentrated force acting a distance $b$ from free end. (a) Coordinate system and important points shown; (b) fictitious force shown along with concentrated force.

From Castigliano's theorem

$$\delta_A = \frac{\partial U}{\partial Q} = \frac{1}{2EI}\left(\frac{2Ql^3}{3} + 2P\int_0^l x\,\langle x - b\rangle\,dx\right) \qquad \text{(a)}$$

Setting $Q = 0$ and integrating give

$$\delta_A = \frac{P}{6EI}[l^2\,(2l - 3b) + b^3] \qquad \text{(b)}$$

---

**EXAMPLE 5.9**

**Given:** The linkage assembly shown in Figure 5.9(a) can be made of different materials and have different cross-sectional areas in its two equal parts, denoted by subscripts 1 and 2.

**Find:** Calculate the horizontal displacement at the point of vertical force application by using Castigliano's theorem.

**Solution:** Note from the free-body diagram in Figure 5.9(b) that since no force was acting horizontally, a fictitious force $Q$ has to be created. From force equilibrium of the vertical and horizontal forces

$$\sum P_v = 0 \rightarrow -P + P_1\sin\theta + P_2\sin\theta = 0$$

$$\sum P_H = 0 \rightarrow Q - P_1\cos\theta + P_2\cos\theta = 0$$

Solving for $P_1$ and $P_2$ gives

$$P_1 = \frac{1}{2}\left(\frac{P}{\sin\theta} + \frac{Q}{\cos\theta}\right) \qquad \text{and} \qquad \frac{\partial P_1}{\partial Q} = \frac{1}{2\cos\theta} \qquad \text{(a)}$$

$$P_2 = \frac{1}{2}\left(\frac{P}{\sin\theta} - \frac{Q}{\cos\theta}\right) \qquad \text{and} \qquad \frac{\partial P_2}{\partial Q} = -\frac{1}{2\cos\theta} \qquad \text{(b)}$$

From Table 5.2 the total strain energy for axial loading can be written as

$$U = \frac{P_1^2 l}{2A_1 E_1} + \frac{P_2^2 l}{2A_2 E_2}$$

(a)                         (b)

**Figure 5.9**   Linkage system arrangement.
(a) Entire assembly; (b) free-body diagram of forces acting at point A.

From Castigliano's theorem the horizontal displacement at point A is

$$\delta_{A,H} = \frac{\partial U}{\partial Q} = \frac{l}{2} \left( \frac{2P_1\, \partial P_1/\partial Q}{A_1 E_1} + \frac{2P_2\, \partial P_2/\partial Q}{A_2 E_2} \right)$$

Substituting Equations (a) and (b) into the equation above gives

$$\delta_{A,H} = \frac{l}{4A_1 E_1} \left( \frac{P}{\sin\theta} + \frac{Q}{\cos\theta} \right) \frac{1}{\cos\theta} - \frac{l}{4A_2 E_2} \left( \frac{P}{\sin\theta} - \frac{Q}{\cos\theta} \right) \frac{1}{\cos\theta}$$

Setting $Q = 0$ gives

$$\delta_{A,H} = \frac{lP}{4\sin\theta\cos\theta} \left( \frac{1}{A_1 E_1} - \frac{1}{A_2 E_2} \right) \tag{c}$$

Note that if the linkage sections 1 and 2 have the same cross-sectional area and are made of the same material (hence the modulus of elasticity is the same), Equation (c) gives $\delta_{A,H} = 0$. In that case there would not be any horizontal displacement.

---

**Given:** Figure 5.10 shows a cantilevered bar with a 90° bend acted upon by a horizontal force $P$ at the free end.

**EXAMPLE 5.10**

**Find:** Calculate the vertical deflection at the free end, assuming that transverse shear is neglected. Use Castigliano's theorem.

**Solution:** Note from Figure 5.10 that since no vertical force exists at the free end, a fictitious force is created. Thus, the four components used to define the total strain energy are

(a) Bending in AB, where $M_{AB} = Py$
(b) Bending in BC, where $M_{BC} = Qx + Ph$
(c) Axial load in AB of magnitude $Q$
(d) Axial load in BC of magnitude $P$

From Table 5.2 the total strain energy can be written as

$$U = \int_0^h \frac{P^2 y^2}{2EI}\, dy + \int_0^l \frac{(Qx + Ph)^2}{2EI}\, dx + \int_0^h \frac{Q^2 dy}{2EA} + \int_0^l \frac{P^2 dx}{2EA}$$

**Figure 5.10** Cantilevered bar with 90° bend acted upon by horizontal force at free end.

The bar's material and cross sections are the same in sections AB and BC. From Castigliano's theorem

$$\delta_{A,V} = \frac{\partial U}{\partial Q} = \int_0^l \frac{(Qx + Ph)x\,dx}{EI} + \int_0^h \frac{Q\,dy}{EA} = \frac{Ql^3}{3EI} + \frac{Phl^2}{2EI} + \frac{Qh}{EA}$$

Setting $Q = 0$ gives

$$\delta_{A,V} = \frac{Phl^2}{2EI}$$

These examples demonstrate the wide range of deflection problems to which Castigliano's theorem can be applied.

## 5.7 SUMMARY

The three main modes in which a machine element will fail are from being overstressed, lack of a tribological film, and excessive elastic deformations. This chapter described the deformations that machine elements may undergo, and deformations due to distributed and concentrated loads were both considered. For a distributed load, four major approaches to describing the deformations were presented: the moment-curvature relation, singularity functions, the method of superposition, and Castigliano's theorem. Each has its particular advantages. The type of load being applied (normal, bending, shear, or transverse shear) determines the approach. Castigliano's theorem is the most versatile of the four approaches considered since it can be applied to a wide range of deflection problems.

## KEY WORDS

**Castigliano's theorem**   when a body is elastically deformed by a system of loads, deflection at any point $p$ in any direction $a$ is equal to the partial derivative of strain energy (with system of loads acting) with respect to the load at $p$ in direction $a$.

**method of superposition**   principle that deflection at any point in a body is equal to sum of deflections caused by each load acting separately.

**moment-curvature relation**   relationship between beam curvature and bending moment, given by (for small bend angles)

$$\frac{d^2y}{dx^2} = \frac{M}{EI}$$

**radius of curvature**   geometrically this is represented in Cartesian coordinates as

$$\frac{1}{r} = \frac{d^2y/dx^2}{\left[1 + (dy/dx)^2\right]^{3/2}}$$

**strain energy**   internal work that was converted from external work done by applying load.

# RECOMMENDED READINGS

Beer, F. P., Johnson, E. R., and Dewolf, J. T. (2001) *Mechanics of Materials*, McGraw-Hill, 3rd ed., New York.

Craig, R. R. (1996) *Mechanics of Materials*, Wiley, New York.

Crandall, S. H., and Dahl, H. C. (1954) *An Introduction to the Mechanics of Solids*, McGraw-Hill, New York.

Fung, Y. C. (1965) *Foundations of Solid Mechanics*, Prentice-Hall, Englewood Cliffs, NJ.

Juvinall, R. C. (1967) *Stress, Strain, and Strengths*, McGraw-Hill, New York.

Lardner, T. J., and Archer, R. R. (1994) *Mechanics of Solids: An Introduction*, McGraw-Hill, New York.

Norton, R. L. (1996) *Machine Design—An Integrated Approach*, Prentice-Hall, Englewood Cliffs, NJ.

Popov, E. P. (1968) *Introduction to Mechanics of Solids*, Prentice-Hall, Englewood Cliffs, NJ.

Shigley, J., and Mischke, C. (2004) *Mechanical Engineering Design*, McGraw-Hill, 7th ed., New York.

Timoshenko, S. P., and Goodier, J. N. (1951) *Theory of Elasticity*, McGraw-Hill, New York.

# PROBLEMS

## Section 5.2

**5.1** A beam is loaded by a concentrated bending moment $M$ at the free end. Find the vertical and angular deformations along the beam by using the equation of the elastic line, Equation (5.3). *Ans.* $y = Mx^2/(2EI)$.

★ **5.2** A simply supported beam of length $l$ carries a force $P$. Find the ratio between the bending stresses in the beam when $P$ is concentrated in the middle of the beam and evenly distributed along it. Use the moment-curvature relation given in Equation (5.3). Also, calculate the ratio of the deformations at the middle of the beam. *Ans.* $y_{conc}/y_{dist} = 1.6$.

**5.3** A simply supported beam with length $l$ is centrally loaded with a force $P$. How large a moment needs to be applied at the ends of the beam

*a)*     To maintain the slope angle of zero at the supports?

*b)*     To maintain the midpoint of the beam without deformation when the load is applied? Use the equation of the elastic line, Equation (5.3). *Ans.* a) $M_0 = Pl/8$, b) $M_0 = Pl/6$.

★ **5.4** Find the relation between $P$ and $w_0$ so that the slope of the deflected beam is zero at the supports for the loading conditions shown in sketch $a$. Assume that $E$ and $A$ are constant. *Ans.* $P = 2w_0l/3$.

Sketch *a*, for Problem 5.4

* Indicates problem of greater difficulty

## Section 5.3

**5.5**  Given a simply supported beam with two concentrated forces acting on it, as shown in sketch *b*, determine the expression for the elastic deformation of the beam for any $x$ by using singularity functions. Assume that $E$ and $I$ are constant. Also determine the location of maximum deflection and derive an expression for it.

(b)

Sketch *b*, for Problem 5.5

★ **5.6**  For the loading condition described in sketch *c*, obtain the internal shear force $V(x)$ and the internal moment $M(x)$ by using singularity functions. Draw $V(x)$, $M(x)$, $q(x)$, and $y(x)$ as a function of $x$. Assume that $w_0 = 9$ kN/m and $l = 3$ m.

Sketch *c*, for Problem 5.6

**5.7**  A simply supported bar is shown in sketch *d* with $w_0 = 4$ kN/m and $l = 12$ m.

*a)*    Draw the free-body diagram of the bar.

*b)*    Use singularity functions to determine shear force, bending moment, slope, and deflection.

Provide both tabular results and a diagram of the shear force, the bending moment, the slope, and the deflection.

Sketch *d*, for Problem 5.7

★ **5.8**  The simply supported beam in Problem 5.5 is altered so that instead of a concentrated force $P$ a concentrated moment $M$ is applied at the same location. The moments are positive and act parallel with each other. Determine the deformation of the beam for any position $x$ along it by using singularity functions. Assume that $E$ and $I$ are constant. Also, determine the location of maximum deflection.

**5.9** The simply supported beam considered in Problem 5.8 has moments applied in opposite directions so that the moment at $x = a$ is $M_0$ and that at $x = l - a$ is $-M_0$. Find the elastic deformation of the beam by using singularity functions. Also, determine the location and size of the maximum deflection.

**5.10** Given the loading condition shown in sketch $e$, find the deflection at the center and ends of the beam. Assume that $EI = 1100$ kN $\cdot$ m$^2$. Ans. $y_{end} = -0.0373$ m, $y_{mid} = 0.00927$ m.

Sketch e, for Problem 5.10

★ **5.11** Given the loading condition shown in sketch $f$, obtain an expression for the deflection at any location on the beam. Assume that $EI$ is constant.

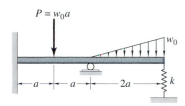

Sketch $f$, for Problem 5.11

**5.12** Given the loading condition and spring shown in sketch $g$, determine the stiffness of the spring so that the bending moment at point B is zero. Assume that $EI$ is constant. Ans. $k = 2EI/l^3$.

Sketch g, for Problem 5.12

★ **5.13** When there is no load acting on the cantilevered beam shown in sketch $h$, the spring has zero deflection. When there is a spring and a force of 20 kN is applied at point C, a deflection of 50 mm occurs at the spring. When there is a spring, if a 50-kN load is applied at the location shown in sketch $h$, what will be the deflection of the bar at point C? Assume that the stiffness of the spring is 450 kN/m.

Sketch *h*, for Problem 5.13

**5.14** Determine the deflection at point A and the maximum moment for the loading shown in sketch *i*. Consider only bending effects and assume that *EI* is constant.

Sketch *i*, for Problem 5.14

**5.15** Determine the maximum deflection of the beam shown in sketch *j*.

Sketch *j*, for Problem 5.15

**5.16** Determine the deflection at any point in the beam shown in sketch *k*. Use singularity functions.

Sketch *k*, for Problem 5.16

## Section 5.4

**5.17** Determine the deformation of a cantilevered beam with loading shown in sketch $l$ as a function of $x$. Also determine the maximum bending stress in the beam and the maximum deflection. Assume that $E = 207$ GPa, $I = 250$ cm$^4$, $P = 2000$ N, $w_0 = 3000$ N/m, $a = 0.5$ m, $b = 0.15$ m, and $c = 0.45$ m. The distance from the neutral axis to the outermost fiber of the beam is 0.040 m. *Ans.* $y_{max} = -1.664$ mm, $\sigma_{max} = 43.84$ MPa.

Sketch $l$, for Problem 5.17

**5.18** Given the loading shown in sketch $m$, let $a = 0.6$ m, $b = 0.7$ m, $M = 13,000$ N · m, and $w_0 = 40,000$ N/m. The beam has a square cross section with sides of 75 mm, and the beam material has a modulus of elasticity of 207 GPa. Determine the beam deformation by using the method of superposition. Also, calculate the maximum bending stress and maximum beam deformation. *Ans.* $y_{max} = -3.988$ mm, $\sigma_{max} = 193.4$ MPa.

Sketch $m$, for Problem 5.18

**5.19** The cantilevered beam shown in sketch $n$ has both a concentrated force and a moment acting on it. Let $a = 1$ m, $b = 0.7$ m, $P = 4350$ N, and $M = 2000$ N · m. The beam cross section is rectangular with a height of 80 mm and a width of 35 mm. Also, $E = 207$ GPa. Calculate the beam deformation by using the method of superposition. Find how large $M$ has to be to give zero deformation at $x = a$. *Ans.* $M = 2900$ N · m.

Sketch $n$, for Problem 5.19

**★ 5.20** A simply supported beam has loads as shown in sketch $o$. Calculate the beam deformation by using the method of superposition; also calculate the maximum bending stress and the maximum beam deformation and their locations. Assume that $E = 207$ GPa and that the beam has a rectangular cross section with a height of 30 mm and a width of 100 mm. Also, $P = 2400$ N, $w_0 = 20{,}000$ N/m, $a = 0.2$ m, $b = 0.1$ m, $c = 0.4$ m, and $d = 0.2$ m. *Ans.* $\sigma_{max} = 107.6$ MPa, $y_{max} = -2.830$ mm.

Sketch $o$, for Problem 5.20

**★ 5.21** The beam shown in sketch $p$ is fixed at both ends and center-loaded with a force of 4000 N. The beam is 3.2 m long and has a square tubular cross section with an outside width of 130 mm and a wall thickness of 10 mm. The tube material is AISI 1080 high-carbon steel. Calculate the deformation at any point along the beam by using the method of superposition.

Sketch $p$, for Problem 5.21

**5.22** Member A shown in sketch $q$ is a 13-mm-diameter aluminum bar; member B is an 8-mm-diameter steel bar. The lower member is of uniform cross section and is assumed to be rigid. Find the distance $x$ if the lower member is to remain horizontal. Assume that the modulus of elasticity for steel is 3 times that for aluminum. *Ans.* $x = 0.564$ m.

Sketch $q$, for Problem 5.22

**★ 5.23** An aluminum rod with $\frac{3}{4}$-in diameter that is 48 in long and a nickel-steel rod with $\frac{1}{2}$-in diameter and 32 in long are spaced 60 in apart and fastened to a horizontal beam that carries a 2000-lb load, as shown in sketch $r$. The beam is to remain horizontal after the load is applied. Assume that the beam is weightless and infinitely rigid. Find the location $x$ of the load and determine the stresses in each rod. Ans. $x = 40.0$ in, $\sigma_A = 1510$ psi, $\sigma_B = 6741$ psi.

3/4-in-diam aluminum rod

1/2-in-diam steel rod

48 in

60 in

32 in

$x$

2000 lb

Sketch $r$, for Problem 5.23

**5.24** Find the force on each of the vertical bars shown in sketch $s$. The 5000-lb member is assumed to be rigid and horizontal, implying that the three vertical bars are connected at the weight in a straight line. Also, assume that the support at the top of the bars is rigid. The bar materials and their circular cross-sectional area are given in the sketch. Ans. $P_s = 1.25 P_b = 1785$ lb.

Steel (0.2 in²)  36 in  Bronze (0.3 in²)  Steel (0.2 in²)

5000 lb

3 in  10 in  10 in  3 in

Sketch $s$, for Problem 5.24

**5.25** Two solid spheres, one made of aluminum alloy 2014 and the other made of AISI 1040 medium-carbon steel, are lowered to the bottom of the sea at a depth of 8000 m. Both spheres have a diameter of 0.3 m. Calculate the elastic energy stored in the two spheres when they are at the bottom of the sea if the density of water is 1000 kg/m³ and the acceleration of gravity is 9.807 m/s². Also calculate how large the steel sphere has to be to have the same elastic energy as the 0.3-m-diameter aluminum sphere. Ans. $U_{al} = 616.4$ N · m, $U_s = 252.2$ N · m.

### Section 5.6

★ **5.26** Use Castigliano's approach instead of singularity functions to solve Problem 5.5. Assume that transverse shear is negligible.

**5.27** Using Castigliano's theorem, find the maximum deflection of the two-diameter cantilevered bar shown in sketch $t$. Neglect transverse shear. *Ans.* $y = 3Pl^3/(16EI)$.

Sketch $t$ for Problem 5.27

★ **5.28** The right-angle cantilevered bracket shown in sketch $u$ is loaded with force $P$ in the $z$ direction. Derive an expression for the deflection of the free end in the $z$ direction by using Castigliano's theorem. Neglect transverse shear effects.

Sketch $u$ for Problem 5.28

**5.29** A triangular cantilevered plate is shown in sketch $v$. Use Castigliano's theorem to derive an expression for the deflection at the free end, assuming that transverse shear is negligible.

Sketch $v$ for Problem 5.29

**5.30** A right-angle cantilevered bracket with concentrated load and torsional loading at the free end is shown in sketch w. Using Castigliano's theorem, find the deflection at the free end in the z direction. Neglect transverse shear effects.

Solid round rod of properties $E, G, A, I,$ and $J$

Sketch w for Problem 5.30

★ **5.31** A cantilevered I-beam has a concentrated load applied to the free end as shown in sketch x. What upward force at point S is needed to reduce the deflection at S to zero? Use Castigliano's theorem. Transverse shear can be neglected. *Ans. $S_y = 10$ kN.*

Sketch x for Problem 5.31

★ **5.32** Using Castigliano's theorem, calculate the horizontal and vertical deflections at point A shown in sketch y. Assume that $E$ and $A$ are constant.

Sketch y for Problem 5.32

**5.33** Calculate the deflection at the point of load application and in the load direction for a load applied as shown in sketch $z$. Assume that $E$ and $I$ are constant.

Sketch $z$, for
Problem 5.33

**5.34** Using Castigliano's theorem, determine the horizontal and vertical deflections at point A of sketch $aa$. Assume that $E$ and $I$ are constant.

Sketch $aa$ for Problem 5.34

★ **5.35** For the structure shown in sketch $bb$, find the force in each member and determine the deflection at point A. Assume that $E$ and $A$ are the same for each member.

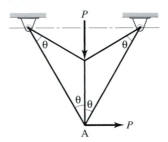

Sketch $bb$ for Problem 5.35

# FAILURE PREDICTION FOR STATIC LOADING

The Liberty Bell, a classic case of brittle fracture. (© R-F Website/Corbis)

*The concept of failure is central to the design process, and it is by thinking in terms of obviating failure that successful designs are achieved.*

Henry Petroski, *Design Paradigms*

## SYMBOLS

| | |
|---|---|
| $A$ | area, m$^2$ |
| $a$ | half of crack length, m |
| $b$ | plate width, m |
| $c$ | distance from neutral axis to outer fiber, m |
| $D$ | major diameter, m |
| $d$ | diameter, m |
| $H$ | major height, m |
| $h$ | minor height, m |
| $I$ | area moment of inertia, m$^4$ |
| $J$ | polar area moment of inertia, m$^4$ |
| $K_c$ | stress concentration factor |
| $K_{ci}$ | fracture toughness, MPa$\sqrt{\text{m}}$ |
| $K_i$ | stress intensity factor, MPa$\sqrt{\text{m}}$ |
| $l$ | length, m |
| $M$ | bending moment, N · m |
| $n_s$ | safety factor |
| $P$ | force, N |
| $q$ | volume flow, m$^3$/s |
| $r$ | radius, m |
| $S_{uc}$ | ultimate stress in compression, Pa |
| $S_{ut}$ | ultimate stress in tension, Pa |
| $S_y$ | yield stress, Pa |
| $S_{yt}$ | yield stress in tension, Pa |
| $T$ | torque, N · m |
| $V$ | shear force, N |
| $u$ | velocity, m/s |
| $Y$ | dimensionless correction factor that accounts for geometry of part containing a crack |
| $\tau$ | shear stress, Pa |
| $\sigma$ | normal stress, Pa |
| $\sigma_e$ | von Mises stress, Pa |
| $\sigma_1, \sigma_2, \sigma_3,$ | principal normal stresses, Pa, where $\sigma_1 \geq \sigma_2 \geq \sigma_3$ |

## SUBSCRIPTS

| | |
|---|---|
| all | allowable |
| avg | average |
| $d$ | design |
| max | maximum |
| nom | nominal |
| oct | octahedral |
| $x, y, z$ | Cartesian coordinates |
| 1,2,3 | principal axes |

## 6.1    INTRODUCTION

A machine element may fail at sites of local stress concentration caused by geometric or microstructural discontinuities. Stress concentration, stress raisers, and stress concentration factors are investigated in this chapter. The presence of cracks within a microstructure is also an important feature in understanding the failure of machine elements. Fracture mechanics is a technique of analysis used to determine the stress level at which preexisting cracks of known size will propagate, leading to fracture. Materials, stress levels, crack-producing flaws, and crack propagation mechanisms are considered while studying fracture toughness and critical crack length. This chapter ends with failure prediction theories for both uniaxial and multiaxial stress states. Various theories are presented for which there are experimental data. Each theory has its own strengths and shortcomings and is best suited for a particular class of material. The loading throughout this chapter is assumed to be static, thus implying that the load is gradually applied and equilibrium is reached in a relatively short time; thus, load is not a function of time.

## 6.2    STRESS CONCENTRATION

Stresses at or near a discontinuity, such as a hole in a plate, are higher than if the discontinuity did not exist. Figure 6.1 shows a rectangular bar with a hole under an axial load. The stress is largest near the hole; therefore, failure will occur first at the hole. The same can be deduced for any other discontinuity, such as a fillet (a narrowing in the width of a plate), a

**Figure 6.1** Rectangular plate with hole subjected to axial load. (a) Plate with cross-sectional plane, (b) one-half of plate with stress distribution, (c) plate with elliptical hole subjected to axial load.

notch (a sharp groove or cut especially intended to initiate failure), an inclusion (such as a discontinuous fiber in a polymer matrix), or an area of load application.

A **stress raiser** is any discontinuity in a part that alters the stress distribution near the discontinuity so that the elementary stress equation no longer describes the state of stress in the part. **Stress concentration** is the region in which stress raisers are present. **Stress concentration factor** $K_c$ is the factor used to relate the actual maximum stress at the discontinuity to the average stress without the discontinuity:

$$K_c = \frac{\text{actual maximum stress}}{\text{average stress}} \tag{6.1}$$

The stress concentration factor assumes that the stress distribution shown in Figure 6.1(b) can be represented by an average stress and that the change to the stress-strain equation can be obtained by using the stress concentration factor. Static loading conditions are assumed. The maximum stress occurs at the smallest cross-sectional area. The value of $K_c$ is difficult to calculate and is usually determined by some experimental technique, such as photoelastic analysis of a plastic model of a part, or by numerical simulation of the stress field.

Some feel for stress concentrations can be obtained by considering the example of an elliptical hole in a plate loaded in tension, as depicted in Figure 6.1(c). The theoretical stress concentration at the edge of the hole is given by

$$K_c = 1 + 2\left(\frac{a}{b}\right) \tag{6.2}$$

where $a$ is the half-length of the ellipse transverse to the stress direction and $b$ is the half-width in the direction of applied stress. Note that as $b$ approaches zero, the ellipse becomes sharper, and the situation approaches that of a very sharp crack. For this case, the stress at the edge of the crack is very large (infinite in the limiting case of $b \to 0$). As either $a$ approaches zero or $b$ becomes very large, the effects of the stress concentration become smaller. Thus the *size* and *orientation* of geometric discontinuities with respect to applied stress play a large role in determining the stress concentration. For more complicated situations, additional factors will play a role, as discussed in Sec. 6.2.1.

**Figure 6.2**  Stress concentration factors for rectangular plate with central hole. (a) Uniform tension; (b) pin-loaded hole; (c) bending. [Adapted from Collins (1981) and Frocht and Hill (1940).]

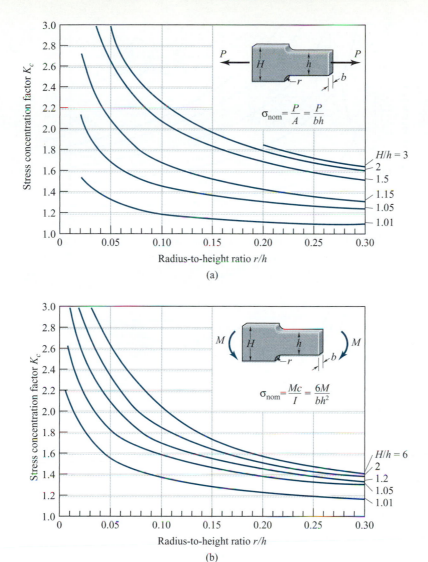

**Figure 6.3** Stress concentration factors for rectangular plate with fillet. (a) Axial load; (b) bending. [Source: Collins (1981).]

## 6.2.1 CHARTS

As stated, the stress concentration factor is a function of the type of discontinuity (hole, fillet, or groove), the geometry of the discontinuity, and the type of loading being experienced. Consideration here is limited to only two geometries, a flat plate and a round bar. Figures 6.2 to 6.4, respectively, determine the stress concentration factor due to bending and axial load for a flat plate with a hole, a fillet, or a groove. Note from Figure 6.2(a) that a small

(a)

(b)

**Figure 6.4**    Stress concentration factors for rectangular plate with groove.
(a) Axial load; (b) bending. [Source: Collins (1981).]

hole in a plate loaded in tension ($d/b \to 0$) leads to $K_c = 3.0$, which is consistent with Equation (6.1). Figures 6.5 and 6.6, respectively, determine the stress concentration factor due to bending and due to torsion for a round bar with a fillet and a groove axially loaded. Figure 6.7 shows the effect of a radial hole in a shaft. These are by no means all the possible geometries but are those most often used in practice. For other geometries refer to Peterson (1974) or Young (1989).

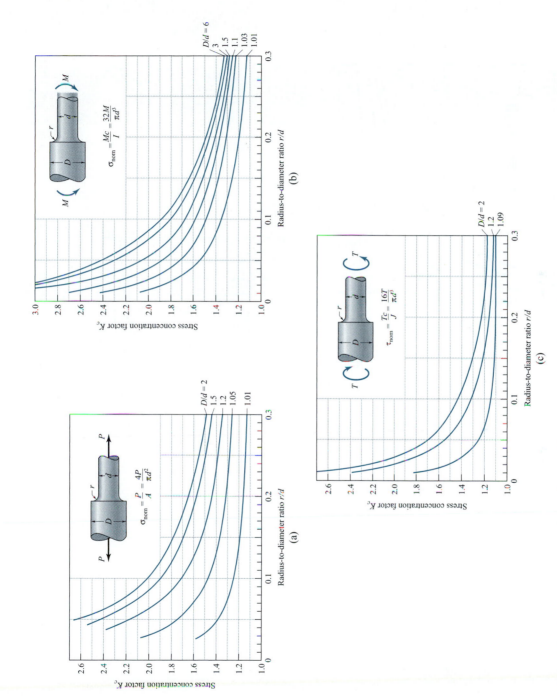

**Figure 6.5**   Stress concentration factors for round bar with fillet. (a) Axial load; (b) bending; (c) torsion. [Source: Collins (1981).]

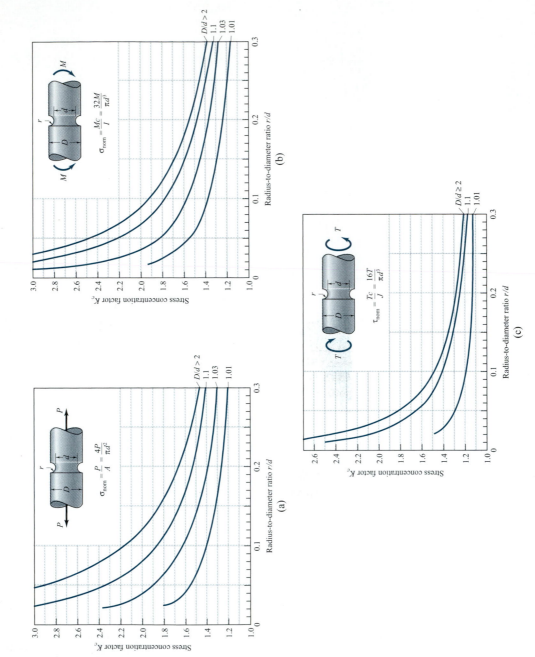

**Figure 6.6** Stress concentration factors for round bar with groove. (a) Axial load; (b) bending; (c) torsion. [Source: Collins (1981).]

**Nominal stresses:**

Axial load:

$$\sigma_{\text{nom}} = \frac{P}{A} = \frac{P}{(\pi D^2/4) - Dd}$$

Bending (plane shown is critical):

$$\sigma_{\text{nom}} = \frac{Mc}{I} = \frac{M}{(\pi D^3/32) - (dD^2/6)}$$

Torsion:

$$\tau_{\text{nom}} = \frac{Tc}{J} = \frac{T}{(\pi D^3/16) - (dD^2/6)}$$

**Figure 6.7**   Stress concentration factors for round bar with hole.

From these figures a number of observations can be made about the stress concentration factor:

1.   The stress concentration factor $K_c$ is independent of the part's material properties.

2.   It is significantly affected by geometry. Note from Figure 6.5 that as the radius $r$ of the discontinuity is decreased, the stress concentration increases.

3.   The stress concentration factor is also affected by the type of discontinuity; the stress concentration factor is considerably lower for a fillet (Fig. 6.3) than for a hole (Fig. 6.2).

These observations are relevant in reducing stresses in a part.

The stress concentration factors given in Figures 6.2 to 6.7 were determined on the basis of static loading, with the additional assumption that the stress in the material does not exceed the proportional limit. If the material is brittle, the proportional limit is the rupture stress, and failure for this material will begin at the point of stress concentration when the proportional limit is reached. It is thus important to apply stress concentration factors when using brittle materials. On the other hand, if the material is ductile and subjected to a static load, designers often ignore stress concentration factors, since a stress that exceeds the proportional limit will not result in a crack. Instead, the ductile material will have reserve strength due to yielding and strain hardening. Further, as a material yields near a stress concentration, the deformation results in blunting of notches, so that the stress concentration is relieved. In applications where stiff designs and tight tolerances are essential, stress concentration will be considered regardless of material ductility.

**Given:**   A flat plate made of a brittle material and with a major height $H$ of 4.5 in, a minor height $h$ of 2.5 in, and a fillet radius $r$ of 0.5 in.

**EXAMPLE 6.1**

**Find:**   The stress concentration factor and the maximum stress for the following conditions:

(a)   Axial loading

(b)   Pure bending
(c)   Axial loading but with fillet radius reduced to 0.25 in

**Solution:**

(a)   Axial loading

$$\frac{H}{h} = \frac{4.5}{2.5} = 1.80 \qquad \frac{r}{h} = \frac{0.5}{2.5} = 0.2$$

From Figure 6.3(a), $K_c = 1.8$. From Equation (6.1) the maximum stress is

$$\sigma_{max} = 1.8 \left( \frac{P}{A} \right) = \frac{1.8P}{bh}$$

where $b$ = width of flat plate.

(b)   Pure bending. From Figure 6.3(b), $K_c = 1.5$. The maximum stress is

$$\sigma_{max} = 1.5 \frac{6M}{bh^2} = \frac{9M}{bh^2}$$

(c)   Axial loading but with fillet radius reduced to 0.25 in is

$$\frac{r}{h} = \frac{0.25}{2.5} = 0.1$$

From Figure 6.3(a), $K_c = 2.2$. The maximum stress is

$$\sigma_{max} = \frac{2.2P}{bh}$$

Thus, reducing the fillet radius by one-half increased the maximum stress by 22%.

---

**EXAMPLE 6.2**

**Given:**   A 50-mm-wide, 5-mm-high rectangular plate has a 5-mm-diameter central hole. The allowable stress due to applying a tensile force is 700 MPa.

**Find:**

(a)   The maximum tensile force that can be applied
(b)   The maximum bending moment that can be applied to reach the maximum stress
(c)   The maximum tensile force and the maximum bending moment if the hole is not put in the plate. Express the results as a ratio when compared to parts $a$ and $b$.

**Solution:**

(a)   The diameter-to-width ratio is $d/b = 5/50 = 0.1$. The cross-sectional area with the hole is

$$A = (b - d)h = (50 - 5)5 = 225 \text{ mm}^2 = 0.225(10^{-3}) \text{ m}^2$$

From Figure 6.2(a) for $d/b = 0.1$ the stress concentration factor $K_c = 2.70$ for axial loading. The maximum tensile force is

$$P_{max} = \frac{\sigma_{all} A}{K_c} = \frac{700(10^6)(0.225)(10^{-3})}{2.70} = 58,330 \text{ N} = 58.33 \text{ kN}$$

(b)  From Figure 6.2(b) for bending when $d/b = 0.1$ and $d/h = 5/5 = 1$, the stress concentration factor $K_c = 2.04$. The maximum bending moment is

$$M_{max} = \frac{Ah\sigma_{all}}{6K_c} = \frac{0.225(10^{-3})5(10^{-3})700(10^6)}{6(2.04)} = 64.34 \text{ N} \cdot \text{m}$$

(c)  The cross-sectional area without the hole is

$$A = bh = (50)5 = 250 \text{ mm}^2 = 0.250(10^{-3}) \text{ m}^2$$

$$\therefore \quad P_{max} = \sigma_{all} A = (700)(10^6)(0.250)(10^{-3}) = 175 \text{ kN}$$

The force ratio is

$$\frac{P_{max}}{(P_{max})_{hole}} = \frac{175}{58.33} = 3.00$$

For bending

$$M_{max} = \frac{\sigma_{all}bh^2}{6} = \frac{\sigma_{all}Ah}{6} = \frac{700(10^6)0.25(10^{-3})5(10^{-3})}{6} = 145.8 \text{ N} \cdot \text{m}$$

The bending moment ratio is

$$\frac{M_{max}}{(M_{max})_{hole}} = \frac{145.8}{64.34} = 2.266$$

## 6.2.2  FLOW ANALOGY

Good design practice drives the mechanical engineer to reduce stress concentration as much as possible. Recommending a means of reducing the stress concentration requires a better understanding of what occurs at the discontinuity to increase the stress. One way of achieving this understanding is to observe similarity between the flow velocity of fluid in a channel and the stress distribution of an axially loaded plate when the channel dimensions are comparable to the size of the plate. The analogy is accurate, since the equations of flow potential in fluid mechanics and stress potential in solid mechanics are of the same form.

If the channel has constant dimensions throughout, the velocities are uniform and the streamlines are equally spaced. For a bar of constant dimensions under axial load, the stresses are uniform and equally spaced. At any point within the channel the flow $q$ must be constant, where the volume flow is

$$q = \int u \, dA \tag{6.3}$$

From solid mechanics the force $P$ must be constant at any location in the plate,

$$P = \int \sigma \, dA \tag{6.4}$$

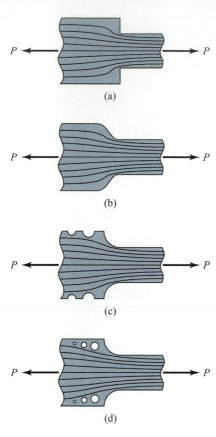

(a)

(b)

(c)

(d)

**Figure 6.8**   Flat plate with fillet axially
loaded showing stress contours for
(a) square corners, (b) rounded corners,
(c) small grooves, and (d) small holes.

If the channel section changes sharply, the flow velocity increases near the shape change, and to maintain equal flow, the streamlines must narrow and crowd together. In a stressed member of the same cross-section, the increase in stress is analogous to the increase in velocity, or inversely to the change in space between the streamlines. Figure 6.8(a) shows the stress distribution around the sharp corners of an axially loaded flat plate; this situation produces a stress concentration factor greater than 3. Recall from Equation (6.1) that this implies that the maximum stress is more than 3 times greater than the average stress. However, this stress concentration factor can be reduced, typically from 3 to 1.5, by rounding the corners as shown in Figure 6.8(b). A still further reduction in the stress concentration factor can be achieved by introducing small grooves or holes, as shown in Figure 6.8(c) and (d), respectively. In Figure 6.8(b) to (d) the design helps to reduce the rigidity of the material at the corners, so that the stress and strain are more evenly distributed throughout the flat plate. The improvements in these designs can be gleaned from the flow analogy.

## 6.3 FRACTURE MECHANICS

Structural studies that consider crack extension as a function of applied load are performed in fracture mechanics. A **crack** is a microscopic flaw that always exists under normal conditions on the surface and within the body of the material. These cracks (or dislocations) on and within the surface are like a dropped stitch in knitting. Under applied stress the crack moves easily through the material, causing a small slip in the plane in which it moves. Materials come undone more easily at these locations. No materials or manufacturing processes yield defect-free crystal structures, so such microscopic imperfections are always present.

Less stress is required to propagate a crack than to initiate it. Propagating a crack is like tearing a cloth. Once you start a little tear, it will propagate rather easily right across the cloth. However, the tear stops at a seam or other interruption of the regular weave of the fabric. So, too, crack propagation can be prevented by introducing discontinuities to act as a seam does.

Fracture failures can occur at stress levels well below the yield stress of a solid material. Fracture mechanics investigates the critical crack length that will make the part fail. **Fracture control** consists of keeping the combination of nominal stress and existing crack size below a critical level for the material being used in a given machine element.

## 6.4 MODES OF CRACK DISPLACEMENT

As shown in Figure 6.9, there are three fundamental **modes of crack propagation,** and each will effect a different crack surface displacement:

1. **Mode I, opening.**  The opening (or tensile) mode [Fig. 6.9(a)] is the most often encountered mode of crack propagation. The crack faces separate symmetrically with respect to the crack plane.

2. **Mode II, sliding.**  The sliding (or in-plane shearing) mode occurs when the crack faces slide relative to each other symmetrically with respect to the normal to the crack plane but asymmetrically with respect to the crack plane [Fig. 6.9(b)].

3. **Mode III, tearing.**  The tearing (or antiplane) mode occurs when the crack faces slide asymmetrically with respect to both the crack plane and its normal [Fig. 6.9(c)].

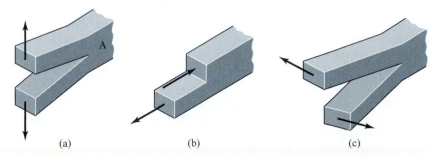

(a)                              (b)                              (c)

**Figure 6.9**   Three modes of crack displacement. (a) Mode I, opening; (b) mode II, sliding, (c) mode III, tearing.

The crack propagation modes are known by their Roman numeral designations given previously (e.g., mode I). Although mode I is the easiest to visualize as a crack-propagating mechanism, applying the discussion of stress raisers in Chapter 5 to geometries such as Figure 6.9 suggests that crack propagation will occur when stresses are higher at the crack tip than elsewhere in the solid.

## 6.5    FRACTURE TOUGHNESS

In this book, considerations of fracture toughness are restricted to mode I crack displacement. But first we need to determine what is meant by **stress intensity factor.** The stress intensity factor $K_i$ specifies the stress intensity at the crack tip [see point A in Fig. 6.9(a)]. The SI unit of $K_i$ is megapascals times meters$^{1/2}$ (MPa$\sqrt{m}$). **Fracture toughness,** however, is the critical value of stress intensity at which crack extension occurs. Fracture toughness is used as a design criterion in fracture prevention for brittle materials, just as yield strength is used as a design criterion in yielding prevention for ductile materials under static loading. Because the stresses near a crack tip [point A in Fig. 6.9(a)] can be defined in terms of the stress intensity factor, a critical value of fracture toughness $K_{ci}$ exists that can be used to determine the condition for brittle fracture. In general, the equation for fracture toughness is

$$K_{ci} = Y\sigma_{nom}\sqrt{\pi a} \tag{6.5}$$

where $Y$ = dimensionless correction factor that accounts for geometry of part containing a crack

$\sigma_{nom}$ = nominal stress at fracture, MPa

$a$ = one-half of crack length, m

Some assumptions imposed in deriving Equation (6.5) are that the load is applied far from the crack and that the crack length $2a$ is small relative to the plate width. The factor $Y$ is summarized for a number of geometries in Appendix C.

The differences between the stress intensity factor $K_i$ and fracture toughness $K_{ci}$ need to be made clear. The stress intensity factor $K_i$ represents the stress level at the crack tip [point A in Fig. 6.9(a)] in a part containing a crack. Fracture toughness $K_{ci}$ is the highest stress intensity that the part can withstand without fracturing. Thus, stress intensity $K_i$ has many values, whereas fracture toughness $K_{ci}$ is a particular value. That the units of stress intensity factor and fracture toughness are the same may seem somewhat strange, but they should be viewed as a combination of units of stress and the square root of crack length.

Table 6.1 shows room-temperature yield stress and fracture toughness (mode I) data for selected engineering materials. Note that the fracture toughness $K_{ci}$ depends on many factors, with the most important being temperature, strain rate, and microstructure. The magnitude of $K_{ci}$ diminishes with increasing strain rate and decreasing temperature. Furthermore, enhancing yield strength by a material process, such as strain hardening, produces a corresponding decrease in $K_{ci}$. Recall that mode I, the opening mode, is assumed.

**Table 6.1** Yield stress and fracture toughness data for selected engineering materials at room temperature.

| Material | Yield stress $S_y$ | | Fracture toughness $K_{ci}$ | |
|---|---|---|---|---|
| | ksi | MPa | ksi $\sqrt{\text{in}}$ | MPa $\sqrt{\text{m}}$ |
| **Metals** | | | | |
| Aluminum alloy 2024–T351 | 47 | 325 | 33 | 36 |
| Aluminum alloy 7075–T651 | 73 | 505 | 26 | 29 |
| Alloy steel 4340 tempered at 260°C | 238 | 1640 | 45.8 | 50.0 |
| Alloy steel 4340 tempered at 425°C | 206 | 1420 | 80.0 | 87.4 |
| Titanium alloy Ti-6Al-4V | 130 | 910 | 40–60 | 44–66 |
| **Ceramics** | | | | |
| Aluminum oxide | — | — | 2.7–4.8 | 3.0–5.3 |
| Soda-lime glass | — | — | 0.64–0.73 | 0.7–0.8 |
| Concrete | — | — | 0.18–1.27 | 0.2–1.4 |
| **Polymers** | | | | |
| Polymethyl methacrylate | — | — | 0.9 | 1.0 |
| Polystyrene | — | — | 0.73–1.0 | 0.8–1.1 |

| SOURCE: Adapted from ASM International (1989).

---

**EXAMPLE 6.3**

**Given:** The following two materials:

(a) AISI 4340 steel tempered at 260°C (500°F)
(b) Aluminum alloy 7075–T651

**Find:** The critical crack length at room temperature. Use the assumptions imposed in deriving Equation (6.5). Also, assume that the fracture stress is 0.8 times the yield stress and that the dimensionless correction factor is unity.

**Solution:**

(a) From Table 6.1 for AISI 4340

$$S_y = 238 \text{ ksi} \rightarrow \sigma_{\text{nom}} = 0.8 S_y = 190.4 \text{ ksi} \quad \text{and} \quad K_{ci} = 45.8 \text{ ksi}\sqrt{\text{in}}$$

From Equation (6.5)

$$a = \frac{1}{\pi}\left(\frac{K_{ci}}{Y\sigma_{\text{nom}}}\right)^2 = \frac{1}{\pi}\left[\frac{(45.8)(10^3)}{(1)(190.4)(10^3)}\right]^2 = 0.01842 \text{ in}$$

(b) From Table 6.1 for aluminum alloy 7075–T651,

$$S_y = 73 \text{ ksi} \rightarrow \sigma_{\text{nom}} = 0.8 S_y = 58.4 \text{ ksi} \quad \text{and} \quad K_{ci} = 26 \text{ ksi}\sqrt{\text{in}}$$

From Equation (6.5)

$$a = \frac{1}{\pi}\left(\frac{K_{ci}}{Y\sigma_{\text{nom}}}\right)^2 = \frac{1}{\pi}\left[\frac{(26)(10^3)}{(1)(58.4)(10^3)}\right]^2 = 0.06309 \text{ in}$$

The stronger material (the steel) will fail first since it has a smaller critical crack length. Thus, the weaker material (the aluminum) is a tougher material as far as crack propagation is concerned.

---

**EXAMPLE 6.4**

**Given:** A container used for compressed air is made of aluminum alloy 2024–T351. The required safety factor against yielding is 1.6, and the largest crack allowed through the thickness of the material is 6 mm. The form of the cracks gives the dimensionless correction factor $Y = 1$.

**Find:**

(a) The stress intensity factor and the safety factor guarding against brittle fracture.
(b) Whether a higher safety factor will be achieved if the material is changed to the stronger aluminum alloy 7075–T651. Assume the same crack exists.

**Solution:**

(a) From Table 6.1 for aluminum alloy 2024–T351

$$S_y = 325 \text{ MPa} \qquad \text{and} \qquad K_{ci} = 36 \text{ MPa}\sqrt{\text{m}}$$

The nominal stress is

$$\sigma_{\text{nom}} = \frac{S_y}{n_s} = \frac{325}{1.6} = 203.1 \text{ MPa}$$

The crack half-length is 3 mm. The stress intensity factor from Equation (6.5) is

$$K_i = Y\sigma_{\text{nom}}\sqrt{\pi a} = 1(203.1)(10^6)\sqrt{3\pi(10^{-3})} = 19.72 \text{ MPa}\sqrt{\text{m}}$$

The safety factor for brittle fracture is

$$n_{s,f} = \frac{K_{ci}}{K_i} = \frac{36}{19.72} = 1.826$$

(b) From Table 6.1 for stronger aluminum alloy 7075–T651

$$S_y = 505 \text{ MPa} \qquad \text{and} \qquad K_{ci} = 29 \text{ MPa}\sqrt{\text{m}}$$

The safety factor guarding against yielding is

$$n_{s,y} = 1.6\left(\frac{505}{325}\right) = 2.49$$

Thus, the increased strength of 7075–T651 results in a higher safety factor guarding against yielding. The safety factor guarding against crack propagation is

$$\frac{K_{ci}}{K_i} = \frac{29}{19.72} = 1.47$$

Thus, the stronger material will fail more easily from crack propagation.

## 6.6 FAILURE PREDICTION FOR UNIAXIAL STRESS STATE

Normal experimental data exist for the axial loading in a uniaxial stress state. Failure is predicted if the design stress $\sigma_d$ is greater than the allowable stress $\sigma_{all}$. Recall from Chapter 3 [Eqs. (3.13) to (3.16)] that the allowable stress depends on the type of load being imposed as well as on whether the material is ductile or brittle. Thus, when either the yield strength or the ultimate strength is known, the allowable stress can be determined for the uniaxial stress state.

The safety factor from Equation (1.1) in a uniaxial stress state can be expressed as

$$n_s = \frac{\sigma_{all}}{\sigma_d} \tag{6.6}$$

The type of loading and the type of material need to be incorporated into Equation (6.6). Therefore, for a uniaxial stress state the allowable stress is divided by the design stress; if it is greater than 1, the design is adequate. Of course, the larger $n_s$ is, the safer is the design, and values of $n_s < 1$ mean that a redesign is necessary. For a redesign, depending on the machine element, it is important to recognize what would reduce the design stress.

---

**Given:** The leaf springs of a truck's rear wheels are loaded in pure bending. The 8-ton axle load is taken up by the two springs, resulting in a bending moment of 9800 N · m in each spring at the load application point. The steel used for the spring is AISI 4340 tempered at 260°C. The dimensions of the leaf spring are such that the width is 10 times the thickness. Assume a safety factor of 5.

**EXAMPLE 6.5**

**Find:** The cross-section of the leaf spring.

**Solution:** Table 6.1 for AISI 4340 steel tempered at 260°C gives $S_y = 1640$ MPa. Using the lower limit of Equation (3.15) (to be conservative) yields

$$\sigma_{all} = 0.6 S_y = 0.6(1640) = 984 \text{ MPa}$$

The design stress from Equation (1.1) is

$$\sigma_d = \frac{\sigma_{all}}{n_s} = \frac{984}{5} = 196.8 \text{ MPa} \tag{a}$$

From Table 4.1 for a rectangular section

$$I = \frac{bh^3}{12} \quad \text{and} \quad c = \frac{h}{2}$$

We are given that $b = 10h$. Substituting the above into Equation (4.48) gives the bending design stress as

$$\sigma_d = \frac{Mc}{I}$$

Making use of Equation ($a$) gives

$$196.8(10^6) = \frac{5880}{h^3}$$

or

$$h^3 = \frac{5880}{196.8(10^6)} = 29.88(10^{-6})$$

$$h = 0.031 \text{ m} = 31.0 \text{ mm}$$

The cross-section of the leaf spring is thus 31 by 310 mm.

## 6.7    FAILURE PREDICTION FOR MULTIAXIAL STRESS STATE

A multiaxial stress state can be either a biaxial or a triaxial stress state. In practice, it is difficult to devise experiments to cover every possible combination of critical stresses because each test is expensive and a large number of them are required. Therefore, a theory is needed that compares the normal and shear stresses $\sigma_x$, $\sigma_y$, $\sigma_z$, $\tau_{xy}$, $\tau_{yz}$, and $\tau_{xz}$ with the uniaxial stress, for which experimental data are relatively easy to obtain. Several failure prediction theories are presented here for a multiaxial stress state while assuming static loading.

### 6.7.1    DUCTILE MATERIALS

Ductile materials include most metals and polymers. In general, metal castings are not as ductile as wrought or cold-worked metal parts. Ductile materials typically have the same tensile strength as compressive strength and are not as susceptible to stress raisers as are brittle materials. For the purposes of this text a ductile material is considered to have failed when it yields. Although in some applications a certain amount of plastic deformation may be acceptable, this is rarely the case in machinery elements. Two commonly used theories of yield criteria are presented: the maximum-shear-stress theory and the distortion energy theory.

*Maximum-Shear-Stress Theory*

The **maximum-shear-stress theory** (MSST) was first proposed by Coulomb (1773) but was independently discovered by Tresca (1868) and is therefore often called the **Tresca yield criterion.** Tresca noted that platinum exhibited bright shear bands under small strains, indicating that metals deformed under shear in all circumstances and that the shear was localized on well-defined planes. His observations led to the MSST, which states that a part subjected to any combination of loads will fail (by yielding or fracturing) whenever the maximum shear stress exceeds a critical value. The critical value can be determined from standard uniaxial tension tests. Experimental evidence verifies that the MSST is a good theory for predicting the yielding of ductile materials, and it is a common approach in design. If the nomenclature $\sigma_1 \geq \sigma_2 \geq \sigma_3$ is used for the principal stresses, the maximum shear stress says that yielding will occur when

$$\sigma_1 - \sigma_3 = \frac{S_y}{n_s} \tag{6.7}$$

where $S_y$ = yield stress of material
$\quad n_s$ = safety factor

For a three-dimensional stress state, the maximum-shear-stress theory provides an envelope describing the stress combinations that cause yielding, as illustrated in Figure 6.10. The curve defined by the yield criterion is known as the *yield locus*. Any stress state in the interior of the yield locus results in the material being distorted elastically. Points outside the yield locus are not possible because any such stress state would cause yielding in the solid before these stresses could be attained. Therefore, any calculations that predict a stress state outside the yield locus predict failure. If a situation arises that would increase the material strength (such as strain rate effects or work hardening), the yield locus expands, so that fracture may not necessarily result.

It is helpful to present the yield criterion in a plane stress circumstance, for which there will be two principal stresses in the plane as well as a principal stress equal to zero

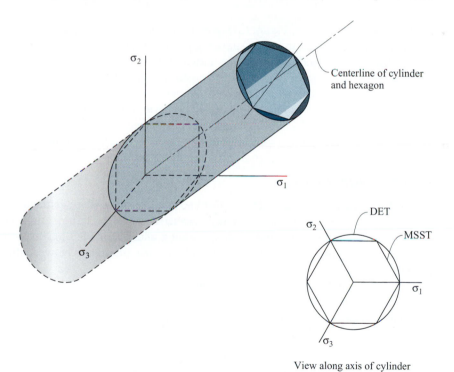

**Figure 6.10**   Three-dimensional yield locus for MSST and DET. [Adapted from Popov (1968).]

**Figure 6.11**    Graphical representation of maximum-shear-stress theory for biaxial stress state ($\sigma_z = 0$).

perpendicular to the plane. Figure 6.11 graphically depicts failure prediction in the plane stress state by the maximum-shear-stress theory. The principal stresses used in the figure are labeled $\sigma_1$ and $\sigma_2$, indicating that the normal ordering of stresses ($\sigma_1 \geq \sigma_2 \geq \sigma_3$) is not being enforced. Remember that the principal stress normal to the page is zero. In the first quadrant, where $\sigma_1$ and $\sigma_2$ are positive, this means that the value of $\sigma_3$ in Equation (6.7) would be zero and that yielding would occur whenever $\sigma_1$ or $\sigma_2$ reached the uniaxial yield strength $S_y$. In the second quadrant, where $\sigma_1$ is guaranteed to be negative and $\sigma_2$ is positive, Equation (6.7) would result in a line as shown in Figure 6.11. The third and fourth quadrants of the curve follow the same reasoning in their development.

### Distortion Energy Theory

The **distortion energy theory** (DET), also known as the **von Mises criterion,** postulates that failure is caused by the elastic energy associated with shear deformation. This theory is valid for ductile materials and predicts yielding under combined loading with greater accuracy than any other recognized theory (although the differences between the DET and the MSST are small).

The DET can be derived mathematically in a number of ways, but one of the more straightforward is to use the concept of octahedral stresses from Chapter 2. From Equation (2.28) the octahedral shear stress produced by uniaxial tension ($\sigma_2 = \sigma_3 = 0$) is

$$\tau_{\text{oct}} = \frac{\sqrt{2}}{3}\sigma_1 \tag{6.8}$$

The maximum octahedral shear stress occurs at

$$(\tau_{\text{oct}})_{\text{limit}} = \frac{\sqrt{2}}{3}\sigma_e \tag{6.9}$$

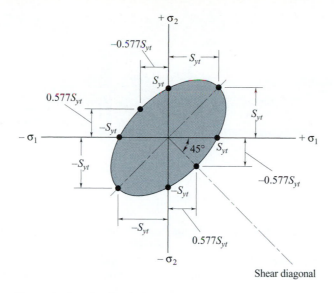

**Figure 6.12** Graphical representation of distortion energy theory for biaxial stress state ($\sigma_z = 0$).

where $\sigma_e =$ **von Mises stress**, Pa. For a triaxial stress state

$$\sigma_e = \frac{1}{\sqrt{2}}\left[(\sigma_2 - \sigma_1)^2 + (\sigma_3 - \sigma_1)^2 + (\sigma_3 - \sigma_2)^2\right]^{1/2} \qquad (6.10)$$

For a biaxial stress state, assuming $\sigma_3 = 0$,

$$\sigma_e = \left(\sigma_1^2 + \sigma_2^2 - \sigma_1\sigma_2\right)^{1/2} \qquad (6.11)$$

Thus, the DET predicts failure if

$$\sigma_e \geq \frac{S_y}{n_s} \qquad (6.12)$$

The DET yield locus is shown in Figure 6.10 for a three-dimensional stress state and in Figure 6.12 for plane stress loading (biaxial stress state). Relative to the MSST the DET has the advantage that the yield criterion is continuous in its first derivative, an important consideration for application in plasticity.

---

**Given:** In the rear wheel suspension of the Volkswagen Beetle, the spring motion is provided by a torsion bar fastened to an arm on which the wheel is mounted. See Figure 6.13 for more details. The torque in the torsion bar is created by the 2500-N force acting on the wheel from the ground through a 300-mm lever arm. Because of space limitations the bearing holding the torsion bar is situated 100 mm from the wheel shaft. The diameter of the torsion bar is 28 mm.

**EXAMPLE 6.6**

**Figure 6.13**    Rear wheel suspension used in Example 6.6.

**Find:**   The stresses in the torsion bar at the bearing by using the distortion energy theory.

**Solution:**   The stresses acting on the torsion bar are a shear stress due to the torsion and a perpendicular tensile/compressive stress from the bending. Using Equation (4.34) gives the shear stress from torsion as

$$\tau = \frac{Tc}{J} = \frac{2500(0.3)(0.014)(32)}{\pi(0.028)^4} \text{ Pa} = 174.0 \text{ MPa}$$

Using Equation (4.48) gives the tensile stress from bending as

$$\sigma = \frac{Mc}{I} = \frac{2500(0.1)(0.014)(64)}{\pi(0.028)^4} \text{ Pa} = 116.0 \text{ MPa}$$

From Equation (2.16) the principal normal stresses are

$$\sigma_1, \sigma_2 = \frac{\sigma_x + \sigma_y}{2} \pm \sqrt{\tau_{xy}^2 + \left(\frac{\sigma_x - \sigma_y}{2}\right)^2} = \frac{116.0}{2} \pm \sqrt{(174.0)^2 + \left(\frac{116.0}{2}\right)^2}$$

so                   $\sigma_1 = 241.4 \text{ MPa}$     and     $\sigma_2 = -125.4 \text{ MPa}$

From Equation (6.11) the von Mises stress is

$$\sigma_e = \left(\sigma_1^2 + \sigma_2^2 - \sigma_1\sigma_2\right)^{0.5} = [(241.4)^2 + (-125.4)^2 - 241.4(-125.4)]^{0.5} = 322.6 \text{ MPa}$$

This is the stress produced as given by the distortion energy theory.

---

**EXAMPLE 6.7**

**Given:**   A round, cantilevered bar made of a ductile material experiences torsion applied to the free end.

**Find:**   Determine when yielding will occur by using (*a*) the MSST and (*b*) the DET.

**Solution:**   Figure 6.14 shows the cantilevered bar, the stresses acting on an element, and Mohr's circle representation of the stress state. The principal stresses are

$$\sigma_1 = \tau_1 \quad \text{and} \quad \sigma_3 = -\tau_1$$

**Figure 6.14** Cantilevered, round bar with torsion applied to free end (used in Example 6.7). (a) Bar with coordinates and load; (b) stresses acting on an element; (c) Mohr's circle representation of stresses.

(a) Using Equation (6.7), the MSST predicts failure if

$$|\sigma_1 - \sigma_3| = 2\tau_{max} \geq \frac{S_y}{n_s}$$

$$\therefore \quad \tau_{max} \geq \frac{S_y}{2n_s} \qquad (a)$$

(b) Using Equation (6.11), yields

$$\sigma_e = \left(\sigma_1^2 - \sigma_1\sigma_2 + \sigma_2^2\right)^{1/2} = \left[\tau_1^2 - \tau_1(-\tau_1) + (-\tau_1)^2\right]^{1/2} = \sqrt{3}\tau_1 = \sqrt{3}\tau_{max} \qquad (b)$$

Therefore, using Equation (6.12), the DET predicts failure if

$$\sigma_e > \frac{S_y}{n_s} \qquad (c)$$

$$\therefore \quad \tau_{max} \geq \frac{S_y}{\sqrt{3}n_s} = 0.5774\frac{S_y}{n_s} \qquad (d)$$

Equations (a) to (d) show that the MSST and the DET are in good agreement. This circumstance, a loading of pure shear, is actually the greatest difference between the MSST and the DET, suggesting that both theories will give close to the same results.

---

**Given:** A round, cantilevered bar, similar to that considered in Example 6.7, is subjected not only to torsion but also to a transverse load at the free end, as shown in Figure 6.15(a). The bar is made of a ductile material having a yield stress of 50,000 psi. The transverse force is 500 lb, and the torque is 1000 lb · in applied to the free end. The bar is 5 in long, and a safety factor of 2 is assumed. Transverse shear can be neglected.

**EXAMPLE 6.8**

**Find:** Determine what the minimum diameter should be to avoid yielding by using both (a) the MSST and (b) the DET.

**Solution:**　Figure 6.15(b) shows the stresses acting on an element on the top of the bar at the wall. Note that in this example $\sigma_z = 0$. The critical section occurs at the wall. By using Equations (4.48) and (4.34) the normal and shear stresses can be obtained as

$$\sigma_x = \frac{Mc}{I} = \frac{Pl(d/2)}{\pi d^4/64} = \frac{32Pl}{\pi d^3}$$

$$\tau_{xy} = \frac{Tc}{J} = \frac{T(d/2)}{\pi d^4/32} = \frac{16T}{\pi d^3}$$

From Equation (2.16) the principal normal stresses in a biaxial stress field can be written as

$$\sigma_1, \sigma_2 = \frac{\sigma_x}{2} \pm \sqrt{\left(\frac{\sigma_x}{2}\right)^2 + \tau_{xz}^2} = \frac{16Pl}{\pi d^3} \pm \sqrt{\left(\frac{16Pl}{\pi d^3}\right)^2 + \left(\frac{16T}{\pi d^3}\right)^2}$$

$$\therefore \quad \sigma_1, \sigma_2 = \frac{16}{\pi d^3}\left[Pl \pm \sqrt{(Pl)^2 + T^2}\right]$$

Putting all the information given into the above equation yields

$$\sigma_1, \ \sigma_2 = \frac{16}{\pi d^3}\left[(500)(5) \pm \sqrt{(500)^2(5)^2 + (1000)^2}\right]$$

$$\therefore \quad \sigma_1 = \frac{26{,}450}{d^3} \quad \text{and} \quad \sigma_2 = -\frac{980.8}{d^3}$$

We can see that the stresses are in the wrong order; to ensure $\sigma_1 \geq \sigma_2 \geq \sigma_3$, we rearrange them so that $\sigma_1 = 26{,}450/d^3$, $\sigma_2 = 0$, and $\sigma_3 = -980.8/d^3$.

From Equation (2.18) the maximum and principal shear stresses can be written as

$$\tau_1, \tau_2 = \pm\sqrt{\tau_{xy}^2 + \frac{(\sigma_x - \sigma_z)^2}{4}}$$

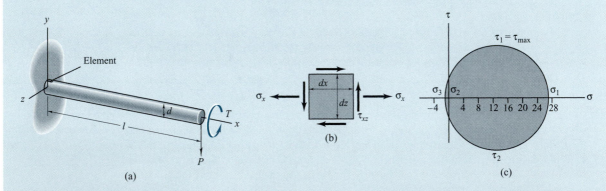

(a)　　　　　　　　　　　　　　　　　(b)　　　　　　　　　　　　　　　(c)

**Figure 6.15**　Cantilevered, round bar with torsion and transverse force applied to free end (used in Example 6.8). (a) Bar with coordinates and loads; (b) stresses acting on element at top of bar and at wall; (c) Mohr's circle representation of stresses.

$$\tau_{max} = \tau_1 = \frac{16}{\pi d^3}\sqrt{(Pl)^2 + T^2} = \frac{16}{\pi d^3}\sqrt{(500)^2(5)^2 + (1000)^2}$$

$$\tau_{max} = \tau_1 = \frac{13,710}{d^3}$$

(a)  Using Equation (6.7), the MSST predicts that failure will be avoided if

$$|\sigma_1 - \sigma_3| = 2\tau_1 = 2\tau_{max} < \frac{S_y}{n_s}$$

$$\therefore \quad \left| \frac{26,450}{d^3} + \frac{980.8}{d^3} \right| < \frac{50,000}{2}$$

or

$$d^3 > \frac{27,430}{25,000} = 1.097 \text{ in}^3$$

$$d > 1.031 \text{ in}$$

(b)  Using Equation (6.11), we have

$$\sigma_e = \left(\sigma_1^2 - \sigma_1\sigma_2 + \sigma_2^2\right)^{1/2} = \left[ \left(\frac{26,450}{d^3}\right)^2 + \left(\frac{26,450}{d^3}\right)\left(\frac{980.8}{d^3}\right) + \left(\frac{980.8}{d^3}\right)^2 \right]^{1/2}$$

$$= \frac{26,950}{d^3}$$

Thus, using Equation (6.12), the DET predicts that failure will be avoided if

$$\sigma_e < \frac{S_y}{n_s}$$

$$\frac{26,950}{d^3} < \frac{50,000}{2}$$

$$\therefore \quad d^3 > \frac{26,950}{25,000} = 1.078 \text{ in}^3$$

$$d > 1.025 \text{ in}$$

Here, both theories give approximately the same solution.

## 6.7.2 BRITTLE MATERIALS

As discussed in Chapter 3, brittle materials do not yield—they fracture. Thus, a failure criterion applied to brittle materials really addresses the circumstances under which the material will literally break. One important consideration with brittle materials is that their strengths in compression are usually much higher than their strengths in tension. Therefore, the failure criterion will show a difference in tensile and compressive behaviors. Three failure criteria are presented: the maximum-normal-stress theory, the internal friction theory, and the modified Mohr theory.

### Maximum-Normal-Stress Theory

The **maximum-normal-stress theory** (MNST) states that a part subjected to any combination of loads will fail whenever the greatest positive principal stress exceeds the tensile strength or whenever the greatest negative principal stress exceeds the compressive strength. This theory works best for fibrous brittle materials and some glasses, and it works reasonably well for brittle materials in general and is therefore common. Failure will occur according to the MNST theory if

$$\sigma_1 \geq \frac{S_{ut}}{n_s} \tag{6.13}$$

$$\sigma_3 \leq \frac{S_{uc}}{n_s} \tag{6.14}$$

where $\sigma_1 \geq \sigma_2 \geq \sigma_3$ = principal normal stresses
$\quad\quad S_{ut}$ = ultimate stress in uniaxial tension
$\quad\quad S_{uc}$ = ultimate stress in uniaxial compression
$\quad\quad n_s$ = safety factor

Note that failure is predicted to occur if either Equation (6.13) or Equation (6.14) is satisfied.

Figure 6.16 graphically presents failure prediction in the biaxial stress state by the maximum-normal-stress theory. The $\sigma_1 - \sigma_2$ plot for biaxial stresses (that is, $\sigma_3 = 0$) shows that the MNST predicts failure for all combinations of $\sigma_1$ and $\sigma_2$ falling outside the shaded area of Figure 6.16.

### Internal Friction Theory

The MNST is most useful for fibrous brittle materials and glasses where the microstructure would orient in the direction of maximum normal stress before fracture could occur. Many brittle materials, such as ceramics and cast metals, do not have this capability, and a different fracture theory is needed. Further, the maximum-shear-stress theory is difficult

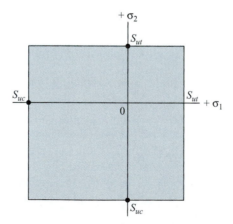

**Figure 6.16**    Graphical representation of maximum-normal-stress theory (MNST) for biaxial stress state ($\sigma_z = 0$).

to apply to brittle materials, since the strength in compression is so much higher than the tensile strength of most brittle materials, such as ceramics and metal castings. A logical extension to the MSST is to separate compressive and tensile strengths, or in mathematical terms,

If $\sigma_1 > 0$ and $\sigma_3 < 0$,
$$\frac{\sigma_1}{S_{ut}} - \frac{\sigma_3}{S_{uc}} = \frac{1}{n_s} \tag{6.15}$$

If $\sigma_3 > 0$,
$$\sigma_1 = \frac{S_{ut}}{n_s} \tag{6.16}$$

If $\sigma_1 < 0$,
$$\sigma_3 = \frac{S_{uc}}{n_s} \tag{6.17}$$

where $\sigma_1 \geq \sigma_2 \geq \sigma_3 =$ ordered principal stresses, Pa
$S_{ut} =$ fracture strength in tension, Pa
$S_{uc} =$ fracture strength in compression, Pa
$n_s =$ safety factor

Although this would appear to be an arbitrary extension of the MSST, Equations (6.15) to (6.17) can be derived analytically if internal friction is considered [Marin (1962)]. For this reason, this fracture criterion is known as the **internal friction theory** (IFT) and is also known as the **Coulomb–Mohr theory.** The IFT has the advantage of being more accurate for brittle materials that have a pronounced difference in tensile and compressive strengths than either the MNST or MSST. Figure 6.17 depicts the IFT for a two-dimensional stress state ($\sigma_3 = 0$).

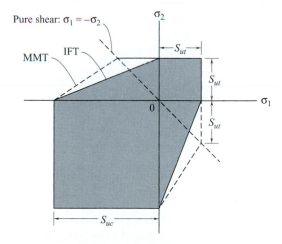

**Figure 6.17**    Internal friction theory and modified Mohr theory for failure prediction of brittle materials.

*Modified Mohr Theory*

The IFT has an analytical basis, but the **modified Mohr theory** arose through efforts at fitting test data. The modified Mohr theory (MMT) predicts brittle material behavior best, especially in the fourth quadrant. The MMT can be expressed as follows:

If $\sigma_1 > 0$ and $\sigma_3 < -S_{ut}$,
$$\sigma_1 - \frac{S_{ut}\sigma_3}{S_{uc} - S_{ut}} = \frac{S_{uc}S_{ut}}{n_s S_{uc} - S_{ut}} \tag{6.18}$$

If $\sigma_3 > -S_{ut}$,
$$\sigma_1 = \frac{S_{ut}}{n_s} \tag{6.19}$$

If $\sigma_1 < 0$,
$$\sigma_3 = \frac{S_{uc}}{n_s} \tag{6.20}$$

Figure 6.17 depicts the MMT, along with the IFT, to demonstrate that there is only a slight difference between the two criteria.

---

**EXAMPLE 6.9**

**Given:** Repeat Example 6.7, but with the cantilever constructed from a brittle material.

**Find:** Determine when fracture will occur by using (*a*) the MNST, (*b*) the IFT, and (*c*) the MMT. Assume that the compressive strength is twice the tensile strength.

**Solution:** Just as in Example 6.7, the stress state is $\sigma_1 = \tau_1$, $\sigma_2 = 0$, and $\sigma_3 = -\tau_1$.

(a) Using Equation (6.13), the MNST predicts failure if

$$\sigma_1 > \frac{S_{ut}}{n_s} \quad \text{or} \quad \tau_1 = \tau_{\max} \geq \frac{S_{ut}}{n_s} \tag{a}$$

Tensile fracture stress was used in Equation (*a*) because the material will fail in tension before it fails in compression.

(b) When the IFT is used, $\sigma_1$ is positive and $\sigma_3$ is negative. Thus, Equation (6.14) yields

$$\frac{\sigma_1}{S_{ut}} - \frac{\sigma_3}{S_{uc}} = \frac{1}{n_s} \quad \frac{\tau_1}{S_{ut}} - \frac{(-\tau_1)}{2S_{ut}} = \frac{1}{n_s} \quad \tau_1 = \frac{2}{3}\frac{S_{ut}}{n_s} \tag{b}$$

(c) Figure 6.17 shows that for pure shear there is no difference between the MMT and the MNST, so that the prediction using the MMT is the same as in part *a*.

---

### 6.7.3    SELECTING FAILURE CRITERION

Selecting a failure criterion to use in design is somewhat of an art. Figure 6.18 shows some available test data on biaxial failures or yielding for several materials. The DET fits ductile materials slightly better than does the MSST, but most of the data fall between the two curves. Although the DET works well for ductile solids and results in a firm mathematical foundation for investigating problems in plasticity, the MSST is often applied. The MSST

**Figure 6.18** Experimental verification of yield and fracture criteria for several materials. (a) Brittle fracture; (b) ductile yielding. [Data from Grassi and Cornet (1949), Dowling (1993), and Murphy (1964).]

is mathematically simple and conservative; for a given circumstance the MSST predicts yielding at lower loads than does the DET. Thus, because there is an added safety factor in using the MSST, it is often used in design. On the other hand, many commercial stress analyses and finite element codes use a von Mises stress to obtain stress profiles, and there is a natural tendency to use a DET in such circumstances.

Ductile materials have high repeatability in terms of strength, so designers can use either failure criterion with a fair amount of confidence. Brittle materials are more difficult to analyze because their properties vary much more than those of ductile materials. Owing to the statistical nature of brittle material properties, experimental strength verification as in Figure 6.18 is usually preferred over theoretical predictions regardless of the test data.

When a ductile material is tested, as in a tension test, calculated yield stresses deviate little among specimens, especially if the specimens have been obtained from the same batch in a manufacturing process. Thus, only a few experiments are needed to specify the strength of a ductile material with good certainty. For most brittle materials, however, many tests must be performed to accurately assess the strength distribution. Test specimens do not fail at the same stress, even if they were manufactured in the same batch with the same processes. It is difficult to specify a strength value for a brittle material with any great certainty. This distinction leads to an entirely different field of study, that of probabilistic design. In deterministic design—the approach followed in this text—it is assumed that an assured minimum strength can be defined (i.e., that all samples of the material in question will have at least this strength). This is certainly possible with metals (even internal flaws are limited in size or else the manufacturing process will be unsuccessful, such as when a bar breaks during extrusion because of a large void). With brittle materials, such as ceramics and castings, a guaranteed minimum strength is difficult, and at times impossible, to specify. Thus, the engineer needs to apply concepts of probability in design. Since failure criteria are deterministic but brittle materials are inherently probabilistic, it could be argued that these failure criteria should never be used with brittle materials. At the very least they should be applied with great caution. The greatest difficulty in using these deterministic criteria lies in determining the strengths (especially the tensile strength) to be used in the equations.

A student of engineering design often wonders why the IFT is presented at all if the MMT fits brittle material data better, as is suggested by Figure 6.18. Given the variation in strengths between brittle test specimens, the difference in the three failure criteria presented here is insignificant. To the question "Which theory is better or best?" no absolute answer can be given, and a designer or organization should use the one with which it has the greatest experience and history.

The IFT and the MSST are identical for metals because the yield strengths of metals in compression and tension are approximately equal. Thus, applying the IFT at all times would ensure a conservative solution. Designers are often reluctant to follow this procedure since the IFT is the most complex criterion from a mathematical standpoint.

The failure criteria given in this chapter are difficult to apply to composite materials and polymers. The behavior of polymers is complex, including their viscoplastic behavior, where a specific yield point is difficult to define. Composite materials require more complex failure theories to account for fiber length and orientation with respect to load, and the interested reader is referred to Jones (1975), Reddy (1996), and Kaw (1997), among others.

For the purposes of this text, ductile materials can properly be analyzed by using the DET or the MSST. Brittle materials should be analyzed by using the MNST, the IFT, or the MMT. A wide variety of additional failure and yield criteria are given in the technical literature. However, those presented in this chapter are by far the most often applied, and a suitable yield criterion can usually be chosen from those presented.

**EXAMPLE 6.10**

**Given:** The following materials and loadings:

(a) Pure aluminum: $S_y = 30$ MPa, $\sigma_x = 10$ MPa, $\sigma_y = -10$ MPa, $\tau_{xy} = 0$
(b) 0.2% carbon steel: $S_y = 65$ ksi, $\sigma_x = -5$ ksi, $\sigma_y = -35$ ksi, $\tau_{xy} = 10$ ksi
(c) Gray cast iron: $S_{ut} = 30$ ksi, $S_{uc} = 120$ ksi, $\sigma_x = -35$ ksi, $\sigma_y = 10$ ksi, $\tau_{xy} = 0$

**Find:** The safety factors for these materials.

**Solution:** The results are summarized in Table 6.2.

**Table 6.2**     Safety factors from different criteria for three different materials, used in Example 6.10

| Part | Criterion | Equation used | Safety factor |
|------|-----------|---------------|---------------|
| a | MSST | (6.7) | 1.5 |
|   | DET | (6.12) | 1.73 |
| b | MSST | (6.7) | 1.71 |
|   | DET | (6.12) | 1.75 |
| c | MNST | (6.14) | 3 |
|   | IFT | (6.15) | 1.6 |
|   | MMT | (6.18) | 1.634 |

## Case Study 6.1 | SAFETY FACTOR FOR A TOTAL HIP REPLACEMENT

**Given:** Figure 6.19 shows a total hip replacement inserted into a human femur and hip. Such devices are commonly used to treat painful arthritic conditions that result in loss of mobility. The operation consists of sawing off portions of the femur, reaming the femoral cavity to allow for implant insertion, and hammering the implant into the femur. The hip portion of the implant is similarly installed, often with screws as shown. The femur portion of the implant includes a stem onto which the highly polished ball is attached. All components are available in a number of sizes so that the operating physician can optimize the components for a particular patient. Figure 6.20 is a drawing of a femoral implant, complete with typical dimensions. The commonly used implant materials are cast cobalt chromium, forged stainless steel, and Ti-6Al-4V (titanium alloy). The allowable stresses in these materials are as follows:

| Material | Allowable stress, ksi |
|---|---|
| Cast cobalt chromium (CC) | 80 (170 in compression) |
| Forged stainless steel (SS) | 120 |
| Ti-6Al-4V titanium alloy (Ti) | 160 |

**Figure 6.19**
Inserted total hip
replacement.

**Find:** Ascertain whether the implant is safe for application from a static load standpoint.

**Solution:** The most difficult part of the problem lies in obtaining the stresses; applying a failure criterion is a relatively short part of the analysis. The hip was analyzed at the sections indicated in Figure 6.21. These sections were selected because they have geometric features that act as stress raisers and because their locations maximize the stresses associated with the applied loads. The following dimensions were obtained for the three sections analyzed, and the stress concentrations were obtained from Figure 6.5(a) and (b).

| Section | $d$ | $D$ | $r$ | $K_c$ (axial) | $K_c$ (bending) |
|---|---|---|---|---|---|
| A-A | 0.50 | 1.255 | 0.25 | 1.8 | 1.2 |
| B-B | **1.0** | 1.255 | 0.06 | 2.0 | 1.8 |
| C-C | **0.75** | **0.8** | 4.54 | 1 | 1 |

Numbers shown in boldface were interpolated from the drawing or extrapolated from the charts. To obtain the stress concentration factor, it was assumed that the geometry is circular in cross-section and that it can be taken as a straight rod with fillets at each section. Although this assumption is not strictly true, the resultant stress concentration factors should be close to those actually seen in the hip. Also, St.-Venant's principle was liberally applied. That is, the stresses at the cross-sections were assumed to be free of end effects, and the stress concentrations not to overlap. A more detailed analysis could be done by methods such as finite element analysis, but this approach will get a fairly accurate result.

The actual load applied to a hip joint is extremely complicated and varies from person to person. Gait, step length, etc., all play a role in the biomechanics of walking. Given that the loading can be complex, the load was taken as 4 times the user's body weight, a peak force measured during a walking step. Running was not considered for this implant, although it can be a factor for some recipients (note that most implant recipients are elderly or have a less active lifestyle). Further, although the direction of the

*(continued)*

**Case Study**          (Continued)

applied force can vary by as much as 30°, it was assumed that the load direction is vertical and centered on the stem (this is also a worst-case assumption—any inclination of the load reduces the bending moment at any section). Load direction could be varied to obtain a better understanding of the effect of stresses on the hip implant. A 200-lb user was considered, or an 800-lb load on the implant. This load, too, is high; most implant recipients weigh less, and obese patients are required to lose weight before receiving an implant.

**Figure 6.20**    Dimension of femoral implant (in inches).

**Figure 6.21**    Sections of femoral stem analyzed for static failure.

## Case Study (Concluded)

The weight leads to a normal force, a shear force, and a bending moment at each section. These values were directly obtained through statics and by using the measured dimensions from Figure 6.20. The resultant loads on each cross-section were as follows:

| Section | Normal force, lb | Shear force, lb | Moment, in · lb |
|---------|------------------|-----------------|-----------------|
| A-A | $800 \cos 36.5°$ $= 640$ | $800 \sin 36.5°$ $= 475$ | $800(1.06/2) \sin 36.5°$ $= 252$ |
| B-B | 640 | 475 | 345 |
| C-C | 800 | 0 | 680 |

Note that section C-C was actually loaded far less severely than is suggested here; part of the load is supported by bone, and the implant is never exposed to these stresses. Regardless, the resultant maximum stresses were calculated from

$$\sigma_{max} = -K_{axial} \frac{P}{A} \pm K_{bending} \frac{Mc}{I}$$

$$\tau = \frac{4V}{3A}$$

yielding the following:

| Section | Maximum normal stress, psi | Maximum shear stress, psi |
|---------|---------------------------|---------------------------|
| A-A | $-30,500$ or $+18,700$ | 3220 |
| B-B | $-7955$ or $+4690$ | 1100 |
| C-C | $-18,200$ or $+14,608$ | 0 |

Clearly, section A-A has the most severe loading, so it was analyzed with respect to failure.

At the extreme ends of the cross-section there is no shear stress, so the stresses given are principal stresses. For uniaxial stress states, the safety factor is merely the ratio of allowable stress to applied stress. For the location of maximum tensile stress, the safety factor is given by $n_s = S_{ut}/\sigma_1$ and equals 4.2 for CC, 6.4 for SS, and 8.5 for Ti. In compression the safety factor is $n_s = S_{uc}/\sigma_3$ and equals 5.6 for CC, 3.9 for SS, and 5.2 for Ti. At the center of the cross-section there is a normal stress component $P/A$ in addition to the maximum shear. Thus, at section A-A the principal stresses (from Mohr's circle analysis) are $\sigma_1 = 9700$ psi, $\sigma_2 = 0$, and $\sigma_3 = -3800$ psi. For cobalt chromium a brittle fracture theory must be chosen, so that the IFT is arbitrarily selected. From Equation (6.15)

$$\frac{\sigma_1}{S_{ut}} - \frac{\sigma_3}{S_{uc}} = \frac{1}{n_s} \qquad \therefore \qquad n_s = 7.0$$

For stainless steel and titanium either the MSST or the DET can be used. Using the MSST gives

$$\sigma_1 - \sigma_3 = \frac{S_y}{n_s}$$

so that $n_s = 8.8$ for SS and 11.8 for Ti.

These safety factors may seem high, but they are quite reasonable for such critical applications, where failure requires immediate surgery and should be avoided whenever possible. Further, this confirms the field experience that most implanted hips do not fail statically; rather they need to be replaced because of fatigue, biological, or tribological issues.

## 6.8 SUMMARY

The main emphasis of this chapter was on failures due to static loading. It was discovered that stress concentrations, such as holes, grooves, and fillets, reduce the safety factor for the part. Using the stress concentration factor enables more accurate depiction of stresses for components with holes, grooves, and fillets. A flow analogy was also presented that suggested some design changes that might reduce the stress near the discontinuity.

Fracture mechanics was discussed in considering crack extension as a function of applied load. Cracks were viewed as microscopic flaws that always exist on the surface and within the body of a solid material. Fracture failures due to crack propagation can occur at stress levels well below the yield stress of the material. There was interest in establishing the critical crack length that will make the part fracture; to do so required an understanding of fracture toughness. Fracture toughness is used as a design criterion in preventing fracture, just as yield strength is used as a design criterion in preventing yielding in a ductile material.

Failure predictions for uniaxial and multiaxial stress states were presented. Failure theories for multiaxial stress states were considered for ductile and brittle materials. The maximum-shear-stress theory states that a part subjected to any combination of loads will yield whenever the maximum shear stress exceeds the shear strength of the material. The distortion energy theory postulates that yielding of ductile materials is caused by the elastic energy associated with shear deformation. The maximum-normal-stress theory states that a brittle material subjected to any combination of loads will fail whenever the greatest positive (or negative) principal stress exceeds the tensile (or compressive) yield strength of the material. Another failure criterion for brittle materials was the internal friction theory, which is similar to the maximum-shear-stress theory but takes into account the higher strength of brittle materials in compression versus tension. The modified Mohr theory deviates from the internal friction theory slightly to better account for the behavior of some brittle materials. In each of these theories a criterion was proposed such that a critical stress was established that was then compared with a critical uniaxial stress for which experimental data are known. This approach may be incompatible with some materials, suggesting that probabilistic methods, or at the very least generous safety factors, may be called for.

## KEY WORDS

**Coulomb–Mohr theory**    theory identical to internal friction theory.

**crack**    microscopic flaw, always present, that can reduce material strength.

**distortion energy theory**    postulate that failure is caused by elastic energy associated with shear deformation.

**fracture control**    maintenance of nominal stress and crack size below a critical level.

**fracture toughness**    critical value of stress intensity at which crack extension occurs.

**internal friction theory**    failure criterion accounting for difference between compressive and tensile strengths of brittle materials.

**maximum-normal-stress theory**    theory that yielding will occur whenever largest positive principal stress exceeds the tensile yield strength or whenever greatest negative principal stress exceeds the compressive yield strength.

**maximum-shear-stress theory**    theory that yielding will occur when the largest shear stress exceeds a critical value.

**modes of crack propagation**    principal mechanisms for cracks to grow: mode I, opening through tension; mode II, sliding or in-plane shearing; mode III, tearing.

**modified Mohr theory**    failure postulate similar to Coulomb–Mohr theory, except that the curve is altered in quadrants II and IV of plane stress plot of principal stresses.

**stress concentration**    region where stress raiser is present.

**stress concentration factor**    factor used to relate actual maximum stress at discontinuity to nominal stress without discontinuity.

**stress intensity factor**    stress intensity at crack tip.

**stress raiser**    discontinuity that alters stress distribution so as to increase maximum stress.

**Tresca yield criterion**    same as maximum-shear-stress theory.

**von Mises criterion**    same as distortion energy theory.

**von Mises stress**    effective stress based on von Mises criterion, given by

$$\sigma_e = \frac{\sqrt{2}}{2} \left[ (\sigma_2 - \sigma_1)^2 + (\sigma_3 - \sigma_1)^2 + (\sigma_3 - \sigma_2)^2 \right]^{1/2}$$

## RECOMMENDED READINGS

Caddell, R. M. (1980) *Deformation and Fracture of Solids*, Prentice Hall, Englewood Cliffs, NJ.

Calladine, C. R. (1969) *Engineering Plasticity*, Pergamon Press, Elmsford, NY.

Dowling, N. E. (1993) *Mechanical Behavior of Materials*, Prentice Hall, Englewood Cliffs, NJ.

Norton, R. L. (2000) *Machine Design: An Integrated Approach*, Prentice Hall, Englewood Cliffs, NJ.

Popov, E. P. (1968) *Introduction to Mechanics of Solids*, Prentice Hall, Englewood Cliffs, NJ.

Shigley, J. E., Mischke, C. R., and Budynas, R. G. (2003) *Mechanical Engineering Design*, 7th ed., McGraw-Hill.

## REFERENCES

ASM International (1989) *Guide to Selecting Engineering Materials*, American Society for Metals, Materials Park, OH.

Collins, J. A. (1981) *Failure of Materials in Mechanical Design*, Wiley, New York.

Coulomb, C. A. (1773) *Sur une Application des Regles de maximis et minimus à quelques problèmes de statique relatifs a l'architecture.*

Dowling, N. E. (1993) *Mechanical Behavior of Materials*, Prentice Hall, Englewood Cliffs, NJ.

Frocht, M. M., and Hill, H. N. (1940) "Stress Concentration Factors around a Central Circular Hole in a Plate Loaded through a Pin in Hole," *Journal of Applied Mechanics*, vol. 7, p. A-5.

Grassi, R. C., and Cornet, I. (1949) "Fracture of Gray Cast Iron Tubes under Biaxial Stresses," *Journal of Applied Mechanics*, vol. 16, pp. 178–183.

Jones, R. M. (1975) *Mechanics of Composite Materials*, Hemisphere Publishing Co., New York.

Kaw, A. K. (1997) *Mechanics of Composite Materials*, CRC Press, Boca Raton, FL.

Marin, J. (1962) *Mechanical Behavior of Engineering Materials*, Prentice Hall, Englewood Cliffs, NJ.

Murphy, G. (1964) *Advanced Mechanics of Materials*, McGraw-Hill, New York.

Peterson, R. E. (1974) *Stress Concentration Factors*, Wiley, New York.

Popov, E. P. (1968) *Introduction to Mechanics of Solids*, Prentice Hall, Englewood Cliffs, NJ.

Reddy, J. N. (1996) *Mechanics of Laminated Composite Plates*, CRC Press, Boca Raton, FL.

Tresca, H. (1868) "Mem. prenetes par divers savants," *Comptes Rendus Acad. Sci.*, vol. 59, p. 754, Paris.

Young, W. C. (1989) *Roark's Formulas for Stress and Strain*, 6th ed., McGraw-Hill, New York.

## PROBLEMS

### Section 6.2

**6.1**  Given that the stress concentration factor is 2.81 for a machine element made of steel with a modulus of elasticity of 207 GPa, find the stress concentration factor for an identical machine element made of aluminum instead of steel. The modulus of elasticity for aluminum is 69 GPa.

**6.2**  A flat part with constant thickness $b$ is loaded in tension as shown in Figure 6.3(a). The height changes from 50 to 100 mm with a radius $r = 10$ mm. Find how much higher a load can be transmitted through the bar if the height is increased from 50 to 87 mm and the radius is decreased from 10 to 4 mm. *Ans.* 42% higher load.

**6.3**  A flat steel plate axially loaded as shown in sketch $a$ has two holes for electric cables. The holes are situated beside each other, and each has a diameter $d$. To make it possible to draw more cables, the two holes are replaced with one hole having twice the diameter $2d$, as shown in sketch $b$. Assume that the ratio of diameter to width is $d/b = 0.2$ for the two-hole plate. Which plate will fail first?

(a)                                        (b)

Sketches *a* and *b*, for Problem 6.3

**6.4**  A 5-mm-thick, 100-mm-wide AISI 1020 steel rectangular plate has a central elliptical hole 6 mm in length transverse to the applied stress and 2 mm in diameter along the stress. Determine the applied load that causes yielding at the edge of the hole.

**6.5**  A round bar has a fillet with $r/d = 0.15$ and $D/d = 1.5$. The bar transmits both bending moment and torque. A new construction is considered to make the shaft stiffer and stronger by making it equally thick on each side of the fillet or groove. Determine whether this is a good idea.

**6.6**  A 10-in-wide plate loaded in tension contains a 2-in-long, $\frac{1}{2}$-in-wide slot. Estimate the stress concentration by

  *a)*     Approximating the slot as an ellipse that is inscribed within the slot.
  *b)*     Obtaining the stress concentration at the edge of the slot by taking a section through the slot and approximating the geometry as a rectangular plate with a groove.

**6.7** A machine has three circular shafts, each with fillets giving stress concentrations. The ratio of fillet radius to shaft diameter is 0.1 for all three shafts. One of the shafts transmits a tensile force, one transmits a bending torque, and one transmits torsion. Because they are stressed exactly to the stress limit ($n_s = 1$), a design change is proposed, doubling the notch radii to get a safety factor greater than 1. What are the safety factors for the three shafts if the diameter ratio is $2$ $(D/d = 2)$? *Ans.* $n_{s,tension} = 1.21$, $n_{s,bending} = 1.19$, $n_{s,torsion} = 1.17$.

**6.8** The shaft shown in sketch $c$ is subjected to tensile, torsional, and bending loads. Determine the principal stresses at the location of stress concentration. *Ans.* $\sigma_1 = 52.99$ MPa, $\sigma_2 = 0$, $\sigma_3 = -12.27$ MPa.

Sketch $c$, for Problem 6.8

**★ 6.9** A steel plate with dimensions shown in sketch $d$ is subjected to 150-kN tensile force and 300-N · m bending moment. The plate is made of AISI 1080 steel. A hole is to be punched in the center of the plate. What is the maximum diameter of the hole for a safety factor of 1.5? *Ans.* $d = 120$ mm.

Sketch $d$, for Problem 6.9

## Section 6.3

**6.10** A Plexiglas plate with dimensions 1 m × 1 m × 10 mm is loaded by a nominal tensile stress of 55 MPa in one direction. The plate contains a small crack perpendicular to the load direction. At this stress level a safety factor of 2 against crack propagation is obtained. Find how much larger the crack can get before it grows catastrophically. *Ans.* $a_2 = 4a_1$.

**6.11** A pressure vessel is made of AISI 4340 steel. The wall thickness is such that the tensile stress in the material is 1100 MPa. The dimensionless geometry correction factor $Y = 1$ for the given geometry. Find the size of the largest allowable crack without failure if the steel is tempered

a)    At 260°C. *Ans.* 1.3 mm.
b)    At 425°C. *Ans.* 4.0 mm.

* Indicates problem of greater difficulty.

**6.12** Two tensile test rods are made of AISI 4340 steel tempered at 260°C and aluminum alloy 2024-T351. The dimensionless geometry correction factor $Y = 1$. Find how high a stress each rod can sustain if there is a crack of 2-mm half-length in each of the rods. *Ans.* AISI 4340: $\sigma = 631$ MPa.

**6.13** A plate made of Ti-6Al-4V titanium alloy has the dimensionless correction factor $Y = 1$. How large can the largest crack in the material be if it still should be possible to plastically deform the plate in tension? *Ans.* 1.488 mm.

**6.14** A Plexiglas model of a gear has a 1-mm half-length crack formed in its fillet curve (where the tensile stress is maximum). The model is loaded until the crack starts to propagate. Let $Y = 1.5$. How much higher a load can a gear made of AISI 4340 steel tempered to 425°C carry with the same crack and the same geometry? *Ans.* $\sigma_{steel}/\sigma_{Plexiglas} = 87.4$

**6.15** A pressure vessel made of aluminum alloy 2024-T351 is manufactured for a safety factor of 2.5 guarding against yielding. The material contains cracks through the wall thickness with a crack half-length less than 3 mm. Let $Y = 1$. Find the safety factor when considering crack propagation.

★ **6.16** The clamping screws holding the top lid of a nuclear reactor are made of AISI 4340 steel tempered at 260°C. They are stressed to a maximum level of 1250 MPa during a pressurization test before starting of the reactor. Find the safety factor guarding against yielding and the safety factor guarding against crack propagation if the initial cracks in the material have $Y = 1$ and $a = 1$ mm. Also, perform the calculations for the same material but tempered to 425°C. *Ans.* AISI 4340 tempered at 260°C: $n_s = 0.714$.

**6.17** A glass tube used in a pressure vessel is made of aluminum oxide (sapphire) to make it possible to apply 30-MPa pressure and still have a safety factor of 2 guarding against fracture. For a soda-lime glass of the same geometry only 7.5-MPa pressure can be allowed if a safety factor of 2 is to be maintained. Determine the size of the cracks the glass tube can tolerate at 7.5-MPa pressure and a safety factor of 2. Let $Y = 1$ for both tubes. *Ans.* Sapphire: $2a < 75.2\,\mu$m, glass: $2a < 65.6\,\mu$m.

**6.18** A stress optic model used for demonstrating the stress concentrations at the ends of a crack is made of polymethyl methacrylate. An artificially made crack 100 mm long is perpendicular to the loading direction. $Y = 1$. Calculate the highest tensile stress that can be applied to the model without propagating the crack. *Ans.* $\sigma_{nom} = 1.78$ MPa.

★ **6.19** A passengerless airplane requires wings that are lightweight and the prevention of cracks more than 2 mm long. The dimensionless geometry correction factor $Y$ is 2.5 for a safety factor of 2.

    *a)*      What is the appropriate alloy for this application? *Ans.* Either aluminum 2020-T351 or alloy steel 4340 tempered at 425°C.

    *b)*      If $Y$ is increased to 2.5, what kind of alloy from Table 6.1 should be used? *Ans.* Al 2020-T351.

## Section 6.6

**6.20** The anchoring of the cables carrying a suspension bridge is made of cylindrical AISI 1080 steel bars 210 mm in diameter. The force transmitted from the cable to the steel bar is 3.5 MN. Calculate the safety factor range guarding against yielding based on the allowable stresses from Equation (3.13). *Ans.* $1.69 < n_s < 2.25$.

**6.21** The arm of a crane has two steel plates connected with a rivet that transfers the force in pure shear. The rivet is made of AISI 1040 steel and has a circular cross-section with a diameter of 25 mm. The load on the rivet is 20 kN. Calculate the safety factor. *Ans.* $n_s = 3.44$.

## Section 6.7

**6.22** A machine element is loaded so that the principal normal stresses at the critical location for a biaxial stress state are $\sigma_1 = 20$ ksi and $\sigma_2 = -15$ ksi. The material is ductile with a yield strength of 60 ksi. Find the safety factor according to

a)    The maximum-shear-stress theory (MSST). *Ans.* $n_s = 1.714$.
b)    The distortion energy theory (DET). *Ans.* $n_s = 1.97$.

**6.23** A bolt is tightened, subjecting its shank to a tensile stress of 80 ksi and a torsional shear stress of 50 ksi at a critical point; all the other stresses are zero. Find the safety factor at the critical point by the DET and the MSST. The material is high-carbon steel (AISI 1080). Will the bolt fail because of the static loading? *Ans.* $n_{s,\text{DET}} = 0.47$, $n_{s,\text{MSST}} = 0.43$.

**6.24** A torque is applied to a piece of chalk used in a classroom until the chalk cracks. Using the maximum-normal-stress theory (MNST) and assuming the tensile strength of the chalk to be small relative to its compressive strength, determine the angle of the cross-section at which the chalk cracks. *Ans.* $45°$.

★ **6.25** A cantilevered bar 500 mm long with square cross-section has 25-mm sides. Two perpendicular forces are applied to its free end: A 1000-N force is applied in the $x$ direction, a 100-N force is applied in the $y$ direction, and an equivalent force of 100 N is applied in the $z$ direction. Calculate the equivalent stress at the clamped end of the bar by using the DET when the sides of the square cross-section are parallel with the $y$ and $z$ directions.

**6.26** A shaft transmitting torque from the gearbox to the rear axle of a truck is unbalanced, so that a centrifugal load of 500 N acts at the middle of the 3-m-long shaft. The AISI 1040 tubular steel shaft has an outer diameter of 70 mm and an inner diameter of 58 mm. Simultaneously, the shaft transmits a torque of 6000 N · m. Use the DET to determine the safety factor guarding against yielding. *Ans.* $n_s = 1.196$.

**6.27** The right-angle cantilevered bracket used in Problem 5.30, sketch $w$, is subjected to a concentrated force of 1000 N and a torque of 300 N · m. Calculate the safety factor. Use the DET and neglect transverse shear. Assume that the bracket is made of AISI 1040 steel, and use the following values: $a = 0.5$ m, $b = 0.3$ m, $d = 0.035$ m, $E = 205$ GPa, and $v = 0.3$. *Ans.* $n_s = 1.76$.

**6.28** A 100-mm-diameter shaft is subjected to a 10-kN·m steady bending moment, an 8-kN · m steady torque, and a 150-kN axial force. The yield strength of the shaft material is 600 MPa. Use the MSST and the DET to determine the safety factors for the various types of loading. *Ans.* $n_s = 4.28$.

★ **6.29** Use the MSST and the DET to determine the safety factor for 2024 aluminum alloy for each of the following stress states:

a)    $\sigma_x = 10$ MPa, $\sigma_y = -60$ MPa. *Ans.* $n_{s,\text{MSST}} = 4.64$.
b)    $\sigma_x = \sigma_y = \tau_{xy} = -30$ MPa. *Ans.* $n_{s,\text{DET}} = 5.42$.
c)    $\sigma_x = -\sigma_y = 20$ MPa and $\tau_{ry} = 10$ MPa. *Ans.* $n_{s,\text{MSST}} = 7.27$.
d)    $\sigma_x = 2\sigma_y = -70$ MPa and $\tau_{xy} = 40$ MPa. *Ans.* $n_{s,\text{DET}} = 3.53$.

**6.30** Four different stress elements, each made of the same material, are loaded as shown in sketches *e*, *f*, *g*, and *h*. Use the MSST and the DET to determine which element is the most critical. *Ans.* Sketch *e* is most critical.

Sketches *e*, *f*, *g*, and *h*, for Problem 6.30          Sketch *i*, for Problem 6.31

★ **6.31** The rod shown in sketch *i* is made of AISI 1040 steel and has two 90° bends. Use the MSST and the DET to determine the minimum rod diameter for a safety factor of 2 at the most critical section. *Ans.* Diameter is acceptable.

**6.32** The shaft shown in sketch *j* is made of AISI 1020 steel. Determine the most critical section by using the MSST and the DET. Dimensions of the various diameters shown in sketch *j* are $d = 30$ mm, $D = 45$ mm, and $d_2 = 40$ mm. *Ans.* Critical location 80 mm.

Sketch *j*, for Problem 6.32

# FAILURE PREDICTION FOR CYCLIC AND IMPACT LOADING

Aloha Airlines Flight 243, a Boeing 737-200, taken April 28, 1988. The midflight fuselage failure was caused by corrosion-assisted fatigue. (Steven Minkowski/ Gamma Liaison)

*All machine and structural designs are problems in fatigue because the forces of Nature are always at work and each object must respond in some fashion.*

Carl Osgood, *Fatigue Design*

## SYMBOLS

| | |
|---|---|
| $A$ | area, $m^2$ |
| $A_a$ | amplitude ratio, $\sigma_a/\sigma_m$ |
| $a$ | fatigue strength exponent |
| $b$ | width, m |
| $b_s$ | slope |
| $C$ | constant used in Equation (7.47) |
| $C_1, C_2$ | integration constants |
| $\bar{C}$ | intercept |
| $c$ | distance from neutral axis to outer fiber of solid, m |
| $d$ | diameter, m |
| $E$ | modulus of elasticity, Pa |
| $e, f$ | factors used in Equation (7.21) |
| $f$ | exponent |
| $g$ | gravitational acceleration, 9.807 $m/s^2$ (386.1 $in/s^2$) |
| $H$ | height including two notch radii, m |
| $h$ | height without two notch radii, m |
| $I$ | area moment of inertia, $m^4$ |
| $I_m$ | impact factor |
| $K$ | stress intensity factor, MPa$\sqrt{m}$ |
| $K_c$ | stress concentration factor |
| $K_f$ | fatigue stress concentration factor |
| $k$ | spring rate, N/m |
| $\Delta K$ | stress intensity range, MPa$\sqrt{m}$ |
| $k_f$ | surface finish factor |
| $k_m$ | miscellaneous factor |
| $k_r$ | reliability factor |
| $k_s$ | size factor |
| $k_t$ | temperature factor |
| $k_0$ | stress concentration factor |
| $l$ | length, m |
| $l_c$ | crack length, m |
| $M$ | bending moment, N · m |
| $m$ | constant used in Equation (7.47) |
| $N$ | fatigue life in cycles |
| $N_c$ | number of stress cycles |
| $N'$ | number of cycles to failure at a specific stress |
| $N'_t$ | total number of cycles to failure |
| $n'$ | number of cycles at a specific stress when $n' < N'$ |
| $n_s$ | safety factor |

| | |
|---|---|
| $P$ | force, N |
| $P_{max}$ | maximum force, N |
| $q_n$ | notch sensitivity factor |
| $R_a$ | arithmetic average roughness, m |
| $R_s$ | stress ratio, $\sigma_{min}/\sigma_{max}$ |
| $r$ | notch radius, m |
| $S_e$ | modified endurance limit, Pa |
| $S'_e$ | endurance limit, Pa |
| $S_f$ | modified fatigue strength, Pa |
| $S'_f$ | fatigue strength, Pa |
| $S'_i$ | strength at $10^3$ cycles for ductile material, Pa |
| $S'_l$ | fatigue strength at which high-cycle fatigue begins, Pa |
| $S_u$ | ultimate strength, Pa |
| $S_{uc}$ | ultimate strength in compression, Pa |
| $S_{ut}$ | ultimate strength in tension, Pa |
| $S_y$ | yield strength, Pa |
| $S_{yt}$ | yield strength in tension, Pa |
| $u$ | sliding velocity, m/s |
| $V$ | shear force, N |
| $W$ | weight, N |
| $Y$ | plate size correction factor |
| $x, y$ | Cartesian coordinates, m |
| $\alpha$ | fatigue ductility exponent |
| $\alpha_i$ | cyclic ratio $n'_i/N'_i$ |
| $\delta$ | deflection, m |
| $\delta_{max}$ | maximum impact deflection, m |
| $\delta_{st}$ | static deflection, m |
| $\epsilon'_f$ | fatigue ductility coefficient |
| $\Delta\epsilon$ | total strain |
| $\sigma$ | stress, Pa |
| $s'_f$ | stress at fracture, Pa |
| $\sigma_r$ | stress range, Pa |

## SUBSCRIPTS

| | |
|---|---|
| $a$ | alternating |
| $f$ | final |
| $i$ | integer |
| $m$ | mean |
| max | maximum |
| min | minimum |
| ref | reference |

## 7.1     INTRODUCTION

Most machine element failures involve load conditions that fluctuate with time. However, the static load conditions covered in Chapter 6 are important, because they provide the foundation for understanding this chapter. This chapter focuses on understanding and predicting component failure under cyclic and impact loading. Fluctuating loads induce fluctuating stresses that often result in failure due to cumulative damage. To better understand failures due to fluctuating stresses, consider the yielding caused by the back-and-forth bending of a paper clip. Bending results in compressive and tensile stresses on opposite sides of the paper clip, and these stresses reverse when the bending direction is changed. Thus, the stress at any point around the wire of the paper clip will vary as a function of time. Repeated bending of the paper clip will eventually exhaust the ductility of the material, resulting in failure.

Stresses and deflections in impact loading are in general much greater than those found in static loading; thus, dynamic loading effects are important. A material's physical properties are a function of loading speed. Fortunately, the more rapid the load, the higher the yield and ultimate strengths. Some examples in which impact loading have to be considered are colliding automobiles, driving a nail with a hammer, and an air hammer breaking up concrete.

## 7.2    FATIGUE

In the 19th century the first industrial revolution was in full swing. Coal was mined, converted to coke, and used to smelt iron from ore. The iron was used to manufacture bridges, railroads, and trains that brought more coal to the cities, allowing more coke to be produced, etc., in a spiral of ever-increasing production. But then bridges, the monuments to engineering, began failing. Even worse, they would fail in extremely confusing ways. A 50-ton locomotive could pass over a bridge without incident; a farmer driving a horse-drawn wagon full of hay would subsequently cause the bridge to collapse. Fear gripped the populace, as people believed death awaited on the bridges (Fig. 7.1).

Today, we no longer fear our "aging infrastructure." The road construction and improvements that delay traffic from time to time make collapses of civil engineering works, such as bridges, truly rare. But fatigue failures, as we now know them to be, are not at all rare in machine elements. Fatigue is the single largest cause of failure in metals, estimated to be the cause of 90% of all metallic failures. Fatigue failures, especially in structures, are catastrophic and insidious, occurring suddenly and often without warning. For that reason engineers must apply fatigue considerations in design.

Fatigue has both microscopic and macroscopic aspects. That is, although a rolling-element bearing failing by surface spalling and a ship breaking in two are quite different events, both the bearing and the ship have failed due to fatigue. The failure stresses were considerably lower than the yield strengths of the materials. The rolling-element bearing suffered surface fatigue failure; the ship, structural fatigue failure. Thus, **fatigue** failure occurs at relatively low stress levels to a component or structure subjected to fluctuating or cyclic stresses. Fatigue failure looks brittle even in ductile metals; little, if any, gross plastic deformation is associated with it.

Some of the basic concepts associated with fatigue are the following:

1.  Fatigue is a complex phenomenon, and no universal theories to describe the behavior of materials subjected to cyclic loadings exist; instead, there are a large number of theories to describe the behavior of particular materials.

2.  Most of the engineering design experience in fatigue is based on an experimental understanding of the behavior of carbon steels. Much effort has been directed toward

ON THE BRIDGE!

**Figure 7.1**    "On the Bridge," an illustration from *Punch* magazine in 1891 warning the populace that death was waiting for them on the next bridge. Note the cracks in the iron bridge. (*Punch* magazine, 1891)

extending these semiempirical rules to other ferrous and nonferrous metals, as well as ceramics, polymers, and composite materials.

3.  For the most part, fatigue involves the accumulation of damage within a material. Damage usually consists of cracks that can grow by a small distance with each stress cycle.

4.  Experimenters have found that fatigue cracks generally begin at a surface and propagate through the bulk. Therefore, much attention is paid to the quality of surfaces in fatigue-susceptible machine elements. (However, if large subsurface flaws or stress raisers exist in the substrate, fatigue cracks can initiate below the surface.)

5.  Fatigue cracks begin at several sites simultaneously and propagate when one flaw becomes dominant and grows more rapidly than the others.

6.  Fatigue testing is imperative to confirm safe mechanical design.

The last of these points cannot be overemphasized, especially since this book concentrates on theoretical approaches to fatigue design. This is not intended to suggest that theoretical approaches are sufficient; experimental verifications are essential, but that is not the focus of this book.

The total life of a component or structure is the time it takes a crack to initiate plus the time it needs to propagate through the cross-section.

## METHODS TO MAXIMIZE DESIGN LIFE:

1. **By minimizing initial flaws, especially surface flaws.** Great care is taken to produce fatigue-insusceptible surfaces through processes, such as grinding or polishing, that leave exceptionally smooth surfaces. These surfaces are then carefully protected before being placed into service.

2. **By maximizing crack initiation time.** Surface residual stresses are imparted (or at least tensile residual stresses are relieved) through manufacturing processes, such as shot peening or burnishing, or by a number of surface treatments.

3. **By maximizing crack propagation time.** Substrate properties, especially those that retard crack growth, are also important. For example, fatigue cracks usually propagate more quickly along grain boundaries than through grains (because grains have much more efficient atomic packing). Thus, using a material that does not present elongated grains in the direction of fatigue crack growth can extend fatigue life (e.g., by using cold-worked components instead of castings).

4. **By maximizing the critical crack length.** Fracture toughness is an essential ingredient. (The material properties that allow for larger internal flaws are discussed in Chap. 6.)

Given a finite number of resources, which one of these approaches should the designer emphasize? Ever-smoother surfaces can be manufactured at ever-increasing costs but at ever-decreasing payoff as the surface valleys become smaller. Maximizing initiation time allows parts to function longer with little or no loss in performance. Maximizing propagation time can allow cracks to be detected before they become catastrophic, the approach previously called the *doctrine of manifest danger*. Maximizing the critical crack length can lead to long life with a greater likelihood of recognizing an imminent failure. The proper emphasis, just as with the safety factor, is product-specific and is best decided through experience.

The designer's job is to select a material and a processing route that will lead to successful products. However, fatigue is an extremely complex subject. This chapter introduces approaches to design for cyclic loading. In practice, extreme caution should be used in applying any of these approaches without generous safety factors.

## 7.3 CYCLIC STRESSES

**Cyclic stress** is a function of time, but the variation is such that the stress sequence repeats itself. The stresses may be axial (tensile or compressive), flexural (bending), or torsional (twisting). Figure 7.2 shows the cyclic variation of nonzero mean stress with time. Also shown are several parameters used to characterize fluctuating cyclic stress. A stress cycle ($N_c = 1$) constitutes a single application and removal of a load and then another application and removal of the load in the opposite direction. Thus, $N_c = \frac{1}{2}$ means that the load is applied once and then released. The stress amplitude alternates about a **mean stress** $\sigma_m$,

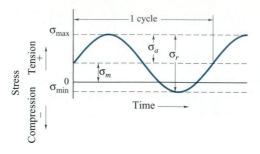

**Figure 7.2**    Variation in nonzero cyclic mean stress.

defined as the average of the maximum and minimum stresses in the cycle, or

$$\sigma_m = \frac{\sigma_{max} + \sigma_{min}}{2} \qquad (7.1)$$

The **stress range** $\sigma_r$ is the difference between $\sigma_{max}$ and $\sigma_{min}$, namely,

$$\sigma_r = \sigma_{max} - \sigma_{min} \qquad (7.2)$$

The **stress amplitude** $\sigma_a$ is one-half of the stress range, or

$$\sigma_a = \frac{\sigma_r}{2} = \frac{\sigma_{max} - \sigma_{min}}{2} \qquad (7.3)$$

The **stress ratio** $R_s$ is the ratio of minimum to maximum stress amplitudes, or

$$R_s = \frac{\sigma_{min}}{\sigma_{max}} \qquad (7.4)$$

Finally, the **amplitude ratio** $A_a$ is the ratio of the stress amplitude to the mean stress, or

$$A_a = \frac{\sigma_a}{\sigma_m} = \frac{\sigma_{max} - \sigma_{min}}{\sigma_{max} + \sigma_{min}} = \frac{1 - R_s}{1 + R_s} \qquad (7.5)$$

The four most frequently used patterns of constant-amplitude cyclic stress are

1.  Completely reversed ($\sigma_m = 0$, $R_s = -1$, $A_a = \infty$)
2.  Nonzero mean (as shown in Fig. 7.2)
3.  Released tension ($\sigma_{min} = 0$, $R_s = 0$, $A_a = 1$, $\sigma_m = \sigma_{max}/2$)
4.  Released compression ($\sigma_{max} = 0$, $R_s = \infty$, $A_a = -1$, $\sigma_m = \sigma_{min}/2$)

---

**EXAMPLE 7.1**

**Given:**   Before the thread-formed additions at the top third of high chimneys were invented, chimneys swayed perpendicularly to the motion of the wind. This swaying motion appeared when the wind speed and direction were constant for some time. In a certain chimney the bending stress was $\pm 100$ MPa.

**Find:**  The mean stress, range of stress, stress amplitude, stress ratio, and amplitude ratio for that swaying steel chimney.

**Solution:**  The drag force acting on the chimney is in the wind direction, and it gives a maximum tensile stress on the front side and a maximum compressive stress on the back side of the chimney. The stress perpendicular to the wind is zero. The swinging motion, which gives the highest stress perpendicular to the wind direction, will thus be symmetric relative to the equilibrium position.

The mean stress is

$$\sigma_m = \frac{\sigma_{max} + \sigma_{min}}{2} = \frac{100 - 100}{2} = 0$$

The range of stress is

$$\sigma_r = \sigma_{max} - \sigma_{min} = 100 - (-100) = 200 \text{ MPa}$$

The alternating stress is

$$\sigma_a = \frac{\sigma_r}{2} = \frac{200}{2} = 100 \text{ MPa}$$

The stress ratio is

$$R_s = \frac{\sigma_{min}}{\sigma_{max}} = \frac{-100}{100} = -1$$

The amplitude ratio is

$$A_a = \frac{\sigma_a}{\sigma_m} = \frac{\sigma_{max} - \sigma_{min}}{\sigma_{max} + \sigma_{min}} = \frac{1 - R_s}{1 + R_s} = \frac{1 - (-1)}{1 + (-1)} = \frac{2}{0} = \infty$$

## 7.4    STRAIN LIFE THEORY OF FATIGUE

Fatigue is a damage accumulation process that manifests itself through crack propagation, but no crack propagation is possible without plastic deformation at the crack tip. Although the volume stressed highly enough for plastic deformation can be extremely small, if the stress fields remain elastic, no crack propagation is possible. Using material properties, such as yield strength or ultimate strength, presents difficulties because cyclic loadings can change these values near a crack tip. They may increase or decrease depending on the material and its manufacturing history. Thus, the material strength at the location where cracks are propagating can differ from the bulk material strength listed in handbooks or obtained from tension tests.

Given the difficulties in expressing material strengths near crack tips, several approaches have been suggested for dealing with the strain encountered at a crack tip. One of the better known is the **Manson–Coffin relationship,** which gives the total strain amplitude as the sum of the elastic and plastic strain amplitudes

$$\frac{\Delta \epsilon}{2} = \frac{\sigma'_f}{E} \left(2N'\right)^a + \epsilon'_f \left(2N'\right)^\alpha \tag{7.6}$$

**Table 7.1**     Cyclic properties of some metals

| Material | Condition | Yield strength $S_y$, MPa | Fracture strength $\sigma_f'$, MPa | Fatigue ductility coefficient $\epsilon_f'$ | Fatigue strength exponent $a$ | Fatigue ductility exponent $\alpha$ |
|---|---|---|---|---|---|---|
| **Steel** | | | | | | |
| 1015 | Normalized | 228 | 827 | 0.95 | −0.110 | −0.64 |
| 4340 | Tempered | 1172 | 1655 | 0.73 | −0.076 | −0.62 |
| 1045 | Q&T* 80°F | — | 2140 | — | −0.065 | −1.00 |
| 1045 | Q&T 306°F | 1720 | 2720 | 0.07 | −0.055 | −0.60 |
| 1045 | Q&T 500°F | 1275 | 2275 | 0.25 | −0.080 | −0.68 |
| 1045 | Q&T 600°F | 965 | 1790 | 0.35 | −0.070 | −0.69 |
| 4142 | Q&T 80°F | 2070 | 2585 | — | −0.075 | −1.00 |
| 4142 | Q&T 400°F | 1720 | 2650 | 0.07 | −0.076 | −0.76 |
| 4142 | Q&T 600°F | 1340 | 2170 | 0.09 | −0.081 | −0.66 |
| 4142 | Q&T 700°F | 1070 | 2000 | 0.40 | −0.080 | −0.73 |
| 4142 | Q&T 840°F | 900 | 1550 | 0.45 | −0.080 | −0.75 |
| **Aluminum** | | | | | | |
| 1100 | Annealed | 97 | 193 | 1.80 | −0.106 | −0.69 |
| 2014 | T6 | 462 | 848 | 0.42 | −0.106 | −0.65 |
| 2024 | T351 | 379 | 1103 | 0.22 | −0.124 | −0.59 |
| 5456 | H311 | 234 | 724 | 0.46 | −0.110 | −0.67 |
| 7075 | T6 | 469 | 1317 | 0.19 | −0.126 | −0.52 |

*Quenched and tempered.
SOURCE: Shigley and Mitchell (1983) and Suresh (1998).

where $\Delta\epsilon$ = total strain

$\sigma_f'$ = stress at fracture in one stress cycle, Pa

$E$ = elastic modulus of material, Pa

$N'$ = number of cycles that will occur before failure

$\epsilon_f'$ = fatigue ductility coefficient (true strain corresponding to fracture in one stress cycle)

$a$ = fatigue strength exponent

$\alpha$ = fatigue ductility exponent

Table 7.1 gives some typical values of the material properties $a$ and $\alpha$.

Landgraf (1968) and Mitchell (1978) have surveyed the industrial applications of the strain life approach to fatigue, but as pointed out by Shigley and Mitchell (1983), the Manson–Coffin relationship is difficult to use because the total strain $\Delta\epsilon$ is difficult to determine. Stress concentration factors, such as those presented in Chapter 6, are readily available in the technical literature, but strain concentration factors, especially in the plastic range, are nowhere to be found. The advantage of the Manson–Coffin equation is that it gives insight into important properties in fatigue strength determination. It shows the importance of strength as well as ductility, and it leads to the conclusion that as long as there is a cyclic plastic strain, no matter how small, eventually there will be failure.

## 7.5    FATIGUE STRENGTH

### 7.5.1    ROTATING-BEAM EXPERIMENTS

Fatigue is inherently probabilistic; that is, there is a great range of performance within samples prepared from the same materials. In previous problems and case studies, a valuable approach called the worst-case scenario was described. To apply this approach to fatigue, a designer would select surface finishes, notch sizes, initial flaw size, etc., that minimize the fatigue strength of the candidate specimen. However, this process would necessarily result in fatigue specimens with zero strength, a situation that does nothing to aid designers. Thus, data on fatigue often reflect the best-case scenario, and the designer is strongly cautioned that great care must be taken in applying fatigue design theories to critical applications.

Because fatigue is basically a damage accumulation phenomenon, initial flaws have a great effect on performance. No manufacturing process produces defect-free parts; indeed, no parts are possible without thousands, even millions, of flaws per cubic inch. Analytical approaches that derive fatigue strengths from first principles are thus very difficult, and most knowledge on material fatigue is experimentally based.

Experimental approaches to fatigue use either exemplars or idealized, standard specimens. The former are more reliable and best for critical specimens. The latter are often used when a direct simulation of the loading environment is cost-prohibitive.

To establish the fatigue strength of an exemplar, a series of tests is performed. The test apparatus duplicates as nearly as possible the stress conditions (stress level, time frequency, stress pattern, etc.) in practice. The exemplar duplicates as nearly as possible any manufacturing and treatment processes. Such experiments give the most direct indication of a component's survivability in the actual loading environment.

To test idealized, standard specimens, a rotating-beam fatigue testing machine is used, most commonly the Moore rotating-beam machine. The specimen is subjected to pure bending, and no transverse shear is imposed. The specimen has specific dimensions (Fig. 7.3) and a highly polished surface, with a final polishing in the axial direction to avoid circumferential scratches. If the specimen breaks into two equal pieces, the test is indicative of the material's **endurance limit** or fatigue strength. If the pieces are unequal, a material or surface flaw has skewed the results. The test specimen is subjected to completely reversed ($\sigma_m = 0$) stress cycling at a relatively large maximum stress amplitude, usually two-thirds of the static ultimate strength, and the cycles to failure are counted. Thus, for each specimen

**Figure 7.3**    R. R. Moore machine fatigue test specimen. Dimensions in inches.

at a specific stress level, the test is conducted until failure occurs. The procedure is repeated on other identical specimens, progressively decreasing the maximum stress amplitude.

### 7.5.2    REGIMES OF FATIGUE CRACK GROWTH

Figure 7.4(a) shows the size of a fatigue crack as a function of number of cycles for two stress ratios. Figure 7.4(b) illustrates the rate of crack growth and more clearly shows three different regimes of crack growth:

1. *Regime A* is a period of very slow crack growth. Note that the crack growth rate can be smaller than an atomic spacing of the material per cycle, a situation that would be corrected by the preferred lattice structure of the metal. Regime A should be recognized as a period of noncontinuum failure processes. The fracture surfaces are faceted or serrated in this regime, indicating crack growth is primarily due to shear deformations within a grain. The crack growth rate is so small that cracks may be negligible over the life of the component if this regime is dominant. Regime A is strongly affected by material microstructure, environmental effects, and stress ratio $R_s$.

2. *Regime B* is a period of moderate crack growth rate, often referred to as the **Paris regime** or Paris power law. In this regime, the rate of crack growth is influenced by several processes, involving material microstructure, mechanical load variables, and the environment. Thus it is not surprising that crack propagation rates cannot be determined for a given material or alloy from first principles, and testing is required to quantify the growth rate.

3. *Regime C* is a period of high growth rate, where the maximum stress intensity factor for the fatigue cycle approaches the fracture toughness of the material. Material microstructural effects and loadings have a large influence on crack growth, and additional static modes such as cleavage and intergranular separation can occur.

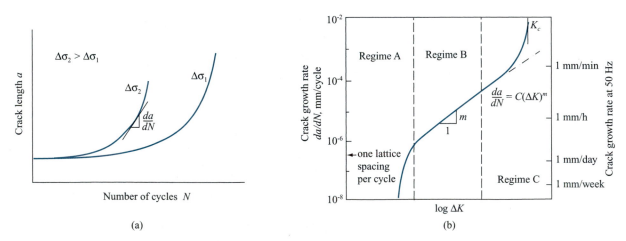

(a)

(b)

**Figure 7.4**    Illustration of fatigue crack growth. (a) Size of a fatigue crack for two different stress ratios as a function of the number of cycles; (b) rate of crack growth, illustrating three regimes. [Source: Suresh (1998).]

### 7.5.3   MICROSTRUCTURE OF FATIGUE FAILURES

As discussed, even the most ductile materials can exhibit brittle behavior in fatigue and will fracture with little or no plastic deformation. The reasons for this are not at all obvious, but an investigation of fatigue fracture microstructure can help explain this behavior.

A typical fatigue fracture surface is shown in Figure 7.5 and has the following features:

1. Near the origin of the fatigue crack, the surface is *burnished,* or very smooth. In the early stages of fatigue, the crack grows slowly and elastic deformations result in microscopic sliding between the two surfaces, resulting in a rubbing of the surfaces.

2. Near the final fracture location, *striations* or *beachmarks* are clearly visible to the naked eye. During the last few cycles of a fatigue failure, the crack growth is very rapid, and these striations are indicative of fast growth and growth-arrest processes.

3. Microscopic striations can exist between these two extremes, as shown in Figure 7.5, and are produced by the slower growth of fatigure cracks at this location in the part.

4. The final fracture surface often looks rough and is indicative of brittle fracture, but it can also appear ductile depending on the material.

The actual pattern of striations depends on the particular geometry, material, and loading (Fig. 7.6) and can require experience to evaluate a failure cause.

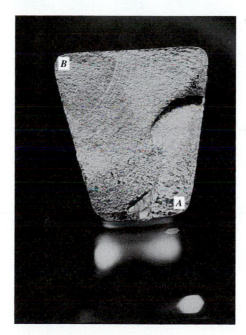

**Figure 7.5**   Cross-section of a fatigued section, showing fatigue striations or beachmarks originating from a fatigue crack at *B*. (From Rimnac et al, 1986. Copyright American Society for Testing and Materials. Reprinted by permission.)

**Figure 7.6**   Typical fatigue fracture surfaces of smooth and notched cross-sections under different loading conditions and stress levels. [From *Metals Handbook*, American Society for Metals (1975).]

### 7.5.4   *S–N* DIAGRAMS

Data from reversed-bending experiments are plotted as the fatigue strength versus the logarithm of the total number of cycles to failure $N_t'$ for each specimen. These plots are called **S–N diagrams** or **Wöhler diagrams,** after August Wöhler, a German engineer who published his fatigue research in 1870. They are a standard method of presenting fatigue data and are common and informative. Two general patterns for two classes of material, those with and those without endurance limits, emerge when the fatigue strength is plotted versus the logarithm of the number of cycles to failure. Figure 7.7 shows typical results for several materials. Figure 7.7(a) presents test data for wrought steel. Note the large amount of scatter in the data, even with the great care in preparing test specimens. Thus, life predictions based on curves such as those in Figure 7.7 are all subject to a great deal of error. Figure 7.7(a) also shows a common result. For some materials with **endurance limits,** such as ferrous and titanium alloys, a horizontal straight line occurs at low stress levels, implying that an endurance limit $S_e'$ is reached below which failure will not occur. This endurance limit $S_e'$

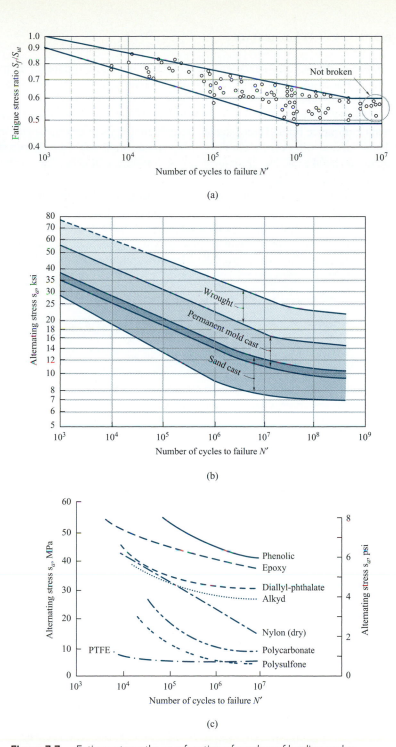

**Figure 7.7** Fatigue strengths as a function of number of loading cycles. (a) Ferrous alloys, showing clear endurance limit [adapted from Lipson and Juvinall (1963)]; (b) aluminum alloys, with less pronounced knee and no endurance limit [adapted from Juvinall and Marshek (1991); (c) selected properties of assorted polymer classes [adapted from Norton (1996).].

represents the largest fluctuating stress that will *not* cause failure for an infinite number of cycles. For many steels the endurance limit ranges between 35% and 60% of the material's ultimate strength.

Most nonferrous alloys (e.g., aluminum, copper, and magnesium) *do not* have a significant endurance limit. Their fatigue strength continues to decrease with increasing cycles. Thus, fatigue will occur regardless of the stress amplitude. The fatigue strength for these materials is taken as the stress level at which failure will occur for some specified number of cycles (e.g., $10^7$ cycles).

Determining the endurance limit experimentally is lengthy and expensive. The Manson–Coffin relationship [Eq. (7.6)] demonstrates that the fatigue life will depend on the material's fracture strength during a single load cycle, suggesting a possible relationship between static material strength and strength in fatigue. Such a relationship has been noted by several researchers (e.g., Fig. 7.8). The stress endurance limits of steel for three types of loading can be approximated as follows:

$$\begin{array}{lll}
\text{Bending} & S'_e = 0.5S_u & \\
\text{Axial} & S'_e = 0.45S_u & \text{(7.7)} \\
\text{Torsion} & S'_e = 0.29S_u &
\end{array}$$

These equations can be used to approximate the endurance limits for other ferrous alloys, but it must be recognized that the limits can differ significantly from experimentally determined endurance limits. Since the ultimate stress and the type of loading are known for various materials, their endurance limits can be approximated.

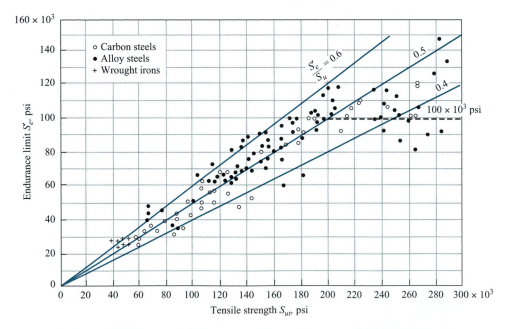

**Figure 7.8**   Endurance limit as function of ultimate strength for wrought steels. [Adapted from Shigley and Mitchell (1983).]

**Table 7.2**    Approximate endurance limit for various materials

| Material | Number of cycles | Relation |
|----------|------------------|----------|
| Magnesium alloys | $10^8$ | $S_e' = 0.35 S_u$ |
| Copper alloys | $10^8$ | $0.25 S_u < S_e' < 0.5 S_u$ |
| Nickel alloys | $10^8$ | $0.35 S_u < S_e' < 0.5 S_u$ |
| Titanium | $10^7$ | $0.45 S_u < S_e' < 0.65 S_u$ |
| Aluminum alloys | $5 \times 10^8$ | $S_e' = 0.40 S_u \ (S_u < 48 \text{ ksi})$ |
| | | $S_e' = 19 \text{ ksi} \ (S_u \geq 48 \text{ ksi})$ |

SOURCE: Adapted from Juvinall and Marshek (1991).

Other materials, for which there is much less experience, are finding increasing uses in fatigue applications. Table 7.2 gives the approximate strengths in fatigue for various material classes. Figure 7.7(c) gives some stress–life curves for common polymers. Because polymers have a much greater variation in properties than metals do, Figure 7.7(c) should be viewed as illustrative of fatigue properties and not used for quantitative data.

## 7.6    FATIGUE REGIMES

The *S–N* diagram [Fig. 7.7(a)] shows different types of behavior as the number of cycles to failure increases. Two basic regimes are **low-cycle fatigue** (generally below $10^3$ stress cycles) and **high-cycle fatigue** (more than $10^3$ but less than $10^6$ stress cycles). The slope of the line is much lower in low-cycle fatigue than in high-cycle fatigue.

Another differentiation can be made between **finite life** and **infinite life.** Figure 7.7(a) shows a clear endurance limit for ferrous alloys, below which any repeating stress will lead to infinite life in the component. Although a distinction between the finite-life and infinite-life portions of the curve is not always clear, for steels it occurs between $10^6$ and $10^7$ cycles. Thus, finite-life classification is considered for any loading below $10^7$ cycles.

### 7.6.1    LOW-CYCLE FATIGUE

Low-cycle fatigue is any loading that causes failure below approximately 1000 cycles. This type of loading is common. A number of devices, such as latches on automotive glove compartments, studs on truck wheels, and setscrews fixing gear locations on shafts, cycle fewer than 1000 times during their service lives. Surviving 1000 cycles means that these devices will last as long as intended.

For components in the low-cycle range, either designers ignore fatigue effects entirely, or they reduce the allowable stress level. Ignoring fatigue seems to be a poor approach. However, the low-cycle portion of the curve in Figure 7.7(a) has a small slope (i.e., the strength at 1000 cycles has not been reduced a great deal). Further, the *y* intercept for the curve is the ultimate strength, not the yield strength. Since static design often uses the yield strength and not the ultimate strength in defining allowable stresses, static approaches are acceptable for designing low-cycle components. In fact, the safety factor compensates for the uncertainty in material strength due to cyclic loading.

Taking low-cycle effects into account allows modifying the material strength based on experimental data. The fatigue strength for steel at which high-cycle fatigue begins can be approximated as follows:

$$
\begin{array}{lll}
\text{Bending} & S'_l = 0.9S_u & \\
\text{Axial} & S'_l = 0.75S_u & \text{(7.8)} \\
\text{Torsion} & S'_l = 0.72S_u &
\end{array}
$$

Again, the fatigue strengths of other ferrous alloys can be approximated by these equations, but they may differ from what is experienced in practice.

### 7.6.2   HIGH-CYCLE, FINITE-LIFE FATIGUE

In many applications the number of stress cycles placed on a component during its service life is between $10^3$ and $10^7$. Examples include car door hinges, aircraft body panels, and aluminum softball bats. Because the strength drops rapidly in this range (Fig. 7.7), an approach not accounting for this drop is inherently flawed.

The fatigue strength at any location between $S'_l$ and $S'_e$ can generally be expressed as

$$ \log S'_f = b_s \log N'_t + \bar{C} \tag{7.9} $$

where $b_s$ = slope
      $\bar{C}$ = intercept

At the endpoints, Equation (7.9) becomes

$$ \log S'_l = b_s \log 10^3 + \bar{C} = 3b_s + \bar{C} \tag{7.10} $$

$$ \log S'_e = b_s \log 10^6 + \bar{C} = 6b_s + \bar{C} \tag{7.11} $$

Subtracting Equation (7.11) from Equation (7.10) gives

$$ b_s = -\frac{1}{3} \log \frac{S'_l}{S'_e} \tag{7.12} $$

Substituting Equation (7.12) into Equation (7.11) gives

$$ \bar{C} = 2 \log \frac{S'_l}{S'_e} + \log S'_e = \log \frac{(S'_l)^2}{S'_e} \tag{7.13} $$

Thus, by using Equations (7.7) and (7.8) the slope $b_s$ and the intercept $\bar{C}$ can be determined for a specific type of loading. Knowing the slope and the intercept from Equation (7.9) yields the fatigue strength as

$$ S'_f = 10^{\bar{C}}(N'_t)^{b_s} \qquad \text{for } 10^3 \leq N'_t \leq 10^6 \tag{7.14} $$

If the fatigue strength is given and the number of cycles until failure is desired, then

$$ N'_t = \left(S'_f 10^{-\bar{C}}\right)^{1/b_s} \qquad \text{for } 10^3 \leq N'_t \leq 10^6 \tag{7.15} $$

Thus, by using Equations (7.7), (7.8), (7.12), and (7.13) the fatigue strength can be obtained from Equation (7.14), or the number of cycles to failure can be determined from Equation (7.15).

---

**Given:** The pressure vessel lids of nuclear power plants are bolted down to seal the high pressure exerted by the water vapor (in a boiler reactor) or the pressurized water (in a pressurized water reactor). The bolts are so heavily stressed that they are replaced after the reactors are opened 25 times. A 20% decrease in stress would give 1000 cycles of life. The ultimate strength is 1080 MPa.

**EXAMPLE 7.2**

**Find:** How low the stress has to be for a life of 10,000 cycles.

**Solution:** Equations (7.7) and (7.8) for axial loading give $S'_e = 0.45S_u$ and $S'_l = 0.75S_u$. Note that $S'_l$ is for 1000 cycles and $S'_e$ is for a life of $10^6$ cycles. Equation (7.12) gives the slope as

$$b_s = -\frac{1}{3} \log \frac{S'_l}{S'_e} = -\frac{1}{3} \log \frac{0.75S_u}{0.45S_u} = -0.07395$$

From Equation (7.13) the intercept is

$$\bar{C} = \log \frac{(S'_l)^2}{S'_e} = \log \frac{(0.75)^2(1080)^2}{(0.45)(1080)} = 3.130$$

Given the slope and intercept, Equation (7.14) gives the fatigue strength as

$$S'_f = 10^{\bar{C}}(N'_t)^{b_s} = 10^{3.13}(10,000)^{-0.07395} = 682.7 \text{ MPa}$$

Thus, the stress has to be decreased to 682.7 MPa for 10,000 cycles.

---

### 7.6.3 HIGH-CYCLE, INFINITE-LIFE FATIGUE

A number of applications call for infinite life, defined for steels as the number of cycles above which an endurance limit can be defined, usually taken as $10^6$ cycles. If a material does not have an endurance limit, it cannot be designed for infinite life. Thus, aluminum alloys, for example, will always be designed for finite life (using the approach given in Sec. 7.6.2).

For ferrous and titanium alloys, however, an infinite-life design approach can be followed. Basically, the designer determines an endurance limit and uses this strength as the allowable stress. Then sizing and selection of components can proceed just as in static design. This approach, which is fairly complex, is described next.

---

## 7.7 ENDURANCE LIMIT MODIFICATION FACTORS

Fatigue experiments assume that the best circumstances exist for promoting long fatigue lives. However, this situation cannot be guaranteed for design applications, so the component's endurance limit must be modified or reduced from the material's best-case endurance limit. The endurance limit modification factors covered in this text are for completely re-versed loading ($\sigma_m = 0$). The **modified endurance limit** can be expressed as

$$S_e = k_f k_s k_r k_t k_m S_e'$$

(7.16)

where $S_e'$ = endurance limit from experimental apparatus under idealized conditions, Pa
$k_f$ = surface finish factor
$k_s$ = size factor
$k_r$ = reliability factor
$k_t$ = temperature factor
$k_m$ = miscellaneous factor

The type of loading has been incorporated into $S_e'$, as presented in Equation (7.7). Many effects can influence the fatigue strength of a part. Some of those most frequently encountered are described in this section.

### 7.7.1   STRESS CONCENTRATION EFFECTS

The Manson–Coffin equation [Eq. (7.6)] showed that the life of a component has a direct correlation with the strain to which it is subjected. Because locations of stress concentration are also locations of strain concentration, these locations can be seen as prime candidates for the promotion of fatigue crack initiation and growth. However, the stress concentration factor developed in Chapter 6 cannot be directly applied to fatigue applications since many materials will relieve stresses near a crack tip through plastic flow. That is, because some materials flow plastically near crack tips, fracture is avoided and the crack's growth is retarded.

For *static* loading the stress concentration factor $K_c$ is used, and for *fatigue* loading the fatigue stress concentration factor $K_f$ is used, where

$$K_f = \frac{\text{endurance limit for notch-free specimen}}{\text{endurance limit for notched specimen}}$$

(7.17)

A notch or stress concentration may be a hole, fillet, or groove. Recall from Section 6.3 that the stress concentration factor is simply a function of geometry. In this section the fatigue stress concentration factor is not only a function of geometry but also a function of the material and type of loading. The consideration of the material is often dealt with by using a **notch sensitivity factor** $q_n$, defined as

$$q_n = \frac{K_f - 1}{K_c - 1}$$

(7.18)

or

$$K_f = 1 + (K_c - 1)q_n$$

(7.19)

From Equation (7.18) note that the range of $q_n$ is between 0 (when $K_f = 1$) and 1 (when $K_f = K_c$). From Equation (7.19) observe that obtaining the fatigue stress concentration factor requires knowing the material's notch sensitivity and the type of loading.

Figure 7.9 is a plot of notch sensitivity versus notch radius for some commonly used materials and for various types of loading. For all the materials considered, the notch sensitivity approaches zero as the notch radius approaches zero. Also, the harder and stronger

**Figure 7.9** Notch sensitivity as function of notch radius for several materials and types of loading. [Adapted from Sines and Waisman (1959).]

steels tend to be more notch-sensitive (have a large value of $q_n$). This is not too surprising, since notch sensitivity is a measure of material ductility and the hardest steels have limited ductility. Figure 7.9 also shows that a given steel is slightly more notch-sensitive for torsional loading than for bending and axial loading.

The endurance limit modification factor taking stress concentrations into account is

$$k_0 = \frac{1}{K_f} \tag{7.20}$$

where $K_f$ is obtained from Equation (7.19).

To apply the effects of stress concentrations, designers can either reduce the endurance limit by $k_0$ or increase the stress by $K_f$; either approach is equally valid. In this text neither factor is included in the modified endurance limit equation [Eq. (7.16)]. It is more convenient to deal with these factors separately, since in later chapters, such as that on shafting (Chap. 11), these effects are treated differently for ductile and brittle materials.

**EXAMPLE 7.3**

**Given:** The driveshaft for a Formula One racing car has a diameter of 30 mm and a half-circular notch with a 1-mm radius. The shaft was dimensioned for a coefficient of friction between the tires and the ground of 1.5 for equal shear and bending stresses. By mounting spoilers and a wing on the car, the load on the tires can be doubled at high speed without increasing the car's mass. The shaft material has an ultimate tensile strength of 965 MPa. Assume from the distortion energy theory that the equivalent stress is proportional to the square root of the bending stress squared plus 3 times the shear stress squared ($\sigma_e = \sqrt{\sigma^2 + 3\tau^2}$).

**Find:** Determine the fatigue stress concentration factors for bending and torsion of the driveshaft. Also, determine if increased acceleration or increased curve handling will give the higher risk of driveshaft failure.

**Solution:** From Figure 7.9 for a notch radius of 1 mm and ultimate strength of 965 MPa, the notch sensitivity is 0.825 for bending and 0.853 for torsion. From Figure 6.9(b) when $r/d = 1/28 = 0.0357$ and $D/d = 30/28 = 1.0714$, the stress concentration factor is 2.2 for bending, and from Figure 6.9(c) the stress concentration factor is 1.8 for torsion.

From Equation (7.19) the fatigue stress concentration factor due to bending is

$$K_f = 1 + (K_c - 1)q_n = 1 + (2.2 - 1)(0.825) = 1.99$$

The fatigue stress concentration factor due to torsion is

$$K_f = 1 + (1.8 - 1)(0.853) = 1.68$$

Let $\sigma_{e1}$ be the equivalent stress for increased curve handling and $\sigma_{e2}$ be the equivalent stress for increased acceleration. Doubling the load and using the distortion energy theory result in

$$\frac{\sigma_{e1}}{\sigma_{e2}} = \frac{\sqrt{(2\sigma)^2 + 3\tau^2}}{\sqrt{\sigma^2 + 3(2\tau)^2}} = \frac{2\sqrt{1 + 0.75(\tau/\sigma)^2}}{\sqrt{1 + 12(\tau/\sigma)^2}}$$

Recall that the shaft was dimensioned such that the shear and bending stresses are equal ($\tau = \sigma$).

$$\therefore \quad \frac{\sigma_{e1}}{\sigma_{e2}} = \frac{2\sqrt{1 + 0.75}}{\sqrt{1 + 12}} = 0.7338$$

$$\sigma_{e2} = \frac{\sigma_{e1}}{0.7338} = 1.363\sigma_{e1}$$

Therefore, increased acceleration gives a higher risk of driveshaft failure than increased curve handling.

### 7.7.2    SURFACE FINISH FACTOR

The specimen shown in Figure 7.3 has a highly polished surface finish with final polishing in the axial direction to smooth any circumferential scratches. Most machine elements do not usually have such a high-quality finish. The modification factor to incorporate the finish effect depends on the process used to generate the surface and on the ultimate strength. Once that process is known, Figure 7.10(a) enables the surface finish factor to be obtained when the ultimate strength in tension is known, or else the coefficients from Table 7.3 can

(a)

(b)

**Figure 7.10**    Surface finish factors for steel. (a) As function of ultimate strength in tension for different manufacturing processes [adapted from Shigley and Mitchell (1983)]. (b) As function of ultimate strength and surface roughness as measured with a stylus profilometer [adapted from Johnson (1967)].

be used with the equation

$$k_f = e S_{ut}^f \qquad\qquad (7.21)$$

where $k_f$ = surface finish factor

$S_{ut}$ = ultimate tensile strength of material

$e, f$ = coefficients defined in Table 7.3

**Table 7.3**    Surface finish factor

| Manufacturing process | Factor e | | Exponent $f$ |
|---|---|---|---|
| | MPa | ksi | |
| Grinding | 1.58 | 1.34 | −0.085 |
| Machining or cold drawing | 4.51 | 2.70 | −0.265 |
| Hot rolling | 57.7 | 14.4 | −0.718 |
| As forged | 272.0 | 39.9 | −0.995 |

I  SOURCE: Shigley and Mitchell (1983).

If the process used to obtain the surface finish is not known but the quality of the surface is known from the arithmetic average surface roughness $R_a$, the surface finish factor can be obtained from Figure 7.10(b).

These approaches are all approximate and are based on the processes given in the charts and used only for well-controlled processes. It is misleading to apply Table 7.3 for other circumstances or operations. For example, plasma spray operations tend to provide an extremely rough surface, but the fatigue properties are determined by the surface layer beneath the plasma-sprayed coating. Further, the data in Table 7.3 are dated. With modern numerically controlled machine tools and improvements in tooling materials, superior finishes are routinely produced today that will give slightly better performance from a fatigue standpoint.

### 7.7.3    SIZE FACTOR

The high-cycle fatigue apparatus used to obtain the endurance limit $S_e'$ was for a specific diameter, namely, 0.30 in, and uses extruded or drawn bar stock for metals. For metals such extrusions have pronounced grain elongation in the direction opposite to the fatigue crack growth. Also, the degree of cold work is high, and the likelihood of large flaws is low. Similar effects are seen for ceramics and castings but for different reasons (smaller shrinkage pores, etc.). Regardless of the manufacturing process, larger parts are more likely to contain flaws and will not demonstrate the strength of the 0.3-in-diameter shaft. The size factor for a round bar is affected by the method of loading. For bending or torsion the size factor is

$$k_s = \begin{cases} 0.869d^{-0.112} & 0.3 \text{ in} < d < 10 \text{ in} \\ 1 & d < 0.3 \text{ in or } d \leq 8 \text{ mm} \\ 1.189d^{-0.112} & 8 \text{ mm} < d \leq 250 \text{ mm} \end{cases} \tag{7.22}$$

For axial loading $k_s = 1$.

For components that are not circular in cross-section, the size factor is difficult to determine. However, because the reasons for strength reductions hold for such components, it is reasonable to define an effective diameter based on an equivalent circular cross-section. The effective diameter can be approximated by using equivalent cross-sections defined from stress distributions as proposed by Shigley and Mischke (1989). A simpler approach is merely to define the diameter from the effective circular cross-section with an area equal to that of the cross-section in question. However, both approaches are too simple. They do not differentiate between material and manufacturing processes and could cause a mistaken belief that the correction factor due to size effects can be calculated with accuracy.

### 7.7.4    RELIABILITY FACTOR

Table 7.4 shows the reliability factor for various percentages of the probability of survival, the probability of surviving to the life indicated at a particular stress. This table is based on the fatigue limit's having a standard deviation of 8%, generally the upper limit for steels. The reliability factor as obtained from Table 7.4 can be considered only as a guide because the actual law of distribution is more complicated than the simple approach used here. We will also assume that the table can be applied to materials other than steel, for which it was originally determined.

### 7.7.5    TEMPERATURE FACTOR

Many high-cycle fatigue applications take place under extremely high temperatures, such as in aircraft engines, where the material is much weaker than at room temperature. Conversely, in some applications, such as automobile axles in Alaska during January, the metal is generally less ductile than at room temperature.

In either case, the Manson–Coffin relationship [Eq. (7.6)] would suggest that the major factor affecting fatigue life is the strength $\sigma_f'$ in one loading cycle. It is therefore reasonable to follow one of two approaches. The designer can either (1) modify the ultimate strength of the material based on its properties at the temperature of interest before determining a material endurance limit in Equation (7.16) or (2) use a temperature factor:

$$k_t = \frac{S_{ut}}{S_{ut,\text{ref}}} \tag{7.23}$$

where $S_{ut}$ = ultimate tensile strength of material at desired temperature, Pa
$S_{ut,\text{ref}}$ = ultimate tensile strength at reference temperature, usually room temperature, Pa

**Table 7.4**    Reliability factors for six probabilities of survival

| Probability of survival, % | Reliability factor $k_r$ |
|:---:|:---:|
| 50 | 1.00 |
| 90 | 0.90 |
| 95 | 0.87 |
| 99 | 0.82 |
| 99.9 | 0.75 |
| 99.99 | 0.70 |

### 7.7.6   MISCELLANEOUS EFFECTS

Several other phenomena can affect a component's strength. Whereas the preceding sections have outlined methods for numerically approximating some effects, other considerations defy quantification. Among these are the following:

1. **Manufacturing history.**   Manufacturing processes play a major role in determining the fatigue life characteristics of engineering materials. This role is manifested in the size factor discussed in Section 7.7.3, but there are other effects as well. Because fatigue crack growth is more rapid along grain boundaries than through grains, any manufacturing process affecting grain size and orientation can affect fatigue. Because some forming operations, such as rolling, extrusion, and drawing, lead to elongated grains, the material's fatigue strength will vary in different directions (anisotropy). With extrusion and drawing this effect is usually beneficial, since the preferred direction of crack propagation becomes the axial direction and crack propagation through the thickness is made more difficult by grain orientation and elongation in metals. Annealing a metal component relieves residual stresses, causes grains to become equiaxed, and may cause grain growth. Relieving tensile residual stresses at a surface is generally beneficial, but equiaxed or larger grains are detrimental from a fatigue standpoint.

2. **Residual stresses.**   These can result from manufacturing processes. A **residual stress** is caused by elastic recovery after nonuniform plastic deformation across a component's thickness. Compressive residual stress in a surface retards crack growth; tensile residual stress can encourage crack growth.

   Compressive residual stresses can be obtained through shot peening and roller burnishing and may be obtained in forging, extrusion, or rolling. **Shot peening** is a cold-working process in which the surface of a part is bombarded with small spherical media called *shot*. Each impact leads to plastic deformation at the workpiece surface, leading to compressive residual stress after elastic recovery. The layer under compressive residual stress is usually less than 1 mm (0.04 in) thick, and the material bulk properties are unaffected. Crack development and propagation is severely retarded by compressive residual stresses; for this reason shot peening is a common surface treatment for fatigue-susceptible parts such as gears, springs, shafts (especially at stress concentrations), and connecting rods.

   The beneficial effect of shot peening on fatigue life can be seen in Figure 7.11. Similar behavior can be found for other materials. This is an important tool for fatigue design, because it represents one of the only strategies that *increases* the fatigue strength of materials, and this increase can be very large. For example, consider an aircraft landing gear, produced from steel with a 2068 MPa (300 ksi) strength. Figure 7.11(a) shows that shot peening can increase the fatigue strength by a factor of 3 over a polished surface. Similar benefits are possible with other materials, but as seen in the figure, the more typical fatigue strength improvement is 15% to 30%.

3. **Coatings.**   These can greatly affect fatigue. Some operations, such as carburizing, lead to a high carbon content in steel surface layers (and thus a high fracture strength) and impart a compressive residual stress on the surface. Electroplated surfaces can be extremely porous and promote crack growth, reducing fatigue strengths by as much as 50%. Zinc plating is the main exception where the fatigue strength is not seriously

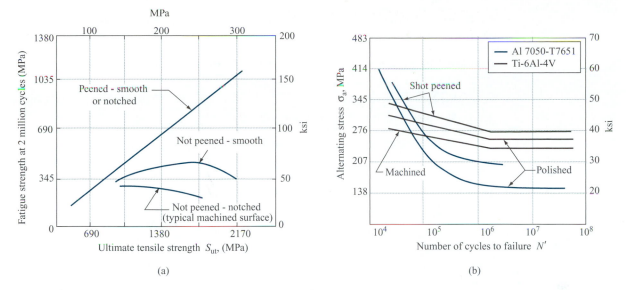

(a)

(b)

**Figure 7.11** The use of shot peening to improve fatigue properties. (a) Fatigue strength at $2 \times 10^6$ cycles for high-strength steel as a function of ultimate strength; (b) typical S–N curves for nonferrous metals. [Source: From Electronics, Inc.]

affected. Anodized oxide coatings are also usually porous, reducing fatigue strength. Coatings applied at high temperatures, such as in chemical vapor deposition processes or hot dipping, may thermally induce tensile residual stresses at the surface.

4. **Corrosion.** It is not surprising that materials operating in corrosive environments have lowered fatigue strengths. The main culprits in corroding metals are hydrogen and oxygen. Hydrogen diffuses into a material near a crack tip, aided by large tensile stresses at the tip, embrittling the material and aiding crack propagation. Oxygen causes coatings to form that are brittle or porous, aiding crack initiation and growth. High temperatures in corrosive environments speed diffusion-based processes.

**Given:** Figure 7.12 shows a tensile-loaded bar. In Figure 7.12(a) the bar is not notched, and in Figure 7.12(b) the bar is notched; but the smallest height is the same in both figures. The bar is machine-made of low-carbon steel (AISI 1020).

**EXAMPLE 7.4**

(a)

(b)

**Figure 7.12** Tensile loaded bar. (a) Unnotched; (b) notched.

**Find:** The modified endurance limits for the notched and unnotched bars as well as the fatigue strengths at $10^4$ cycles.

**Solution:** From Table A.1 for low-carbon steel (AISI 1020) the ultimate strength is

$$S_u = 57 \text{ ksi}$$

The loading is axial, so from Equations (7.7) and (7.8)

$$S'_e = 0.45 S_u = (0.45)(57)(10^3) \text{ psi} = 25.65 \text{ ksi}$$

$$S'_l = 075 S_u = (0.75)(57)(10^3) \text{ psi} = 42.75 \text{ ksi}$$

From Equations (7.12) and (7.13) the slope and intercept are, respectively,

$$b_s = -\frac{1}{3} \log \frac{S'_l}{S'_e} = -\frac{1}{3} \log \frac{42.75}{25.65} = -0.07400$$

$$\bar{C} = \log \frac{(S'_l)^2}{S'_e} = \log[(1.667)(42.75)(10^3)] = 4.853$$

From Figure 7.10(a) for machined surfaces the surface finish factor $k_f = 0.84$. For axial loading the size factor $k_s = 1$. For the unnotched bar the fatigue stress concentration factor $K_f = 1$. For the notched bar with a groove Figure 6.4 is to be used. From Figure 7.12

$$H = 1.2 \text{ in} \qquad r = 0.1 \text{ in} \qquad b = 1 \text{ in} \qquad h = 1 \text{ in}$$

$$\therefore \qquad \frac{H}{h} = \frac{1.2}{1} = 1.2 \qquad \text{and} \qquad \frac{r}{h} = \frac{0.1}{1} = 0.10$$

From Figure 6.4(a) for $H/h = 1.2$ and $r/h = 0.1$, the stress concentration factor $K_c$ is 2.35. From Figure 7.9 for $S_u = 57$ ksi and $r = 2.5$ mm, the notch sensitivity is 0.73. From Equation (7.19) the fatigue stress concentration can be expressed as

$$K_f = 1 + (K_c - 1)q_n = 1 + (2.35 - 1)(0.73) = 1.986$$

From Equation (7.16) the modified endurance limit for the notched bar is

$$\frac{S_e}{K_f} = \frac{k_f k_s S'_e}{K_f} = \frac{(0.84)(1)(25.65)(10^3)}{1.986} \text{ psi} = 10.85 \text{ ksi}$$

For the unnotched bar, the modified endurance limit is

$$\frac{S_e}{K_f} = \frac{(0.84)(1)(25.65)(10^3)}{1} \text{ psi} = 21.55 \text{ ksi}$$

Thus, even though the notched and unnotched bars have the same height, width, and length, the presence of the notch decreases the modified endurance limit by one-half.

The fatigue strength at $10^4$ cycles from Equation (7.14) is

$$S'_f = 10^{\bar{C}} (N'_t)^{b_s} = 10^{4.853}(10^4)^{-0.074} = 10^{4.557} \text{ psi} = 36.06 \text{ ksi}$$

Thus, for the notched bar

$$\frac{S_f}{K_f} = \frac{k_f k_s S_f'}{K_f} = \frac{(0.84)(1)(36.06)(10^3)}{1.986} \text{ psi} = 15.25 \text{ ksi}$$

and for the unnotched bar

$$\frac{S_f}{K_f} = \frac{(0.84)(1)(36.06)(10^3)}{1} \text{ psi} = 30.29 \text{ ksi}$$

## 7.8   CUMULATIVE DAMAGE

In constructing the solid curve in Figure 7.7(c), it was assumed that the cyclic variation was completely reversed ($\sigma_m = 0$). Furthermore, for any stress level between the strengths $S_l'$ and $S_e'$, say $S_1'$, the maximum stress level in the completely reversed variation was kept constant until failure occurred at $N_1'$ cycles. Operating at stress amplitude $S_1'$ for a number of cycles $n_1' < N_1'$ produced a smaller damage fraction. Because cyclic variations are not constant in practice, engineers must deal with several different levels of completely reversed stress cycles. Operating over stress levels between $S_l'$ and $S_e'$, say $S_i'$, at a number of cycles $n_i' < N_i'$ resulted in the damage fraction $n_i'/N_i'$. When the damage fraction due to different levels of stress exceeds unity, failure is predicted. Thus, *failure is predicted* if

$$\frac{n_1'}{N_1'} + \frac{n_2'}{N_2'} + \cdots \geq 1 \tag{7.24}$$

This formulation is frequently called the **linear damage rule** (sometimes called **Miner's rule**), since it states that the damage at any stress level is directly proportional to the number of cycles (assuming that each cycle does the same amount of damage). The rule also assumes that the stress sequence does not matter and that the rate of damage accumulation at a particular stress level is independent of the stress history. Despite these shortcomings, the linear damage rule remains popular, largely because it is so simple.

   If $N_t'$ is the total number of cycles to failure when there are different cyclic patterns (all of which are completely reversed), the ratio of the number of cycles at a specific stress level to the total number of cycles to failure is

$$\alpha_i = \frac{n_i'}{N_t'} \qquad \text{or} \qquad n_i' = \alpha_i N_t' \tag{7.25}$$

Substituting Equation (7.25) into Equation (7.24) gives that failure will occur if

$$\sum \frac{\alpha_i}{N_i'} \geq \frac{1}{N_t'} \tag{7.26}$$

**EXAMPLE 7.5**

**Given:** For the unnotched bar in Example 7.4, the fatigue stress is 25 ksi for 20% of the time, 30 ksi for 30%, 35 ksi for 40%, and 40 ksi for 10%.

**Find:** The number of cycles until cumulative failure.

**Solution:** Note that $S'_1 = 25$ ksi is less than $S'_e$, which was found to be 25.65 ksi. Therefore, $N'_1 = \infty$, implying that at this stress level, failure will not occur. From Equation (7.14) for the other three fatigue stress levels

$$N'_2 = \left(S'_f 10^{-\bar{C}}\right)^{1/b_s} = [(30,000)(10)^{-4.853}]^{-1/0.074} = 1.201 \times 10^5 \text{ cycles}$$

$$N'_3 = [(35,000)(10)^{-4.853}]^{-1/0.074} = 1.495 \times 10^4 \text{ cycles}$$

$$N'_4 = [(40,000)(10)^{-4.853}]^{-1/0.074} = 2.461 \times 10^3 \text{ cycles}$$

Making use of Equation (7.26) gives

$$\frac{\alpha_1}{N'_1} + \frac{\alpha_2}{N'_2} + \frac{\alpha_3}{N'_3} + \frac{\alpha_4}{N'_4} = \frac{1}{N'_t}$$

$$\frac{0.2}{\infty} + \frac{0.3}{1.201 \times 10^5} + \frac{0.4}{1.495 \times 10^4} + \frac{0.1}{2.461 \times 10^3} = \frac{1}{N'_t}$$

$$\frac{1}{N'_t} = 0 + 2.498 \times 10^{-6} + 2.676 \times 10^{-5} + 4.063 \times 10^{-5}$$

$$N'_t = 14,310 \text{ cycles}$$

The cumulative number of cycles to failure is 14,310.

## 7.9    INFLUENCE OF NONZERO MEAN STRESS

Other than in classifying cyclic behavior, completely reversed ($\sigma_m = 0$) stress cycles have been assumed. Many machine elements involve fluctuating stresses about a nonzero mean. The experimental apparatus used to generate the results shown in Figure 7.7 could not apply mean and alternating stresses. Because data are not available, the influence of nonzero mean stress must be estimated by using one of several empirical relationships that determine failure at a given life when alternating and mean stresses are both nonzero.

### 7.9.1    DUCTILE MATERIALS

Figure 7.13 shows how four empirical relationships estimate the influence of nonzero mean stress on fatigue life for ductile materials loaded in tension. On the ordinate, the yield strength $S_{yt}$ is plotted, reminding us that yielding rather than fatigue might be the failure criterion. On the abscissa, the yield strength in tension $S_{yt}$ and the ultimate strength in tension $S_{ut}$ are plotted. Failure is predicted for values of $\sigma_a$ and $\sigma_m$ above the curves.

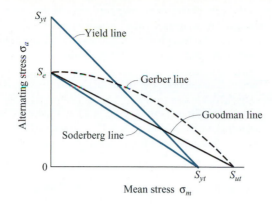

**Figure 7.13**    Influence of nonzero mean stress on fatigue life for tensile loading as estimated by four empirical relationships.

### Gerber Line

The **Gerber line** is sometimes called the Gerber parabolic relationship because the equation is

$$\frac{K_f n_s \sigma_a}{S_e} + \left(\frac{n_s \sigma_m}{S_{ut}}\right)^2 = 1 \tag{7.27}$$

where $S_e$ = modified endurance limit, Pa
  $S_{ut}$ = ultimate strength in tension, Pa
  $n_s$ = safety factor
  $\sigma_a$ = alternating stress, Pa
  $\sigma_m$ = mean stress, Pa
  $K_f$ = fatigue stress concentration factor

This line passes through the central portion of the experimental failure points and hence should be the best predictor of failure. The parabolic nature of Equation (7.27) poses problems for the practical implementation of this equation.

### Goodman Line

The **Goodman line** proposes connecting the modified endurance limit on the alternating stress axis with the ultimate strength in tension on the mean stress axis in Figure 7.13 by a straight line, or

$$\frac{K_f \sigma_a}{S_e} + \frac{\sigma_m}{S_{ut}} = \frac{1}{n_s} \tag{7.28}$$

Note the linearization of Equation (7.28) relative to Equation (7.27). Equation (7.28) fits experimental data reasonably well and is simpler to use than Equation (7.27). The starting and ending points for the Goodman and Gerber lines are the same in Figure 7.13, but between these points the Goodman line is linear and the Gerber line is parabolic.

**EXAMPLE 7.6**

**Given:** A straight, circular rotating beam with a 30-mm diameter and a 1-m length has an axial load of 30,000 N applied at the end and a stationary radial load of 400 N. The beam is cold-drawn, and the material is AISI 1040 steel. Assume that $k_s = k_r = k_t = k_m = 1$.

**Find:** The safety factor for infinite life by using the Goodman line.

**Solution:** From Table A.1 for AISI 1040 steel $S_u = 520$ MPa. From Figure 7.10(a) for the cold-drawn process and $S_u = 520$ MPa, the surface finish factor is 0.78. From Equation (7.16) the modified endurance limit while making use of Equation (7.7) is

$$S_e = k_f k_s k_r k_t k_m S_e' = (0.78)(1)(1)(1)(1)(0.45)(520) = 182.5 \text{ MPa}$$

The bending stress from Equation (4.48) gives the alternating stress that the beam experiences as

$$\sigma_a = \frac{Mc}{I} = \frac{(64)(400)(1)(0.03/2)}{\pi(0.03)^4} \text{ Pa} = 150.9 \text{ MPa}$$

The mean stress due to the axial load is

$$\sigma_m = \frac{P_a}{A} = \frac{(30,000)(4)}{\pi(0.03)^2} \text{ Pa} = 42.44 \text{ MPa}$$

For an unnotched beam $K_f = 1$. From Equation (7.28)

$$\frac{K_f \sigma_a}{S_e} + \frac{\sigma_m}{520} = \frac{1}{n_s}$$

$$\therefore \quad \frac{(1)(150.9)}{182.5} + \frac{42.44}{520} = \frac{1}{n_s}$$

$$n_s = \frac{1}{0.9084} = 1.101$$

Using the Goodman line, the safety factor for infinite life is 1.101.

### Soderberg Line

The **Soderberg line** is conservative and is given as

$$\frac{K_f \sigma_a}{S_e} + \frac{\sigma_m}{S_{yt}} = \frac{1}{n_s} \tag{7.29}$$

Note from Figure 7.13 and Equations (7.28) and (7.29) that the ultimate strength in the Goodman relationship has been replaced with the yield strength in the Soderberg relationship.

### Yield Line

To complete the possibilities, the **yield line** is given. It is used to define yielding on the first cycle, or

$$\frac{\sigma_a}{S_{yt}} + \frac{\sigma_m}{S_{yt}} = \frac{1}{n_s} \tag{7.30}$$

This completes the description of the theories presented in Figure 7.13.

## Modified Goodman Diagram

The Goodman relationship given in Equation (7.28) is modified by combining fatigue failure with failure by yielding. The complete **modified Goodman diagram** is shown in Figure 7.14. Thus, all points inside a modified Goodman diagram (ABCDEFGH) correspond to fluctuating stresses that should cause neither fatigue failure nor yielding. The word *complete* is used to indicate that the diagram is valid for both tension and compression. The word *modified* designates that the Goodman line shown in Figure 7.13 has been modified in Figure 7.14; that is, in Figure 7.13 the Goodman line extends from the endurance limit on the alternating stress ordinate to the ultimate strength on the mean stress abscissa. In the modified Goodman diagram in Figure 7.14, the Goodman line AB is modified such that for stresses larger than the yield strength the yield line BC is used. Thus, the modified Goodman diagram combines fatigue criteria as represented by the Goodman line and yield criteria as represented by the yield line. Note in Figure 7.14 that lines AB, DE, EF, and HA are Goodman lines and that lines BC, CD, FG, and GH are yield lines. The static load is represented by line CG. Table 7.5 gives the equations and range of applicability for the construction of the Goodman and yield lines of the complete modified Goodman diagram. Note that when the ultimate and yield strengths are known for a specific material, as well as the corresponding endurance limit for a particular part made of that material, the modified Goodman diagram can be constructed.

**Figure 7.14**  Complete modified Goodman diagram, plotting stress as ordinate and mean stress as abscissa.

**Table 7.5**        Equations and range of applicability for construction of complete modified Goodman diagram

| Line | Equation | Range |
|------|----------|-------|
| AB | $\sigma_{max} = \dfrac{S_e}{K_f} + \sigma_m\left(1 - \dfrac{S_e}{S_u K_f}\right)$ | $0 \le \sigma_m \le \dfrac{S_y - S_e/K_f}{1 - S_e/(K_f S_u)}$ |
| BC | $\sigma_{max} = S_y$ | $\dfrac{S_y - S_e/K_f}{1 - S_e/(K_f S_u)} \le \sigma_m \le S_y$ |
| CD | $\sigma_{min} = 2\sigma_m - S_y$ | $\dfrac{S_y - S_e/K_f}{1 - S_e/(K_f S_u)} \le \sigma_m \le S_y$ |
| DE | $\sigma_{min} = \left(1 + \dfrac{S_e}{K_f S_u}\right)\sigma_m - \dfrac{S_e}{K_f}$ | $0 \le \sigma_m \le \dfrac{S_y - S_e/K_f}{1 - S_e/(K_f S_u)}$ |
| EF | $\sigma_{min} = \sigma_m - \dfrac{S_e}{K_f}$ | $\dfrac{S_e}{K_f} - S_y \le \sigma_m \le 0$ |
| FG | $\sigma_{min} = -S_y$ | $-S_y \le \sigma_m \le \dfrac{S_e}{K_f} - S_y$ |
| GH | $\sigma_{max} = 2\sigma_m + S_y$ | $-S_y \le \sigma_m \le \dfrac{S_e}{K_f} - S_y$ |
| HA | $\sigma_{max} = \sigma_m + \dfrac{S_e}{K_f}$ | $\dfrac{S_e}{K_f} - S_y \le \sigma_m \le 0$ |

As an example of the way the modified Goodman diagram aids in visualizing the various combinations of fluctuating stress, consider the mean stress indicated by point L in Figure 7.14. The Goodman criterion indicates that this stress can fluctuate between points M and N. The nonzero mean cyclic fluctuation is sketched on the right of the figure.

Also shown in Figure 7.14 are the four regions of mean stress on the abscissa. Table 7.6 gives the failure equation for each of these regions as well as the validity limits for each equation.

**Table 7.6**        Failure equations and validity limits of equations for four regions of complete modified Goodman relationship

| Region in Figure 7.14 | Failure equation | Validity limits of equation |
|-----------------------|------------------|-----------------------------|
| a | $\sigma_{max} - 2\sigma_m = \dfrac{S_y}{n_s}$ | $-S_y \le \sigma_m \le \dfrac{S_e}{K_f} - S_y$ |
| b | $\sigma_{max} - \sigma_m = \dfrac{S_e}{n_s K_f}$ | $\dfrac{S_e}{K_f} - S_y \le \sigma_m \le 0$ |
| c | $\sigma_{max} + \sigma_m\left(\dfrac{S_e}{K_f S_u} - 1\right) = \dfrac{S_e}{n_s K_f}$ | $0 \le \sigma_m \le \dfrac{S_y - S_e/K_f}{1 - S_e/(K_f S_u)}$ |
| d | $\sigma_{max} = \dfrac{S_y}{n_s}$ | $\dfrac{S_y - S_e/K_f}{1 - S_e/(K_f S_u)} \le \sigma_m \le S_y$ |

**Given:** A steel alloy has an ultimate tensile strength of 90 ksi, a yield strength of 60 ksi, and a modified endurance limit of 30 ksi; and no stress concentration is present.

**EXAMPLE 7.7**

**Find:** Determine the numerical coordinates of the critical points on the modified Goodman diagram. Limit the results to positive mean stresses. Also indicate the nonfailure region of the diagram.

**Solution:** Line AF (Goodman line) is

$$\sigma_{max} = b_s\sigma_m + \bar{C} \qquad (a)$$

where $b_s$ = slope and $\bar{C}$ = intercept. But

$$\sigma_m = 0 \qquad \sigma_{max} = 30 \text{ ksi} \qquad \therefore \qquad \bar{C} = 30 \text{ ksi}$$

$$\therefore \qquad \sigma_{max} = b_s\sigma_m + 30 \text{ ksi} \qquad (b)$$

But $\qquad \sigma_{max} = 90 \text{ ksi} \qquad$ when $\sigma_m = 90 \text{ ksi}$

Substituting this into Equation (b) gives $b_s = \frac{2}{3}$. Therefore, line AB is given as

$$\sigma_{max} = \frac{2}{3}\sigma_m + 30 \text{ ksi} \qquad (c)$$

$$\therefore \qquad \text{Point A} = (0, 30 \text{ ksi})$$

$$\text{Point B} \qquad \sigma_{max} = 60 \text{ ksi}$$

From Equation (c)

$$60 \text{ ksi} = \frac{2}{3}\sigma_m + 30 \text{ ksi}$$

$$\therefore \qquad \sigma_m = 45 \text{ ksi}$$

$$\text{Point B} = (45 \text{ ksi}, 60 \text{ ksi})$$

$$\text{Point C} = (60 \text{ ksi}, 60 \text{ ksi})$$

Line EF (Goodman line) is

$$\sigma_{min} = b_s\sigma_m + \bar{C} \qquad (d)$$

Given that

$$\sigma_m = 0 \qquad \sigma_{min} = -30 \text{ ksi}$$

$$\therefore \qquad \bar{C} = -30 \text{ ksi} \qquad (e)$$

and $\qquad \sigma_{min} = b_s\sigma_m - 30 \text{ ksi}$

Given that

$$\sigma_m = 90 \text{ ksi} \qquad \sigma_{min} = 90 \text{ ksi}$$

$$\therefore \qquad 90 \text{ ksi} = b_s(90) \text{ ksi} - 30 \text{ ksi} \qquad (f)$$

$$\therefore \qquad b_s = \frac{4}{3}$$

Substituting Equations ($e$) and ($f$) into Equation ($d$) gives

$$\sigma_{min} = \frac{4}{3}\sigma_m - 30 \text{ ksi}$$

$$\therefore \qquad \text{Point E} = (0, -30 \text{ ksi})$$

$$\text{Point D} = (45 \text{ ksi}, 30 \text{ ksi})$$

Figure 7.15 shows a plot of these points as well as the shaded nonfailure region.

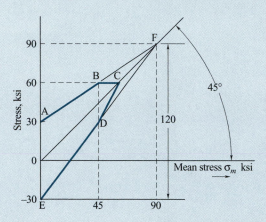

**Figure 7.15**    Modified Goodman diagram for Example 7.7.

---

**EXAMPLE 7.8**

**Given:**   For the same two bars given in Example 7.4, the load is cyclically varying between 2 and 10 ksi.

**Find:**   Using the modified Goodman relationship, determine the safety factor guarding against fatigue failure.

**Solution:**   From Table A.1 for low-carbon steel (AISI 1020)

$$S_u = 57 \text{ ksi} \qquad \text{and} \qquad S_y = 43 \text{ ksi} \tag{a}$$

From Example 7.4 the modified endurance limit is

$$\frac{S_e}{K_f} = 10.85 \text{ ksi} \qquad \text{for notched bar} \tag{b}$$

$$\frac{S_e}{K_f} = 21.55 \text{ ksi} \qquad \text{for unnotched bar} \tag{c}$$

The mean stress from Equation (7.1) is

$$\sigma_m = \frac{\sigma_{\max}}{2} = 6 \text{ ksi} \qquad\qquad \textbf{(d)}$$

Because the loading is tensile, the region in Figure 7.14 is either c or d. To establish which, evaluate as follows:

Notched bar:
$$\frac{S_y - S_e/K_f}{1 - S_e/(S_u K_f)} = \frac{(43 - 10.85)10^3}{1 - 10.85/57} \text{ psi} = 39.71 \text{ ksi}$$

$$\therefore \qquad 0 \le \sigma_m \le \frac{S_y - S_e/K_f}{1 - S_e/(S_u K_f)} \qquad \text{or} \qquad 0 \le 6 \text{ ksi} \le 39.71 \text{ ksi}$$

$$\therefore \qquad \text{Region c}$$

Unnotched bar:
$$\frac{S_y - S_e/K_f}{1 - S_e/(S_u K_f)} = \frac{(43 - 21.55)10^3}{1 - 21.55/57} \text{ psi} = 34.49 \text{ ksi}$$

$$\therefore \qquad 0 \le \sigma_m \le \frac{S_y - S_e/K_f}{1 - S_e/(S_u K_f)} \qquad \text{or} \qquad 0 \le 6 \text{ ksi} \le 34.49 \text{ ksi}$$

$$\therefore \qquad \text{Region c}$$

From Table 7.5 for region c failure will occur if

$$\sigma_{\max} - \sigma_m = \frac{S_e}{K_f}\left(\frac{1}{n_s} - \frac{\sigma_m}{S_u}\right)$$

Notched bar:
$$10 - 6 = 10.85\left(\frac{1}{n_s} - \frac{6}{57}\right)$$
$$n_s = 2.110$$

Unnotched bar:
$$10 - 6 = 21.55\left(\frac{1}{n_s} - \frac{6}{57}\right)$$
$$n_s = 3.438$$

Neither bar has a safety factor that indicates an infinite life cannot be achieved. However, if there are additional effects (such as corrosion) that apply equally to the two bars, then it is expected that the notched bar will fail before the unnotched bar.

## 7.9.2 BRITTLE MATERIALS

Figure 7.16 shows the alternating stress ratio as a function of the mean stress ratio for axially loaded cast iron. The figure is skewed since the compressive strength is typically several times greater than the tensile strength. Until recently, the use of brittle materials in a fatigue

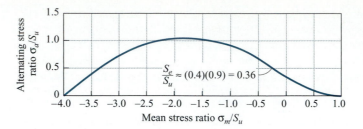

**Figure 7.16**    Alternating stress ratio as function of mean stress ratio for axially loaded cast iron.

environment has been limited to gray cast iron in compression. Now, however, carbon fibers and ceramics have had significant usage in fatigue environments.

In Figure 7.16 the dimensionalization of the alternating and mean stresses is with respect to the ultimate strength rather than to the yield strength, as done for ductile materials. Also, the compressive mean stress permits large increases in alternating stress.

For *brittle* materials a stress raiser increases the likelihood of failure under either steady or alternating stresses, and it is customary to apply a stress concentration factor to both. Thus, designers apply the fatigue stress concentration factor $K_f$ to the alternating component of stress for ductile materials but apply the stress concentration factor $K_c$ to both the alternating and mean components of stress for brittle materials.

For a single normal stress in brittle materials, the equation for the safety factor with a steady stress $\sigma_m$ and with the ultimate tensile strength $S_{ut}$ as the basis of failure is

$$n_s = \frac{S_{ut}}{K_c \sigma_m} \tag{7.31}$$

With an alternating stress $\sigma_a$ and a modified endurance limit $S_e$

$$n_s = \frac{S_e}{K_c \sigma_a} \tag{7.32}$$

For a single shear stress on a brittle component and with a *steady* shear stress $\tau_m$, the safety factor is

$$n_s = \frac{S_{ut}}{K_{cs} \tau_m (1 + S_{ut}/S_{uc})} \tag{7.33}$$

With an alternating shear stress $\tau_a$ and a modified endurance limit $S_e$,

$$n_s = \frac{S_e}{K_{cs} \tau_a (1 + S_{ut}/S_{uc})} \tag{7.34}$$

## 7.10    INFLUENCE OF MULTIAXIAL STRESS STATES

The previous sections have considered fatigue failures for uniaxial stress states. Most machine element applications encounter more complicated loading conditions. Two special cases are important. The first situation exists where the applied stresses are in phase, a situation referred to as **simple multiaxial stress.** For example, a cylindrical pressure vessel that is periodically pressurized will have a hoop and axial stress that are both directly related to the pressure, so that they are subjected to their maximum and minimum values at the same time. On the other hand, a shaft with two gears mounted on it will see a periodic variation in normal forces and torques applied by the gears; these may be caused by forcing functions in the driven machinery. Clearly, the normal forces and torques do not have to be in phase in this circumstance; this situation is called **complex multiaxial stress.**

Caution should be given that the current theoretical approaches for multiaxial stress states are even less developed than uniaxial fatigue approaches, so that experimental confirmation of designs is even more imperative.

## 7.10.1   SIMPLE MULTIAXIAL STRESS

### Fully Reversing Stresses

For *fully reversing, simple multiaxial stresses,* the mean stress is zero for all applied normal and shear stresses. For such a circumstance, experimental evidence suggests that a combination of an equivalent von Mises effective stress, defined from the alternating principal stress components, can be used in conjunction with the uniaxial fatigue failure criterion. That is, an effective stress can be defined from

$$\sigma'_e = \sqrt{\sigma^2_{a,1} + \sigma^2_{a,2} + \sigma^2_{a,3} - \sigma_{a,1}\sigma_{a,2} - \sigma_{a,2}\sigma_{a,3} - \sigma_{a,1}\sigma_{a,3}} \qquad (7.35)$$

If the stress state is two-dimensional (with $\sigma_{a,3} = 0$), this equation can be written as

$$\sigma'_e = \sqrt{\sigma^2_{a,1} + \sigma^2_{a,2} - \sigma_{a,1}\sigma_{a,2}} \qquad (7.36)$$

The safety factor can then be calculated from

$$n_s = \frac{S'_f}{\sigma'_e} \qquad \text{or} \qquad n_s = \frac{S'_e}{\sigma'_e} \qquad (7.37)$$

Here $S'_f$ is used for finite-life applications and is the fatigue strength at the desired life; $S'_e$ is used for infinite-life applications.

All fatigue strength reduction factors, including stress concentration effects, should be used in the application of Equations (7.35) to (7.37).

### Simple Multiaxial Stresses with Nonzero Mean

A number of studies have addressed the situation where the applied stresses are in phase, but the mean stress is non-zero. Two of the more common theories are referred to as the **Sines method** and the **von Mises method.** These theories use the common approach of defining an effective alternating and mean stress and then inserting these effective stresses in the modified Goodman failure criterion given in Table 7.5.

The approach of Sines (1955) uses the following equivalent stresses:

$$\sigma_a' = \sqrt{\frac{(\sigma_{a,x} - \sigma_{a,y})^2 + (\sigma_{a,y} - \sigma_{a,z})^2 + (\sigma_{a,z} - \sigma_{a,x})^2 + 6\left(\tau_{a,xy}^2 + \tau_{a,yz}^2 + \tau_{a,zx}^2\right)}{2}} \tag{7.38}$$

$$\sigma_m' = \sigma_{m,x} + \sigma_{m,y} + \sigma_{m,z} \tag{7.39}$$

For a two-dimensional stress state ($\sigma_z = \tau_{xz} = \tau_{yz} = 0$), these equations can be simplified to, respectively,

$$\sigma_a' = \sqrt{\sigma_{a,x}^2 + \sigma_{a,y}^2 - \sigma_{a,x}\sigma_{a,y} + 3\tau_{a,xy}^2} \tag{7.40}$$

$$\sigma_m' = \sigma_{m,x} + \sigma_{m,y} \tag{7.41}$$

Note that the mean component of shear stress does not appear in these equations. This is acceptable for some circumstances, but is nonconservative for situations where a stress concentration such as a notch or fillet is present. For this reason, another approach using the von Mises effective stresses defines the effective alternating and mean stresses as

$$\sigma_a' = \sqrt{\frac{(\sigma_{a,x} - \sigma_{a,y})^2 + (\sigma_{a,y} - \sigma_{a,z})^2 + (\sigma_{a,z} - \sigma_{a,x})^2 + 6(\tau_{a,xy}^2 + \tau_{a,yz}^2 + \tau_{a,zx}^2)}{2}} \tag{7.42}$$

$$\sigma_m' = \sqrt{\frac{(\sigma_{m,x} - \sigma_{m,y})^2 + (\sigma_{m,y} - \sigma_{m,z})^2 + (\sigma_{m,z} - \sigma_{m,x})^2 + 6(\tau_{m,xy}^2 + \tau_{m,yz}^2 + \tau_{m,zx}^2)}{2}} \tag{7.43}$$

or, for a two-dimensional stress state with $\sigma_{a,z} = \sigma_{m,z} = \tau_{a,xz} = \tau_{a,yz} = \tau_{m,xz} = \tau_{m,yz} = 0$,

$$\sigma_a' = \sqrt{\sigma_{a,x}^2 + \sigma_{a,y}^2 - \sigma_{a,x}\sigma_{a,y} + 3\tau_{a,xy}^2} \tag{7.44}$$

$$\sigma_m' = \sqrt{\sigma_{m,x}^2 + \sigma_{m,y}^2 - \sigma_{m,x}\sigma_{m,y} + 3\tau_{m,xy}^2} \tag{7.45}$$

One of the difficulties in applying these theories is that conflicts may arise in determining stress concentration factors and fatigue strengths depending on the loading condition selected. Recall from Equation (7.8) that $S_l'$ will vary depending on whether the loading is bending, axial, or torsional. If the loading is a combination of these loadings, it is not clear how to calculate $S_l'$. Similarly, stress concentration factors can be defined based on the loading; thus it is not obvious which value to use in Table 7.5. Recognizing that an experimental verification of a design is imperative for critical applications, it is reasonable to follow any of the following approximations:

1. Perform a worst-case scenario, using the smallest resulting strengths and largest stress concentrations that result from the loading.

2. Since mode I failure is usually most critical, calculate strengths and stress concentrations based on axial loads when they are present. If normal stresses are present due to

bending moments, calculate material strengths and stress concentration effects based on bending.

3. An experienced engineer can evaluate the applied stresses to determine which is the most likely failure mode. That is, if the torsional stresses are dominant, it is reasonable to calculate strengths and concentration factors based on torsion. However, this is a very subjective approach that all but guarantees experimental confirmation of the design.

### 7.10.2 COMPLEX MULTIAXIAL STRESSES

For complex multiaxial stress states, where the normal and shear stress maxima and minima do not occur at the same time, failure theories are not well accepted. True asynchronous situations cannot be analyzed with existing failure theories, and an experimental program may be required. A designer's goal in such circumstances is to obtain estimates of machine element dimensions for use in the experimental program. The following approaches have been applied for complex multiaxial stress states:

1. It has been shown that the fatigue strengths of some metals are not less than their strengths in a simple multiaxial stress state. Thus, it may be reasonable to approximate the stresses as synchronous and to analyze the situation as a simple multiaxial stress state, using Equations (7.38) and (7.39) or (7.42) and (7.43).

2. For situations where the loading consists of bending and torsion (such as is commonly encountered in shafts), an approach in the American Society of Mechanical Engineers Boiler Code can be used. Referred to as SEQA, it defines an effective stress given by

$$\sigma_{SEQA} = \frac{\sigma}{\sqrt{2}} \left[ 1 + 3 \left( \frac{\tau}{\sigma} \right)^2 + \sqrt{1 + 6\frac{\tau}{\sigma} \cos 2\phi + 9 \left( \frac{\tau}{\sigma} \right)^4} \right]^{1/2} \tag{7.46}$$

where $\sigma$ = bending stress amplitude including stress concentration effects
$\tau$ = torsional stress amplitude including stress concentration effects
$\phi$ = phase angle between bending and torsion

Equation (7.46) can be used to obtain both mean and alternating components of stress.

It can be shown by comparing Equations (7.46) and (7.42) to (7.43) that the von Mises approach is conservative for any phase difference or stress ratio. However, this is true only for high-cycle fatigue and can be nonconservative for finite-life applications. Since shafts are usually designed for infinite life, this is rarely a concern.

3. Behavior for a given loading and material combination may be well quantified within an organization or can be found in the technical literature.

## 7.11 FRACTURE MECHANICS APPROACH TO FATIGUE

With the increasing interest in materials without clear endurance limits, special attention must be paid to damage accumulation and replacing fatigued components before catastrophic failure can occur. Indeed, this is a main design challenge in the aircraft industry, where aluminum alloys, although they have no endurance limit, are used because of their high strength/weight ratios. Routine nondestructive evaluation to determine the size of flaws in stress-bearing members is conducted to identify and remove suspect components. This approach is called **fault-tolerant design** since it recognizes the presence of defects and allows the use of material as long as the defects remain smaller than a critical size.

Parris et al. (1961) hypothesized that crack growth in a cyclic loading should follow the rule

$$\frac{dl_c}{dN} = C(\Delta K)^m \tag{7.47}$$

where $dl_c/dN$ is the change in crack length per load cycle ($l_c$ is the crack length and $N$ is the number of stress cycles), $C$ and $m$ are empirical constants, and $\Delta K$ is the stress intensity range, defined as

$$\Delta K = K_{max} - K_{min} \tag{7.48}$$

where $K_{max}$ and $K_{min}$ are the maximum and minimum stress intensity factors, respectively, around a crack during a loading cycle. For a center-cracked plate with crack length $2l_c$

$$K_{max} = Y\sigma_{max}\sqrt{\pi l_c} \tag{7.49}$$

where $\sigma_{max}$ is the maximum far-field stress and $Y$ is a correction factor to account for finite plate sizes. Appendix C gives values of $Y$. Equation (7.47) is known as the **Paris power law** and is the most widely used equation in fracture mechanics approaches to fatigue problems. Suresh (1991) has derived the life of a component based on the Paris power law as

$$N = \frac{2}{(m-2)CY^m(\Delta\sigma)^m\pi^{m/2}}\left[\frac{1}{(l_{c0})^{(m-2)/2}} - \frac{1}{(l_{cf})^{(m-2)/2}}\right] \tag{7.50}$$

unless $m = 2$, when the fatigue life is

$$N = \frac{1}{CY^2(\Delta\sigma)^2\pi} \ln\frac{l_{cf}}{l_{c0}} \tag{7.51}$$

where $N$ = fatigue life in cycles
$C, m$ = material constants
$Y$ = correction factor to account for finite plate sizes
$\Delta\sigma$ = range of far-field stresses to which component is subjected
$l_{c0}$ = initial crack size
$l_{cf}$ = critical crack size based on fracture mechanics

---

**EXAMPLE 7.9**

**Given:** An aluminum alloy aircraft component in the form of a 100-mm-wide plate is subjected to a 100-MPa stress during pressurization of the aircraft cabin. Superimposed on this stress is a fluctuation arising from vibration, with an amplitude of 10 MPa and a frequency of 45 Hz. Nondestructive crack detection techniques do not detect any flaws, but the smallest detectable flaw is 0.2 mm.

**Find:**  Determine the minimum expected life of the component. If upon reinspection a centered 1.1-mm-long crack is found, what is the expected life from the time of inspection? Use a fracture toughness of 29 MPa$\sqrt{\text{m}}$ and a fracture stress of 260 MPa, and assume $m = 2.5$ and $C = 6.9 \times 10^{-12}$ m/cycle for $\Delta\sigma$ in megapascals.

**Solution:**  The critical crack length $l_{cf}$ is given by Equation (6.5) and is found to be 1.60 mm. A worst-case scenario would occur if the largest undetectable flaw resulting in the largest value of $Y$ were located in the geometry. This flaw occurs for the double-edge cracked tension specimen, where $Y$ equals 2.0. Therefore, the life is found from Equation (7.50) as

$$N = \frac{2}{(m-2)CY^m(\Delta\sigma)^m \pi^{m/2}} \left[ \frac{1}{(l_{c0})^{(m-2)/2}} - \frac{1}{(l_{cf})^{(m-2)/2}} \right] = 324 \text{ h}$$

If $l_{c0} = 1.1$ mm, the life until fracture is approximately 47 h.

# 7.12  LINEAR IMPACT STRESSES AND DEFORMATIONS

Thus far this chapter has focused on cyclic load variation; this section focuses on **impact loading.** Throughout most of this text, a load is applied to a body gradually such that when the load reaches a maximum, it remains constant or static. For the material covered thus far in this chapter, the loading is **dynamic** (i.e., it varies with time). When loads are rapidly applied to a body as in impact loading, the stress levels and deformations induced are often much larger than those in static or cyclic loading.

The properties of the material are a function of the loading speed; the more rapid the loading, the higher both the yield strength and the ultimate strength of the material. Figure 7.17 shows the variation of the mechanical properties with loading speed for a typical mild steel. For average strain rates from $10^{-1}$ to $10^3$ s$^{-1}$ the yield strength increases significantly.

If no energy is assumed to be lost during impact, the mechanics of impact can be studied by using conservation of energy. Consider a simple block falling a distance $h$ and striking a spring compressed a distance $\delta_{max}$ before momentarily coming to rest. If the mass of the spring is neglected and it is assumed that the spring responds elastically, conservation of energy requires that the kinetic energy be completely transformed to elastic strain energy:

$$W(h + \delta_{max}) = \frac{1}{2}(k\delta_{max})\delta_{max} \qquad (7.52)$$

where $k$ = spring constant, N/m, and $W$ = weight of block, N. Equation (7.52) can be expressed as a quadratic equation:

$$\delta_{max}^2 - \frac{2W}{k}\delta_{max} - 2\left(\frac{W}{k}\right)h = 0 \qquad (7.53)$$

Solving for $\delta_{max}$ gives

$$\delta_{max} = \frac{W}{k} + \sqrt{\left(\frac{W}{k}\right)^2 + 2\left(\frac{W}{k}\right)h} \qquad (7.54)$$

**Figure 7.17**    Mechanical properties of mild steel at room temperature as a function of average strain rate [Manjoine (1944)].

If the weight is applied statically (or gradually) to the spring, the static displacement is

$$\delta_{st} = \frac{W}{k} \qquad (7.55)$$

Substituting Equation (7.55) into Equation (7.54) gives

$$\delta_{max} = \delta_{st} + \sqrt{(\delta_{st})^2 + 2\delta_{st}h}$$

or

$$\delta_{max} = \delta_{st}\left(1 + \sqrt{1 + \frac{2h}{\delta_{st}}}\right) \qquad (7.56)$$

From the maximum displacement in Equation (7.56) the maximum force is

$$P_{max} = k\delta_{max} \qquad (7.57)$$

Recall that in dropping the block from some distance $h$ the maximum force $P_{max}$ on impact is instantaneous. The block will continue to oscillate until the motion dampens and the block assumes the static position. This analysis assumes that when the block first makes contact with the spring, the block does not rebound from the spring (does not separate from the spring). Making use of the spring constant $k$ in Equations (7.55) and (7.57) relates the static and dynamic effects.

The impact factor can be expressed as

$$I_m = \frac{\delta_{max}}{\delta_{st}} = \frac{P_{max}}{W} = 1 + \sqrt{1 + \frac{2h}{\delta_{st}}} \qquad (7.58)$$

Note that once the impact factor is known, the impact load, stresses, and deflections can be calculated. The impact stress is

$$\sigma = \frac{P_{max}}{A} \qquad (7.59)$$

where $A$ = area of surface impacting spring, m$^2$.

If, instead of the block dropping vertically, it slides with a velocity $u$ on a surface that provides little frictional resistance (so that it can be neglected) and the block impacts the spring, the block's kinetic energy $Wu^2/(2g)$ is transformed to stored energy in the spring, or

$$\frac{1}{2}\left(\frac{W}{g}\right)u^2 = \frac{1}{2}k\delta_{max}^2 \qquad (7.60)$$

where $g$ = gravitational acceleration, 9.807 m/s$^2$. Note that the right side of Equation (7.52) is identical to the right side of Equation (7.60). Solving for $\delta_{max}$ in Equation (7.60) gives

$$\delta_{max} = \sqrt{\frac{Wu^2}{gk}} \qquad (7.61)$$

By using Equation (7.55), Equation (7.61) becomes

$$\delta_{max} = \sqrt{\frac{\delta_{st}u^2}{g}} \qquad (7.62)$$

In both situations the moving body (the block) is assumed to be rigid, and the stationary body (the spring) is assumed to be deformable. The material is assumed to behave in a linear elastic manner. Thus, whether a block falls a distance $h$ and impacts on a spring, or a block moves at a velocity $u$ and impacts on a spring, the formulation for the deformation, the impact force, or the impact stress can be determined.

---

**EXAMPLE 7.10**

**Given:** A diver jumps up 2 ft on the free end of a diving board before diving into the water. Figure 7.18 shows a sketch of the diver and the dimensions of the diving board. The supported end of the diving board is fixed. The modulus of elasticity is $10.3 \times 10^6$ psi, and the yield strength is 30 ksi. The weight of the diver is 200 lb.

**Find:** The safety factor for impact loading based on yielding.

**Solution:** Equations (7.54) and (7.55) should be used to determine the maximum deflection at impact at the end of the diving board. The spring rate is not given and will need to be determined.

**Figure 7.18** Diver impacting diving board, used in Example 7.10. (a) Side view; (b) front view; (c) side view showing forces and coordinates.

The forces acting are shown in Figure 7.18(c). From Equation (5.3)

$$\frac{d^2y}{dx^2} = \frac{M}{EI} = -\frac{Px}{EI}$$

Integrating this equation gives

$$EI\frac{dy}{dx} = -\frac{Px^2}{2} + C_1$$

where $C_1 =$ integration constant. Integrating again gives

$$EIy = -\frac{Px^3}{6} + C_1 x + C_2$$

where $C_2 =$ integration constant. The boundary conditions are

1.  $x = l, y' = 0 \qquad \rightarrow C_1 = \dfrac{Pl^2}{2}$

2.  $x = l, y = 0 \qquad \rightarrow C_2 = -\dfrac{Pl^3}{3}$

The deflection at the end of the board is of interest when $x = 0$.

$$\therefore \qquad EIy = -\frac{Px^3}{6} + \frac{Pl^2 x}{2} - \frac{Pl^3}{3}$$

or

$$\frac{6EIy}{P} = -x^3 + 2l^2 x - 2l^3$$

$$\therefore \qquad \delta = y|_{x=0} = -\frac{Pl^3}{3EI}$$

The spring constant is

$$k = \frac{-P}{\delta} = \frac{3EI}{l^3}$$

where $I$ = area moment of inertia = $\dfrac{bh^3}{12} = \dfrac{18(1.5)^3}{12} = 5.063 \text{ in}^4$

$l$ = length = 5 ft = 60 in

$E$ = modulus of elasticity = $10.3 \times 10^6$ psi

$k = \dfrac{3(10.3)(10^6)(5.063)}{(60)^3} = 724.3 \text{ lb/in}$

From Equation (7.55)

$$\delta_{st} = \frac{W}{k} = \frac{200 \text{ lb}}{724.3 \text{ lb/in}} = 0.2761 \text{ in}$$

From Equation (7.58) the impact factor is

$$I_m = \frac{\delta_{max}}{\delta_{st}} = 1 + \sqrt{1 + \frac{2h}{\delta_{st}}} = 1 + \sqrt{1 + \frac{2(24)}{0.2761}} = 14.22$$

$$\therefore \quad \delta_{max} = I_m \delta_{st} = (14.22)(0.2761) = 3.927 \text{ in}$$

From Equation (7.57)

$$P_{max} = k\delta_{max} = (724.3)(3.927) = 2844 \text{ lb}$$

The maximum bending stress from Equation (4.48) is

$$\sigma_{max} = \frac{Mc}{I} = \frac{(2844)(5)(12)(0.75)}{5.063} \text{ psi} = 25.28 \text{ ksi}$$

The yield stress is given as 30 ksi. The safety factor for yielding is

$$n_s = \frac{S_y}{\sigma_{max}} = \frac{(30)(10^3)}{(25.28)(10^3)} = 1.187$$

Thus, $n_s > 1$ and thus failure should not occur; however because the safety factor is just above 1, the margin of safety is a minimum.

---

## Case Study 7.1 | POWER PRESS BRAKE STUD DESIGN REVIEW

**Given:** An engineer working for a manufacturer of mechanical power presses is assisting in the development of a new size of machine. When starting to design a new brake stud, the engineer finds the information in Figure 7.19 from an existing computer-assisted drawing (CAD) of a machine with the closest capacity to the new project. The brake stud supports the brake on the camshaft of the mechanical power press. The brake actuates with every cycle of the press, stopping the ram at the top-dead-center position so that an operator can remove a workpiece from the dies and insert another workpiece for the next cycle. If the stud fails, the press could continue to coast downward and could result in a serious injury.

Focusing on section A-A, the engineer recognizes that no fillet radius had been specified for the brake stud for machines sold previously. Conversations with machinists who routinely worked on the part led to the conclusion that the common practice was to undercut the fillet to make sure assembly was complete; the fillet radius in effect was the radius of the machine tool insert, a value as low as $\frac{1}{8}$ in. The immediate concern is whether a product recall is in

*(continued)*

**Case Study**     (Continued)

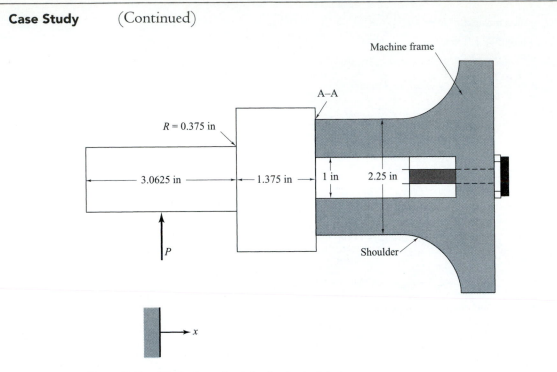

**Figure 7.19**    Dimensions of existing brake stud design.

order, since a brake stud failure could result in a machine operator's hands being in a die while the machine fails to stop after a cycle. The supervisor shares the engineer's fears but assigns one immediate task.

**Find:**   Is the machine safe as manufactured? What concerns exist for this machine in service? Is a recall necessary? The carbon steel used for the stud has a minimum ultimate strength of 74.5 ksi, and no yield strength was prescribed in the drawing.

**Solution:**

(a) *Applied stresses.* Figure 7.19 shows that the power press frame and the brake stud share the loading if the stud-retaining nut is tight. If the nut is loose, the stud can conceivably carry the entire loading. Existing brake design calculations reveal that the brake

acts over one-quarter of the camshaft revolution and has a relatively constant peak load of almost 1000 lb. Figure 7.20 shows the shear and bending moment diagrams for the stud and the loading with respect to time. The critical loading is bending (shear must also be considered, but in this case, as with most shafts, bending stresses are the critical consideration). The maximum stress in the absence of stress raisers is given by

$$\sigma_{\max} = \frac{Mc}{I} = \frac{(1000)(3.0625/2 + 1.375)(0.5)}{(\pi/4)(0.5)^4} \text{ psi}$$

$$= 29.6 \text{ ksi}$$

The loading is not completely reversing; the mean stress is 14.8 ksi, and the stress amplitude is 14.8 ksi (Fig. 7.20).

The stress raiser cannot be ignored in this case. From Figure 6.5 the theoretical stress concentration

**Case Study** (Continued)

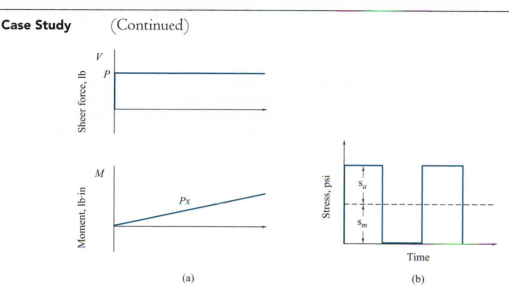

**Figure 7.20** Press brake stud loading. (a) Shear and bending-moment diagrams for applied load; (b) stress cycle.

factor is approximately 1.7. The notch sensitivity is obtained from Figure 7.9 as 0.85. Thus, from Equation (7.19)

$$K_f = 1 + (1.7 - 1)0.85 = 1.6$$

Recall that for ductile materials a stress concentration factor is not used for static loads, with the rationale that plastic flow (even at microscopic scales) will relieve stress concentrations.

The brake actuates every time the press cycles. Some mechanical presses operate at more than 100 cycles/min, but the design speed for the subject machine is 40 cycles/min. Still, even when used for only one 8-h shift per day, the number of loading cycles on the stud would be $10^7$ after 20 years. Many mechanical presses have expected lives of twice this. Clearly, the stud must be designed for infinite life.

(b) *Fatigue strength calculation.* The steel has an ultimate strength of 74.5 ksi, but no yield strength was given. Thus, the Goodman line will be used for fatigue analysis. Although this may seem conservative, if a plastic deformation in the stud were possible, it would manifest itself in trials conducted on the machines before shipping. Therefore, the design reviewer feels reasonably confident that the main failure mode is fatigue.

From Equation (7.7) the approximate value of an unmodified endurance limit is $S'_e = 0.5S_{ut} = 37.25$ ksi. The following correction factors are obtained:

- Surface finish: A machined finish so that from Equation (7.21)

$$k_f = 2.70(74.5)^{-0.265} = 0.86$$

- Size factor: From Equation (7.22)

$$k_s = 0.869(1)^{-0.112} = 0.869$$

- Reliability factor (chosen from Table 7.3 for a 99% reliability)

$$k_r = 0.82$$

The endurance limit for the steel brake stud is then given by Equation (7.16) as

$$S_e = (37.25)(0.86)(0.869)(0.82) = 22.8 \text{ ksi}$$

(c) *Fatigue criterion.* Using the Goodman line [Eq. (7.28)] gives the safety factor as

$$\frac{K_f \sigma_a}{S_e} + \frac{\sigma_m}{S_{ut}} = \frac{1.6(14.8)}{22.8} + \frac{14.8}{74.5} = \frac{1}{n_s}$$

$$\therefore \quad n_s = 0.8$$

*(continued)*

**Case Study**    (Concluded)

Thus, the design is inadequate. However, if there were no stress concentration at all, the safety factor would increase to 1.17, certainly a low safety factor regardless.

**Discussion:**    The machine manufacturer had not experienced failures in the brake stud, even though the existing design had been on the market for over three decades. Through laboratory testing it was determined that if the brake-stud-retaining bolt were loose, the brake would

make a noticeably loud squeal and the stud would clank. Undoubtedly, such problems were promptly identified by maintenance personnel, and the stud bolt was retightened whenever it became loose in service. As long as the brake stud was snug against the machine support, or stud shoulder, the safety factor for infinite life was high. It was ultimately determined that a recall was not in order, but a service memorandum was issued to all previous customers informing them of the need to properly maintain the tightness of the stud-retaining bolt. Subsequent service manuals also mentioned this important consideration.

## 7.13    SUMMARY

Failures in components or structures are often caused by fluctuating stresses. If the stress variation sequence repeats itself, it is called cyclic. The various cyclic patterns were described. The fatigue strength versus the logarithm of the number of cycles to failure was presented for materials with and without an endurance limit. Ferrous materials tend to have endurance limits, and nonferrous materials tend not to have endurance limits. The endurance limit for ferrous materials is a function of the type of loading the component is subjected to. Also, *low*-cycle fatigue failure has been classified as occurring at less than $10^3$ cycles and *high*-cycle fatigue failure as occurring above $10^3$ cycles.

The endurance limit has been experimentally determined for a specimen of specific size with a mirrorlike surface finish. The specimen was precisely prepared and tested under controlled conditions. In practice, conditions differ significantly from those in a test situation. To more accurately characterize the conditions that prevail in practice, the modified endurance limit is used. Correction factors for the endurance limit include effects of surface finish, size, stress concentration, and other common effects.

Various approaches for estimating when failure will occur under nonzero mean cyclic stresses were also considered. The Soderberg, Gerber, Goodman, and modified Goodman theories were presented. These theories allow the variation of mean and alternating stresses. The construction of a complete modified Goodman diagram was shown, and fatigue and yield criteria were used. Failure equations were given for specific regions of the modified Goodman diagram.

Most of the results presented were for ductile materials, but the behavior of brittle materials was briefly described. The major difference between brittle and ductile materials is that the compressive and tensile strengths are nearly identical for ductile materials whereas for brittle materials the compressive strength is several times greater than the tensile strength.

Impact loading has to be considered when loads are rapidly applied. The stress levels and deformations induced are much larger than with static or cyclic loading. The more rapid the loading, the higher the yield and ultimate strengths of a material. By equating the kinetic and elastic strain energies, two types of impact loading were considered: a block falling from a given height onto a spring and a block sliding downward into a spring. An impact factor, maximum and static deformations, and maximum and static loads were also obtained.

## KEY WORDS

**amplitude ratio**   ratio of stress amplitude to mean stress.

**complex multiaxial stress**   multiaxial fatigue loading where the forces and torques do not need to be in phase.

**cyclic stress**   stress sequence that repeats over time.

**dynamic**   adjective indicating variation with time.

**endurance limit**   stress level below which infinite life can be realized.

**fatigue**   failure, at relatively low stress levels, of structures that are subject to fluctuating and cyclic stresses.

**fault tolerant design**   design that allows the presence of defects and the use of materials as long as the defects remain smaller than a critical size.

**finite life**   life until failure due to fatigue.

**Gerber line**   parabolic relationship taking mean and alternating stresses into account.

**Goodman line**   theory connecting modified endurance limit and ultimate strength on plot of alternating stress versus mean stress.

**high-cycle fatigue**   fatigue failure that occurs above $10^3$ cycles but below $10^6$ cycles.

**impact loading**   load rapidly applied to body.

**infinite life**   stress levels that do not cause fatigue failure.

**linear damage rule**   theory that damage at any stress level is proportional to number of cycles.

**low-cycle fatigue**   fatigue failure that occurs below $10^3$ cycles.

**Manson–Coffin relationship**   theoretical approach to fatigue based on strain.

**mean stress**   average of minimum and maximum stresses in cycle.

**Miner's rule**   same as **linear damage rule.**

**modified endurance limit**   corrections for endurance limit based on surface finish, material, specimen size, loading type, temperature, etc.

**modified Goodman diagram**   diagram that defines all stress states not resulting in fatigue failure or yielding.

**notch sensitivity factor**   measure of material property that reflects ability of ductile materials to be less susceptible to stress raisers in fatigue.

**Paris power law**   postulate that crack growth in cyclic loading follows the power law.

**residual stress**   internal stress usually caused by manufacturing process.

**shot peening**   cold working process to surfaces that severely retards crack development and propagation.

**simple multiaxial stress**   multiaxial fatigue loading where the forces and torques are in phase.

**Sines method**   non-conservative method used for simple multiaxial stresses with non-zero mean.

**S–N diagram**   plot of stress level versus number of cycles before failure.

**Soderberg line**   theory connecting modified endurance limit and yield strength on plot of alternating stress versus mean stress.

**stress amplitude**   one-half of stress range.

**stress range**   difference between maximum and minimum stresses in cycle.

**stress ratio**   ratio of minimum to maximum stresses.

**von Mises method**   conservative method used for simple multiaxial stresses with non-zero mean.

**Wöhler diagram**   same as *S–N diagram.*

**yield line**   failure criterion that postulates yielding on first cycle of cyclic loading with nonzero mean.

## RECOMMENDED READINGS

Juvinall, R. C. (1967) *Engineering Considerations of Stress, Strain, and Strengths*, McGraw-Hill, New York.

Madayag, A. F. (1969) *Metal Fatigue: Theory and Design*, Wiley, New York.

Rice, R. C., ed. (1988) *Fatigue Design Handbook,* Society of Automotive Engineers, Warrendale, PA.

Suresh, S. (1998) *Fatigue of Materials,* 2nd ed., Cambridge University Press, Cambridge, United Kingdom.

Zahavi, E., and Torbilo, V. (1996) *Fatigue Design*, CRC Press, Boca Raton, FL.

## REFERENCES

American Society of Materials International, (1975), Metals Handbook, vol. 11: Failure Analysis and Prevention, Metals Park, Ohio.

Dowling, N. E. (1993) *Mechanical Behavior of Materials,* Prentice-Hall, Englewood Cliffs, NJ.

Johnson, R. C. (1967) Predicting Part Failures, Part I, *Machine Design,* vol. 45, p. 108.

Juvinall, R. C. (1967) *Engineering Considerations of Stress, Strain, and Strength*, McGraw-Hill, New York.

Juvinall, R. C., and Marshek, K. M. (1991) *Fundamentals of Machine Component Design*, Wiley, New York.

Landgraf, R. W. (1968) Cyclic Deformation and Fatigue Behavior of Hardened Steels, Report no. 330, Department of Theoretical and Applied Mechanics, University of Illinois, Urbana.

Lipson, C., and Juvinall, R. C. (1963) *Handbook of Stress and Strength*, Macmillan, New York.

Manjoine, M. J. (1944) "Influence of Rate of Strain and Temperature on Yield Stresses of Mild Steel," *Journal of Applied Mechanics*, vol. 66, pp. A-211 to A-218.

Mitchell, M. (1978) "Fundamentals of Modern Fatigue Analysis for Design," *Fatigue and Microstructure*, M. Meschii, ed., American Society for Metals, Metals Park, OH, pp. 385–437.

Norton, R. L. (1996) *Machine Design*, Prentice-Hall, Englewood Cliffs, NJ.

Parris, P. C., Gomez, M. P., and Anderson, W. P. (1961) "A Rational Analytic Theory of Fatigue," *Trend Engineering,* University of Washington, vol. 13, no. 1, pp. 9–14.

Petroski, H. (1992) *To Engineer Is Human*, Cambridge University Press, Cambridge, United Kingdom.

Shigley, J. E., and Mischke, C. R. (1989) *Mechanical Engineering Design*, 5th ed., McGraw-Hill, New York.

Shigley, J. E., and Mitchell, L. D. (1983) *Mechanical Engineering Design*, 4th ed., McGraw-Hill, New York.

Sines (1959), Behavior of Metals under Complex Static and Alternating Stresses. In Sines, G., and Waisman, J. L. eds. Metal Fatigue, McGraw-Hill, New York.

Sines, G., and Waisman, J. L. (1959) *Metal Fatigue,* McGraw-Hill, New York.

Suresh, S. (1998) *Fatigue of Materials*, 2nd ed., Cambridge University Press, Cambridge, United Kingdom.

## PROBLEMS

### Section 7.3

**7.1** A tuning fork is hit with a pencil and starts to vibrate with a frequency of 440 Hz. The maximum bending stress in the tuning fork is 2 MPa at the end positions. Calculate the mean stress, range of stress, stress amplitude, stress ratio, and amplitude ratio. Also calculate how much stress the tuning fork can sustain without being plastically deformed if it is made of AISI 1080 steel. *Ans.* $\sigma_m = 0, \sigma_r = 4$ MPa, $R_s = -1$.

### Section 7.5

**7.2** The jack for a Volvo consists of a mechanism in which the lift screw extends horizontally through two corners of the mechanism while the other two corners apply a force between the ground and the car to be lifted. The maximum compressive stress in the jack is 190 MPa when the car is jacked up so high that both wheels on one side of the car are in the air and the load on the jack is 8000 N. How many times can the jack be used for a small truck that weighs 6 tons and loads the jack to 17,000 N before it fails from fatigue? The jack material is AISI 1080 steel. *Ans.* 6050 cycles.

### Section 7.6

**7.3** The shaft shown in sketch *a* rotates at high speed while the imposed loads remain static. The shaft is machined from ground, high-carbon steel (AISI 1080). If the loading were sufficiently large to produce a fatigue failure after 1 million cycles, where would the failure most likely occur? Show all necessary computations and reasoning.

Sketch *a*, for Problem 7.3

## Section 7.7

**7.4**   For each of the AISI 1040 cold-drawn steel bars shown in sketches *b* and *c*, determine

   *a)*     The static tensile load causing fracture
   *b)*     The alternating (completely reversing) axial load $\pm P$ that would be just on the verge of producing eventual fatigue failure

Sketches *b* and *c*, for Problem 7.4

★ **7.5**   A stepped shaft, as shown in sketch *d*, was machined from high-carbon steel (AISI 1080). The loading is one of completely reversed torsion. During a typical 30 s of operation under overload conditions, the nominal stress in the 1-in-diameter section was calculated to be as shown in sketch *e*. Estimate the life of the shaft when it is operating continually under these conditions. *Ans.* 4.79 h.

Sketch *d*, for Problem 7.5

*Indicates problem of greater difficulty.

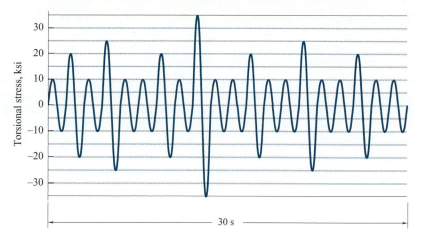

Sketch e, for Problem 7.5

**7.6** A flood protection dam gate is supposed to operate only once per week for 100 years, but after 30 years of use it needs to be operated twice per day (each time the high tide comes in). Find how much lower the bending stress must be from then on to still give a total life of 100 years. The material being fatigued is medium-carbon steel (AISI 1040). *Ans.* 20% lower stress.

★ **7.7** A hydraulic cylinder has a piston diameter $D = 100$ mm and a piston rod diameter $d = 33$ mm. The hydraulic pressure alternately applied to each side of the piston is $p = 300$ bars. This pressure induces in the rod a compressive force of $\pi p D^2 / 4$ or a tensile force of $\pi p (D^2 - d^2)/4$, depending on which side of the piston the pressure is applied. The piston rod material is martensitic stainless steel (AISI 410). Find the endurance life of the piston rod. *Ans.* 87,290 cycles.

**7.8** A 20-mm-diameter shaft transmits a variable torque of $500 \pm 400$ N·m. The frequency of the torque variation is $0.1$ s$^{-1}$. The shaft is made of high-carbon steel (AISI 1080). Find the endurance life of the shaft. *Ans.* 62,900 cycles, or 174 h.

**7.9** For the shaft in Problem 7.8 determine how large the shaft diameter has to be for infinite life. *Ans.* $d = 22.5$ mm.

**7.10** A notched bar has diameter $D = 25$ mm, and notch radius $r = 0.5$ mm, and the bottom diameter of the notch is 24 mm. It experiences rotational bending fatigue. Determine which of the steels AISI 1020, 1040, 1080, or 316 will give the highest allowable bending moment for infinite life. Calculate that moment.

★ **7.11** A straight, circular rotating beam has a diameter of 30 mm and a length of 1 m. At the end of the beam, a stationary load of 600 N is applied perpendicular to the beam, giving a bending deformation. Find which surface machining method can be used to give infinite life to the rotating beam if it is made of

*a)* AISI 1020 steel
*b)* AISI 1080 steel

Note that $k_s = k_r = k_m = 1$.

**7.12** The pedals on a bicycle are screwed into the crank with opposite threads, one left-hand thread and one right-hand thread, to ensure that the pedals do not accidentally unscrew. Just outside the thread is a 0.75-mm-radius fillet connecting the 12.5-mm-diameter threaded part with an

11-mm-diameter central shaft in the pedal. The shaft is made of AISI 4340 alloy steel tempered at a low temperature to give an ultimate stress of 2 GPa. Find the fatigue stress concentration factor for the fillet, and calculate the maximum allowable pedal force for infinite life if the force from the foot is applied 70 mm from the fillet. *Ans.* 1090 N.

**7.13** During the development of a new submarine a 1-to-20 model was used. The model was tested to find if there was any risk of fatigue failure in the full-size submarine. To be on the safe side, the stresses in the model were kept 25% higher than the stresses in the full-size submarine. Is it possible to conclude that the full-size submarine will be safe if the model was safe at a 25% higher stress level?

★ **7.14** During the development of a new car it was found that the power from the motor was too high for the gearbox. When maximum torque was transmitted through the gearbox, 50% of the gearboxes failed within the required lifetime of 800 h. How much stronger should the gearbox be made to ensure that only one gearbox out of 1000 would fail within 800 h?

**7.15** A tension member in service has been inspected, and a fatigue crack has been discovered, as shown in sketch *f*. A proposed solution is to drill a hole at the tip of the crack, the intent being to reduce the stress concentration at the crack tip. The material is AISI 1040 medium-carbon steel and was produced through forging. The load is a completely reversing 3500 lb, and the original design is as shown.

    *a)*    What was the original factor of safety?

    *b)*    What is the smallest drilled hole that restores the safety factor to that of the original design? (Use the nearest $\frac{1}{32}$-in increment that is satisfactory.) Use a reliability of 90%.

Sketch *f*, for Problem 7.15

## Section 7.8

**7.16** Truck gearboxes are dimensioned for infinite life at maximum torque regarding contact stresses and bending stresses for all gear steps. Car gearboxes are dimensioned for finite life according to typical running conditions. Maximum torque for the first gear can typically be maintained only 3 to 6 s for each acceleration before the maximum speed is reached. If a driver accelerates at full power 20 times per day for 20 years on the first gear, the required life is only 60 to 120 h. The normal load spectrum gives a life of 200,000 km for 99% of the gearboxes. A driver uses the car twice a year to move his 10-ton boat 50 km (the distance between his home and the harbor) during the 10-year life of the car. Calculate how much of the 200,000-km nominal life of the gearbox is consumed by the boat moving if the life is inversely proportional to the load raised to the 3.5 power. Assume that during the move the gearbox load is 4 times higher than normal.

**7.17** The hand brakes of a bicycle are often the same type for the front and rear wheels. The high center of gravity (compared with the distance between the wheels) will increase the contact force between the front wheel and the ground and will decrease the contact force between the rear wheel and the ground when the brakes are applied. If equal force is applied to each of the two brakes, the rear wheel will start sliding while the front wheel is still rolling. The manufacturer of the brake wires did not know about this difference between the front and rear wheels, so the wires were dimensioned as if the wheels had equal contact force and thus needed equal force in the two brake wires. The wires were originally dimensioned to withstand 20 years' use with brake applications at a force necessary to lock the two (equally loaded) wheels. How long will the life of the wires be if the friction between the front wheel and the ground is just enough to lift the real wheel from the ground? Assume the endurance life is proportional to the stress raised to the $-10$ power.

★ **7.18** Ball bearings often run at varying loads and speeds. Bearing life is inversely proportional to the contact stress raised to the $31/3$ power for a number of bearing types. For ball bearings the contact stresses are approximately proportional to the third root of the load. A ball bearing in a gearbox is dimensioned to have a life of 1752 million revolutions, one-half of which are at one-half the maximum motor torque, one-quarter at full motor torque, and one-quarter at 75% of full motor torque. Calculate the bearing life if the motor torque is kept constant at the maximum level. *Ans.* 681 million revolutions.

**7.19** A round cold-drawn steel bar with a solid cross-section is subjected to a cyclic force that ranges from a maximum of 80,000 lb in tension to a minimum of 30,000 lb in compression. The ultimate strength of the steel is 158,000 psi, and the yield strength is 133,000 psi. The critical safety factor is 2. Determine the following:

a) The modified endurance limit. *Ans.* 50.20 ksi.
b) The cross-sectional area that will produce fatigue failure and the corresponding diameter *Ans.* $d=1.79$ in, $A = 2.51$ in$^2$.
c) The region (a, b, c, or d) in Figure 7.14, assuming modified Goodman criteria, and why.

★ **7.20** Both bars used in Problem 7.4 and shown in sketches *b* and *c* are made from cold-drawn, medium-carbon steel (AISI 1040). Determine the safety factor for each bar. The load varies from 10 to 40 kN. Use the modified Goodman relationship. *Ans.* $n_{s,\text{notched}} = 2.8$.

★ **7.21** The $\frac{7}{16}$-in-thick component in sketch *g* is designed with a fillet and a hole. The load varies from 12,000 to 2000 lb. The following strengths are given: $S_u = 56$ ksi and $S_y = 41$ ksi. Using the modified Goodman failure theory, determine the safety factor for the hole as well as for the fillet. At which location will failure first occur? Will the component fail? Explain.

Sketch g, for Problem 7.21

★ **7.22** The flat bar shown in sketch *h* is made of cold-drawn, high-carbon steel (AISI 1080). The cyclic, nonzero, mean axial load varies from a minimum of 2 kN to a maximum of 10 kN. Using the Goodman failure theory, determine the safety factors for the hole, fillet, and groove. Also, indicate where the flat bar will first fail. *Ans.* $n_{s,\text{hole}} = 2.97$, $n_{s,\text{fillet}} = 3.25$.

Sketch *h*, for Problem 7.22

**7.23**  A straight, circular rotating beam with a diameter of 30 mm and a length of 1 m has an axial load of 3000 N applied at the end and a stationary radial load of 400 N. The beam is cold-drawn, and the material is titanium. Note that $k_s = k_r = k_m = 1$. Find the safety factor $n_s$ for infinite life by using the Goodman line.

**7.24**  The cold-drawn AISI 1040 steel bar shown in sketch *i* is subjected to a tensile load fluctuating between 800 and 3000 lb. Estimate the factor of safety, using the Goodman criterion.

Sketch *i*, for Problem 7.24

**7.25**  The cantilever shown in sketch *j* carries a downward load $F$ that varies from 300 to 700 lb.

*a)*    Compute the resulting safety factor for static and fatigue failure if the bar is made from AISI 1040 steel.

*b)*    What fillet radius is needed for a fatigue failure safety factor of 3.0 (use a constant notch sensitivity)?

Sketch *j*, for Problem 7.25

**7.26** The $\frac{7}{16}$-in-thick component in Problem 7.21 is constructed from AISI 1020 steel, but is used at elevated temperature so that its yield strength is 70% of its room-temperature value, and its ultimate strength is 60% of its room-temperature value.

    *a)*    If the desired reliability is 90%, what completely reversing load can be supported without failure?

    *b)*    If the load varies from 1000 to 2000 lb, what is the safety factor guarding against fatigue failure? Use the Goodman failure criterion.

★ **7.27** The stepped rod shown in sketch *k* is subjected to a tensile force that varies between 5000 and 8000 lb. The rod has a machined surface finish everywhere except in the shoulder area, where a grinding operation has been performed to improve the fatigue resistance of the rod. Using a 99% probability of survival, determine the safety factor for infinite life if the rod is made of AISI 1080 steel. Use the Goodman line. Does the part fail at the fillet? Explain.

Sketch *k*, for Problem 7.27

## Section 7.12

★ **7.28** A 10,000-lb elevator is supported by a stranded steel cable having a 2.5-in$^2$ cross-section and an effective modulus of elasticity of $12 \times 10^6$ psi. As the elevator is descending at a constant velocity of 400 ft/min, an accident causes the cable, 60 ft above the elevator, to suddenly stop. Determine the static elongation of the cable, impact factor, maximum elongation, and maximum tensile stress developed in the cable.

**7.29** A person is planning a bungee jump from a 40-m-high bridge. Under the bridge is a river with crocodiles, so the person does not want to be submerged in the water. The rubber rope fastened to the ankles has a spring constant of 3600 N, divided by the length of the rope. The distance from the ankles to the top of the head is 1.75 m, and the person has a mass of 80 kg. Calculate how long the rope should be. *Ans.* 21.29 m.

**7.30** A toy with a bouncing 50-mm-diameter steel ball has a compression spring with a spring constant of 100,000 N/m. The ball falls from a 3-m height down onto the spring (which can be assumed to be weightless) and bounces away and lands in a hole. Calculate the maximum force on the spring and the maximum deflection during the impact. The steel ball density is 7840 kg/m$^3$. *Ans.* 1743 N.

**7.31** At a building site a 1-ton container hangs from a crane wire and is then placed on the floor so that the wire becomes unloaded. The container is pushed to the elevator shaft where it is to be lowered as shown in sketch *l*. By mistake there is a 1-m slack in the wire from the crane when the container falls into the elevator shaft. Calculate the maximum force in the wire if it has a cross-sectional steel surface of 500 mm$^2$ and an effective modulus of elasticity of 70 GPa and is 25 m long from the crane to the container. *Ans.* 175.8 kN.

25 m

500 mm²

Elevator shaft

Sketch I, for Problem 7.31

★ **7.32** Modern kitchen drawers have small rubber springs mounted onto the sides of the inside of the front plate to take up the force and stop the drawer when it is being closed. The spring constant for each of the two rubber springs is 400 kN/m. The drawer is full of cutlery, which weighs 5 kg, and is closed with a speed of 0.5 m/s. Calculate the maximum force in each rubber spring if the drawer itself weighs 1 kg and

*a)*     The cutlery is in a container that is fixed to the drawer so that it moves with the drawer.

*b)*     The cutlery is in a plastic container that can slide 80 mm with a coefficient of friction of 0.25 inside the drawer.

**7.33** Car doors are easy to slam shut but difficult to press shut by hand force. The door lock has two latches, the first easily engaged and the second requiring the rubber seal around the door to be quite compressed before it can engage. The rubber seal has a spring constant of 50,000 N/m at the locked position for the door, and the mass moment of inertia for the door around its hinges is 2.5 kg · m². The distance between the lock and the hinges is 1 m. Calculate the force needed to press the car door shut at the lock if a speed of 0.8 m/s at the lock slams it shut. *Ans.* 282.8 N.

# LUBRICATION, FRICTION, AND WEAR

Greases are a necessary lubricant for many applications, including rolling element bearings, for the reduction of friction and wear. (Courtesy of SKF.)

" . . . among all those who have written on the subject
of moving forces, probably not a single one has given
sufficient attention to the effect of friction in machines . . . "

Guillaume Amontons (1699)

## SYMBOLS

| | |
|---|---|
| $A$ | area, m$^2$ |
| $A_o$ | sum of projected areas in sliding, m$^2$ |
| $A_r$ | sum of real areas in contact, m$^2$ |
| $b^*$ | contact semiwidth, m |
| $C_1, C_2$ | constants used in Equation (8.33) |
| $D_x$ | diameter of contact ellipse along $x$ axis, m |
| $D_y$ | diameter of contact ellipse along $y$ axis, m |
| $d$ | diameter, m |
| $E$ | modulus of elasticity, Pa |
| $E'$ | effective elastic modulus, Pa |
| $\mathcal{E}$ | complete elliptic integral of second kind |
| $F$ | friction force, N |
| $\mathcal{F}$ | complete elliptic integral of first kind |
| $H$ | hardness of softer material, Pa |
| $h$ | film thickness, m |
| $h_{\min}$ | minimum film thickness, m |
| $k_e$ | ellipticity parameter, $D_y/D_x$ |
| $k_1$ | adhesive wear constant |
| $k_2$ | abrasive wear constant |
| $L$ | sliding distance, m |
| $l$ | length, m |
| $m_g$ | gravitational mass, kg |
| $m_p$ | mass of person, kg |
| $N$ | number of measurements of $z_i$ |
| $P_1, P_2$ | forces, N |
| $p$ | pressure, Pa |
| $p_a$ | ambient pressure, Pa |
| $p_H$ | maximum Hertzian contact pressure, Pa |
| $p_{\max}$ | maximum pressure, Pa |
| $p_s$ | supply pressure, Pa |
| $R$ | radius, m; curvature sum, m |
| $R_a$ | arithmetic average surface roughness, m |
| $R_q$ | root-mean-square surface roughness, m |
| $R_d$ | curvature difference |

| | |
|---|---|
| $r$ | radius, m |
| $S_y$ | yield stress in tension, Pa |
| $s$ | shear strain rate, s$^{-1}$ |
| $t_m$ | temperature, °C |
| $u$ | velocity, m/s |
| $u_b$ | velocity of upper surface, m/s |
| $v$ | wear volume, m$^3$ |
| $v_{abr}$ | abrasive wear volume, m$^3$ |
| $W$ | normal load, N |
| $W'$ | dimensionless load for rectangular contact |
| $w'$ | load per unit length, N/m |
| $w_a$ | squeeze velocity, m/s |
| $x, y, z$ | Cartesian coordinate system, m |
| $z$ | coordinate in direction of film, m |
| $z_i$ | height from reference line, m |
| $\alpha_r$ | radius ratio, $R_y/R_x$ |
| $\delta_{\max}$ | maximum deformation, m |
| $\eta$ | absolute viscosity, Pa · s |
| $\eta_k$ | kinematic viscosity, m$^2$/s |
| $\eta_0$ | absolute viscosity at $p = 0$ and at constant temperature, Pa · s |
| $\theta$ | cone angle, rad |
| $\Lambda$ | dimensionless film parameter |
| $\mu$ | coefficient of sliding friction |
| $\mu_d$ | coefficient of dynamic friction |
| $\mu_r$ | coefficient of rolling friction |
| $\nu$ | Poisson's ratio |
| $\xi$ | pressure-viscosity coefficient, m$^2$/N |
| $\rho$ | density, kg/m$^3$ |
| $\sigma$ | normal stress, Pa |
| $\tau$ | shear stress, Pa |

## SUBSCRIPTS

| | |
|---|---|
| $a$ | solid $a$ |
| $b$ | solid $b$ |
| $x, y, z$ | Cartesian coordinate system |

## 8.1   INTRODUCTION

**Tribology** is generally defined as the study of the lubrication, friction, and wear, and it plays a significant role in machine element life. Not only must the design stress be less than the allowable stress and must the deformation not exceed some maximum value, but also tribological considerations must be properly understood for machine elements to be successfully designed. The interaction of surfaces in relative motion considered in tribology should not be regarded as a special subject; like the strength of materials, tribology is basic to every engineering design of machine elements. There are few machine elements that do not depend on tribological considerations.

Fluid film lubrication occurs when opposing surfaces are completely separated by a lubricant film and no asperities are in contact. The applied load is carried by pressure generated within the fluid, and frictional resistance to motion arises entirely from the shearing of the viscous fluid.

This chapter presents the lubrication, friction, and wear considerations that are important in the successful design of machine elements. It includes the concept of conformal and nonconformal surfaces, or geometric conformity, which is used throughout the text. This chapter also considers the various lubrication regimes, surface and film parameters, lubricant viscosity, deformation due to concentrated loading, friction, and wear.

## 8.2 CONFORMAL AND NONCONFORMAL SURFACES

**Conformal surfaces** fit snugly into each other with a high degree of geometric conformity so that the load is carried over a relatively large area; for example, the lubrication area of a journal bearing would be $2\pi$ times the radius times the length. The load-carrying surface area remains essentially constant even if the load is increased. Fluid film journal bearings (Fig. 8.1) and slider bearings have conformal surfaces. In journal bearings, the radial clearance between the journal and the sleeve is typically one-thousandth of the journal diameter; in slider bearings, the inclination of the bearing surface to the runner is typically one part in a thousand.

Many fluid-film-lubricated machine elements have surfaces that do not conform to each other well. The full burden of the load must then be carried by a small area. The lubrication area of a nonconformal conjunction is typically three orders of magnitude less than that of a conformal conjunction. In general, the lubrication area between **nonconformal surfaces** enlarges considerably with increasing load, but it is still smaller than the lubrication area between conformal surfaces. Some examples of nonconformal surfaces are mating gear teeth, cams and followers, and rolling-element bearings (Fig. 8.2).

**Figure 8.1** Conformal surfaces. [From Hamrock and Anderson (1983).]

**Figure 8.2** Nonconformal surfaces. [From Hamrock and Anderson (1983).]

## 8.3   CONCENTRATED LOADING; DEFORMATIONS AND STRESSES

The text up to this point has focused on distributed loading, with the exception of Section 2.3 which pointed out the differences between distributed and concentrated loading. This section describes not only the deformation due to concentrated loading but also the surface and subsurface stresses. The focus is mostly on elliptical contacts, but at the end of this section rectangular contact situations are also presented.

### 8.3.1   ELLIPTICAL CONTACTS

The undeformed geometry of nonconformal contacting solids can be represented in general terms by two ellipsoids, as shown in Figure 8.3. Two solids with different radii of curvature in a pair of principal planes ($x$ and $y$) passing through the conjunction between the solids make contact at a single point under the condition of zero applied load. Such a condition is called *point contact* and is shown in Figure 8.3, where the radii of curvature are denoted by $r$'s. It is assumed throughout this book that convex surfaces such as those in Figure 8.3 exhibit positive curvature, and concave surfaces exhibit negative curvature. Therefore, if the center of curvature lies within the solid, the radius of curvature is positive; if the center of curvature lies outside the solid, the radius of curvature is negative. Figure 8.4 shows the elements and bearing races. The importance of the sign of the radius of curvature presents itself later in the chapter.

Note that if coordinates $x$ and $y$ are chosen such that

$$\frac{1}{r_{ax}} + \frac{1}{r_{bx}} \geq \frac{1}{r_{ay}} + \frac{1}{r_{by}}$$

(8.1)

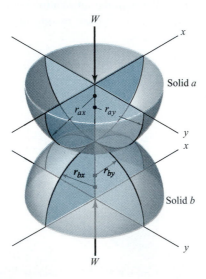

**Figure 8.3**   Geometry of contacting elastic solids.

(a)

(b)

(c)

**Figure 8.4** Sign designations for radii of curvature of various machine elements. (a) Rolling elements; (b) ball bearing races; (c) rolling-bearing races.

coordinate $x$ then determines the direction of the semiminor axis of the contact area when a load is applied, and $y$ determines the direction of the semimajor axis. The direction of the entraining motion is always considered to be along the $x$ axis.

## Curvature Sum and Difference

The curvature sum and difference, which are important in analyzing contact stresses and deformation, are

$$\frac{1}{R} = \frac{1}{R_x} + \frac{1}{R_y} \tag{8.2}$$

and

$$R_d = R\left(\frac{1}{R_x} - \frac{1}{R_y}\right) \tag{8.3}$$

where

$$\frac{1}{R_x} = \frac{1}{r_{ax}} + \frac{1}{r_{bx}} \tag{8.4}$$

$$\frac{1}{R_y} = \frac{1}{r_{ay}} + \frac{1}{r_{by}} \tag{8.5}$$

Equations (8.4) and (8.5) effectively redefine the problem of two ellipsoidal solids approaching each other in terms of an equivalent solid of radii $R_x$ and $R_y$ approaching a plane. Note that the curvature difference expressed in Equation (8.3) is dimensionless.

The radius ratio $\alpha_r$ is

$$\alpha_r = \frac{R_y}{R_x} \tag{8.6}$$

Thus, if Equation (8.1) is satisfied, $\alpha_r > 1$; and if it is not satisfied, $\alpha_r < 1$.

Machine elements (considered later in the text) with nonconformal surfaces (concentrated loading) have a range of radius ratios from 0.03 to 100. For example, for a traction drive simulated by a disk rolling on a plane, $\alpha_r = 0.03$; for a ball-on-plane contact, $\alpha_r = 1.0$; and for a contact approaching a nominal line contact, $\alpha_r \to 100$, such as in a barrel-shaped roller bearing against a plane. Some further examples of machine elements with radius ratios less than 1 are Novikov gear contacts, locomotive wheel-rail contacts, and roller-flange contacts in a radially loaded roller bearing. Some further examples of machine elements with radius ratios greater than 1 are rolling-element bearings and most gears. These topics are considered later in the text, but the general information on contact geometry of nonconformal surfaces is an important feature to consider here.

### Ellipticity Parameter

The **ellipticity parameter** $k_e$ is defined as the ratio of elliptical contact diameter in the $y$ direction (transverse direction) to the elliptical contact diameter in the $x$ direction (direction of entraining motion), or

$$k_e = \frac{D_y}{D_x} \tag{8.7}$$

If $\alpha_r \geq 1$, the contact ellipse will be oriented with its major diameter transverse to the direction of motion, and consequently $k_e \geq 1$; otherwise, the major diameter will lie along the direction of motion with both $\alpha_r < 1$ and $k_e < 1$. To avoid confusion, the commonly used solutions to the surface deformation and stresses are presented only for $\alpha_r > 1$.

Note that the ellipticity parameter is a function only of the solid's radii of curvature

$$k_e = f(r_{ax}, r_{bx}, r_{ay}, r_{by}) \tag{8.8}$$

That is, as the load increases, the semiaxes in the $x$ and $y$ directions of the contact ellipse increase proportionately to each other so that the ellipticity parameter remains constant.

### Contact Pressure

When an elastic solid is subjected to a load, stresses are produced that increase as the load is increased. These stresses are associated with deformations, which are defined by strains. Unique relationships exist between stresses and their corresponding strains, as shown in Appendix B. For elastic solids the stresses are linearly related to the strains, with the modulus of elasticity (or the proportionality constant) being an elastic constant that acquires different values for different materials, as covered in Section 3.7. The modulus of elasticity $E$ and Poisson's ratio $\nu$ are two important parameters (defined in Chapter 3) that are used in this chapter to describe contacting solids.

As the stresses increase within the material, elastic behavior is replaced by plastic flow in which the material is permanently deformed. The stress state at which the transition from elastic to plastic behavior occurs, known as the *yield stress,* has a definite value for a given material at a given temperature. In this book only elastic behavior is considered.

When two elastic nonconformal solids are brought together under a load, a contact area develops whose shape and size depend on the applied load, the elastic properties of the materials, and the curvature of the surfaces. When the two solids shown in Figure 8.3 have a normal load applied to them, the contact area is elliptical. It has been common to refer to elliptical contacts as point contacts; but since under load these contacts become elliptical, they are referred to here as such. For the special case where $r_{ax} = r_{ay}$ and $r_{bx} = r_{by}$, the resulting contact is a circle rather than an ellipse. When $r_{ay}$ and $r_{by}$ are both infinite, the initial line contact develops into a rectangle when load is applied.

Hertz (1881) considered the stresses and deformations in two perfectly smooth, ellipsoidal contacting solids much like those shown in Figure 8.3. His application of the classical elasticity theory to this problem has formed the basis of stress calculations for machine elements, such as ball and roller bearings, gears, and cams and followers. Hertz made the following assumptions:

1. The materials are homogeneous, and the yield stress is not exceeded.
2. No tangential forces are induced between the solids.
3. Contact is limited to a small portion of the surface such that the dimensions of the contact region are small relative to the radii of the ellipsoids.
4. The solids are at rest and in equilibrium.

Making use of these assumptions, Hertz was able to develop the following expression for the pressure distribution within an ellipsoidal contact (shown in Figure 8.5):

$$p_H = p_{max} \left[ 1 - \left( \frac{2x}{D_x} \right)^2 - \left( \frac{2y}{D_y} \right)^2 \right]^{1/2}$$

(8.9)

where $D_x$ = diameter of contact ellipse in $x$ direction, m
$\quad\quad\ D_y$ = diameter of contact ellipse in $y$ direction, m

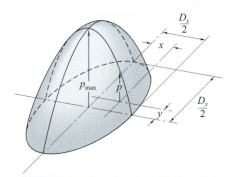

**Figure 8.5**  Pressure distribution in ellipsoidal contact.

The maximum pressure is given by

$$p_{\max} = \frac{6W}{\pi D_x D_y}$$

(8.10)

where $W$ is the normal applied load. Equation (8.9) determines the distribution of pressure or compressive stress on the common interface, which is clearly a maximum at the contact center and decreases to zero at the periphery.

### Simplified Solutions

The classical Hertzian solution requires the calculation of the ellipticity parameter $k_e$ as well as the complete elliptic integrals of the first and second kinds $\mathcal{F}$ and $\mathcal{E}$. This calculation entails finding a solution to a transcendental equation relating $k_e$, $\mathcal{F}$, and $\mathcal{E}$ to the geometry of the contacting solids and is usually accomplished by some iterative numerical procedure. Hamrock and Brewe (1983) used a linear regression by the method of least squares to power-fit the pairs of data and arrived at the simplified equations shown in Table 8.1. From this table the ellipticity parameter for the complete radius ratio range can be expressed as

$$k_e = \alpha_r^{2/\pi}$$

(8.11)

Figure 8.6 shows the ellipticity parameter and the elliptic integrals of the first and second kinds for a radius ratio ($\alpha_r = R_y/R_x$) range usually encountered in nonconformal conjunctions. Note from Figure 8.6 that $\mathcal{F} = \mathcal{E} = \pi/2$ when $\alpha_r = 1$. Also, both $\mathcal{F}$ and $\mathcal{E}$ have discontinuous derivatives at $\alpha_r = 1$, thus the reason for the two columns presented in Table 8.1.

When the ellipticity parameter $k_e$, the normal applied load $W$, Poisson's ratio $v$, and the modulus of elasticity $E$ of the contacting solids are known, the major and minor axes

**Table 8.1**    Simplified equations. [From Hamrock and Brewe (1983).]

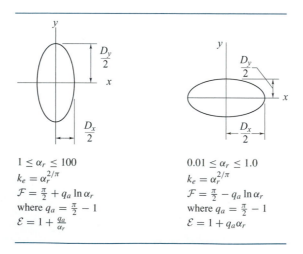

| $1 \le \alpha_r \le 100$ | $0.01 \le \alpha_r \le 1.0$ |
|---|---|
| $k_e = \alpha_r^{2/\pi}$ | $k_e = \alpha_r^{2/\pi}$ |
| $\mathcal{F} = \frac{\pi}{2} + q_a \ln \alpha_r$ | $\mathcal{F} = \frac{\pi}{2} - q_a \ln \alpha_r$ |
| where $q_a = \frac{\pi}{2} - 1$ | where $q_a = \frac{\pi}{2} - 1$ |
| $\mathcal{E} = 1 + \frac{q_a}{\alpha_r}$ | $\mathcal{E} = 1 + q_a \alpha_r$ |

**Figure 8.6** Variation of ellipticity parameter and elliptic integrals of first and second kinds as function of radius ratio.

of the contact ellipse ($D_y$ and $D_x$) and the maximum deformation at the contact center can be written as

$$D_y = 2 \left( \frac{6 k_e^2 \mathcal{E} W R}{\pi E'} \right)^{1/3}$$ (8.12)

$$D_x = 2 \left( \frac{6 \mathcal{E} W R}{\pi k_e E'} \right)^{1/3}$$ (8.13)

$$\delta_{\max} = \mathcal{F} \left[ \frac{9}{2 \mathcal{E} R} \left( \frac{W}{\pi k_e E'} \right)^2 \right]^{1/3}$$ (8.14)

$$E' = \frac{2}{\left(1 - v_a^2\right)/E_a + \left(1 - v_b^2\right)/E_b}$$ (8.15)

In these equations $D_y$ and $D_x$ are proportional to $W^{1/3}$, and $\delta_{\max}$ is proportional to $W^{2/3}$.

---

**Given:** A solid sphere of 20-mm radius ($r_{ax} = r_{ay} = 20$ mm) rolls on the inside of an outer race with a 100-mm internal radius ($r_{bx} = -100$ mm) and a large width in the axial direction. The sphere is made of silicon nitride, and the outer race is made of stainless steel. The normal applied load is 1000 N.

**EXAMPLE 8.1**

**Find:**

(a) Curvature sum $R$
(b) Ellipticity parameter $k_e$
(c) Elliptic integrals of the first and second kinds $\mathcal{F}$ and $\mathcal{E}$
(d) Effective modulus of elasticity $E'$
(e) Dimensions of elliptical contact $D_x$ and $D_y$
(f) Maximum Hertzian pressure or stress on surface $p_{\max}$
(g) Maximum deformation $\delta_{\max}$

**Solution:**

(a)  Note that $r_{bx}$ is negative, so the curvature sum is

$$\frac{1}{R_x} = \frac{1}{r_{ax}} + \frac{1}{r_{bx}} = \frac{1}{0.02} - \frac{1}{0.10} = 40 \text{ m}^{-1}$$

$$\therefore \qquad R_x = 0.025 \text{ m}$$

$$\frac{1}{R_y} = \frac{1}{r_{ay}} + \frac{1}{r_{by}} = \frac{1}{0.02} + \frac{1}{\infty} = \frac{1}{0.02} \text{ m}^{-1}$$

$$\therefore \qquad R_y = 0.02 \text{ m}$$

$$\frac{1}{R} = \frac{1}{R_x} + \frac{1}{R_y} = \frac{1}{0.025} + \frac{1}{0.020} = 90 \text{ m}^{-1}$$

$$\therefore \qquad R = 0.0111 \text{ m}$$

(b)  The ellipticity parameter, where $\alpha_r = R_y/R_x = 0.020/0.025 = 0.80$, is

$$k_e = \alpha_r^{2/\pi} = (0.80)^{2/\pi} = 0.8676$$

(c)  From Table 8.1 the elliptical integrals of the first and second kinds are

$$\mathcal{F} = \frac{\pi}{2} - \left(\frac{\pi}{2} - 1\right) \ln 0.8 = 1.698$$

$$\mathcal{E} = 1 + \left(\frac{\pi}{2} - 1\right)(0.8) = 1.457$$

(d)  From Table 3.2, $E_a = 314$ GPa and $E_b = 193$ GPa; and from Table 3.3, $\nu_a = 0.26$ and $\nu_b = 0.30$. The effective modulus of elasticity is

$$E' = \frac{2}{(1 - \nu_a^2)/E_a + (1 - \nu_b^2)/E_b}$$

$$= \frac{2}{(1 - 0.26^2)/3.14 \times 10^{11} + (1 - 0.32^2)/1.93 \times 10^{11}} = 2.603 \times 10^{11} \text{ N/m}^2$$

(e)  The dimensions of the elliptical contact, from Equations (8.12) and (8.13), are

$$D_y = 2\left(\frac{6k_e^2 \mathcal{E} W R}{\pi E'}\right)^{1/3} = 2\left[\frac{6(0.8676)^2(1.457)(1000)(0.0111)}{\pi(2.603)(10)^{11}}\right]^{1/3} = 0.8940 \times 10^{-3} \text{ m}$$

$$D_x = 2\left(\frac{6\mathcal{E} W R}{\pi k_e E'}\right)^{1/3} = 2\left[\frac{6(1.457)(1000)(0.0111)}{\pi(0.8676)(2.603)(10)^{11}}\right]^{1/3} = 1.0304 \times 10^{-3} \text{ m}$$

(f)  The maximum pressure, from Equation (8.10), is

$$p_{max} = \frac{6W}{\pi D_x D_y} = \frac{6(1000)}{\pi(0.8943)(1.0308)(10)^{-6}} \text{ Pa} = 2.073 \text{ GPa}$$

(g) From Equation (8.14) the maximum deformation at the center of the conjunction is

$$\delta_{max} = \mathcal{F} \left[ \frac{9}{2\mathcal{E}R} \left( \frac{W}{\pi k_e E'} \right)^2 \right]^{1/3}$$

$$= 1.6981 \left\{ \frac{4.5}{(1.4566)(0.0111)} \left[ \frac{1000}{\pi (0.8676)(2.603)(10)^{11}} \right]^2 \right\}^{1/2} = 13.94 \times 10^{-6} \text{ m}$$

---

**EXAMPLE 8.2**

**Given:** The balls in a deep-groove ball bearing have a 17-mm diameter. The groove in the inner race has an 8.84-mm radius; the radius from the center of the race to the bottom of the groove is 27.5 mm. The load being applied is 20,000 N. The ball and race material is steel.

**Find:**

(a) The dimensions of the contact ellipse
(b) The maximum deformation

**Solution:**

(a) From Equations (8.4) and (8.5) the curvature sums in the $x$ and $y$ directions are, respectively,

$$\frac{1}{R_x} = \frac{1}{r_{ax}} + \frac{1}{r_{bx}} = \frac{1}{8.5 \times 10^{-3}} + \frac{1}{27.5 \times 10^{-3}} = 154.0 \text{ m}^{-1}$$

$$\frac{1}{R_y} = \frac{1}{r_{ay}} + \frac{1}{r_{by}} = \frac{1}{8.5 \times 10^{-3}} - \frac{1}{8.84 \times 10^{-3}} = 4.525 \text{ m}^{-1}$$

From Equations (8.2) and (8.3) the curvature sum and difference are, respectively,

$$\frac{1}{R} = \frac{1}{R_x} + \frac{1}{R_y} = 158.5 \text{ m}^{-1}$$

$$R_d = R \left( \frac{1}{R_x} - \frac{1}{R_y} \right) = \frac{1}{158.5}(154.0 - 4.525) = 0.9431$$

Using Equation (8.6) gives the radius ratio as

$$\alpha_r = \frac{R_y}{R_x} = \frac{154.0}{4.525} = 34.03$$

From Table 8.1 the elliptical parameter and the elliptic integrals of the first and second kinds are

$$k_e = \alpha_r^{2/\pi} = 34.03^{2/\pi} = 9.446$$

$$\mathcal{F} = \frac{\pi}{2} + \left( \frac{\pi}{2} - 1 \right) \ln \alpha_r = \frac{\pi}{2} + \left( \frac{\pi}{2} - 1 \right) \ln 34.03 = 3.584$$

$$\mathcal{E} = 1 + \frac{\pi/2 - 1}{34.03} = 1.017$$

The effective modulus of elasticity is

$$E' = \frac{E}{1 - \nu^2} = \frac{207 \times 10^9}{1 - 0.3^2} \text{ Pa} = 227.5 \text{ GPa}$$

From Equations (8.12) and (8.13) the dimensions of the contact ellipse are

$$D_y = 2\left(\frac{6k_e^2 \mathcal{E} W R}{\pi E'}\right)^{1/3} = 2\left[\frac{6(9.446)^2(1.017)(20,000)}{158.5\pi(227.5)(10^9)}\right]^{1/3} \text{ m} = 9.162 \text{ mm}$$

$$D_x = \frac{D_y}{k_e} = \frac{9.162}{9.446} = 0.9699 \text{ mm}$$

(b) From Equation (8.14) the maximum deformation in the center of contact is

$$\delta_{max} = \mathcal{F}\left[\frac{9}{2\mathcal{E}R}\left(\frac{W}{\pi k_e E'}\right)^2\right]^{1/3} = 3.584\left[\frac{9(158.5)}{2(1.017)}\left(\frac{20,000}{\pi(9.446)(227.5)(10^9)}\right)^2\right]^{1/3}$$

$$= 65.68 \times 10^{-6} \text{ m} = 65.68 \,\mu\text{m}$$

### 8.3.2    RECTANGULAR CONTACTS

For rectangular conjunctions the contact ellipse discussed throughout this chapter has infinite width in the transverse direction ($D_y \to \infty$). This type of contact is exemplified by a cylinder loaded against a plane, a groove, or another parallel cylinder or by a roller loaded against an inner or outer race. In these situations the contact semiwidth is given by

$$b^* = R_x\left(\frac{8W'}{\pi}\right)^{1/2} \tag{8.16}$$

where the dimensionless load is

$$W' = \frac{w'}{E'R_x} \tag{8.17}$$

and $w'$ is the load per unit length along the contact. The maximum deformation for a rectangular conjunction can be written as

$$\delta_{max} = \frac{2W'R_x}{\pi}\left(\ln\frac{2\pi}{W'} - 1\right) \tag{8.18}$$

The maximum Hertzian contact pressure in a rectangular conjunction can be written as

$$p_{max} = E'\left(\frac{W'}{2\pi}\right)^{1/2} \tag{8.19}$$

---

**EXAMPLE 8.3**

**Given:** A solid cylinder of 20-mm radius ($r_{ax} = 20$ mm and $r_{ay} = \infty$) rolls around the inside of an outer race with a 100-mm internal radius ($r_{bx} = -100$ mm) and a large width in the axial ($y$) direction ($r_{by} = \infty$). The cylinder is made of silicon nitride, and the outer race is made of stainless steel. The normal applied load per unit length is 1000 N/m.

**Find:**

(a) Curvature sum $R$
(b) Semiwidth of contact $b^*$
(c) Maximum Hertzian contact pressure $p_{max}$
(d) Maximum deformation $\delta_{max}$

Also, compare the results with those of Example 8.2.

**Solution:**

(a) From Example 8.2, $E' = 2.603 \times 10^{11}$ N/m². The curvature sum is

$$\frac{1}{R_x} = \frac{1}{r_{ax}} + \frac{1}{r_{bx}} = \frac{1}{0.02} - \frac{1}{0.10} = 40 \text{ m}^{-2} \qquad \therefore \qquad R_x = 0.025 \text{ m}$$

The dimensionless load is

$$W' = \frac{w'}{E'R_x} = \frac{1000}{(2.603)(10^{11})(2.5)(10^{-2})} = 1.538 \times 10^{-7}$$

(b) The semiwidth of the contact is

$$b^* = R_x \left(\frac{8W'}{\pi}\right)^{1/2} = 0.025 \left[\frac{8(1.538)(10^{-7})}{\pi}\right]^{1/2} = 15.64 \times 10^{-6} \text{ m} = 15.64 \,\mu\text{m}$$

(c) The maximum contact pressure is

$$p_{max} = E' \left(\frac{W'}{2\pi}\right)^{1/2} = 2.603(10^{11}) \left[\frac{(1.538)(10^{-7})}{2\pi}\right]^{1/2} = 40.68 \times 10^6 \text{ Pa}$$

(d) The maximum deformation at the center of the contact is

$$\delta_{max} = \frac{2W'R}{\pi} \left(\ln \frac{2\pi}{W'} - 1\right)$$

$$= \frac{2(1.538)(10^{-7})(2.5)(10^{-2})}{\pi} \left(\ln \frac{2\pi}{0.1538 \times 10^{-6}} - 1\right) \qquad \text{m}$$

$$= 0.0405 \,\mu\text{m}$$

# 8.4    LUBRICATION

Features that distinguish the four lubrication regimes present in machine elements are considered here.

## 8.4.1    HYDRODYNAMIC LUBRICATION

**Hydrodynamic lubrication** (HL) is generally characterized by conformal surfaces with fluid-film lubrication. A positive pressure develops in a hydrodynamically lubricated journal or thrust bearing because the bearing surfaces converge and their relative motion and the

Conformal surfaces
$p_{max} \approx 5$ MPa
$h_{min} = f(W, u_b, \eta_0, R_x, R_y) > 1$ μm
No elastic effect

**Figure 8.7**    Characteristics of hydrodynamic lubrication. [From Hamrock, et al. (2004).]

viscosity of the fluid separate the surfaces. The existence of this positive pressure implies that a normal applied load may be supported. The magnitude of the pressure developed (usually less than 5 MPa) is not generally high enough to cause significant elastic deformation of the surfaces.

The minimum film thickness in a hydrodynamically lubricated bearing is a function of normal applied load $W$, velocity $u_b$, lubricant absolute viscosity $\eta_0$, and geometry ($R_x$ and $R_y$). Figure 8.7 shows some of these characteristics of hydrodynamic lubrication. Minimum film thickness $h_{min}$ as a function of $u_b$ and $W$ for sliding motion is given as

$$(h_{min})_{HL} \propto \left(\frac{u_b}{W}\right)^{1/2} \tag{8.20}$$

In practice, the minimum film thickness normally exceeds 1 μm.

In hydrodynamic lubrication the films are generally thick, so that opposing solid surfaces are prevented from coming into contact. This condition is often called the *ideal form* of lubrication because it provides low friction and very low wear. Lubrication of the solid surfaces is governed by the bulk physical properties of the lubricant, notably the viscosity; the frictional characteristics arise purely from the shearing of the viscous lubricant.

For a normal load to be supported by a bearing, positive-pressure profiles must be developed over the bearing length. Figure 8.8 illustrates three ways of developing positive pressure in hydrodynamically lubricated bearings. For positive pressure to be developed in a slider bearing [Fig. 8.8(a)] the lubricant film thickness must be decreasing in the sliding direction. In a squeeze film bearing [Fig. 8.8(b)] the bearing surfaces approach each other with velocity $w_a$, providing a valuable cushioning effect; positive pressures will be generated only when the film thickness is diminishing. In an externally pressurized bearing, sometimes called a *hydrostatic bearing* [Fig. 8.8(c)], the pressure drop across the bearing supports the

**Figure 8.8** Mechanism of pressure development for hydrodynamic lubrication. (a) Slider bearing; (b) squeeze film bearing; (c) externally pressurized bearing. [From Hamrock et al. (2004).]

load. The load-carrying capacity is independent of bearing motion and lubricant viscosity. There is no surface contact wear at starting and stopping as there is with the slider bearing.

## 8.4.2 ELASTOHYDRODYNAMIC LUBRICATION

**Elastohydrodynamic lubrication** (EHL) is a form of hydrodynamic lubrication where elastic deformation of the lubricated surfaces becomes significant. Historically, elastohydrodynamic lubrication may be viewed as one of the major developments in the field of lubrication in the 20th century. Not only did it reveal the existence of a previously unsuspected regime of lubrication in highly stressed nonconforming machine elements, but also it brought order to the complete spectrum of lubrication regimes, ranging from boundary to hydrodynamic. The features important in a hydrodynamically lubricated slider bearing [Fig. 8.8(a)]—converging film thickness, sliding motion, and a viscous fluid between the surfaces—are also important here. Elastohydrodynamic lubrication is normally associated with nonconformal surfaces and fluid-film lubrication. There are two distinct forms of EHL.

**Hard EHL** (HEHL) relates to materials of high elastic modulus, such as metals and ceramics. In this form of lubrication, the elastic deformation and the pressure-viscosity effects are equally important. Figure 8.9 gives the characteristics of hard elastohydrodynamically lubricated conjunctions. The maximum pressure is typically between 0.5 and 4 GPa; the minimum film thickness normally exceeds 0.1 $\mu$m. These conditions are dramatically different from those found in a hydrodynamically lubricated conjunction (Fig. 8.7). At loads normally experienced in nonconformal machine elements, the elastic deformations are 2 orders of magnitude larger than the minimum film thickness. Furthermore, the lubricant viscosity can vary by as much as 20 orders of magnitude within the lubricated conjunction. The minimum film thickness is a function of the same parameters as for hydrodynamic lubrication (Fig. 8.7) but with the additions of the effective elastic modulus $E'$ and the pressure-viscosity coefficient $\xi$ of the lubricant. The effective modulus of elasticity is

$$E' = \frac{2}{\left(1 - v_a^2\right)/E_a + \left(1 - v_b^2\right)/E_b} \tag{8.21}$$

where $v$ = Poisson's ratio
$E$ = modulus of elasticity, Pa

Nonconformal surfaces
High-elastic-modulus material (e.g., steel)
$p_{max} \approx 0.5$ to 4 GPa
$h_{min} = f(W, u_b, \eta_0, R_x, R_y, E', \xi) > 0.1\ \mu m$
Elastic and viscous effects both important

Nonconformal surfaces (e.g., nitrite rubber)
$p_{max} \approx 0.5$ to 4 MPa
$h_{min} = f(W, u_b, \eta_0, R_x, R_y, E') \approx 1\ \mu m$
Elastic effects predominate

**Figure 8.9**   Characteristics of hard
elastohydrodynamic lubrication. [From
Hamrock, et al. (2004).]

**Figure 8.10**   Characteristics of soft
elastohydrodynamic lubrication. [From
Hamrock et al. (2004).]

The relationships between the minimum film thickness $h_{min}$ and the normal applied load $W$ and upper surface velocity $u_b$ for hard EHL as obtained from Hamrock and Dowson (1977) are

$$(h_{min})_{HEHL} \propto W^{-0.073} \tag{8.22}$$

$$(h_{min})_{HEHL} \propto u_b^{0.68} \tag{8.23}$$

Comparing the results for hard EHL [Eqs. (8.22) and (8.23)] with those for hydrodynamic lubrication [Eq. (8.20)] results in the following conclusions:

1.  The exponent on the normal applied load is nearly 7 times larger for hydrodynamic lubrication than for hard EHL. This difference implies that the film thickness is only slightly affected by load for hard EHL but is significantly affected for hydrodynamic lubrication.

2.  The exponent on mean velocity is slightly higher for hard EHL than for hydrodynamic lubrication.

Engineering applications in which elastohydrodynamic lubrication is important for high-elastic-modulus materials include gears, rolling-element bearings, and cams.

**Soft EHL** relates to materials of low elastic modulus, such as rubber and polymers. Figure 8.10 shows the characteristics of soft EHL materials. In soft EHL the elastic distortions are large, even with light loads. The maximum pressure for soft EHL is 0.5 to 4 MPa (typically 1 MPa), in contrast to 0.5 to 4 GPa (typically 1 GPa ) for hard EHL (Fig. 8.9). This low pressure has a negligible effect on the viscosity variation throughout the conjunction. The minimum film thickness is a function of the same parameters as in hydrodynamic lubrication with the addition of the effective elastic modulus. The minimum film thickness for soft EHL is typically 1 $\mu$m. Engineering applications in which elastohydrodynamic lubrication is important for low-elastic-modulus materials include seals, natural human joints, tires, and a number of lubricated machine elements that use rubber as a material.

The common features of hard and soft EHL are that the local elastic deformation of the solids provides coherent fluid films and that asperity interaction is largely prevented. Lack of asperity interaction implies that the frictional resistance to motion is due to lubricant shearing.

### 8.4.3  BOUNDARY LUBRICATION

In **boundary lubrication** the solids are not separated by the lubricant; thus fluid-film effects are negligible, and there is significant asperity contact. The contact lubrication mechanism is governed by the physical and chemical properties of thin surface films of molecular proportions. The properties of the bulk lubricant are of minor importance, and the coefficient of friction is essentially independent of fluid viscosity. The frictional characteristics are determined by the properties of the solids and the lubricant film at the common interfaces. The surface films vary in thickness from 1 to 10 nm, depending on the lubricant's molecular size.

Figure 8.11 illustrates the film conditions existing in fluid film and boundary lubrication. The surface slopes and film thicknesses in this figure are greatly distorted for purposes of illustration. To scale, real surfaces would appear as gently rolling hills rather than sharp peaks. The surface asperities are not in contact for fluid-film lubrication but are in contact for boundary lubrication.

Figure 8.12 shows the behavior of the coefficient of friction in the different lubrication regimes. In boundary lubrication, although the friction is much higher than in the hydrodynamic regime, it is still much lower than for unlubricated surfaces. The mean coefficient of friction increases a total of three orders of magnitude in going from the hydrodynamic to the elastohydrodynamic to the boundary to the unlubricated regime.

Figure 8.13 shows the wear rate in the various lubrication regimes as determined by the operating load. In the hydrodynamic and elastohydrodynamic regimes, there is little or no wear because there is no asperity contact. In the boundary lubrication regime, the degree of asperity interaction and the wear rate increase as the load increases. The transition from boundary lubrication to an unlubricated condition is marked by a drastic change in wear

Boundary film

Bulk lubricant

(a)                    (b)                    (c)

**Figure 8.11**   Regimes of lubrication. (a) Fluid film lubrication—surfaces separated by bulk lubricant film. This regime is sometimes further classified as thick- or thin-film lubrication. (b) Partial lubrication—both bulk lubricant and boundary film play a role. (c) Boundary lubrication—performance depends essentially on boundary film. [From Hamrock et al. (2004).]

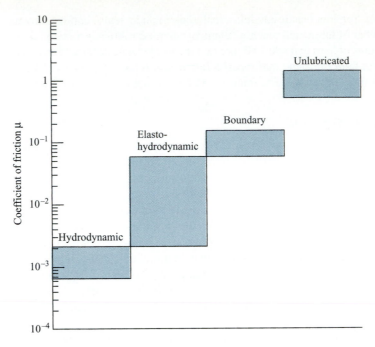

**Figure 8.12**    Bar diagram showing coefficient of friction for various lubrication conditions. [From Hamrock et al. (2004).]

rate. As the relative load is increased in the unlubricated regime, the wear rate increases until scoring or seizure occurs and the machine element can no longer operate successfully. Most machine elements cannot operate long with unlubricated surfaces. Together Figures 8.12 and 8.13 show that the friction and wear of unlubricated machine element surfaces can be greatly decreased by providing boundary lubrication.

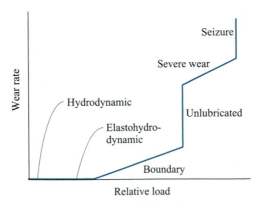

**Figure 8.13**    Wear rate for various lubrication regimes. [From Beerbower (1972).]

Boundary lubrication is encountered for machine elements with heavy loads and low running speeds, where fluid-film lubrication is difficult to attain. Mechanisms such as door hinges operate under conditions of boundary lubrication; other applications where low cost is of primary importance use boundary lubrication as in rubbing sleeve bearings.

### 8.4.4 PARTIAL LUBRICATION

If the pressures in elastohydrodynamically lubricated machine elements are too high or the running speeds are too low, the lubricant film will be penetrated. Some contact will then take place between the asperities, and **partial lubrication** (sometimes called **mixed lubrication**) will occur. The behavior of the conjunction in a partial lubrication regime is governed by a combination of boundary and fluid-film effects. Interaction takes place between one or more molecular layers of boundary lubricating films. A partial fluid-film lubrication action develops in the bulk of the space between the solids. The average film thickness in a partially lubricated conjunction is between 0.001 and 1 $\mu$m.

It is important to recognize that the transition from elastohydrodynamic to partial lubrication does not take place instantaneously as the severity of loading is increased; rather, a decreasing proportion of the load is carried by pressures within the fluid that fills the space between the opposing solids. As the load increases, a larger part of the load is supported by the contact pressure between the asperities of the solids. Furthermore, for conformal surfaces the regime of lubrication goes directly from hydrodynamic to partial lubrication.

## 8.5 SURFACE PARAMETERS

Designing machine elements is ultimately a problem of two surfaces that are either in contact or separated by a thin fluid film. In either case the surface texture is important in ensuring long component life. In Figure 8.14 the magnification in the vertical direction is 1000 and that in the horizontal is 20, so that the ratio of vertical to horizontal magnification is 50 : 1. It is therefore important to take into account this difference in magnification. This is often not done, and thus a false impression of the nature of surfaces has become prevalent.

Figure 8.14 is a graph of a surface profile generated by a stylus profilometer. The surface profile shows the surface height variation relative to a mean reference line. The mean, or M system, reference line is based on selecting the mean line as the centroid of the profile. Thus, the areas above and below this line are equal, so that the average of $z_i$ is zero. The light area below the reference line equals the dark shaded area above the reference line.

**Figure 8.14** Surface profile showing surface height variation relative to mean reference line.

Two different surface parameters may be computed:

1. **Centerline average** or arithmetic average surface roughness, denoted by $R_a$,

$$R_a = \frac{1}{N} \sum_{i=1}^{N} |z_i|$$

(8.24)

where $z_i$ = height from reference line, m
$N$ = number of height measurements taken

2. **Root-mean-square** (rms) surface roughness, denoted by $R_q$,

$$R_q = \left( \frac{1}{N} \sum_{i=1}^{N} z_i^2 \right)^{1/2}$$

(8.25)

If a Gaussian height distribution is assumed, $R_q$ has the advantage of being the standard deviation of the profile. For a simple sinusoidal distribution, the ratio of $R_q$ to $R_a$ is

$$\frac{R_q}{R_a} = \frac{\pi}{2\sqrt{2}} = 1.11$$

(8.26)

**Table 8.2**  Typical arithmetic average surface roughness for various manufacturing processes and machine components. [From Hamrock, et al. (2004).]

| | Arithmetic average surface roughness, $R_a$ | |
|---|---|---|
| | $\mu$m | $\mu$in |
| **Processes** | | |
| Sand casting; hot rolling | 12.5–25 | 500–1000 |
| Sawing | 3.2–25 | 128–1000 |
| Planing and shaping | 0.8–25 | 32–1000 |
| Forging | 3.2–12.5 | 128–500 |
| Drilling | 1.6–6.3 | 64–250 |
| Milling | 0.8–6.3 | 32–250 |
| Boring; turning | 0.4–6.3 | 16–250 |
| Broaching; reaming; cold rolling; drawing | 0.8–3.2 | 32–128 |
| Die casting | 0.8–1.6 | 32–64 |
| Grinding, coarse | 0.4–1.6 | 16–64 |
| Grinding, fine | 0.1–0.4 | 4–16 |
| Honing | 0.03–0.4 | 1.2–16 |
| Polishing | 0.02–0.2 | 0.8–8 |
| Lapping | 0.005–0.1 | 0.2–4 |
| **Components** | | |
| Gears | 0.25–10 | 10–400 |
| Plain bearings—journal (runner) | 0.12–0.5 | 5–20 |
| Plain bearings—bearing (pad) | 0.25–1.2 | 10–50 |
| Rolling bearings—rolling elements | 0.015–.12 | 0.6–5 |
| Rolling bearings—tracks | 0.1–0.3 | 4–12 |

Table 8.2 gives typical values of the arithmetic average for various processes and components. Note from this table that as higher-precision processes are applied, the $R_a$ values decrease significantly. The finest process shown is lapping, which produces an $R_a$ between 0.005 and 0.1 $\mu$m. This table contains $R_a$ values for a number of machine elements considered in Chapters 9 through 20.

## 8.6 FILM PARAMETER

When machine elements such as rolling-element bearings, gears, hydrodynamic journal and thrust bearings, and seals (all of which are considered later in this text) are adequately designed and fluid-film-lubricated, the lubricated surfaces are completely separated by a lubricant film. Endurance testing of ball bearings, for example, as reported by Tallian et al. (1967), has demonstrated that when the lubricant film is thick enough to separate the contacting bodies, bearing fatigue life is greatly extended. Conversely, when the film is not thick enough to provide full separation between the asperities in the contact zone, bearing life is adversely affected by the high shear stresses resulting from direct metal-to-metal contact.

The four lubrication regimes are defined in Section 8.4, and calculation methods for determining the rms surface finish are introduced in Section 8.5. This section introduces a **film parameter** and describes its range of values for the four lubrication regimes. The relationship between the dimensionless film parameter $\Lambda$ and the minimum film thickness $h_{\min}$ is

$$\Lambda = \frac{h_{\min}}{\left(R_{qa}^2 + R_{qb}^2\right)^{1/2}} \qquad (8.27)$$

where $R_{qa}$ = rms surface roughness of surface $a$
$\phantom{where }R_{qb}$ = rms surface roughness of surface $b$

The film parameter is used to define the four important lubrication regimes. The range for these four regimes is

1. Hydrodynamic lubrication, $5 \leq \Lambda < 100$
2. Elastohydrodynamic lubrication, $3 \leq \Lambda < 10$
3. Partial lubrication, $1 \leq \Lambda < 5$
4. Boundary lubrication, $\Lambda < 1$

These values are rough estimates. The major differences in geometric conformity in going from hydrodynamically lubricated conjunctions to elastohydrodynamically lubricated conjunctions make it difficult for clear distinctions to be made.

**Running in** is a process that affects the film parameter. This process allows wear to occur so that the mating surfaces can adjust to each other to provide for smooth running; this type of wear can be viewed as beneficial. The film parameter will increase with running in, since the composite surface roughness will decrease. Running in also has a significant effect on the shape of the asperities that is not captured by the composite surface roughness. With running in, the tips of the asperities in contact become flattened.

**EXAMPLE 8.4**

**Given:** Gears for an excavator are manufactured by sand casting. The as-cast surface roughness is measured to have a centerline average of 18 $\mu$m. This high surface roughness makes the gear wear rapidly. The film thickness for the grease-lubricated gears is calculated to be 1.6 $\mu$m.

**Find:** How should the sand-cast gears be machined to give a film parameter of 1?

**Solution:** Using Equation (8.27) while assuming that the roughnesses are equal on the two surfaces gives

$$\Lambda = \frac{h_{min}}{R_q \sqrt{2}} \quad \text{or} \quad R_q = \frac{h_{min}}{\Lambda \sqrt{2}} = \frac{1.6}{1 \sqrt{2}} = 1.131 \, \mu m$$

Table 8.2 shows that for a surface roughness of 1 $\mu$m, grinding is the fastest and cheapest method of achieving these surface finishes. Smoother surfaces can be manufactured by honing, polishing, and lapping, but these processes are considerably more expensive.

## 8.7   LUBRICANT VISCOSITY

A **lubricant** is any substance that reduces friction and wear and provides smooth running and a satisfactory life for machine elements. Most lubricants are liquids (such as mineral oils, synthetic esters, silicone fluids, and water), but they may also be solids (such as polytetrafluoroethylene, or Teflon) for use in dry bearings, greases for use in rolling-element bearings, or gases (such as air) for use in gas bearings. The physical and chemical interactions between the lubricant and the lubricating surfaces must be understood in order to provide the machine elements with satisfactory life.

A lubricant is a solid, liquid, or gas that is interposed between solid surfaces to facilitate their relative sliding or rolling, thus controlling friction and wear. Separation of the surfaces is, however, not the only function of a lubricant. Liquid lubricants have desirable secondary properties and characteristics:

1. They can be drawn in between moving parts by hydrodynamic action.
2. They have relatively high heat-sink capacity to cool the contacting parts.
3. They are easily blended with chemicals to give a variety of properties, such as corrosion resistance, detergency, or surface-active layers.
4. They can remove wear particles.

Liquid lubricants can be divided into those of petroleum origin, known as mineral oils, and those of animal or vegetable origin, known as fatty oils; synthetic oils are often grouped with the latter. For a lubricant to be effective, it must be viscous enough to maintain a lubricant film under operating conditions but should be as fluid as possible to remove heat and to avoid power loss due to viscous drag. The most important lubricant property, the viscosity, is considered in the following sections.

## 8.7.1    ABSOLUTE VISCOSITY

**Absolute** or dynamic **viscosity** can be defined in terms of the simple model shown in Figure 8.15, which depicts two parallel flat plates separated by a distance $h$, with the upper plate moving with velocity $u_b$ and the lower plate stationary. Oil molecules are visualized as small spheres that roll along in layers between the flat plates. To move the upper plate of area $A$ at a constant velocity $u_b$ across the surface of the oil and cause adjacent layers to flow past each other, a tangential force must be applied. Since the oil will "wet" and cling to the two surfaces, the bottommost layer will not move at all, the topmost layer will move with a velocity equal to the velocity of the upper plate, and the layers between the plates will move with velocities directly proportional to their distances $z$ from the stationary plate. This type of orderly movement in parallel layers is known as *streamline, laminar,* or *viscous flow.*

The shear stress on the oil causing relative movement of the layers is equal to $F/A$. The *shear strain rate $s$* of a particular layer is defined as the ratio of its velocity to its perpendicular distance from the stationary surface $z$ and is constant for each layer.

$$\therefore \qquad s = \frac{u}{z} = \frac{u_b}{h} \tag{8.28}$$

The shear strain rate has the unit of $s^{-1}$.

Newton correctly deduced that the force required to maintain a constant velocity $u_b$ of the upper surface is proportional to the area and the shear strain rate, or

$$F = \eta \frac{Au_b}{h} \tag{8.29}$$

where $\eta$ = absolute viscosity, Pa · s. By rearranging Equation (8.29) the absolute viscosity can be expressed as

$$\eta = \frac{F/A}{u_b/h} = \frac{\text{shear stress}}{\text{shear strain rate}} \tag{8.30}$$

It follows from Equation (8.30) that the unit of viscosity must be the unit of shear stress, divided by the unit of shear strain rate. The units of viscosity $\eta$ for three different systems

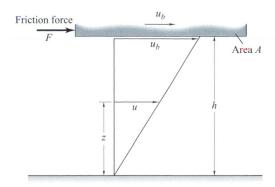

**Figure 8.15**    Slider bearing illustrating absolute viscosity.

**Table 8.3**    Absolute viscosity conversion factors

| To convert from | To cP | kgf · s/m² | N · s/m² | lb · s/in² |
|---|---|---|---|---|
| | | Multiply by | | |
| cP | 1 | $1.02 \times 10^{-4}$ | $10^{-3}$ | $1.45 \times 10^{-7}$ |
| kgf · s/m² | $9.807 \times 10^3$ | 1 | 9.807 | $1.422 \times 10^{-3}$ |
| N · s/m² | $10^3$ | $1.02 \times 10^{-1}$ | 1 | $1.45 \times 10^{-4}$ |
| reyn, or lb · s/in² | $6.9 \times 10^6$ | $7.03 \times 10^2$ | $6.9 \times 10^3$ | 1 |

are as follows:

1.  SI units: newton-seconds per square meter or, since a newton per square meter is also called a pascal, a pascal-second.

2.  The cgs units: dyne-seconds per square centimeter, or centipoise (cP), where $1 \text{ cP} = 10^{-2} \text{ P}$.

3.  English units: pound-seconds per square inch, called a *reyn* in honor of Osborne Reynolds.

Conversion of absolute viscosity from one system to another can be facilitated by Table 8.3. To convert from a unit in the column on the left side of the table to a unit at the top of the table, multiply by the corresponding value given in the table. Figure 8.16 gives the absolute viscosities of a number of fluids for a wide range of temperature.

---

**EXAMPLE 8.5**

**Given:** An absolute viscosity of $0.04 \text{ N} \cdot \text{s/m}^2$.

**Find:** The absolute viscosity in reyn, centipoise, and poise.

**Solution:** Using Table 8.3 gives

$$\eta = 0.04 \text{ N} \cdot \text{s/m}^2 = 0.04 \text{ Pa} \cdot \text{s} = 5.8 \times 10^{-6} \text{ lb} \cdot \text{s/in}^2$$

Note also that

$$\eta = 0.04 \text{ N} \cdot \text{s/m}^2 = 0.04 \text{ Pa} \cdot \text{s} = 5.8 \times 10^{-6} \text{ reyn}$$

and

$$\eta = 0.04 \text{ N} \cdot \text{s/m}^2 = (0.04)(10^3) \text{ cP} = 40 \text{ cP} = 0.4 \text{ P}$$

---

**EXAMPLE 8.6**

**Given:** An absolute viscosity of 1 reyn.

**Find:** The absolute viscosity in centipoise.

**Solution:** From Table 8.3,

$$1 \text{ reyn} = 1 \text{ lb} \cdot \text{s/in}^2 = 6.9 \times 10^6 \text{ cP}$$

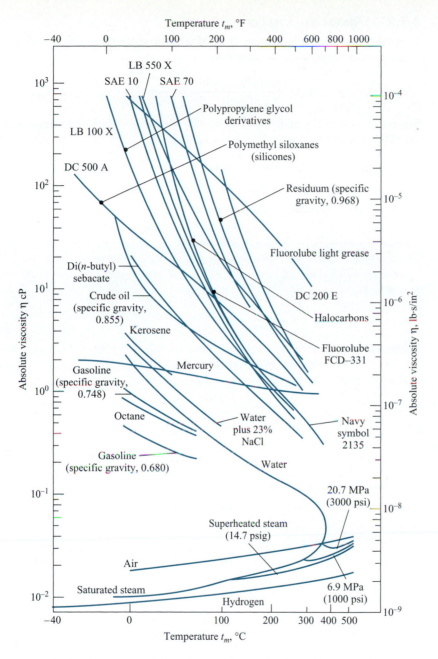

**Figure 8.16**    Absolute viscosities of a number of fluids for a wide range of temperatures.

### 8.7.2   KINEMATIC VISCOSITY

In many situations it is convenient to use **kinematic viscosity** rather than absolute viscosity. The kinematic viscosity $\eta_k$ is defined as

$$\eta_k = \frac{\text{absolute viscosity}}{\text{density}} = \frac{\eta}{\rho} = \frac{N \cdot s/m^2}{N \cdot s^2/m^4} = m^2/s \tag{8.31}$$

The ratio given in Equation (8.31) is literally kinematic; all trace of force or mass cancels out. The units of kinematic viscosity are as follows:

1.   SI units: square meters per second.
2.   The cgs units: square centimeters per second, called a *stoke*.
3.   English units: square inches per second.

---

**EXAMPLE 8.7**

**Given:**   Both mercury and water have an absolute viscosity of 1.5 cP at 5°C, but mercury has 13.6 times higher density than water. Assume that no change in density occurs when the temperature is increased.

**Find:**   The kinematic viscosities of mercury and water at 5°C and at 90°C.

**Solution:**   Figure 8.16 gives the absolute viscosities of various fluids as a function of temperature. At 90°C the absolute viscosity of water is 0.32 cP, and for mercury it is 1.2 cP. The kinematic viscosity of *mercury* at 5°C is

$$\eta_k = \frac{\eta}{\rho} = \frac{1.5(10^{-3})}{13,600} = 0.110(10^{-6}) \text{ m}^2/s$$

and at 90°C it is

$$\eta_k = \frac{\eta}{\rho} = \frac{1.2(10^{-3})}{13,600} = 0.0882(10^{-6}) \text{ m}^2/s$$

The kinematic viscosity of *water* at 5°C is

$$\eta_k = \frac{\eta}{\rho} = \frac{1.5(10^{-3})}{1000} = 1.50(10^{-6}) \text{ m}^2/s$$

and at 90°C it is

$$\eta_k = \frac{\eta}{\rho} = \frac{0.32(10^{-3})}{1000} = 0.32(10^{-6}) \text{ m}^2/s$$

Although mercury has the same absolute viscosity as water at 5°C and 3.75 times higher absolute viscosity at 90°C, the kinematic viscosities for mercury are much lower than those for water because of mercury's high density.

---

### 8.7.3   VISCOSITY-PRESSURE EFFECTS

In highly loaded contacts such as ball bearings, gears, and cams, the pressure is high enough to increase the lubricant viscosity significantly. The increase of a lubricant's viscosity with pressure is known as a **viscosity-pressure effect** or **piezoviscous effect,** and it is

**Table 8.4**     Absolute and kinematic viscosities of various fluids at atmospheric pressure and different temperatures

| | Temperature $t_m$, °C | | | | | |
|---|---|---|---|---|---|---|
| | 38 | 99 | 149 | 38 | 99 | 149 |
| Fluid | Absolute viscosity at $p = 0$, $\eta_0$, cP | | | Kinematic viscosity at $p = 0$, $\eta_k$, m²/s | | |
| Advanced ester | 25.3 | 4.75 | 2.06 | $2.58 \times 10^{-5}$ | $0.51 \times 10^{-5}$ | $0.23 \times 10^{-5}$ |
| Formulated advanced ester | 27.6 | 4.96 | 2.15 | $2.82 \times 10^{-5}$ | $0.53 \times 10^{-5}$ | $0.24 \times 10^{-5}$ |
| Polyalkyl aromatic | 25.5 | 4.08 | 1.80 | $3.0 \times 10^{-5}$ | $0.50 \times 10^{-5}$ | $0.23 \times 10^{-5}$ |
| Synthetic paraffinic oil (lot 3) | 414 | 34.3 | 10.9 | $49.3 \times 10^{-5}$ | $4.26 \times 10^{-5}$ | $1.4 \times 10^{-5}$ |
| Synthetic paraffinic oil (lot 4) | 375 | 34.7 | 10.1 | $44.7 \times 10^{-5}$ | $4.04 \times 10^{-5}$ | $1.3 \times 10^{-5}$ |
| Synthetic paraffinic oil (lot 2) plus antiwear additive | 370 | 32.0 | 9.93 | $44.2 \times 10^{-5}$ | $4.00 \times 10^{-5}$ | $1.29 \times 10^{-5}$ |
| Synthetic paraffinic oil (lot 4) plus antiwear additive | 375 | 34.7 | 10.1 | $44.7 \times 10^{-5}$ | $4.04 \times 10^{-5}$ | $1.3 \times 10^{-5}$ |
| C-ether | 29.5 | 4.67 | 2.20 | $2.5 \times 10^{-5}$ | $0.41 \times 10^{-5}$ | $0.20 \times 10^{-5}$ |
| Superrefined naphthenic mineral oil | 68.1 | 6.86 | 2.74 | $7.8 \times 10^{-5}$ | $0.82 \times 10^{-5}$ | $0.33 \times 10^{-5}$ |
| Synthetic hydrocarbon (traction fluid) | 34.3 | 3.53 | 1.62 | $3.72 \times 10^{-5}$ | $0.40 \times 10^{-5}$ | $0.19 \times 10^{-5}$ |
| Fluorinated polyether | 181 | 20.2 | 6.68 | $9.66 \times 10^{-5}$ | $1.15 \times 10^{-5}$ | $0.4 \times 10^{-5}$ |

| SOURCE: From Jones et al. (1975).

especially pronounced in mineral oils and other fluids with a large molecular chain length. Barus proposed the following formula in 1893 for the isothermal viscosity-pressure dependence of liquids:

$$\ln \frac{\eta}{\eta_0} = \xi p \qquad (8.32)$$

where  ln = natural, or Napierian, logarithm to base $e$
   $\eta_0$ = absolute viscosity at $p = 0$ and at a constant temperature, cP
   $\xi$ = pressure-viscosity coefficient of lubricant dependent on temperature, m²/N
   $p$ = pressure, Pa

   Table 8.4 lists the kinematic viscosities in square meters per second and the absolute viscosities in centipoise of various fluids at zero pressure and different temperatures. These values of the absolute viscosity correspond to $\eta_0$ in Equation (8.32) for the particular fluid and temperature used. The pressure-viscosity coefficients $\xi$ for these fluids, expressed in square meters per newton, are given in Table 8.5. The values correspond to $\xi$ in Equation (8.32).

---

**Given:**  A synthetic paraffin has an absolute viscosity of 32 cP at 99°C, and a fluorinated polyether has an absolute viscosity of 20.2 cP at 99°C.

**EXAMPLE 8.8**

**Find:**

(a)  The pressure at which the two oils have the same absolute viscosity.
(b)  The pressure at which the paraffin is 100 times less viscous than the fluorinated polyether.

**Solution:**

(a)   From Table 8.5, the pressure-viscosity coefficient at 99°C for the paraffin (lot 2) is $1.37 \times 10^{-8}$ m²/N, and for the polyether it is $3.24 \times 10^{-8}$ m²/N. Using Equation (8.32) and equating the viscosity give

$$32e^{1.37 \times 10^{-8}p} = 20.2e^{3.24 \times 10^{-8}p}$$

$$1.58e^{1.37 \times 10^{-8}p} = e^{3.24 \times 10^{-8}p}$$

$$1.58 = e^{0.4574}$$

$$\therefore \quad e^{0.4574 + 1.37 \times 10^{-8}p} = e^{3.24 \times 10^{-8}p}$$

$$p = 24.46 \times 10^{6} \text{ Pa} = 24.46 \text{ MPa}$$

(b)   Using the Barus equation [Eq. (8.32)] while letting the paraffin be 100 times less viscous than the polyether gives

$$1.58(100)e^{1.37 \times 10^{-8}p} = e^{3.24 \times 10^{-8}p}$$

$$158 = e^{5.063}$$

$$\therefore \quad e^{5.063 + 1.37 \times 10^{-8}p} = e^{3.24 \times 10^{-8}p}$$

$$p = 2.71 \times 10^{8} \text{ Pa} = 0.271 \text{ GPa}$$

## 8.7.4    VISCOSITY-TEMPERATURE EFFECTS

The viscosity of mineral and synthetic oils decreases with increasing temperature; therefore, the temperature at which the viscosity was measured must be quoted with every viscosity reported. Figures 8.16 and 8.17 show how absolute viscosity varies with temperature. Figure 8.16 presents the absolute viscosity of several fluids for a wide temperature range.

**Table 8.5**    Pressure-viscosity coefficients of various fluids at different temperatures

| | Temperature $t_m$, °C | | |
|---|---|---|---|
| | 38 | 99 | 149 |
| **Fluid** | Pressure-viscosity coefficient $\xi$, m²/N | | |
| Advanced ester | $1.28 \times 10^{-8}$ | $0.987 \times 10^{-8}$ | $0.851 \times 10^{-8}$ |
| Formulated advanced ester | $1.37 \times 10^{-8}$ | $1.00 \times 10^{-8}$ | $0.874 \times 10^{-8}$ |
| Polyalkyl aromatic | $1.58 \times 10^{-8}$ | $1.25 \times 10^{-8}$ | $1.01 \times 10^{-8}$ |
| Synthetic paraffinic oil (lot 3) | $1.77 \times 10^{-8}$ | $1.51 \times 10^{-8}$ | $1.09 \times 10^{-8}$ |
| Synthetic paraffinic oil (lot 4) | $1.99 \times 10^{-8}$ | $1.51 \times 10^{-8}$ | $1.29 \times 10^{-8}$ |
| Synthetic paraffinic oil (lot 2) plus antiwear additive | $1.81 \times 10^{-8}$ | $1.37 \times 10^{-8}$ | $1.13 \times 10^{-8}$ |
| Synthetic paraffinic oil (lot 4) plus antiwear additive | $1.96 \times 10^{-8}$ | $1.55 \times 10^{-8}$ | $1.25 \times 10^{-8}$ |
| C-ether | $1.80 \times 10^{-8}$ | $0.980 \times 10^{-8}$ | $0.795 \times 10^{-8}$ |
| Superrefined naphthenic mineral oil | $2.51 \times 10^{-8}$ | $1.54 \times 10^{-8}$ | $1.27 \times 10^{-8}$ |
| Synthetic hydrocarbon (traction fluid) | $3.12 \times 10^{-8}$ | $1.71 \times 10^{-8}$ | $0.939 \times 10^{-8}$ |
| Fluorinated polyether | $4.17 \times 10^{-8}$ | $3.24 \times 10^{-8}$ | $3.02 \times 10^{-8}$ |

SOURCE: From Jones et al. (1975).

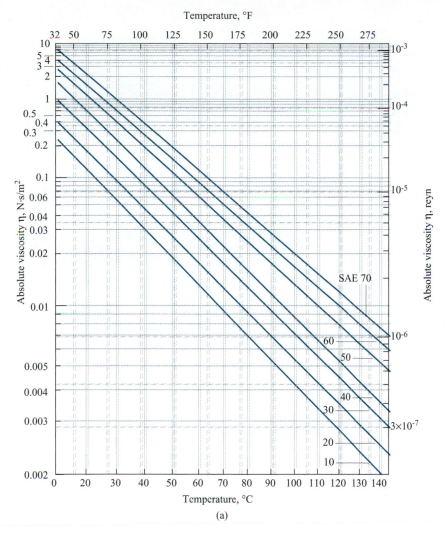

**Figure 8.17(a)** Absolute viscosities of SAE lubricating oils at atmospheric pressure. (a) Single-grade oils.

The interesting point of this figure is how drastically the slope and level of viscosity change for different fluids. The viscosity varies by five orders of magnitude, with the slope being highly negative for the Society of Automotive Engineers (SAE) oils and positive for gases.

Figure 8.17 gives the viscosity of SAE oils as a function of temperature. The SAE standards allow for a range of values and specify a kinematic viscosity, which results in the approximate absolute viscosity curves shown in Figure 8.17. Further, SAE specifies viscosity only at one temperature for the lubricants in Figure 8.17(a), namely, 100°C (212°F). Multigrade oils, shown in Figure 8.17(b), have a low-temperature viscosity requirement defined at −18°C (0°F), which they must meet in addition to the high-temperature viscosity

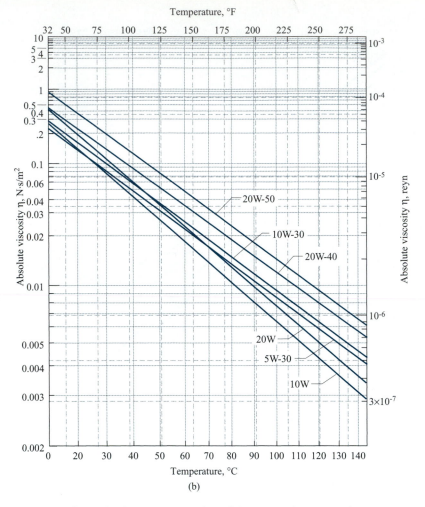

**Figure 8.17(b)**    Absolute viscosities of SAE lubricating oils at atmospheric pressure. (b) Multigrade oils.

requirement. This demanding set of requirements can be met through the use of viscosity index modifiers.

The viscosities of SAE oils can also be calculated by using the data in Table 8.6. These use the curve fit

$$
\eta = \begin{cases} C_1 \exp \dfrac{C_2}{t_F + 95} & \text{English units} \\[3mm] C_1 \exp \dfrac{C_2}{1.8 t_C + 127} & \text{SI units} \end{cases}
\tag{8.33}
$$

where $C_1, C_2$ = curve–fit constants

$\quad\quad t_F$ = lubricant temperature, °F

$\quad\quad t_C$ = lubricant temperature, °C

**Table 8.6** Curve fit data for SAE single-grade oils for use in Equation (8.33)

| SAE grade | Constant $C_1$ | | Constant $C_2$ |
|---|---|---|---|
| | reyn | $N \cdot s/m^2$ | |
| 10 | $1.58 \times 10^{-8}$ | $1.09 \times 10^{-4}$ | 1157.5 |
| 20 | $1.36 \times 10^{-8}$ | $9.38 \times 10^{-5}$ | 1271.6 |
| 30 | $1.41 \times 10^{-8}$ | $9.73 \times 10^{-5}$ | 1360.0 |
| 40 | $1.21 \times 10^{-8}$ | $8.35 \times 10^{-5}$ | 1474.4 |
| 50 | $1.70 \times 10^{-8}$ | $1.17 \times 10^{-4}$ | 1509.6 |
| 60 | $1.87 \times 10^{-8}$ | $1.29 \times 10^{-4}$ | 1564.0 |

| SOURCE: [From Seirig and Dandage (1982).]

**EXAMPLE 8.9**

**Given:** An SAE 10 oil is used in a car motor where the working temperature is 110°C.

**Find:** How many times more viscous would the oil be if the motor were operating in arctic conditions where the working temperature was −30°C?

**Solution:** Figure 8.17 gives the viscosity at 110°C to be 0.0033 N · s/m$^2$ and by extrapolating to be 1.04 N · s/m$^2$ at −30°C. The viscosity ratio is 1.04/0.0033 = 315.2. The oil is thus 315.2 times more viscous at −30°C than at 110°C.

## 8.8 FRICTION

**Friction** may be viewed as the force resisting relative movement between surfaces in contact. Two main classes of friction, sliding and rolling, are shown in Figure 8.18. Whereas sliding surfaces are often conformal, rolling friction involves nonconformal surfaces. However, most rolling contacts also experience some sliding.

In both rolling and sliding contacts, a tangential force $F$ in the direction of motion is needed to move the upper body over the stationary lower body. The ratio between the tangential force and the normal applied load $W$ is known as the *coefficient of friction* $\mu$

$$\mu = \frac{F}{W} \tag{8.34}$$

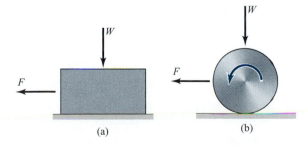

**Figure 8.18** Friction force in (a) sliding and (b) rolling.

**Table 8.7**    Typical coefficients of friction for combinations of unlubricated metals in air

|  | Coefficient of friction $\mu$ |
|---|---|
| **Self-mated metals in air** | |
| Gold | 2.5 |
| Silver | 0.8–1 |
| Tin | 1 |
| Aluminum | 0.8–1.2 |
| Copper | 0.7–1.4 |
| Indium | 2 |
| Magnesium | 0.5 |
| Lead | 1.5 |
| Cadmium | 0.5 |
| Chromium | 0.4 |
| **Pure metals and alloys sliding on steel (0.13% carbon) in air** | |
| Silver | 0.5 |
| Aluminum | 0.5 |
| Cadmium | 0.4 |
| Copper | 0.8 |
| Chromium | 0.5 |
| Indium | 2 |
| Lead | 1.2 |
| Copper—20% lead | 0.2 |
| White metal (tin-based) | 0.8 |
| White metal (lead-based) | 0.5 |
| $\alpha$-Brass (copper—30% zinc) | 0.5 |
| Leaded $\alpha/\beta$ brass (copper—40% zinc) | 0.2 |
| Gray cast iron | 0.4 |
| Mild steel (0.13% carbon) | 0.8 |

I  SOURCE: Hutchings (1992).

Throughout this text $\mu$ designates sliding friction and $\mu_r$ designates rolling friction. Also, the term *coefficient of friction* refers to the coefficient of sliding friction; whenever the coefficient of rolling friction is used, it will be so designated. Rolling friction is much smaller (by at least one order of magnitude) than sliding friction. Rolling and sliding friction varies from 0.001 in lightly loaded rolling-element bearings to greater than 10 for clean metals sliding against themselves in vacuum. For most common materials sliding in air, the value of $\mu$ lies in a narrow range from approximately 0.1 to 2.0. Table 8.7 gives typical coefficients of sliding friction for unlubricated metals in air; note that the range is 0.2 to 2.5, one order of magnitude.

## 8.8.1    LOW FRICTION

In a number of situations low friction is desirable, for example,

1.  Turbines and generators in electric or hydroelectric power stations, which use oil-lubricated hydrodynamic bearings

2.  Gyroscopes, spinning at very high speeds, which use gas-lubricated hydrodynamic bearings

Friction can be reduced by using special low-friction materials, by lubricating the surface if it is not already being done, or by using clever designs that convert sliding motion to rolling. Reducing the wear and heat produced by friction is one of the most important factors in extending machine element life.

## 8.8.2  HIGH FRICTION

In some situations high controlled friction is desirable, for example,

1. Brakes
2. Interaction between shoe and floor
3. Interaction between tire and road
4. Interaction between a nail and the wood into which it is hammered
5. Grip between a nut and a bolt

It must be emphasized that in these high-friction applications the friction must be controlled.

## 8.8.3  LAWS OF DRY FRICTION

The three **laws of friction** may be stated as follows:

1. The friction force is proportional to the normal load.
2. The friction force is not dependent on the apparent area of the contacting solids; that is, it is independent of the size of the solid bodies.
3. The friction force is independent of the sliding velocity.

These laws are applicable for most sliding conditions in the absence of a lubricant. The first two laws are normally called *Amonton's law*; he rediscovered them in 1699, with Leonardo da Vinci usually given the credit for discovering them some 200 years earlier.

The first two laws are found today to be generally satisfied for metals but are violated when polymers are the solid materials in contact. The third law of friction is less well founded than the first two laws. The friction force needed to begin sliding is usually greater than that necessary to maintain it; thus, the coefficient of sliding friction $\mu$ is greater than the coefficient of dynamic friction $\mu_d$. However, once sliding is established, $\mu_d$ is often nearly independent of sliding velocity.

The phenomenon important to machine element design is stick-slip. When stick-slip occurs, the friction force fluctuates between two extreme positions. All stick-slip processes are caused by the fact that the friction force does not remain constant as a function of some other variable, such as distance, time, or velocity.

## 8.8.4  SLIDING FRICTION OF METALS

Bowden and Tabor (1973) recognized that surfaces in contact touch only at points of asperity interaction, and that very high stresses induced in such small areas would lead readily to local plastic deformation. The penetration of an asperity into the opposing surface can be likened to a miniature hardness test, and the mean normal stress $\sigma$ over the real areas of asperity contact $A_r$ can be represented to all intents by the hardness $H$ of the softer material. Likewise, if $\tau$ represents the shear stress of the asperity junctions, the normal applied load $W$, the friction force $F$, and the coefficient of friction $\mu$ can be expressed as

$$\mu = \frac{F}{W} = \frac{A_r \tau}{A_r H} = \frac{\tau}{H}$$

(8.35)

This expression is an important step in understanding friction, although it is incomplete because it neglects the more complex nature of asperity interactions and deformations and accounts only for the adhesive element of friction. These limitations are apparent when it is recognized that for metals $\tau \approx 0.5 S_y$ and $H \approx 3 S_y$, where $S_y$ is the yield strength. Thus, from Equation (8.35) all clean metals should have a universal coefficient of friction of $\frac{1}{6}$, which is not representative of more sensitive experimental findings.

Another explanation for friction is that a force is required to move hard asperities through or even over another surface and that the resulting microcutting motion represents the friction process. The idea is that if the sum of the projected areas of the indenting asperities perpendicular to the sliding direction is $A_o$ and if the mean stress resisting plastic deformation of the softer material that is being cut is equal to the hardness, then the friction force $F$ can be expressed as

$$F = A_o H$$

(8.36)

The normal applied load $W$ carried by a number of asperities can be expressed as

$$W = AH$$

where $A$ = contact area, m². The coefficient of friction then becomes

$$\mu = \frac{F}{W} = \frac{A_o}{A}$$

(8.37)

Figure 8.19 shows a conical asperity having a mean angle $\theta$. The coefficient of friction for these considerations is

$$\mu = \frac{A_o}{A} = \frac{2}{\pi} \tan \theta$$

(8.38)

Only the front end of the conical asperity shown in Figure 8.19 is in contact.

Both molecular (adhesive) and deformation (plowing) actions are active, and the coefficient of friction is a function of both, or

$$\mu = \frac{\tau}{H} + \frac{2}{\pi} \tan \theta$$

(8.39)

**Figure 8.19**    Conical asperity having mean angle $\theta$ plowing through a softer material. This action simulates abrasive wear.

Experiments have demonstrated that the adhesion term [first term on the right side of Equation (8.39)] plays a major role in determining the friction between metals.

### 8.8.5 SLIDING FRICTION OF POLYMERS

The coefficient of friction is lower for machine elements made of polymers, such as polyethylene, acrylics, polystyrene, and nylon, than for those made of metals. For clean copper on steel in air, the coefficient of friction is around 1.0, but for most plastics it is typically 0.4. Adhesion at the interface is so strong that shear actually occurs in the polymer itself, and the strength properties of the polymer determine the friction.

One major difference between the frictional behavior of polymers and that of metals is the effect of load and geometry. Geometry refers to the shape of the surfaces, whether they are flat or curved and, if they are curved, how sharp the curvature is. With metals, the area of *true* contact is determined only by the load and the yield pressure, and not by the shape of the surfaces. With polymers it is otherwise. These materials deform viscoelastically, implying that deformation depends on the normal applied load $W$, the geometry, and the time of loading. For a fixed time of loading and fixed geometry, for example, a sphere on a flat surface, the area of true contact is not proportional to the load, as it is with metals. It is proportional to $W^n$, where $n$ is less than 1 and usually near $\frac{3}{4}$.

$$F \propto W^{3/4} \qquad \text{for polymers} \tag{8.40}$$

$$F \propto W \qquad \text{for metals} \tag{8.41}$$

Thus, the difference in friction behavior is quite significant.

### 8.8.6 SLIDING FRICTION OF RUBBER

Rubber is an extreme example of a material that deforms elastically. Regardless of the amount that rubber becomes distorted, it will return to its original shape when the deforming force is removed, as long as the rubber is not torn or cut. In contrast, polymers deform elastically and also flow viscously, and metals deform elastically over a limited range and then flow plastically. Thus, the friction of rubber often deviates appreciably from the laws of friction. The friction depends markedly on the load and the geometry of the contacting surfaces.

For rubber the relationship between friction force and the normal applied load is

$$F \propto W^{2/3} \tag{8.42}$$

Rubber is truly an elastic solid, and this relationship is valid over a wide load range.

## 8.9 WEAR

**Wear** has long been recognized as very important and is usually considered detrimental in machine elements, yet scientific studies of the wear phenomenon began only relatively recently. The historical order of tribological studies has been friction, lubrication, and then wear. Indeed, most research on wear has been conducted in the last 50 years.

This chronology is all the more remarkable in view of the importance of wear to machine element life. In a more general sense, practically everything made by humans wears out, yet the fundamental actions that govern the process remain elusive. Initially, in the tribologic chronology attention was focused on friction, since it governs the ability of machine elements to function properly. The role of lubricants in controlling friction and wear for some machine elements was then explored, with little attention to the detailed mechanism of wear. In fact, it remains true that today's engineer is better equipped to design a machine element to withstand known loads than to specify a given life of the machine element.

Wear is the progressive loss of material from the operating surface of a body occurring as a result of loading and relative motion at the surface. Wear can be classified by the physical nature of the underlying process, such as abrasion, adhesion, and fatigue. These three main types of wear are considered in this section. Other types of wear, such as erosion and corrosion, are beyond the scope of this text.

### 8.9.1    ADHESIVE WEAR

Figure 8.20 shows the mechanism of **adhesive wear,** where material is transferred from one surface to the other by solid-phase welding. Adhesive wear is the most common type of wear and the least preventable. Observe from Figure 8.20 that as the moving asperities pass each other, the microscopic weld often breaks and material is removed from the surface having the lower strength.

The volume of material removed by wear $v$ is directly proportional to the sliding distance $L$ and the normal applied load $W$ and is inversely proportional to the hardness $H$ of the softer of the two materials, or

$$v \propto \frac{WL}{H} \tag{8.43}$$

However, there are many exceptions to this formula. After a dimensionless adhesive wear coefficient $k_1$ is introduced, Equation (8.43) becomes

$$v = \frac{k_1 WL}{3H} \tag{8.44}$$

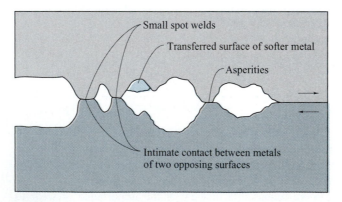

**Figure 8.20**    Adhesive wear model.

**Table 8.8** Coefficients of rubbing friction and adhesive wear constant for selected rubbing materials

| Rubbing materials | Coefficient of friction $\mu$ | Adhesive wear coefficient $k_1$ |
|---|---|---|
| Gold on gold | 2.5 | 0.1–1 |
| Copper on copper | 1.2 | 0.01–0.1 |
| Mild steel on mild steel | 0.6 | $10^{-2}$ |
| Brass on hard steel | 0.3 | $10^{-3}$ |
| Lead on steel | 0.2 | $2 \times 10^{-5}$ |
| Polytetrafluoroethylene (Teflon) on steel | 0.2 | $2 \times 10^{-5}$ |
| Stainless steel on hard steel | 0.5 | $2 \times 10^{-5}$ |
| Tungsten carbide on tungsten carbide | 0.35 | $10^{-6}$ |
| Polyethylene on steel | 0.5 | $10^{-8}$–$10^{-7}$ |

The dimensionless adhesive wear coefficient $k_1$ is a measure of the probability that any interaction between asperities of the two surfaces in contact will produce a wear particle due to any of the wear mechanisms. Thus, if $k_1$ is 1, every junction involving surface contact will produce a wear fragment; if $k_1 = 0.1$, one-tenth of the contacts between surfaces will produce a wear fragment. Studies have shown that, for unlubricated surfaces, the lowest value of $k_1$ is obtained for polyethylene sliding on steel. For this situation $k_1 = 1 \times 10^{-7}$, which means that 1 in 10 million contacts will produce a wear fragment.

Table 8.8 shows some typical values of $k_1$ for several materials in contact. As presented in Equation (8.44), the wear volume is a function not only of $k_1$ but also of the hardness of the softer of the two materials in contact. For example, although the $k_1$ value of polyethylene on steel is one-tenth that for tungsten carbide sliding on itself, the wear volume is 10 times larger because tungsten carbide wears 10 times less than the polymer. Therefore, the volume of the wear fragments is considerably smaller because tungsten is several orders of magnitude harder than polyethylene.

Table 8.8 also gives the coefficient of friction as well as the adhesive wear constant. The wear constant varies by eight orders of magnitude for the various rubbing materials, but the coefficient of friction varies by only one order of magnitude.

**EXAMPLE 8.10**

**Given:** A journal bearing in a dam gate moves slowly and operates under a high load. The entire bearing is made of AISI 1040 steel. To increase the bearing wear life, a material change is considered for one of the surfaces.

**Find:** Which of the following materials will give the longest wear life: brass, lead, polytetrafluoroethylene (PTFE), or polyethylene? The hardness for lead is 30 MPa, for PTFE it is 50 MPa, and for polyethylene it is 70 MPa.

**Solution:** From Equation (8.44) the wear volume is

$$v = \frac{k_1 W L}{3H}$$

The wear volume is a minimum when $H/k_1$ is a maximum. Note that $H \approx 3S_y$. Making use of Table 8.8 for the materials being considered gives the following:

Brass $\qquad \dfrac{H}{k_1} = \dfrac{3S_y}{k_1} = \dfrac{3(75)(10^6)}{10^{-3}} = 225(10^9)\text{ Pa} = 225\text{ GPa}$

Lead $\qquad \dfrac{H}{k_1} = \dfrac{30(10^6)}{2(10^{-5})} = 1500(19^9)\text{ Pa} = 1500\text{ GPa}$

PTFE $\qquad \dfrac{H}{k_1} = \dfrac{50(10^6)}{2(10^{-5})} = 2500(10^9)\text{ Pa} = 2500\text{ GPa}$

Polyethylene $\qquad \dfrac{H}{k_1} = \dfrac{70(10^6)}{10^{-7}} = 700{,}000(10^9)\text{ Pa} = 700{,}000\text{ GPa}$

Thus, polyethylene will give much longer life than any of the other materials.

### 8.9.2   ABRASIVE WEAR

**Abrasive wear** arises when two interacting surfaces are in direct physical contact and one is significantly harder than the other. Under a normal load the asperities of the harder surface penetrate the softer surface, producing plastic deformations and indentation and implying that permanent deformation is occurring. Figure 8.19 illustrates abrasive wear model. A hard conical asperity with slope $\theta$ under a normal load $W$ plows through the softer surface, removing material and producing a groove. The amount of material lost by abrasive wear is

$$v_{\text{abr}} \approx \frac{2k_1}{\pi}\frac{\tan\theta}{H}WL = \frac{k_1 k_2 WL}{H} \qquad (8.45)$$

where $\qquad k_2 = \dfrac{2\tan\theta}{\pi} = \text{abrasive wear coefficient} \qquad (8.46)$

$k_1 = \text{adhesive wear coefficient, given in Table 8.8}$

For most asperities $\theta$ is small, typically between $5°$ and $10°$.

A similarity of form is evident between Equations (8.44) and (8.45), and it is clear that the general **laws of wear** can be stated as follows:

1.  Wear increases with sliding distance.

2.  Wear increases with normal applied load.

3.  Wear decreases as the hardness of the sliding surface increases.

These laws, along with Equations (8.44) and (8.45), reveal the role of the principal variables in abrasive wear, but there are sufficient exceptions to the predicted behavior to justify a measure of caution in their use. Furthermore, wear rates, or the life of rubbing components, cannot be determined without some knowledge of the wear coefficients, which are experimentally determined.

### 8.9.3 FATIGUE WEAR

For nonconformal contacts in such machine elements as rolling-element bearings, gears, friction drives, cams, and tappets, a prevalent form of failure is fatigue wear. In these situations the removal of material results from a cyclic load variation. Figure 8.21 illustrates the stresses developed on and below the surface that deform and weaken the metal. **Fatigue wear** is caused by the propagation of subsurface damage to the surface due to cyclic loadings.

As cyclic loading continues, defects or cracks develop below the surface. Eventually, the defects coalesce near the surface. Material at the surface is then easily broken away, degrading the component's surface and releasing work-hardened particulate contaminants.

Fatigue wear occurs in nonconformal machine elements, even in well-lubricated situations. Cyclic loading of nonconformal surfaces can cause extremely high stresses to the solid material, although the surfaces may never be in direct contact, due to the loading and

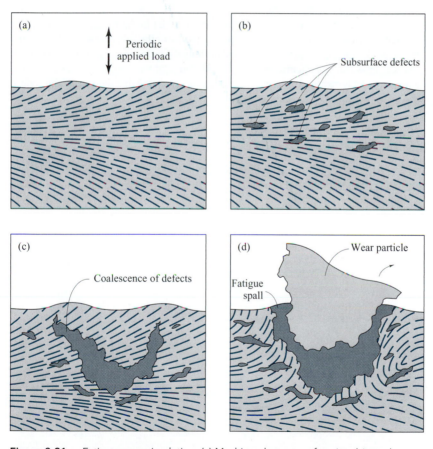

**Figure 8.21**    Fatigue wear simulation. (a) Machine element surface is subjected to cyclic loading; (b) defects and cracks develop near the surface; (c) the cracks grow and coalesce, eventually extending to the surface until (d) a wear particle is produced, leaving a fatigue spall in the material.

their nonconformal nature. After a few million such encounters the asperities will fatigue, and a piece of material (wear particle) may be lost from the surface.

A group of apparently identical nonconformal machine elements subjected to identical loads, speeds, lubrication, and environmental conditions may exhibit wide variation in failure times. For this reason the fatigue wear process must be treated statistically. Thus, the fatigue life of a bearing is normally defined in terms of its statistical ability to survive a certain period of time or by the allowable wear before the component ceases to function adequately. The only requirement for fatigue failure is that the surface material be loaded. The other wear mechanisms require not only loading but also physical contact between the surfaces. If the surfaces are separated by a lubricant film, adhesive wear and abrasive wear are virtually eliminated, but fatigue wear can still occur.

---

**Case Study 8.1**    PIN-ON-DISK WEAR TESTS

**O**ften, engineers need to determine the tribological performance of mating components. In theory, this can be done by manufacturing prototypes of candidate materials and evaluating them in service. However, this approach is time-consuming, laborious, and costly, especially for a large number of materials. The alternative is to perform a simulation in a simplified environment and infer the performance. This difficult practice usually requires significant experience to make inferences with confidence.

**Given:** A common test procedure is the pin-on-disk method illustrated in Figure 8.22. The pin is constructed

from one material in the tribological pair, and the disk is constructed from the other material. By rotating the disk and pushing the pin against the disk, a sliding situation can be simulated. The wear on the pin can be measured with a microscope by measuring the size of the wear scar, and it is given by

$$v = \frac{k_1}{3H} WL = \frac{\pi}{64} \frac{d^4}{R} \qquad (8.47)$$

where $d$ = diameter of wear flat [see Fig. 8.22(b)], m
$\quad\quad\;\; R$ = radius of pin, m

(a)                                    (b)

**Figure 8.22**    (a) Illustration of a pin-on-disk wear test; (b) detail of pin, with dashed line showing the worn surface.

---

## Case Study (Concluded)

**Table 8.9** Test data for aircraft braking system used in Case Study 8.1

| Test | Material | Rotation speed of disk, rpm | Contact force, N | Test duration, min | Wear scar, mm |
|------|----------|------------------------------|-------------------|---------------------|----------------|
| 1 | A | 30 | 100 | 350 | 9.70 |
| 2 | A | 30 | 200 | 100 | 8.81 |
| 3 | A | 60 | 200 | 60 | 16.01 |
| 4 | B | 30 | 100 | 480 | 12.63 |
| 5 | B | 60 | 100 | 480 | 15.27 |
| 6 | C | 60 | 100 | 480 | 15.89 |
| 7 | D | 60 | 200 | 480 | 20.83 |
| 8 | E | 60 | 100 | 240 | 13.02 |

The test data from the pin-on-disk apparatus are shown in Table 8.9.

**Find:** A material recommendation regarding a low-wear component for an aircraft braking system. Use a distance of 150 mm from the pin to the center of disk rotation and an initial pin radius of 100 mm. Because the pin diameter is approximately 30 mm, all tests are valid.

**Solution:** Let $W = 100$ N, $R = 100$ mm, $d = 9.7$ mm, and $L = \pi(300)$ mm. Note that $L$ is per revolution. Substituting the data into Equation (8.47) gives

$$k_1 = 1.317 \times 10^{-8} H \qquad \mathrm{Pa}^{-1}$$

Performing the calculation for all trials gives the results shown in Table 8.10.

From these results the lowest wear coefficient was for material A in test 1. However, the extremely large wear coefficient for the same material in test 3 is clearly a concern. A dramatic increase in wear rate such as this is common for materials with a protective oxide surface. When the speed or stress becomes large enough, the oxide is removed and does not have time to become reestablished. If the application can be reasonably assumed to avoid this condition, material A may be the best; otherwise, material B is superior.

This case study demonstrates typical problems encountered by tribologists. Wear is a complicated phenomenon, and to apply test data, one must be certain that the test properly simulates the actual loading environment. The consequences cannot be overstated: If the test does not simulate the actual loading environment, it has little use because there can be no certainty that the wear mode will be the same as that in the test. With experience such assurances are possible. Another problem is that decisions must often be made with limited experimental data.

**Table 8.10** Test results of wear coefficient divided by hardness used in Case Study 8.1

| Test | 1 | 2 | 3 | 4 | 5 | 6 | 7 | 8 |
|------|-----|------|-------|------|------|------|------|------|
| $k_1/H \times 10^{-8}$ | 1.317 | 1.57 | 14.26 | 2.76 | 2.95 | 3.44 | 5.11 | 3.12 |

---

## 8.10 SUMMARY

In this chapter conformal and nonconformal surfaces were defined. Conformal surfaces fit snugly into each other with a high degree of geometric conformity so that the load is carried over a relatively large area and the area remains essentially constant as the load is increased. Nonconformal surfaces do not geometrically conform to each other well and have small load-carrying surface areas that enlarge with increasing load but are still small relative to those of conformal surfaces.

A lubricant's physical and chemical actions within a lubricated conjunction were described for the four lubrication regimes: hydrodynamic, elastohydrodynamic, partial, and boundary. Hydrodynamic lubrication is characterized by conformal surfaces. The lubricating film is thick enough to prevent the opposing solids from coming into contact. Friction arises only from the shearing of the viscous lubricant. The pressures developed in hydrodynamic lubrication are low (usually less than 5 MPa), so that the surfaces may generally be considered rigid and the pressure-viscosity effects are small. Three modes of pressure development within hydrodynamic lubrication were presented: slider, squeeze, and external pressurization. For hydrodynamic lubrication with sliding motion, the minimum film thickness is quite sensitive to load, being inversely proportional to the square root of the normal applied load.

Elastohydrodynamic lubrication (EHL) is characterized by nonconformal surfaces, and again there is no asperity contact of the solid surfaces. Two modes of elastohydrodynamic lubrication exist: hard and soft. Hard EHL is characterized by metallic surfaces, and soft EHL, by surfaces made of elastomeric materials. The pressures developed in hard EHL are high (typically between 0.5 and 4 GPa), so that elastic deformation of the solid surfaces becomes important as do the pressure-viscosity effects of the lubricant. As with hydrodynamic lubrication, friction is due to the shearing of the viscous lubricant. The minimum film thickness for hard EHL is relatively insensitive to load because the contact area increases with increasing load, thereby providing a larger lubricated area to support the load. For soft EHL the elastic distortions are large, even for light loads, and the viscosity within the conjunction varies little with pressure because the pressures are relatively low and the elastic effect predominates. Both hydrodynamic and elastohydrodynamic lubrication are fluid-film lubrication phenomena in that the film is thick enough to prevent opposing solid surfaces from coming into contact.

In boundary lubrication considerable asperity contact occurs, and the lubrication mechanism is governed by the physical and chemical properties of thin surface films that are of molecular size (from 1 to 10 nm). The friction characteristics are determined by the properties of the solids and the lubricant film at the common interfaces. Partial lubrication (sometimes called mixed lubrication) is governed by a mixture of boundary and fluid-film effects. Most of the scientific unknowns lie in this lubrication regime.

Because in many lubricated machine elements the solids are separated by thin films, the physical nature of the solids' surface topographies must be understood. In computing the parameters that define surface texture, the height measurements were made from a defined reference line. The mean, or M system, reference line was used; it is based on selecting the mean line as the centroid of the profile. The area of peaks above this line equals the area of valleys below this line. Two ways of computing surface parameters, namely, centerline average and root-mean-squared roughness, were also discussed.

For concentrated loads both elliptical and rectangular contacts were discussed. It was emphasized that a number of machine elements considered later in the text, such as rolling-element bearings and gears, experience concentrated loading. The major features of concentrated loading (first presented in Chap. 2) are that the load is applied to an extremely small area and that the load per unit area is much larger than normally found in distributed loading situations. Equations for curvature sum and maximum contact pressure were presented, as well as simplified expressions that enable the dimensions of the contact, maximum pressure, and maximum deformation to be determined.

Features of rolling and sliding friction were presented as well as the laws of friction. The coefficient of friction was defined as the ratio of the friction force to the normal applied load. Sliding friction of metals, polymers, plastics, and rubber was also discussed. For these various classes of material the relationship between the friction force and the normal applied load varies significantly.

The useful life of engineering machine components is limited by breakage, obsolescence, and *wear*. Practically everything made will wear out. Wear is the progressive loss of substance from the operating surface of a body occurring as a result of loading and relative motion at the surface. The most common forms of wear were discussed:

1. In adhesive wear, material transfers from one surface to another due to solid-phase welding.
2. In abrasive wear, material is displaced by hard particles.
3. In fatigue wear, material is removed by cyclic stress variations even though the surfaces may not be in contact.

## KEY WORDS

**abrasive wear**   wear caused by physical damage from penetration of a hard surface into a softer one.

**absolute viscosity**   shear stress divided by shear strain rate, having SI units of pascal-seconds.

**adhesive wear**   wear caused by solid-state weld junctions of two surfaces.

**boundary lubrication**   lubrication condition where considerable asperity interaction occurs between solids and lubrication mechanism is governed by properties of thin surface films that are of molecular proportion.

**conformal surfaces**   surfaces that fit snugly into each other with high degree of conformity, as in journal bearings.

**elastohydrodynamic lubrication**   lubrication condition where nonconformal surfaces are completely separated by lubricant film and no asperities are in contact.

**ellipticity parameter**   diameter of contact ellipse in $y$ direction divided by diameter of contact ellipse in $x$ direction.

**fatigue wear**   wear caused by propagation of subsurface damage to surface due to cyclic loading.

**film parameter**   minimum film thickness divided by composite surface roughness.

**fluid-film lubrication**   lubrication condition where lubricated surfaces are completely separated by a lubricant film and no asperities are in contact.

**friction**   force resisting relative motion between surfaces in contact.

**hard EHL**   elastohydrodynamic lubrication of surfaces with high elastic modulus materials, such as metals and ceramics.

**hydrodynamic lubrication**   fluid-film lubrication of conformal surfaces, as in journal bearings.

**kinematic viscosity**   absolute viscosity divided by density, with SI units of meters squared per second.

**laws of friction**   these can be summarized as follows:

1. Friction force is proportional to normal load.
2. Friction force is not dependent on apparent area of contact solids; that is, it is independent of size of solid bodies.
3. Friction force is independent of sliding velocity.

**laws of wear**    these can be summarized as follows:

1.    Wear increases with sliding distance.

2.    Wear increases with normal applied load.

3.    Wear decreases as hardness of sliding surface increases.

**lubricant**    any substance that reduces friction and wear and provides smooth running and satisfactory life for machine elements.

**mixed lubrication**    same as **partial lubrication.**

**nonconformal surfaces**    surfaces that do not conform to each other very well as in rolling-element bearings.

**partial lubrication**    lubrication condition where the load between two surfaces in contact is transmitted partially through lubricant film and partially through asperity contact.

centerline average or arithmetic average surface roughness.

root-mean-square surface roughness.

**running in**    process through which beneficial wear causes surfaces to adjust to each other and improve performance.

**soft EHL**    elastohydrodynamic lubrication of surfaces with low elastic modulus materials, such as rubber and polymers.

**tribology**    study of lubrication, friction, and wear of moving or stationary parts.

**wear**    progressive loss of material from operating surface of body occurring as result of loading and relative motion of surface.

---

# RECOMMENDED READINGS

Arnell, R. D., Davies, P. B., Halling, J., and Wholmes, T. L. (1991) *Tribology: Principles and Design Applications,* Springer-Verlag, New York.

Bhushan, B. (2002) *Introduction to Tribology,* Wiley, New York.

Bhushan, B., ed. (2001) *Modern Tribology Handbook,* CRC Press, Boca Raton, FL.

Dowson, D. (1998) *History of Tribology,* 2nd ed., Professional Engineering Publishing, London.

Hamrock, B. J., Schmid, S. R., and Jacobson, B. O. (2004) *Fluid Film Lubrication,* 2nd ed., Marcel Dekker, New York.

Johnson, K. L. (1985) *Contact Mechanics,* Cambridge University Press, Cambridge, United Kingdom.

Ludema, K. (1996) *Friction, Wear, Lubrication: A Textbook in Tribology,* CRC Press, Boca Raton, FL.

Rabinowicz, E. (1995) *Friction and Wear of Materials,* 2nd ed., Wiley, New York.

Szeri, A. Z. (1998) *Fluid Film Lubrication,* Cambridge University Press, Cambridge, United Kingdom.

---

# REFERENCES

Beerbower, A. (1972) "Boundary Lubrication," GRU.IGBEN.72, Report on Scientific and Technical Application Forecasts (available NTIS AD-747336).

Bowden, F. P., and Tabor, D. (1973) *Friction—An Introduction to Tribology,* Anchor Press/Doubleday, New York.

Hamrock, B. J., and Anderson, W. J. (1983) *Rolling-Element Bearings,* NASA Reference Publication 1105.

Hamrock, B. J., and Brewe, D. E. (1983) "Simplified Solution for Stresses and Deformations," *Journal of Lubrication Technology,* vol. 105, no. 2, pp. 171–177.

Hamrock, B. J., and Dowson, D. (1977) "Isothermal Elastohydrodynamic Lubrication of Point Contacts, Part III—Fully Flooded Results," *Journal of Lubrication Technology,* vol. 99, no. 2, pp. 264–276.

Hamrock, B. J., Schmid, S. R., and Jacobson, B. O. (2004) *Fundamentals of Fluid Film Lubrication,* 2nd ed., Marcel Dekker, New York.

Hertz, H. (1881) "The Contact of Elastic Solids," *J. Reine Angew. Math.,* vol. 92, pp. 156–171.

Hutchings, I. M. (1992) *Tribology—Friction and Wear of Engineering Materials,* Edward Arnold, London.

Jones, W. R., et al. (1975) "Pressure-Viscosity Measurements for Several Lubricants to $5.5 \times 10^8$ Pa · s and 149°C," *ASLE Transactions,* vol. 18, no. 4, pp. 249–262.

Kalpakjian, S., and Schmid, S. R. (2003) *Manufacturing Processes for Engineering Materials,* 4th ed., Prentice-Hall, Upper Saddle River, NJ.

Seireg, A. S., and Dandage, S. (1982) "Empirical Design Procedure for the Thermodynamic Behavior of Journal Bearings," *Journal of Lubrication Technology,* vol. 104, pp. 135–148.

Tallian, T. E., et al. (1967) "On Computing Failure Modes in Rolling Contacts," *ASLE Transaction,* vol. 10, no. 4, pp. 418–435.

## PROBLEMS

### Section 8.2

**8.1** Describe the difference between conformal and nonconformal surfaces.

**8.2** Determine which of the following contact geometries is conformal and which is nonconformal.

    *a)*     Meshing gear teeth
    *b)*     Ball and inner race of a ball bearing
    *c)*     Journal bearing
    *d)*     Railway wheel and rail contact
    *e)*     Car making contact with the road
    *f)*     Egg and egg cup
    *g)*     Human knee

### Section 8.3

★ **8.3** A single ball rolling in a groove has a 10-mm diameter and a 4-N normal force acting on it. The ball and the groove have a 200-GPa modulus of elasticity and a Poisson's ratio of 0.3. Assuming a 6.08-mm-radius groove in a semi-infinite steel block, determine the following:

    *a)*     Contact zone dimensions. *Ans.* $D_y = 0.227$ mm, $D_x = 0.0757$ mm.

*Indicates problem of greater difficulty

      *b)*      Maximum elastic deformation. *Ans.* $\delta_{max} = 0.391$ $\mu$m.

      *c)*      Maximum pressure. *Ans.* 0.445 GPa.

★ **8.4**   A solid cylinder rolls with a load against the inside of an outer cylindrical race. The solid cylinder radius is 20 mm, and the race internal radius is 150 mm. The race and the roller have the same axial length. What is the radius of a geometrically equivalent cylinder near a plane? The cylinder is made of silicon nitride ($E = 314$ GPa, $v = 0.26$), and the race is made of stainless steel ($E = 193$ GPa, $v = 0.30$). If the normal applied load per unit width is 12,000 N/m, determine

      *a)*      Contact semiwidth. *Ans.* $b^* = 50.93\mu$m.

      *b)*      Maximum surface stress. *Ans.* $p_{max} = 150$ MPa.

      *c)*      Maximum elastic deflection. *Ans.* $\delta_{max} = 0.393$ $\mu$m.

   Also, indicate what these values would be if the silicon nitride cylinder were replaced with a stainless steel cylinder.

**8.5**   A 100-mm-diameter shaft has a 20-mm-diameter ball rolling around the outside. Find the curvature sum, the maximum contact stress, the maximum deflection, and the contact dimensions if the ball load is 1000 N. The ball is made of silicon nitride ($E = 314$ GPa, $v = 0.26$), and the shaft is made of steel ($E = 206$ GPa, $v = 0.30$). Also, determine these values if both ball and shaft are made of steel. *Ans.* For SiN on steel, $D_y = 7.80 \times 10^{-4}$ m, $D_x = 6.95 \times 10^{-4}$ m.

**8.6**   The ball–outer race contact of a ball bearing has a 17-mm ball diameter, an 8.84-mm outer-race groove radius, and a 44.52-mm radius from the bearing axis to the bottom of the groove. The load on the most highly loaded ball is 25,000 N. Calculate the dimensions of the contact ellipse and the maximum deformation at the center of the contact. The race and the ball are made of steel. *Ans.* $D_y = 9.42$ mm, $D_x = 1.35$ mm.

## Section 8.4

**8.7**   Describe three applications for each of the four lubrication regimes: hydrodynamic lubrication, elastohydrodynamic lubrication, boundary lubrication, and partial lubrication.

**8.8**   A hydrodynamic journal bearing is loaded with a normal load $W$ and is rotating with a surface velocity $u_b$. Find how much higher the rotational speed the bearing needs to be to maintain the same minimum film thickness if the load $W$ is doubled.

★ **8.9**   A ball bearing lubricated with a mineral oil runs at 4000 rpm. The bearing is made of AISI 5210 steel and is loaded to a medium-high level. Find how much thinner the oil film will be if the load is increased 5 times. How much must the speed be increased to compensate for the higher load, while keeping the oil film thickness constant? *Ans.* $h_{min,1} = 1.125h_{min,2}$, 19% increase in speed.

## Section 8.5

★ **8.10**  Two bearing types, one hydrodynamic journal bearing and one ball bearing, operate with a film parameter of 5. Calculate how much lower the speed for each bearing has to be to decrease the film parameter to 3. *Ans.* Ball bearing: $u_{b2} = 0.4718u_{b1}$.

**8.11**  A machined steel surface has roughness peaks and valleys. The roughness wavelength is such that 400 peaks and 400 valleys are found during one roughness measurement. Find how much the arithmetic average surface roughness $R_a = 0.2\mu$m will change if one of the surface roughness peaks and one of the valleys increase from 0.25 to 2.5 $\mu$m. *Ans.* 0.2056 $\mu$m.

★ **8.12** A surface with a triangular sawtooth roughness pattern has a peak-to-valley height of 4 $\mu$m. Find the $R_a$ and $R_q$ values. *Ans. $R_a = 1.00$ $\mu$m.*

**8.13** A precision ball bearing has a race rms surface roughness of 0.07 $\mu$m and a ball rms surface roughness of 0.02 $\mu$m. Changing the roughness of the components may give a higher $\Lambda$ value. Determine which components to smooth if it costs equally to halve the roughness of the race and to halve the roughness of the balls. Note that $h_{min} = 0.2$ $\mu$m.

**8.14** The minimum film thickness in a particular application is 10 $\mu$m. Assume that the surface roughnesses of the two surfaces being lubricated are identical. What lubrication regime would you expect the application to be operating in? Also, what is the maximum surface roughness allowed?

## Section 8.6

**8.15** Two equally rough surfaces with rms surface roughness $R_q$ are lubricated and loaded together. The oil film thickness is such that $\Lambda = 2$. Find the $\Lambda$ value if one of the surfaces is polished so that $R_q \to 0$ (i.e., the surface becomes absolutely smooth). *Ans. $\Lambda = 2.83$.*

★ **8.16** Two lubricated surfaces have rms surface roughnesses $R_q$ of 0.23 and 0.04 $\mu$m, respectively. By using a new honing machine either surface can be made twice as smooth. For good lubrication, which surface roughness is it more important to decrease?

## Section 8.7

**8.17** The absolute viscosity of a fluid at atmospheric conditions is $6 \times 10^{-3}$ kgf · s/m². Give this absolute viscosity in

    *a)*    Reyn
    *b)*    Poise
    *c)*    Pound-seconds per square inch
    *d)*    Newton-seconds per square meter

**8.18** Given a fluid with an absolute viscosity $\eta$ between two 1-m² surfaces spaced 1 mm apart, find how fast the surfaces will move relative to each other if a 10-N force is applied in the direction of the surfaces when $\eta$ is

    *a)*    0.001 N · s/m² (water). *Ans. $u_b = 10$ m/s.*
    *b)*    0.100 N · s/m² (a thin oil at room temperature).
    *c)*    10.0 N · s/m² (syrup; cold oil). *Ans. $u_b = 1$ mm/s.*
    *d)*    $10^8$ N · s/m² (asphalt).

## Section 8.8

★ **8.19** A polymer box stands on a slope. The box weighs 10 kg, and the slope angle is 25°. To pull the box down the slope, a force $F = 5$ N is required. The friction force is proportional to the load raised to 0.75. Find the additional weight needed in the box to make it slide down the slope. *Ans. 5.60 kg.*

**8.20** A moose suddenly jumps out onto a dry asphalt road 80 m in front of a car running at 108 km/h. The maximum coefficient of friction between the rubber tires and the road is 1.0 when they just start to slide and 0.8 when the locked wheels slide along the road. It takes 1 s for the driver to apply the brakes after the moose jumps onto the road. Find whether the car can stop in time before it hits the moose

a)     With locked wheels sliding along the road
b)     With an automatic braking system (ABS) that keeps the sliding speed so low that maximum friction is maintained

★ **8.21** Given a block on an incline as shown in sketch *a*, find the force $P$ required

a)     To prevent motion downward. *Ans.* 198 lb.
b)     To cause motion upward. *Ans.* 760 lb.

The coefficient of friction is 0.30. Draw a free-body diagram for both situations showing the forces involved.

$P \longrightarrow$   $W = 600$ lb

35°

Sketch a, for Problem 8.21.

### Section 8.9

★ **8.22** Three equal blocks of plastic are carrying the weight of a steel slider. At first each block carries one-third of the weight. But the central block is made of polytetrafluoroethylene (PTFE), and the two outer blocks are made of polyethylene; so they have different adhesive wear constants (PTFE, $k_1 = 2 \times 10^{-5}$; polyethylene, $k_1 = 2 \times 10^{-8}$). Determine how the load redistributes between the blocks if the hardnesses of the plastics are assumed to be the same.

**8.23** Given the plastic blocks in Problem 8.22 and sharing the load so that the wear is equal on each block, find the coefficient of friction.

★ **8.24** A 50-kg copper piece is placed on a flat copper surface, as shown in sketch *b*. Assuming that copper has a hardness of 275 MPa, calculate the shear stress, abrasive wear volume, and adhesive wear volume. *Ans.* $\tau = 312$ MPa, $k_2 = 0.0669$.

6°

7.5

Sketch b, for Problem 8.24.

# MACHINE ELEMENTS

**Outline**

# COLUMNS

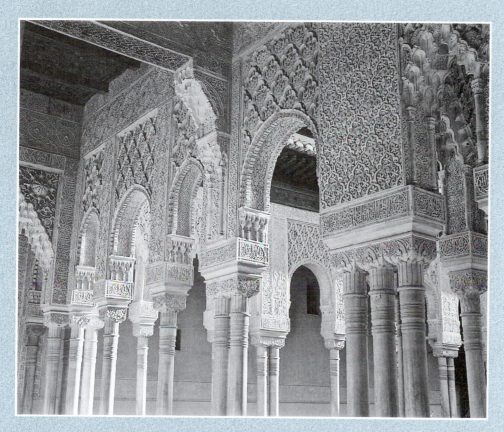

Columns from the
Alhambra in Granada,
Spain. (Art Resource)

*And as imagination bodies forth the forms of things unknown,*
*The poet's pen turns them to shapes*
*And gives to airy nothingness a local habitation and a name*
William Shakespeare, *A Midsummer Night's Dream*

## SYMBOLS

| | |
|---|---|
| $A$ | cross-sectional area, $m^2$ |
| $a$ | side of square, m |
| $a_o$ | outer dimension of hollow square, m |
| $a_i$ | inner dimension of hollow square, m |
| $b$ | base dimension of rectangular section, m |
| $C_c$ | slenderness ratio |
| $C_1, C_2$ | integration constants |
| $c$ | distance from neutral axis to outer fiber of column, m |
| $d$ | diameter, m |
| $E$ | modulus of elasticity, Pa |
| $E_t$ | tangent modulus, Pa |
| $e$ | eccentricity, m |
| $g$ | gravitational acceleration, $9.807$ m/s$^2$ |
| $h$ | height dimension of rectangular section, m |
| $I$ | area moment of inertia, $m^4$ |
| $l$ | length of column, m |
| $l_e$ | effective length of column, m |
| $M$ | moment, N · m |
| $M'$ | statically equivalent moment, N · m |
| $m_a$ | mass, kg |
| $n$ | integer, 1, 2, . . . |
| $n_s$ | stress reduction factor in AISC criteria |
| $P$ | force, N |

| | |
|---|---|
| $R_e$ | resultant force, N |
| $r$ | radius of column, m |
| $r_g$ | radius of gyration, m |
| $S_y$ | yield strength, Pa |
| $s$ | arc length, m |
| $t$ | time, s |
| $t_h$ | thickness, m |
| $V$ | shear force, N |
| $W_e$ | weight, N |
| $x$ | length dimension of column, m |
| $y$ | transverse dimension or deflection of column, m |
| $\theta$ | position angle, deg |
| $\rho$ | density, kg/m$^3$ |
| $\sigma$ | normal stress, Pa |
| $\sigma_{all}$ | allowable normal stress, Pa |

## SUBSCRIPTS

| | |
|---|---|
| cr | critical |
| $E$ | Euler |
| $i$ | inner |
| $J$ | Johnson |
| max | maximum |
| $o$ | outer |
| T | tangency point |

## 9.1 INTRODUCTION

The basic understanding of loads, stresses, and deformations obtained in the preceding chapters is related here to columns. A **column** is a straight and long (relative to its cross section) bar that is subjected to compressive, axial loads. The reason for this special consideration of columns is that failures due to yielding, determined from Equation (4.24), and due to deformation, determined from Equation (4.25), are not correct in predicting failures of long columns. Specific failure theories for columns need to be developed. Because of their slender shape, columns tend to deform laterally upon loading; and if the deflection becomes larger than their respective critical values, they fail catastrophically. This situation is known as **buckling,** which can be defined as a sudden, large deformation of a structure due to a slight increase of the applied load, under which the structure had exhibited little, if any, deformation before this increase.

This chapter first describes the meaning of elastic stability and end conditions. It then establishes the various failure criteria for concentrically loaded columns and describes the nature of the instability that can occur, thus predicting when buckling will take place. It also establishes the failure criteria for eccentrically loaded columns, so that proper design of concentric and eccentric columns is ensured.

# 9.2    EQUILIBRIUM REGIMES

To understand why columns buckle, it is first necessary to understand the equilibrium regimes. An important question to be answered is, When an equilibrium position is disturbed slightly, does the component tend to return to the equilibrium position or does it tend to depart even farther? To visualize what is happening, consider Figure 9.1, which shows the three equilibrium regimes: stable, neutral, and unstable.

## 9.2.1    STABLE EQUILIBRIUM

Figure 9.1(a) attempts to illustrate what occurs in stable equilibrium. Assume that the surfaces are frictionless and that the sphere has a light weight. The forces on the sphere (gravity and normal surface reaction) are in balance whenever the surface is horizontal. These balanced positions are indicated by a zero in the figure. Figure 9.1(a) shows the sphere displaced slightly from its equilibrium position. The forces on it no longer balance, but the resultant imbalance is a restoring force (i.e., gravity is accelerating the sphere back toward the equilibrium position). Such a situation is called *stable equilibrium*.

## 9.2.2    NEUTRAL EQUILIBRIUM

Figure 9.1(b) considers the sphere in neutral equilibrium. After the sphere has been slightly displaced from the equilibrium position, it is still in equilibrium at the displaced position, and there is no tendency either to return to the previous position or to move to some other position. Equilibrium is always satisfied. Because the surface is flat, the sphere does not move after being placed in another position; this is *neutral equilibrium*.

## 9.2.3    UNSTABLE EQUILIBRIUM

Figure 9.1(c) shows the case of unstable equilibrium, which is the opposite situation from that presented in Figure 9.1(a). When the sphere is displaced from its equilibrium position (to either the right or the left), the resultant imbalance is a disturbing force (i.e., it accelerates the sphere away from its equilibrium position). Such a situation is called *unstable equilibrium*. Gravity and convex surfaces cause the sphere to move farther from the balanced position.

   To generalize from this example, unstable equilibrium occurs if for small displacements from the equilibrium position the disturbing forces tend to accelerate the part away from the equilibrium position. Columns in unstable equilibrium are unreliable and hazardous; a small displacement can cause a catastrophic change in the configuration of a column. Thus, as the load on a column increases, a critical load is reached at which unstable equilibrium occurs and the column will not return to its straight configuration. The load cannot be increased

    (a)                        (b)                        (c)

**Figure 9.1**    Depiction of equilibrium regimes. (a) Stable; (b) neutral; (c) unstable.

beyond this value unless the column is laterally restrained. Thus, for long, slender columns a critical buckling load exists. This critical buckling load (when defined by the cross-sectional area) gives a critical buckling stress. For columns, this critical buckling stress is much lower than the yield strength of the material. Thus, columns will generally fail from buckling long before they fail from yielding.

The shape as well as the load establishes if buckling will occur. A column may be viewed as a straight bar with a large slenderness ratio $l/r$ (typically 100) subjected to axial compression. Buckling occurs in a column with a large slenderness ratio when the column is loaded to a critical load and marked changes in deformation occur that do not result from the material yielding.

**EXAMPLE 9.1**

**Given:** A simple pendulum (Fig. 9.2) in which a ball is hung by a thin wire and is acted on by gravitational acceleration.

**Find:** Is the equilibrium neutral, stable, or unstable?

**Solution:** The force restoring the ball to the center position ($\theta = 0$) is

$$P = m_a g \sin \theta \qquad \text{(a)}$$

From simple harmonic motion

$$P = -m_a l \frac{d^2\theta}{dt^2} \qquad \text{(b)}$$

Combining Equations ($a$) and ($b$) gives

$$\frac{d^2\theta}{dt^2} = -\frac{g}{l} \sin \theta \qquad \text{(c)}$$

**Figure 9.2**     Pendulum used in Example 9.1.

For any angle $\theta$ the angular acceleration $d^2\theta/dt^2$ has the opposite sign from $\theta$ for $-\pi \le \theta \le \pi$. Thus, the ball will always return to the center position ($\theta = 0$), implying that stable equilibrium prevails.

## 9.3 CONCENTRICALLY LOADED COLUMNS

### 9.3.1 LINEAR ELASTIC MATERIALS

Figure 9.3(a) shows a concentrically loaded column with pinned ends; thus the ends are kept in position but are free to rotate. Assume that the column is initially straight and that the load is concentric, as depicted in Figure 9.3(b). Figure 9.3(c) shows a free-body diagram of the loads acting on the column.

Equation (5.3) describes the moment as

$$M = EI\frac{d^2y}{dx^2}$$

Equilibrium of the section cut from the bar requires that $M = -Py$. Substituting this into Equation (5.3) gives

$$\frac{d^2y}{dx^2} + \frac{P}{EI}y = 0 \qquad (9.1)$$

(a)  (b)  (c)

**Figure 9.3** Column with pinned ends. (a) Assembly; (b) deformation shape; (c) load acting.

This is a homogeneous, second-order, linear differential equation with constant coefficients. The general solution of this equation is

$$y = C_1 \sin\left(x\sqrt{\frac{P}{EI}}\right) + C_2 \cos\left(x\sqrt{\frac{P}{EI}}\right) \tag{9.2}$$

where $C_1$ and $C_2$ are integration constants.

The boundary conditions are

1.  $y = 0$ at $x = 0 \rightarrow C_2 = 0$
2.  $y = 0$ at $x = l$, so that

$$C_1 \sin\left(l\sqrt{\frac{P}{EI}}\right) = 0 \tag{9.3}$$

However, $C_1$ cannot be equal to zero; otherwise, a trivial solution of $y = 0$ results and the column will always remain straight, which is contrary to experience. The other possibility of satisfying Equation (9.3) is for

$$\sin\left(l\sqrt{\frac{P}{EI}}\right) = 0 \tag{9.4}$$

which is satisfied when

$$l\sqrt{\frac{P}{EI}} = n\pi \tag{9.5}$$

$$P = \frac{n^2\pi^2 EI}{l^2} \tag{9.6}$$

where $n = 1, 2, \ldots$. The smallest value of $P$ is obtained for $n = 1$. Thus, the critical load $P_{cr}$ for a column with pinned ends is

$$P_{cr} = \frac{\pi^2 EI}{l^2} \tag{9.7}$$

This load is sometimes called the **Euler load,** named after the Swiss mathematician Leonhard Euler, who originally solved this problem in 1757.

An interesting aspect of Equation (9.7) is that the critical load is independent of the material's strength; rather, it depends only on the column's dimensions (expressed in $I$ and $l$) and the material's modulus of elasticity $E$. For this reason, as far as elastic buckling is concerned, columns made, for example, of high-strength steel offer no advantage over those made of lower-strength steel, since both steels have approximately the same modulus of elasticity.

Another interesting aspect of Equation (9.7) is that the critical load capacity of a column will increase as the moment of inertia of the cross-sectional area increases. Thus, columns are designed so that most of their cross-sectional area is located as far as possible from the section's principal centroidal axes. This implies that a hollow tube is preferred over a solid section.

It is also important to realize that a column will buckle about the principal axis of the cross section having the least moment of inertia (the weakest axis). For example, a column having a rectangular cross section, such as a meter stick, as shown in Figure 9.4, will buckle about the $x$ axis and not the $y$ axis.

Substituting Equation (9.5) and $C_2 = 0$ into Equation (9.2) gives the buckling shape as

$$y = C_1 \sin \frac{\pi x}{l} \qquad (9.8)$$

when $x = l/2$, $y = y_{max}$, and $C_1 = y_{max}$,

$$\therefore \qquad y = y_{max} \sin \frac{\pi x}{l} \qquad (9.9)$$

Thus, the buckling shape varies sinusoidally, with zero at the ends and a maximum at half-length.

Engineers are often interested in defining the critical stress of a column. The radius of gyration $r_g$, given in Equation (4.14), is substituted into Equation (9.7), giving the critical stress for the Euler equation as

$$(\sigma_{cr})_E = \frac{P_{cr}}{A} = \frac{\pi^2 E}{(l/r_g)^2} \qquad (9.10)$$

Thus, the smallest radius of gyration for a section should be used. Also, the critical stress $\sigma_{cr}$ is an average stress in the column just before the column buckles. This stress is an elastic stress and is therefore less than or equal to the material's yield strength.

**Figure 9.4**
Buckling of rectangular section.

### 9.3.2 INELASTIC BUCKLING

Section 9.3.1 determined the stress at which a column will buckle if the material in the column is linear elastic. As presented in Chapter 3, this is the case for most materials below the proportional limit. Above the proportional limit and up to the yield point, the material may still be elastic, but it will not behave linearly. The effect is that the elastic modulus at the buckling stress in Equation (9.10) may be significantly lower than one would expect from published values of elastic modulus. Equation (9.10) is often modified as

$$\sigma_{cr} = \frac{\pi^2 E_t}{(l/r_g)^2} \qquad (9.11)$$

where $E_t$ is the **tangent modulus,** defined as the elastic modulus at the stress level in the column.

Equation (9.11) is called the **tangent modulus equation** or the **Essenger equation.** It is extremely difficult to apply in design, since the tangent modulus is never well quantified. However, it is mentioned here because of the importance of the tangent modulus effect.

## 9.4 END CONDITIONS

Table 9.1 shows four end conditions. End conditions affect the effective length of the column; the effective length is the overall column length minus the portion that takes into account the end conditions. The critical load and stress given in Equations (9.7) and (9.10), respectively,

**Table 9.1**     Effective length for four end conditions

| End condition description | Both ends pinned | One end pinned and one end fixed | Both ends fixed | One end fixed and one end free |
|---|---|---|---|---|
| Illustration of end condition | $l = l_e$ | $l_e = 0.7l$ | $l_e = 0.5l$ | $l_e = 2l$ |
| Theoretical effective column length | $l_e = l$ | $l_e = 0.7l$ | $l_e = 0.5l$ | $l_e = 2l$ |
| AISC (1989)–recommended effective column length | $l_e = l$ | $l_e = 0.8l$ | $l_e = 0.65l$ | $l_e = 2.1l$ |

are modified simply by replacing $l$ with the effective length $l_e$ for the corresponding end condition. Substituting $l_e$ for $l$ in Equations (9.7) and (9.10) gives

$$(P_{\text{cr}})_E = \frac{\pi^2 EI}{l_e^2} \tag{9.12}$$

$$(\sigma_{\text{cr}})_E = \frac{(P_{\text{cr}})_E}{A} = \frac{\pi^2 E}{(l_e/r_g)^2} \tag{9.13}$$

These equations for the critical load and stress for the Euler criterion (see Sec. 9.5) are valid for any end condition. The minimum American Institute of Steel Construction (AISC) recommendations given in Table 9.1 apply to end constructions where "ideal conditions are approximated."

EXAMPLE 9.2

**Given:** A column has a square tubular cross section. The wall thickness is 10 mm, and the column length is 12 m. The column is axially loaded in compression and has pinned ends. The mass of the tubular column must be no larger than 213.3 kg.

**Find:** Determine which metal in Tables A.1 and A.2 gives the highest buckling load; also calculate the critical load.

**Solution:** The area of a square tubular cross section with outside dimension $h$ is

$$A = h^2 - (h - 0.02)^2 \qquad \text{(a)}$$

The mass of the column is

$$m_a = \rho A l = \rho l[h^2 - (h - 0.02)^2] \qquad \text{(b)}$$

The area moment of inertia for a square tube is

$$I = \frac{h^4 - (h - 0.02)^4}{12} = \frac{[h^2 - (h - 0.02)^2][h^2 + (h - 0.02)^2]}{12} = \frac{m_a[h^2 + (h - 0.02)^2]}{12\rho l} \qquad \text{(c)}$$

Substituting Equation (c) into Equation (9.7) gives the critical load as

$$P_{cr} = \frac{\pi^2 E m_a}{12\rho l^3}[h^2 + (h - 0.02)^2] \qquad \text{(d)}$$

Thus, the largest $P_{cr}$ would result from the largest value of $E/\rho$. From Tables A.1 and A.2 observe that magnesium would give the largest $E/\rho$ and that $E = 45$ GPa and $\rho = 1740$ kg/m³. In Equation (b), given that $m_a = 213.3$ kg, $l = 12$ m, and $\rho = 1740$ kg/m³,

$$\frac{m_a}{\rho l} = 0.04(h - 0.01)$$

$$\frac{213.3}{(1740)(12)(0.04)} + 0.01 = h$$

$$h = 0.2654 \text{ m} = 265.4 \text{ mm}$$

From Equation (d) the critical load is

$$P_{cr} = \frac{\pi^2(45)(10^9)(213.3)}{12(1740)(12)^3}[(0.2654)^2 + (0.2454)^2] = 3.431 \times 10^5 \text{ N}$$

## 9.5 EULER'S BUCKLING CRITERION

One approach to determining the buckling criteria for columns is simply to state that the critical stress is equivalent to the Euler critical stress as long as it is less than the allowable stress. The American Institute of Steel Construction (1989) assumes that the proportional limit of a material exists at one-half the yield strength, or that the allowable stress for elastic buckling is

$$\sigma_{\text{all}} = 0.5 S_y \qquad \text{(9.14)}$$

where $S_y$ = yield strength of material in Table A.1 for several ferrous alloys. Substituting Equation (9.14) into Equation (9.13) gives the **slenderness ratio** $C_c$ from Euler's formula as

$$C_c = \left(\frac{l_e}{r_g}\right)_E = \sqrt{\frac{2E\pi^2}{S_y}}$$  (9.15)

## 9.6    JOHNSON'S BUCKLING CRITERION

Figure 9.5 shows curves for normal stress as a function of slenderness ratio plotted from the Euler and Johnson equations. The abrupt change in the Euler curve [Eq. (9.7)] as it approaches the yield strength (see point A in Fig. 9.5) results in empirical modification at this location. Perhaps the most widely used modification is the parabola proposed by J. B. Johnson around the turn of the century. The **Johnson equation** is

$$(\sigma_{cr})_J = \frac{(P_{cr})_J}{A} = S_y - \frac{S_y^2}{4\pi^2 E}\left(\frac{l_e}{r_g}\right)^2$$  (9.16)

Also shown in Figure 9.5 is the tangency point T, the point at which the Euler and Johnson equations are equal. This tangency point distinguishes between intermediate columns (Johnson range) and long columns (Euler range). To determine the value of $(l_e/r_g)_T$, equate Equations (9.13) and (9.16). Thus,

$$\left(\frac{l_e}{r_g}\right)_T^4 - \frac{4\pi^2 E}{S_y}\left(\frac{l_e}{r_g}\right)_T^2 + \frac{4\pi^4 E^2}{S_y^2} = 0$$  (9.17)

Solving for $(l_e/r_g)_T$ gives

$$\left(\frac{l_e}{r_g}\right)_T = \sqrt{\frac{2\pi^2 E}{S_y}} = \sqrt{\frac{\pi^2 EA}{P_{cr}}}$$  (9.18)

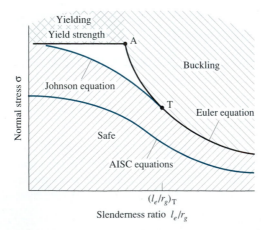

**Figure 9.5**    Normal stress as function of slenderness ratio obtained by Euler, Johnson, and AISC equations, as well as yield strength.

Note from Equations (9.15) and (9.18) that $(l_e/r_g)_E = (l_e/r_g)_T$. If $l_e/r_g \leq (l_e/r_g)_T$, the Johnson equation given in Equation (9.16) should be used; if $l_e/r_g \geq (l_e/r_g)_T$, the Euler equation given in Equation (9.13) should be used. Also shown in Figure 9.5 are three regions: safe, buckling, and yielding. Thus, by knowing the normal stress and the slenderness ratio, one can quickly determine the active buckling region for a particular design.

## DETERMINATION OF BUCKLING LOAD FOR A COLUMN

1. Given the material, cross-section and dimensions for a column, calculate the critical slenderness ratio $C_c$ from Eq. (9.15). Be sure to use the effective length for the end constraints as given in Table 9.1.

2. Calculate the slenderness ratio for the column, $l_e/r_g$.

3. If $l_e/r_g \leq C_c$, the columns will buckle inelastically, and the Johnson equation [Eq. (9.16)] should be used to obtain the stress at buckling. The load can be obtained by multiplying this stress by the cross-sectional area.

4. If the $l_e/r_g \geq C_c$, the column will buckle elastically, and the buckling load can be calculated from the Euler equation [Eq. (9.12)].

---

**EXAMPLE 9.3**

**Given:** A column with one end fixed and the other free is to be made of aluminum alloy 2014. The column's cross-sectional area is 600 mm², and its length is 2.5 m.

**Find:** The critical column buckling load for the following shapes:

    (a) A solid round bar
    (b) A cylindrical tube with a 50-mm outer diameter
    (c) A square tube with a 50-mm outer dimension
    (d) A square bar

**Solution:**

    (a) The cross-sectional area of a solid round bar is

$$A = \frac{\pi d^2}{4} \quad \text{or} \quad d = \sqrt{\frac{4A}{\pi}} = \sqrt{\frac{4(600)}{\pi}} = 27.64 \text{ mm}$$

The area moment of inertia for a circular section is

$$I = \frac{\pi d^4}{64} = \frac{\pi (27.64)^4}{64} = 28,650 \text{ mm}^4$$

The radius of gyration is

$$r_g = \sqrt{\frac{I}{A}} = \sqrt{\frac{28,650}{600}} = 6.910 \text{ mm}$$

From Table A.2 for aluminum alloy 2014, $E = 72$ GPa and $S_y = 97$ MPa. From Equation (9.18), $(l_e/r_g)_T = 121.0$. From Table 9.1 for one end fixed and one end free, for the recommended consideration,

$$l_e = 2.1l = 2.1(2.5) = 5.25 \text{ m} = 5250 \text{ mm}$$

$$\frac{l_e}{r_g} = \frac{5250}{6.910} = 759.8$$

Since $l_e/r_g > (l_e/r_g)_T$, the Euler formula in Equation (9.12) gives the critical load at the onset of buckling as

$$P_{cr} = \frac{\pi^2 EI}{l_e^2} = \frac{\pi^2 (72)(10^9)(28,650)(10^{-12})}{(5.25)^2} = 738.6 \text{ N}$$

(b)   The cross-sectional area of a cylindrical tube is $A = (\pi/4)(d_o^2 - d_i^2)$. Given that

$$d_o = 50 \text{ mm} \quad \text{and} \quad A = 600 \text{ mm}^2$$

$$d_i = \sqrt{d_o^2 - \frac{4A}{\pi}} = \sqrt{50^2 - \frac{4(600)}{\pi}} = 41.67 \text{ mm}$$

The area moment of inertia for a hollow circular cross section is

$$I = \frac{\pi}{64}(d_o^4 - d_i^4) = \frac{\pi}{64}[(50)^4 - (41.67)^4] = 1.588 \times 10^5 \text{ mm}^4 = 1.588 \times 10^{-7} \text{ m}^4$$

The radius of gyration is

$$r_g = \sqrt{\frac{I}{A}} = \sqrt{\frac{(1.588)(10^5)}{600}} = 16.27 \text{ mm} = 0.01627 \text{ m}$$

$$\therefore \quad \frac{l_e}{r_g} = \frac{5.25}{0.01627} = 322.7$$

Since $l_e/r_g > (l_e/r_g)_T$, the Euler equation should be used. Thus,

$$P_{cr} = \frac{\pi^2 EI}{l_e^2} = \frac{\pi^2 (72)(10^9)(1.588)(10^{-7})}{(5.25)^2} = 4094 \text{ N}$$

In parts $a$ to $d$, the same amount of material is used for the construction of the column. Also, for the cylindrical tube, the area moment of inertia and the critical buckling load are 5.5 times larger than found in part $a$ for a solid round bar.

(c)   The cross-sectional area of a square tube is $A = a_1^2 - a_2^2$. The inner dimension is

$$a_2 = \sqrt{a_1^2 - A} = \sqrt{50^2 - 600} = 43.59 \text{ mm}$$

The area moment of inertia is

$$I = \frac{a_1^4 - a_2^4}{12} = \frac{(50)^4 - (43.59)^4}{12} = 220,000 \text{ mm}^4 = 2.2 \times 10^{-7} \text{ m}^4$$

The radius of gyration is

$$r_g = \sqrt{\frac{I}{A}} = 10^2 \sqrt{\frac{22}{600}} = 19.15 \text{ mm} = 0.01915 \text{ m}$$

$$\therefore \quad \frac{l_e}{r_g} = \frac{5.25}{0.01915} = 274.2$$

Since $l_e/r_g > (l_e/r_g)_T$, the Euler formula should be used to calculate the critical load:

$$P_{cr} = \frac{\pi^2 EI}{l_e^2} = \frac{\pi^2 (72)(10^9)(0.22)(10^{-6})}{(5.25)^2} = 5672 \text{ N}$$

Comparing results from parts $b$ and $c$ shows that the hollow square had a critical load that was 39% greater than found in part $b$ for a hollow circular section.

(d)   The cross-sectional area of a square column is $A = a^2$, or $a = \sqrt{A} = \sqrt{600} = 24.49$ mm. The area moment of inertia is

$$I = \frac{a^4}{12} = \frac{(24.5)^4}{12} = 30,000 \text{ mm}^4 = 3 \times 10^{-8} \text{ m}^4$$

The radius of gyration is

$$r_g = \sqrt{\frac{I}{A}} = \sqrt{\frac{30{,}000}{600}} = 7.071 \text{ mm}$$

$$\frac{l_e}{r_g} = \frac{5.25}{(7.071)(10^{-3})} = 742.5$$

Since $l_e/r_g > (l_e/r_g)_T$, the Euler formula should be used:

$$P_{cr} = \frac{\pi^2 EI}{l_e^2} = \frac{\pi^2(72)(10^9)(3)(10^{-8})}{(5.25)^2} = 773.5 \text{ N}$$

Figure 9.6 shows the cross sections of the four shapes considered as well as the critical buckling loads. This figure clearly illustrates why in designing columns the column's cross-sectional area should be located as far as possible from the section's principal centroidal axes.

**Figure 9.6** Cross-sectional areas, drawn to scale, from results of Example 9.3, as well as critical buckling load for each cross-sectional area.

**Given:** A column with one end fixed and the other end pinned is made of a low-carbon steel. The column's cross section is rectangular with $h = 0.5$ in and $b = 1.5$ in.

**EXAMPLE 9.4**

**Find:** The buckling load for the following three lengths:

(a) 0.5 ft
(b) 2 ft
(c) 4 ft

**Solution:** From Table A.1 for low-carbon steel, $S_y = 43$ ksi and $E = 30 \times 10^6$ psi. The cross-sectional area is

$$A = bh = (1.5)(0.5) = 0.75 \text{ in}^2$$

For a rectangular cross section, the area moment of inertia is

$$I = \frac{bh^3}{12} = \frac{(1.5)(0.5)^3}{12} = 0.0156 \text{ in}^4$$

The radius of gyration is

$$r_g = \sqrt{\frac{I}{A}} = \sqrt{\frac{0.0156}{0.75}} = 0.1442 \text{ in}$$

(a)   From Table 9.1 for one end pinned and the other end fixed and $l = 0.5$ ft,

$$l_e = 0.8l = 0.8(6) = 4.8 \text{ in}$$

$$\therefore \quad \frac{l_e}{r_g} = \frac{4.8}{0.1442} = 33.28$$

From Equation (9.18)

$$\left(\frac{l_e}{r_g}\right)_T = \sqrt{\frac{2\pi^2 E}{S_y}} = \sqrt{\frac{2\pi^2(30)(10^6)}{43,000}} = 117.4$$

Since $l_e/r_g < (l_e/r_g)_T$, the Johnson equation [Eq. (9.16)] should be used:

$$P_{cr} = A\left[S_y - \frac{1}{E}\left(\frac{S_y}{2\pi}\frac{l_e}{r_g}\right)^2\right]$$

$$= 0.75\left\{(4.3)(10^4) - \frac{1}{(30)(10^6)}\left[\frac{(4.3)(10^4)(33.28)}{2\pi}\right]^2\right\} = 30,950 \text{ lb}$$

(b)   From Table 9.1 for one end pinned and one end fixed and $l = 2$ ft,

$$l_e = 0.80l = 0.8(24) = 19.2 \text{ in}$$

$$\frac{l_e}{r_g} = \frac{19.2}{0.1442} = 133.1$$

Since $l_e/r_g > (l_e/r_g)_T$, the Euler equation should be used. From Equation (9.12)

$$P_{cr} = \frac{\pi^2 EI}{l_e^2} = \frac{\pi^2(30)(10^6)(0.0156)}{(19.2)^2} = 12,530 \text{ lb}$$

(c)   The Euler equation is definitely valid for $l = 4$ ft:

$$l_e = 0.8(4)(12) = 38.4 \text{ in}$$

The critical load is

$$P_{cr} = \frac{\pi^2 EI}{l_e^2} = \frac{\pi^2(30)(10^6)(0.0156)}{(38.4)^2} = 3132 \text{ lb}$$

Observe that in part $a$ the column was short, so that yield strength predominated and the Johnson equation was used; but in parts $b$ and $c$ the columns were long enough that the Euler equation was used. Also note that as the column becomes longer, the critical load decreases significantly.

## 9.7   AISC CRITERIA

The American Institute of Steel Construction [AISC (1989)] has produced design guidelines for elastic stability conditions. A number of subtleties are associated with these recommendations.

As shown by the tangent modulus equation, the elastic modulus based on the linear elastic portion of a stress–strain curve will lead to erroneous results when applied to buckling problems. However, tangent modulus data are not readily available in the technical literature and are difficult to obtain experimentally. Thus, the **AISC equations** correct for reductions in elastic modulus as the column stress exceeds the proportional limit of the material.

Further, long columns are more difficult to design because they are extremely susceptible to defects in straightness or to eccentricity in loading. Therefore, a weighted reduction in the allowable stress is prescribed. The allowable normal stress for elastic buckling is given by

$$\sigma_{\text{all}} = \frac{12\pi^2 E}{23(l_e/r_g)^2} \tag{9.19}$$

and for inelastic buckling

$$\sigma_{\text{all}} = \frac{\left[1 - (l_e/r_g)^2/(2C_c^2)\right] S_y}{n_\sigma} \tag{9.20}$$

where $C_c$ = slenderness ratio for Euler buckling defined by Equation (9.15)

$n_\sigma$ = reduction in allowable stress given by

$$n_\sigma = \frac{5}{3} + \frac{3(l_e/r_g)}{8C_c} - \frac{(l_e/r_g)^3}{8C_c^3} \tag{9.21}$$

Figure 9.5 compares the AISC as well as the Euler and Johnson equations. The American Association of State Highway and Transportation Officials (AASHTO) uses equations identical to Equations (9.19) and (9.20) but requires a constant stress reduction $n_\sigma = 2.12$ for both elastic and inelastic buckling.

Note that $n_\sigma$ is not a safety factor but a mandatory reduction in a material's allowable stress. Buckling load can be predicted by using these equations, but designs where safety is paramount require the use of an additional safety factor.

## DESIGN OF COLUMNS

These rules are intended to guide a designer in determining the safe buckling load according to AISC criteria. This approach predicts a load lower than that needed to buckle a column under ideal conditions.

1.  Given the material, cross-section and dimensions for a column, calculate the critical slenderness ratio from Eq. (9.15). Be sure to use the AISC-defined effective length for the end constraints as given in Table 9.1.
2.  Calculate the slenderness ratio for the column, $l_e/r_g$.
3.  if $l_e/r_g \leq C_c$ the buckling is inelastic and Eq. (9.20) gives the allowable stress for the column.
4.  If $l_e/r_g \geq C_c$ the allowable stress is given by Eq. (9.19).
5.  Multiply the allowable stress by the cross-sectional area to obtain the allowable load.

## 9.8   ECCENTRICALLY LOADED COLUMNS

Columns used in applications rarely have the applied load aligned coincidentally with the centroidal axis of the cross section. The distance between the two axes is called the *eccentricity* and is designated by *e*. It is assumed that the load is always parallel with the centroid of the columns. Just as for concentrically loaded columns (in Sec. 9.3) the analysis of eccentricity is restricted to columns with pinned ends. Figure 9.7(a) shows a pinned-end column subjected to forces acting at a distance *e* from the centerline of the undeformed column. It is assumed that the load is applied to the column at a short eccentric distance from the centroid of the cross section. This loading on the column is statically equivalent to the axial load and bending moment $M' = -Pe$ shown in Figure 9.7(b). As when one is considering concentrically loaded columns, small deflections and linear elastic material behavior are assumed. The $xy$ plane is a plane of symmetry for the cross-sectional area.

From a free-body diagram of an arbitrary section shown in Figure 9.7(c), the internal moment in the column is

$$M = -P(e + y) \tag{9.22}$$

The differential equation for the deflection curve is obtained from Equations (5.3) and (9.22), or

$$\frac{d^2y}{dx^2} + \frac{Py}{EI} = -\frac{Pe}{EI} = \text{constant} \tag{9.23}$$

If the eccentricity is zero, Equations (9.23) and (9.1) are identical.

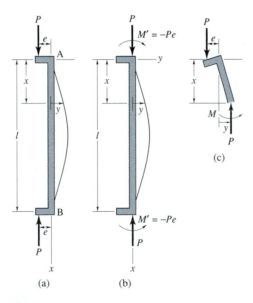

**Figure 9.7**   Eccentrically loaded column.
(a) Eccentricity; (b) statically equivalent bending moment; (c) free-body diagram through arbitrary section.

The general solution to Equation (9.23) is

$$y = C_1 \sin\left(x\sqrt{\frac{P}{EI}}\right) + C_2 \cos\left(x\sqrt{\frac{P}{EI}}\right) - e \tag{9.24}$$

The boundary conditions are

1.  At $x = 0$, $y = 0 \rightarrow C_2 = e$
2.  At $x = l/2$, $\partial y/\partial x = 0$

Taking the derivative of Equation (9.24) gives

$$\frac{dy}{dx} = C_1\sqrt{\frac{P}{EI}} \cos\left(x\sqrt{\frac{P}{EI}}\right) - e\sqrt{\frac{P}{EI}} \sin\left(x\sqrt{\frac{P}{EI}}\right) \tag{9.25}$$

Using Equation (9.25) and boundary condition 2 gives

$$C_1 = e \tan\left(\frac{l}{2}\sqrt{\frac{P}{EI}}\right) \tag{9.26}$$

Substituting $C_2 = e$ and Equation (9.26) into Equation (9.24) gives

$$y = e\left[\tan\left(\frac{l}{2}\sqrt{\frac{P}{EI}}\right) \sin\left(x\sqrt{\frac{P}{EI}}\right) + \cos\left(x\sqrt{\frac{P}{EI}}\right) - 1\right]$$

The maximum deflection occurs at $x = l/2$.

$$\therefore \quad y_{max} = e\left[\frac{\sin^2\left(\frac{l}{2}\sqrt{\frac{P}{EI}}\right)}{\cos\left(\frac{l}{2}\sqrt{\frac{P}{EI}}\right)} + \cos\left(\frac{l}{2}\sqrt{\frac{P}{EI}}\right) - 1\right] \tag{9.27}$$

$$y_{max} = e\left[\sec\left(\frac{l}{2}\sqrt{\frac{P}{EI}}\right) - 1\right] \tag{9.28}$$

The maximum stress on the column is caused by the axial load and the moment. The maximum moment occurs at the column's midheight and has a magnitude of

$$M_{max} = |P(e + y_{max})|$$

$$M_{max} = Pe \sec\left(\frac{l}{2}\sqrt{\frac{P}{EI}}\right) \tag{9.29}$$

The maximum stress in the column is compressive and is

$$\sigma_{max} = \frac{P}{A} + \frac{M_{max}c}{I}$$

Making use of Equation (9.29) gives

$$\sigma_{max} = \frac{P}{A} + \frac{Pec}{I} \sec\left(\frac{l}{2}\sqrt{\frac{P}{EI}}\right)$$

Since the radius of gyration is $r_g^2 = I/A$, the preceding equation becomes

$$\sigma_{max} = \frac{P}{A}\left[1 + \frac{ec}{r_g^2} \sec\left(\frac{l}{2r_g}\sqrt{\frac{P}{EA}}\right)\right] \tag{9.30}$$

where $P$ = critical load where buckling will occur in eccentrically loaded column, N

$A$ = cross-sectional area of column, m$^2$

$e$ = eccentricity of load measured from neutral axis of column's cross-sectional area to load's line of action, m

$c$ = distance from neutral axis to outer fiber of column, m

$r_g$ = radius of gyration, m

$l$ = length before load is applied, m

$E$ = modulus of elasticity of column material, Pa

For end conditions other than pinned, the length is replaced with the effective length (using Table 9.1), and Equations (9.28) and (9.30) become

$$y_{\max} = e\left[\sec\left(\frac{l_e}{2}\sqrt{\frac{P}{EI}}\right) - 1\right]$$   (9.31)

$$\sigma_{\max} = \frac{P}{A}\left[1 + \frac{ec}{r_g^2}\sec\left(\frac{l_e}{2r_g}\sqrt{\frac{P}{EA}}\right)\right]$$   (9.32)

Equation (9.31) is known as the **secant equation,** and the parameter $ec/r_g^2$ is called the **eccentricity ratio.** Note from Equation (9.32) that it is not convenient to calculate the load explicitly.

Figure 9.8 shows the effect of slenderness ratio $l_e/r_g$ on normal stress $\sigma$ for an eccentrically loaded column. These results are for structural-grade steel where the modulus of elasticity is $29 \times 10^6$ psi and the yield strength is 36,000 psi. Note that as $e \to 0$, $\sigma \to P/A$, where $P$ is the critical column load as established by the Euler formula [Eq. (9.12)].

The curves in Figure 9.8 indicate that differences in eccentricity ratio have a significant effect on column stress when the slenderness ratio is small. On the other hand, columns with large slenderness ratios tend to fail at the Euler (or critical) load regardless of the eccentricity ratio.

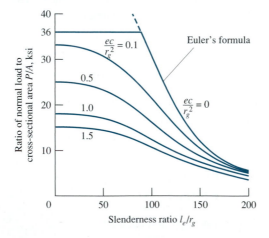

**Figure 9.8**   Stress variation with slenderness ratio for five eccentricity ratios, for structural steel with $E = 29 \times 10^6$ psi and $S_y = 36 \times 10^3$ psi.

If $\sigma_{max} = \sigma_{all} = S_y/2$, Equation (9.32) becomes

$$P = \frac{S_y A/2}{1 + (ec/r_g^2)\sec[(l_e/2r_g)\sqrt{P/(EA)}]} = f(P) \tag{9.33}$$

An initial guess for $P$ is the value obtained from the concentric loading condition, or

$$P = P_{cr} \tag{9.34}$$

Using Equation (9.34) in Equation (9.33) gives a new value of $P$. This iterative process is continued until a desired accuracy is achieved, such as

$$\left| \frac{P - f(P)}{f(P)} \right| \le 1 \times 10^{-4} \tag{9.35}$$

---

EXAMPLE 9.5

**Given:** A 36-in-long hollow aluminum alloy 2014 tube with a 3.0-in outside diameter, a 0.03-in wall thickness, and with both ends pinned.

**Find:** The critical buckling load for

(a) Concentric loading
(b) Eccentric loading, with an eccentricity of 0.15 in

**Solution:**

(a) The inside diameter of the column is

$$d_i = d_o - 2t_h = 3.0 - 2(0.03) = 2.94 \text{ in}$$

The cross-sectional area is

$$A = \frac{\pi}{4}(d_o^2 - d_i^2) = \frac{\pi}{4}[(3)^2 - (2.94)^2] = 0.280 \text{ in}^2$$

The area moment of inertia is

$$I = \frac{\pi}{64}(d_o^4 - d_i^4) = \frac{\pi}{64}[(3)^4 - (2.94)^4] = 0.309 \text{ in}^4$$

The radius of gyration is

$$r_g = \sqrt{\frac{I}{A}} = \sqrt{\frac{0.309}{0.280}} = 1.050 \text{ in}$$

From Table 9.1 for a column with both ends pinned, the effective length is equal to the actual length ($l = l_e$). From Table A.2 for aluminum alloy 2014

$$E = 10.5 \times 10^6 \text{ psi} \quad \text{and} \quad S_y = 14 \times 10^3 \text{ psi}$$

From Equation (9.18)

$$\left(\frac{l_e}{r_g}\right)_T = \sqrt{\frac{2\pi^2 E}{S_y}} = \sqrt{\frac{2\pi^2(10.5)(10^6)}{14(10^3)}} = 121.7$$

and

$$\frac{l_e}{r_g} = \frac{36}{1.05} = 34.29$$

Since $l_e/r_g < (l_e/r_g)_T$, the Johnson formula [Eq. (9.16)] should be used:

$$P_{cr} = AS_y\left[1 - \frac{S_y}{4\pi^2 E}\left(\frac{l_e}{r_g}\right)^2\right] = (0.28)(14)(10^3)\left[1 - \frac{(14)(10^3)(34.29)^2}{4\pi^2(10.5)(10^6)}\right] = 3764 \text{ lb}$$

(b) The distance from the neutral axis to the outer fiber is $c = d_o/2 = 1.5$ in. The eccentricity ratio is

$$\frac{ce}{r_g^2} = \frac{(1.5)(0.15)}{(1.05)^2} = 0.2041$$

$$P = P_{cr} = 3764 \text{ lb}$$

From Equation (9.33), and noting that $S_y A/2 = 1960$ lb,

$$P_1 = \frac{1960}{1 + 0.2041 \sec\{(34.29/2)\sqrt{3764/[2.940(10^6)]}\}} = 1569 \text{ lb}$$

$$P_2 = \frac{1960}{1 + 0.2041 \sec\{(34.29/2)\sqrt{1569/[2.940(10^6)]}\}} = 1628 \text{ lb}$$

The critical load for the eccentric loading is 1628 lb, and that for the concentric loading is 3764 lb.

---

## Case Study 9.1    |    DESIGN OF TELESCOPING BOOM FOR HYDRAULIC CRANE

**A** mobile hydraulic boom crane (Fig. 9.9) is used in various materials handling activities. The boom is a set of telescoping cylinders driven by hydraulic pressure. The load is applied eccentrically from the boom axis. At low boom angles the crane's capacity is defined by its tipping stability, but at high boom angles boom buckling is a major concern. In the manufacture of the telescoping boom, great care is taken to ensure that the boom sections are extremely straight; the deviation from straightness is no more than 0.02 in per 10-ft section. This tolerance is ensured by clamping the boom section in two rigid clamps, 10 ft apart, and tracing a dial indicator along the entire exposed length of the boom. Any section found to deviate from straightness is carefully bent plastically with a hydraulic jack until the 0.02-in limit is achieved.

**Given:** The inner diameter of the tube must be at least 5 in based on the hydraulic requirements, and the thickness must be a convenient dimension so that standard extrusions can be used (i.e., an integer multiple of eighths of an inch). The bucket must telescope at least 60 ft and must be constructed from three sections to ensure a reasonable contracted length. The eccentricity is a constant value of 3 ft and cannot be modified, as this dimension is determined by rigging requirements. Use Tri-ten steel ($E = 29 \times 10^6$ psi; $S_y = 30 \times 10^3$ psi) as the boom material. Assume the theoretical effective column length.

**Figure 9.9**    Mobile hydraulic boom crane in Case Study 9.1. (AP/Wide World Photos)

**Case Study** (Concluded)

**Find:** A tubular cross section that will deform less than 2 ft under a 10,000-lb load for a fully extended vertical boom.

**Solution:** From Table 9.1 the effective boom length for the relevant buckling mode is 60 ft times 2, or 120 ft. The boom consists of three sections, each more slender nearer to the load end. As a worst-case analysis the entire boom length is assumed to have a constant cross section equal to the most slender section of the boom. Also, the deviation from straightness of each section is assumed to be continuous and cumulative. Thus, if the ultimate design is still acceptable based on deformation and stress requirements, the final boom will be even stronger.

The initial deformation can be taken as that caused by a fictitious load $P_i$ for the load eccentricity. Since the cumulative deviation from straightness is 0.120 in, the fictitious load is determined by solving the secant equation [Eq. (9.31)] for the load $P_i$, or

$$0.120 = 36 \left[ \sec \left( \frac{120 \times 12}{2} \sqrt{\frac{P_i}{EI}} \right) - 1 \right]$$

Solving for $P_i$ gives

$$\sqrt{\frac{P_i}{EI}} = \frac{\cos^{-1}(36/36.12)}{720}$$

$$\therefore \quad P_i = EI(1.28 \times 10^{-8})$$

The load that causes an additional 24-in deflection (for a total of 24.12 in) is $P_i + 10,000$ lb. Again applying the secant equation gives

$$24.120 = 36 \left[ \sec \left( \frac{120 \times 12}{2} \sqrt{\frac{P_i + 10,000}{EI}} \right) - 1 \right]$$

Numerically solving for $I$, we have $I = 208.8$ in$^4$. Since

$$I = \frac{\pi}{64} \left[ (5 + 2t_h)^4 - (5)^4 \right]$$

the minimum thickness is 1.68 in. The next-largest multiple of $\frac{1}{8}$ in is thus 1.75 in, but we will use $I = 208.8$ in$^4$, recognizing a larger, commercially available cross section will give better performance.

Before this tube wall thickness can be accepted, we must be assured that the allowable boom stress is not exceeded, since the tangent modulus can be much less than the elastic modulus reported for the steel. The iterative approach described in Equations (9.33) to (9.35) is used. The cross-sectional area of the tube is $A = \pi[(2.5 + 1.75)^2 - (2.5)^2] = 37.1$ in$^2$. The tube's radius of gyration is calculated from

$$r_g = \sqrt{\frac{I}{A}} = \sqrt{\frac{208.8}{37.1}} = 2.466 \text{ in}$$

Equation (9.34) yields

$$P = P_{cr} = 7200 \text{ lb}$$

Substituting into Equation (9.33) yields

$$P = \frac{S_y A/2}{1 + \dfrac{ec}{r_g^2} \sec \left( \dfrac{l_e}{3 r_g} \sqrt{\dfrac{P}{EA}} \right)} = 14,000 \text{ lb}$$

Using this value in Equation (9.33) gives 9400 lb, which then gives 12,700 lb, etc., until eventually a solution of 11,300 lb is reached. Thus, this value also ensures that the beam remains linearly elastic under the design load.

# 9.9 SUMMARY

In an attempt to better understand column buckling, three equilibrium regimes were studied. Stable, neutral, and unstable equilibrium were explained by observing what happens to a sphere or cylinder on concave, flat, and convex surfaces. It was concluded that when columns are in unstable equilibrium, they are unreliable and can be hazardous. A small displacement can cause a catastrophic change in the configuration of the column. Thus, as the load on a column is increased, a critical load is reached at which unstable equilibrium will occur and the column will not return to its straight configuration. Thus, for long, slender columns there is a critical buckling load. The column's critical buckling stress was also determined to be much less than the yield strength of the material. End conditions were found to affect the critical buckling load. The effective length was used to handle four types of end condition. The Euler and Johnson buckling criteria were developed for concentrically loaded columns. Eccentrically loaded columns were also studied, and the secant formulas were derived.

## KEY WORDS

**AISC equations**    estimations of allowable stress for prevention of buckling in structures (corrections for reductions in elastic modulus as column stress exceeds proportional limit).

**buckling**    sudden, large deformation of a structure due to a slight increase of applied load.

**column**    straight and long (relative to cross-sectional dimension) member subjected to compressive axial loads.

**eccentricity ratio**    measure of how far off-center the load is applied, given by $ec/r_g^2$.

**Essenger equation**    critical buckling load in nonlinear elastic buckling, given by $\sigma_{cr} = \pi^2 E_t/(l/r_g)^2$.

**Euler load**    critical load of elastic column, given by $P_{cr} = \pi^2 EI/l^2$.

**Johnson equation**    critical load for inelastic buckling, given by

$$(\sigma_{cr})_J = \frac{(P_{cr})_J}{A} = S_y - \frac{S_y^2}{4\pi^2 E}\left(\frac{l_e}{r_g}\right)^2$$

**secant equation**    deflection due to eccentric loading, given by

$$y_{max} = e\left[\sec\left(\frac{l_e}{2}\sqrt{\frac{P}{EI}}\right) - 1\right]$$

**slenderness ratio**    measure of column slenderness, given by

$$C_c = \left(\frac{l_e}{r_g}\right)_E = \sqrt{\frac{2E\pi^2}{S_y}}$$

**tangent modulus**    elastic modulus at stress level in column.

**tangent modulus equation**    same as **Essenger equation.**

## RECOMMENDED READINGS

Juvinall, R. C., and Marshek, K. M. (2003) *Fundamentals of Machine Component Design,* 3rd ed., Wiley, New York.

Ketter, R. L., Lee, G. C., and Prawel, S. P. (1979) *Structural Analysis and Design,* McGraw-Hill, New York.

*Manual of Steel Construction* (1980) 8th ed., American Institute of Steel Construction, Chicago.

Norton, R. L. (2003) *Machine Design,* 3rd ed., Prentice-Hall, Englewood Cliffs, NJ.

Popov, I. (1968) *Introduction to Mechanics of Solids,* Prentice-Hall, Englewood Cliffs, NJ.

Shigley, J. E., Mischke, C. R., and Budynas, R. G. (2003) *Mechanical Engineering Design,* 7th ed., McGraw-Hill, New York.

Timoshenko, S. P., and Goodier, J. N. (1970) *Theory of Elasticity,* McGraw-Hill, New York.

Ugural, A. C., and Fenster, S. K. (2003) *Advanced Strength and Applied Elasticity,* 4th ed., Prentice-Hall, Englewood Cliffs, NJ.

Willems, N., Easley, J., and Rolfe, S. (1981) *Strength of Materials,* McGraw-Hill, New York.

## REFERENCE

American Institute of Steel Construction (AISC) (1989) *Manual of Steel Construction,* 9th ed., Chicago.

## PROBLEMS

### Section 9.2

**9.1**  A person rides a bike on a flat, level road. Is this a neutral, stable, or unstable equilibrium position?

**9.2**  A golf ball is placed

  *a)*     On top of a small hill
  *b)*     On a horizontal flat plane
  *c)*     In a shallow groove

Describe the type of equilibrium for each condition.

### Section 9.3

**9.3**  A column has pinned ends and is axially loaded in compression. The length is 6 m, and the weight is 1962 N, but the form of the cross section can be changed. The column is made of steel (AISI 1040). Find the Euler buckling load for

  *a)*     A solid circular section. *Ans. $P_{cr} = 81.7$ kN.*
  *b)*     A solid square section
  *c)*     A circular tube with outer diameter of 100 mm. *Ans. $P_{cr} = 220$ kN.*
  *d)*     A square tube with outside dimension of 100 mm

**9.4**  A column with both ends pinned is 3 m long and has a tubular section with an outer diameter of 30 mm and wall thickness of 5 mm. Find which material given in Tables A.1 through A.4

produces the highest buckling load. Also, give the buckling loads for AISI 1080 steel, aluminum alloy 2014, and molybdenum. *Ans.* $P_{cr, \text{AISI } 1090} = 7245$ N.

**9.5**   Determine the critical stresses for the four columns considered in Problem 9.3. Let the column be 8 m long and the cross-sectional area be 5000 mm². The column material is AISI 1080 steel. *Ans.* For a solid square cross section, $\sigma_{cr} = 13.3$ MPa.

## Section 9.4

**9.6**   A column is axially loaded in compression. The ends were specified to be fixed, but because of a manufacturing error they had

*a)*      One end fixed and the other pinned
*b)*      Both ends pinned

Find out how much the critical elastic buckling load is decreased because of the errors. Also, calculate the buckling loads for the 4-m-long column made of AISI 1040 steel having a solid, square cross section with 30-mm sides for the three end conditions. Assume theoretical effective column length. *Ans.* $P_{cr, \text{fixed-fixed}} = 34.5$ kN.

**9.7**   An elastic column has one end pinned and the other end fixed in a bushing so that the values given in Table 9.1 do not apply. Instead, the effective length is $l_e = 0.83l$. The cross section of the column is a circular tube with an outer diameter of 73 mm and 3.2-mm wall thickness; the column is 11.2 m long. Calculate the elastic buckling load if the column is made of

*a)*      AISI 1080 steel. *Ans.* 10.1 kN.
*b)*      Polycarbonate. *Ans.* 116 N.

**9.8**   An elastic AISI 1020 steel column has both ends pinned. It is 12.5 m long and has a square, tubular cross section with an outside dimension of 160 mm and 4-mm wall thickness. Its compressive axial load is 130 kN.

*a)*      Determine the safety factors guarding against buckling and yielding. *Ans.* $n_{s, \text{buckling}} = 1.02$, $n_{s, \text{yielding}} = 5.66$.
*b)*      If the ends are changed to fixed and the material is changed to aluminum alloy 2014, calculate the safety factors guarding against buckling and yielding.

★ **9.9**   A beam  has both ends mounted in stiff rubber bushings, giving bending moments in the beam ends proportional to the angular displacements at the beam ends. Calculate the effective beam length if the angular spring constant at the ends is $\partial M / \partial \theta = 10^5$ N · m/rad in all directions. The beam is 3 m long and has a solid, circular cross section of AISI 1080 steel with a 24-mm diameter. *Ans.* 1.534 m.

## Section 9.5

**9.10**   Two solid, circular columns are made of different materials, steel and aluminum. The cross-sectional areas are the same, and the moduli of elasticity are 207 GPa for the steel and 72 GPa for the aluminum. Find the ratio of the critical buckling lengths for the columns, assuming that the same buckling load is applied to both columns. *Ans.* $l_{es} / l_{ea} = 1.696$.

| *Indicates problem of greater difficulty

**★ 9.11** A television mast shown in sketch *a* consists of a circular tube with an outer diameter $d_o$ and a wall thickness $t_h$. Calculate how long the distance $l$ between the anchoring points for the guy wires can be if the mast should deform plastically rather than buckle.

Sketch *a*, for Problem 9.11

## Section 9.6

**★ 9.12** A 3-m-long column of square, solid cross section with both ends fixed must sustain a critical load of $3 \times 10^5$ N. The material is steel with a modulus of elasticity of 207 GPa and a yield stress of 700 MPa.

*a)* What minimum dimensions of the cross section are permitted without having failure? Should the Euler or the Johnson equation be used? *Ans. w = 0.045 m.*

*b)* If the critical load is increased by two orders of magnitude to $P_{cr} = 3 \times 10^7$ N, what minimum dimensions of the cross section are permitted without failure? Also, indicate whether the Euler or the Johnson equation should be used.

**9.13** A solid, round column with a length of 2 m and a diameter of 50 mm is fixed at one end and is free at the other end. The material's yield strength is 300 MPa, and its modulus of elasticity is 207 GPa. Assuming concentric loading of the column, determine

*a)* The critical load. *Ans. $P_{cr} = 39.2$ kN.*

*b)* The critical load if the free end is also fixed (i.e., both ends are now fixed). *Ans. $P_{cr} = 450$ kN.*

**9.14** A circular-cross-section bar with a diameter of 2 in and a length of 40 in is axially loaded. The bar is made of medium-carbon steel. Both ends are pinned. Determine

*a)* Whether the Johnson or the Euler formula should be used.

*b)* The critical load. *Ans. 116 kips.*

★ **9.15** A low-carbon-steel pipe, as shown in sketch *b*, has an outer diameter of 2 in and a thickness of 0.5 in. If the pipe is held in place by a guy wire, determine the largest horizontal force *P* that can be applied without causing the pipe to buckle. Assume that the ends of the pipe are pin-connected. *Ans.* 4.38 kips.

P ◄

12 ft

|◄——— 5 ft ———►|

Sketch *b*, for Problem 9.15

**9.16** A 20-mm-diameter, medium-carbon steel (AISI 1040) rod is loaded as a column with rounded ends. If the critical load is 110 kN, how long can the rod be and still carry the following percentages of the critical load? (*a*) 90%, (*b*) 50%, and (*c*) 2%. Also, indicate whether the Johnson or the Euler equation should be used. *Ans. a*) 0.241 m; *b*) 0.540 m; *c*) 2.7 m.

**9.17** Most types of steel have similar moduli of elasticity but can have widely different yield strength properties depending on the alloying and the heat treatment. For a column having a solid, circular cross section with a diameter of 20 mm, the yield strengths for three steels are 300, 600, and 1000 MPa. Find for each steel the critical column length where buckling would not be a problem. *Ans.* For $S_y = 300$ MPa, $l_e = 0.5835$ m.

**9.18** For a fixed cross-sectional area of a tubular steel column, find the geometry that would give the highest buckling load. The minimum column wall thickness is 5 mm.

★ **9.19** A rectangular-cross-section tube is axially loaded in compression. The tube's outside dimensions are 70 and 90 mm; it is 3.0 m long with both ends pinned; the wall thickness is 9 mm. Determine which of the steels in the table in Appendix A.1 should be used so that plastic buckling is not a problem.

**9.20** The column considered in Problem 9.19 is made of AISI 1020 steel. Find the critical buckling load. *Ans.* 392 kN.

**9.21** A 2.5-m-long column with one end fixed and the other end free is made of aluminum alloy 2014 and has a solid, round cross section. Determine the diameter of the column for the following loads:

*a)*      $P = 500$ kN
*b)*      $P = 800$ kN

## Section 9.8

★ **9.22** A 1-m-long column with a 60-mm by 100-mm rectangular cross section and made of aluminum alloy 2014 is subjected to a compressive axial load. Determine the critical buckling load

   *a)*   If both ends are fixed. *Ans.* 565 kN.
   *b)*   If one end is fixed and the other is free. *Ans.* 316 kN.
   *c)*   If the load is applied eccentrically at a distance of 10 mm for case *b. Ans.* 109.6 kN.

★ **9.23** A column with pinned ends and solid rectangular cross section with dimensions of 35 mm by 60 mm is made of AISI 1040 steel. The 4-m-long column is loaded with an eccentricity of 15 mm. Find the elastic deflection if a 100-N compressive load is applied; also, calculate how large a load can be applied without permanent deformation occurring.

★ **9.24** The column considered in Problem 9.23 is subjected to a varying load in the range of 0 to 24,000 N. Calculate and plot the deformation as a function of the axial load.

# STRESSES AND DEFORMATIONS IN CYLINDERS

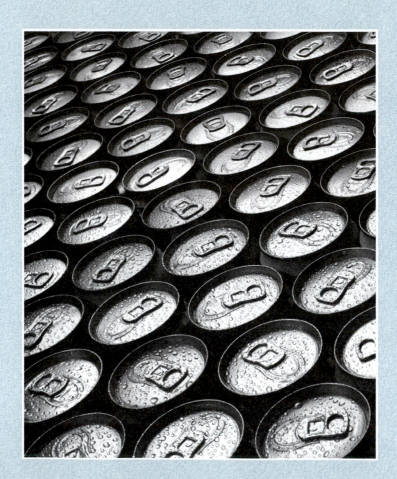

Common beverage cans. Along with food containers, these are the most common pressure vessels. (© R-F Website/Getty)

*In all things, success depends on previous preparation.*
*And without such preparation there is sure to be failure.*

Confucius, *Analects*

## SYMBOLS

| | |
|---|---|
| $A$ | cross-sectional area, $m^2$ |
| $\bar{a}$ | coefficient of linear thermal expansion, $(°C)^{-1}$ |
| $C_1, C_2$ | integration constants |
| $c$ | radial clearance, m |
| $d$ | diameter, m |
| $E$ | modulus of elasticity, Pa |
| $l$ | length, m |
| $n_s$ | safety factor |
| $P$ | force, N |
| $p$ | pressure, Pa |
| $r$ | radius, m |
| $S_y$ | yield strength, Pa |
| $T$ | torque, N $\cdot$ m |
| $t_h$ | thickness, m |
| $t_l$ | tolerance, m |
| $\Delta t_m$ | temperature change, °C |
| $x, y$ | Cartesian coordinates, m |
| $z$ | axial direction in cylindrical polar coordinates, m |
| $\alpha$ | cone angle, deg |
| $\beta$ | body force per volume, $N/m^3$ |
| $\delta$ | interference or displacement, m |
| $\epsilon$ | strain |
| $\theta$ | circumferential direction in cylindrical polar coordinates, rad |
| $\mu$ | coefficient of friction |
| $\nu$ | Poisson's ratio |
| $\rho$ | density, $kg/m^3$ |
| $\sigma$ | normal stress, Pa |
| $\omega$ | angular velocity, rad/s |

## SUBSCRIPTS

| | |
|---|---|
| $a$ | axial |
| avg | average |
| $c$ | circumferential |
| $e$ | von Mises |
| $f$ | fit |
| $h$ | hub |
| $i$ | inner or internal |
| max | maximum |
| min | minimum |
| $o$ | outer or external |
| $r$ | radial |
| $s$ | shaft |
| $t$ | tangential |
| $t_m$ | temperature change |
| $x, y$ | Cartesian coordinates |
| $z$ | axial direction in cylindrical polar coordinates |
| $\theta$ | circumferential direction in cylindrical polar coordinates |
| $\sigma$ | normal stress |
| 1,2,3 | principal axes |

# 10.1    INTRODUCTION

Just as Chapter 9 dealt with a specific shape [a straight and long (relative to its radius) bar] and the unique stresses and strains acting on columns, this chapter deals with a specific shape in the form of cylinders. Many important engineering applications rely on this shape. This chapter begins with the consideration of tolerancing in the design of cylindrical members, which will prove to be important for a number of machine elements considered later in the text.

Stresses and deformations of thin-walled, thick-walled, internally pressurized, externally pressurized, and rotating cylinders are considered as well as press and shrink fits. The material developed in this chapter is important to shafting (Chap. 11) and to a number of machine elements, such as rolling-element bearings (Chap. 13) and hydrodynamic bearings (Chap. 12).

## 10.2   TOLERANCES AND FITS

A number of definitions about tolerancing need to be given:

1. **Tolerance** $t_l$ is the maximum variation in the size of a part. Two types of tolerance are

    (a) **Bilateral tolerance.**   A part is permitted to vary both above and below the nominal size, such as $1.5 \pm 0.003$.

    (b) **Unilateral tolerance.**   A part is permitted to vary either above or below the nominal size, but not both, such as $1.5^{+0.000}_{-0.003}$.

2. **Nominal diameter** $d$ is the approximate size chosen and the one to which allowances and tolerances are applied.

3. **Allowance** $a$ is the difference between the nominal diameters of mating parts.

4. **Interference** $\delta$ is the actual difference in the size of mating parts.

Note that the tolerances of the shaft $t_{ls}$ and the hub $t_{lh}$ may be different.

Because it is impossible to make parts to an exact size, tolerances are used to control the variation between mating parts. Two examples illustrate the importance of tolerancing:

1. In hydrodynamic bearings (Chap. 12) a critical part of the design is the specification of the radial clearance between the journal and the bearing. A typical value is 0.001 in. However, variations in the journal's outside diameter and the bearing's inside diameter cause additional or smaller clearances. Such variations must be accounted for in analyzing bearing performance. Too small a clearance could cause failure; too large a clearance would reduce the precision of the machine and adversely affect the lubrication. Thus, tolerancing and accuracy of the dimensions can have a significant effect on the performance of hydrodynamic bearings.

2. Rolling-element bearings (Chap. 13) are generally designed to be installed on a shaft (Chap. 11) with an interference fit. The inside diameter of the bearing is smaller than

**Table 10.1**     Classes of fit

| Class | Description | Type | Applications |
|---|---|---|---|
| 1 | Loose | Clearance | Where accuracy is not essential, such as in building and mining equipment |
| 2 | Free | Clearance | In rotating journals with speeds of 600 rpm or greater, such as in engines and some automotive parts |
| 3 | Medium | Clearance | In rotating journals with speeds under 600 rpm, such as in precision machine tools and precise automotive parts |
| 4 | Snug | Clearance | Where small clearance is permissible and where mating parts are not intended to move freely under load |
| 5 | Wringing | Interference | Where light tapping with a hammer is necessary to assemble the parts |
| 6 | Tight | Interference | In semipermanent assemblies suitable for drive or shrink fits on light sections |
| 7 | Medium | Interference | Where considerable pressure is required to assemble and for shrink fits of medium sections; suitable for press fits on generator and motor armatures and for automotive wheels |
| 8 | Heavy force or shrink | Interference | Where considerable bonding between surfaces is required, such as locomotive wheels and heavy crankshaft disks of large engines |

the outside diameter of the shaft where the bearing is to be seated. A significant force is required to press the bearing onto the shaft, thus imposing significant stresses on both the shaft and the bearing. What should be the shaft diameter relative to the bearing bore size to ensure that failure due to overstressing the members does not occur? Proper tolerancing of the members will greatly contribute to successful design. Table 10.1 describes the eight classes of fit, the type of fit, and their applications. Note that for classes 1 to 4 the members are interchangeable, but for classes 5 to 8 the members are not interchangeable.

Once the classes of fit have been established, the next task will be to indicate the tolerance applicable for a specific class. Tables 10.2 and 10.3 give the tolerances for various classes of fit in inches and millimeters, respectively. In these tables the nominal diameter is designated by $d$.

After the allowance, interference, and tolerance have been established, the next step is to establish the maximum and minimum diameters of the hub and shaft. Table 10.4 gives these dimensions for the two types of fit (clearance and interference). Tables 10.2 through 10.4 thus establish the upper and lower limits of the shaft and hub diameters for the eight classes of fit.

**Table 10.2**   Recommended tolerances in **inches** for classes of fit

| Class | Allowance $a$ | Interference $\delta$ | Hub tolerance $t_{lh}$ | Shaft tolerance $t_{ls}$ |
|-------|---------------|-----------------------|------------------------|---------------------------|
| 1 | $0.0025d^{2/3}$ | — | $0.0025d^{1/3}$ | $0.0025d^{1/3}$ |
| 2 | $0.0014d^{2/3}$ | — | $0.0013d^{1/3}$ | $0.0013d^{1/3}$ |
| 3 | $0.0009d^{2/3}$ | — | $0.0008d^{1/3}$ | $0.0008d^{1/3}$ |
| 4 | $0.000$ | — | $0.0006d^{1/3}$ | $0.0004d^{1/3}$ |
| 5 | — | $0.000$ | $0.0006d^{1/3}$ | $0.0004d^{1/3}$ |
| 6 | — | $0.00025d$ | $0.0006d^{1/3}$ | $0.0006d^{1/3}$ |
| 7 | — | $0.0005d$ | $0.0006d^{1/3}$ | $0.0006d^{1/3}$ |
| 8 | — | $0.0010d$ | $0.0006d^{1/3}$ | $0.0006d^{1/3}$ |

**Table 10.3**   Recommended tolerances in **millimeters** for classes of fit

| Class | Allowance $a$ | Interference $\delta$ | Hub tolerance $t_{lh}$ | Shaft tolerance $t_{ls}$ |
|-------|---------------|-----------------------|------------------------|---------------------------|
| 1 | $0.0073d^{2/3}$ | — | $0.0216d^{1/3}$ | $0.0216d^{1/3}$ |
| 2 | $0.0041d^{2/3}$ | — | $0.0112d^{1/3}$ | $0.0112d^{1/3}$ |
| 3 | $0.0026d^{2/3}$ | — | $0.0069d^{1/3}$ | $0.0069d^{1/3}$ |
| 4 | $0.000$ | — | $0.0052d^{1/3}$ | $0.0035d^{1/3}$ |
| 5 | — | $0.000$ | $0.0052d^{1/3}$ | $0.0035d^{1/3}$ |
| 6 | — | $0.00025d$ | $0.0052d^{1/3}$ | $0.0052d^{1/3}$ |
| 7 | — | $0.0005d$ | $0.0052d^{1/3}$ | $0.0052d^{1/3}$ |
| 8 | — | $0.0010d$ | $0.0052d^{1/3}$ | $0.0052d^{1/3}$ |

**Table 10.4**    Maximum and minimum diameters of shaft and hub for two types of fit

| | Hub diameter | | Shaft diameter | |
|---|---|---|---|---|
| | **Maximum** | **Minimum** | **Maximum** | **Minimum** |
| **Type of fit** | $d_{h,\max}$ | $d_{h,\min}$ | $d_{s,\max}$ | $d_{s,\min}$ |
| Clearance | $d + t_{lh}$ | $d$ | $d - a$ | $d - a - t_{ls}$ |
| Interference | $d + t_{lh}$ | $d$ | $d + \delta + t_{ls}$ | $d + \delta$ |

---

**EXAMPLE 10.1**

**Given:** A medium-force fit is applied to a shaft with a nominal diameter of 3 in.

**Find:** Determine the interference, the hole and shaft tolerances, and the hub and shaft diameters.

**Solution:** From Table 10.1 the class of fit is 7, and the type of fit is interference. From Table 10.2 for a class 7 fit

$$\delta = 0.0005d = 0.0005(3) = 0.0015 \text{ in}$$

$$t_{lh} = t_{ls} = 0.0006d^{1/3} = 0.0006(3)^{1/3} = 0.865 \times 10^{-3} \text{ in} \approx 0.009 \text{ in}$$

From Table 10.4 for interference fit, the hub diameter is

$$d_{h,\max} = d + t_{lh} = 3 + 0.0009 = 3.0009 \text{ in}$$

$$d_{h,\min} = d = 3.000 \text{ in}$$

and the shaft diameter is

$$d_{s,\max} = d + \delta + t_{ls} = 3 + 0.0015 + 0.0009 = 3.0024 \text{ in}$$

$$d_{s,\min} = d + \delta = 3 + 0.0015 = 3.0015 \text{ in}$$

---

## 10.3    PRESSURIZATION EFFECTS

This section presents thin-walled and thick-walled cylinders with internal and external pressurization. The material in this section is used for a wide range of applications, some of which are

1. Pressure vessels loaded internally and/or externally
2. Storage tanks
3. Gun barrels
4. Hydraulic and pneumatic cylinders
5. Transmission pipelines
6. Machine elements, such as rolling-element bearings or gears pressed onto shafts

A distinction must be made between **thin-walled cylinders** and **thick-walled cylinders**. Generally, when a cylinder's inner diameter $d_i$ is 40 times larger than its thickness $t_h$, thin-wall analysis can be safely used. For smaller ratios of cylinder inner diameter to

(a)                                                    (b)

**Figure 10.1**    Internally pressurized thin-walled cylinder. (a) Stress element on cylinder; (b) stresses acting on element.

thickness, thick-wall analysis should be used. Mathematically expressing gives

$$\frac{d_i}{t_h} > 40 \qquad \rightarrow \qquad \text{thin-walled cylinders} \qquad (10.1)$$

$$\frac{d_i}{t_h} < 40 \qquad \rightarrow \qquad \text{thick-walled cylinders} \qquad (10.2)$$

Thin-wall analysis becomes more accurate as $d_i/t_h$ increases. Equation (10.1) must be adhered to, but Equation (10.2) may be valid for a larger range than stated.

## 10.3.1    THIN-WALLED CYLINDERS

Figure 10.1(a) shows a thin-walled cylinder subjected to internal pressure $p_i$. It is assumed that the stress distribution is uniform throughout the wall thickness. The radial stress is small relative to the circumferential stress because $t_h/d_i \ll 1$. Thus, a small element can be considered to be in plane stress with the principal stresses shown in Figure 10.1(b).

Figure 10.2, the front view of the cylinder shown in Figure 10.1, shows the forces acting on a small element due to the internal pressure. This element also has a length $dl$ coming out of the paper. Summing forces in the radial direction gives

$$p_i r_i \, d\theta \, dl = 2\sigma_{\theta,\text{avg}} \sin\left(\frac{d\theta}{2}\right) t_h \, dl$$

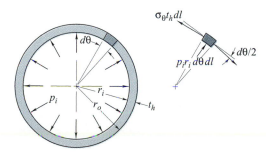

**Figure 10.2**    Front view of internally pressurized, thin-walled cylinder.

Since $d\theta/2$ is small, $\sin(d\theta/2) = d\theta/2$ and thus

$$\sigma_{\theta,avg} = \frac{p_i r_i}{t_h} \tag{10.3}$$

The $\theta$ component of stress is called the *tangential stress* or **hoop stress;** the term originates from the manufacture of wooden barrels by shrinking a hot iron ring, or hoop, around closely fitting planks to achieve a seal. The maximum hoop stress can be obtained by modifying Equation (10.3) such that $r_i$ is replaced with $r_{avg}$:

$$r_{avg} = r_i + \frac{t_h}{2} = \frac{d_i + t_h}{2} \tag{10.4}$$

$$\therefore \qquad \sigma_{\theta,max} = \frac{p_i r_{avg}}{t_h} = \frac{p_i(d_i + t_h)}{2t_h} \tag{10.5}$$

Thus, from Equations (10.3) and (10.5) for thin-walled cylinders, where $d_i/t_h > 40$,

$$\sigma_\theta = \sigma_{\theta,avg} = \sigma_{\theta,max} = \frac{p_i r}{t_h} \tag{10.6}$$

$$\sigma_r = 0 \tag{10.7}$$

The area subjected to axial stress is

$$A = \pi\left(r_o^2 - r_i^2\right) = 2\pi r_{avg} t_h \tag{10.8}$$

Thus, the average axial tensile stress is

$$\sigma_{z,avg} = \frac{p_i r_i^2}{r_o^2 - r_i^2} = \frac{p_i r_i^2}{2 r_{avg} t_h}$$

For thin-walled cylinders $r_i \approx r_{avg} \approx r$

$$\therefore \qquad \sigma_z = \frac{p_i r}{2t_h} \tag{10.9}$$

In summary, the stresses in a thin-walled cylinder are

$$\sigma_r = 0 \tag{10.7}$$

$$\sigma_\theta = \frac{p_i r}{t_h} \tag{10.6}$$

$$\sigma_z = \frac{p_i r}{2t_h} \tag{10.9}$$

Note that the circumferential (hoop) stress is twice the axial stress. Also, the principal stresses for thin-walled cylinders are

$$\sigma_1 = \sigma_\theta = \frac{p_i r}{t_h} \tag{10.10}$$

$$\sigma_2 = \sigma_z = \frac{p_i r}{2t_h} \tag{10.11}$$

**Given:** A 100-in-inner-diameter cylinder made of a material having a yield strength of 60 ksi is subjected to an internal pressure of 300 psi. Use a safety factor of 3, and assume that thin-wall analysis is adequate.

**EXAMPLE 10.2**

**Find:** The wall thickness required to prevent yielding, based on

    (a) Maximum-shear-stress theory (MSST)
    (b) Distortion energy theory (DET)

**Solution:** From Equations (10.10) and (10.11)

$$\sigma_1 = \sigma_\theta = \frac{p_i r}{t_h} = \frac{(300)(50)}{t_h}$$

$$\sigma_2 = \sigma_z = \frac{p_i r}{2t_h} = \frac{(300)(50)}{2t_h}$$

$$\sigma_3 = \sigma_r = 0$$

Because $\sigma_1 > \sigma_2 > \sigma_3$, the principal stresses are ordered properly.

    (a) Using MSST, from Equation (6.7), yielding will not occur if

$$\sigma_1 - \sigma_3 < \frac{S_y}{n_s} \rightarrow \frac{p_i r}{t_h} < \frac{S_y}{n_s}$$

$$\frac{n_s p_i r}{S_y} < t_h$$

To avoid yielding,

$$t_h > \frac{p_i r n_s}{S_y} = \frac{(300)(50)(3)}{(60)(10^3)} = 0.75 \text{ in}$$

    (b) Using DET, from Equation (6.11), the von Mises stress for a biaxial stress state is

$$\sigma_e = \left(\sigma_1^2 + \sigma_2^2 - \sigma_1 \sigma_2\right)^{1/2} = \left(\frac{3\sigma_1^2}{4}\right)^{1/2} = \sqrt{\frac{3}{4}}\sigma_1$$

$$\therefore \quad \sigma_e = (0.8660)\frac{(300)(50)}{t_h} = \frac{(1.299)(10^4)}{t_h}$$

From Equation (6.12) yielding will not occur if

$$\sigma_e < \frac{S_y}{n_s} \rightarrow \frac{(1.299 \times 10^4)n_s}{S_y} < t_h$$

Thus, to avoid yielding,

$$t_h > \frac{(1.299)(10^4)n_s}{S_y} = \frac{(1.299)(10^4)(3)}{(6)(10^4)} = 0.6495 \text{ in}$$

## 10.3.2 THICK-WALLED CYLINDERS

It was indicated in Section 10.3.1 that if the cylinder walls are thin, the circumferential or hoop stress $\sigma_\theta$ can be assumed to be uniform throughout the wall thickness. This assumption cannot be made for thick-walled cylinders.

**Figure 10.3**     Complete front view of thick-walled cylinder internally and externally pressurized. (a) With stresses acting on cylinder; (b) with stresses acting on element.

Figure 10.3 shows a radially loaded, thick-walled cylinder subjected to internal pressure $p_i$ and external pressure $p_o$. Because the body and the loading are symmetric about the axis, shear stresses in the circumferential and radial directions are not present, and only normal stresses $\sigma_\theta$ and $\sigma_r$ act on the element. The loading is two-dimensional; therefore, only plane stresses are involved. If an axial loading is superimposed [Eq. (10.9) developed in Sec. 10.3.1], the third principal stress is merely changed from zero to $\sigma_z$. Recall that Equation (10.9) is valid regardless of cylinder thickness.

Figure 10.4 shows a cylindrical polar element before and after deformation. The radial and circumferential (hoop) displacements are given by $\delta_r$ and $\delta_\theta$, respectively. The radial strain is the change in the element thickness in the radial direction, divided by the element thickness:

$$\epsilon_r = \frac{\delta_r + (\partial \delta_r / \partial r)\, dr - \delta_r}{dr} = \frac{\partial \delta_r}{\partial r} \tag{10.12}$$

The circumferential strain is affected by the displacement to the new radius $r + \delta_r$, or

$$\epsilon_\theta = \frac{(r + \delta_r)\, d\theta - r\, d\theta}{r\, d\theta} = \frac{\delta_r}{r} \tag{10.13}$$

The circumferential variations shown in Figure 10.4 are not present in the press-fit equations because of circular symmetry. From Hooke's law the stress–strain relationship for the biaxial stress state [see Eq. (B.58)], while making use of Equations (10.12) and (10.13), gives

$$\epsilon_r = \frac{\partial \delta_r}{\partial r} = \frac{1}{E}(\sigma_r - \nu \sigma_\theta) \tag{10.14}$$

$$\epsilon_\theta = \frac{\delta_r}{r} = \frac{1}{E}(\sigma_\theta - \nu \sigma_r) \tag{10.15}$$

The expressions for the radial and circumferential stresses are obtained from force equilibrium of the element. Figure 10.3(b) gives details of the stresses acting on the element.

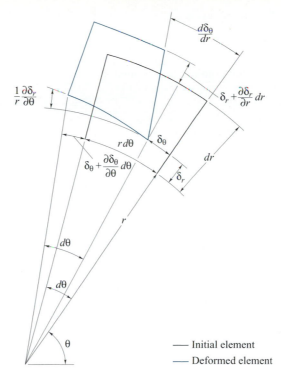

**Figure 10.4** Cylindrical polar element before and after deformation.

Summing the forces in the radial direction gives

$$(\sigma_r + d\sigma_r)(r + dr)\, d\theta\, dz - \sigma_r r\, d\theta\, dz - 2\sigma_\theta \sin\left(\frac{d\theta}{2}\right) dr\, dz = 0 \qquad \text{(10.16a)}$$

For small $d\theta$, $\sin(d\theta/2) = d\theta/2$ and Equation (10.16a) reduces to

$$\sigma_\theta = r\frac{d\sigma_r}{dr} + \sigma_r \qquad \text{(10.16b)}$$

Equations (10.14), (10.15), and (10.16b) are three equations with three unknowns $\delta_r$, $\sigma_r$, and $\sigma_\theta$. Substituting Equation (10.16b) into Equations (10.14) and (10.15) and then differentiating Equation (10.15) with respect to $r$ while equating Equations (10.14) and (10.15) give

$$0 = 3\frac{d\sigma_r}{dr} + r\frac{d^2\sigma_r}{dr^2} \qquad \text{(10.17)}$$

Equation (10.17) can be rewritten as

$$0 = 2\frac{d\sigma_r}{dr} + \frac{d}{dr}\left(r\frac{d\sigma_r}{dr}\right) \qquad \text{(10.18)}$$

Integrating gives

$$0 = 2\sigma_r + r\frac{d\sigma_r}{dr} + C_1$$

This equation can be rewritten as

$$0 = \frac{d}{dr}\left(r^2\sigma_r\right) + C_1 r$$

Integrating gives

$$\sigma_r = -\frac{C_1}{2} - \frac{C_2}{r^2} \tag{10.19}$$

The boundary conditions for thick-walled cylinders pressurized both internally and externally are

1.  $\sigma_r = -p_i$ at $r = r_i$
2.  $\sigma_r = -p_o$ at $r = r_o$

Note that because the pressurization is compressive, the minus signs appear in the boundary conditions. Applying the boundary conditions gives

$$\sigma_r = \frac{p_i r_i^2 - p_o r_o^2 + (p_o - p_i)(r_o r_i/r)^2}{r_o^2 - r_i^2} \tag{10.20}$$

$$\frac{d\sigma_r}{dr} = -\frac{2(p_o - p_i)(r_o r_i)^2}{r^3\left(r_o^2 - r_i^2\right)} \tag{10.21}$$

Substituting Equations (10.20) and (10.21) into Equation (10.16$b$) gives

$$\sigma_\theta = \frac{p_i r_i^2 - p_o r_o^2 - (r_i r_o/r)^2(p_o - p_i)}{r_o^2 - r_i^2} \tag{10.22}$$

### Internally Pressurized

If, as in many applications, the outer pressure $p_0$ is zero, Equations (10.20) and (10.22) reduce to

$$\sigma_r = \frac{p_i r_i^2\left(1 - r_o^2/r^2\right)}{r_o^2 - r_i^2} \tag{10.23}$$

$$\sigma_\theta = \frac{p_i r_i^2\left(1 + r_o^2/r^2\right)}{r_o^2 - r_i^2} \tag{10.24}$$

Figure 10.5 shows the radial and circumferential (hoop) stresses in an internally pressurized cylinder. The circumferential stress is tensile, and the radial stress is compressive. Further, the maximum stresses occur at $r = r_i$ and are

$$\sigma_{r,\text{max}} = -p_i \tag{10.25}$$

$$\sigma_{\theta,\text{max}} = p_i\left(\frac{r_o^2 + r_i^2}{r_o^2 - r_i^2}\right) \tag{10.26}$$

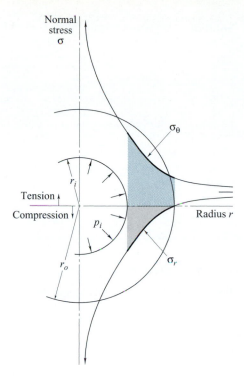

**Figure 10.5**   Internally pressurized, thick-walled cylinder showing circumferential (hoop) and radial stress for various radii. [Juvinall (1967)]

From Equation (10.15) the circumferential strains for internal pressurization as evaluated at the location of maximum stress are

$$\epsilon_\theta = \frac{\delta_r}{r_i} = \frac{p_i}{E}\left(\frac{r_o^2 + r_i^2}{r_o^2 - r_i^2} + \nu\right) \qquad (10.27)$$

where $\nu$ = Poisson's ratio of solid material. The radial displacement is outward:

$$\delta_r = \frac{p_i r_i}{E}\left(\frac{r_o^2 + r_i^2}{r_o^2 - r_i^2} + \nu\right) \qquad (10.28)$$

---

**Given:**   A hydraulic cylinder connected to a high-pressure hydraulic circuit can be pressurized to 100 MPa. The inner radius of the cylinder is 100 mm, and the wall thickness is 35 mm.

**EXAMPLE 10.3**

**Find:**   The tangential and axial stresses in the cylinder wall at the inner diameter for both thin- and thick-wall analysis.

**Solution:** From the *thin*-wall theory given in Equations (10.10) and (10.11), the following can be written:

$$\sigma_1 = \frac{p_i r}{t_h} = \frac{(100)(10^6)(0.1)}{0.035} \text{ Pa} = 285.7 \text{ MPa}$$

$$\sigma_2 = \frac{p_i r}{2t_h} = 142.9 \text{ MPa}$$

From *thick*-wall theory with internal pressurization, Equation (10.26) gives

$$\sigma_{\theta,\max} = p_i \left( \frac{r_o^2 + r_i^2}{r_o^2 - r_i^2} \right) = (100)(10^6) \left( \frac{0.135^2 + 0.10^2}{0.135^2 - 0.10^2} \right) \text{ Pa} = 343.2 \text{ MPa}$$

and the axial stress is

$$\sigma_z = \frac{\pi r_i^2 p_i}{\pi (r_o^2 - r_i^2)} = \frac{(0.10)^2 (10^8)}{0.135^2 - 0.10^2} \text{ Pa} = 121.6 \text{ MPa}$$

### Externally Pressurized

If the internal pressure is zero and the external pressure is not zero, Equations (10.20) and (10.22) reduce to

$$\sigma_r = \frac{p_o r_o^2}{r_o^2 - r_i^2} \left( \frac{r_i^2}{r^2} - 1 \right) \tag{10.29}$$

$$\sigma_\theta = -\frac{p_o r_o^2}{r_o^2 - r_i^2} \left( \frac{r_i^2}{r^2} + 1 \right) \tag{10.30}$$

Figure 10.6 shows the radial and circumferential stresses in an externally pressurized cylinder. Note that both stresses are compressive. Furthermore, the maximum circumferential stress occurs at $r = r_i$, and the maximum radial stress occurs at $r = r_o$. These expressions are

$$\sigma_{r,\max} = -p_o \tag{10.31}$$

$$\sigma_{\theta,\max} = -\frac{2r_o^2 p_o}{r_o^2 - r_i^2} \tag{10.32}$$

---

**EXAMPLE 10.4**

**Given:** A thick-walled cylinder with 0.3-m internal diameter and 0.4-m external diameter has a maximum circumferential (hoop) stress of 250 MPa. The material has a Poisson's ratio of 0.3 and a modulus of elasticity of 207 GPa.

**Find:** Determine the following:

(a) For internal pressurization ($p_o = 0$) the maximum pressure to which the cylinder may be subjected
(b) For external pressurization ($p_i = 0$) the maximum pressure to which the cylinder may be subjected
(c) The radial displacement of a point on the inner surface for the situation presented in part *a*

**Solution:**

(a) Internal pressurization. From Equation (10.26)

$$p_i = \frac{\sigma_{\theta,\max}\left(r_o^2 - r_i^2\right)}{r_o^2 + r_i^2} = \frac{(250)(10^6)(0.2^2 - 0.15^2)}{0.2^2 + 0.15^2} \text{ Pa} = 70 \text{ MPa}$$

(b) External pressurization. From Equation (10.32)

$$p_o = -\frac{\sigma_{\theta,\max}\left(r_o^2 - r_i^2\right)}{2r_o^2} = \frac{(250)(10^6)(0.2^2 - 0.15^2)}{2(0.2)^2} \text{ Pa} = 54.69 \text{ MPa}$$

(c) Radial displacement for internally pressurized cylinders. From Equation (10.28)

$$\delta_{r,\max} = \frac{p_i r_i}{E}\left(\frac{r_o^2 + r_i^2}{r_o^2 - r_i^2} + v\right) = \frac{(70)(10^6)(0.15)}{(2.07)(10^{11})}\left(\frac{0.2^2 + 0.15^2}{0.2^2 - 0.15^2} + 0.3\right)$$

$$= 1.96 \times 10^{-4} \text{ m} = 0.196 \text{ mm}$$

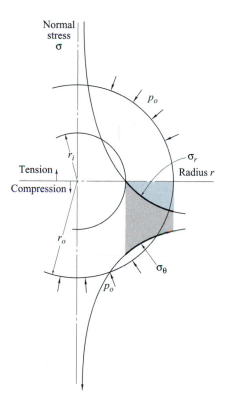

**Figure 10.6** Externally pressurized, thick-walled cylinder showing circumferential (hoop) and radial stress for various radii. [Juvinall (1967)]

## 10.4    ROTATIONAL EFFECTS

Rotating cylinders are encountered in a number of machine elements such as flywheels, gears, pulleys, and sprockets. This section considers rotation of the cylinder while assuming no pressurization ($p_i = p_o = 0$).

If a body force is included when one is considering the stresses acting on the elements shown in Figure 10.3, Equation (10.16$b$) becomes

$$\frac{\sigma_\theta - \sigma_r}{r} - \frac{d\sigma_r}{dr} - \beta = 0 \qquad\qquad \text{(10.33)}$$

where $\beta$ = body force per volume, N/m$^3$. Here, the body force is the rotating inertia force that acts radially and is given as

$$\beta = r\omega^2\rho \qquad\qquad \text{(10.34)}$$

where $\omega$ = rotational speed, s$^{-1}$
$\quad\rho$ = density, kg/m$^3$
$\quad r$ = radius of cylinder, m

Two special cases are considered: a cylinder with a central hole and a solid cylinder. Section 10.3 presented a complete derivation of the equations; however, in this section, since the procedures are similar, only the resulting equations are presented.

### 10.4.1    CYLINDER WITH CENTRAL HOLE

The circumferential (hoop) and radial stresses while considering rotation but neglecting pressurization in a cylinder with a central hole are

$$\sigma_\theta = \frac{3+\nu}{8}\rho\omega^2\left(r_i^2 + r_o^2 + \frac{r_i^2 r_o^2}{r^2} - \frac{1+3\nu}{3+\nu}r^2\right) \qquad\qquad \text{(10.35)}$$

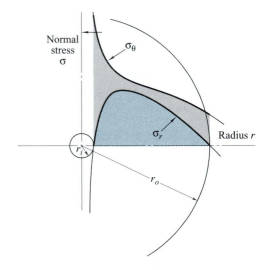

**Figure 10.7**    Stresses in rotating cylinder with central hole and no pressurization [Juvinall (1967)].

$$\sigma_r = \frac{3+\nu}{8}\rho\omega^2\left(r_i^2 + r_o^2 - \frac{r_i^2 r_o^2}{r^2} - r^2\right) \tag{10.36}$$

where $\nu$ = Poisson's ratio. Figure 10.7 shows how these stresses vary with radius. Note that both stresses are tensile. The radial stress is zero at the inner and outer radii, with a maximum occurring between $r_i$ and $r_o$. The circumferential stress is a maximum at $r = r_i$ and exceeds the radial stress for any value of radius $r$. The maximum circumferential stress is

$$\sigma_{\theta,max} = \frac{3+\nu}{4}\rho\omega^2\left[r_o^2 + \frac{r_i^2(1-\nu)}{3+\nu}\right] \tag{10.37}$$

The maximum radial stress can be obtained by differentiating Equation (10.36) with respect to $r$ and setting the result equal to zero. Thus,

$$\sigma_{r,max} = \frac{3+\nu}{8}\rho\omega^2(r_i - r_o)^2 \tag{10.38}$$

$$r = \sqrt{r_i r_o} \tag{10.39}$$

---

**Given:** Two concentric AISI 1040 steel tubes are press-fit together at 110 MPa. The nominal sizes of the tubes are 100-mm outer diameter and 80-mm inner diameter and 80-mm outer diameter and 60-mm inner diameter, respectively.

**EXAMPLE 10.5**

**Find:** How fast does the combined tube have to rotate to decrease the press-fit pressure to zero?

**Solution:** By assuming linear elasticity the press-fit pressure of 110 MPa has to be compensated for by an equally large radial stress at the location of the fit caused by the rotation. Equation (10.36) thus gives

$$\sigma_r = \frac{3+\nu}{8}\rho\omega^2\left(r_i^2 + r_o^2 - \frac{r_i^2 r_o^2}{r^2} - r^2\right)$$

$$\sigma_r = \frac{3+0.3}{8}(7850)\omega^2\left[(0.03)^2 + (0.05)^2 - \frac{(0.03)^2(0.05)^2}{(0.04)^2} - (0.04)^2\right] = (110)(10^6)$$

$$\omega = 9288 \text{ rad/s} = 88{,}700 \text{ rpm}$$

Thus, at 88,700 rpm the press-fit pressure becomes zero. This very high rotational speed clearly shows that press fits of tubes do not loosen due to inertial effects in such machine elements.

---

## 10.4.2 SOLID CYLINDER

Setting $r_i = 0$ in Equations (10.35) and (10.36) gives the circumferential and radial stresses when one is considering rotation of a solid cylinder but neglecting pressurization effects.

$$\sigma_\theta = \frac{3+\nu}{8}\rho\omega^2\left[r_o^2 - \frac{r^2(1+3\nu)}{3+\nu}\right] \tag{10.40}$$

$$\sigma_r = \frac{3+\nu}{8}\rho\omega^2\left(r_o^2 - r^2\right) \tag{10.41}$$

Figure 10.8 shows the stress distribution for both stress components as a function of radius for a solid cylinder. Note that both stresses are tensile, with the maximum occurring at $r = 0$

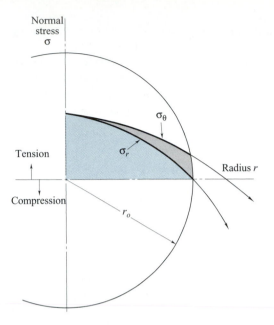

**Figure 10.8**    Stresses in rotating solid cylinder with no pressurization [Juvinall (1967)].

and both having the same value at that location. The maximum stress is

$$\sigma_{\theta,\max} = \sigma_{r,\max} = \frac{3+v}{8}\rho(r_o\omega)^2 \qquad \textbf{(10.42)}$$

Some interesting observations can be made in comparing Equations (10.37) and (10.38) with Equation (10.42). As $r_i$ becomes very small, Equations (10.38) and (10.42) are in complete agreement. However, as $r_i \to 0$, Equations (10.37) and (10.42) differ by a factor of 2 in circumferential stress. This difference is due to the radial stress in the inner portion of the solid cylinder, which decreases the circumferential stress by one-half.

---

**EXAMPLE 10.6**

**Given:** A cylindrical flywheel is press-fit onto a solid shaft. Both are made of AISI 1080 steel. The press-fit pressure is 185 MPa, the shaft diameter is 100 mm, and the outside diameter of the flywheel is 550 mm.

**Find:** The shaft speed when the press-fit pressure disappears due to centrifugal effects.

**Solution:** When the radial stress due to centrifugal acceleration at the shaft surface is equal to the original press-fit pressure, the flywheel will start to separate from the shaft. From Table A.1 and Equation (10.41)

$$\sigma_r = \frac{(3+v)\rho\omega^2\left(r_o^2 - r^2\right)}{8} = \frac{(3+0.3)(7850)\omega^2(0.275^2 - 0.050^2)}{8} = 185 \times 10^6$$

$$\omega = 883.9 \text{ rad/s} = 8441 \text{ rpm}$$

Thus, at 8441 rpm the press-fit pressure becomes zero. This speed is high, but not unreasonable for a flywheel-mounted shaft; thus press fits can fail in this manner for overspeeding shafts with flywheels.

## 10.5 PRESS FITS

In a **press fit,** the pressure $p_f$ is caused by the radial interference between the shaft and the hub. This pressure increases the radius of the hole and decreases the radius of the shaft. Section 10.2 described shaft and hub dimensions in terms of tolerance, which results in specific fits. This section focuses on the stress and strain found in press fits and uses material developed in Section 10.3.2 for thick-walled cylinders.

Figure 10.9 shows a side view of interference in a press fit; there is a radial displacement of the hub $\delta_{rh}$ and a radial displacement of the shaft $\delta_{rs}$. Figure 10.10 shows the front view of an interference fit. In Figure 10.10(a) the cylinders are assembled with an interference fit; in Figure 10.10(b) the hub and shaft are disassembled, and the dimensions of each are clearly shown. This figure also shows the interference pressure being internal for the hub and external for the shaft. The shaft is shown as hollow in order to present the most general case.

### 10.5.1 HUB

By using Equation (10.15) the hub displacement is

$$\delta_{rh} = \frac{r_f}{E_h}(\sigma_\theta - \nu_h\sigma_r) \qquad (10.43)$$

where $E_h$ = modulus of elasticity of hub material, Pa
$\nu_h$ = Poisson's ratio of hub material

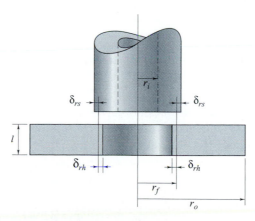

**Figure 10.9** Side view showing interference in press fit of hollow shaft to hub.

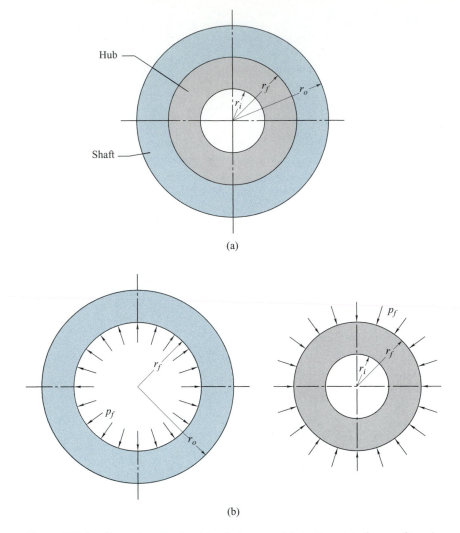

**Figure 10.10**     Front view showing (a) cylinder assembled with an interference fit and (b) hub and hollow shaft disassembled (also showing interference pressure).

For internally pressurized, thick-walled cylinders and from Equations (10.23) and (10.24), the radial and circumferential stresses for the hub while letting $p_i = p_f, r = r_f$, and $r_i = r_f$ are

$$\sigma_r = \frac{p_f r_f^2 \left(1 - r_o^2/r_f^2\right)}{r_o^2 - r_f^2} = -p_f \tag{10.44}$$

$$\sigma_\theta = \frac{p_f r_f^2 \left(1 + r_o^2/r_f^2\right)}{r_o^2 - r_f^2} = \frac{p_f \left(r_o^2 + r_f^2\right)}{r_o^2 - r_f^2} \tag{10.45}$$

Substituting Equations (10.44) and (10.45) into Equation (10.43) gives

$$\delta_{rh} = \frac{r_f p_f}{E_h} \left( \frac{r_o^2 + r_f^2}{r_o^2 - r_f^2} + v_h \right) \tag{10.46}$$

The positive sign of $\delta_{rh}$ indicates that the radial displacement of the hub is outward.

## 10.5.2 SHAFT

From Equation (10.15) the displacement of the shaft is

$$\delta_{rs} = \frac{r_f}{E_s} (\sigma_\theta - v_s \sigma_r) \tag{10.47}$$

where $E_s$ = modulus of elasticity of shaft material, Pa
$\quad\quad v_s$ = Poisson's ratio of shaft material

The circumferential and radial stresses for externally pressurized, thick-walled cylinders can be obtained from Equations (10.29) and (10.30) by letting $p_o = p_f$, $r_o = r_f$, and $r = r_f$, to give

$$\sigma_r = \frac{p_f r_f^2}{r_f^2 - r_i^2} \left( \frac{r_i^2}{r_f^2} - 1 \right) = -p_f \tag{10.48}$$

$$\sigma_\theta = -\frac{p_f r_f^2}{r_f^2 - r_i^2} \left( \frac{r_i^2}{r_f^2} + 1 \right) = -\frac{p_f (r_f^2 + r_i^2)}{r_f^2 - r_i^2} \tag{10.49}$$

Substituting Equations (10.48) and (10.49) into Equation (10.47) gives

$$\delta_{rs} = -\frac{r_f p_f}{E_s} \left( \frac{r_f^2 + r_i^2}{r_f^2 - r_i^2} - v_s \right) \tag{10.50}$$

Because the first term in parentheses in Equation (10.50) is greater than unity and Poisson's ratio is less than or equal to 0.5, $\delta_{rs}$ is negative; hence, shaft displacement is directed inward toward the center of the shaft.

## 10.5.3 INTERFERENCE FIT

The total radial displacement in an interference fit is shown in Figure 10.9. Recall that outward deflection (expansion of the inside diameter of the hub) is positive in sign and inward deflection (reduction of the outside diameter of the shaft) is negative. Thus, the total radial interference is

$$\delta_r = \delta_{rh} - \delta_{rs} = r_f p_f \left[ \frac{r_o^2 + r_f^2}{E_h (r_o^2 - r_f^2)} + \frac{v_h}{E_h} + \frac{r_f^2 + r_i^2}{E_s (r_f^2 - r_i^2)} - \frac{v_s}{E_s} \right] \tag{10.51}$$

If the shaft and the hub are made of the same material, $E = E_s = E_h$ and $v = v_s = v_h$ and Equation (10.51) reduces to

$$\delta_r = \frac{2 r_f^3 p_f (r_o^2 - r_i^2)}{E (r_o^2 - r_f^2)(r_f^2 - r_i^2)} \tag{10.52}$$

Furthermore, if the shaft is solid rather than hollow, $r_i = 0$ and Equation (10.52) further reduces to

$$\delta_r = \frac{2 r_f p_f r_o^2}{E \left( r_o^2 - r_f^2 \right)} \tag{10.53}$$

From these equations, it can be seen that if displacement is known, the interference pressure may be an unknown and these equations can readily be used in its evaluation.

### 10.5.4     FORCE AND TORQUE

The maximum force $P_{max}$ to assemble a press fit varies directly with the thickness of the outer member, the length of the outer member, the difference in diameters of the mating shaft and hub, and the coefficient of friction $\mu$. The maximum stress is

$$\tau_{max} = p_f \mu = \frac{P_{max}}{A} = \frac{P_{max}}{2 \pi r_f l} \tag{10.54}$$

The torque is

$$T = P_{max} r_f = 2 \pi \mu r_f^2 l p_f \tag{10.55}$$

The axial and circumferential stresses are related to the maximum stress by

$$\tau_a^2 + \tau_c^2 = \tau_{max}^2$$

where $\tau_a = \dfrac{P_a}{2 \pi r_f l} = $ axial stress

$\tau_c = \dfrac{P_c}{2 \pi r_f l} = $ circumferential stress

---

**EXAMPLE 10.7**

**Given:**   A 6-in-diameter steel shaft is to have a press fit with a 12-in-outside-diameter cast iron hub. Both the hub and the shaft are 10 in long. The maximum circumferential stress is to be 5000 psi. The moduli of elasticity are $30 \times 10^6$ psi for steel and $15 \times 10^6$ psi for cast iron. Poisson's ratio for both steel and cast iron is 0.3, and the coefficient of friction for the two materials is 0.12. That is,

$$r_f = 3 \text{ in} \qquad r_i = 0 \qquad r_o = 6 \text{ in}$$
$$E_s = 30 \times 10^6 \text{ psi} \qquad E_h = 15 \times 10^6 \text{ psi} \qquad \nu_s = \nu_h = 0.3$$
$$\mu = 0.12 \qquad l = 10 \text{ in} \qquad \sigma_{\theta,max} = 5000 \text{ psi}$$

**Find:**   Determine

    (a)   The interference
    (b)   The axial force required to press the hub on the shaft
    (c)   The torque that this press fit can transmit

**Solution:**

(a) From Equation (10.45) the interference pressure is

$$p_f = \frac{\sigma_{\theta,\max}\left(r_o^2 - r_f^2\right)}{r_o^2 + r_f^2} = \frac{5000(6^2 - 3^2)}{6^2 + 3^2} = 3000 \text{ psi}$$

From Equation (10.51) the maximum permissible radial interference is

$$\delta_r = r_f p_f \left[ \frac{r_o^2 + r_f^2}{E_h\left(r_o^2 - r_f^2\right)} + \frac{\nu_h}{E_h} + \frac{r_f^2 + r_i^2}{E_s\left(r_f^2 - r_i^2\right)} - \frac{\nu_s}{E_s} \right]$$

$$= \frac{3(3000)}{15(10^6)}\left(\frac{6^2 + 3^2}{6^2 - 3^2} + 0.3 + \frac{1}{2} - \frac{0.3}{2}\right) = 1.390 \times 10^{-3} \text{ in}$$

(b) From Equation (10.54) the force required for the press fit is

$$P_{\max} = 2\pi \mu r_f l p_f = (2)(\pi)(0.12)(3)(10)(3000) = 67{,}860 \text{ lb}$$

(c) From Equation (10.55) the torque is

$$T = P_{\max} r_f = (67{,}860)(3) = 203{,}600 \text{ lb} \cdot \text{in}$$

---

**Given:** A wheel hub is press-fit onto a 105-mm-diameter solid shaft. The coefficient of friction is 0.11, and the hub and shaft material is AISI 1080 steel. The hub's outer diameter is 160 mm, and its width is 120 mm. The radial interference between the shaft and the hub is 65 $\mu$m (the shaft diameter is 130 $\mu$m larger than the inside diameter of the hub).

**EXAMPLE 10.8**

**Find:** The axial force necessary to dismount the hub.

**Solution:** Equation (10.53) gives the relationship between radial displacement and pressure.

$$\delta_r = \frac{2r_f p_f r_o^2}{E\left(r_o^2 - r_f^2\right)} = (65)(10^{-6}) = \frac{(2)(0.0525)p_f(0.080)^2}{(2.07)(10^{11})[(0.080)^2 - (0.0525)^2]}$$

$$\therefore \quad p_f = 72.96 \text{ MPa}$$

The axial force necessary to dismount the hub is

$$P = \mu p_f A = (0.11)(72.96)(10^6)\pi(0.105)(0.120) = 317{,}700 \text{ N} = 317.7 \text{ kN}$$

---

## 10.6  SHRINK FITS

In producing a **shrink fit,** it is common to heat the outer component (hub) in order to expand it beyond the interference, and then slip it over the inner component (shaft); cooling then contracts the outer component. Temperature change produces a strain, called *thermal strain,* even in the absence of stress. Although thermal strain is not exactly linear with temperature change, for temperature changes of a few hundred degrees Fahrenheit the actual variation can be closely described by a linear approximation. According to this linear relationship, the temperature difference, through which the outer component must be heated to obtain the required expansion over the undeformed solid shaft, is

$$\Delta t_m = \frac{\delta_r}{\bar{a} r_f} \tag{10.56}$$

where $\bar{a}$ = coefficient of thermal expansion (see Table 3.5 and Fig. 3.14). Equation (10.56) can be expressed in terms of radial strain as

$$\epsilon_r = \frac{\delta_r}{r_f} = \bar{a}\Delta t_m \tag{10.57}$$

The deformation is

$$\delta_r = \epsilon_r r_f = \bar{a}\Delta t_m r_f \tag{10.58}$$

These equations are valid not only for shrink fits of shaft and hub but also for a wide range of thermal problems.

The strain due to a temperature change may be added algebraically to a local strain by using the *principle of superposition*. The principle states that stresses and strains (at a point on a given plane) due to different loads may be computed separately and added algebraically, provided that the sum does not exceed the proportionality limit of the material and that the structure remains stable. The method of superposition for different types of loading was covered in Section 5.4. Thus, the normal strain due to normal load and temperature effects is

$$\epsilon = \epsilon_\sigma + \epsilon_{tm} \tag{10.59}$$

where $\epsilon_\sigma$ = strain due to normal stress
$\epsilon_{tm}$ = strain due to temperature change

Thus, the general (triaxial stress state) stress–strain relationship developed in Appendix B, Equation (B.44), may be expressed while considering thermal strain as

$$\epsilon_x = \frac{1}{E}[\sigma_x - \nu(\sigma_y + \sigma_z)] + \bar{a}\Delta t_m$$

$$\epsilon_y = \frac{1}{E}[\sigma_y - \nu(\sigma_z + \sigma_x)] + \bar{a}\Delta t_m \tag{10.60}$$

$$\epsilon_z = \frac{1}{E}[\sigma_z - \nu(\sigma_x + \sigma_y)] + \bar{a}\Delta t_m$$

---

**EXAMPLE 10.9**

**Given:**    A 10-in-long steel tube has a cross-sectional area of 1 in$^2$ that expands by 0.008 in from a stress-free condition at 80°F when the tube is heated to 480°F.

**Find:**    The load and stress acting on the steel tube.

**Solution:**   From Table 3.5 for steel alloy $\bar{a} = 6.1 \times 10^{-6}$ °F, and from Equation (10.57),

$$\epsilon = \bar{a}\,\Delta t_m = (6.1)(10^{-6})(400) = 2.44 \times 10^{-3}$$

$$\Delta l = l\epsilon = (10)(2.44)(10^{-3}) = 0.0244 \text{ in}$$

Because the measured expansion was only 0.008 in, the constraint due to compressive normal loading must apply a force sufficient to deflect the tube axially by the following amount:

$$\delta = \frac{Pl}{AE} \rightarrow P = \frac{AE\,\delta}{l}$$

From Table A.1 the modulus of elasticity is $30 \times 10^6$ psi. Therefore,

$$P = \frac{(1)(30)(10^6)(0.0164)}{10} = 4.92 \times 10^4 \text{ lb} = 49,200 \text{ lb}$$

This, then, is the compressive, normal, axial load being exerted on the steel tube. The axial stress is

$$\sigma = -\frac{P}{A} = -49,200 \text{ psi}$$

---

**Given:**   A block of aluminum alloy is placed between two rigid jaws of a clamp, and the jaws are tightened to a snug state. The temperature of the entire assembly is raised by 250°C in an oven. The cross-sectional areas are 65 mm² for the block and 160 mm² for the stainless steel screws.

**EXAMPLE 10.10**

**Find:**   The stresses induced in the screws and the block.

**Solution:**   Figure 10.11 shows the block-and-screw assembly and the forces acting on these components. From force equilibrium

$$P_a = 2P_s \qquad\qquad\qquad \textbf{(a)}$$

Here subscript $a$ refers to the aluminum block, and subscript $s$ refers to the stainless steel screws. Compatibility requires that the length changes of the block and the screws be the same, or

$$\delta_a = \delta_s \qquad\qquad\qquad \textbf{(b)}$$

(a)                                                          (b)

**Figure 10.11**   (a) Block placed between two rigid jaws of clamp and (b) forces acting.

Thermal expansion will induce an axial force, as shown in Figure 10.11. The displacements of the block and screws are

$$\delta_a = \bar{a}_a l \Delta t_m - \frac{P_a l}{E_a A_a} \tag{c}$$

$$\delta_s = \bar{a}_s l \Delta t_m + \frac{P_s l}{E_s A_s} \tag{d}$$

Substituting Equations $(a)$, $(c)$, and $(d)$ into Equation $(b)$ gives

$$P_s = \frac{\Delta t_m (\bar{a}_a - \bar{a}_s)}{1/(E_s A_s) + 2/(E_a A_a)} \tag{e}$$

From Tables 3.2 and 3.5

$$E_a = 70 \text{ GPa} \qquad E_s = 193 \text{ GPa}$$

$$\bar{a}_a = 24 \times 10^{-6} (°C)^{-1} \qquad \bar{a}_s = 17 \times 10^{-6} (°C)^{-1}$$

It is given that $A_a = 65 \text{ mm}^2$ and $A_s = 160 \text{ mm}^2$; substituting these values and the above into Equation $(e)$ gives the force acting on each screw as

$$P_s = \frac{(250)(24 - 17)(10^{-6})}{1/(193)(10^9)(160)(10^{-6}) + 2/(70)(10^9)(65)(10^{-6})} = 3708 \text{ N}$$

The force acting on the aluminum block is

$$P_a = 2P_s = 7416 \text{ N}$$

The axial stresses of the block and screw are

$$\sigma_a = \frac{P_a}{A_a} = -\frac{7416}{(65)(10^{-6})} \text{ Pa} = -114.1 \text{ MPa}$$

$$\sigma_s = \frac{P_s}{A_s} = \frac{3708}{(160)(10^{-6})} \text{ Pa} = 23.18 \text{ MPa}$$

Note that the stress acting on the aluminum block is compressive, and that acting on the screws is tensile.

---

**Case Study 10.1** | DESIGN OF A SHOT SLEEVE FOR A DIE CASTING MACHINE

**Given:** Die casting is a common and important manufacturing process. A wide variety of products are produced through die casting, such as personal computer and camera frames, automotive structural components, fasteners, and toy cars. In die casting, molten metal is placed in a shot sleeve and injected into a metal die outfitted with cooling lines to extract the heat and cause the cast metal to solidify quickly. A schematic illustration of a die casting machine is shown in Figure 10.12.

For the process to be economically viable, a number of features are essential:

•  The tooling is very expensive, and therefore a fairly large production run is required to justify the use of this process. Therefore, the cycle time (and cooling time) must be short to achieve required production runs.

•  Given that the cooling time should be very short, it is essential that the molten metal be injected into the die under high pressure and velocity to ensure that the mold is filled completely before local solidification results in an underfill. Injection pressures can be as high as 70 MPa (10 ksi), although 15 to 20 MPa (2 to 3 ksi) is more typical.

**Case Study** (Continued)

**Figure 10.12** Schematic illustration of a die casting machine [From Kalpakjian and Schmid (2003)].

- The molten metal is transferred to the die casting machine from a furnace and held temporarily in the *shot sleeve*. The shot sleeve must have sufficient volume to fill the die cavity.

- When the metal is injected into the cavity, a high pressure is generated in the shot sleeve through the compressive action of a hydraulic cylinder.

- The die obviously must have a higher melting temperature than the metal being cast; this usually limits the tooling to die steels, and the workpiece metal to aluminum, magnesium, or copper alloys.

A die casting machine is being designed for an operating pressure of 20 MPa, typical for the *hot chamber* process where the dies are maintained at a high temperature to aid in filling of the cavity. The hydraulic cylinder that has been prescribed can generate the required pressure for an inside diameter of the shot sleeve of 170 mm. A material typically used for this application is H13 steel, which has an ultimate strength of 768 MPa at room temperature and 650 MPa at casting temperatures for aluminum.

**Find:** Evaluate the design of a shot sleeve with a thickness of 60 mm, allowing for $8 \times 10^{-3}$ m³ to be cast in one shot. Assume the sleeve is produced from H13 steel. Part of the proposed instrumentation for the machine will include

placing strain measurement devices and thermocouples on the sleeve. Calculate the hoop strain at the exterior of the sleeve as a function of pressure, and evaluate whether this can be easily measured.

**Solution:** The fatigue strength of the H13 steel needs to be estimated. The loading in this case consists of hoop and radial stresses only—significant axial stresses are not developed in the sleeve. On one end, the hydraulic cylinder applies a force that compresses the fluid, and on the nozzle end the cylinder is mounted so that it does not see such loads. We approximate the loading as axial and apply Equation (7.7) to obtain

$$S'_e = 0.45 S_u = 0.45(768) = 345 \text{ MPa}$$

Of the correction factors that need to be considered, a surface finish correction factor will be first taken for a machined surface. From Figure 7.7, the value of $k_f$ is approximated as 0.7. A reliability of 50% will be assumed, so that $k_r = 1.0$. The temperature factor is calculated from Equation (7.23) as

$$k_t = \frac{S_{ut}}{S_{ut,\text{ref}}} = \frac{650}{768} = 0.846$$

All other fatigue strength correction factors will be ignored. Therefore, the endurance limit is estimated from

*(continued)*

**Case Study**        (Concluded)

Equation (7.16) as

$$S_e = k_f k_t S'_e = (0.70)(0.846)(345) = 204 \text{ MPa}$$

To obtain the stresses, we apply Equations (10.25) and (10.26), using $r_i = 85$ mm $= 0.085$ m, $r_o = 145$ mm $= 0.145$ m, and $p_i = 20$ MPa. Therefore,

$$\sigma_{r,max} = -p_i = -20 \text{ MPa}$$

$$\sigma_{\theta,max} = p_i \left( \frac{r_o^2 + r_i^2}{r_o^2 - r_i^2} \right) = (20 \times 10^6) \frac{(0.145)^2 + (0.085)^2}{(0.145)^2 - (0.085)^2}$$

$$= 40.9 \text{ MPa}$$

The von Mises stress for this case is given by Equation (6.9), using $\sigma_1 = 40.9$ MPa, $\sigma_2 = 0$, and $\sigma_3 = -20$ MPa, yielding

$$\sigma_e = \frac{1}{\sqrt{2}} \left[ (\sigma_2 - \sigma_1)^2 + (\sigma_3 - \sigma_1)^2 + (\sigma_3 - \sigma_2)^2 \right]^{1/2}$$

$$= \frac{1}{\sqrt{2}} \left[ (0 - 40.9)^2 + (-20 - 40.9)^2 + (-20 - 0)^2 \right]^{1/2}$$

$$= 53.77 \text{ MPa}$$

Compared to the endurance limit of the material, these stresses are very low. Note that thermal stresses, especially during start-up, can have a significant influence on the

stresses in the sleeve, and fatigue failures of such components are not uncommon.

The stresses at the outside of the sleeve must be determined to obtain the circumferential strain at this location. Using $r = r_o = 145$ mm in Equations (10.23) and (10.24) yields the following stresses as a function of internal pressure:

$$\sigma_r = \frac{p_i r_i^2 \left( 1 - r_o^2 / r^2 \right)}{r_o^2 - r_i^2} = 0$$

$$\sigma_\theta = \frac{p_i r_i^2 \left( 1 + r_o^2 / r^2 \right)}{r_o^2 - r_i^2}$$

$$= \frac{(p_i)(0.085)^2 \left( 1 + 0.145^2 / 0.145^2 \right)}{(0.145)^2 - (0.085)^2} = 1.047 p_i$$

Young's modulus and Poisson's ratio for steel are obtained from the inside front cover as 207 GPa and 0.3, respectively. Therefore, applying Equation (10.15) gives

$$\epsilon_\theta = \frac{1}{E} (\sigma_\theta - \nu \sigma_r) = \frac{1}{207 \times 10^9} (1.047 p_i - 0)$$

$$= 5.058 \times 10^{-12} p_i$$

where $p_i$ is in pascals. Note that for a 20-MPa internal pressure, the hoop strain at the outer radius is $101 \times 10^{-6}$, which is low but certainly measurable.

## 10.7    SUMMARY

In this chapter it was shown that fits must be specified to ensure the proper mating assembly of a shaft and a hub because it is impossible to manufacture these parts with exactly the desired dimensions. A system was devised to tolerate small dimensional variations of the mating shaft and hub without sacrificing their proper functioning.

Pressurization effects on cylinders were considered. Thin-wall and thick-wall analyses were described for both internal and external pressurization. These effects are important in a large number of applications, ranging from pressure vessels to gun barrels. A major assumption made in the thin-wall analysis was that the circumferential (hoop) stress is uniform throughout the wall thickness. This assumption is not valid for thick-wall analysis. Ranges of diameter to thickness were discussed for thin- and thick-wall analyses. For thick-wall situations, radial and circumferential variations with radius for both internal and external pressurization were shown. For internal pressurization, the maximum stress occurred at the inner radius for both radial and circumferential components, with the radial stress being compressive and the circumferential stress tensile. For external pressurization, the maximum radial stress occurred at the outer radius, and the maximum circumferential stress occurred at the inner radius, with both stress components being compressive.

Rotational effects while assuming no pressurization of cylinders were also considered. Rotational effects are important in such machine elements as flywheels, gears, and pulleys. The rotating inertial force was considered in establishing the radial and circumferential stresses for both a cylinder with a central hole and a solid cylinder. The chapter ended with consideration of press and shrink fits of a shaft and a hub. The interference, axial force, and torque were developed for these situations.

## KEY WORDS

**allowance**   difference between nominal diameters of mating parts.

**bilateral tolerance**   variation above and below nominal size.

**hoop stress**   circumferential stress in a pressure vessel.

**interference**   difference in size of mating parts.

**nominal diameter**   approximate size of element.

**press fit**   connections where interfacial pressure is due to interference between mating parts and assembly is accomplished by elastic deformation due to large forces.

**shrink fit**   connections where interfacial pressure is due to interference between mating parts and assembly is accomplished by heating the outer component and then cooling it over the inner component.

**thick-walled cylinder**   cylinder whose ratio of diameter to thickness is less than 40.

**thin-walled cylinder**   cylinder where radial stress is negligible, approximately true for diameter-to-thickness ratios greater than 40.

**tolerance**   maximum variation in part size.

**unilateral tolerance**   variation above or below nominal size, but not both.

## RECOMMENDED READINGS

Beer, F. P., et al. (2001) *Mechanics of Materials,* 3rd ed., McGraw-Hill, New York.

Chuse, R., and Carson, B. E. (1993) *Pressure Vessels,* 7th ed., McGraw-Hill, New York.

Craig, R. R. (1966) *Mechanics of Materials,* Wiley, New York.

Fung, Y. C. (1965) *Foundations of Solid Mechanics,* Prentice-Hall, Englewood Cliffs, NJ.

Harvey, J. F. (1991) *Theory and Design of Pressure Vessels,* 2nd ed., Van Nostrand Reinhold, New York.

Hibbeler, R. C. (2002) *Mecahnics of Materials,* 5th ed., Prentice Hall, Upper Saddle River, NJ.

Juvinall, R. C. (1967) *Stress, Strain, and Strength,* McGraw-Hill, New York.

Juvinall, R. C., and Marshek, K. M. (2003) *Fundamentals of Machine Component Design,* 3rd ed., Wiley, New York.

Lardner, T. J., and Archer, R. R. (1994) *Mechanics of Solids: An Introduction,* McGraw-Hill, New York.

Megyesy, E. F. (1986) *Pressure Vessel Handbook,* 7th ed., Pressure Vessel Handbook Publishing, Inc., Tulsa, OK.

Mott, R. L. (1998) *Machine Elements in Mechanical Design,* 3rd ed., Prentice-Hall, Upper Saddle River, NJ.

Popov, E. P. (1968) *Introduction to Mechanics of Solids,* Prentice-Hall, Englewood Cliffs, NJ.

Shigley, J. E., Mischke, C. R., and Budynas, R. (2003) *Mechanical Engineering Design,* 6th ed., McGraw-Hill, New York.

Timoshenko, S., and Goodier, J. (1970) *Theory of Elasticity,* McGraw-Hill, New York.

## REFERENCES

Juvinall, R. C. (1967) *Stress, Strain, and Strength,* McGraw-Hill, New York.

Kalpakjian, S., and Schmid, S. R. (2003) *Manufacturing Engineering and Technology,* 4th ed., Prentice Hall, New Jersey.

## PROBLEMS

### Section 10.2

**10.1**   A journal bearing is to be manufactured with optimum geometry for minimum power loss for a given load and speed. The relative clearance $c/r = 0.001$; the journal diameter is 100 mm. Find the accuracy to which the bearing parts have to be manufactured so that there is no more than $\pm 10\%$ error in the relative clearance.

**10.2**   A press fit between a solid steel shaft and a steel housing is dimensioned to be of class 7. By mistake the shaft is ground at 22°C higher temperature than originally anticipated, so that when the shaft cools, the diameter is slightly too small. The shaft material is AISI 1040 steel. What is the class of fit between the shaft and the housing because of this mistake? Also, if the grinding temperature were 50°C higher rather than 22°C higher, what would be the class of fit?

### Section 10.3

★ **10.3**   A cylinder with a 0.30-m inner diameter and a 0.40-m outer diameter is internally pressurized to 140 MPa. Determine the maximum shear stress at the outer surface of the cylinder. *Ans.* $\tau = 180$ MPa

**10.4**   A rubber balloon has the shape of a cylinder with spherical ends. At low pressure the cylindrical part is 250 mm long with a diameter of 20 mm. The rubber material has constant thickness and is linearly elastic. How long will the cylindrical part of the balloon be when it is inflated to a diameter of 100 mm? Assume that the rubber's modulus of elasticity is constant and that Poisson's ratio is 0.5.

| * Indicates problem of greater difficulty.

★ **10.5** A thin-walled cylinder containing pressurized gas is fixed by its two ends between rigid walls. Obtain an expression for the wall reactions in terms of cylinder length $l$, thickness $t_h$, radius $r$, and internal pressure $p_i$.

★ **10.6** A pressurized cylinder has an internal pressure of 1 MPa, a thickness of 8 mm, a length of 4 m, and a diameter of 1.08 m. The cylinder is made of AISI 1080 steel. What is the volume increase of the cylinder due to the internal pressure? *Ans.* $\Delta V = 0.00227$ m³

**10.7** A 1-m-diameter cylindrical container with two hemispherical ends is used to transport gas. The internal pressure is 5 MPa, and the safety factor is 2.0. Using the MSST, determine the container thickness. Assume that the tangential strains of the cylinder and the sphere are equal, since they are welded joints. The material is such that $E = 200$ GPa, $S_y = 430$ MPa, and $v = 0.3$. *Ans.* $t_h > 11.6$ mm

## Section 10.4

**10.8** A solid cylindrical shaft made of AISI 1020 steel rotates at a speed that produces a safety factor of 3 against the stress causing yielding. To instrument the shaft, a small hole is drilled in its center for electric wires. At the same time the material is changed to AISI 1080 steel. Find the safety factor against yielding of the new shaft. *Ans.* $n_s = 1.92$

★ **10.9** A flywheel is mounted on a tubular shaft. The shaft's inner diameter is 30 mm, and its outer diameter is 50 mm. The flywheel is a cylindrical disk with an inner diameter of 50 mm, an outer diameter of 300 mm, and a thickness of 35 mm. Both the shaft and the flywheel are made of AISI 1080 steel. The flywheel is used to store energy, so that the rotational acceleration is proportional to the angular speed $\omega$. The angular acceleration $\partial\omega/\partial t$ is to be held to within $\pm0.2\omega$. Calculate the shaft speed $\omega$ at which the flywheel starts to slide on the shaft if the press-fit pressure at $\omega = 0$ is 127 MPa and the coefficient of friction between the shaft and the flywheel is 0.13. *Ans.* $\omega = 15,750$ rpm

**10.10** Assume that the flywheel and the shaft given in Problem 10.9 are made of aluminum alloy 2014 instead of steel. Find the speed at which the flywheel will start to slide if the coefficient of friction is 0.14 and the press-fit pressure is 30 MPa. *Ans.* $\omega = 12,800$ rpm

**10.11** The flywheel and the shaft given in Problem 10.9 are axially loaded by a force of 50,000 N. Calculate the shaft speed at which the flywheel starts to slide on the shaft. *Ans.* $\omega = 10,700$ rpm

## Section 10.5

**10.12** A 6-in-diameter solid steel shaft is to have a press fit with a 12-in-outer-diameter by 10-in-long hub made of cast iron. The maximum allowable hoop stress is 5000 psi. The moduli of elasticity are $30 \times 10^6$ psi for steel and $20 \times 10^6$ psi for cast iron. Poisson's ratio for steel and cast iron is 0.3. The coefficient of friction for both steel and cast iron is 0.11. Determine the following:

*a)* Total radial interference. *Ans.* $\delta_r = 0.001095$ in
*b)* Axial force required to press the hub on the shaft.
*c)* Torque transmitted with this fit. *Ans.* 186.6 kip-in

**10.13** A flat, 0.5-m-outer-diameter, 0.1-m-inner-diameter, 0.08-m-thick steel disk shown in sketch *a* is shrink-fit onto a shaft. If the assembly is to transmit a torque of 100 kN · m, determine the fit pressure and the total radial interference. The coefficient of friction is 0.25.

Sketch *a*, for Problem 10.13

**10.14** A flat, 0.1-m-outer-diameter, 0.05-m-inner-diameter, 0.12-m-thick steel disk is shrink-fit onto a shaft with a 0.02-m inner diameter. If the assembly is to transmit a torque of 12 kN · m, determine the fit pressure and the total radial interference. The coefficient of friction is 0.25.

★ **10.15** A supported cantilevered beam is loaded as shown in sketch *b*. If at the time of assembly the load was zero and the beam was horizontal, determine the stress in the round rod after the load is applied and the temperature is lowered by 70°C. The beam moment of inertia is $I_{\text{beam}} = 9.8 \times 10^7$ mm$^4$. The beam material and rod material are both high-carbon steel.

Sketch *b*, for Problem 10.15

**10.16** A 10-in-long tube (with properties $E = 30 \times 10^6$ psi and $\bar{a} = 7 \times 10^{-6}/°$F) having a cross-sectional area of 1 in$^2$ is installed with fixed ends so that it is stress-free at 80°F. In operation the tube is heated throughout to a uniform 480°F. Measurements indicate that the fixed ends separate by 0.008 in. What loads are exerted on the ends of the tube, and what are the resultant stresses?

10.17 A solid square bar is constrained between two fixed supports, as shown in sketch c. The square bar has a cross section of 5 in by 5 in and is 2 ft long. The bar just fits between fixed supports at the initial temperature of 60°F. If the temperature is raised 120°F, determine the average thermal stress developed in the bar. Assume that $E = 30 \times 10^6$ psi and $\bar{a} = 6.5 \times 10^{-6}/°F$. Ans. $\sigma = -11.7$ ksi

5 in by 5 in

2 ft

Sketch c, for Problem 10.17

★ 10.18 Two stiff shafts are connected by a thin-walled elastic tube to a press-fit connection, as shown in sketch d. The contact pressure between the shafts and the tube is p. The coefficient of friction is $\mu$. Calculate the maximum torque $T_{max}$ that can be transmitted through the press fit. Describe and calculate what happens if the torque decreases from $T_{max}$ to $\theta T_{max}$ where $0 < \theta < 1$.

l       a       l

Sketch d, for Problem 10.18

10.19 A 50-mm-diameter steel shaft and a 60-mm-long cylindrical bushing of the same material with an outer diameter of 80 mm have been incorrectly shrink-fit together and have to be dismounted. What axial force is needed for this operation if the diametral interference is 50 mm and the coefficient of friction is 0.2? Ans. $P_{max} = 119$ kN

10.20 A bushing press-fit on a shaft shown in sketch e is going to be dismounted. What axial force $P_a$ is needed to dismount the bushing if, at the same time, the bushing transmits a torque $T = 500$ N · m? The diametral interference $\delta = 30$ $\mu$m, the coefficient of friction $\mu = 0.1$, and the modulus of elasticity $E = 210$ GPa.

b = 100 mm

$P_a$

$P_a$

d = 50 mm

D = 70 mm

Sketch e, for Problem 10.20

**10.21** To help in dismounting the wheel in Example 10.8, both an axial force and a moment are applied at the shaft-flywheel junction. How large a moment is needed to decrease the axial force to 50 kN when the wheel is dismounted?

★ **10.22** Two AISI 1040 steel cylinders are press-fit on each other. The inner diameter is 200 mm, the common diameter is 300 mm, and the outer diameter is 400 mm. The radial interference of the two cylinders is 0.1 mm.

a)    Draw the radial and tangential stress distributions due to the press fit.
b)    If the internal pressure is 207 MPa and the external pressure is 50 MPa, determine the radial and tangential stress distributions due only to these pressures.
c)    Superimpose the stress distributions from (a) and (b) to obtain the total radial and tangential stress distribution.

★ **10.23** A thick-walled cylinder is placed freely inside another thick-walled cylinder. What pressure is induced on the surfaces between the two cylinders by internally pressurizing the inner cylinder? *Hint:* Equate the radial displacement of the outer wall of the inner cylinder with that of the inner wall of the outer cylinder.

## Section 10.6

**10.24** A 0.5-m-outer-diameter, 0.1-m-inner-diameter, 0.1-m-thick flat disk is shrink-fit onto a solid shaft. Both the shaft and the disk are made of high-carbon steel. The assembly transmits 10 MW at 1000 rpm. Calculate the minimum temperature to which the disk must be heated for this shrink fit. The coefficient of friction is 0.25. *Ans.* $\Delta t_m = 222.5°C$

**10.25** Two shafts are connected by a shrink-fit bushing with an outer diameter of 120 mm. The diameter of each shaft is 80 mm, and each shaft is 2 m long. Before the shrink-fit bushing was mounted, the diametral interference was $\delta = 80\,\mu$m. The bushing and the shaft are made of steel with a modulus of elasticity $E = 210$ GPa. Find the allowable temperature increase without slip in the press fit if a power of 250 kW is transmitted and the far ends of the shafts cannot move axially. The coefficient of thermal expansion is $11.5 \times 10^{-6}/°C$, and the coefficient of friction is 0.1. The axial length of the bushing press-fit on each shaft is 80 mm. The rotational speed is 1500 rpm. *Ans.* $\Delta t_m = 58°C$

**10.26** A 10-mm-thick, 100-mm-wide ring is shrunk onto a 100-mm-diameter shaft. The diametral interference is 75 $\mu$m. Find the surface pressure in the shrink fit and the maximum torque that can be transmitted if the coefficient of friction $\mu = 0.10$. The modulus of elasticity $E = 210$ GPa. *Ans.* $T = 3780$ N-m

★ **10.27** A railway car buffer has a spring consisting of 11 outer rings and 11 inner rings where one of each ring type is a half-ring at the end of the spring. On one occasion a railway car with a total mass of 10,000 kg rolls until it hits a rigid stop. The force stopping the car is equal in the outer and the inner rings. The speed of the car just before the stop is 18 km/h. Find the compression of the outer and inner rings and the stresses in the rings if the coefficient of friction $\mu = 0.10$ and the modulus of elasticity $E = 210$ GPa. The ring dimensions are shown in sketch $f$. When a spring of this type is compressed, the spring rate $k$ (i.e., force divided by deformation) is

$$k = \frac{P}{\delta} = \frac{\pi E \tan \alpha \tan(\alpha + \gamma)}{n(r_y/A_y + r_i/A_i)}$$

Sketch $f$, for Problem 10.27

where $\alpha$ = cone angle for spring
     $\gamma$ = friction angle
   $\tan \gamma = \mu$ = coefficient of friction
   $r_y, r_i$ = radii to surface center of gravity
   $A_y, A_i$ = cross-sectional areas of rings

★ **10.28** A railway car weighing 20 tons has in each end two buffers of the type described in Problem 10.27. The data are $A_y = A_i = 200\,\text{mm}^2, r_y = 70\,\text{mm}, r_i = 60\,\text{mm}, n = 22, \alpha = 14°, \gamma = 7°$, and $E = 206$ GPa. The car hits a rigid stop with a speed of 1 m/s. Find the maximum force in each buffer and the energy absorbed. At what speed will the car bounce back? *Ans.* $P_{max} = 208$ kN, $v = 0.565$ m/s

# CHAPTER

# II SHAFTING AND ASSOCIATED PARTS

Bedplate with crankshaft.
Sulzer RTA96C (Courtesy
of Wärtsilä).

*When a man has a vision, he cannot get the*
*power from the vision until he has performed*
*it on the Earth for the people to see.*
   Black Elk, Oglala Sioux visionary, as told to
      John Niedhart

## SYMBOLS

| | |
|---|---|
| $A$ | area, m$^2$ |
| $\tilde{A}$ | constant defined in Equation (11.27) |
| $A_1$ | integration constant |
| $a$ | length from $x = 0$ to force, m |
| $\tilde{B}$ | constant defined in Equation (11.28) |
| $b$ | length from force to end of beam, m |
| $C_f$ | coefficient of fluctuation, Equation (11.70) |
| $c$ | distance from neutral axis to outer fiber, m |
| $d$ | diameter, m |
| $E$ | modulus of elasticity, Pa |
| $g$ | gravitational acceleration, 9.807 m/s$^2$ |
| $h$ | height of key, m |
| $h_p$ | power, W |
| $I$ | area moment of inertia, m$^4$ |
| $I_m$ | mass moment of inertia, kg · m$^2$ |
| $J$ | polar area moment of inertia, m$^4$ |
| $K_c$ | stress concentration factor |
| $K_e$ | kinetic energy, N · m |
| $K_f$ | fatigue stress concentration factor |
| $k$ | spring rate, N/m |
| $k_f$ | surface finish factor |
| $k_r$ | reliability factor |
| $k_s$ | size factor |
| $l$ | length, m |
| $M$ | moment, N · m |
| $M_f$ | performance index, J/kg |
| $m_a$ | mass, kg |
| $n_s$ | safety factor |
| $P$ | normal force, N |
| $p$ | pressure, Pa |
| $p_f$ | interference pressure, Pa |
| $q_n$ | notch sensitivity factor |
| $R$ | reaction force, N |
| $r$ | radius, m |
| $S_e$ | modified endurance limit, Pa |
| $S_e'$ | endurance limit, Pa |
| $S_{se}$ | shear modified endurance limit, Pa |
| $S_{sy}$ | shear yield stress, Pa |
| $S_u$ | ultimate stress, Pa |
| $S_{ut}$ | ultimate tensile stress, Pa |
| $S_y$ | yield stress, Pa |
| $T$ | torque, N · m |
| $T_l$ | load torque, N · m |
| $T_m$ | mean torque, N · m |
| $t$ | time, s |
| $t_h$ | thickness, m |
| $U$ | potential energy, N · m |
| $u$ | velocity, m/s |
| $W$ | load, N |
| $w$ | width of key, m |
| $x, y, z$ | Cartesian coordinates, m |
| $\delta$ | deflection, m |
| $\epsilon$ | normal strain |
| $\theta$ | cylindrical polar coordinate |
| $\theta_{\omega\max}$ | location within a cycle where speed is maximum, deg |
| $\theta_{\omega\min}$ | location within a cycle where speed is minimum, deg |
| $\nu$ | Poisson's ratio |
| $\rho$ | density, kg/m$^3$ |
| $\sigma$ | normal stress, Pa |
| $\sigma_e$ | critical stress using distortion energy theory, Pa |
| $\sigma_\phi$ | normal stress acting on oblique plane, Pa |
| $\tau$ | shear stress, Pa |
| $\tau_\phi$ | shear stress acting on oblique plane, Pa |
| $\phi$ | oblique angle, deg |
| $\omega$ | angular speed, rad/s |
| $\omega_\phi$ | fluctuation speed, rad/s |

## SUBSCRIPTS

| | |
|---|---|
| $a$ | alternating |
| all | allowable |
| avg | average |
| $c$ | compression |
| cr | critical |
| $e$ | endurance limit |
| $f$ | flywheel |
| $i$ | inner |
| $m$ | mean |
| max | maximum |
| min | minimum |
| $o$ | outer |
| $p$ | pressure |
| $r$ | radial |
| $s$ | shear |
| $x, y, z$ | Cartesian coordinates |
| $\theta$ | circumferential |
| $\omega$ | speed |
| 1,2,3 | principal axes |

## 11.1    INTRODUCTION

A **shaft** is a rotating or stationary member usually having a circular cross section much smaller in diameter than the shaft length and having mounted on it such power-transmitting elements as gears, pulleys, belts, chains, cams, flywheels, cranks, sprockets, and rolling-element bearings. The loading on the shaft can be various combinations of bending (almost always fluctuating); torsion (may or may not be fluctuating); shock; or axial, normal, or transverse shear. All these types of loading are separately considered in Chapter 4. The geometry of the shaft is such that the diameter will generally be the variable used to satisfy the design. In practical applications the shaft may often be stepped instead of having a constant diameter. Some of the main considerations in designing a shaft are strength, using yield or fatigue (or both) as a criterion; deflection; or the dynamics established by the critical speeds.

This chapter on shafts is placed before the chapters on the various machine elements mounted on a shaft mainly because these elements impose a force that can be considered in general and is equally applicable for whatever machine element it represents. This chapter makes extensive use of the failure prediction techniques developed in Chapters 6 and 7. Here combinations of loading are presented, whereas in Chapter 4 each type of loading was considered by itself. The dynamics and the first critical speed are important, since the rotating shaft becomes dynamically unstable and large vibrations are likely to develop. The chapter ends by considering the design of flywheels.

## 11.2    SHAFT DESIGN PROCEDURE

In the process of transmitting power at a given rotational speed, a shaft may be subjected to a torsional movement, or torque. Thus, a torsional shear stress is developed in the shaft. Also, some machine elements when mounted on a shaft exert forces on it in the transverse direction (perpendicular to the shaft axis). Thus, bending moments are developed in the shaft. A shaft carrying one or more of the various machine elements must be supported by bearings (considered in Chaps. 12 and 13). If two bearings can provide radial support to limit shaft bending and deflection to acceptable values, this is highly desirable and simplifies the design. If, on the other hand, three or more bearings must be used to provide adequate support and stiffness, precise alignment of the bearings must be maintained.

### PROCEDURE FOR SHAFT DESIGN

1. Develop a free-body diagram by replacing the various machine elements mounted on the shaft by their statically equivalent load or torque components. To illustrate this, Figure 11.1(a) shows two gears exerting forces on a shaft, and Figure 11.1(b) then shows a free-body diagram of the gears acting on the shaft.

2. Draw a bending moment diagram in the $xy$ and $xz$ planes as shown in Figure 11.1(c). The resultant internal moment at any section along the shaft may be expressed as

$$M_x = \sqrt{M_{xy}^2 + M_{xz}^2} \qquad (11.1)$$

3. Develop a torque diagram as shown in Figure 11.1(d). Torque developed from one power-transmitting element must balance the torque from other power-transmitting elements.

4. Establish the location of the critical cross section, or the $x$ location where the torque and moment are the largest.

5. For ductile materials use the maximum-shear-stress theory (MSST) or the distortion energy theory (DET) covered in Section 6.7.1.

6. For brittle materials use the maximum-normal-stress theory (MNST), the internal friction theory (IFT), or the modified Mohr theory (MMT), covered in Section 6.7.2.

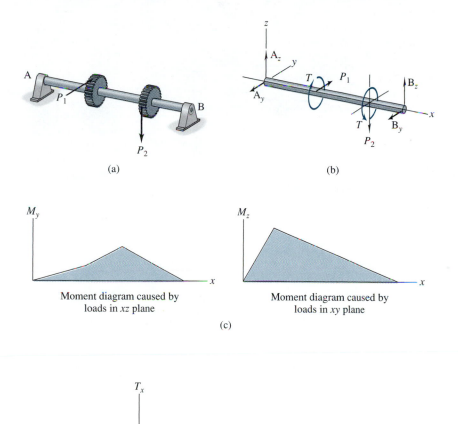

(a)

(b)

Moment diagram caused by loads in $xz$ plane

Moment diagram caused by loads in $xy$ plane

(c)

(d)

**Figure 11.1** Shaft assembly. (a) Shaft with two bearings at A and B and two gears with resulting forces $P_1$ and $P_2$; (b) free-body diagram of torque and forces resulting from assembly drawing; (c) moment diagram in $xz$ and $xy$ planes; (d) torque diagram.

**EXAMPLE 11.1**

**Given:** The power from the motor of a front-wheel-drive car is transmitted to the gearbox by a chain drive [Fig. 11.2(a)]. The two chain wheels are the same size. The chain is not prestressed, so the loose chain exerts no force. The safety factor is 4. The shaft is to be made of AISI 1080 steel. The chain transmits 100 kW of power at the chain speed of 50 m/s when the motor speed is 6000 rpm.

**Find:** The appropriate shaft diameter by using the DET.

**Solution:** Equation (4.41) gives the force transmitted through the chain as

$$P = \frac{h_p}{u} = \frac{100,000}{50} = 2000 \text{ N}$$

(a)

(b)

(c)

**Figure 11.2**    Illustration for Example 11.1. (a) Chain drive assembly; (b) free-body diagram of forces; (c) bending moment diagram.

The chain wheel radius is

$$r = \frac{u}{\omega} = \frac{50}{(6000)\,(2\pi/60)} = \frac{1}{4\pi} = 0.07958 \text{ m} = 79.58 \text{ mm}$$

Figure 11.2(b) shows a free-body diagram of the forces acting on the shaft. The torque being applied by the chain is

$$T = Pr = (2000)(0.07958) = 159.15 \text{ N} \cdot \text{m}$$

Force equilibrium gives

$$-R_A + R_B - P = 0 \tag{a}$$

Summation of moments about point A gives

$$0.2P - R_B(0.160) = 0$$

$$\therefore \quad R_B = 2500 \text{ N}$$

Substituting this into Equation (a) gives

$$R_A = R_B - P = 500 \text{ N}$$

Figure 11.2(c) shows the bending moment diagram. The maximum bending moment occurs at $x = 0.160$ m and has a value of 80 N · m. The torque is constant along the shaft, thus the critical section occurs at $x = 0.160$ m. From Equations (4.34) and (4.48)

$$\tau_{xy} = \frac{cT}{J} = \frac{(d/2)(159.15)}{\pi d^4/32} = \frac{2546.4}{\pi d^3}$$

and

$$\sigma_x = \frac{cM}{I} = \frac{(d/2)(80)}{\pi d^4/64} = \frac{2560}{\pi d^3}$$

From Equation (2.16) the principal stresses when $\sigma_y = 0$ are

$$\sigma_1, \sigma_2 = \frac{\sigma_x}{2} \pm \sqrt{\tau_{xy}^2 + \frac{\sigma_x^2}{4}} = \frac{1280}{\pi d^3} \pm \frac{1}{\pi d^3}\sqrt{(2546.4)^2 + \frac{(2560)^2}{4}}$$

$$\sigma_1 = \frac{4130}{\pi d^3} \quad \text{and} \quad \sigma_2 = -\frac{1570}{\pi d^3}$$

From Equation (6.11) the von Mises stress for a biaxial stress state is

$$\sigma_e = \frac{1}{\pi d^3}\left[(4130)^2 + (1570)^2 + (1570)(4130)\right]^{1/2} = \frac{5100}{\pi d^3}$$

From Equation (6.12) the DET predicts that failure will not occur if

$$\sigma_e = \frac{5100}{\pi d^3} < \frac{S_y}{n_s}$$

$$d^3 \geq \frac{(5100)(4)}{\pi(380)(10^6)} = (1.7087)(10^{-5})$$

$$\therefore \quad d \geq 0.02576 \text{ m}$$

The shaft diameter should be 26 mm.

## 11.3   STATIC LOADING

A number of different loading conditions are considered here. The shaft designer must establish either the minimum shaft diameter to successfully support the loads acting on the shaft or the safety factor for a specific design. Equations for both are presented below.

### 11.3.1   BENDING MOMENT AND TORSION

The force exerted on a shaft in the transverse direction (perpendicular to the shaft axis) produces a maximum stress, from Equation (4.48), of

$$\sigma_x = \frac{Mc}{I} \tag{11.2}$$

Similarly, from Equation (4.34)

$$\tau_{xy} = \frac{Tc}{J} \tag{11.3}$$

where $\quad c = \dfrac{d}{2} \quad I = \dfrac{\pi d^4}{64} \quad$ and $\quad J = \dfrac{\pi d^4}{32} \tag{11.4}$

for a circular cross section. Substituting Equation (11.4) into Equations (11.2) and (11.3) gives

$$\sigma_x = \frac{64Md}{2\pi d^4} = \frac{32M}{\pi d^3} \tag{11.5}$$

$$\tau_{xy} = \frac{Td/2}{\pi d^4/32} = \frac{16T}{\pi d^3} \tag{11.6}$$

For the plane stress state, the principal normal stresses, from Equation (2.16) when $\sigma_y = 0$, are

$$\sigma_1, \sigma_2 = \frac{\sigma_x}{2} \pm \sqrt{\left(\frac{\sigma_x}{2}\right)^2 + \tau_{xy}^2} \tag{11.7}$$

Substituting Equations (11.5) and (11.6) into Equation (11.7) gives

$$\sigma_1, \sigma_2 = \frac{16M}{\pi d^3} \pm \sqrt{\left(\frac{16M}{\pi d^3}\right)^2 + \left(\frac{16T}{\pi d^3}\right)^2} = \frac{16}{\pi d^3}\left(M \pm \sqrt{M^2 + T^2}\right) \tag{11.8}$$

From Equation (2.18) the principal shear stresses are

$$\tau_1, \tau_2 = \pm\sqrt{\tau_{xy}^2 + \left(\frac{\sigma_x}{2}\right)^2} \tag{11.9}$$

Substituting Equations (11.5) and (11.6) into Equation (11.9) gives

$$\tau_1, \tau_2 = \pm\frac{16}{\pi d^3}\sqrt{M^2 + T^2} \tag{11.10}$$

### Distortion Energy Theory

As shown in Section 6.7.1 and by Equations (6.11) and (6.12), the DET predicts failure if the von Mises stress satisfies the condition

$$\sigma_e = \left(\sigma_1^2 + \sigma_2^2 - \sigma_1\sigma_2\right)^{1/2} \geq \frac{S_y}{n_s} \tag{11.11}$$

where $S_y$ = yield strength of shaft material, Pa
$\quad n_s$ = safety factor

From Equation (11.8) the DET predicts failure if

$$\frac{16}{\pi d^3}\left(4M^2 + 3T^2\right)^{1/2} \geq \frac{S_y}{n_s} \tag{11.12}$$

Thus, the DET predicts the smallest diameter where failure will first start to occur as

$$d = \left(\frac{32n_s}{\pi S_y}\sqrt{M^2 + \frac{3}{4}T^2}\right)^{1/3} \tag{11.13}$$

Often the torque is not explicitly given in Equation (11.13), and the power transfer equations [Eqs. (4.40) and (4.42)] developed in Chapter 4 must be used.

If the shaft diameter is known and the safety factor is an unknown, Equation (11.13) becomes

$$n_s = \frac{\pi d^3 S_y}{32\sqrt{M^2 + \frac{3}{4}T^2}} \tag{11.14}$$

### Maximum-Shear-Stress Theory

As shown in Section 6.7.1 and by Equation (6.7), the MSST predicts failure for a plane or biaxial stress state ($\sigma_3 = 0$) if

$$|\sigma_1 - \sigma_2| \geq \frac{S_y}{n_s} \tag{11.15}$$

Equation (11.8) gives

$$\frac{32\sqrt{M^2 + T^2}}{\pi d^3} \geq \frac{S_y}{n_s} \tag{11.16}$$

Thus, the MSST predicts the smallest diameter where failure will first start to occur as

$$d = \left(\frac{32n_s}{\pi S_y}\sqrt{M^2 + T^2}\right)^{1/3} \tag{11.17}$$

If the shaft diameter is known and the safety factor is an unknown, Equation (11.17) becomes

$$n_s = \frac{\pi d^3 S_y}{32(M^2 + T^2)^{1/2}} \tag{11.18}$$

**EXAMPLE 11.2**

**Given:**    An assembly of belts has tensile forces applied as shown in Figure 11.3(a) and friction-less journal bearings at locations A and B. The yield strength of the shaft material is 500 MPa, and the safety factor is 2.

**Figure 11.3**    Figure used for Example 11.2. (a) Assembly drawing; (b) free-body diagram; (c) moment diagram in *xz* plane; (d) moment diagram in *xy* plane; (e) torque diagram.

**Find:** Determine the smallest safe shaft diameter by using both the DET and the MSST. Also, provide a free-body diagram as well as moment and torque diagrams.

**Solution:** A free-body diagram is shown in Figure 11.3(b); a moment diagram in the $xz$ plane, in Figure 11.3(c); and a moment diagram in the $xy$ plane, in Figure 11.3(d). From the moment diagrams the maximum moment is

$$M_{max} = \sqrt{(118.75)^2 + (37.5)^2} = 124.5 \text{ N} \cdot \text{m}$$

Figure 11.3(e) gives the torque diagram.
    Using the DET [Eq. (11.13)], the smallest safe diameter is

$$d = \left( \frac{32n_s}{\pi S_y} \sqrt{M^2 + \frac{3T^2}{4}} \right)^{1/3} = \left\{ \frac{32(2)}{\pi (500)(10^6)} \left[ 124.5^2 + \frac{3(7.5)^2}{4} \right]^{1/2} \right\}^{1/3}$$

$$= 0.0172 \text{ m} = 17.2 \text{ mm}$$

Using the MSST [Eq. (11.17)] gives

$$d = \left( \frac{32n_s}{\pi S_y} \sqrt{M^2 + T^2} \right)^{1/3} = \left[ \frac{32(2)}{\pi (500)(10^6)} \left( 124.5^2 + 7.5^2 \right)^{1/2} \right]^{1/3}$$

$$= 0.0172 \text{ m} = 17.2 \text{ mm}$$

Since the torque is small relative to the moment, little difference exists between the DET and MSST predictions.

## 11.3.2    BENDING MOMENT, TORSION, AND AXIAL LOADING

If axial loading is added to what was considered in Section 11.3.1 [see Eq. (11.5)], the normal stress is

$$\sigma_x = \frac{32M}{\pi d^3} + \frac{4P}{\pi d^2} \tag{11.19}$$

The shear stress is expressed by Equation (11.6); and the principal normal stresses, by Equation (11.7). Substituting Equations (11.19) and (11.6) into Equation (11.7) gives

$$\sigma_1, \sigma_2 = \frac{16M}{\pi d^3} + \frac{2P}{\pi d^2} \pm \sqrt{\left( \frac{16M}{\pi d^3} + \frac{2P}{\pi d^2} \right)^2 + \left( \frac{16T}{\pi d^3} \right)^2}$$

$$= \frac{2}{\pi d^3} \left[ 8M + Pd \pm \sqrt{(8M + Pd)^2 + (8T)^2} \right] \tag{11.20}$$

Substituting Equations (11.19) and (11.6) into Equation (11.9) gives the principal shear stresses as

$$\tau_1, \tau_2 = \pm \frac{2}{\pi d^3} \sqrt{(8M + Pd)^2 + (8T)^2} \tag{11.21}$$

### Distortion Energy Theory

Substituting Equation (11.20) into Equation (11.11) shows that the DET predicts failure if

$$\frac{4}{\pi d^3}\sqrt{(8M + Pd)^2 + 48T^2} \geq \frac{S_y}{n_s} \tag{11.22}$$

Because axial loading was added, an explicit expression for the diameter $d$ cannot be obtained.

### Maximum-Shear-Stress Theory

Substituting Equation (11.20) into Equation (11.15) shows that the MSST predicts failure if

$$\frac{4}{\pi d^3}\sqrt{(8M + Pd)^2 + 64T^2} \geq \frac{S_y}{n_s} \tag{11.23}$$

Again, because axial loading was added, an explicit expression for the diameter $d$ cannot be obtained.

## 11.4   CYCLIC LOADING

In cyclic loading the loads vary throughout a cycle rather than remain constant as in static loading. Here, a general analysis is derived for the fluctuating normal and shear stresses for ductile materials, and appropriate equations are then given for brittle materials.

### 11.4.1   DUCTILE MATERIALS

Figure 11.4 shows the normal and shear stresses acting on a shaft. In Figure 11.4(a) the stresses act on a rectangular element, and in Figure 11.4(b) they act on an oblique plane at an angle $\phi$. The normal stresses are denoted by $\sigma$ and the shear stresses by $\tau$. Subscript $a$ designates alternating and subscript $m$ designates mean or steady stress. Also, $K_f$ designates the fatigue stress concentration factor due to normal loading, and $K_{fs}$ designates the fatigue concentration factor due to shear loading. On a rectangular element [Fig. 11.4(a)] the normal stress is $\sigma_m \pm K_f \sigma_a$ and the shear stress is $\tau_m \pm K_{fs}\tau_a$. The largest stress occurs when $\sigma_a$ and $\tau_a$ are in phase, or when the frequency of one is an integral multiple of the frequency of the other. Summing the forces tangent to the diagonal gives

$$-\tau_\phi A + (\tau_m + K_{fs}\tau_a)A\cos\phi\cos\phi - (\tau_m + K_{fs}\tau_a)A\sin\phi\sin\phi$$
$$+ (\sigma_m + K_f\sigma_a)A\cos\phi\sin\phi = 0$$

Making use of double angles reduces the above equation to

$$\tau_\phi = (\tau_m + K_{fs}\tau_a)\cos 2\phi + \tfrac{1}{2}(\sigma_m + K_f\sigma_a)\sin 2\phi$$

Separating the mean and alternating components of stress gives the stress acting on the oblique plane as

$$\tau_\phi = \tau_{\phi m} + \tau_{\phi a} = \left(\frac{\sigma_m}{2}\sin 2\phi + \tau_m\cos 2\phi\right) + \left(\frac{K_f\sigma_a}{2}\sin 2\phi + K_{fs}\tau_a\cos 2\phi\right)$$

$$\tag{11.24}$$

Recall the Soderberg line in Figure 7.9 for tensile loading. For shear loading the endpoints of the Soderberg line are $S_{se} = S_e/(2n_s)$ and $S_{sy} = S_y/(2n_s)$. Figure 11.5 shows the

(a)

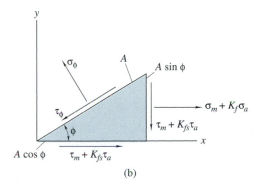

(b)

**Figure 11.4** Fluctuating normal and shear stresses acting on shaft. (a) Stresses acting on rectangular element; (b) stresses acting on oblique plane at angle $\phi$.

**Figure 11.5** Soderberg line for shear stress.

Soderberg line for shear stress. From the proportional triangles GHF and D0F of Figure 11.5,

$$\frac{\text{HF}}{\text{0F}} = \frac{\text{HG}}{\text{0D}} \quad \text{or} \quad \frac{S_y/(2n_s) - \tau_{\phi m}}{S_y/(2n_s)} = \frac{\tau_{\phi a}}{S_e/(2n_s)}$$

$$\frac{1}{n_s} = \frac{\tau_{\phi a}}{S_e/2} + \frac{\tau_{\phi m}}{S_y/2} \tag{11.25}$$

Note the similarities between Equation (7.29) and Equation (11.25). Substituting into Equation (11.25) the expressions for $\tau_{\phi a}$ and $\tau_{\phi m}$ gives

$$\frac{1}{n_s} = \frac{(K_f\sigma_a/2)\sin 2\phi + K_{fs}\tau_a \cos 2\phi}{S_e/2} + \frac{(\sigma_m/2)\sin 2\phi + \tau_m \cos 2\phi}{S_y/2}$$

$$= \tilde{A}\sin 2\phi + 2\tilde{B}\cos 2\phi \tag{11.26}$$

where

$$\tilde{A} = \frac{\sigma_m}{S_y} + \frac{K_f\sigma_a}{S_e} \tag{11.27}$$

$$\tilde{B} = \frac{\tau_m}{S_y} + \frac{K_{fs}\tau_a}{S_e} \tag{11.28}$$

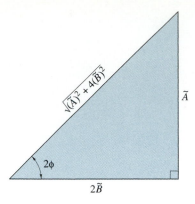

**Figure 11.6**    Illustration of relationship given in Eq. (11.29).

We are interested in the stress combination that produces the smallest $n_s$, since this corresponds to a maximum-stress situation. The minimum value of $n_s$ corresponds to a maximum value of $1/n_s$. Differentiating $1/n_s$ in Equation (11.26) and equating the result to zero give

$$\frac{d}{d\phi}\left(\frac{1}{n_s}\right) = 2\tilde{A}\cos 2\phi - 4\tilde{B}\sin 2\phi = 0$$

$$\therefore \qquad \frac{\sin 2\phi}{\cos 2\phi} = \tan 2\phi = \frac{\tilde{A}}{2\tilde{B}} \tag{11.29}$$

This relationship is illustrated in Figure 11.6, which shows that

$$\sin 2\phi = \frac{\tilde{A}}{\sqrt{(\tilde{A})^2 + 4(\tilde{B})^2}} \qquad \text{and} \qquad \cos 2\phi = \frac{2\tilde{B}}{\sqrt{(\tilde{A})^2 + 4(\tilde{B})^2}} \tag{11.30}$$

Substituting these into Equation (11.26) gives

$$\frac{1}{n_s} = \frac{(\tilde{A})^2}{\sqrt{(\tilde{A})^2 + 4(\tilde{B})^2}} + \frac{4(\tilde{B})^2}{\sqrt{(\tilde{A})^2 + 4(\tilde{B})^2}} = \sqrt{(\tilde{A})^2 + 4(\tilde{B})^2}$$

Substituting Equations (11.27) and (11.28) into the previous equation gives

$$\frac{1}{n_s} = \sqrt{\left(\frac{\sigma_m}{S_y} + \frac{K_f \sigma_a}{S_e}\right)^2 + 4\left(\frac{\tau_m}{S_y} + \frac{K_{fs}\tau_a}{S_e}\right)^2}$$

$$\frac{S_y}{n_s} = \sqrt{\left(\sigma_m + \frac{S_y}{S_e}K_f\sigma_a\right)^2 + 4\left(\tau_m + \frac{S_y}{S_e}K_{fs}\tau_a\right)^2} \tag{11.31}$$

By setting $\sigma_y = 0$, $\sigma_x = \sigma$, and $\tau_{xy} = \tau$ in Equation (2.18) for biaxial stresses, the maximum shear stress is

$$\tau_{max} = \sqrt{\left(\frac{\sigma}{2}\right)^2 + \tau^2} \tag{11.32}$$

and the safety factor is

$$n_s = \frac{S_y/2}{\tau_{max}} = \frac{S_y/2}{\sqrt{(\sigma/2)^2 + \tau^2}} = \frac{S_y}{\sqrt{\sigma^2 + 4\tau^2}}$$

$$\frac{S_y}{n_s} = \sqrt{\sigma^2 + 4\tau^2} \tag{11.33}$$

Equations (11.33) and (11.31) have the same form, and

$$\sigma = \sigma_m + \frac{S_y}{S_e} K_f \sigma_a \quad \text{and} \quad \tau = \tau_m + \frac{S_y}{S_e} K_{fs} \tau_a$$

Note that the normal and shear stresses each contain a steady and an alternating component, the latter weighted for the effect of fatigue and stress concentration.

By making use of Equations (11.5) and (11.6), Equation (11.31) becomes

$$n_s = \frac{\pi d^3 S_y}{32\sqrt{\left[M_m + (S_y/S_e)K_f M_a\right]^2 + \left[T_m + (S_y/S_e)K_{fs} T_a\right]^2}} \tag{11.34}$$

If, instead of the safety factor, we want to determine the smallest safe diameter for a specified safety factor, Equation (11.34) can be rewritten as

$$d = \left[\frac{32n_s}{\pi S_y} \sqrt{\left(M_m + \frac{S_y}{S_e}K_f M_a\right)^2 + \left(T_m + \frac{S_y}{S_e}K_{fs} T_a\right)^2}\right]^{1/3} \tag{11.35}$$

The shear fatigue stress concentration factor $K_{fs}$ can be estimated by using torsion instead of axial loading in the stress concentration factor.

Equations (11.34) and (11.35) represent the general form of the MSST. Note in Equation (11.35) that $S_y$, $S_y/S_e$, $K_f$, and $K_{fs}$ depend on the shaft diameter $d$. The approach to be used is to estimate the range of the shaft diameter and iterate the results if the final design is outside the estimated range.

Peterson (1974) modified Equation (11.31) by changing the coefficient of the shear stress term from 4 to 3, such that the DET is satisfied and gives

$$\frac{S_y}{n_s} = \sqrt{\left(\sigma_m + \frac{S_y}{S_e}K_f \sigma_a\right)^2 + 3\left(\tau_m + \frac{S_y}{S_e}K_{fs} \tau_a\right)^2} \tag{11.36}$$

By making use of Equations (11.5) and (11.6), Equation (11.36) becomes

$$n_s = \frac{\pi d^3 S_y}{32\sqrt{\left[M_m + (S_y/S_e)K_f M_a\right]^2 + \frac{3}{4}\left[T_m + (S_y/S_e)K_{fs} T_a\right]^2}} \tag{11.37}$$

The smallest safe diameter corresponding to a specific safety factor can then be expressed as

$$d = \left[ \frac{32n_s}{\pi S_y} \sqrt{\left( M_m + \frac{S_y}{S_e} K_f M_a \right)^2 + \frac{3}{4} \left( T_m + \frac{S_y}{S_e} K_{fs} T_a \right)^2} \right]^{1/3} \tag{11.38}$$

The distinction between Equations (11.34) and (11.35) and Equations (11.37) and (11.38) needs to be recognized. Equations (11.34) and (11.35) assume that the MSST is valid; Equations (11.37) and (11.38) assume that the DET is valid. All four equations are general equations applicable for ductile materials.

---

**EXAMPLE 11.3**

**Given:** When a rear-wheel-drive car accelerates around a bend at high speeds, the driveshafts are subjected to both bending and torsion. The acceleration torque $T$ is reasonably constant at $400 \, \text{N} \cdot \text{m}$ while the bending moment is varying due to cornering and is expressed in newton-meters as

$$M = 250 + 800 \sin \omega t$$

Thus, the mean and alternating moments are, respectively, $M_m = 250 \, \text{N} \cdot \text{m}$ and $M_a = 800 \, \text{N} \cdot \text{m}$. Assume there is no notch giving a stress concentration. The reliability must be 99%, and the safety factor is 4.5. The shaft is forged from AISI 1080 steel.

**Find:** The shaft diameter while using the MSST.

**Solution:** From Equation (7.7) and Table A.1 the bending endurance limit for AISI 1080 steel is

$$S'_e = 0.5 S_u = 0.5(615 \, \text{MPa}) = 307.5 \, \text{MPa}$$

From Figure 7.10 the surface finish factor for an as-forged surface at $S_{ut} = 615 \, \text{MPa}$ is $k_f = 0.42$. In evaluating the size factor, the shaft diameter needs to be chosen. From Equation (7.22), and assuming $d = 30$ mm,

$$k_s = 1.189 d^{-0.112} = 1.189(30)^{-0.112} = 0.8124$$

From Table 7.4 for 99% probability of survival, the reliability factor is 0.82. Substituting the above into Equation (7.16) gives the endurance limit as

$$S_e = k_f k_s k_r S'_e = (0.42)(0.8124)(0.82)(307.5)(10^6) = 86.03 \, \text{MPa}$$

Equation (11.35) with $T_a = 0$ gives

$$d = \left[ \frac{32 n_s}{\pi S_y} \sqrt{\left( M_m + \frac{S_y}{S_e} K_f M_a \right)^2 + T_m^2} \right]^{1/3}$$

$$= \left\{ \frac{32(4.5)}{\pi(380)(10^6)} \sqrt{\left[ 250 + \frac{(380)(10^6)}{(86.03)(10^6)} (1)(800) \right]^2 + (400)^2} \right\}^{1/3}$$

$$= 0.07714 \, \text{m} = 77.14 \, \text{mm}$$

A new calculation has to be made for the size factor, since initially the shaft diameter was chosen as 30 mm. From Equation (7.22)

$$k_s = 1.189 d^{-0.112} = 1.189(77.14)^{-0.112} = 0.7308$$

$$\therefore \quad S_e = (0.42)(0.7308)(0.82)(307.5)(10^6) = 77.39 \text{ MPa}$$

From Equation (11.35)

$$d = \left\{ \frac{32(4.5)}{\pi(380)(10^6)} \sqrt{\left[250 + \frac{(380)(10^6)}{(77.39)(10^6)}(1)(800)\right]^2 + (400)^2} \right\}^{1/3} = 0.0797 \text{ m}$$

The shaft diameter should be at least 80 mm.

## 11.4.2   BRITTLE MATERIALS

Although shafts are usually cold-worked metals machined to final desired dimensions, there are applications where castings, which are often brittle materials, are used as shafts. As discussed in Chapter 6, this requires a slightly different analysis approach than for ductile materials.

For brittle materials, the forces in Figure 11.4(b) are assumed to be *normal* rather than tangent to the diagonal. Also, the design line as dictated by the maximum-normal-stress theory (MNST) extends from $S_e/n_s$ to $S_u/n_s$ instead of from $S_e/(2n_s)$ to $S_y/(2n_s)$, as was true for the MSST. Following procedures similar to those used in obtaining Equation (11.31) gives

$$\frac{2S_u}{n_s} = K_c\left(\sigma_m + \frac{S_u}{S_e}\sigma_a\right) + \sqrt{K_c^2\left(\sigma_m + \frac{S_u}{S_e}\sigma_a\right)^2 + 4K_{cs}^2\left(\tau_m + \frac{S_u}{S_e}\tau_a\right)^2} \tag{11.39}$$

where $K_c$ = stress concentration factor.

By making use of Equations (11.5) and (11.6), Equation (11.39) can be written as

$$n_s = \frac{\pi d^3 S_u/16}{K_c[M_m + (S_u/S_e)M_a] + \sqrt{K_c^2[M_m + (S_u/S_e)M_a]^2 + K_{cs}^2[T_m + (S_u/S_e)T_a]^2}} \tag{11.40}$$

If we want the minimum safe diameter of the shaft for a specific safety factor,

$$d = \left\{ \frac{16n_s}{\pi S_u}\left[K_c\left(M_m + \frac{S_u}{S_e}M_a\right) + \sqrt{K_c^2\left(M_m + \frac{S_u}{S_e}M_a\right)^2 + K_{cs}^2\left(T_m + \frac{S_u}{S_e}T_a\right)^2}\right]\right\}^{1/3} \tag{11.41}$$

The important difference in the equations developed above for the safety factor and the smallest safe diameter is that Equations (11.34) and (11.35) are applicable for ductile materials while assuming the MSST, Equations (11.37) and (11.38) are also applicable for ductile materials but while assuming the DET, and Equations (11.40) and (11.41) are applicable for brittle materials while assuming the MNST. Note the major differences between the equations developed for brittle and ductile materials. For brittle materials [Eqs. (11.40) and (11.41)] the stress concentration factor $K_c$ and the ultimate stress $S_u$ are used, whereas for ductile materials [Eqs. (11.34), (11.35), (11.37), and (11.38)] the fatigue stress concentration factor $K_f$ and the yield stress $S_y$ are used. Having these general expressions, we will now consider specific fluctuating stresses.

A specific cyclic stress variation that occurs in practical applications is reversed bending and steady torsion. From Section 7.3 note that reversed bending implies that either $\sigma_m = 0$ or $M_m = 0$. Also, steady torsion implies that either $\tau_a = 0$ or $T_a = 0$. Thus, to apply the condition of reversed bending and steady torsion ($M_m = 0$ and $T_a = 0$), and knowing whether the material is ductile or brittle and which failure theory should be applied, the reduced forms of Equations (11.34) and (11.35), (11.37) and (11.38), or (11.40) and (11.41) should be used.

**EXAMPLE 11.4**

**Given:** The shaft made of high-carbon steel shown in Figure 11.7 is subjected to completely reversed bending and steady torsion. A standard ball bearing is to be placed on diameter $d_2$, and this surface will therefore be machined to form a good seat for the bearing. The groove between the sections is present to make sure that the large-diameter section is not damaged by the grinding operation, and is called a *grinding relief.* Assume that standard ball bearing bore sizes are in 5-mm increments in the range of 15 to 50 mm. Design so that the relative sizes are approximately $d_2 = 0.75d_3$ and $d_1 = 0.65d_3$. The completely reversed bending moment is 70 N · m, and the steady torsion is 45 N · m. Assume a safety factor of 2.5, and size the shaft for infinite life.

**Find:** Determine the diameter $d_2$ by both the MSST and the DET.

**Solution:** From Table A.1 for high-carbon steel, $S_u = 615$ MPa and $S_y = 380$ MPa. From Equation (7.7) for bending

$$S'_e = 0.5S_u = 307.5 \text{ MPa}$$

From Figure 7.10 the surface finish factor $k_f$ for machined surfaces is 0.75. Assume that the diameter $d_1 = 20$ mm. From Equation (7.22) the size factor is

$$k_s = 1.189d^{-0.112} = 1.189(20)^{-0.112} = 0.8501$$

**Figure 11.7**   Section of shaft in Example 11.4.

Given that

$$d_3 = \frac{d_2}{0.75} = \frac{d_1}{0.65} \rightarrow \frac{d_2}{d_1} = \frac{0.75}{0.65} = 1.154$$

from Figure 6.6(b) for

$$\frac{r}{d} = \frac{d_2 - d_1}{2d_1} = \frac{1}{2}\left(\frac{d_2}{d_1} - 1\right) = 0.0769$$

the stress concentration factor $K_c = 1.9$. Recall that

$$r = \frac{d_2 - d_1}{2} = \frac{d_2}{2}\left(1 - \frac{1}{1.154}\right) = 0.06672 d_2$$

$$\therefore \quad r = (0.06672)(1.154)(20) = 1.540 \text{ mm}$$

From Figure 7.9 for $S_u = 615$ MPa, bending, and $r = 1.335$ mm, the notch sensitivity factor is $q_n = 0.75$. From Equation (7.19)

$$K_f = 1 + (K_c - 1)q_n = 1 + (1.9 - 1)0.75 = 1.675$$

From Equation (7.16) the modified endurance limit is

$$S_e = k_f k_s S_e' = (0.75)(0.85)(307.5) = 196.0 \text{ MPa}$$

Recall that for completely reversed bending and steady torsion $M_m = 0$ and $T_a = 0$.
Using Equation (11.35) and the MSST gives

$$d_1 = \left[\frac{32(2.5)}{\pi(380)(10^6)}\sqrt{\left(\frac{380}{196}\right)^2 (1.675)^2(70)^2 + (45)^2}\right]^{1/3} = 0.02495 \text{ m} = 24.95 \text{ mm}$$

$$\therefore \quad d_1 = 25 \text{ mm}$$

$$d_3 = \frac{d_1}{0.65} = \frac{25}{0.65} = 38.46 \text{ mm}$$

$$d_2 = 0.75 d_3 = 0.75(38.46) = 28.85 \text{ mm}$$

Having established that $d_1 = 25$ mm instead of the 20 mm initially guessed, we need to recalculate the size factor:

$$k_s = 1.189 d^{-0.112} = (1.189)(25)^{-0.112} = 0.8291$$

$$\frac{r}{d} = \frac{d_2 - d_1}{2d_1} = \frac{28.85 - 25}{2(25)} = 0.0770$$

Figure 6.6(b) for $r/d = 0.077$ and $d_2/d_1 = 30/26 = 1.154$ gives a stress concentration of 1.9. Also,

$$r = \frac{d_2 - d_1}{2} = \frac{28.85 - 25}{2} = 1.925 \text{ mm}$$

From Figure 7.9 for $S_u = 615$ MPa, bending, and $r = 2.0$ mm, the notch sensitivity $q_n = 0.76$. The fatigue stress concentration factor can be expressed as

$$K_f = 1 + (K_c - 1)q_n = 1 + (1.9 - 1)(0.76) = 1.684$$

From Equation (7.16) the modified endurance limit is

$$S_e = k_f k_s S_r' = (0.75)(0.8291)307.5(10^6) \text{ Pa} = 191.2 \text{ MPa}$$

Using Equation (11.35) and the MSST gives

$$d_1 = \left[ \frac{32(2.5)}{\pi(380)(10^6)} \sqrt{\left(\frac{380}{191.2}\right)^2 (1.684)^2(70)^2 + (45)^2} \right]^{1/3} = 0.02519 \text{ m} = 25.19 \text{ mm}$$

$$\therefore \qquad d_1 = 25 \text{ mm}, \qquad d_3 = 38.5 \text{ mm} \qquad d_2 = 28.9 \text{ mm}$$

Using Equation (11.38) and the DET gives

$$d_1 = \left[ \frac{32(2.5)}{\pi(380)(10^6)} \sqrt{\left(\frac{380}{191.2}\right)^2 (1.684)^2(70)^2 + 0.75(45)^2} \right]^{1/3} = 0.02515 \text{ m} = 25.15 \text{ mm}$$

$$\therefore \qquad d_1 = 26 \text{ mm} \qquad d_3 = 40 \text{ mm} \qquad d_2 = 30 \text{ mm}$$

Thus, the MSST and the DET give the same results.

## 11.5    CRITICAL SPEED OF ROTATING SHAFTS

All shafts deflect during rotation. The magnitude of the deflection depends on the stiffness of the shaft and its supports, the total mass of the shaft and its attached parts, and the amount of system damping. The **critical speed** of a rotating shaft, sometimes called the **natural frequency,** is the speed at which the rotating shaft becomes dynamically unstable and large vibration amplitudes are likely to develop. For any shaft there are an infinite number of critical speeds, but only the lowest (first) and occasionally the second are generally of interest to designers. The others are usually so high as to be well out of the operating range of shaft speed. This text considers only the first critical speed of the shaft. Two approximate methods of finding the first critical speed (or lowest natural frequency) of a system are given in this section, one attributed to Rayleigh and the other to Dunkerley.

### 11.5.1    SINGLE-MASS SYSTEM

The **first critical speed** (or **lowest natural frequency**) can be obtained by observing the rate of interchange between the kinetic (energy of motion) and potential (energy of position) energies of the system during its cyclic motion. A single mass on a shaft can be represented by the simple spring and mass shown in Figure 11.8. The dashed line indicates the static equilibrium position. The potential energy of the system is

$$U = \int_0^\delta (m_a g + k\delta)\, d\delta - m_a g\delta$$

where $m_a$ = mass, kg
$g$ = gravitational acceleration, 9.807 m/s$^2$
$k$ = spring rate, N/m
$\delta$ = deflection, m

**Figure 11.8**
Simple single-mass system.

Integrating gives

$$U = \tfrac{1}{2}k\delta^2 \qquad (11.42)$$

The kinetic energy of the system with the mass moving with a velocity of $\dot\delta$ is

$$K_e = \tfrac{1}{2}m_a(\dot\delta)^2 \qquad (11.43)$$

Observe the following about Equations (11.42) and (11.43):

1. As the mass passes through the static equilibrium position, the potential energy $U$ is zero and the kinetic energy $K_e$ is at a maximum and equal to the total mechanical energy of the system.

2. When the mass is at the position of maximum displacement and is on the verge of changing direction, its velocity is zero. At this point the potential energy $U$ is at a maximum and is equal to the total mechanical energy of the system.

Thus, the total mechanical energy is the sum of the potential and kinetic energies and is constant at any time.

$$\therefore \qquad \frac{d}{dt}(U + K_e) = 0 \qquad (11.44)$$

Substituting Equations (11.42) and (11.43) into Equation (11.44) gives

$$\frac{d}{dt}\left[\frac{1}{2}k\delta^2 + \frac{1}{2}m_a\left(\dot\delta\right)^2\right] = k\delta\dot\delta + m_a\dot\delta\ddot\delta = 0$$

$$\therefore \qquad \dot\delta(m_a\ddot\delta + k\delta) = 0 \qquad (11.45)$$

$$\ddot\delta + \omega^2\delta = 0$$

where

$$\omega = \sqrt{\frac{k}{m_a}} \qquad \text{rad/s} \qquad (11.46)$$

The general solution to this differential equation is

$$\delta = A_1 \sin(\omega t + \phi) \qquad (11.47)$$

where $A_1$ = integration constant. The first critical speed (or lowest natural frequency) is $\omega$.

Substituting Equation (11.47) into Equations (11.42) and (11.43) gives

$$U = \frac{k}{2} A_1^2 \sin^2(\omega t + \phi) \tag{11.48}$$

$$K_e = \frac{m_a}{2} A_1^2 \omega^2 \cos^2(\omega t + \phi) \tag{11.49}$$

Note from Equation (11.47) that for static deflection if $k = W/\delta$ and $m_a = W/g$, then

$$\omega = \sqrt{\frac{k}{m_a}} = \sqrt{\frac{W/y}{W/g}} = \sqrt{\frac{g}{\delta}} \tag{11.50}$$

## 11.5.2    MULTIPLE-MASS SYSTEM

From Equation (11.43) the kinetic energy for $n$ masses is

$$K_e = \tfrac{1}{2} m_{a1}(\dot{\delta}_1)^2 + \tfrac{1}{2} m_{a2}(\dot{\delta}_2)^2 + \cdots + \tfrac{1}{2} m_{an}(\dot{\delta}_n)^2 \tag{11.51}$$

If the deflection is represented by Equation (11.47), then $y_{max} = A_1$. Also, $\dot{y}_{max} = A_1 \omega = y_{max}\omega$. Therefore, the maximum kinetic energy is

$$K_{e,max} = \frac{\omega^2}{2} \left[ m_{a1}(\delta_{1,max})^2 + m_{a2}(\delta_{2,max})^2 + \cdots + m_{an}(\delta_{n,max})^2 \right] \tag{11.52}$$

From Equation (11.42) the potential energy for $n$ masses is

$$U = \tfrac{1}{2} k_1 \delta_1^2 + \tfrac{1}{2} k_2 \delta_2^2 + \cdots + \tfrac{1}{2} k_n \delta_n^2 \tag{11.53}$$

and the maximum potential energy is

$$U_{max} = \tfrac{1}{2} k_1 (\delta_{1,max})^2 + \tfrac{1}{2} k_2 (\delta_{2,max})^2 + \cdots + \tfrac{1}{2} k_n (\delta_{n,max})^2 \tag{11.54}$$

The Rayleigh method assumes that $K_{e,max} = U_{max}$, or

$$\frac{1}{2} \omega^2 \sum_{i=1,\ldots,n} m_{ai}(\delta_{i,max})^2 = \frac{1}{2} \sum_{i=1,\ldots,n} k_i(\delta_{i,max})^2$$

$$\omega^2 = \frac{\displaystyle\sum_{i=1,\ldots,n} k_i(\delta_{i,max})^2}{\displaystyle\sum_{i=1,\ldots,n} m_{ai}(\delta_{i,max})^2} \tag{11.55}$$

but

$$k_i = \frac{W_i}{\delta_{i,max}} \quad \text{and} \quad m_{ai} = \frac{W_i}{g} \tag{11.56}$$

where $W_i = i$th weight placed on shaft, N
$g$ = gravitation acceleration, 9.807 m/s$^2$

Substituting Equation (11.56) into Equation (11.55) gives

$$\omega_{cr} = \sqrt{\frac{g \displaystyle\sum_{i=1,\ldots,n} W_i \delta_{i,max}}{\displaystyle\sum_{i=1,\ldots,n} W_i \delta_{i,max}^2}} \tag{11.57}$$

This, then, is the first critical speed (first natural frequency) of a multiple-mass system when the Rayleigh method is used. Equation (11.57), known as the **Rayleigh equation,** *overestimates* the first critical speed. Because the actual displacements are larger than the static displacements used in Equation (11.57), the energies in both the denominator and the numerator will be underestimated by the Rayleigh formulation. However, the error in the underestimate will be larger in the denominator, since it involves the square of the approximated displacements. Thus, Equation (11.57) overestimates (provides an upper bound on) the first critical speed.

The **Dunkerley equation** is another approximation to the first critical speed of a multiple-mass system; it is given as

$$\frac{1}{\omega_{cr}^2} = \frac{1}{\omega_1^2} + \frac{1}{\omega_2^2} + \cdots + \frac{1}{\omega_n^2} \tag{11.58}$$

where $\omega_1$ = critical speed if only mass 1 exists
$\omega_2$ = critical speed if only mass 2 exists
$\omega_n$ = critical speed if only the $n$th mass exists

Recall from Equation (11.50) that $\omega_i = \sqrt{g/\delta_i}$.

The Dunkerley equation *underestimates* (provides a lower bound on) the first critical speed. The major difference between the Rayleigh and Dunkerley equations is in the deflections. In the Rayleigh equation the deflection at a specific mass location takes into account the deflections due to all the masses acting on the system; in the Dunkerley equation the deflection is due only to the individual mass being evaluated.

---

**Given:** Figure 11.9 shows a simply supported shaft arrangement. A solid shaft of 2-in diameter made of low-carbon steel is used. The following are given: $x_1 = 30$ in, $x_2 = 40$ in, $x_3 = 20$ in, $P_A = 80$ lb, and $P_B = 120$ lb.

**EXAMPLE 11.5**

**Find:** Determine the first critical speed by using

    (a) The Rayleigh method
    (b) The Dunkerley method

**Solution:** From Table 5.1(b) for simply supported ends:
For $0 \leq x \leq a$

$$\delta_x = \frac{Pbx}{6l\,EI}(l^2 - x^2 - b^2) \tag{a}$$

**Figure 11.9** Simply supported shaft arrangement for Example 11.5.

For $a \leq x \leq l$

$$\delta_x = \frac{Pa(l-x)}{6lEI}(2lx - a^2 - x^2) \tag{b}$$

From Table A.1 the modulus of elasticity for carbon steel is $3 \times 10^7$ psi. For a solid, round shaft the area moment of inertia is

$$I = \frac{\pi d^4}{64} = \frac{\pi(2)^4}{64} = \frac{\pi}{4} = 0.7854 \text{ in}^4$$

The deflection at location A due to load $P_A$ from Equation (a) and $a = x_1 = x = 30$ in and $b = x_2 + x_3 = 60$ in is

$$\delta_{AA} = \frac{P_A bx}{6lEI}(l^2 - x^2 - b^2) = -\frac{80(60)(30)(90^2 - 30^2 - 60^2)}{6(90)(3)(10^7)(0.7854)} = -0.04074 \text{ in}$$

Note that in Figure 11.9 the $y$ direction is upward; thus, $\delta_{AA}$ is negative. Also, the first subscript in $\delta_{AA}$ designates the location where the deflection occurs, and the second subscript designates the loading that contributes to the deflection.

The deflection at location A due to load $P_B$ from Equation (a) and $a = x_1 + x_2$, $x = x_1 = 30$ in, and $b = 20$ in is

$$\delta_{AB} = \frac{P_B bx}{6lEI}(l^2 - x^2 - b^2) = -\frac{120(20)(30)(90^2 - 30^2 - 20^2)}{6(90)(3)(10^7)(0.7854)} = -0.03848 \text{ in}$$

The total deflection at location A is

$$\delta_A = \delta_{AA} + \delta_{AB} = -0.04074 - 0.03848 = -0.07922 \text{ in}$$

The deflection at location B due to load $P_B$ from Equation (a) and $x = a = x_1 + x_2 = 70$ in and $b = 20$ in is

$$\delta_{BB} = \frac{P_B bx}{6lEI}(l^2 - x^2 - b^2) = -\frac{120(20)(70)(90^2 - 70^2 - 20^2)}{6(90)(3)(10^7)(0.7854)} = -0.03697 \text{ in}$$

The deflection at location B due to load $P_A$ from Equation (b) and $a = x_1 = 30$ in, $b = x_2 + x_3 = 60$ in, and $x = x_1 + x_2 = 70$ in is

$$\delta_{BA} = \frac{P_A a(l-x)}{6lEI}(2lx - a^2 - x^2)$$

$$= -\frac{(80)(30)(90 - 70)[2(90)(70) - 30^2 - 70^2]}{6(90)(3)(10^7)(0.7854)} = -0.02565 \text{ in}$$

Thus, the total deflection at location B is

$$\delta_B = \delta_{BA} + \delta_{BB} = -0.02565 - 0.03697 = -0.06262 \text{ in}$$

(a)  Using the Rayleigh equation [Eq. (11.57)] gives the first critical speed as

$$\omega_{cr} = \sqrt{\frac{g(P_A \delta_A + P_B \delta_B)}{P_A \delta_A^2 + P_B \delta_B^2}} = \sqrt{\frac{(386)[80(0.07912) + 120(0.06262)]}{80(0.07912)^2 + 120(0.06262)^2}}$$

$$= 74.14 \text{ rad/s} = 11.8 \text{ rps} = 708 \text{ rpm}$$

(b)  Using the Dunkerley equation [Eq. (11.58)] gives

$$\frac{1}{\omega_{cr}^2} = \frac{1}{\omega_{cr,A}^2} + \frac{1}{\omega_{cr,B}^2}$$

where

$$\omega_{cr,A} = \sqrt{\frac{g}{\delta_{AA}}} = \sqrt{\frac{386}{0.04074}} = 97.34 \text{ rad/s} = 15.49 \text{ rps} = 929.5 \text{ rpm}$$

$$\omega_{cr,B} = \sqrt{\frac{g}{\delta_{BB}}} = \sqrt{\frac{386}{0.03692}} = 102.2 \text{ rad/s} = 16.27 \text{ rps} = 976.4 \text{ rpm}$$

$$\therefore \quad \omega_{cr} = \frac{\omega_{cr,A}\omega_{cr,B}}{\sqrt{\omega_{cr,A}^2 + \omega_{cr,B}^2}} = \frac{(929.5)(976.4)}{\sqrt{(929.5)^2 + (976.4)^2}} = 11.22 \text{ rps} = 673 \text{ rpm}$$

The Rayleigh equation gives $\omega_{cr} = 708$ rpm, which overestimates the first critical speed; the Dunkerley equation gives $\omega_{cr} = 673$ rpm, which underestimates the first critical speed. Therefore, the actual first critical speed is between 673 and 708 rpm, and the shaft design should avoid this range of operation.

## 11.6 KEYS AND PINS

A variety of machine elements, such as gears, pulleys, and cams, are mounted on rotating shafts. The portion of the mounted member in contact with the shaft is called the **hub.** This section describes methods of attaching the hub to the shaft.

There are a great variety of **keys** and **pins,** as shown in Figure 11.10. The simplest and most commonly used is the **parallel key,** which has a constant cross section across its length. A **tapered key** has a constant width, but the height varies with a taper of 0.125 in per 12 in (0.597°), and is driven into a tapered slot on the hub until it locks. A gib head can be included on the key to facilitate removal. A **Woodruff key** has a constant width and a semicircular profile, and it fits into a semicircular keyseat machined onto the shaft. A Woodruff key is light-duty because it has a deep seat in the shaft, which can compromise strength, but it aligns itself readily against the hub. Selected standard key sizes are summarized in Tables 11.1 through 11.3; the complete list of standard sizes is contained in ANSI B17.1-1967. Keys are usually manufactured from low-carbon steel, but can be heat-treated to obtain higher strengths. However, this is rare, as ductility in the key material is usually important.

**Roll pins** or **grooved pins** are alternatives to keys, as shown in Figure 11.10(h) and (i). Roll and grooved pins are used for light-duty service, and they remain in place because of elastic deformation of the pin in the hole. **Setscrews** are threaded fasteners that are driven radially through a threaded hole in the hub. The point (which can be of a number of different geometries or materials) bears against the shaft, and torque is transmitted by frictional resistance of the setscrew against the shaft and the shaft against the hub. Note that setscrews can develop only moderate forces, as summarized in Table 11.4.

Consider the behavior of a simple square key, as shown in Figure 11.10a. The main purpose of the key is to prevent motion between the shaft and the connected machine element through which torque is being transmitted. The purpose of using the key is to transmit the *full* torque.

**Figure 11.10** Illustration of keys and pins. (a) Dimensions of shaft with keyway in shaft and hub; (b) square parallel key; (c) flat parallel key; (d) tapered key; (e) tapered key with gib head, or gib head key—the gib head assists in removal of the key; (f) round key; (g) Woodruff key with illustration of mounting; (h) pin, which is often grooved—the pin is slightly larger than the hole so that friction holds the pin in place; (i) roll pin. Elastic deformation of the pin in the smaller hole leads to friction forces that keep the pin in place.

**Table 11.1**    Dimensions of selected square plain parallel stock keys

| Shaft diameter, in | Key width, in | Distance from keyseat to opposite side of shaft, in | Shaft diameter, in | Key width, in | Distance from keyseat to opposite side of shaft, in |
|---|---|---|---|---|---|
| 0.500 | 0.125 | 0.430 | 2.000 | 0.50 | 1.718 |
| 0.625 | 0.1875 | 0.517 | 2.250 | 0.50 | 1.972 |
| 0.750 | 0.1875 | 0.644 | 2.500 | 0.625 | 2.148 |
| 0.875 | 0.1875 | 0.771 | 2.750 | 0.625 | 2.402 |
| 1.000 | 0.25 | 0.859 | 3.000 | 0.75 | 2.577 |
| 1.125 | 0.25 | 0.956 | 3.250 | 0.75 | 2.831 |
| 1.250 | 0.25 | 1.112 | 3.500 | 0.875 | 3.007 |
| 1.375 | 0.3125 | 1.201 | 3.750 | 0.875 | 3.261 |
| 1.500 | 0.375 | 1.289 | 4.000 | 1.00 | 3.437 |
| 1.675 | 0.375 | 1.416 | 4.500 | 1.00 | 3.944 |
| 1.750 | 0.375 | 1.542 | 5.000 | 1.25 | 4.296 |
| 1.875 | 0.50 | 1.591 | 6.000 | 1.50 | 5.155 |

SOURCE: From ANSI B17.1-1967, Keys and Keyseats, available from the American Society of Mechanical Engineers.

**Table 11.2**    Dimensions of square and flat taper stock keys

| Shaft diameter, in | Square Type Width $w$, in | Height[a] $h$, in | Flat Type Width $w$, in | Height[a] $h$, in | Available Lengths $l$ Minimum, in | Maximum, in | Available increments, in |
|---|---|---|---|---|---|---|---|
| 0.5–0.5625 | 0.125 | 0.125 | 0.125 | 0.09375 | 0.50 | 2.00 | 0.25 |
| 0.625–0.875 | 0.1875 | 0.1875 | 0.1875 | 0.125 | 0.75 | 3.00 | 0.375 |
| 0.9375–1.25 | 0.25 | 0.25 | 0.25 | 0.1875 | 1.00 | 4.00 | 0.50 |
| 1.3125–1.375 | 0.3125 | 0.3125 | 0.3125 | 0.25 | 1.25 | 5.25 | 0.625 |
| 1.4375–1.75 | 0.375 | 0.375 | 0.375 | 0.25 | 1.50 | 6.00 | 0.75 |
| 1.8125–2.25 | 0.5 | 0.5 | 0.5 | 0.375 | 2.00 | 8.00 | 1.00 |
| 2.3125–2.75 | 0.625 | 0.625 | 0.625 | 0.4375 | 2.50 | 10.00 | 1.25 |
| 2.875–3.25 | 0.75 | 0.75 | 0.75 | 0.50 | 3.00 | 12.00 | 1.50 |
| 3.375–3.75 | 0.875 | 0.875 | 0.875 | 0.625 | 3.50 | 14.00 | 1.75 |
| 3.875–4.5 | 1.00 | 1.00 | 1.00 | 0.75 | 4.00 | 16.00 | 2.00 |
| 4.75–5.5 | 1.25 | 1.25 | 1.25 | 0.875 | 5.00 | 20.00 | 2.50 |
| 5.75–6 | 1.50 | 1.50 | 1.50 | 1.00 | 6.00 | 24.00 | 3.00 |

[a] Measured at a distance equal to the key width $w$ from the large end.

SOURCE: From ANSI B17.1-1967, Keys and Keyseats, available from the American Society of Mechanical Engineers.

A key is also intended as a safety system. Most machines have an operating speed and torque that define the required size of the key. However, in the event of a drastic increase in the load conditions, the key will shear before the shaft or machine element (gear, cam, pulley, etc.) fails. Since keys are inexpensive and can be quickly replaced, designers use them to protect more expensive machine components.

**Table 11.3**     Dimensions of selected Woodruff keys

| Key no. | Suggested shaft sizes, in | Nominal key size[a] $w \times l$, in | Height $h$ of key, in | Shearing area, in$^2$ |
|---------|---------------------------|--------------------------------------|------------------------|-----------------------|
| 204     | 0.3125–0.375              | 0.062 × 0.500                        | 0.203                  | 0.030                 |
| 305     | 0.4375–0.50               | 0.094 × 0.625                        | 0.250                  | 0.052                 |
| 405     | 0.6875–0.75               | 0.125 × 0.625                        | 0.250                  | 0.072                 |
| 506     | 0.8125–0.9375             | 0.156 × 0.750                        | 0.313                  | 0.109                 |
| 507     | 0.875–0.9375              | 0.156 × 0.875                        | 0.375                  | 0.129                 |
| 608     | 1.00–1.1875               | 0.188 × 1.000                        | 0.438                  | 0.178                 |
| 807     | 1.25–1.3125               | 0.250 × 0.875                        | 0.375                  | 0.198                 |
| 809     | 1.25–1.75                 | 0.250 × 1.125                        | 0.484                  | 0.262                 |
| 810     | 1.25–1.75                 | 0.250 × 1.250                        | 0.547                  | 0.296                 |
| 812     | 1.25–1.75                 | 0.250 × 1.500                        | 0.641                  | 0.356                 |
| 1012    | 1.8125–2.5                | 0.312 × 1.500                        | 0.641                  | 0.438                 |
| 1212    | 1.875–2.5                 | 0.375 × 1.500                        | 0.641                  | 0.517                 |

[a]The key extends into the hub a distance of $w/2$.

SOURCE:  From ANSI B17.2-1967, Woodruff Keys and Keyseats, available from the American Society of Mechanical Engineers.

**Table 11.4**     Holding force generated by setscrews

| Screw diameter, in | Holding force, lb |
|--------------------|-------------------|
| 0.25               | 100               |
| 0.375              | 250               |
| 0.50               | 500               |
| 0.75               | 1300              |
| 1.0                | 2500              |

SOURCE: From Oberg and Jones (2000).

Note in Figure 11.10 that if $h = w$, the key is square, a special case of a flat key. Keys are usually made of low-carbon steel (AISI 1020) and have a cold-drawn finish. Keys should fit tightly so that key rotation is not possible. In some applications setscrews may be required to restrict motion.

Key design is successful if failure from shear or compression is prevented, both of which are considered here:

1.   Failure due to shear:

$$P = \frac{T}{d/2} = \frac{2T}{d} \tag{11.59}$$

The area for a key as shown in Figure 11.10(d) is $A_s = wl$. The design shear stress is

$$\tau_{\text{design}} = \frac{P}{A_s} = \frac{2T}{dwl} \tag{11.60}$$

The right side of Equation (11.60) is independent of the height $h$. To avoid failure due to shear,

$$\tau_{\text{design}} \leq \frac{S_{sy}}{n_s} = \frac{\text{yield stress in shear}}{\text{safety factor}} \qquad (11.61)$$

Recall from Equation (3.14) that $\tau_{\text{all}} = S_{sy} = 0.40S_y$.

2. Failure due to compressive or bearing stress. The compression or bearing area of the key is

$$A_c = \frac{lh}{2} \qquad (11.62)$$

The compressive or bearing design stress is

$$\sigma_{\text{design}} = \frac{P}{A_c} = \frac{2T}{dlh/2} = \frac{4T}{dlh} \qquad (11.63)$$

The right side of Equation (11.63) is independent of the width $w$. While use is made of Equation (3.16), failures due to compressive or bearing stress can be avoided if

$$\sigma_{\text{design}} \leq \frac{0.90S_y}{n_s} \qquad (11.64)$$

---

**Given:** A 4-in-diameter shaft with a hub is made of high-carbon steel. A square key made of low-carbon steel has a width and height of 1 in. Assume a torque value at a 2-in radius.

**EXAMPLE 11.6**

**Find:** The critical length of the key while assuming a safety factor of 2 and considering both compression and shear.

**Solution:** From Table A.1 the yield strength for the shaft and hub is $S_y = 55$ ksi. From Equations (3.14) and (3.16)

$$\tau_{\text{all}} = S_{sy} = 0.4S_y = 22 \text{ ksi}$$

$$\sigma_{\text{all}} = S_{cy} = 0.9S_y = 49.5 \text{ ksi}$$

Thus, the design stresses are

$$\tau_{\text{design}} = \frac{S_{sy}}{n_s} = \frac{22(10^3)}{2} = 11,000 \text{ psi} = 11 \text{ ksi}$$

$$\sigma_{\text{design}} = \frac{S_{cy}}{n_s} = \frac{(49.5)(10^3)}{2} = 24,750 \text{ psi} = 24.75 \text{ ksi}$$

For a shaft with a circular cross section

$$J = \frac{\pi d^4}{32} = \frac{\pi (4)^4}{32} = 25.13 \text{ in}^4$$

The maximum torque acting on the shaft at a radius of 2 in is

$$T_{\text{max}} = \frac{\tau_{\text{design}} J}{r} = \frac{11(10^3)(23.13)}{2} = 138,200 \text{ lb} \cdot \text{in}$$

The maximum force is

$$P_{max} = \frac{T_{max}}{r} = 69,100 \text{ lb}$$

From Table A.1 the yield strength for the key is $S_y = 43$ ksi. From Equations (3.14) and (3.16)

$$\tau_{all} = S_{sy} = 0.4S_y = 17.20 \text{ ksi}$$

$$\sigma_{all} = S_{cy} = 0.9S_y = 38.70 \text{ ksi}$$

The design stresses are

$$\tau_{design} = \frac{S_{sy}}{n_s} = \frac{17,200}{2} = 8600 \text{ psi}$$

$$\sigma_{design} = \frac{S_{cy}}{n_s} = \frac{38,700}{2} = 19,350 \text{ psi}$$

From Equation (11.60)

$$l_{cr} = \frac{2T_{max}}{dw\tau_{design}} = \frac{2(138,200)}{4(1)(8600)} = 8.035 \text{ in}$$

Thus, to avoid key shear failure, the key should be at least 8.035 in long. From Equation (11.63)

$$l_{cr} = \frac{4T_{max}}{\sigma_{design}dh} = \frac{4(138,200)}{(19,350)(4)(1)} = 7.142 \text{ in}$$

To avoid key compressive or bearing failure, the key should be at least 7.142 in long. Therefore, failure will first occur from key shearing, and the key must be at least 8.035 in long.

## 11.7   FLYWHEELS

Large variations in acceleration within a mechanism can cause large oscillations in the torque. The peak torque can be so high as to require an overly large motor. However, the average torque over the cycle, due mainly to losses and external work done, may often be much lower than the peak torque. To smooth out the velocity changes and stabilize the back-and-forth flow in energy of rotating equipment, a **flywheel** is often attached to a shaft. The use of a flywheel will allow the following to occur:

1. Reduced amplitude of speed fluctuation
2. Reduced maximum torque required
3. Energy stored and released when needed during cycle

A flywheel with driving (mean) torque $T_m$ and load torque $T_l$ is shown in Figure 11.11. The flywheel in this figure is a flat circular disk. A motor supplies a torque $T_m$, which for design purposes should be as constant as possible. The load torque $T_l$, for such applications as crushing machinery, punch presses, and machine tools, varies considerably.

**Figure 11.11**    Flywheel with driving (mean) torque $T_m$ and load torque $T_l$.

## 11.7.1    DYNAMICS

The kinetic energy in a rotating system like that shown in Figure 11.11 is

$$K_e = \frac{I_m \omega^2}{2} \tag{11.65}$$

where $I_m$ = mass moment of inertia, kg $\cdot$ m$^2$
$\omega$ = angular speed, rad/s

The mass moments of inertia for six solids are given in Table 4.2, and mass moment of inertia is discussed in Section 4.2.6. From Newton's second law of motion as applied to the flywheel shown in Figure 11.11,

$$-T_l + T_m = I_m \frac{d\omega}{dt} \tag{11.66}$$

The design motor torque should be equivalent to the average torque, or $T_m = T_{avg}$. But

$$\frac{d\omega}{dt} = \frac{d\omega}{dt} \frac{d\theta}{d\theta} = \frac{d\theta}{dt} \frac{d\omega}{d\theta} = \omega \frac{d\omega}{d\theta} \tag{11.67}$$

Substituting Equation (11.67) into Equation (11.66) gives

$$-T_l + T_{avg} = I_m \omega \frac{d\omega}{d\theta}$$

$$-(T_l - T_{avg})\, d\theta = I_m \omega\, d\omega \tag{11.68}$$

This equation can be written in terms of a definite integral as

$$-\int_{\theta_{\omega_{min}}}^{\theta_{\omega_{max}}} (T_l - T_{avg})\, d\theta = \int_{\omega_{min}}^{\omega_{max}} I_m \omega\, d\omega = \frac{I_m}{2} \left( \omega_{max}^2 - \omega_{min}^2 \right) \tag{11.69}$$

The *left side* of Equation (11.69) represents the change in kinetic energy between the maximum and minimum shaft speeds and is equal to the area under the torque-angle diagram between the extreme values of speed. The *far right side* of Equation (11.69) describes the change in energy stored in the flywheel. The only way to extract energy from the flywheel is to slow it down; adding energy will speed it up. The best that can be done is to minimize the speed variation ($\omega_{max} - \omega_{min}$) by providing a flywheel with a sufficiently large mass moment of inertia $I_m$.

The location of minimum angular speed $\theta_{\omega_{min}}$ within a cycle of operation occurs after the maximum positive energy has been delivered from the rotor to the load. The location of maximum angular speed $\theta_{\omega_{max}}$ within a cycle of operation occurs after the maximum negative energy has been returned to the load, the point where the ratio of the energy summation to the area in the torque pulse is the largest negative value.

## 11.7.2    FLYWHEEL SIZING

The size of a flywheel required to absorb the energy with an acceptable change in speed needs to be determined. The change in shaft speed during a cycle of operation is called the *fluctuation speed $\omega_f$* and is expressed as

$$\omega_f = \omega_{max} - \omega_{min}$$

The **coefficient of fluctuation** is defined as

$$C_f = \frac{\omega_{max} - \omega_{min}}{\omega_{avg}} = \frac{2(\omega_{max} - \omega_{min})}{\omega_{max} + \omega_{min}} \qquad (11.70)$$

Table 11.5 gives coefficients of fluctuation $C_f$ for various types of equipment. The smaller $C_f$ is, the larger the diameter of the flywheel used should be. A larger flywheel will add increased cost and weight to the system, factors that have to be weighed against the smoothness of the operation desired.

**Table 11.5**    Coefficient of fluctuation for various types of equipment

| Type of equipment | Coefficient of fluctuation, $C_f$ |
| --- | --- |
| Crushing machinery | 0.200 |
| Electrical machinery | 0.003 |
| Electrical machinery, direct driven | 0.002 |
| Engines with belt transmissions | 0.030 |
| Flour milling machinery | 0.020 |
| Gear wheel transmission | 0.020 |
| Hammering machinery | 0.200 |
| Machine tools | 0.030 |
| Paper-making machinery | 0.025 |
| Pumping machinery | 0.030–0.050 |
| Shearing machinery | 0.030–0.050 |
| Spinning machinery | 0.010–0.020 |
| Textile machinery | 0.025 |

SOURCE: Kent (1969).

The far right side of Equation (11.69) describes the kinetic energy of the flywheel and can be rewritten as

$$K_e = \frac{I_m}{2} (\omega_{max} + \omega_{min})(\omega_{max} - \omega_{min})$$

Making use of Equation (11.70) in the above equation gives

$$K_e = I_m \omega_{avg}^2 C_f \tag{11.71}$$

$$I_m = \frac{K_e}{C_f \omega_{avg}^2} \tag{11.72}$$

By knowing the desired coefficient of fluctuation for a specific application, obtaining the change in kinetic energy $K_e$ from the integration of the torque curve, and knowing the average angular velocity, the mass moment of inertia $I_m$ required can be determined. By knowing $I_m$, the dimension (the diameter) of the flywheel can be determined.

The most efficient flywheel design is obtained by maximizing $I_m$ for the minimum volume of flywheel material used. Ideally, this design would take the shape of a ring supported by a thin disk, but thin cross sections can be difficult to manufacture. Efficient flywheels have their masses concentrated at the largest radius (or in the rim) and their hubs supported on spokes, as bicycle wheels are. This configuration places the greatest part of the mass at the largest radius possible and minimizes the weight for a given mass moment of inertia while making manufacturing problems tractable.

### DESIGN PROCEDURE FOR SIZING A FLYWHEEL

1. Plot the load torque $T_l$ versus $\theta$ for one cycle.

2. Determine $T_{l,avg}$ over one cycle.

3. Find the locations $\theta_{\omega max}$ and $\theta_{\omega min}$.

4. Determine kinetic energy by integrating the torque curve.

5. Determine $\omega_{avg}$.

6. Determine $I_m$ from Equation (11.72).

7. Find the dimensions of the flywheel.

---

**Given:** The output, or load torque, of a flywheel used in a punch press for each revolution of the shaft is 12 N · m from 0 to $\pi$ and from $3\pi/2$ to $2\pi$ and 144 N · m from $\pi$ to $3\pi/2$. The coefficient of fluctuation is 0.05 about an average speed of 600 rpm. Assume that the flywheel's solid disk is made of low-carbon steel of constant 1-in thickness.

**EXAMPLE 11.7**

**Find:** Determine the following:

(a) The average load or output torque
(b) The locations $\theta_{\omega max}$ and $\theta_{\omega min}$
(c) The energy required
(d) The outside diameter of the flywheel

**Figure 11.12**    Load or output torque variation for one cycle used in Example 11.7.

## Solution:

(a)  By using the load or output torque variation for one cycle from Figure 11.12, the average load torque is

$$2\pi\, T_{l,\mathrm{avg}} = \pi(12) + \frac{\pi}{2}(144) + \frac{\pi}{2}(12)$$

$$T_{l,\mathrm{avg}} = 6 + 36 + 3 = 45 \text{ N} \cdot \text{m}$$

(b)  From Figure 11.12

$$\theta_{\omega\max} = \pi \qquad \text{and} \qquad \theta_{\omega\min} = \frac{3\pi}{2}$$

(c)  From the left side of Equation (11.69) the kinetic energy for one cycle is

$$K_e = -\int_{\theta_{\omega\min}}^{\theta_{\omega\max}} (T_l - T_{\mathrm{avg}})\, d\theta = (144 - 45)\frac{\pi}{2} = 155.5 \text{ N} \cdot \text{m}$$

(d)  The average angular speed can be expressed as

$$\omega_{\mathrm{avg}} = 600 \frac{\mathrm{r}}{\mathrm{min}} \left(\frac{1 \text{ min}}{60 \text{ s}}\right)\left(\frac{2\pi}{1 \text{ r}}\right) = 62.83 \text{ rad/s}$$

From Equation (11.72) the mass moment of inertia is

$$I_m = \frac{K_e}{C_f \omega_{\mathrm{avg}}^2} = \frac{155.5}{(0.05)(62.83)^2} = 0.7879 \text{ kg} \cdot \text{m}^2 \qquad \text{(a)}$$

From Table 4.2 (also given on the inside back cover) the mass moment of inertia for a solid round disk is

$$I_m = \frac{m_a d^2}{8} = \frac{\pi d^2 t_h \rho}{4}\left(\frac{d^2}{8}\right) = \frac{\pi}{32}\rho d^4 t_h$$

For low-carbon steel from Table A.1, $\rho = 7860$ kg/m³.

$$\therefore \qquad I_m = \frac{\pi}{32}(7860)(0.0254)d^4 = 19.60d^4 \qquad \text{(b)}$$

Equating Equations (a) and (b) gives

$$d^4 = \frac{0.7879}{19.60} = 0.0402 \text{ m}^4$$

$$d = 0.4478 \text{ m} = 447.8 \text{ mm}$$

### 11.7.3 STRESSES

The total stresses in the flywheel are the sum of the rotational effects and the stresses associated with internal pressurization. Material covered in Chapter 10 will be used here, in particular Section 10.4 for rotational effects and Section 10.3.2 for internal pressurization. The flywheel is assumed to be a disk with a central hole and constant thickness. The stresses in the flywheel due to rotational and press-fit effects can be written as

$$\sigma_\theta = \sigma_{\theta\omega} + \sigma_{\theta p} \tag{11.73}$$

$$\sigma_r = \sigma_{r\omega} + \sigma_{rp} \tag{11.74}$$

Making use of Equations (10.35) and (10.36) and Equations (10.23) and (10.24), respectively, gives the tangential (hoop) and radial stresses as

$$\sigma_\theta = \frac{3+\nu}{8}\rho\omega^2 \left( r_i^2 + r_o^2 + \frac{r_i^2 r_o^2}{r^2} - \frac{1+3\nu}{3+\nu}r^2 \right) + \frac{p_i r_i^2 \left( 1 + r_o^2/r^2 \right)}{r_o^2 - r_i^2} \tag{11.75}$$

$$\sigma_r = \frac{3+\nu}{8}\rho\omega^2 \left( r_i^2 + r_o^2 - \frac{r_i^2 r_o^2}{r^2} - r^2 \right) + \frac{p_i r_i^2 \left( 1 - r_o^2/r^2 \right)}{r_o^2 - r_i^2} \tag{11.76}$$

Note that $\sigma_\theta$ and $\sigma_r$ are principal stresses since there is no shear stress associated with them. Usually, $\sigma_\theta$ is larger than $\sigma_r$, implying that the maximum principal stress equals the tangential (hoop) stress ($\sigma_1 = \sigma_\theta$). Also, the circumferential stress is largest at $r = r_i$ for the flywheel. Thus, for *brittle* materials the design of the flywheel should be such that

$$\sigma_1 < \frac{S_u}{n_s} \tag{11.77}$$

For *ductile* materials a multiaxial failure theory, such as the DET, should be used. If $\sigma_1 = \sigma_\theta$, $\sigma_2 = \sigma_r$, and $\sigma_z = 0$, from Equation (6.11)

$$\sigma_e = \sqrt{\sigma_\theta^2 + \sigma_r^2 - \sigma_\theta \sigma_r}$$

From Equation (6.12) the DET predicts failure if

$$\sigma_e \geq \frac{S_y}{n_s}$$

---

**Given:** A flywheel made of low-carbon steel has an outside radius of 6 in and an inside radius of 1 in. A shaft made of low-carbon steel has a radius of 1.003 in. The flywheel is to be assembled onto the shaft and will operate at a speed of 5000 rpm.

**EXAMPLE 11.8**

**Find:** Determine the following:

(a) Circumferential stresses on the flywheel
(b) Radial stresses on the flywheel
(c) Speed at which the flywheel will break loose

**Solution:** From Table A.1 for low-carbon steel

$$\rho = 7860 \text{ kg/m}^3 = 0.2840 \text{ lbm/in}^3 = 7.358 \times 10^{-4} \text{ lb} \cdot \text{s}^2/\text{in}^4$$

$$E = 3.0 \times 10^7 \text{ lb/in}^2 \quad \text{and} \quad \nu = 0.30$$

The angular speed is

$$\omega = \frac{2\pi N_a}{60} = \frac{2\pi (5000)}{60} = 523.6 \text{ rad/s}$$

(a) Stresses and deformation of the flywheel at the shaft outside radius: For $r = r_o$ and $r_i = 0$ from Equations (11.75) and (11.76)

$$\sigma_r = 0$$

$$\sigma_\theta = \frac{\rho\omega^2 r_o^2(1-\nu)}{4} = \frac{(7.358)(10^{-4})(523.6)^2(1.003)^2(1-0.3)}{4} = 35.51 \text{ psi}$$

When the stress is known, the strain can be expressed from Equation (B.45) as

$$\epsilon_\theta = \frac{\sigma_\theta}{E} - \frac{\nu\sigma_r}{E} = \frac{\sigma_\theta}{E} = \frac{35.51}{(30)(10^6)} = 1.184 \times 10^{-6}$$

The deflection can be expressed as

$$\delta = \epsilon_\theta r = (1.184)(10^{-6})(1.003) = 1.187 \times 10^{-6} \text{ in}$$

It can be concluded that rotational effects do not increase the shaft radius since $\delta$ is so small.

(b) Stresses and deformation of the flywheel at the shaft inside radius: The stresses on the inside radius of the flywheel due to rotational effects can be obtained from the first term on the right-hand side of Equations (11.75) and (11.76).

$$\sigma_{r\omega} = 0$$

$$\sigma_{\theta\omega} = \frac{(3+\nu)\rho\omega^2}{8}\left(r_i^2 + r_o^2 + \frac{r_i^2 r_o^2}{r^2} - \frac{1+3\nu}{3+\nu}r^2\right)$$

But $r = r_i = 1$ in and $r_o = 6$ in. Therefore,

$$\sigma_{\theta\omega} = \frac{(3+0.3)(7.358)(10^{-4})(523.6)^2}{8}\left[1 + 36 + 36 - \frac{1+0.9}{3+0.3}(1)\right] = 6027 \text{ psi}$$

The expansion of the inside radius of the flywheel due to rotational effects is

$$\delta_\omega = \frac{\sigma_{\theta\omega}r_i}{E} = \frac{(6027)(1)}{(3)(10^7)} = 2.009 \times 10^{-4} \text{ in}$$

Rotational effects expand the inside radius of the flywheel to

$$r_1 = 1.000 + 2.009 \times 10^{-4} = 1.0002 \text{ in}$$

The interference between the flywheel inside radius and the shaft outside radius is 0.00280 in ($\delta_i = 0.0028$ in). From Equation (10.53) the interference pressure of placing a flywheel on a solid shaft is

$$p_f = \frac{E\delta_r\left(r_o^2 - r_i^2\right)}{2r_i r_o^2} = \frac{3(10^7)(2.8)(10^{-3})(36 - 1)}{2(1)(36)} = 4.083 \times 10^4 \text{ lb/in}^2$$

Substituting the above into Equations (11.75) and (11.76) for the dimensions, materials, and flywheel speed given results in the following equations:

$$\sigma_\theta = 4245 + \frac{49{,}992}{r^2} - 47.91r^2$$

$$\sigma_r = 4245 - \frac{44{,}992}{r^2} - 83.21r^2$$

The circumferential stress is tensile and largest at the inside radius, and the radial stress is compressive and largest at the inside radius.

(c) At what speed does the flywheel separate from the shaft? Separation occurs when there is no interference ($\delta = 0$). Recall that the initial interference was $\delta_o = 0.003$ in. Therefore, for the conditions given, the following equations must be satisfied:

$$\delta_f - \delta_s = \delta_o$$

$$\left(\frac{\sigma_\theta}{E}\right)_{\omega f} - \left(\frac{\sigma_\theta}{E}\right)_{\omega s} = \delta_o$$

Making use of Equation (11.75) (only the $\sigma$ term) and Equation (10.35) while recognizing that for the flywheel $r_o = 6$ in and $r_i = r = 1$ in and that for the shaft $r_i = 0$ and $r_o = r = 1$ in gives

$$\omega^2 = \frac{8(0.003)(3)(10^7)}{(72)(7.358)(10^{-4})(3 + 0.3)}$$

$$\omega = 2029 \text{ rad/s} = 19{,}379 \text{ rpm}$$

## 11.7.4 MATERIALS

The main purpose of a flywheel is to store energy and then use it efficiently. Small flywheels, the type found in children's toys, are often made of lead, but steel and iron are preferred for health reasons. Old steam engines have cast iron flywheels. More recently, flywheels have been proposed for vehicle power storage and regenerative braking systems; high-strength steels and composites will be used for these vehicular flywheels. Thus, a great diversity of materials are being used for flywheels. The obvious question is, what is the best material for use in flywheels?

Recall from Equations (11.75) and (11.76) that the speed effect is more significant than the pressurization effect. Thus, the maximum stress is

$$\sigma_{\max} = \frac{3 + \nu}{8}\rho\omega^2 f(r) \tag{11.78}$$

**Table 11.6**    Materials for flywheels

| Material | Performance index, $M_f$, kJ/kg | Comment |
|---|---|---|
| Ceramics | 200–2000 (compression only) | Brittle and weak in tension. Use is usually discouraged. |
| Composites: | | |
|   Ceramic-fiber-reinforced polymer | 200–500 | The best performance; a good choice. |
|   Graphite-fiber-reinforced polymer | 100–400 | Almost as good as CFRP and cheaper; an excellent choice. |
| Beryllium | 300 | Good but expensive, difficult to work, and toxic. |
| High-strength steel | 100–200 | All about equal in performance; |
| High-strength aluminum (Al) alloys | 100–200 | steel and Al alloys less expensive |
| High-strength magnesium (Mg) alloys | 100–200 | than Mg and Ti alloys. |
| Titanium alloys | 100–200 | |
| Lead alloys | 3 | High density makes these a good |
| Cast iron | 8–10 | (and traditional) selection when performance is velocity-limited, not strength-limited. |

| SOURCE: Ashby (1992).

From Equation (11.65) the kinetic energy is

$$K_e = \tfrac{1}{2} I_m \omega^2 \tag{11.79}$$

The mass moment of inertia $I_m$ from Table 4.2 can be represented as

$$I_m = m_a g(r) \tag{11.80}$$

$$\therefore \qquad K_e = \tfrac{1}{2} m_a g(r) \omega^2 \tag{11.81}$$

Assume that $\omega^2 f(r) \approx g(r)\omega^2$. From Equations (11.81) and (11.78)

$$\frac{K_e}{m_a} \propto \frac{\sigma_{max}}{\rho}$$

Thus, the best materials for a flywheel are those with high values of the performance index

$$M_f = \frac{\sigma_{max}}{\rho} \qquad \text{J/kg} \tag{11.82}$$

Recall that 1 joule (J) is 1 newton-meter (N · m).

Table 11.6 gives performance indices for various materials. From this table observe that the best materials are ceramics, beryllium, and composites. Lead and cast iron are the worst flywheel materials.

## 11.8    SUMMARY

A shaft is usually a circular-cross-section rotating member that has such power-transmitting elements as gears, pulleys, flywheels, sprockets, and rolling-element bearings mounted on it. The loading on the shaft can be one or various combinations of bending, torsion, or axial or transverse shear. Furthermore, these types of loading can be either static or cyclic. A design procedure was presented that established the appropriate shaft diameter for specific conditions. If the shaft diameter is known, the safety factor (or the smallest diameter where failure first occurs) is often an important consideration. Three failure prediction theories considering important combinations of loading were presented. These theories are by no means all-inclusive but should provide the essential understanding from which any other considerations can be easily obtained.

Shaft dynamics and in particular the first critical speed are important to design, since the rotating shaft becomes dynamically unstable and large vibrations are likely to develop. Both the Rayleigh and the Dunkerley equations for determining the first critical speed of a multiple-mass system were presented. The Rayleigh overestimates and the Dunkerley underestimates the exact solution, thus providing a range of operating speeds that the design should avoid.

The last section was on flywheels. In some applications large variations in acceleration occur that can cause large oscillations in torque. The peak torque can be so high as to require an overly large motor. A flywheel is often used to smooth out the velocity changes and stabilize the back-and-forth energy flow of rotating equipment. A procedure for designing a flywheel was presented as well as flywheel dynamics, sizing, stresses, and material selection.

# KEY WORDS

**coefficient of fluctuation**   dimensionless speed variation $\dfrac{\omega_{max} - \omega_{min}}{\omega_{avg}}$.

**critical speed**   speed at which a rotating shaft becomes dynamically unstable.

**Dunkerley equation**   relation for first critical speed that underestimates frequency.

**first critical speed**   lowest frequency at which dynamic instability occurs.

**flywheel**   element that stores energy through rotational inertia.

**grooved pin**   a pin having longitudinal grooves with raised edges that give a tight fit in a hole when the pin is hammered in.

**hub**   portion of member mounted onto shaft that directly contacts shaft.

**key**   element that transmits power from shaft to hub.

**lowest natural frequency**   same as **first critical speed.**

**natural frequency**   same as **critical speed.**

**pin**   element that transmits power from shaft to hub.

**Rayleigh equation**   relation for first critical speed that overestimates frequency.

**roll pin**   a tube made of spring material with a longitudinal slit that is driven into a hole whose size tends to close the slit, thus producing a tight fit.

**set screw**   threaded fastener driven radially into a hub.

**shaft**    rotating or stationary member, usually of circular cross section with small diameter relative to length and used to transmit power through such elements as gears, sprockets, pulleys, and cams.

**tapered key**    a rectangular key with parallel sides slightly tapered in thickness along its depth.

**Woodruff key**    constant width and a semicircular profile key.

## RECOMMENDED READINGS

Juvinall, R. C., and Marshek, K. M. (2003) *Fundamentals of Machine Component Design,* 3rd ed., Wiley, New York.

Mott, R. L. (1998) *Machine Elements in Mechanical Design,* 3rd ed. Prentice-Hall, Upper Saddle River, NJ.

Norton, R. L. (2000) *Machine Design,* Prentice Hall, Englewood Cliffs, NJ.

Oberg, E., and Jones, F. D. (2000) *Machinery's Handbook,* 26th ed., Industrial Press, New York.

Piotrowski, J. (1995) *Shaft Alignment Handbook,* 2nd ed., Marcel Dekker, New York.

Shigley, J. E., Mischke, C. R., and Budynas, R. G. (2003) *Mechanical Engineering Design,* 7th ed., McGraw-Hill, New York.

## REFERENCES

Ashby, M. F. (1992) *Materials Selection in Mechanical Design,* Pergamon Press, Oxford, England.

*Kent's Mechanical Engineers Handbook* (1969), 12th ed., "Design and Production," p. 740.

Oberg, E., and Jones, F. D. (2000) *Machinery's Handbook,* 26th ed., Industrial Press, New York.

Peterson, R. E. (1974) *Stress Concentration Factors,* Wiley, New York.

## PROBLEMS

### Section 11.2

★ **11.1**  A shaft assembly shown in sketch *a* is driven by a flat belt at location A and drives a flat belt at location B. The drive belt pulley diameter is 300 mm; the driven belt pulley diameter is 500 mm. The distance between sheaves is 800 mm, and the distance from each sheave to the nearest bearing is 200 mm. The belts are horizontal and load the shaft in opposite directions. Determine the size of the shaft and the types of steel that should be used. Assume a safety factor of 5. *Ans.* For AISI 1080 steel, $d = 30$ mm.

### Section 11.3

**11.2**  The gears in the shaft assembly shown in Figure 11.1(a) transmit 100 kW of power and rotate at 3600 rpm. Gear wheel 1 is loaded against another gear such that the force $P_1$ acts in a 45° direction at a radius of 80 mm from the shaft center. The force $P_2$ acts vertically downward at a radius of 110 mm from the shaft center. The distance from bearing A to gear 1 is 100 mm,

|    * Indicates problem of greater difficulty.

Sketch a, for Problem 11.1

that from gear 1 to gear 2 is 85 mm, and that from gear 2 to bearing B is 50 mm. Perform the following:

a)   Draw a free-body diagram with forces acting on the shaft when bearings A and B transmit only radial forces.
b)   Find values of force components as well as the resultant force at locations A and B.
c)   Find the transmitted torque. *Ans. T* = 265 N · m.
d)   Draw a bending moment diagram in the $xy$ and $xz$ planes along with a torque diagram; also, indicate the maximum bending moment and the maximum torque.
e)   Find the safety factor according to the distortion-energy theory (DET) and the maximum-shear-stress theory (MSST) if the shaft has a diameter of 35 mm and is made of high-carbon steel (AISI 1080). *Ans. $n_{s,\text{MSST}}$* = 5.18, $n_{s,\text{DET}}$ = 5.73.

**11.3** The shaft assembly given in Problem 11.2 has an extra loading from thermal expansion of the shaft. The bearings are assumed to be rigid in the axial direction, so that when the shaft heats up, it cannot elongate and instead compressive stress builds up. Determine the thermal stress, and find the safety factor by using the DET if the shaft heats up by

a)   5°C *Ans. $n_s$* = 5.17.
b)   15°C *Ans. $n_s$* = 4.24.

**11.4** Given the shaft assembly in Problem 11.2 but calculating as if the AISI 1080 steel were brittle [using the maximum-normal-stress theory (MNST)], find the safety factor $n_s$ for fatigue by using the information in Problem 11.2. *Ans. $n_{s,\text{MNST}}$* = 5.0.

## Section 11.4

**11.5**   Gears 3 and 4 act on the shaft shown in sketch *b*. The resultant gear force, $P_A$ = 600 lb, acts at an angle of 20° from the $y$ axis. The yield stress for the shaft, which is made of cold-drawn steel, is 71,000 psi, and the ultimate stress is 85,000 psi. The shaft is solid and of constant diameter. The safety factor is 2.6. Assume the DET throughout. Also, for fatigue loading conditions assume completely reversed bending with a bending moment amplitude equal to that used for static conditions. The alternating torque is zero. Determine the safe shaft diameter due to static and fatigue loading. Show shear and moment diagrams in the various planes. *Ans.* $d_{static} =$ 1.8 in, $d_{fatigue} = 2.25$ in.

Sketch *b*, for Problem 11.5

**11.6**    Derive Equation (11.39) in Section 11.4.2. Start by showing the stresses acting on an oblique plane at angle $\phi$. The MNST should be used, implying that the critical stress line extends from $S_e/n_s$ to $S_u/n_s$.

**11.7**    The shaft shown in sketch *c* rotates at 1000 rpm and transfers 6 kW of power from input gear A to output gears B and C. The spur gears A and C have pressure angles of 20°. The helical gear has a pressure angle of 20° and a helix angle of 30° and transfers 70% of the input power. All important surfaces are ground. All dimensions are in millimeters. The shaft is made of annealed carbon steel with $S_{ut} = 636$ MPa and $S_y = 365$ MPa.

Sketch *c*, for Problem 11.7

*a)*    Draw a free-body diagram as well as the shear and moment diagrams of the shaft.
*b)*    Which bearing should support the thrust load and why? *Ans.* The left bearing.
*c)*    Determine the minimum shaft diameter for a safety factor of 2 and 99% reliability. *Ans. d* = 20 mm.

**11.8** In Problem 11.7 if the shaft diameter is 30 mm and it is made of AISI 1030 cold-drawn steel, what is the safety factor while assuming 90% reliability? *Ans.* $n_s = 16.5$.

**11.9** The shaft shown in sketch *d* supports two gears. The shaft is made from high-carbon steel (AISI 1080), and it is to be designed with a safety factor of 2.0. The gears transmit a constant torque caused by $P_A = 2000$ N acting vertically as shown. The shaft has a constant machined cross section. (a) What is the reaction force on gear C? (b) What is the critical location in the shaft? (c) Using the Soderberg line, obtain the required shaft diameter. (*Note:* Ignore all endurance limit modification factors except the surface finish factor.) (d) One of the gears has been attached with a keyway, the other with a shrink-fit. Which gear was attached with a keyway, and why?

Sketch *d*, for Problem 11.9

## Section 11.5

**11.10** The rotor shown in sketch *e* has a stiff bearing on the left. Find the critical speed when the shaft is made of steel with $E = 210$ GPa. *Ans.* $N = 3040$ rpm.

Sketch *e*, for Problem 11.10

**11.11** Determine the critical speed in bending for the shaft assembly shown in sketch *f*. The modulus of elasticity of the shaft $E = 207$ GPa, its length $l = 350$ mm, its diameter $d = 8$ mm, and the rotor mass $m_a = 2.3$ kg. *Ans.* $N = 1530$ rpm.

Sketch $f$, for Problem 11.11

**11.12** Calculate the diameter of the shaft in the assembly shown in sketch $g$ so that the first critical speed is 9000 rpm. The shaft is made of steel with $E = 207$ GPa. The distance $a = 300$ mm, and the mass $m_a = 100$ kg; the mass of the shaft is neglected. Use both the Rayleigh and Dunkerley methods. *Ans.* $d_R = 131$ mm, $d_D = 133$ mm.

Sketch g, for Problem 11.12

**★ 11.13** The simply supported shaft shown in sketch $h$ has two weights on it. Neglecting the shaft weight and using the Rayleigh method for the stainless steel ($E = 26 \times 10^6$ psi) shaft, determine the safe diameter to ensure that the first critical speed is no less than 3600 rpm. *Ans.* $d_R = 2.23$ in, $d_D = 2.26$ in.

Sketch h, for Problem 11.13

| * Indicates problem of greater difficulty.

★ **11.14** Determine the first critical speed by the Dunkerley and Rayleigh methods for the steel shaft shown in sketch *i*. Neglect the shaft mass. The area moment of inertia is $I = \pi r^4/4$, where $r$ = shaft radius. The method of superposition may be used with the following given:

$$\delta = \frac{P}{6EI}\left[\frac{bx^3}{l} - \langle x-a\rangle^3 - \frac{xb(l^2 - b^2)}{l}\right]$$

*Ans.* $N_R = 708$ rpm, $N_D = 673$ rpm.

120 lbm    80 lbm    2-in-diam shaft

20 in    40 in    30 in

Sketch *i*, for Problem 11.14

**11.15** Calculate the diameter $d$ of the shaft in sketch *j* so that the lowest critical speed in bending becomes 9000 rpm. The modulus of elasticity $E = 107$ GPa, the distance $a = 300$ mm, and the mass $m_a = 100$ kg. Neglect the mass of the shaft. Use the Rayleigh and Dunkerley methods. *Ans.* $d_R = 155$ mm, $d_D = 156$ mm.

y   10 kg   10 kg   d   x

375    600    375

Sketch *j*, for Problem 11.15

**11.16** The rotor in sketch *k* has a shaft diameter of 32 mm, and the disk mass is 170 kg. Calculate the critical speed if the spring rates are 1668 and 3335 N/mm, respectively, and are the same in all directions. The elastic modulus $E = 206$ GPa. For a shaft without springs the influence coefficient $\alpha_{11} = l^3/(6EI)$. *Ans.* $N = 1360$ rpm.

$l = 160$ mm    $l = 160$ mm

Sketch *k*, for Problem 11.16

★ **11.17** Calculate the critical speed for a rotor shown in sketch *l* that has two moment-free bearings. The shaft has a diameter of 20 mm and is made of steel with $E = 206$ GPa. *Ans.* $N = 1990$ rpm.

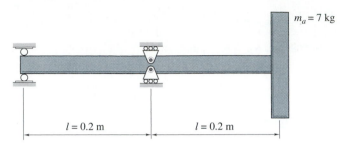

$m_a = 7$ kg

$l = 0.2$ m        $l = 0.2$ m

Sketch *l*, for Problem 11.17

**11.18** The deflection at the center of the steel shaft shown in sketch *k* is equal to 0.35 mm. Find the diameter of the solid circular shaft and the first critical speed, using both the Rayleigh and Dunkerly methods. All length dimensions are in millimeters.

## Section 11.6

★ **11.19** A flywheel has a hub made of aluminum alloy 2014. The hub is connected to a 2-in-diameter AISI 1040 steel shaft with a flat tapered key made of AISI 1020 steel. Determine the maximum torque that can be transmitted with a safety factor of 3.

★ **11.20** A jaw crusher is used to crush iron ore down to the particle size needed for iron ore pellet production. The pellets are later used in a blast furnace to make steel. If the ore pieces fed into the crusher are too large, the crusher is protected by a torque-limiting key made of copper. The torque on the 400-mm-diameter shaft connecting the flywheel with the jaw mechanism should never go above 700 kN · m. Dimension a copper square tapered key so that the shaft is not damaged when an oversized ore particle comes into the crusher.

**11.21** Given the sketch of the Woodruff key in Figure 11.10, show that the shear area is given by

$$A_t = 2w\sqrt{c(D-c)}$$

where $c$ = diametral clearance, m
$\quad\quad D$ = key diameter, m

## Section 11.7

**11.22** A punch press is to be driven by a constant-torque electric motor that operates at 1200 rpm. A flywheel of constant thickness cut from 2-in-thick steel plate is to be used to ensure smooth operation. The punch press torque steps from 0 to 10,000 ft · lb, remaining constant for the first 45° of camshaft rotation, back to 0 for the next 45°, up to 6000 ft · lb for the next 45°, and back to 0 for the remainder of the cycle. What is the minimum diameter of the flywheel?

**11.23** The output torque of a flywheel for each revolution of a shaft is 10 N · m from 0 to $\pi$ and 120 N · m from $\pi$ to $2\pi$. The coefficient of fluctuation is 0.04 at an average speed of 2000 rpm. Assume that the flywheel disk is an AISI 1040 steel plate of 30-mm constant thickness. Determine the following:

a) Average load torque
b) Locations $\theta_{\omega max}$ and $\theta_{\omega min}$
c) Energy required
d) Outside diameter of flywheel

**11.24** The output torque of a flywheel for each revolution of the shaft is 200 in · lb from 0 to $\pi/3$, 1600 in · lb from $\pi/3$ to $2\pi/3$, 400 in · lb from $2\pi/3$ to $\pi$, 900 in · lb from $\pi$ to $5\pi/3$, and 200 in · lb from $5\pi/3$ to $2\pi$. Input torque is assumed to be constant. The average speed is 860 rpm, and the coefficient of fluctuation is 0.10. Find the diameter of the flywheel if it is to be cut from a 1-in-thick steel plate. *Ans. d = 12 in.*

**11.25** A 20-mm-thick flywheel is made of aluminum alloy 2014 and runs at 9000 rpm in a racing car motor. What is the safety factor if the aluminum alloy 2014 is stressed to one-quarter of its yield strength at 9000 rpm? To decrease the flywheel outer diameter, higher-density exotic materials are investigated. Find the best material from Table A.2 that can be substituted for aluminum alloy 2014 but with the same safety factor. The flywheel is machined from a solid piece of aluminum alloy 2014 with no central hole, and the thickness cannot be larger than 20 mm.

★ **11.26** The aluminum alloy flywheel considered in Problem 11.25 has a mass moment of inertia of $0.0025$ kg · m². By mistake the motor is accelerated to 7000 rpm in neutral gear, and the throttle sticks in the fully open position. The only way to stop the motor is to disconnect the electric lead from the spark plug, and it takes 6 s before the motor is motionless. Neglecting the inertia of all movable parts in the motor except the flywheel, calculate the internal friction moment in the motor and the (mechanical part of the) friction losses in the motor at 5000 rpm. The friction moment is assumed to be constant at all speeds. *Ans. $T_f = -0.305$ N · m.*

**11.27** A one-cylinder ignition bulb motor to an old fishing boat has a flywheel that gives the motor a coefficient of fluctuation of 25% when it idles at 180 rpm. The mass moment of inertia for the flywheel is 1.9 kg · m². Determine the coefficient of fluctuation at 500 rpm if the compression stroke consumes equally large energy at all speeds. Also, calculate the mass moment of inertia needed to obtain a 20% coefficient of fluctuation at 500 rpm. *Ans. $C_f = 0.032$, $I_m = 0.31$ kg · m².*

**11.28** A flywheel for a city bus drive should store as much energy as possible for a given flywheel weight. The diameter must be less than 1.5 m, and the mass must be smaller than 250 kg. Find which material from Tables A.1 and A.2 gives the highest possible stored energy for a safety factor of 4, assuming that the flywheel has constant thickness.

**11.29** A flywheel on an AISI 1080 steel shaft is oscillating due to disturbances from a combustion engine. The engine is a four-stroke engine with six cylinders, so that the torque disturbances occur 3 times per revolution. The flywheel shaft has a diameter of 20 mm and is 1 m long, and the flywheel moment of inertia is 0.5 kg · m². Find the engine speed at which the large torsional vibrations begin to appear. The shaft material is AISI 1080 steel. *Ans. N = 160 rpm.*

# 12
# HYDRODYNAMIC AND
# HYDROSTATIC BEARINGS

A Kingsbury Bearing.
(Albert Kingsbury's
Company, 1926)

*"A cup of tea, standing in a dry saucer, is apt to slip about in an
awkward manner, for which a remedy is found in introduction of
a few drops of water, or tea, wetting the parts in contact."*

Lord Rayleigh (1918)

## SYMBOLS

| | |
|---|---|
| $A$ | area, $m^2$ |
| $\tilde{A}$ | integration constant |
| $B_j$ | bearing number for journal slider bearing, |

$$\left(\frac{\eta_0 \omega_b r_b w_t}{\pi W_r}\right)\left(\frac{r_b}{c}\right)^2$$

| | |
|---|---|
| $B_t$ | bearing number for thrust slider bearing, |

$$\left(\frac{\eta_0 u_b w_t}{W_z}\right)\left(\frac{l}{s_h}\right)^2$$

| | |
|---|---|
| $\bar{B}$ | integration constant |
| $\tilde{B}$ | integration constant |
| $\tilde{C}$ | integration constant |
| $C_p$ | specific heat of lubricant at constant pressure, J/(kg · °C) |
| $C_s$ | volumetric specific heat, J/(m³·°C) |
| $C_1, \ldots, C_4$ | integration constants |
| $c$ | radial clearance in journal bearing, m |
| $c_b$ | bearing clearance at pad minimum film thickness, m |
| $c_s$ | stiffness coefficient |
| $\tilde{D}$ | integration constant |
| $d$ | diameter, m |
| $e$ | eccentricity of journal bearing, m |
| $F$ | friction force, N |
| $g$ | acceleration due to gravity, 9.807 m/s² (386.1 in/s²) |
| $H$ | dimensionless film thickness, $h/s_h$ |
| $H_o$ | outlet film thickness ratio, $h_o/s_h$ |
| $H_p$ | dimensionless total power loss |
| $h$ | film thickness, m |
| $h_o$ | outlet (minimum) film thickness, m |
| $h_p$ | rate of working against viscous stress (power loss), W |
| $k$ | slope of film thickness |
| $L_p$ | dimensionless pivot location, $l_p/l$ |
| $l$ | length in sliding direction, m |
| $l_p$ | length to pivot from inlet, m |
| $m_p$ | preload factor |
| $N_a$ | rotational speed, rps |
| $N_p$ | number of pads |
| $n_s$ | step location |
| $P$ | dimensionless pressure, |
| $P_{max}$ | dimensionless maximum film pressure, |

$$\frac{W_r}{2 r_b w_t p_{max}}$$

| | |
|---|---|
| $p$ | pressure, Pa |
| $p_l$ | lift pressure, Pa |
| $p_{max}$ | maximum film pressure, Pa |
| $p_r$ | recess pressure, Pa |
| $p_s$ | supply pressure, Pa |
| $Q$ | dimensionless volumetric flow rate |
| $q$ | circumferential volumetric flow rate, m³/s |
| $q_s$ | side-leakage volumetric flow rate, m³/s |
| $q_x$ | volumetric flow rate in sliding direction, m³/s |
| $q'_r$ | radial flow rate per unit width, m²/s |
| $q'_x$ | volumetric flow rate per unit width in sliding direction, m²/s |
| $q'_y$ | volumetric flow rate per unit width in transverse direction, m²/s |
| $R_a$ | centerline average surface roughness, m |
| $\bar{R}$ | gas constant, m²/s² · °C |
| $r$ | radius, m |
| $r_b$ | radius of journal bearing, m |
| $s_h$ | shoulder height, m |
| $T$ | friction torque, N · m |
| $t$ | time, s |
| $\Delta t$ | time change, s |
| $t_m$ | absolute temperature, °C |
| $\tilde{t}_m$ | mean temperature, °C |
| $\Delta t_m$ | temperature change, °C |
| $t_w$ | thickness, m |
| $u$ | fluid velocity in sliding direction, m/s |
| $\tilde{u}$ | average velocity in sliding direction, m/s |
| $V$ | volume, m³ |
| $v$ | fluid velocity in transverse direction, m/s |
| $W_r$ | radial load, N |
| $W_r^*$ | radial load per area, $W_r/(2 r_b w_t)$, Pa |
| $W_t$ | total load, N |
| $W_z$ | normal load, N |
| $W'_z$ | normal load per unit width, N/m |
| $\bar{W}_z$ | dimensionless normal load |
| $w$ | squeeze velocity, m/s |
| $w_t$ | width (in side-leakage direction) of bearing, m |
| $X$ | dimensionless coordinate |
| $x$ | Cartesian coordinate in direction of sliding, m |
| $y$ | Cartesian coordinate in side-leakage direction, m |
| $z$ | Cartesian coordinate across film, m |
| $\alpha$ | inclination angle of fixed-incline slider bearing, deg |
| $\alpha_a$ | offset factor |
| $\Delta$ | step height, m |

| | |
|---|---|
| $\epsilon$ | eccentricity ratio, $e/c$ |
| $\eta$ | absolute viscosity, Pa·s |
| $\eta_0$ | absolute viscosity at $p = 0$ and constant temperature, Pa·s |
| $\lambda$ | length-to-width ratio, $l/w_t$ |
| $\lambda_j$ | diameter-to-width ratio, $2r/w_t$ |
| $\mu$ | coefficient of sliding friction |
| $\rho$ | density, kg/m³ |
| $\tau$ | shear stress, Pa |
| $\Phi$ | attitude angle, deg |
| $\phi$ | cylindrical polar coordinate |
| $\phi_{max}$ | location of maximum pressure, deg |
| $\phi_0$ | location of terminating pressure, deg |
| $\omega$ | angular velocity, $2\pi N_a$, rad/s |

**SUBSCRIPTS**

| | |
|---|---|
| $a$ | solid $a$ (upper surface) |
| all | allowable |
| atm | atmospheric |
| $b$ | solid $b$ (lower surface) |
| $d$ | disk |
| $i$ | inlet; inner |
| $m$ | mean |
| max | maximum |
| min | minimum |
| $o$ | outlet; outer |
| $p$ | punch |
| $x, y, z$ | Cartesian coordinates |
| 1,2 | regions in thrust slider bearing |

## 12.1  INTRODUCTION

The characterization of **hydrodynamic lubrication** was presented in Section 8.4.1 and will not be repeated here. For the bearings considered in this chapter it will be assumed that **fluid film lubrication** occurs, or that the lubricated surfaces are separated by a fluid film. Also, the bearings are assumed to have conformal surfaces, which were considered in Section 8.2. These three topics (characterization of hydrodynamic lubrication, fluid film lubrication, and conformal surfaces), which were covered in Chapter 8, should be understood before proceeding with this chapter.

The history of hydrodynamic lubrication begins with the classical experiments of Tower (1883), who detected the existence of a film from measurements of pressure within the lubricant, and of Petrov (1883), who reached the same conclusion from friction measurements. This work was closely followed by Reynolds' (1886) celebrated analytical paper in which he used a reduced form of the Navier-Stokes equations in association with the continuity equation to generate a second-order differential equation for the pressure in the narrow, converging gap (or conjunction) between bearing surfaces. This pressure enables a load to be transmitted between the surfaces with extremely low friction, since the surfaces are completely separated by a fluid film. In such a situation the physical properties of the lubricant, notably the dynamic viscosity, dictate the behavior in the conjunction.

Designing hydrodynamic bearings is an excellent opportunity to use the basic knowledge obtained from fluid mechanics. Making the connection between bearing design and fluid mechanics is extremely important in the development of an engineer. This chapter hopes to provide such an experience.

This chapter presents the most important aspects of hydrodynamic and hydrostatic bearings. The derivation of the Reynolds equation is presented, and the physical significance of its various terms is explained. The important aspects of self-lubricated thrust and journal bearings and seals are then presented. Useful design charts are given, and how these design charts were obtained is explained. The chapter ends with brief considerations of normal squeeze film and hydrostatic bearings.

## 12.2 THE REYNOLDS EQUATION

The full Navier-Stokes equation covered in most fluid mechanics texts [e.g., see Currie (1993)] contains inertia, body, pressure, and viscous terms. These equations are sufficiently complicated to prohibit analytical solutions to most practical problems. There is, however, a class of flow conditions, known as *slow viscous motion,* in which the pressure and viscous terms predominate. Fluid film lubrication problems belong to this class.

The differential equation governing the pressure distribution in fluid film lubrication is the **Reynolds equation,** first derived in a remarkable paper by Osborne Reynolds in 1886. This paper not only contained the basic differential equation of fluid film lubrication but also directly compared his theoretical predictions with the experimental results obtained earlier by Tower (1883).

### 12.2.1 DERIVATION OF THE REYNOLDS EQUATION

From an order-of-magnitude analysis the general Navier-Stokes equations [see any fluid mechanics text, e.g., Currie (1993)] can be considerably reduced. If $h_{min}$ is the minimum film thickness and is in the $z$ direction, if $l$ is the length and is in the $x$ direction (the direction of motion), and if $w_t$ is the width and is in the $y$ direction, then in hydrodynamic and hydrostatic bearing situations $l$ and $w_t$ are several (usually 3) orders of magnitude larger than the minimum film thickness. Thus, if terms of order $(h/l)^2$, $(h/w_t)^2$, $h/l$, and $h/w_t$ and smaller are neglected and only first-order terms are considered, the reduced Navier-Stokes equations reduce to

$$\frac{\partial p}{\partial x} = \frac{\partial}{\partial z}\left(\eta \frac{\partial u}{\partial z}\right) \tag{12.1}$$

$$\frac{\partial p}{\partial y} = \frac{\partial}{\partial z}\left(\eta \frac{\partial v}{\partial z}\right) \tag{12.2}$$

$$\frac{\partial p}{\partial z} = 0 \qquad \rightarrow \qquad p = f(x, y, t) \tag{12.3}$$

where $\eta$ = absolute viscosity, Pa·s
$u$ = fluid velocity in $x$ direction, m/s
$v$ = fluid velocity in $y$ direction, m/s
$p$ = pressure, Pa

Equations (12.1) and (12.2) can be directly integrated to give the general expressions for the velocity gradients

$$\frac{\partial u}{\partial z} = \frac{z}{\eta}\frac{\partial p}{\partial x} + \frac{\tilde{A}}{\eta} \tag{12.4}$$

$$\frac{\partial v}{\partial z} = \frac{z}{\eta}\frac{\partial p}{\partial y} + \frac{\tilde{C}}{\eta} \tag{12.5}$$

where $\tilde{A}$ and $\tilde{C}$ are integration constants.

The lubricant viscosity might change considerably across the thin film ($z$ direction) as a result of temperature variations that arise in some bearing problems. In this case, progress toward developing a simple Reynolds equation is considerably complicated. An approach that is satisfactory in most fluid film applications is to treat $\eta$ as the average viscosity across the film. This approach, which does not restrict variation in the $x$ and $y$ directions, is pursued here.

With $\eta$ representing the average viscosity across the film, integrating Equations (12.4) and (12.5) gives the velocity components as

$$u = \frac{z^2}{2\eta}\frac{\partial p}{\partial x} + \tilde{A}\frac{z}{\eta} + \tilde{B} \tag{12.6}$$

and

$$v = \frac{z^2}{2\eta}\frac{\partial p}{\partial y} + \tilde{C}\frac{z}{\eta} + \tilde{D} \tag{12.7}$$

If zero slip at the fluid-solid interface is assumed, the boundary values for the velocity are

1.  $z = 0$    $u = u_b$    $v = v_b$
2.  $z = h$    $u = u_a$    $v = v_a$

The subscripts $a$ and $b$ refer to conditions at the upper and lower surfaces, respectively. Therefore, $u_a$, $v_a$, and $w_a$ refer to the velocity components of the upper surface in the $x$, $y$, and $z$ directions, respectively, and $u_b$, $v_b$, and $w_b$ refer to the velocity components of the lower surface in the same directions.

With the boundary conditions applied to Equations (12.6) and (12.7), the velocity components are

$$u = -z\left(\frac{h-z}{2\eta}\right)\frac{\partial p}{\partial x} + \frac{u_b(h-z)}{h} + \frac{u_a z}{h} \tag{12.8}$$

and

$$v = -z\left(\frac{h-z}{2\eta}\right)\frac{\partial p}{\partial y} + \frac{v_b(h-z)}{h} + \frac{v_a z}{h} \tag{12.9}$$

The volume flow rates per unit width in the $x$ and $y$ directions are defined as

$$q'_x = \int_0^h u\,dz \tag{12.10}$$

$$q'_y = \int_0^h v\,dz \tag{12.11}$$

Substituting Equations (12.8) and (12.9) into these equations gives

$$q'_x = -\frac{h^3}{12\eta}\frac{\partial p}{\partial x} + \frac{h(u_a + u_b)}{2} \tag{12.12}$$

$$q'_y = -\frac{h^3}{12\eta}\frac{\partial p}{\partial y} + \frac{h(v_a + v_b)}{2} \tag{12.13}$$

The first term on the right side of Equations (12.12) and (12.13) represents the **Poiseuille (or pressure) flow,** and the second term represents the **Couette (or velocity) flow.**

From a standard fluid mechanics text, the continuity equation can be expressed as

$$\frac{\partial \rho}{\partial t} + \frac{\partial}{\partial x}(\rho u) + \frac{\partial}{\partial y}(\rho v) + \frac{\partial}{\partial z}(\rho w) = 0 \tag{12.14}$$

where $\rho$ = density, kg/m$^3$
$\quad w$ = squeeze velocity, m/s

The density $\rho$ expressed in Equation (12.14) and the absolute viscosity $\eta$ expressed in Equations (12.8) and (12.9) are functions of $x$ and $y$, but not of $z$. It is convenient to express the continuity equation [Eq. (12.14)] in integral form as

$$\int_0^h \left[ \frac{\partial \rho}{\partial t} + \frac{\partial}{\partial x}(\rho u) + \frac{\partial}{\partial y}(\rho v) + \frac{\partial}{\partial z}(\rho w) \right] dz = 0 \tag{12.15}$$

Now, a general rule of integration is that

$$\int_0^h \frac{\partial}{\partial x}[f(x, y, z)]\, dz = -f(x, y, h)\frac{\partial h}{\partial x} + \frac{\partial}{\partial x}\left[ \int_0^h f(x, y, z)\, dz \right] \tag{12.16}$$

Thus, the $u$ component of Equation (12.15) becomes

$$\int_0^h \frac{\partial}{\partial x}(\rho u)\, dz = -(\rho u)_{z=h}\frac{\partial h}{\partial x} + \frac{\partial}{\partial x}\left( \int_0^h \rho u\, dz \right) = -\rho u_a \frac{\partial h}{\partial x} + \frac{\partial}{\partial x}\left( \rho \int_0^h u\, dz \right) \tag{12.17}$$

Similarly, for the $v$ component

$$\int_0^h \frac{\partial}{\partial y}(\rho v)\, dz = -\rho v_a \frac{\partial h}{\partial y} + \frac{\partial}{\partial y}\left( \rho \int_0^h v\, dz \right) \tag{12.18}$$

The $w$ component term can be integrated directly to give

$$\int_0^h \frac{\partial}{\partial z}(\rho w)\, dz = \rho(w_a - w_b) \tag{12.19}$$

After substituting Equations (12.17) to (12.19) into Equation (12.15), the integrated continuity equation becomes

$$h\frac{\partial \rho}{\partial t} - \rho u_a \frac{\partial h}{\partial x} + \frac{\partial}{\partial x}\left( \rho \int_0^h u\, dz \right) - \rho v_a \frac{\partial h}{\partial y} + \frac{\partial}{\partial y}\left( \rho \int_0^h v\, dz \right) + \rho(w_a - w_b) = 0 \tag{12.20}$$

The integrals in this equation represent the volume flow per unit width ($q_x'$ and $q_y'$) given in Equations (12.12) and (12.13). Introducing these flow rate expressions into the integrated continuity equation yields the general Reynolds equation

$$0 = \frac{\partial}{\partial x}\left(-\frac{\rho h^3}{12\eta}\frac{\partial p}{\partial x}\right) + \frac{\partial}{\partial y}\left(-\frac{\rho h^3}{12\eta}\frac{\partial p}{\partial y}\right) + \frac{\partial}{\partial x}\left[\frac{\rho h(u_a + u_b)}{2}\right] + \frac{\partial}{\partial y}\left[\frac{\rho h(v_a + v_b)}{2}\right]$$

$$+ \rho(w_a - w_b) - \rho u_a\frac{\partial h}{\partial x} - \rho v_a\frac{\partial h}{\partial y} + h\frac{\partial \rho}{\partial t} \tag{12.21}$$

---

**EXAMPLE 12.1**

**Given:**  A pump piston without rings pumps an oil with a viscosity of 0.143 Pa · s at a pressure differential of 20 MPa. The piston moves concentrically in the cylinder so that the radial clearance is constant around the piston. The piston is 100 mm in diameter and 80 mm long, and the radial clearance is 50 $\mu$m. The piston moves with a speed of 0.2 m/s.

**Find:**  Determine the leakage past the piston when

(a)  The piston pumps the oil (as a pump)
(b)  The oil drives the piston (as a hydraulic cylinder)

**Solution:**  The pressure gradient is

$$\frac{\partial p}{\partial x} = \frac{\Delta p}{\Delta x} = \frac{(20)(10^6)}{0.080} = 2.5 \times 10^8 \text{ N/m}^3 \tag{a}$$

(a)  When the piston pumps the oil, the cylinder wall passes the piston from high pressure to low pressure. From Equation (12.12) the pressure flow and the motion flow together are

$$q = \pi dq'_x = \pi d\left[\frac{h^3}{12\eta}\frac{\partial p}{\partial x} + \frac{h(u_a + u_b)}{2}\right]$$

$$= \pi(0.100)\left[\frac{(50)^3(10^{-18})(2.5)(10^8)}{(12)(0.143)} + \frac{(50)(10^{-6})(0.2)}{2}\right]$$

$$= 7.292 \times 10^{-6} \text{ m}^3/\text{s}$$

(b)  When the oil drives the piston, the pressure flow is opposite to the flow from the motion.

$$q = \pi d\left[\frac{h^3}{12\eta}\frac{dp}{dx} - \frac{h(u_a + u_b)}{2}\right]$$

$$= \pi(0.100)\left[\frac{(50)^3(10^{-18})(2.5)(10^8)}{12(0.143)} - \frac{(50)(10^{-6})(0.2)}{2}\right]$$

$$= 4.150 \times 10^{-6} \text{ m}^3/\text{s}$$

Thus, when the piston pumps the oil, the leakage is 7.292 cm³/s; and when the oil drives the piston, the leakage is 4.150 cm³/s. Therefore, the cylinder wall motion pulls oil into the high-pressure region.

## 12.2.2 PHYSICAL SIGNIFICANCE OF TERMS IN THE REYNOLDS EQUATION

The first two terms in Equation (12.21) are the Poiseuille terms, and they describe the net flow rates due to pressure gradients within the lubricated area; the third and fourth terms are the Couette terms, and they describe the net entraining flow rates due to surface velocities. The fifth to seventh terms describe the net flow rates due to a squeezing motion, and the last term describes the net flow rate due to local expansion. The flows or "actions" can be considered, without any loss of generality, by eliminating the side-leakage terms $(\partial/\partial y)$ in Equation (12.21).

$$\underbrace{\frac{\partial}{\partial x}\left(\frac{\rho h^3}{12\eta}\frac{\partial p}{\partial x}\right)}_{\text{Poiseuille}} = \underbrace{\frac{\partial}{\partial x}\left[\frac{\rho h(u_a + u_b)}{2}\right]}_{\text{Couette}} + \underbrace{\rho\left(w_a - w_b - u_a\frac{\partial h}{\partial x}\right)}_{\text{squeeze}} + \underbrace{h\frac{\partial \rho}{\partial t}}_{\text{local expansion}}$$

$$\underbrace{\frac{h(u_a + u_b)}{2}\frac{\partial \rho}{\partial x}}_{\text{density wedge}} \qquad \underbrace{\frac{\rho h}{2}\frac{\partial}{\partial x}(u_a + u_b)}_{\text{stretch}} \qquad \underbrace{\frac{\rho(u_a + u_b)}{2}\frac{\partial h}{\partial x}}_{\text{physical wedge}} \qquad (12.22)$$

As can be seen in Equation (12.22), the Couette terms lead to three distinct actions. The physical significance of each term within the Reynolds equation is now discussed in detail.

***Density Wedge Term*** $\dfrac{h(u_a + u_b)}{2}\dfrac{\partial \rho}{\partial x}$

**Density wedge** is concerned with the rate at which lubricant density changes in the sliding direction, as shown in Figure 12.1. If the lubricant density decreases in the sliding direction, the Couette mass flows differ for each location at the three velocity profiles of Figure 12.1. For continuity of mass flow, this discrepancy must be eliminated by generating a balancing Poiseuille flow.

Note from Figure 12.1 that the density must decrease in the sliding direction if positive pressures are to be generated. This effect could be introduced by raising the temperature of the lubricant as it passes through the bearing. The density wedge (sometimes called the *thermal wedge*) mechanism is not important in most bearings; however, it has been suggested that it may play a significant role in the performance of parallel-surface thrust bearings where the major pressure-generating actions are absent.

***Stretch Term*** $\dfrac{\rho h}{2}\dfrac{\partial(u_a + u_b)}{\partial x}$

The **stretch term** concerns the rate at which the surface velocity changes in the sliding direction. This effect is produced if the bounding solids are elastic and the extent to which the surfaces are stretched varies through the bearing. For positive pressures to be developed, the surface velocities have to decrease in the sliding direction, as shown in Figure 12.2. Stretch action is not encountered in conventional bearings.

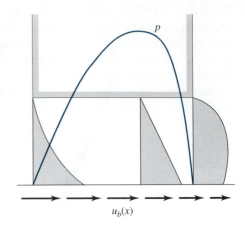

**Figure 12.1**    Density wedge mechanism. [From Hamrock et al. (2004).]

**Figure 12.2**    Stretch mechanism. [From Hamrock et al. (2004).]

*Physical Wedge Term* $\dfrac{\rho(u_a + u_b)}{2}\dfrac{\partial h}{\partial x}$

**Physical wedge** action is very important and is the best known mechanism for generating pressure. Figure 12.3 illustrates this action for a plane slider and a stationary bearing pad. At each of the three sections considered, the Couette volume flow rate is proportional to the area of the triangle of height $h$ and base $u$. Since $h$ varies along the bearing, there is a different Couette flow rate at each section, and flow continuity can be achieved only if a balancing Poiseuille flow is superimposed. For a positive load-carrying capacity the thickness of the lubricant film must decrease in the sliding direction.

*Normal Squeeze Term* $\rho(w_a - w_b)$

**Normal squeeze** action provides a valuable cushioning effect when bearing surfaces tend to be pressed together. Positive pressures will be generated when the film thickness is diminishing. The physical wedge and normal squeeze actions are the two major pressure-generating mechanisms in hydrodynamic or self-acting fluid film bearings. In the absence of sliding the squeeze effect arises directly from the difference in normal velocities $(w_a - w_b)$, as illustrated in Figure 12.4. Positive pressures will clearly be achieved if the film thickness is decreasing $(w_b > w_a)$.

*Translation Squeeze Term* $-\rho u_a \dfrac{\partial h}{\partial x}$

**Translation squeeze** action results from the translation of inclined surfaces. The local film thickness may be reduced by the sliding of the inclined bearing surface, as shown in Figure 12.5. The rate at which the film thickness decreases is shown in the figure. In this case, the pressure profile is moving over the space covered by the fixed coordinate system, the pressure at any fixed point being a function of time.

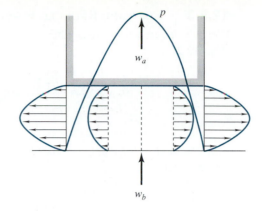

**Figure 12.3**   Physical wedge mechanism.
[From Hamrock et al. (2004).]

**Figure 12.4**   Normal squeeze mechanism.
[From Hamrock et al. (2004).]

## *Local Expansion Term* $h \dfrac{\partial \rho}{\partial t}$

The local time rate of density change governs the **local expansion** term. The pressure-generating mechanism can be visualized by considering the thermal expansion of the lubricant contained between stationary bearing surfaces, as shown in Figure 12.6. If heat is supplied to the lubricant, it will expand and the excess volume will have to be expelled from the space between the bearing surfaces. In the absence of surface velocities, the excess lubricant volume must be expelled by a pressure (Poiseuille) flow action. Pressures are thus generated in the lubricant, and for a positive load-carrying capacity, $\partial \rho / \partial t$ must be negative (i.e., the volume of a given mass of lubricant must increase). Local expansion, which is a transient mechanism of pressure generation, is generally of no significance in bearing analysis.

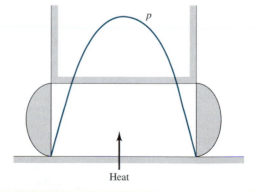

**Figure 12.5**   Translation squeeze mechanism.
[From Hamrock et al. (2004).]

**Figure 12.6**   Local expansion mechanism.
[From Hamrock et al. (2004).]

### 12.2.3    STANDARD REDUCED FORMS

For only tangential motion, where $w_a = u_a\,\partial h/\partial x$ and $w_b = 0$, the Reynolds equation [Eq. (12.21)] becomes

$$\frac{\partial}{\partial x}\left(\frac{\rho h^3}{\eta}\frac{\partial p}{\partial x}\right) + \frac{\partial}{\partial y}\left(\frac{\rho h^3}{\eta}\frac{\partial p}{\partial y}\right) = 12\tilde{u}\frac{\partial(\rho h)}{\partial x} \qquad (12.23)$$

where

$$\tilde{u} = \frac{u_a + u_b}{2} = \text{constant}$$

This equation is applicable for elastohydrodynamic lubrication. For hydrodynamic lubrication the fluid properties do not vary significantly throughout the bearing and may be considered to be a constant. Thus, the corresponding Reynolds equation is

$$\frac{\partial}{\partial x}\left(h^3\frac{\partial p}{\partial x}\right) + \frac{\partial}{\partial y}\left(h^3\frac{\partial p}{\partial y}\right) = 12\tilde{u}\eta_0\frac{\partial h}{\partial x} \qquad (12.24)$$

Equation (12.23) not only allows the fluid properties to vary in the $x$ and $y$ directions but also permits the bearing surfaces to be of finite length in the $y$ direction. Side leakage, or flow in the $y$ direction, is associated with the second term in Equations (12.23) and (12.24). If the pressure in the lubricant film has to be considered as a function of $x$ and $y$, Equation (12.23) can rarely be solved analytically.

The Reynolds equation that is valid for gas-lubricated bearings is discussed next. The equation of state for a perfect gas is

$$p = \rho \bar{R} t_m \qquad (12.25)$$

where $\bar{R}$ = gas constant, $m^2/(s^2 \cdot K)$
$\quad\ t_m$ = absolute temperature, K

Therefore, from Equation (12.25),

$$\rho = \frac{p}{\bar{R} t_m} \qquad (12.26)$$

Substituting this equation into Equation (12.23) yields the Reynolds equation normally used for gas-lubricated bearings for tangential motion only. Because the viscosity of a gas does not vary much, it can be considered to be constant. Thus,

$$\frac{\partial}{\partial x}\left(p h^3\frac{\partial p}{\partial x}\right) + \frac{\partial}{\partial y}\left(p h^3\frac{\partial p}{\partial y}\right) = 12\tilde{u}\eta_0\frac{\partial(p h)}{\partial x} \qquad (12.27)$$

## 12.3    THRUST SLIDER BEARINGS

The surfaces of **thrust bearings** are perpendicular to the axis of rotation. A hydrodynamically lubricated slider bearing develops load-carrying capacity by virtue of the relative motion of the two surfaces separated by a fluid film. A slider bearing may be viewed as a bearing that develops a positive pressure due to the physical wedge mechanism presented in Section 12.2.2. Thrust slider bearings are considered in this section and journal slider bearings in Section 12.4. The processes occurring in a bearing with fluid film lubrication can be better understood by considering qualitatively the development of oil pressure in a thrust slider bearing.

### 12.3.1 MECHANISM OF PRESSURE DEVELOPMENT

An understanding of the development of load-supporting pressures in hydrodynamic bearings can be gleaned by considering, from a purely physical point of view, the conditions of geometry and motion required to develop pressure. An understanding of the physical situation can make the mathematics of hydrodynamic lubrication much more meaningful. By considering only what must happen to maintain continuity of flow, much of what the mathematical equations tell us later in this chapter can be deduced.

Figure 12.7 shows velocity profiles for two plane surfaces separated by a constant lubricating film thickness. The plates are so wide that side-leakage flow (into and out of the paper) can be neglected. The upper plate is moving with a velocity $u_a$, and the bottom plate is held stationary. No slip occurs at the surfaces. The velocity varies uniformly from 0 at surface AB to $u_a$ at surface A'B', thus implying that the rate of shear $du/dz$ throughout the oil film is constant. The volume of fluid flowing across section AA' per unit time has to be equal to that flowing across section BB'. The flow crossing the two boundaries results only from velocity gradients, and since the gradients are equal, the flow continuity requirement is satisfied without any pressure buildup within the film. Because the ability of a lubricating film to support a load depends on the pressure buildup in the film, a slider bearing with parallel surfaces is not able to support a load by a fluid film. If any load is applied to the surface AB, the lubricant will be squeezed out and the bearing will operate under conditions of boundary lubrication.

Consider now the case of two nonparallel plates as shown in Figure 12.8(a). Again, the plates are wide in the direction perpendicular to the motion, so that lubricant flow in this direction is negligibly small. The volume of lubricant that the surface A'B' tends to carry into the space between the surfaces AB and A'B' through section AA' during unit time is AC'A'. The volume of lubricant that the surface tends to discharge from the space through section BB' during the same time is BD'B'. Because the distance AA' is greater than the distance BB', the volume AC'A' is greater than the volume BD'B' by the volume AEC'. From flow continuity the actual volume of lubricant carried into the space must equal the volume discharged from this space.

It can be surmised that there will be a pressure buildup in the lubricating film until flow continuity is satisfied. Figure 12.8(b) shows the velocity profiles due to Poiseuille flow. The flow is outward from both the leading and trailing edges of the bearing because flow is always from a region of higher pressure to a region of lower pressure. The pressure flow at boundary AA' opposes the velocity flow, but the pressure flow at BB' is in the same direction as the velocity flow.

Figure 12.8(c) shows the results of superimposing Couette and Poiseuille flows. The form of the velocity distribution curves obtained in this way must satisfy the condition that the flow rate through section AA' equal the flow rate through section BB'. Therefore, area AHC'A' must equal area BID'B'. The area between the straight, dashed line AC' and

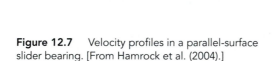

**Figure 12.7**   Velocity profiles in a parallel-surface slider bearing. [From Hamrock et al. (2004).]

**Figure 12.8**   Flow within a fixed-incline slider bearing. (a) Couette flow; (b) Poiseuille flow; (c) resulting velocity profile. [From Hamrock et al. (2004).]

the curve AHC′ in section AA′ and the area between the dashed line BD′ and curve BID′ represent the pressure-induced flow through these areas.

The pressure is maximum in section JJ′, somewhere between sections AA′ and BB′. There is no Poiseuille flow contribution in section JJ′, since the pressure gradient is zero at this location, and all the flow is Couette. Flow continuity is satisfied in that triangle JK′J′ is equal to areas AHC′A′ and BID′B′.

## 12.3.2   GENERAL THRUST SLIDER BEARING THEORY

Solutions of the Reynolds equation presented in Section 12.2 for actual bearing configurations are usually obtained in approximate numerical form. Analytical solutions are possible only for the simplest problems. By restricting the flow to two dimensions (say, the $xz$ plane) analytical solutions for many common bearings become available. The quantitative value of these solutions is limited, since flow in the third dimension $y$, known as side leakage, plays an important part in fluid film bearing performance. The two-dimensional solutions have a definite value because they provide a good deal of information about the general characteristics of bearings, information that leads to a clear physical picture of the performance of the lubricant films. The focus of this text is on design and thus this will not be discussed, but further discussion can be found in Hamrock et al. (2004). This text describes the relevant equations used for finite-width bearings, numerically evaluates them, and then obtains useful design charts.

Although side-leakage effects are considered in the results shown in this section, a simplification is to neglect the pressure and temperature effects on the lubricant properties, namely, viscosity and density. Lubricant viscosity is particularly sensitive to temperature (as demonstrated in Table 8.4 and Fig. 8.17), and since the heat generated in hydrodynamic bearings is often considerable, the limitation imposed by this simplification is at once apparent. The temperature rise within the film can be calculated if it is assumed that all the heat produced by the viscous action is carried away by the lubricant (the adiabatic assumption).

For finite-width bearings having hydrodynamic pressure generated within the lubricant film due to the physical wedge mechanism (see Sec. 12.2.2), some flow will take place in the $y$ direction. By using the adiabatic assumption (implying that the heat produced in the bearing is carried away by the lubricant), the viscosity and density can be considered constant within the $xy$ plane. The appropriate Reynolds equation from Equation (12.24) when the velocity of the top surface is zero ($u_a = 0$) is

$$\frac{\partial}{\partial x}\left(h^3\frac{\partial p}{\partial x}\right) + \frac{\partial}{\partial y}\left(h^3\frac{\partial p}{\partial y}\right) = 6\eta_0 u_b \frac{\partial h}{\partial x} \qquad \textbf{(12.28)}$$

Thus, for a specific film shape $h(x, y)$ the pressure throughout the bearing can be obtained numerically. Knowing the pressure from the Reynolds equation and integrating over the $xy$ plane produce the load-carrying capacity.

### 12.3.3   HYDRODYNAMIC THRUST BEARINGS—NEGLECTING SIDE LEAKAGE

Many loads carried by rotary machinery have components that act in the direction of the shaft's axis of rotation. These thrust loads are frequently carried by self-acting or hydrodynamic bearings of the form shown in Figure 12.9. A thrust plate attached to, or forming part of, the rotating shaft is separated from the sector-shaped bearing pads by a lubricant film. The load-carrying capacity of the bearing arises entirely from the pressures generated by the geometry of the thrust plate over the bearing pads. The $x$ direction (direction of motion) is along a mean arc AA′, as shown in Figure 12.9. The transverse direction (side-leakage direction) would be perpendicular to arc AA′. All pads are identical and separated by deep lubrication grooves, implying that atmospheric pressure exists completely around the pad. Thus, the total thrust load-carrying capacity is simply the number of pads multiplied by the load-carrying capacity of an individual pad.

This section will consider three different slider bearings while neglecting side leakage: parallel slider, fixed-incline slider, and parallel-step slider. These three slider bearings will have three different film shapes across arc AA′ in Figure 12.9.

**Figure 12.9**    Thrust slider bearing geometry. [From Hamrock et al. (2004)]

If side leakage is neglected, Equation (12.28) reduces to

$$\frac{\partial}{\partial x}\left(h^3 \frac{\partial p}{\partial x}\right) = 6\eta_0 u_b \frac{\partial h}{\partial x} \qquad (12.29)$$

This, then, is the Reynolds equation used for the three types of slider bearing being considered.

### Parallel-Surface Slider Bearing

The film thickness for the parallel-surface slider bearing is constant across the arc length AA′ shown in Figure 12.9. Equation (12.29) becomes

$$\frac{d}{dx}\left(\frac{dp}{dx}\right) = 0$$

Integrating twice gives

$$p = \bar{A}x + \bar{B}$$

where $\bar{A}$ and $\bar{B}$ = integration constants. The boundary conditions are

1.  $p = 0$ at $x = 0 \rightarrow \bar{B} = 0$
2.  $p = 0$ at $x = l \rightarrow \bar{A} = 0$

Therefore, the pressure is zero throughout the bearing. The normal load-carrying capacity of the bearing is the integral of the pressure across the length, and since the pressure is zero across the length, the normal applied load would also be zero.

## Fixed-Incline Slider Bearing

The film shape equation for the fixed-incline slider bearing is

$$h = h_0 + s_h\left(1 - \frac{x}{l}\right) \tag{12.30}$$

When $x = 0$, $h = h_o + s_h$; and when $x = l$, $h = h_o$.

Equation (12.29) can be integrated to give

$$\frac{dp}{dx} = \frac{6\eta_0 u_b}{h^2} + \frac{\bar{A}}{h^3}$$

where $\bar{A}$ = integration constant. The boundary condition to be applied is

$$h = h_m \qquad \text{and} \qquad \frac{dp}{dx} = 0 \tag{12.31}$$

which implies that $\bar{A} = -6h_0 u_b h_m$

$$\therefore \qquad \frac{dp}{dx} = \frac{6\eta_0 u_b}{h^3}(h - h_m) \tag{12.32}$$

This equation can be written in dimensionless form if

$$P = \frac{p s_h^2}{\eta_0 u_b l} \qquad H = \frac{h}{s_h} \qquad H_m = \frac{h_m}{s_h} \qquad H_0 = \frac{h_o}{s_h} \qquad X = \frac{x}{l}$$

as

$$\frac{dP}{dX} = 6\frac{H - H_m}{H^3} \tag{12.33}$$

where

$$H = \frac{h}{s_h} = H_o + 1 - X \tag{12.34}$$

$$\frac{dH}{dX} = -1 \rightarrow dH = -dX \tag{12.35}$$

Note that Equations (12.34) and (12.35) are valid only for a fixed-incline bearing.

Integrating Equation (12.33) gives

$$P = 6\int\left(\frac{1}{H^2} - \frac{H_m}{H^3}\right)dX = -6\int\left(\frac{1}{H^2} - \frac{H_m}{H^3}\right)dH$$

$$\therefore \qquad P = 6\left(\frac{1}{H} - \frac{H_m}{2H^2}\right) + \bar{B} \tag{12.36}$$

The boundary conditions are

1.  $P = 0$ when $X = 0 \rightarrow H = H_o + 1$
2.  $P = 0$ when $X = 1 \rightarrow H = H_o$

Making use of these boundary conditions gives

$$H_m = \frac{2H_o(1 + H_o)}{1 + 2H_o} \qquad (12.37)$$

$$\bar{B} = -\frac{6}{1 + 2H_o} \qquad (12.38)$$

Substituting Equations (12.37) and (12.38) into Equation (12.36) gives

$$P = \frac{6X(1 - X)}{(H_0 + 1 - X)^2(1 + 2H_o)} \qquad (12.39)$$

The normal applied load per unit width is

$$W_z' = \frac{W_z}{w_t} = \int_0^l p \, dx = \frac{\eta_0 u_b l^2}{s_h^2} \int_0^1 P \, dX$$

The dimensionless normal load is

$$\therefore \qquad \bar{W}_z = \frac{W_z s_h^2}{\eta_0 u_b l^2 w_t} = \int_0^1 P \, dX \qquad (12.40)$$

Recall that $dH = -dX$, and from Equation (12.34) note that if

$$X = 0 \rightarrow H = H_0 + 1$$

then

$$X = 1 \rightarrow H = H_0$$

Thus, Equation (12.40) becomes

$$\bar{W}_z = \frac{W_z s_h^2}{\eta_0 u_b l^2 w_t} = \int_{H_0 + 1}^{H_0} -P \, dH \qquad (12.41)$$

Substituting Equations (12.36) and (12.38) into Equation (12.41) gives

$$\bar{W}_z = \frac{W_z s_h^2}{\eta_0 u_b l^2 w_z} = 6 \ln \frac{H_0 + 1}{H_0} - \frac{12}{1 + 2H_0} \qquad (12.42)$$

### Parallel-Step Slider Bearing

For the parallel-step slider bearing, the film thickness in each of the two regions is constant, implying that the appropriate Reynolds equation is

$$\frac{d^2 p}{dx^2} = 0$$

Integrating gives

$$\frac{dp}{dx} = \text{constant}$$

However, since the film thicknesses between regions differ, their pressure gradients also differ; thus there will be no discontinuity at the step.

$$p_{max} = n_s l \left(\frac{dp}{dx}\right)_i = -(l - n_s l)\left(\frac{dp}{dx}\right)_o \tag{12.43}$$

The inlet and outlet flow rates at the step must be the same.

$$\therefore \qquad q'_{xo} = q'_{xi}$$

Making use of Equation (12.12) gives

$$-\frac{h_o^3}{12\eta_0}\left(\frac{dp}{dx}\right)_o + \frac{u_b h_o}{2} = -\frac{(h_0 + s_h)^3}{12\eta_0}\left(\frac{dp}{dx}\right)_i + \frac{u_b(h_o + s_h)}{2} \tag{12.44}$$

Equations (12.43) and (12.44) represent a pair of simultaneous equations with unknowns $(dp/dx)_o$ and $(dp/dx)_i$. The solutions are

$$\left(\frac{dp}{dx}\right)_i = \frac{6\eta_0 u_b(1 - n_s)s_h}{(1 - n_s)(h_o + s_h)^3 + n_s h_o^3} \tag{12.45}$$

$$\left(\frac{dp}{dx}\right)_o = -\frac{6\eta_0 u_b n_s s_h}{(1 - n_s)(h_o + s_h)^3 + n_s h_o^3} \tag{12.46}$$

The maximum pressure at the step can be found by substituting Equations (12.45) and (12.46) into Equation (12.43) to give

$$p_{max} = \frac{6\eta_0 u_b l n_s(1 - n_s)s_h}{(1 - n_s)(h_o + s_h)^3 + n_s h_o^3} \tag{12.47}$$

$$P_{max} = \frac{p_{max} s_h^2}{\eta_0 u_b l} = \frac{6n_s(1 - n_s)}{(1 - n_s)(H_o + 1)^3 + n_s H_o^3} \tag{12.48}$$

The pressures in the inlet and outlet regions are

$$P_i = \frac{X P_{max}}{n_s} = \frac{6X(1 - n_s)}{(1 - n_s)(H_o + 1)^3 + n_s H_o^3} \qquad 0 \leq X \leq n_s \tag{12.49}$$

$$P_o = \frac{(1 - X)P_{max}}{1 - n_s} = \frac{6(1 - X)n_s}{(1 - n_s)(H_o + 1)^3 + n_s H_o^3} \qquad n_s \leq X \leq 1 \tag{12.50}$$

The normal applied load per unit width is

$$\frac{W_z}{w_t} = \frac{p_{max} l}{2} = \frac{3\eta_0 u_b l^2 n_s(1 - n_s)s_h}{(1 - n_s)(h_o + s_h)^3 + n_s h_o^3} \tag{12.51}$$

The dimensionless normal applied load is

$$\bar{W}_z = \frac{W_z}{\eta_0 u_b w_t}\left(\frac{s_h}{l}\right)^2 = \frac{3n_s(1 - n_s)}{(1 - n_s)(H_o + 1)^3 + n_s H_o^3} = \frac{P_{max}}{2} \tag{12.52}$$

The largest $\bar{W}_z$ is obtained when $n_s = 0.7182$ and $H_o = 0.5820$.

**EXAMPLE 12.2**

**Given:** A fixed-incline self-acting thrust bearing with the width of the pad being much larger than the length. The viscosity of the lubricant is 0.01 N·s/m², the sliding velocity is 10 m/s, the pad length is 0.3 m, the minimum film thickness is 15 $\mu$m, and the inlet film thickness is twice the outlet film thickness.

**Find:** The magnitude and location of the maximum pressure in the bearing.

**Solution:** For the width of the pad much larger than the length, side leakage can be neglected and the results given in Equations (12.37) and (12.39) are valid. Also the pressure is a maximum when $dP/dX = 0$, $H = H_m$, and $X = X_m$. We are given that $H_o = 1$. From Equation (12.37)

$$H_m = \frac{2H_o(1 + H_o)}{1 + 2H_o} = \frac{(2)(2)}{3} = \frac{4}{3}$$

From Equation (12.34),

$$X_m = H_o + 1 - H_m = 1 + 1 - \frac{4}{3} = \frac{2}{3}$$

From Equation (12.39),

$$P_m = \frac{6X_m(1 - X_m)}{(H_o + 1 - X_m)^2(1 + 2H_o)} = \frac{6\left(\frac{2}{3}\right)\left(1 - \frac{2}{3}\right)}{\left(1 + 1 - \frac{2}{3}\right)^2(1 + 2)} = \frac{4/3}{(4/3)^2(3)} = \frac{1}{4}$$

Thus,

$$P_m = \frac{p_m s_h^2}{\eta_0 u_b l} = \frac{1}{4} \quad \text{or} \quad p_m = \frac{\eta_0 u_b l}{4s_h^2}$$

Given that $\eta_0 = 0.01$ N·s/m², $u_b = 10$ m/s, $l = 0.3$ m, and $s_h = h_o = 15$ $\mu$m,

$$p_m = \frac{(0.01)(10)(0.3)}{4(15)^2(10^{-12})} \text{ Pa} = 33.3 \text{ MPa}$$

### 12.3.4    OPERATING AND PERFORMANCE PARAMETERS

The last section neglected side-leakage effects in the design of slider bearings; this section will now consider these effects. The following general definitions, related to the bearing geometry shown in Figure 12.10, and relationships are employed: The forces acting on the bearing surfaces of an individual pad can be considered in two groups. The loads, which act in the direction normal to the surface, yield normal loads that can be resolved into components $W_x$ and $W_z$. The viscous surface stresses, which act in the direction tangent to the surface, yield shear forces on the solids that have components $F$ in the $x$ direction. The shear forces in the $z$ direction are negligible. Once the pressure is obtained from the Reynolds equation [Eq. (12.28)], the force components acting on the bearing surfaces are

$$W_{za} = W_{zb} = \int_0^{w_t} \int_0^l p \, dx \, dy \qquad (12.53)$$

$$W_{xb} = 0 \qquad (12.54)$$

**Figure 12.10** Force components and oil film geometry in a hydrodynamically lubricated thrust slider bearing. [From Hamrock et al. (2004).]

$$W_{xa} = -\int_0^{w_t} \int_{h_o+s_h}^{h_o} p\,dh\,dy = -\int_0^{w_t} \int_0^l p\frac{dh}{dx}\,dx\,dy$$

$$= -\int_0^{w_t} (ph)_0^l\,dy + \int_0^{w_t}\int_0^l h\frac{dp}{dx}\,dx\,dy = \int_0^{w_t}\int_0^l h\frac{dp}{dx}\,dx\,dy \qquad (12.55)$$

$$W_b = \sqrt{W_{zb}^2 + W_{xb}^2} = W_{zb} \qquad (12.56)$$

$$W_a = \sqrt{W_{za}^2 + W_{xa}^2} \qquad (12.57)$$

$$\Phi = \tan^{-1}\frac{W_{xa}}{W_{za}} \qquad (12.58)$$

The shear force acting on solid $b$ is

$$F_b = \int_0^{w_t}\int_0^l (\tau_{zx})_{z=0}\,dx\,dy \qquad (12.59)$$

but

$$(\tau_{zx})_{z=0} = \left(\eta\frac{du}{dz}\right)_{z=0} \qquad (12.60)$$

Substituting Equation (12.8) into Equation (12.60) gives

$$(\tau_{zx})_{z=0} = -\frac{h}{2}\frac{dp}{dx} - \frac{\eta u_b}{h} \qquad (12.61)$$

Substituting Equation (12.61) into Equation (12.59) gives

$$F_b = \int_0^{w_t}\int_0^l \left(-\frac{h}{2}\frac{dp}{dx} - \frac{\eta u_b}{h}\right)dx\,dy$$

Making use of Equation (12.55) gives

$$F_b = -\frac{W_{xa}}{2} - \int_0^{w_t} \int_0^l \frac{\eta u_b}{h} \, dx \, dy \tag{12.62}$$

Similarly, the shear force acting on solid $a$ is

$$F_a = \int_0^{w_t} \int_0^l (\tau_{zx})_{z=h} \, dx \, dy = -\frac{W_{xa}}{2} + \int_0^{w_t} \int_0^l \frac{\eta u_b}{h} \, dx \, dy \tag{12.63}$$

Note from Figure 12.10 that

$$F_b - F_a + W_{xa} = 0 \tag{12.64}$$

$$W_{zb} - W_{za} = 0 \tag{12.65}$$

These equations represent the condition of static equilibrium.

The viscous stresses generated by the shearing of the lubricant film give rise to a resisting force of magnitude $-F_b$ on the moving surface. The rate of working against the viscous stresses, or the power loss, for one pad is

$$h_p = -F_b u_b \tag{12.66}$$

The work done against the viscous stresses appears as heat within the lubricant. Some of this heat may be transferred to the surroundings by radiation or by conduction, or it may be convected from the clearance space by the lubricant flow.

The bulk temperature rise of the lubricant for the case in which all the heat is carried away by convection is known as the **adiabatic temperature rise.** This bulk temperature increase can be calculated by equating the rate of heat generated within the lubricant to the rate of heat transferred by convection

$$h_p = \rho q_x C_p \, \Delta t_m$$

or the adiabatic temperature rise may be expressed as

$$\Delta t_m = \frac{h_p}{\rho q_x C_p} \tag{12.67}$$

where $\rho$ = density, kg/m$^3$

$\quad q_x$ = volume flow rate in $x$ direction (direction of motion), m$^3$/s

$\quad C_p$ = specific heat of lubricant at constant pressure, J/(kg $\cdot$ °C)

Equations (12.53) to (12.67) are among the relevant equations used in thrust bearing analysis. Recall that the pressure-generating mechanism used in this section and the next (Secs. 12.3 and 12.4) is the physical wedge, as discussed in Section 12.2.2. The general thrust analysis will now be used to design two types of thrust bearing, the fixed-incline and pivoted-pad bearings. A detailed numerical evaluation will not be made of the Reynolds equation and its load and friction force components, since the emphasis of this text is on design. [If the reader is interested, two references are recommended. The original source is Raimondi and Boyd (1955); a more recent text is Hamrock et al. (2004).]

The operating parameters affecting the pressure generation, film thickness, load, and friction components are

1. **Bearing number** for a thrust slider bearing

$$B_t = \left( \frac{\eta_0 u_b w_t}{W_z} \right) \left( \frac{l}{s_h} \right)^2 \tag{12.68}$$

2. Length-to-width ratio

$$\lambda = \frac{l}{w_t} \tag{12.69}$$

3. Location of pivot from inlet

$$X_p = \frac{x_p}{l} \tag{12.70}$$

For a fixed-incline bearing only the first two operating parameters affect the pressure generation, film thickness, load, and friction force components. The operating parameters are viewed as the parameters given in the design of a hydrodynamic thrust bearing. The performance parameters that will result from choosing a given set of operating parameters are

1. Outlet (minimum) film thickness $h_o$
2. Temperature rise due to lubricant shearing $\Delta t_m$
3. Power loss $h_p$
4. Coefficient of sliding friction $\mu$
5. Circumferential and side-leakage volumetric flow rates $q$ and $q_s$

These performance parameters are needed to evaluate the design of a hydrodynamic thrust slider bearing.

## 12.3.5 FIXED-INCLINE SLIDER BEARING

The simplest form of fixed-incline bearing provides only straight-line motion and consists of a flat surface sliding over a fixed pad or land having a profile similar to that shown in Figure 12.11. The fixed-incline bearing depends for its operation on the lubricant being drawn into a wedge-shaped space and, thus, producing pressure that counteracts the load and prevents contact between the sliding parts. Since the wedge action takes place only when the sliding surface moves in the direction in which the lubricant film converges, the fixed-incline bearing (Fig. 12.11) can carry load only in this direction. If reversibility is desired, a combination of two or more pads with their surfaces sloped in opposite directions is required. Fixed-incline pads are used in multiples as in the thrust bearing shown in Figure 12.12.

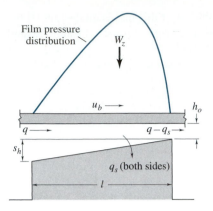

**Figure 12.11**    Side view of fixed-incline slider bearing. [From Raimondi and Boyd (1955).]

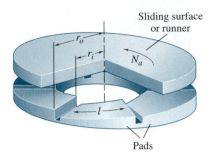

**Figure 12.12**    Configuration of multiple fixed-incline thrust slider bearing. [From Raimondi and Boyd (1955).]

## PROCEDURE FOR DESIGNING A FIXED-INCLINE THRUST BEARING

1.  Choose a pad length-to-width ratio. A square pad ($\lambda = 1$) is generally thought to give good performance. If it is known whether maximum load or minimum power loss is more important in a particular application, the outlet film thickness ratio $H_o$ can be determined from Figure 12.13.

2.  Once $\lambda$ and $H_o$ are known, Figure 12.14 can be used to obtain the bearing number $B_t$.

3.  From Figure 12.15 determine the temperature rise due to shear heating for a given $\lambda$ and $B_t$. The volumetric specific heat $C_s = \rho C_p$, which is the dimensionless temperature rise parameter, is relatively constant for mineral oils and is equivalent to $1.36 \times 10^6$ N/(m² · °C).

4.  Determine lubricant temperature. Mean temperature can be expressed as

$$\tilde{t}_m = t_{mi} + \frac{\Delta t_m}{2} \tag{12.71}$$

where $t_{mi}$ = inlet temperature, °C. The inlet temperature is usually known beforehand. Once the mean temperature $\tilde{t}_m$ is known, it can be used in Figure 8.17 to determine the viscosity of SAE oils, or Figure 8.16 or Table 8.4 can be used. In using Table 8.4 if the temperature is different from the three temperatures given, a linear interpolation can be used.

5.  Make use of Equations (12.34) and (12.68) to get the outlet (minimum) film thickness $h_0$ as

$$h_o = H_o l \sqrt{\frac{\eta_0 u_b w_t}{W_z B_t}} \tag{12.72}$$

Once the outlet film thickness is known, the shoulder height $s_h$ can be directly obtained from $s_h = h_o/H_o$. If in some applications the outlet film thickness is specified and either the velocity $u_b$ or the normal applied load $W_z$ is not known, Eq. (12.72) can be rewritten to establish $u_b$ or $W_z$.

6. Check Table 12.1 to see if the outlet (minimum) film thickness is sufficient for the pressurized surface finish. If $(h_o)_{\text{Eq. (12.72)}} \geq (h_o)_{\text{Table 12.1}}$, go to step 7. If $(h_o)_{\text{Eq. (12.72)}} < (h_o)_{\text{Table 12.1}}$, consider one or both of the following steps:

(a) Increase the bearing speed.

(b) Decrease the load, the surface finish, or the inlet temperature. Upon making this change, return to step 3.

7. Evaluate the other performance parameters. Once an adequate minimum film thickness and a proper lubricant temperature have been determined, the performance parameters can be evaluated. Specifically, from Figure 12.16 the power loss, the coefficient of friction, and the total and side flows can be determined.

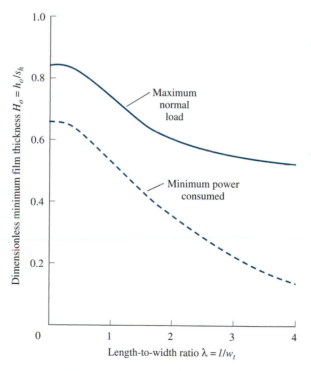

**Figure 12.13**    Chart for determining minimum film thickness corresponding to maximum load or minimum power loss for various pad proportions in fixed-incline bearings. [From Raimondi and Boyd (1955).]

**Figure 12.14**  Chart for determining minimum film thickness for fixed-incline thrust bearings. [From Raimondi and Boyd (1955).]

**Figure 12.15**  Chart for determining dimensionless temperature rise due to viscous shear heating of lubricant in fixed-incline thrust bearings. [From Raimondi and Boyd (1955).]

**Table 12.1** Allowable outlet (minimum) film thickness for a given surface finish

| Surface Finish Centerline (Average) $R_a$ | | Description of surface | Examples of manufacturing methods | Approximate relative costs | Allowable Outlet (minimum) Film Thickness[a] $h_o$ | |
|---|---|---|---|---|---|---|
| $\mu$m | $\mu$in | | | | $\mu$m | $\mu$in |
| 0.1–0.2 | 4–8 | Mirrorlike surface without tool marks; close tolerances | Grind, lap, and superfinish | 17–20 | 2.5 | 100 |
| 0.2–0.4 | 8–16 | Smooth surface without scratches; close tolerances | Grind and lap | 17–20 | 6.2 | 250 |
| 0.4–0.8 | 16–32 | Smooth surfaces; close tolderances | Grind, file, and lap | 10 | 12.5 | 500 |
| 0.8–1.6 | 32–63 | Accurate bearing surface without tool marks | Grind, precision mill, and file | 7 | 25 | 1000 |
| 1.6–3.2 | 63–125 | Smooth surface without objectionable tool marks; moderate tolerances | Shape, mill, grind, and turn | 5 | 50 | 2000 |

[a]The values of film thickness are given only for guidance. They indicate the film thickness required to avoid metal-to-metal contact under clean oil conditions with no misalignment. It may be necessary to take a larger film thickness than that indicated (e.g., to obtain an acceptable temperature rise). It has been assumed that the average surface finish of the pads is the same as that of the runner.
SOURCE: ESDU (1984).

(a)

**Figure 12.16** Chart for determining performance parameters of fixed-incline thrust bearings. (a) Friction coefficient; *(continued)*

(b)

(c)

**Figure 12.16**    (b) power loss; (c) lubricant flow; *(continued)*

Bearing number $B_t$

(d)

**Figure 12.16**    *(concluded)* (d) lubricant side flow. [From Raimondi and Boyd (1955).]

**Given:**  A fixed-incline slider thrust bearing has the following operating parameters: $W_z = 3600$ lb, $u_b = 1200$ in/s, $l = 3$ in, $w_t = 3$ in, SAE 10 oil, and $t_{mi} = 40°C$.

**EXAMPLE 12.3**

**Find:**  For the maximum load condition determine the following performance parameters: $s_h$, $h_o$, $\Delta t_m$, $\mu$, $h_p$, $q$, and $q_s$.

**Solution:**  From Figure 12.13 for $\lambda = l/w_t = 3/3 = 1$ and maximum normal load, the outlet film thickness ratio $H_o = 0.73$. Therefore, from Figure 12.14 the bearing number $B_t = 8.0$, and from Figure 12.15

$$\frac{0.9\rho C_p l w_t \, \Delta t_m}{W_z} = 13$$

Inserting the input parameters and noting the fact that

$$C_s = \rho C_p = 1.36 \times 10^6 \text{ N/(m}^2 \cdot °\text{C)} = 197.2 \text{ lb/(in}^2 \cdot °\text{C)}$$

give

$$\Delta t_m = \frac{13 W_z}{0.9\rho C_p l w_t} = \frac{13(3.6)(10^3)}{(0.9)(197.2)(3)(3)} = 29.30°\text{C}$$

From Equation (12.71)

$$\tilde{t}_m = t_{mi} + \frac{\Delta t_m}{2} = 40 + \frac{29.30}{2} = 54.65°\text{C}$$

From Figure 8.17 for SAE 10 oil at 55°C the absolute viscosity is $2.32 \times 10^{-6}$ lb · s/in². Making use of Equation (12.72) gives

$$h_o = H_o l \sqrt{\frac{\eta_0 u_b w_t}{W_z B_t}} = 0.73(3) \sqrt{\frac{(2.32)(10^{-6})(1.2)(10^3)(3)}{(3.6)(10^3)(8)}} = 1.179 \times 10^{-3} \text{ in}$$

$$s_h = \frac{h_o}{H_o} = \frac{(1.179)(10^{-3})}{0.73} = 1.616 \times 10^{-3} \text{ in}$$

All but the last row of surface finish will ensure that $h_o > (h_o)_{\text{all}}$. Thus, as long as the surface finish is less than $1 \times 10^{-3}$ in, the design is adequate.

The performance parameters for $\lambda = 1$ and $B_t = 8.0$ are as follows:

(a)  From Figure 12.16(a)

$$\frac{\mu l}{s_h} = 8.5$$

$$\therefore \quad \mu = \frac{(8.5)(1.616)(10^{-3})}{3} = 0.0046$$

(b)  From Figure 12.16(b)

$$\frac{1.5 h_p l}{W_z u_b s_h} = 11$$

$$\therefore \quad h_p = \frac{11 W_z u_b s_h}{1.5 l} = \frac{11(3.6)(10^3)(1.2)(10^3)(1.616)(10^{-3})}{1.5(3)} = 17.06 \times 10^3 \text{ W}$$

$$= 17.06 \text{ kW}$$

(c)  From Figure 12.16(c)

$$\frac{q}{w_t u_b s_h} = 0.58$$

$$\therefore \quad q = 0.58 w_t u_b s_h = 0.58(3)(1.2)(10^3)(1.616)(10^{-3} \text{ in}^3/\text{s}) = 3.374 \text{in}^3/\text{s}$$

(d)  $q_s = (0.3)q = 1.012 \text{ in}^3/\text{s}$.

### 12.3.6    THRUST SLIDER BEARING GEOMETRY

This chapter has dealt with the performance of an individual pad of a thrust bearing. Normally, a number of identical pads are assembled in a thrust bearing as shown, for example, in Figures 12.9 and 12.12. The length, width, speed, and load of an individual pad can be related to the geometry of a thrust bearing by the following formulas:

$$w_t = r_o - r_i \tag{12.73}$$

$$l = \frac{r_o + r_i}{2}\left(\frac{2\pi}{N_p} - \frac{\pi}{36}\right) \tag{12.74}$$

$$u_b = \frac{(r_o + r_i)\omega}{2} \tag{12.75}$$

$$W_t = N_p W_z \tag{12.76}$$

where $N_p$ is the number of identical pads placed in the thrust bearing (usually between 3 and 20). The $\pi/36$ term of Equation (12.74) accounts for feed grooves between the pads. These are deep grooves that ensure that ambient pressure is maintained between the pads. Also $W_t$ in Equation (12.76) is the total thrust load on the bearing.

## 12.4    JOURNAL SLIDER BEARINGS

The last section dealt with slider bearing pads as used in thrust bearings. The surfaces of *thrust bearings* are *perpendicular* to the axis of rotation, as shown in Figure 12.9. This section deals with **journal bearings,** where the bearing surfaces are parallel to the axis of rotation, as shown in Figure 12.17. Journal bearings are used to support shafts and to carry radial loads with minimum power loss and minimum wear. The journal bearing can be represented by a plain cylindrical sleeve (bushing) wrapped around the journal (shaft) but can adopt a variety of forms. The lubricant is supplied at some convenient point in the bearing through a hole or a groove. If the bearing extends around the full 360° of the journal, it is described as a *full* journal bearing; if the wrap angle is less than 360°, the term *partial* journal bearing is used.

### 12.4.1    PETROV'S EQUATION

Petrov's equation applies Newton's postulate (Sec. 8.7.1) to a full journal running concentrically with the bearing (Fig. 12.18). As shown in Section 12.4.3, the journal will run

**Figure 12.17**    Pressure distribution around a journal bearing.

**Figure 12.18**    Concentric journal bearing.

concentrically with the bearing only when one of the following conditions prevails: (1) the radial load acting on the bearing is zero, (2) the viscosity of the lubricant is infinite, or (3) the speed of the journal is infinite. None of these conditions are practically possible. However, if the load is light enough, if the viscosity is sufficiently high, and if the journal has a sufficiently high speed, the eccentricity of the journal relative to the bearing may be so small that the oil film around the journal can be considered practically to be of uniform thickness.

Since the oil film in a journal bearing is always thin relative to the bearing radius, the curvature of the bearing surface may be ignored. The film may be considered as an unwrapped body having a thickness equal to the radial clearance, a length equal to $2\pi r$, and width $w_t$. Assume now that the viscosity throughout the oil film is constant. In Figure 12.19 the bottom surface is stationary, and the top surface is moving with constant velocity $u$. Petrov (1883) assumed no slip at the interface between the lubricant and the solids.

Making use of Newton's postulate as expressed in Equation (8.10) gives the friction force in a concentric journal bearing as

$$F = \eta_0 A \frac{u}{c} = \eta_0 2\pi r w_t \frac{2\pi r N_a}{c} = \frac{4\pi^2 \eta_0 r^2 w_t N_a}{c} \tag{12.77}$$

where $N_a$ = rotational speed, rps
 $\eta_0$ = viscosity at $p = 0$ and constant temperature

**Figure 12.19**    Developed journal and bearing surfaces for a concentric journal bearing.

The coefficient of friction for a concentric journal bearing is thus

$$\mu = \frac{F}{W_r} = \frac{4\pi^2 \eta_0 r^2 w_t N_a}{c W_r} \tag{12.78}$$

where $c$ = radial clearance, m
$\quad W_r$ = normal applied load, N

The friction torque for a concentric journal bearing can then be written as

$$T = Fr = \frac{4\pi^2 \eta_0 r^3 w_t N_a}{c} = \frac{2\pi \eta_0 r^3 w_t \omega}{c} \tag{12.79}$$

where $\omega = 2\pi N_a$ = angular velocity, rad/s. Equation (12.79) is generally called **Petrov's equation** (after N. Petrov, who suggested a similar equation for torque in his paper published in 1883).

The power loss is the torque times the angular velocity, and for a concentric (lightly loaded) journal bearing it can be expressed in horsepower as

$$h_p = \frac{8\pi^3}{(12)(550)} \frac{\eta_0 r^3 w_t N_a^2}{c} = 0.03758 \frac{\eta_0 r^3 w_t N_a^2}{c} \tag{12.80}$$

where $\eta_0$ = viscosity at $p = 0$ and constant temperature, lb · s/in$^2$
$\quad r$ = radius of journal, in
$\quad w_t$ = width of journal, in
$\quad N_a$ = rotational speed, rps
$\quad c$ = radial clearance, in

Equation (12.80) is valid only for these units.

## 12.4.2 JOURNAL SLIDER BEARING OPERATION

Journal bearings (Fig. 12.17) rely on shaft sliding motion to generate the load-supporting pressures in the lubricant film. The shaft does not normally run concentric with the bearing. The displacement of the shaft center relative to the bearing center is known as the **eccentricity** (designated by $e$ in Fig. 12.17). The shaft's eccentric position within the bearing clearance is influenced by the load that it carries. The amount of eccentricity adjusts itself until the load is balanced by the pressure generated in the converging lubricating film. The line drawn through the shaft center and the bearing center is called the **line of centers.** The physical wedge pressure-generating mechanism mentioned in Section 12.2.2 and used for thrust slider bearings in Section 12.3.5 is also valid in this section.

The pressure generated and therefore the load-carrying capacity of the bearing depend on the shaft eccentricity, the angular velocity, the effective viscosity of the lubricant, and the bearing dimensions and clearance:

$$W_r = f(e, \omega, \eta_0, r, w_t, c)$$

The load and the angular velocity are usually specified, and the minimum shaft radius is often predetermined. To complete the design, it will be necessary to calculate the bearing width and clearance and to choose a suitable lubricant if one is not already specified.

### 12.4.3   OPERATING AND PERFORMANCE PARAMETERS

The Reynolds equation used for thrust slider bearings [Eq. (12.28)] is modified for a journal slider bearing with $x = r_b \phi$ and $\tilde{u} = u_b/2 = r_b \omega_b / 2$.

$$\therefore \qquad \frac{\partial}{\partial \phi}\left(h^3 \frac{\partial p}{\partial \phi}\right) + r_b^2 \frac{\partial}{\partial y}\left(h^3 \frac{\partial p}{\partial y}\right) = 6\eta_0 \omega_b r_b^2 \frac{\partial h}{\partial \phi} \qquad (12.81)$$

The film thickness around the journal is expressed as

$$h = c(1 + \epsilon \cos \phi) \qquad (12.82)$$

where $\epsilon = e/c$ = eccentricity ratio. Substituting Equation (12.82) into Equation (12.81) gives

$$\frac{\partial}{\partial \phi}\left(h^3 \frac{\partial p}{\partial \phi}\right) + r_b^2 h^3 \frac{\partial^2 p}{\partial y^2} = -6\eta_0 \omega_b r_b^2 e \sin \phi \qquad (12.83)$$

Analytical solutions to Equation (12.83) are not normally available, and numerical methods are thus needed. Equation (12.83) is often solved by using a relaxation method in which the first step is to replace the derivatives in the equation by finite difference approximations. The lubrication area is covered by a mesh, and the numerical method relies on the fact that a function can be represented, with sufficient accuracy, over a small range by a quadratic expression. The Reynolds boundary condition which requires that $p = 0$ and $dp/dx = 0$ at $\phi = \phi^*$ (the outlet boundary) is used. Only the results obtained by using this numerical method are presented in this section.

The *operating* parameters for hydrodynamic journal bearings are

1.  Bearing number for journal bearings (also called the **Sommerfeld number**)

$$B_j = \left(\frac{\eta_0 \omega_b r_b w_t}{\pi W_r}\right)\left(\frac{r_b}{c}\right)^2 \qquad (12.84)$$

2.  Diameter-to-width ratio

$$\lambda_j = \frac{2r_b}{w_t} \qquad (12.85)$$

3.  Angular extent of journal (full or partial)

The *performance* parameters for hydrodynamic journal bearings are

1.  Eccentricity $e$
2.  Location of minimum film thickness, sometimes called *attitude angle* $\Phi$
3.  Coefficient of sliding friction $\mu$
4.  Total and side-leakage volumetric flow rates $q$ and $q_s$
5.  Angle of maximum pressure $\phi_{max}$
6.  Location of terminating pressure $\phi_0$
7.  Temperature rise due to lubricant shearing $\Delta t_m$

The parameters $\Phi$, $\phi_{max}$, and $\phi_0$ are described in Figure 12.17, which gives the pressure distribution around a journal bearing. Note from this figure that if the bearing is concentric ($e = 0$), the film shape around the journal is constant and equal to $c$, and no fluid film pressure is developed. At heavy loads, which is the other extreme, the journal is forced downward, and the limiting position is reached when $h_{\min} = 0$ and $e = c$ (i.e., the journal is in contact with the bearing).

Temperature rise due to lubricant shearing will be considered here as was done in Section 12.3.4 for thrust bearings. In Equation (12.83) the lubricant viscosity corresponds to the viscosity when $p = 0$ but can vary as a function of temperature. Since work is done on the lubricant as the fluid is being sheared, the temperature of the lubricant is higher when it leaves the conjunction than on entry. In Chapter 8 (Figs. 8.17 and 8.18 as well as Table 8.4) it was shown that oil viscosity drops off significantly with rising temperature. This decrease is compensated for by using a mean of the inlet and outlet temperatures

$$\tilde{t}_m = t_{mi} + \frac{\Delta t_m}{2} \tag{12.86}$$

where $t_{mi}$ = inlet temperature
$\Delta t_m$ = temperature rise of lubricant from inlet to outlet

The viscosity used in the bearing number $B_j$ and other performance parameters is at the mean temperature $\tilde{t}_m$. The temperature rise of the lubricant from inlet to outlet $\Delta t_m$ can be determined from the performance charts provided in this section.

## 12.4.4   DESIGN PROCEDURE

Now that the operating and performance parameters have been defined, the design procedure for a hydrodynamic journal bearing can be presented. The results are for a full journal bearing; results for a *partial* journal bearing can be obtained from Raimondi and Boyd (1958).

Most design problems are underconstrained, and one or more dimensions have to be specified before analysis can commence. In such a circumstance, the data in Table 12.2 can help in prescribing a candidate design, which can then be modified if the performance is unsatisfactory. For example, given the load and shaft diameter, a bearing width can be determined through use of typical average pressures from Table 12.2. This table can also be useful in confirming the results of analysis.

Figure 12.20 shows the effect of the bearing number $B_j$ on the minimum film thickness for four diameter-to-width ratios. The following relationship should be observed:

$$h_{\min} = c - e \tag{12.87}$$

In dimensionless form

$$H_{\min} = \frac{h_{\min}}{c} = 1 - \epsilon \tag{12.88}$$

where

$$\epsilon = \frac{e}{c} = \text{eccentricity ratio} \tag{12.89}$$

The bearing number for journal bearings is expressed in Equation (12.84); in a given design it is affected by

1. Absolute lubricant viscosity $\eta_0$
2. Angular shaft speed $\omega_b$
3. Radial load $W_r$
4. Radial clearance $c$
5. Journal dimensions $r_h$ and $w_t$

**Table 12.2**    Typical radial load per area $W_r^*$ in use for journal bearings

| Application | Average Radial Load per Area, $W_r^*$ | |
| --- | --- | --- |
|  | psi | MPa |
| Automotive engines | | |
|    Main bearings | 600–750 | 4–5 |
|    Connecting rod bearing | 1700–2300 | 10–15 |
| Diesel engines | | |
|    Main bearings | 900–1700 | 6–12 |
|    Connecting rod bearing | 1150–2300 | 8–15 |
| Electric motors | 120–250 | 0.8–1.5 |
| Steam turbines | 150–300 | 1.0–2.0 |
| Gear reducers | 120–250 | 0.8–1.5 |
| Centrifugal pumps | 100–180 | 0.6–1.2 |
| Air compressors | | |
|    Main bearings | 140–280 | 1–2 |
|    Crankpin | 280–500 | 2–4 |
| Centrifugal pumps | 100–180 | 0.6–1.2 |

I  SOURCE: (From Juvinall and Marshek 2003)

All these parameters affect the bearing number and thus the design of the journal bearing.

In Figure 12.20 a recommended operating eccentricity ratio, or minimum film thickness, is indicated as well as a preferred operating area. The left boundary of the shaded zone defines the optimum eccentricity ratio for a minimum coefficient of friction; and the right boundary, the optimum eccentricity ratio for maximum load. The recommended operating eccentricity is midway between these two boundaries.

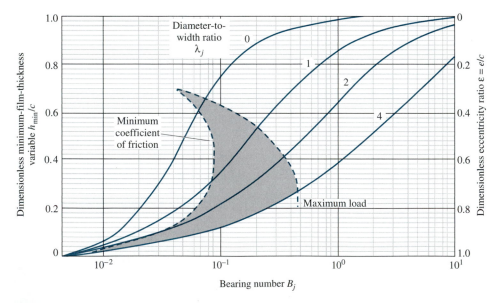

**Figure 12.20**    Effect of bearing number on minimum film thickness for four diameter-to-width ratios. [From Raimondi and Boyd (1958).]

Figure 12.21 shows the effect of the bearing number on the attitude angle $\Phi$ [angle between the load direction and a line drawn through the centers of the bearing and journal (see Fig. 12.17)] for four values of $\lambda_j$. This angle establishes where the minimum and maximum film thicknesses are located within the bearing.

Figure 12.22 shows the effect of the bearing number on the coefficient of friction for four values of $\lambda_j$. The effect is small for a complete range of dimensionless load parameters.

Figure 12.23 shows the effect of bearing number on the dimensionless volumetric flow rate $Q = 2\pi q/(r_b c w_t \omega_b)$ for four values of $\lambda_j$. The dimensionless volumetric flow rate $Q$ that is pumped into the converging space by the rotating journal can be obtained from this figure. Of the volume of oil $q$ pumped by the rotating journal, an amount $q_s$ flows out the ends and hence is called *side-leakage volumetric flow*. This side leakage can be computed from the volumetric flow ratio $q_s/q$ of Figure 12.24.

Figure 12.25 illustrates the maximum pressure developed in a journal bearing; in this figure the maximum film pressure is made dimensionless with the load per unit area. The maximum pressure and its location are shown in Figure 12.17. Figure 12.26 shows the effect of bearing number on the location of the terminating and maximum pressures for four values of $\lambda_j$.

The temperature rise in degrees Celsius of the lubricant from the inlet to the outlet can be obtained from Shigley and Mitchell (1983) as

$$\Delta t_m = \frac{8.3 W_r^* (r_b/c)\mu}{Q(1 - 0.5 q_s/q)} \tag{12.90a}$$

where in Equation (12.90a) $W_r^* = W_r/(2r_b w_t)$ is in megapascals. Therefore, the temperature rise can be directly obtained by substituting the values of $r_b \mu/c$ obtained from Figure 12.22, $Q$ from Figure 12.23, and $q_s/q$ from Figure 12.24 into Equation (12.90a). The temperature

**Figure 12.21** Effect of bearing number on attitude angle for four diameter-to-width ratios. [From Raimondi and Boyd (1958).]

**Figure 12.22**   Effect of bearing number on coefficient of friction for four diameter-to-width ratios. [From Raimondi and Boyd (1958).]

rise in degrees Fahrenheit is given by

$$\Delta t_m = \frac{0.103 W_r^* \, (r_b/c) \, \mu}{Q(1 - 0.5 q_s/q)} \tag{12.90b}$$

where

$$W_r^* = \frac{W}{2 r_b w_t} \tag{12.90c}$$

and $W_r^*$ is in pounds per square inch.

Once the viscosity is known, the bearing number can be calculated and then the performance parameters can be obtained from Figures 12.20 to 12.26 and Equation (12.90).

The results presented thus far have been for $\lambda_j$ of 0, 1, 2, and 4. If $\lambda_j$ is some other value, use the following formula for establishing the performance parameters:

$$y = \frac{1}{(w_t/2 r_b)^2} \left[ -\frac{1}{8} \left( 1 - \frac{w_t}{2 r_b} \right) \left( 1 - \frac{w_t}{r_b} \right) \left( 1 - 2 \frac{w_t}{r_b} \right) y_0 + \frac{1}{3} \left( 1 - \frac{w_t}{r_b} \right) \left( 1 - 2 \frac{w_t}{r_b} \right) y_1 \right.$$

$$\left. - \frac{1}{4} \left( 1 - \frac{w_t}{2 r_b} \right) \left( 1 - 2 \frac{w_t}{r_b} \right) y_2 + \frac{1}{24} \left( 1 - \frac{w_t}{2 r_b} \right) \left( 1 - \frac{w_t}{r_b} \right) y_4 \right] \tag{12.91}$$

**Figure 12.23** Effect of bearing number on dimensionless volumetric flow rate for four diameter-to-width ratios. [From Raimondi and Boyd (1958).]

**Figure 12.24** Effect of bearing number on side-leakage flow ratio for four diameter-to-width ratios. [From Raimondi and Boyd (1958).]

**Figure 12.25**     Effect of bearing number on dimensionless maximum film pressure for four diameter-to-width ratios. [From Raimondi and Boyd (1958).]

**Figure 12.26**     Effect of bearing number on location of terminating and maximum pressures for four diameter-to-width ratios. [From Raimondi and Boyd (1958).]

where $y$ is any one of the performance parameters ($H_{min}$, $\Phi$, $r_b\mu/c$, $Q$, $q_s/q$, $p_{max}$, $\phi_0$, or $\phi_{max}$), and where the subscript on $y$ is the $\lambda_j$ value. For example, $y_1$ is equivalent to $y$ evaluated at $\lambda_j = 1$. All the results presented are valid for a full journal bearing.

---

**Given:**  A full journal bearing has the specification of SAE 60 oil with an inlet temperature of $40°C$, $N_a = 30$ rps, $W_r = 2200$ N, $r_b = 20$ mm, and $w_t = 40$ mm.

**Find:**  From the figures given in this section, establish the operating and performance parameters for this bearing while designing for maximum load.

**Solution:**  The angular speed can be expressed as

$$\omega_b = 2\pi N_a = 2\pi(30) = 60\pi \text{ rad/s}$$

The diameter-to-width ratio is

$$\lambda_j = \frac{2r_b}{w_t} = \frac{2(2)}{4} = 1$$

From Figure 12.20 for $\lambda_j = 1$ and designing for maximum load

$$B_j = 0.2 \qquad \frac{h_{min}}{c} = 0.53 \qquad \text{and} \qquad \epsilon = 0.47 \qquad \text{(a)}$$

For $B_j = 0.2$ and $\lambda_j = 1$ from Figures 12.22 to 12.24

$$\frac{r_b\mu}{c} = 4.9 \qquad Q = 4.3 \qquad \text{and} \qquad \frac{q_s}{q} = 0.6 \qquad \text{(b)}$$

From Equation (12.90c) the radial load per area is

$$W_r^* = \frac{W_r}{2r_b w_t} = \frac{2200}{2(2)(4)(10^{-4})} \text{ Pa} = 1.375 \text{ MPa} \qquad \text{(c)}$$

The lubricant temperature rise obtained by using Equation (12.90a) and the results from Equations (b) and (c) is

$$\Delta t_m = \frac{8.3 W_r^*(r_b/c)\mu}{Q(1 - 0.5q_s/q)} = \frac{8.3(1.375)(4.9)}{4.3[1 - (0.5)(0.6)]} = 18.58°C$$

From Equation (12.86) the mean temperature in the lubricant conjunction is

$$\tilde{t}_m = t_{mi} + \frac{\Delta t_m}{2} = 40 + \frac{18.58}{2} = 49.29°C$$

From Figure 8.17 for SAE 60 oil at $49.3°C$ the absolute viscosity is

$$2.5 \times 10^{-5} \text{ reyn} = 1.70 \times 10^2 \text{ cp} = 0.170 \text{ N} \cdot \text{s/m}^2$$

From Equation (12.84) the radial clearance can be expressed as

$$c = r_b\sqrt{\frac{\eta_0\omega_b r_b w_t}{\pi W_r B_j}} = (2)(10^{-2})\sqrt{\frac{(0.170)60\pi(2)(10^{-2})(4)(10^{-2})}{\pi(2.2)(10^3)(0.2)}} = 0.0861 \times 10^{-3} \text{ m}$$

The coefficient of friction from Equation (b) is

$$\mu = \frac{4.9c}{r_b} = \frac{(4.9)(0.0861)(10^{-3})}{(2)(10^{-2})} = 0.021$$

**EXAMPLE 12.4**

The circumferential volumetric flow rate is

$$q = \frac{Qr_b cw_t \omega_b}{2\pi} = \frac{(4.3)(2)(10^{-2})(0.0861)(10^{-3})(4)(10^{-2})60\pi}{2\pi} = 8.89 \times 10^{-6} \text{ m}^3/\text{s}$$

$$q_s = 5.330 \times 10^{-6} \text{ m}^3/\text{s}$$

From Figure 12.21 for $B_j = 0.2$ and $\lambda_j = 1$ the attitude angle is 61°. From Figure 12.25 for $B_j = 0.2$ and $\lambda_j = 1$ the dimensionless maximum pressure is $P_{max} = 0.46$. The maximum pressure is thus

$$p_{max} = \frac{W_r}{2r_b w_t P_{max}} = \frac{2200}{2(2)(10^{-2})(4)(10^{-2})(0.46)} \text{ Pa} = 2.989 \text{ MPa}$$

From Figure 12.26 for $B_j = 0.2$ and $\lambda_j = 1$ the location of the maximum pressure from the applied load is 18°, and the location of the terminating pressure from the applied load is 86°.

### 12.4.5    OPTIMIZATION TECHNIQUES

The radial clearance $c$ is the most difficult of the performance parameters to control accurately during manufacturing, and it may increase because of wear. Figure 12.27 shows the performance of a particular bearing for a range of radial clearances. If the clearance is too tight, the temperature will be too high and the minimum film thickness too low. High temperature may cause the bearing to fail by fatigue. If the oil film is too thin, dirt particles may not pass through without scoring, or may embed themselves in the bearing. In either event there will be excessive wear and friction, resulting in high temperatures and possible seizing. A large clearance will permit dirt particles to pass through and also permit a large flow of oil, lowering the temperature and lengthening bearing life. However, if the clearance

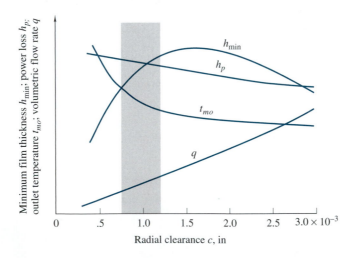

**Figure 12.27**    Effect of radial clearance on some performance parameters for a particular case.

becomes too large, the bearing becomes noisy and the minimum film thickness begins to decrease again.

Figure 12.27 shows the best compromise, when both the production tolerance and the future wear on the bearing are considered, to be a clearance range slightly to the left of the top of the minimum-film-thickness curve. Future wear will move the operating point to the right, thus increasing the film thickness and decreasing the operating temperature.

## 12.5   SQUEEZE FILM BEARINGS

In Sections 12.3 and 12.4 slider bearings were considered where the positive bearing pressure is due to the physical wedge mechanism covered in Section 12.2.2. Here the focus is on the development of positive pressure in a fluid contained between two surfaces moving toward each other. Because a finite time is required to squeeze the fluid out of the gap, this action provides a useful cushioning effect in bearings. The reverse effect, which occurs when the surfaces are moving apart, can lead to cavitation in the fluid film.

The concept of **squeeze film bearings** was introduced in this text in Figure 8.8(b), and the pressure-generating mechanism was covered earlier in Section 12.2.2. For squeeze film bearings a relationship needs to be developed between load and normal velocity at any instant. The time required for the separation of the surfaces to change by a specified amount can be determined by a single integration with respect to time.

The appropriate Reynolds equation for squeeze film bearings is

$$\frac{\partial}{\partial x}\left(h^3\frac{\partial p}{\partial x}\right) + \frac{\partial}{\partial y}\left(h^3\frac{\partial p}{\partial y}\right) = 12\eta_0\frac{\partial h}{\partial t} = -12\eta_0 w \qquad (12.92)$$

where $w =$ squeeze velocity, m/s.

### 12.5.1   PARALLEL-SURFACE SQUEEZE FILM THRUST BEARING

Figure 12.28 shows a simple parallel-surface squeeze film bearing. When side leakage is neglected [flow is in the $y$ direction (the direction out from the paper)], the second term in

**Figure 12.28**   Parallel-surface squeeze film bearing.

Equation (12.92) is neglected. This assumption is valid for bearings whose width $w_t$ in the $y$ direction is much larger than their length $l$, or $w_t/l > 4$. Thus, if side leakage is neglected and a parallel film is assumed, Equation (12.92) reduces to

$$\frac{\partial^2 p}{\partial x^2} = -\frac{12\eta_0 w}{h_o^3} \tag{12.93}$$

In Figure 12.28 the coordinate system origin is at the midpoint of the bearing since symmetric oil film geometry makes it convenient to place it there. Integrating twice gives the pressure as

$$p = -\frac{6\eta_0 w}{h_o^3}x^2 + \tilde{A}x + \tilde{B} \tag{12.94}$$

where $\tilde{A}$ and $\tilde{B}$ = integration constants. The boundary conditions are

$$p = 0 \qquad \text{when } x = \pm\frac{l}{2}$$

Making use of these boundary conditions gives

$$p = \frac{3\eta_0 w}{2h_o^3}(l^2 - 4x^2) \tag{12.95}$$

The normal load-carrying capacity per unit width is

$$W_z' = \int_{-l/2}^{l/2} p\,dx = \frac{3\eta_0 w}{2h_o^3}\int_{-l/2}^{l/2}(l^2 - 4x^2)\,dx = \frac{\eta_0 l^3 w}{h_o^3} \tag{12.96}$$

For time-independent loads ($W_z'$ not a function of $t$) Equation (12.96) can be used to determine the time for the gap between the parallel surfaces to be reduced by a given amount. The squeeze velocity $w$ can be expressed as $-\partial h/\partial t$. Since the film thickness does not vary with $x$ but varies with time, Equation (12.96) becomes

$$W_z' = -\frac{\eta_0 l^3}{h_o^3}\frac{\partial h_o}{\partial t} \tag{12.97}$$

Rearranging terms and integrating give

$$-\frac{W_z'}{\eta_0 l^3}\int_{t_1}^{t_2} dt = \int_{h_{o,1}}^{h_{o,2}}\frac{dh_o}{h_o^3}$$

$$\therefore \qquad \Delta t = t_2 - t_1 = \frac{\eta_0 l^3}{2W_z'}\left(\frac{1}{h_{o,2}^2} - \frac{1}{h_{o,1}^2}\right) \tag{12.98}$$

The final outlet film thickness $h_{o,2}$ can then be expressed in terms of the initial film thickness $h_{o,1}$ and the time interval $\Delta t$ as

$$h_{o,2} = \frac{h_{o,1}}{\left[1 + 2W_z'\Delta t h_{o,1}^2/(\eta_0 l^3)\right]^{1/2}} \tag{12.99}$$

EXAMPLE 12.5

**Given:** Two parallel plates are 0.025 m long and infinitely wide. The lubricant separating the plates is initially 25 $\mu$m thick and has a viscosity of 0.5 Pa·s. The load per unit width is 20,000 N/m.

**Find:** Calculate the time required to reduce the film thickness to (a) 2.5 $\mu$m, (b) 0.25 $\mu$m, and (c) zero.

**Solution:** Making use of Equation (12.98) gives

$$\Delta t = \frac{\eta_0 l^3}{2W_z'} \left( \frac{1}{h_{o,2}^2} - \frac{1}{h_{o,1}^2} \right) = \frac{(0.5)(0.025)^3}{2(20,000)} \left[ \frac{1}{(2.5 \times 10^{-6})^2} - \frac{1}{(25 \times 10^{-6})^2} \right]$$

1. For $h_o = 2.5\ \mu$m

$$\Delta t = 1.95 \times 10^{-10}(0.16 - 0.0016)(10^{12}) = 30.9\ \text{s} \qquad \text{(a)}$$

2. For $h_o = 0.25\ \mu$m

$$\Delta t = 1.95 \times 10^{-10} \left[ \frac{1}{(0.25 \times 10^{-6})^2} - \frac{1}{(25 \times 10^{-6})^2} \right] = 3120\ \text{s} \qquad \text{(b)}$$

3. For $h_o = 0$

$$\Delta t = 1.95 \times 10^{-10} \left[ \frac{1}{0} - \frac{1}{(25 \times 10^{-6})^2} \right] = \infty \qquad \text{(c)}$$

Equation (c) implies that, theoretically, the lubricant will never be squeezed out of the space between the parallel plates.

## 12.5.2 GENERAL COMMENTS ABOUT SQUEEZE FILM BEARINGS

For a thrust bearing subjected to a squeeze velocity, a parallel film shape produces the largest load-carrying capacity of all possible film shapes. In contrast, for a thrust slider bearing a parallel film should produce no positive pressure and no load-carrying capacity. Recall that a slider bearing uses a physical wedge mechanism (see Sec. 12.2.2) to produce a positive pressure. Squeeze film bearings can be applied to a number of bearing configurations other than the parallel-surface thrust bearing considered earlier. Hamrock (1993) applied squeeze films to a number of conformal and nonconformal surfaces, such as journal bearings, parallel circular plates, and an infinitely long cylinder near a plate.

The second major observation about normal squeeze film bearings is that as the bearing surfaces move toward each other, the viscous fluid shows great reluctance to be squeezed out the sides of the bearing. The tenacity of the squeeze film is remarkable, and the survival of many modern bearings depends on this phenomenon.

The third remarkable feature of squeeze film bearings is that a small approach velocity will provide an extremely large load-carrying capacity. The reason is mainly that in Equation (12.96) the normal applied load per unit width is inversely proportional to the film thickness raised to the third power and directly proportional to the squeeze velocity.

# 12.6    HYDROSTATIC BEARINGS

Slider bearings, considered in Sections 12.3 and 12.4, use a physical wedge pressure-generating mechanism to develop pressure within the bearing. Such bearings, in addition to having low frictional drag and hence low power loss, have the great advantage that they are basically simple and therefore are reliable and inexpensive and require little attention. Slider bearings have, however, certain important limitations:

1.  If the design speed is low, it may not be possible to generate sufficient hydrodynamic pressure.

2.  Fluid film lubrication may break down during starting, direction changing, and stopping.

3.  In a journal slider bearing (Sec. 12.4), the shaft runs eccentrically, and the bearing location varies with load, thus implying low stiffness.

In **hydrostatic** (also called **externally pressurized**) **lubricated bearings** the bearing surfaces are separated by a fluid film maintained by a pressure source outside the bearing. Hydrostatic bearings avoid disadvantages 1 and 2 and reduce the variation of bearing location with load mentioned in disadvantage 3. The characteristics of hydrostatically lubricated bearings are

1.  Extremely low friction

2.  Extremely high load-carrying capacity at low speeds

3.  High positional accuracy in high-speed, light-load applications

4.  A lubrication system more complicated than that for slider bearings (considered in Secs. 12.3 and 12.4)

Therefore, hydrostatically lubricated bearings are used when the requirements are demanding, as in large telescopes and radar tracking units, where very heavy loads and low speeds are used; or in machine tools and gyroscopes, where extremely high speeds, light loads, and gas lubricants are used.

Figure 12.29 shows how a fluid film forms in a hydrostatically lubricated bearing system. In a simple bearing system under no pressure [Fig. 12.29(a)] the runner, under the influence of load $W_z$, is seated on the pad. As the source pressure builds up [Fig. 12.29(b)], the pressure in the pad recess also increases. The recess pressure is built up to a point [Fig. 12.29(c)] where the pressure on the runner over an area equal to the pad recess area is just sufficient to lift the load. This is commonly called the *lift pressure* $p_l$. Just after the runner separates from the bearing pad [Fig. 12.29(d)], the recess pressure is less than that required to lift the bearing runner ($p_r < p_l$). After lift, flow commences through the system. Thus, a pressure drop exists between the pressure source and the bearing (across the restrictor) and from the recess to the bearing outlet. If more load is added to the bearing [Fig. 12.29(e)], the film thickness will decrease and the recess pressure will rise until the integrated pressure across the land equals the load. If the load is then reduced to less than the original [Fig. 12.29(f)], the film thickness will increase to some higher value and the recess pressure will decrease accordingly. The maximum load that can be supported by the pad will be reached theoretically when the recess pressure is equal to the source pressure. If a load greater than this is applied, the bearing will seat and remain seated until the load is reduced and can again be supported by the supply pressure.

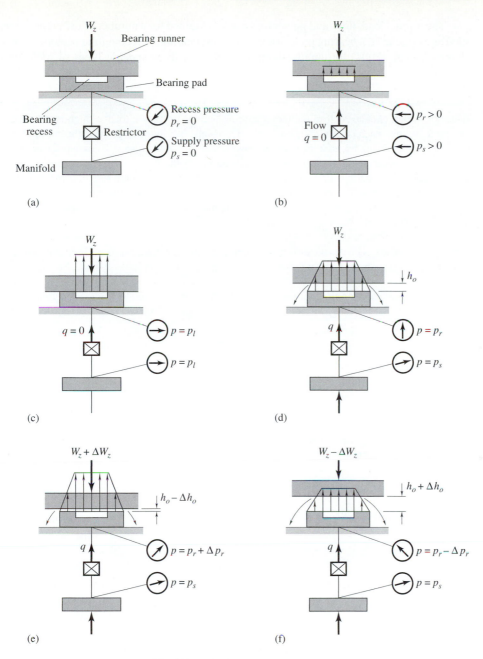

**Figure 12.29** Formation of fluid film in hydrostatic bearing system. (a) Pump off; (b) pressure buildup; (c) pressure times recess area equals normal applied load; (d) bearing operating; (e) increased load; (f) decreased load. [From Rippel (1963).]

The normal load, flow, and power loss can be obtained for a number of different shapes, such as a circular step bearing pad, an annular thrust bearing, or rectangular sectors. Consult Hamrock et al. (2004) for details about the design of these specific types of hydrostatic bearing.

---

**EXAMPLE 12.6**

**Given:**   A flat, circular hydrostatic thrust bearing is used to carry a load as high as 25 tons in a large milling machine. The purpose of the hydrostatic bearing is to position the workpiece accurately without any friction. The vertical positioning of the workpiece is influenced by the oil film thickness and its variation with load. To minimize the variation for different loads, a step is made in the bearing, as shown in Figure 12.30. The pressure in the step is constant, and $s_h/h_o > 10$. Thus, $p = p_r$ through the step or for $0 < r < r_i$. The step height $s_h = 5$ mm and $r_o = 200$ mm. The oil viscosity is $0.25$ N · s/m², and the smallest value of $h_o$ allowed is $50$ $\mu$m. The oil flow through the bearing is constant at $0.1 \times 10^{-3}$ m³/s.

**Find:**   Determine $r_i$ such that the oil outlet (minimum) film thickness $h_o$ varies as little as possible when the load varies between 5 and 25 tons.

**Solution:**   Assuming that the film thickness is the same in any radial or angular position and that the pressure does not vary in the $\theta$ direction, the appropriate Reynolds equation is

$$\frac{\partial}{\partial r}\left( r\frac{\partial p}{\partial r} \right) = 0$$

Integrating once gives

$$r\frac{dp}{dr} = \tilde{A}$$

**Figure 12.30**   Radial-flow hydrostatic thrust bearing with circular step pad.

and integrating again gives

$$p = \tilde{A} \ln r + \tilde{B} \tag{a}$$

The boundary conditions for the circular thrust bearing shown in Figure 12.30 are

1. $p = p_r$ at $r = r_i$
2. $p = 0$ at $r = r_o$

Making use of these boundary conditions gives

$$p = p_r \frac{\ln(r/r_o)}{\ln(r_i/r_o)} \tag{b}$$

$$\frac{dp}{dr} = \frac{p_r}{r \ln(r_i/r_o)} \tag{c}$$

The radial volumetric flow rate per circumference is given as

$$q_r' = -\frac{h_o^3}{12\eta_0} \frac{dp}{dr} = -\frac{h_o^3 p_r}{12\eta_0 r \, \ln(r_i/r_o)} \tag{d}$$

The total circumferential volumetric flow rate is

$$q = 2\pi r q_r' = -\frac{\pi h_o^3 p_r}{6\eta_0 \ln(r_i/r_o)} = \frac{\pi h_o^3 p_r}{6\eta_0 \ln(r_o/r_i)} = \text{constant} \tag{e}$$

Making use of Equation (b) gives the pressure distribution in a radial-flow hydrostatic bearing. The normal load component is balanced by the total pressure force, or

$$W_z = \pi r_i^2 p_r + \int_{r_i}^{r_o} \frac{p_r \ln(r/r_o)}{\ln(r_i/r_o)} 2\pi r \, dr$$

This equation reduces to

$$W_z = \frac{\pi p_r (r_o^2 - r_i^2)}{2 \ln(r_o/r_i)} \tag{f}$$

From Equations (e) and (f)

$$W_z = \frac{(r_o^2 - r_i^2) 6 q \eta_0}{2 h_o^3} \tag{g}$$

From Equation (g), $(W_z)_{max}$ occurs when $(h_o)_{min}$ and $(r_i)_{min}$ whereas $(W_z)_{min}$ occurs when $(h_o)_{max}$ and $(r_i)_{max}$.

$$\therefore \qquad (h_o)_{min}^3 = \frac{6\eta_0 q}{(W_z)_{max}} \frac{r_o^2 - r_{i,max}^2}{2}$$

$$(50)^3 (10^{-6})^3 = \frac{(6)(0.25)(0.1)(10^{-3})}{(25,000)(9.807)} \frac{0.200^2 - r_{i,max}^2}{2}$$

$$\therefore \qquad r_{i,max} = 0.1990 \text{ m}$$

Also,

$$(h_o)^3_{max} = \frac{6\eta_0 q}{(W_z)_{min}} \frac{r_o^2 - r_{i,min}^2}{2} = \frac{(6)(0.25)(10^{-4})}{(5000)(9.807)} \frac{0.200^2 - 0.1990^2}{2}$$

$$(h_o)_{max} = 84.82 \times 10^{-6} \text{ m} = 84.82\,\mu\text{m}$$

The inner radius should therefore be 199 mm, which gives a minimum film thickness variation of $84.82 - 50 = 34.82\,\mu$m.

---

**Case Study 12.1** | DESIGN OF A CRANKSHAFT MAIN BEARING

An automotive crankshaft is shown inside an internal combustion engine in Figure 12.31. The crankshaft is in one of the most difficult-to-access locations of the engine. Therefore, it is imperative that the crankshaft and its bearings be designed so that maintenance will never be needed in the lifetime of the automobile. This requires that the bearings be designed to operate in the full-film lubrication regime.

The load acting on one of the bearings is shown in Figure 12.32, where both magnitude and direction are given as functions of time. The loading is complex and reflects the large dynamic loads associated with ignition within each cylinder. The peak loads are actually softened by the grudgeon pin bearing in the connecting rod, an example of a squeeze film bearing.

**Figure 12.31**    Cross section of internal combustion engine, showing main bearing.

**Case Study**    (Continued)

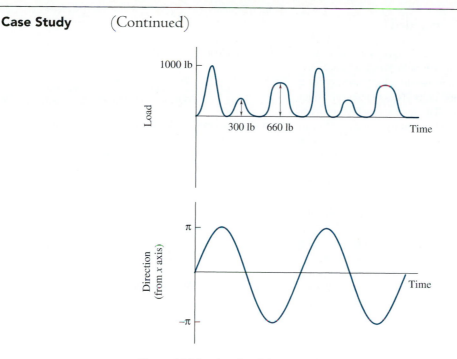

**Figure 12.32**    Load and direction of load as a function of
time at main bearing location under consideration.

It is desired to specify a clearance between the
crankshaft and the bearing, and to estimate the temperature
rise in the lubricant so that a proper cooling system can be
designed. (Obviously, the lubricant is not the only heat
source in the engine.) However, it is an important component, since oil temperatures should not become too high
inside a bearing, or else the lubricant will chemically degrade (break down).

**Given:**    A ground and lapped crankshaft with the loading
shown. The shaft diameter is 2.0 in. The lubricant is SAE
40 at a maximum inlet temperature of 200°F. Shaft speeds
vary from 800 rpm at idle to 4500 rpm maximum. The
greatest challenge is to design a bearing that operates at
high speed and high temperature and is suddenly brought
to idle. Thus, only the idle speed is considered in the problem, recognizing that the inlet temperature specified reflects high operating temperatures. The bearing width dimension must be less than 1.5 in for the design to remain
compact.

**Find:**    A bearing clearance if the surfaces are ground and
lapped, and a bearing width subject to the constraint that
the maximum lubricant temperature in the bearing not exceed 300°F.

**Solution:**    A full solution of the Reynolds equation for
this problem would be difficult because the time-dependent
load would require a time-dependent solution of a highly
nonlinear differential equation. Clearly, the approach of a
worst-case scenario will be useful in making this problem
tractable.

It is recognized that the larger the load, the smaller
the film thickness if a steady state is assumed. Thus, if the
bearing is analyzed by using the constant peak load, the
calculated film thickness will be less than the film thickness
actually encountered in application. If this film thickness
leads to acceptable frictional and thermal properties in the
bearing, the bearing will be at least as robust in practice.
Therefore, the maximum load is taken as 1000 lb.

*(continued)*

**Case Study**    (Concluded)

The bearing dimensions in the designer's direct control are the width and the clearance ratio. Obviously, increasing the width will also increase the film thickness for a given set of operating conditions. These increases would allow an increase in the radial clearance, which is beneficial for manufacturing and assembly reasons. Also, to maintain a full lubricant film, the minimum thickness must be 10 times the surface roughness in the bearing. Since it was specified that the surface is ground and lapped, the achievable surface roughness from Table 12.1 is approximately 10 $\mu$in. Thus, the design film thickness in the bearing must be larger than the allowable film thickness.

Since we are unsure which value of either the width or the clearance to use, we will need to make an assumption. Only one example of the analysis will be explained in depth, as this can be repeated by the student. We examine Table 12.2 and note that for an automotive main bearing, the average radial load per area is between 600 and 700 lb/in². Using the low value of 600 psi, we obtain a reasonable first approximation for a candidate bearing width:

$$p_{axe} = 600 \text{ lb/in}^2 = \frac{W_r}{2r_b w_t} = \frac{1000 \text{ lb}}{2(1.0)w_t}; \quad w_t = 0.833 \text{ in}$$

We could use this value and proceed, but it is convenient to use multiples of eighths or quarters of an inch in specifying dimensions. For illustrative purposes we specify $w_t = 1.0$ in, which leads to a conservative design. This bearing width can be modified as needed, but it complies with the stated requirements for a compact design. The diameter-to-width ratio is then

$$\lambda_j = \frac{2r}{wt} = \frac{2}{1} = 2$$

Since in the statement of the problem there is no mention whether the design is to be optimized for maximum load or minimum coefficient of friction in Figure 12.20, we choose a value midway between these optimum conditions. Thus, we find that $B_j = 0.19$. For $\lambda_j = 2$ and $B_j = 0.19$ from Figures 12.22 to 12.24

$$\frac{r_b\mu}{c} = 6.0 \qquad Q = 5.4 \quad \text{and} \quad \frac{q_s}{q} = 0.82$$

The radial load per area is

$$W_r^* = \frac{W_r}{2r_b w_t} = \frac{1000}{2(1.0)(1.0)} = 500 \text{ psi}$$

From Equation (12.90b) the temperature rise is

$$\Delta t_m = \frac{0.103 W_r^*(r_b\mu/c)}{Q(1 - 0.5q_s/q)} = \frac{0.103(500)(6)}{5.4[1 - 0.5(0.82)]} = 96.99°\text{F}$$

The inlet temperature is 200°F, hence from Equation (12.71) the mean temperature is

$$\tilde{t}_m = t_{mi} + \frac{\Delta t_m}{2} = 200 + \frac{96.99}{2} = 248.5°\text{F}$$

Note that the maximum temperature requirement is met, but just barely. Recognizing that we are performing a worst-case analysis, we realize that under typical operating conditions the temperature will be lower, and we are satisfied that the lubricant will not encounter excessive temperatures.

From Figure 8.17 for SAE 40 oil at 248°F the absolute viscosity is 0.7 $\mu$reyn. From Equation (12.84) solving for the radial clearance gives

$$c = r_b\sqrt{\frac{\eta_0\omega_b r_b w_t}{\pi W_r B_j}} = 1.0\sqrt{\frac{0.7(10^{-6})(83.78)(1.0)(1.0)}{\pi(1000)(0.19)}}$$
$$= 3.134 \times 10^{-4} \text{ in}$$

From Figure 12.20 for $\lambda_j = 2$ and $B_j = 0.19$

$$\frac{h_{\min}}{c} = 0.30 \quad \text{or} \quad h_{\min} = 0.30(3.134)(10^{-4}) = 94.0 \mu\text{in}$$

Although this film thickness is less than 250 $\mu$in, the design may still be adequate. Recall that if the film thickness is 10 times the surface roughness, the lubrication regime is full film lubrication. This is the case here, so even under the worst-case scenario of operating at idle speed under maximum load, the bearing will encounter very low wear. It can be shown through the same approach that the film thickness at high speed will be adequate.

## 12.7 SUMMARY

The chapter began with the derivation of the Reynolds equation by coupling the Navier-Stokes equations with the continuity equation. The Reynolds equation contains Poiseuille, physical wedge, stretch, local expansion, and normal and translation squeeze terms. Each of these terms describes a specific type of physical motion, and the physical significance of each term was explained. Standard forms of the Reynolds equation were also discussed.

Design information was given for thrust and journal bearings. Results were presented for fixed-incline thrust slider bearings from numerical evaluations of the Reynolds equation. A procedure was outlined to assist in designing these bearings. The procedure provided an optimum pad configuration and described performance parameters such as normal applied load, coefficient of friction, power loss, and lubricant flow through the bearing. Similar design information was given for a plain journal bearing.

In a normal squeeze film bearing, a positive pressure was found to occur in a fluid contained between two surfaces when the surfaces are moving toward each other. A finite time is required to squeeze the fluid out of the gap, and this action provides a useful cushioning effect in bearings. The reverse effect, which occurs when the surfaces are moving apart, can lead to cavitation in fluid films. It was found for normal squeeze film bearings that a parallel film shape produces the largest normal load-carrying capacity of all possible film shapes. In contrast, for slider bearings the parallel film was shown to produce no positive pressure and therefore no load-carrying capacity. It was also found that for the normal squeeze action, as the bearing surfaces move toward each other, the viscous fluid shows great reluctance to be squeezed out the sides of the bearing. The tenacity of a squeeze film is remarkable, and the survival of many modern bearings depends on this phenomenon. It was found that a relatively small approach velocity will provide an extremely large load-carrying capacity.

Hydrostatic bearings were briefly considered and were found to offer certain operating advantages over other types of bearings. Probably the most useful characteristics of hydrostatic bearings are their high load-carrying capacity and inherently low friction at any speed, even zero. The principles and basic concepts of film formation were discussed.

## KEY WORDS

**adiabatic temperature rise**     temperature rise in lubricant if all heat is carried away through convection.

**bearing number**     dimensionless operating parameter, specific to each type of bearing, that is important for determining bearing performance.

**Couette flow**     velocity-driven fluid flow.

**density wedge**     term dealing with density changes in bearing lubricant.

**eccentricity**     displacement of shaft center relative to bearing center.

**externally pressurized bearings**     same as **hydrostatic lubricated bearings.**

**fluid film lubrication**   lubrication condition where two surfaces transmitting a load are separated by a pressurized fluid film.

**hydrodynamic lubrication**   lubrication activated by motion of bearing surfaces.

**hydrostatic lubricated bearings**   slider bearings where lubricant is provided at elevated pressure.

**journal bearings**   bearings whose surfaces are parallel to axis of rotation.

**line of centers**   line containing bearing and shaft centers.

**local expansion**   term involving time rate of density change.

**normal squeeze**   term dealing with approach of two bearing surfaces toward each other.

**Petrov's equation**   equation for frictional torque of concentric journal bearings, given by

$$T = Fr = \frac{4\pi^2 \eta_0 r^3 w_t N_a}{c} = \frac{2\pi \eta_0 r^3 w_t \omega}{c}$$

**physical wedge**   term dealing with rate of channel convergence.

**Poiseuille flow**   pressure-driven fluid flow.

**preload factor**   fractional reduction of film clearance when pads are brought in.

**Reynolds equation**   differential equation governing pressure distribution and film thickness in bearings.

**Sommerfeld number**   bearing characteristic number for a journal bearing.

**squeeze film bearings**   bearing where one dominant film-generating mechanism is squeeze effect (see terms in Reynolds equation).

**stretch term**   term dealing with rate of velocity change in sliding direction.

**thrust bearings**   bearings whose surfaces are perpendicular to axis of rotation.

**translation squeeze**   term dealing with translation of two surfaces.

## RECOMMENDED READINGS

Halling, J. (ed.) (1978) *Principles of Tribology,* Macmillan, New York.

Hamrock, B. J., Schmid, S. R., and Jacobson, B. O. (2004) *Fundamentals of Fluid Film Lubrication,* 2nd. ed., Marcel Dekker, New York.

Juvinall, R. C., and Marshek, K. M. (2003) *Fundamentals of Machine Component Design,* 3rd ed. Wiley, New York.

Mott, R. L. (1998) *Machine Elements in Mechanical Design,* 3rd ed. Prentice-Hall, Upper Saddle River, NJ.

Norton, R. L. (1996) *Machine Design,* Prentice-Hall, Englewood Cliffs, NJ.

Radzimovski, E. I. (1959) *Lubrication of Bearings,* Ronald Press, New York.

Shigley, J. E., Mischke, C. R., and Budynas, R. G. (2003), *Mechanical Engineering Design,* 7th ed., McGraw-Hill, New York.

Szeri, A. Z. (ed.) (1980) *Tribology: Friction, Lubrication and Wear,* Hemisphere Publishing Co., Washington.

# REFERENCES

Currie, I. G. (1993) *Fundamental Mechanics of Fluids,* 2nd ed., McGraw-Hill, New York.

Dowson, D. (1998) *History of Tribology,* 2nd ed., Longmans, London.

Engineering Sciences Data Unit (ESDU) (1967) *General Guide to the Choice of Thrust Bearing Type,* Item 67033, Institution of Mechanical Engineers, London.

Hamrock, B. J. (1993) *Fundamentals of Fluid Film Lubrication,* McGraw-Hill, New York.

Hamrock, B. J., Schmid, S. R., and Jacobson, B. O. (2004) *Fundamentals of Fluid Film Lubrication,* 2nd ed., Marcel Dekker, New York.

Petrov, N. P. (1883) Friction in Machines and the Effect of the Lubricant, *Inzh. Zh. St. Petersburg,* vol. 1, pp. 71–140; vol. 2, pp. 227–279; vol. 3, pp. 377–463; vol. 4, pp. 535–564.

Raimondi, A. A., and Boyd, J. (1955) Applying Bearing Theory to the Analysis and Design of Pad-Type Bearings, *ASME Trans.,* vol. 77, no. 3, pp. 287–309.

Raimondi, A. A., and Boyd, J. (1958) A Solution for the Finite Journal Bearing and Its Application to Analysis and Design-I, -II, -III, *ASLE Trans.,* vol. 1, no. 1, I—pp. 159–174; II—pp. 175–193; III—pp. 194–209.

Reynolds, O. (1886) On the Theory of Lubrication and Its Application to Mr. Beauchamp Tower's Experiments, Including an Experimental Determination of the Viscosity of Olive Oil, *Philos. Trans. R. Soc.,* vol. 177, pp. 157–234.

Rippel, H. C. (1963) *Cast Bronze Hydrostatic Bearing Design Manual,* 2nd ed., Cast Bronze Bearing Institute, Inc., Cleveland, OH.

Shigley, J. E., and Mitchell, L. D. (1983) *Mechanical Engineering Design,* 4th ed., McGraw-Hill, New York.

Tower, B. (1883) First Report on Friction Experiments (Friction of Lubricated Bearings), *Proc. Inst. Mech. Eng. (London),* pp. 632–659.

# PROBLEMS

## Section 12.2

**12.1** To decrease the leakage past a seal of the type discussed in Example 12.1, the sealing gap is tapered and the film shape is

$$h(x) = h_o + x\,\frac{dh}{dx} = h_o + kx$$

Because of the roughness of the surfaces, the outlet (minimum) oil film thickness $h_o$ has to be at least 10 $\mu$m. Find the optimum film shape to minimize the total leakage past the 10-mm-wide seal during a full work cycle (back and forth) of the piston when the fluid viscosity is 0.075 Pa · s, the piston speed is 0.8 m/s in both directions, the stroke length is 85 mm, the sealing pressure is 1 MPa, and the piston diameter is 50 mm.

## Section 12.3

★ **12.2** The oil viscosity is 0.01 Pa · s in a very wide slider bearing with an exponential oil film shape

$$h = (h_o + s_h)\,e^{-x/l}$$

* Indicates problem of greater difficulty.

Find the pressure distribution and calculate the load-carrying capacity per unit width when the sliding speed is 10 m/s if $h_0 = 60 \ \mu$m and the bearing is 0.10 m long in the direction of motion. *Ans.* $W_z/w_t = 45$ kN/m.

★ **12.3** A flat strip of metal emerges from an oil bath having viscosity $\eta_0$ and inlet pressure $p_i$ with velocity $u_b$ on passing through a slot of the form shown in sketch *a*. In the initial convergent region (region 1) of the slot, the film thickness decreases linearly from $h_0 + s_h$ to $h_0$ over the length $n_s l$ on each side of the strip. In the final section of the slot, the film on each side has a constant thickness $h_0$ over the length $l(1 - n_s)$. The width of the flat strip is large relative to its length, so that it can be considered infinitely wide. Find the pressure distribution and volume flow rate in the $x$ direction.

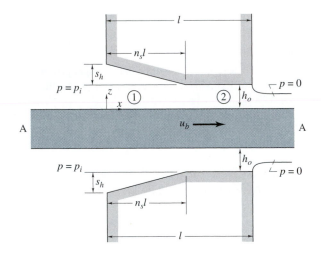

Sketch *a*, for Problem 12.3

**12.4** A fixed-incline slider bearing is optimized to give minimum relative power loss. The pad length $l$ equals the width $w_t$, and the bearing carries the load $W_z = 5$ kN. The viscosity has been chosen to avoid surface contact through the lubricant film. The sliding speed $u_b = 15$ m/s. Find the minimum film thickness for the bearing if $\eta = 0.08$ N $\cdot$ s/m², $\lambda = w_t/l = 1$, and $l = 0.070$ m. *Ans.* $h_0 = 79.0 \ \mu$m.

**12.5** The flat slider bearing in Problem 12.4 is split into two equal halves, each with width $w_t = 0.035$ m. Find the viscosity needed at the running temperature to obtain the same minimum film thickness as in Problem 12.4. The load is equally split between the two halves, with 2.5 kN each. The bearing halves should both be optimized for minimum relative power loss. *Ans.* $\eta = 0.195$ N $\cdot$ s/m².

★ **12.6** A very wide slider bearing lubricated with an oil having a viscosity of 0.05 N $\cdot$ s/m² consists of two parts, each with a constant film thickness (see sketch *b*). Find the pressure distribution in the oil and the position and size of the bearing load per unit width of the bearing. All dimensions are in millimeters. *Ans.* $W_z/w_t = 1.379$ MN/m.

Sketch b, for Problem 12.6

**12.7** Given the infinitely wide slider bearing shown in sketch c, find the pressure distribution by using the Reynolds equation.

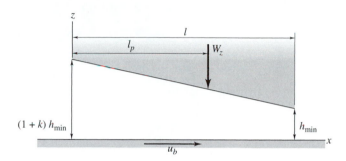

Sketch c, for Problem 12.7

**12.8** A fixed-incline pad thrust bearing carrying the weight of a water turbine for a hydroelectric powerplant is lubricated with water. The design should be optimized to obtain minimum power loss when the roughness of the bearing surfaces is $R_a = 7$ $\mu$m, indicating that the bearing minimum film thickness should be 100 $\mu$m. The load on the bearing from the weight of the turbine and the water is 1 MN. The viscosity of water is 0.001 Pa · s, and the bearing length-to-width ratio is 1, which gives minimum power loss for five pads circumferentially around the bearing with a mean radius equal to the bearing pad width. Determine the bearing dimensions, and calculate the coefficient of friction at a rotating speed of 100 rpm. *Ans.* $w_t = 1.32$ m, $\mu = 0.00067$.

**12.9** For a fixed-incline-pad thrust bearing with a total normal load of 12,000 lb, $r_o = 4$ in, $r_i = 2$ in, $\omega = 30$ r/s, $\ell/b = 1$, SAE 10 oil, and $t_i, = 100°$F, determine the following: $N_0$, $s_h$, $h_o$, $\Delta t_m$, $\mu$, $h_p$, $q$, and $q_s$.

## Section 12.4

**12.10** A plain journal bearing has a diameter of 2 in and a width of 1 in. The full journal bearing is to operate at a speed of 2000 rpm and carries a load of 750 lb. If SAE 20 oil at an inlet temperature of 100°F is to be used, determine the following:

a)  The radial clearance for optimum load-carrying capacity, the temperature rise, and the mean temperature

b)  The performance parameters: coefficient of friction, flow rate, side flow, and attitude angle

c)  The kinematic viscosity of the oil at the mean temperature if the oil density is 0.89 g/cm³

**12.11** An oil pump is designed like a journal bearing with a constant oil film thickness, as shown in sketch $d$. The wrap angle is 320°, the film thickness is 0.2 mm, and the oil viscosity is $0.022$ N·s/m². The shaft diameter $2r = 60$ mm, and the length of the pump in the axial direction is 50 mm. Find the volume pumped per unit time as a function of the resisting pressure at the rotational speed $N_a = 1500$ rpm.

Sketch d, for Problem 12.11

**12.12** A full journal bearing is used in a powerplant generator in the United States; it has four poles and delivers electricity with 60-Hz frequency. The rotor has a 30,000-kg mass and is carried equally by two journal slider bearings. The shaft diameter is 300 mm, and each bearing is 300 mm wide. The bearing and shaft surfaces are ground to a surface finish $R_a = 1$ $\mu$m. Dimension the bearings for minimum power loss, and calculate the coefficient of friction, the side-leakage ratio, and the location of the terminating pressure to find where the oil feed groove can be placed. To ensure that the bearings are dynamically stable, choose an eccentricity ratio of 0.82. Calculate the power loss in the bearing. *Ans.* $h_p = 8.10$ kW.

**12.13** The powerplant generator considered in Problem 12.12 is rebuilt and moved to Europe. The rebuilt generator has two poles, and the frequency of the electricity is 50 Hz. Redimension the bearings for minimum power loss, and calculate the coefficient of friction, the side-leakage ratio, and the location of the terminating pressure. Let the eccentricity ratio be 0.82 to make sure that the bearings are dynamically stable. Calculate the power loss. *Ans.* $h_p = 13.5$ kW.

**12.14** A motor speed of 110 rpm is used in a gearbox to provide a propeller speed of 275 rpm. The propeller shaft diameter is 780 mm, and the shaft is directly coupled to the outgoing shaft of the gearbox. The combined influence of the propeller shaft weight and the gear forces applies a 1.1-MN load on the journal bearing. Dimension the journal bearing for a diameter-to-width ratio of 2, and find the viscosity and radial clearance needed to give a minimum power loss; also calculate the power loss in the bearing. The journal bearing diameter is 780 mm, and the surface roughness $R_a = 3\mu$m. *Ans.* $\eta = 0.039$ N·s/m$^2$, $c = 500$ $\mu$m, $h_p = 24$ kW.

**12.15** A steam turbine rotor for a small powerplant in a paper factory has two journal bearings, each carrying one-half the weight of a 20,000-N rotor. The rotor speed is 3000 rpm, the bearing diameters are both 120 mm, and the bearings are 120 mm wide. Design the bearing with surface roughness $R_a = 0.6$ mm and minimum power loss. Calculate the coefficient of friction, the attitude angle, the oil flow rate, and the side-leakage flow rate. Choose an eccentricity ratio of 0.82 to avoid dynamic instability. *Ans.* $\mu = 0.00231$, $q = 1.20 \times 10^{-4}$ m$^3$/s.

**12.16** A full journal bearing is used in an air compressor to support a 5.2-kN radial load at a speed of 2300 rpm. The diameter-to-width ratio is 2, and SAE 50 oil is used with an inlet temperature of 100°C. Determine the minimum bearing dimensions and the radial clearance for both maximum load and minimum coefficient of friction. The radial load per area is 1.5 MPa. *Ans.* $w_t = 0.0416$ m.

**12.17** A shaft rotates at 4000 rpm and is supported by a journal bearing lubricated with SAE 30 oil at an inlet temperature of 25°C. The bearing has to support a 10-kN load and has a 40-mm diameter and 10-mm width. It operates at a condition which gives the largest possible load support, and the surfaces each have a roughness of 0.1 $\mu$m. Determine the film parameter, the temperature rise in the lubricant, and the coefficient of friction for this bearing.

★ **12.18** A 3.2-kN radial load is applied to a 50-mm-diameter shaft rotating at 1500 rpm. A journal bearing is used to carry the radial load, has a diameter-to-width ratio of 1, and is lubricated with SAE 20 oil with an inlet temperature of 35°C. Determine the following:

*a)* Mean temperature and temperature rise in the bearing. *Ans.* $\Delta t_m = 13.6$°C.
*b)* Minimum film thickness and its location. *Ans.* $h_{min} = 26\mu$m, $\phi = 55°$.
*c)* Maximum pressure and its location. *Ans.* $p_{max} = 2.91$ MPa, $\phi_{max} = 76°$.
*d)* Total and side-leakage flow rates. *Ans.* $q = 8.12 \times 10^{-6}$ m$^3$/s.

**12.19** A 4000-N radial load is applied to a 50-mm-diameter shaft rotating at 2000 rpm. A journal bearing is to be designed to support the radial load. The journal has a diameter-to-width ratio of 1 and is lubricated with SAE 30 oil with an inlet temperature of 35°C. It is to be designed for minimum friction. Determine the following:

*a)* The required radial clearance
*b)* Mean temperature and temperature rise in the bearing
*c)* Minimum film thickness
*d)* Total and side flow leakage rates

**12.20** A journal bearing 100 mm in diameter and 100 mm long has a radial clearance of 0.05 mm. It rotates at 2000 rpm and is lubricated with SAE 10 oil at 100°C.

*a)* Estimate the power loss and the friction torque, using the Petrov equation. Calculate the coefficient of friction if the applied load is 1000 N.
*b)* Calculate the clearance, power loss, friction torque, and coefficient of friction if the applied load is 1000 N, using the charts in Section 12.4.4.

**12.21** A 50-mm-diameter shaft rotating at 5000 rpm is to be supported by a journal bearing. The bearing is lubricated by SAE 40 oil with a mean temperature of 40°C. The bearing supports a load of 10 kN. Using a diameter-to-width ratio of 2.0, obtain the following:

a)     The required radial clearance for maximum load-carrying capability. *Ans.* $c = 47.3 \ \mu$m
b)     The coefficient of friction for the bearing. *Ans.* $\mu = 0.017$
c)     The volume flow rate of oil through the bearing
d)     The side leakage of oil in the bearing

**12.22** A 45-kN radial load is applied to a 100-mm-diameter shaft rotating at 500 rpm. A journal bearing is to be designed to support the radial load. The journal has a diameter-to-width ratio of 1 and is lubricated with SAE 30 oil at a mean temperature of 90°C. It is to be designed for maximum load support. Determine the following:

a)     The required radial clearance. *Ans.* $c = 14.4 \ \mu$m
b)     Coefficient of friction. *Ans.* $\mu = 0.00144$
c)     Minimum film thickness
d)     Total and side flow leakage rates

★ **12.23** A shaft is rotating at 1000 rpm and is supported by two journal bearings at the two ends of the shaft. The bearings are lubricated with SAE 40 oil with an inlet temperature of 25°C. A 5-kN load is applied 0.5 m from the left bearing where the total shaft length is 2.5 m. The bearing width is 25 mm, the diameter is 50 mm. Determine

a)     Radial clearance
b)     Temperature rise. *Ans.* $\Delta t_m = 50$°C.
c)     Minimum film thickness. *Ans.* $h_0 = 10 \ \mu$m.
d)     Maximum pressure. *Ans.* $p_{\max} = 9.7$ MPa.
e)     Side-leakage flow rate.
f)     Bearing power loss. *Ans.* $h_p = 79.6$ W.

**12.24** A plain journal bearing has a diameter of 2 in and a length of 1 in. The full journal bearing is to operate at a speed of 2000 rpm and carries a load of 750 lb. If SAE 10 oil at an inlet temperature of 110°F is to be used, what should the radial clearance be for optimum load-carrying capacity? Describe the surface finish that would be sufficient and yet less costly. Also indicate what the temperature rise, coefficient of friction, flow rate, side flow rate, and attitude angle are.

## Section 12.5

★ **12.25** The infinitely wide slider bearing shown in Figure 12.28 has parallel surfaces, squeeze velocity across the film of $w$, oil film thickness $h$, and bearing length $l$. Find the bearing damping constant $B = -\partial W_z'/\partial w$, where $W_z'$ is the bearing load per unit width. Use the Reynolds equation for squeeze motion. *Ans.* $B = \eta_0 l^3/h_o^3$.

## Section 12.6

★ **12.26** The circular hydrostatic thrust bearing shown in sketch $e$ has inner radius $r_i$ and outer radius $r_o = 3r_i$, a constant oil film thickness $h$, and a constant oil flow rate $q$. The bearing load is $P$, and the oil viscosity is $\eta$. Find the bearing stiffness coefficient $c_s = -\partial P/\partial h$. *Ans.* $c_s = 72\eta_0 q r_i^2/h^4$.

Sketch e, for Problem 12.26

**12.27** A circular hydrostatic thrust bearing, shown in sketch $f$, has a constant oil film thickness $h$. A central lubricant reservoir, whose height is much larger than the oil film thickness, is fed by a liquid with pressure $p_r$. Find the pump power needed if the bearing carries load $P$. Given: $r_i = 27$ mm, $\eta = 0.00452$ N·s/m$^2$, $h = 0.120$ $\mu$m, $r_o = 200$ mm, $P = 3.08 \times 10^4$ N. *Ans.* $h_p = 100$ W.

Sketch f, for Problem 12.27

★ **12.28** A sculptural display uses a 300-mm granite sphere suspended on a pressurized waterbed. Design the pad and pump to support the sphere so that the minimum film thickness is 50 $\mu$m.

# 13

# Rolling-Element Bearings

An assortment of rolling-element bearings. (Courtesy of Timken, Inc.)

*Since there is no model in nature for guiding wheels on axles or axle journals, man faced a great task in designing bearings—a task which has not lost its importance and attraction to this day.*

Rolling Bearings and their contribution to the Progress of Technology, FAG Kugelfischer Georg Schäfer KGaA, Schweinfurt, Germany, Lewis Brooks Ltd., 1986, ISBN 0-7124-1500-9.

## SYMBOLS

| | |
|---|---|
| $A$ | area, m$^2$ |
| $\tilde{A}$ | constant |
| $B$ | total conformity ratio |
| $b_w$ | bearing width, m |
| $b^*$ | contact semiwidth, m |
| $\tilde{b}$ | semiwidth in $y$ direction of uniform pressure, m |
| $C$ | dynamic load rating, N |
| $\bar{C}$ | specific dynamic capacity or load rating of bearing, N |
| $C_0$ | static load rating, N |
| $c$ | radial clearance, m |
| $c_d$ | diametral clearance, m |
| $c_e$ | free endplay, m |
| $c_r$ | distance between race curvature centers, m |
| $D_x$ | diameter of contact ellipse along $x$ axis, m |
| $D_y$ | diameter of contact ellipse along $y$ axis, m |
| $\bar{D}$ | material factor |
| $d$ | diameter of rolling element, m |
| $d_e$ | pitch diameter, m |
| $d_i, d_o$ | inner- and outer-race diameters, m |
| $E'$ | effective elastic modulus, $2\left(\dfrac{1-v_a^2}{E_a}+\dfrac{1-v_b^2}{E_b}\right)^{-1}$, Pa |
| $\bar{E}$ | metallurgical processing factor |
| $\mathcal{E}$ | complete elliptic integral of second kind |
| $\bar{e}$ | plane strain component in solid |
| $e_1$ | constant |
| $\bar{F}_l$ | lubrication factor |
| $\mathcal{F}$ | complete elliptic integral of first kind |
| $f_0$ | dimensionless factor |
| $G$ | dimensionless materials parameter, $\xi E'$ |
| $\bar{G}$ | speed effect factor |
| $H$ | dimensionless film thickness |
| $H_{max}$ | dimensionless maximum film thickness |
| $H_{min}$ | dimensionless minimum film thickness |
| $\bar{H}_m$ | misalignment factor |
| $\tilde{H}_{e,min}$ | dimensionless minimum film thickness for elliptical conjunctions obtained from curve-fitting results, $\tilde{h}_m/R_x$ |
| $h$ | film thickness, m |
| $h_{min}$ | minimum film thickness, m |
| $\tilde{h}_{min}$ | minimum film thickness obtained from curve-fitting results, m |
| $h_0$ | central film thickness, m |
| $K_1$ | constant defined in Equation (13.36) |
| $K_{1.5}$ | constant defined in Equation (13.35) |

| | |
|---|---|
| $k_e$ | ellipticity parameter, $D_y/D_x$ |
| $k_o$ | flow rate constant of orifice, m$^4$/(s $\cdot$ N$^{1/2}$) |
| $\bar{k}_e$ | ellipticity parameter from approximate formula, $\alpha_r^{2/\pi}$ |
| $L_A$ | adjusted life, millions of cycles |
| $\tilde{L}$ | fatigue life, millions of cycles |
| $\tilde{L}_{10}$ | fatigue life for probability of survival of 0.90, millions of cycles |
| $l$ | length of rolling element, m |
| $l_e$ | length of roller land, m |
| $l_l$ | roller effective length, m |
| $l_t$ | roller length, m |
| $l_v$ | length dimension in stressed volume, m |
| $\tilde{M}$ | probability of failure |
| $m_k$ | load-life exponent |
| $\bar{m}$ | slope |
| $N_a$ | rotational speed, rps |
| $N_b$ | rotational speed, rpm |
| $N_c$ | number of cycles to failure |
| $n$ | number of rolling elements per row |
| $n_s$ | static safety factor |
| $P$ | normal applied load, N |
| $P_a$ | dimensionless pressure at inlet, $p_a h_o^2/(\eta_0 l u_b)$; axial component of load, N |
| $P_e$ | dimensionless pressure, $p/E'$ |
| $P_{z_1}$ | normal applied load, N |
| $P_z$ | normal applied load, N |
| $P_z'$ | normal applied load per unit width, N/m |
| $P_r$ | radial component of load, N |
| $P_t$ | system thrust load, N |
| $P_0$ | equivalent static load, N |
| $\bar{P}$ | bearing equivalent load, N |
| $p$ | pressure, Pa |
| $p_a$ | ambient pressure, Pa |
| $p_H$ | maximum Hertzian pressure, Pa |
| $p_{max}$ | maximum pressure, Pa |
| $q_a$ | constant, $\pi/2 - 1$ |
| $R$ | curvature sum, m |
| $R_d$ | curvature difference |
| $R_q$ | root-mean-square (rms) surface roughness, m |
| $R_r$ | race conformity, $r/(2r_r)$ |
| $R_x, R_y$ | effective radii in $x$ and $y$ directions, respectively, m |
| $r$ | radius, m |
| $r_c$ | corner radius, m |
| $r_i$ | inner radius, m |
| $r_o$ | outer radius, m |
| $r_r$ | crown radius, m |
| $\tilde{S}$ | probability of survival |

| | |
|---|---|
| $s_h$ | shoulder height, m |
| $s_1$ | $r/h_0$ |
| $t_0$ | characteristic time, s |
| $U$ | dimensionless speed parameter, $\eta_0\tilde{u}/(E'R_x)$ |
| $u$ | velocity in $x$ direction, m/s |
| $u_a$ | surface $a$ velocity, m/s |
| $u_b$ | surface $b$ velocity, m/s |
| $u_0$ | characteristic velocity in $x$ direction, m/s |
| $\bar{u}$ | dimensionless velocity in $x$ direction, $u/u_0$ |
| $\tilde{u}$ | mean surface velocity in $x$ direction, $(u_a+u_b)/2$, m/s |
| $v_i, v_o$ | linear velocities of inner and outer contacts, m/s |
| $W$ | dimensionless load parameter, $P_z'/(E'R_x)$ |
| $W_t$ | total load, N |
| $W_\psi$ | load at $\psi$, N |
| $w$ | bearing load, N |
| $w_e$ | equivalent bearing load, N |
| $w_0$ | equivalent static bearing load, N |
| $w_t$ | total thrust load of bearing, N |
| $X$ | radial factor |
| $X_0$ | radial factor for static loading |
| $Y$ | thrust factor |
| $Y_0$ | thrust factor for static loading |
| $\tilde{x}, \tilde{y}$ | factors for calculating equivalent load |
| $x$ | Cartesian coordinate in direction of sliding, m |
| $y$ | Cartesian coordinate in side-leakage direction, m |
| $Z_w$ | constant defined in Equation (13.48) |
| $z$ | Cartesian coordinate in direction of film, m |
| $z_o$ | depth of maximum shear stress, m |
| $\alpha_r$ | load angle, deg |
| $\beta$ | contact angle, deg |
| $\beta_f$ | free contact angle, deg |
| $\gamma_d$ | angle defined in Figure 13.9 |
| $\delta$ | elastic deformation, m |
| $\delta_a$ | axial deflection, m |
| $\delta_m$ | elastic deformation, m |

| | |
|---|---|
| $\delta_{max}$ | interference, or total elastic compression on load line, m |
| $\delta_t$ | deflection due to thrust load, m |
| $\delta_\varphi$ | elastic compression of ball, m |
| $\eta_0$ | absolute viscosity at $p=0$ and constant temperature, Pa·s |
| $\theta_s$ | angle used to define shoulder height, deg |
| $\Lambda$ | dimensionless film parameter, $h_{min}/(R_{qa}^2 + R_{qb}^2)^{1/2}$ |
| $\nu$ | Poisson's ratio |
| $\xi$ | pressure-viscosity coefficient, m²/N |
| $\rho$ | density of lubricant, kg/m³ |
| $\rho_0$ | density at $p=0$, kg/m³ |
| $\tau_0$ | shear stress at zero pressure, Pa |
| $\phi_s$ | angle locating ball-spin vector, deg |
| $\psi$ | angle to load line, deg |
| $\psi_t$ | angular extent of bearing loading, deg |
| $\omega$ | angular velocity, rad/s |
| $\omega_b$ | angular velocity of surface $b$ or of rolling element about its own axis, rad/s |
| $\omega_c$ | angular velocity of separator or of ball set, rad/s |
| $\omega_i, \omega_o$ | angular velocities of inner and outer races, rad/s |
| $\omega_l$ | rotational speed of load vector, rad/s |
| $\omega_r$ | angular velocity of race, rad/s |
| $\omega_s$ | angular velocity due to spinning, rad/s |

**SUBSCRIPTS**

| | |
|---|---|
| $a$ | solid a |
| all | allowable |
| $b$ | solid b |
| $i$ | inner; inlet |
| max | maximum |
| $o$ | outer; outlet |
| $x, y, z$ | coordinates |
| $0$ | static loading |

## 13.1  INTRODUCTION

Chapter 12 considered hydrodynamic bearings, which assume that the surfaces are conformal and that the motion is sliding. This chapter considers **rolling-element bearings,** precise, yet simple, machine elements of great utility where the surfaces are nonconformal and the motion is primarily rolling. First, the history of rolling-element bearings is briefly reviewed. Subsequent sections describe the types of rolling-element bearing and their geometry and kinematics. The organization of this chapter is such that unloaded and unlubricated rolling-element bearings are considered in Sections 13.2 through 13.6, loaded but unlubricated rolling-element bearings in Section 13.7; and loaded and lubricated rolling-element bearings in Sections 13.8 and 13.9.

## 13.2   HISTORICAL OVERVIEW

The purpose of a bearing is to provide relative positioning and rotational freedom while transmitting a load between two structures, usually a shaft and a housing. The basic form and concept of the rolling-element bearing are simple. If loads are to be transmitted between surfaces in relative motion in a machine, the action can be facilitated most effectively if rolling elements are interposed between the sliding members. The frictional resistance encountered in sliding is then largely replaced by the much smaller resistance associated with rolling, although the arrangement inevitably involves high stresses in the restricted regions of effective load transmission.

The precision rolling-element bearing of the 20th century, a product of exacting technology and sophisticated science, has been very effective in reducing friction and wear in a wide range of machinery. The rapid development of numerous forms of rolling-element bearing in the 20th century is well known and documented. However, the origins of these vital machine elements can be traced to long before there was a large industrial demand for such devices and certainly long before adequate machine tools for their effective manufacture existed in large quantities. A complete history of rolling-element bearings is given in Hamrock and Dowson (1981); only a brief overview is presented here.

The influence of general technological progress on the development of rolling-element bearings, particularly those used in moving heavy stone building blocks and carvings and in road vehicles, precision instruments, water-raising equipment, and windmills, is discussed in Hamrock and Dowson (1981). The concept of rolling-element bearings emerged in Roman times, faded from memory during the Middle Ages, revived during the Renaissance, developed steadily in the 17th and 18th centuries for various applications, and was firmly established for individual road carriage bearings during the Industrial Revolution. Toward the end of the 19th century, the great merit of ball bearings for bicycles promoted interest in the manufacture of accurate steel balls. Initially, the balls were turned from bars on special lathes, with individual general machine manufacturing companies making their own bearings. Growing demand for both ball and roller bearings encouraged the formation of specialist bearing manufacturing companies at the turn of the century and thus laid the foundations of a great industry. The advent of precision grinding techniques and the availability of improved materials did much to confirm the future of the new industry.

The essential features of most forms of the modern rolling-element bearing were therefore established by the second half of the 19th century, but it was the formation of specialist precision-manufacturing companies in the early years of the 20th century that finally established the rolling-element bearing as a most valuable, high-quality, readily available machine component. The availability of ball and roller bearings in standard sizes has had a large impact on machine design throughout the 20th century. Such bearings still provide a challenging field for further research and development, and many engineers and scientists are currently engaged in various research projects in this area. In many cases new materials or enlightened design concepts have extended the life and range of application of rolling-element bearings. In other respects, much remains to be done in explaining the extraordinary operating characteristics of these bearings, which have served our technology so well for almost a century.

## 13.3     BEARING TYPES

Ball and roller bearings are available to the designer in a great variety of designs and size ranges. Tables 13.1 to 13.5 illustrate some of the more widely used bearing types; numerous types of specialty bearings are also available. Size ranges are generally given in metric units, because traditionally, most rolling-element bearings have been manufactured to metric dimensions, predating the efforts toward a metric standard.

Tables 13.1 to 13.5 also list approximate relative load-carrying capacities, both radial and thrust, and, where relevant, approximate tolerances to misalignment.

Rolling-element bearings are an assembly of several parts: an inner race, an outer race, a set of balls or rollers, and a cage or separator. The cage or separator maintains even spacing of the rolling elements. A cageless bearing, in which the annulus is packed with the maximum rolling-element complement, is called a *full-complement bearing*. Full-complement bearings have high load-carrying capacity but lower speed limits than bearings equipped with cages. Tapered-roller bearings are an assembly of a cup, a cone, a set of tapered rollers, and a cage.

Although rolling-element bearings will function well without a lubricant and are sometimes operated that way, it is often advantageous to provide a lubricant film to extend their life. Since lubricant supply is difficult with rolling-element bearing geometry, bearings are often packed with thick greases and then sealed so that the bearing operates under lubrication. The seals are a critical component, both to keep the lubricant inside the bearing pack and to prevent the lubricant from being washed away (e.g., as in an automobile axle when the car is driven through a puddle of water). Although lubrication is often a critical issue in bearing performance, this subject is not investigated in depth in this chapter. It should be recognized, however, that rolling-element bearings usually operate under elastohydrodynamic lubrication, often with thick greases as the fluid, and that proper lubrication is essential for long service life.

### 13.3.1     BALL BEARINGS

**Ball bearings** are used in greater quantity than any other type of rolling-element bearing. For an application where the load is primarily radial with some thrust load present, one of the types in Table 13.1 can be chosen. A **Conrad,** or **deep-groove, bearing** has a ball complement limited by the number of balls that can be packed into the annulus between the inner and outer **races** with the inner race resting against the inside diameter of the outer race. A stamped and riveted two-piece **cage,** piloted on the ball set, or a machined two-piece cage, either ball-piloted or race-piloted, is almost always used in a Conrad bearing. The only exception is a one-piece cage with open-sided pockets that is snapped into place. A filled-notch bearing most often has both inner and outer races notched so that a ball complement limited only by the annular space between the races can be used. It has low thrust capacity because of the filling notch.

**Table 13.1** Characteristics of representative radial ball bearings

| Type | | Approximate Range of Bore Sizes, mm | | Relative Capacity | | Limiting speed factor | Tolerance to misalignment |
|------|---|---------|---------|--------|--------|---------|---------|
| | | Minimum | Maximum | Radial | Thrust | | |
| Conrad or deep groove | | 3 | 1060 | 1.00 | [a]0.7 | 1.0 | ±0°15′ |
| Maximum capacity or filling notch | | 10 | 130 | 1.2–1.4 | [a]0.2 | 1.0 | ±0°3′ |
| Self-aligning, internal | | 5 | 120 | 0.7 | [b]0.2 | 1.0 | ±2°30′ |
| Self-aligning, external | | — | — | 1.0 | [a]0.7 | 1.0 | High |
| Double row, maximum | | 6 | 110 | 1.5 | [a]0.2 | 1.0 | ±0°3′ |
| Double row, deep groove | | 6 | 110 | 1.5 | [a]1.4 | 1.0 | 0° |

[a] Two directions.
[b] One direction.
SOURCE: From Hamrock and Anderson (1983).

The **self-aligning internal bearing** (Table 13.1) has an outer-race ball path ground in a spherical shape so that it can accept high levels of misalignment. The **self-aligning external bearing** (Table 13.1) has a multipiece outer race with a spherical inner race. It, too, can accept high misalignment and has higher capacity than the self-aligning internal bearing. However, the self-aligning external bearing is somewhat less self-aligning than its internal counterpart because of friction in the multipiece outer race.

Representative **angular-contact ball bearings** are illustrated in Table 13.2. An angular-contact ball bearing has a two-shouldered ball groove in one race and a single-shouldered ball groove in the other race. Thus, it can support only a unidirectional thrust load. The cutaway shoulder allows bearing assembly by snapping the inner race over the ball set after it is positioned in the cage and outer race. It also permits use of a one-piece, machined, race-piloted cage that can be balanced for high-speed operation. Typical contact angles vary from 15° to 40°.

Angular-contact ball bearings are used in **duplex pairs** mounted either back-to-back or face-to-face, as shown in Table 13.2. Duplex bearing pairs are manufactured so that they "preload" each other when clamped together in the housing and on the shaft. Preloading provides stiffer shaft support and helps prevent bearing skidding at light loads. Proper levels of preload can be obtained from the manufacturer. A duplex pair can support bidirectional thrust load. The back-to-back arrangement offers greater resistance to moment or overturning loads than does the face-to-face arrangement.

**Table 13.2**     Characteristics of representative angular-contact ball bearings (minimum bore size 10 mm)

| Type | | Approximate maximum bore size, mm | Relative Capacity | | Limiting speed factor | Tolerance to misalignment |
|---|---|---|---|---|---|---|
| | | | Radial | Thrust | | |
| One-directional thrust | | 320 | [b]1.00–1.15 | [a,b]1.5–2.3 | [b]1.1–3.0 | ±0°2′ |
| Duplex, back to back | | 320 | 1.85 | [c]1.5 | 3.0 | 0° |
| Duplex, face to face | | 320 | 1.85 | [c]1.5 | 3.0 | 0° |
| Duplex, tandem | | 320 | 1.85 | [a]2.4 | 3.0 | 0° |
| Two-directional or split-ring | | 110 | 1.15 | [c]1.5 | 3.0 | ±0°2′ |
| Double row | | 140 | 1.5 | [c]1.85 | 0.8 | 0° |

[a]One direction.
[b]Depends on contact angle.
[c]Two directions.
SOURCE: From Hamrock and Anderson (1983).

Where thrust loads exceed the load-carrying capacity of a simple bearing, two bearings can be used in tandem, with each bearing supporting part of the thrust load. Three or more bearings are occasionally used in tandem, but this is discouraged because of the difficulty in achieving effective load sharing; even slight differences in operating temperature will cause an uneven distribution of load sharing.

The **split-ring bearing** (Table 13.2) offers several advantages. The split ring (usually the inner race) has its ball groove ground as a circular arc with a shim between the ring halves. The shim is then removed when the bearing is assembled so that the split-ring ball groove has the shape of a gothic arch. This shape reduces the axial play for a given radial play and results in more accurate axial positioning of the shaft. The bearing can support bidirectional thrust loads but must not be operated for prolonged times at predominantly radial loads. This restriction results in three-point, ball-race contact and relatively high frictional losses. As with the conventional angular-contact bearing, a one-piece, precision-machined cage is used.

**Thrust ball bearings** (90° contact angle), shown in Table 13.3, are used almost exclusively for machinery with vertically oriented shafts. The flat-race bearing allows eccentricity of the fixed and rotating members; an additional bearing must be used for radial positioning.

**Table 13.3**  Characteristics of representative thrust ball bearings

| Type | Approximate Range of Bore Sizes, mm | | Relative thrust | Limiting speed | Tolerance to misalignment |
|---|---|---|---|---|---|
| | Minimum | Maximum | | | |
| One-directional, flat race | 6.45 | 88.9 | [a]0.7 | 0.10 | [b]0° |
| One-directional, grooved race | 6.45 | 1180 | [a]1.5 | 0.30 | 0° |
| Two-directional, grooved race | 15 | 220 | [c]1.5 | 0.30 | 0° |

[a]One direction.

[b]Accepts eccentricity.

[c]Two directions.

SOURCE: From Hamrock and Anderson (1983).

It has low thrust load-carrying capacity because of its small ball-race contact and consequent high Hertzian stress. Grooved-race bearings have higher thrust load-carrying capacity and can support low-magnitude radial loads. All the pure-thrust ball bearings have modest speed capability because of the 90° contact angle and the consequent high level of ball spinning and frictional losses.

## 13.3.2  ROLLER BEARINGS

**Cylindrical roller bearings** (Table 13.4) provide purely radial load support in most applications. An N- or NU-type bearing will allow free axial movement of the shaft relative to the housing to accommodate differences in thermal growth. An F- or J-type bearing will support a light thrust load in one direction; and a T-type bearing, a light bidirectional thrust load. Cylindrical roller bearings have moderately high radial load-carrying capacity as well as high-speed capability, exceeding those of either spherical or tapered-roller bearings. A commonly used bearing combination for supporting a high-speed rotor is an angular-contact ball bearing (or a duplex pair) and a cylindrical roller bearing. As explained in Section 13.4, the rollers in cylindrical roller bearings are seldom pure cylinders. They are usually crowned, or made slightly barrel-shaped, to relieve stress concentrations on the roller ends if any misalignment of the shaft and housing is present. Cylindrical roller bearings may be equipped with one- or two-piece cages, usually race-piloted. Full-complement bearings can be used for greater load-carrying capacity but at a significant sacrifice in speed capability.

**Spherical roller bearings** (Table 13.5) are made as either single- or double-row bearings. The more popular bearing design (convex) uses barrel-shaped rollers. An alternative design (concave) employs hourglass-shaped rollers. Spherical roller bearings combine

**Table 13.4**    Characteristics of representative cylindrical roller bearings

| Type | Approximate Range of Bore Sizes, mm | | Relative Capacity | | Limiting speed factor | Tolerance to misalignment |
|------|---------|---------|--------|--------|--------|--------|
| | Minimum | Maximum | Radial | Thrust | | |
| Separable outer ring, nonlocating (N) | 10 | 320 | 1.55 | 0 | 1.20 | ±0°5′ |
| Separable inner ring, nonlocating (NU) | 12 | 500 | 1.55 | 0 | 1.20 | ±0°5′ |
| Separable inner ring, one-direction locating (NJ) | 12 | 320 | 1.55 | [a]Locating | 1.15 | ±0°5′ |
| Separable inner ring, two-direction locating | 20 | 320 | 1.55 | [b]Locating | 1.15 | ±0°5′ |

[a]One direction.
[b]Two directions.
SOURCE: Hamrock and Anderson (1983).

**Table 13.5**    Characteristics of representative spherical roller bearings

| Type | Approximate Range of Bore Sizes, mm | | Relative Capacity | | Limiting speed factor | Tolerance to misalignment |
|------|---------|---------|--------|--------|--------|--------|
| | Minimum | Maximum | Radial | Thrust | | |
| Single row, barrel or convex | 20 | 320 | 2.10 | 0.20 | 0.50 | ±2° |
| Double row, barrel or convex | 25 | 1250 | 2.40 | 0.70 | 0.50 | ±1°30′ |
| Thrust | 85 | 360 | [a]0.10 [b]0.10 | [a]1.80 [b]2.40 | 0.35–0.50 | ±3° |
| Double row, concave | 50 | 130 | 2.40 | 0.70 | 0.50 | ±1°30′ |

[a]Symmetric rollers.
[b]Asymmetric rollers.
SOURCE: From Hamrock and Anderson (1983).

extremely high radial load-carrying capacity with modest thrust load-carrying capacity (with the exception of the thrust type) and excellent tolerance to misalignment. They find widespread use in heavy-duty rolling mill and industrial gear drives, where all these bearing characteristics are requisite.

Other types of roller bearing, such as tapered-roller and needle roller bearings, are not considered here but are discussed in Hamrock et al. (2004) and Hamrock and Anderson (1983).

---

**EXAMPLE 13.1**

**Given:** The geometry of the rolling surfaces in rolling-element bearings is determined from the load-carrying capacity of the elements and the type of motion. The load in ball bearings is carried by the balls. Depending on the geometry of the races, radial, axial, or moment loads can be transmitted through a ball bearing.

**Find:** Describe which ball bearing type should be used for

(a) Pure radial load
(b) Pure axial load
(c) Bending moment load
(d) High radial stiffness
(e) High axial stiffness

**Solution:**

(a) For pure radial loading, a radial ball bearing is recommended. Conrad (deep-groove) ball bearings are the least recommended. A filled-notch bearing offers additional radial capacity. If there are no space limitations and the bearing width is not restricted, a double-row bearing should be considered. Table 13.1 indicates bearing type preferences for radially loaded bearings.

(b) For pure axial loading the force transmitted through the balls should be as close as possible to the bearing axis. This location gives the smallest possible internal force acting through the balls. Table 13.3 shows thrust bearings that are appropriate for pure axial loading.

(c) For bending moment loading a bearing must support loads in both the axial and radial directions; deep-groove and double deep-groove ball bearings can be used. If two bearings are mounted so far apart that the center distance between the balls is about the same as the bearing outer diameter, any type of bearing shown in Tables 13.1 and 13.2 can be used.

(d) Equation (8.14) shows that the elastic deformation $\delta_{max}$ in a ball bearing is related to the load $W$ as follows:

$$\delta_{max} \propto W^{2/3}$$

and the contact stiffness is

$$\frac{\partial W}{\partial \delta_{max}} = \frac{1}{\partial \delta_{max}/\partial W} = \frac{1}{\frac{2}{3}W^{-1/3}} = \frac{3}{2}W^{1/3}$$

When $W$ approaches zero, the contact stiffness also approaches zero; thus the only way to get high radial stiffness in a ball bearing is to preload the balls. To obtain high radial and high axial stiffness and high running accuracy, pairs of angular-contact ball bearings are used with axial preloading against each other. The same effect can also be achieved by using a single bearing with a split ring, as shown in Table 13.2.

(e)  High axial stiffness can be achieved by preloading thrust ball bearings, such as deep-groove or angular-contact ball bearings, against each other. The higher the preload, the higher the stiffness becomes, but at the same time the shorter the bearing life becomes and the higher the power loss (temperature) becomes.

## 13.4   GEOMETRY

The operating characteristics of a rolling-element bearing depend greatly on its diametral clearance; this clearance varies for the different types of bearing discussed in Section 13.3. The principal geometric relationships governing the operation of unloaded rolling-element bearings are developed in this section. This information will be of interest when stress, deflection, load-carrying capacity, and life are considered in subsequent sections. Although bearings rarely operate in the unloaded state, an understanding of this section is essential to appreciation of the remaining sections in this chapter.

### 13.4.1   BALL BEARINGS

As described below, the geometry of ball bearings involves pitch diameter and clearance, race conformity, contact angle, endplay, shoulder height, and curvature sum and difference.

#### Pitch Diameter and Clearance

The cross section through a radial single-row ball bearing shown in Figure 13.1 depicts the radial clearance and various diameters. The **pitch diameter** $d_e$ is the mean of the inner- and outer-race contact diameters and is given by

$$d_e = d_i + \frac{d_o - d_i}{2}$$

or
$$d_e = \frac{d_o + d_i}{2} \tag{13.1}$$

Also, from Figure 13.1 the diametral clearance $c_d$ can be written as

$$c_d = d_o - d_i - 2d \tag{13.2}$$

Diametral clearance may, therefore, be thought of as the maximum distance that one race can move diametrally with respect to the other when nonmeasurable force is applied and both races lie in the same plane. Although diametral clearance is generally used in connection with single-row radial bearings, Equation (13.2) is also applicable to angular-contact bearings.

#### Race Conformity

Race conformity is a measure of the geometric conformity of the race and the ball in a plane passing through the bearing axis, which is a line passing through the center of the bearing perpendicular to its plane and transverse to the race. Figure 13.2 is a cross section of a ball

(a)

(b)

**Figure 13.1** (a) Cross section through radial single-row ball bearings [from Hamrock and Anderson (1983)]; (b) examples of radial single-row ball bearings. [Courtesy of SKF.]

bearing showing race conformity, expressed as

$$R_r = \frac{r}{d} \qquad (13.3)$$

In perfect conformity, the race radius is equal to the ball radius; thus $R_r$ is equal to $\frac{1}{2}$. The closer the race conforms to the ball, the greater the frictional heat within the contact. On the other hand, open-race curvature and reduced geometric conformity, which both reduce friction, also increase the maximum contact stresses and consequently reduce the bearing fatigue life. For this reason most ball bearings made today have race conformity ratios in the range $0.51 \le R_r \le 0.54$, with $R_r = 0.52$ being the most common. The race conformity ratio for the outer race is usually made slightly larger than that for the inner race to compensate for the closer conformity in the plane of the bearing between the outer race and the ball than between the inner race and the ball. This larger ratio tends to equalize the contact stresses at the inner- and outer-race contacts. The difference in race conformities does not normally exceed 0.02.

**Figure 13.2** Cross section of ball and outer race, showing race conformity. [From Hamrock and Anderson (1983).]

### Contact Angle

Radial bearings have some axial play since they are generally designed to have a diametral clearance, as shown in Figure 13.3. Axial play implies a free contact angle different from zero. **Angular-contact bearings** are specifically designed to operate under thrust loads. The clearance built into the unloaded bearing, along with the race conformity ratio, determines the bearing free contact angle. Figure 13.3 shows a radial bearing with contact due to the axial shift of the inner and outer races when no measurable force is applied.

Before the free contact angle is discussed, it is important to define the distance between the centers of curvature of the two races in line with the center of the ball in Figure 13.3(a)

**Figure 13.3**   Cross section of radial bearing, showing ball-race contact due to axial shift of inner and outer races. (a) Initial position; (b) shifted position. [From Hamrock and Anderson (1983).]

and (b). This distance—denoted by $x$ in Figure 13.3(a) and by $c_r$ in Figure 13.3(b)—depends on race radius and ball diameter. When quantities referred to the inner and outer races are denoted by the subscripts $i$ and $o$, respectively, Figure 13.3(a) and (b) shows that

$$\frac{c_d}{4} + d + \frac{c_d}{4} = r_o - x + r_i$$

$$x = r_o + r_i - d - \frac{c_d}{2}$$

$$d = r_o - c_r + r_i$$

or
$$c_r = r_o + r_i - d \qquad \textbf{(13.4)}$$

From these equations

$$x = c_r - \frac{c_d}{2}$$

This distance, shown in Figure 13.3(b), will be useful in defining the contact angle. Equation (13.3) can be used to write Equation (13.4) as

$$c_r = Bd \qquad \textbf{(13.5)}$$

where
$$B = R_{ro} + R_{ri} - 1 \qquad \textbf{(13.6)}$$

The quantity $B$ in Equation (13.6) is known as the *total conformity ratio* and is a measure of the combined conformity of both the outer and inner races to the ball. Calculations of bearing deflection in later sections depend on the quantity $B$.

The **free contact angle** $\beta_f$ [Fig. 13.3(b)] is defined as the angle made by a line through the points where the ball contacts both races and a plane perpendicular to the bearing axis of rotation when nonmeasurable force is applied. Note that the centers of curvature of both the outer and inner races lie on the line defining the free contact angle. From Figure 13.3(b) the expression for the free contact angle can be written as

$$\cos \beta_f = 1 - \frac{c_d}{2c_r} \qquad \textbf{(13.7)}$$

Equations (13.2) and (13.4) can be used to write Equation (13.7) as

$$\beta_f = \cos^{-1} \frac{r_o + r_i - (d_o - d_i)/2}{r_o + r_i - d} \tag{13.8}$$

Equation (13.8) shows that if the size of the balls is increased and everything else remains constant, the free contact angle decreases. Similarly, if the ball size is decreased, the free contact angle increases.

From Equation (13.7) the diametral clearance $c_d$ can be written as

$$c_d = 2c_r(1 - \cos \beta_f) \tag{13.9}$$

This is an alternative definition of the diametral clearance given in Equation (13.2).

### Endplay

Free **endplay** $c_e$ is the maximum axial movement of the inner race with respect to the outer race when both are coaxially centered and no measurable force is applied. Free endplay depends on total curvature and contact angle, as shown in Figure 13.3, and can be written as

$$c_e = 2c_r \sin \beta_f \tag{13.10}$$

Figure 13.4 shows the variation of free contact angle and endplay with the diametral clearance ratio $c_d/(2d)$ for four values of the total conformity ratio normally found in

**Figure 13.4** Free contact angle and endplay as function of $c_d/(2d)$ for four values of total conformity ratio. [From Hamrock and Anderson (1983).]

single-row ball bearings. Eliminating $\beta_f$ in Equations (13.9) and (13.10) enables the following relationships between free endplay and diametral clearance to be established:

$$c_d = 2c_r - \left[(2c_r)^2 - c_e^2\right]^{1/2}$$

$$c_e = \sqrt{4c_r c_d - c_d^2}$$

### Shoulder Height

**Shoulder height,** or **race depth,** is the depth of the race groove measured from the shoulder to the bottom of the groove and is denoted by $s_h$ in Figure 13.5. From this figure the equation defining the shoulder height can be written as

$$s_h = r(1 - \cos\theta_s) \qquad (13.11)$$

The maximum possible diametral clearance for complete retention of the ball-race contact within the race under zero thrust load is given by the condition $(\beta_f)_{\max} = \theta_s$. Making use of Equations (13.9) and (13.11) gives

$$(c_d)_{\max} = \frac{2c_r s_h}{r} \qquad (13.12)$$

**Figure 13.5**
Shoulder height in a ball bearing. [From Hamrock and Anderson (1983).]

### Curvature Sum and Difference

The undeformed geometry of contacting solids in a ball bearing can be represented by two ellipsoids, as discussed in Section 8.3.1. This section will apply the information provided there to a ball bearing.

Figure 13.6 shows a cross section of a ball bearing operating at a contact angle. Equivalent radii of curvature for both inner- and outer-race contacts in, and normal to, the direction of rolling can be calculated from this figure. The radii of curvature for the *ball–inner-race*

**Figure 13.6**   Cross section of ball bearing. [From Hamrock and Anderson (1983).]

*contact* are

$$r_{ax} = r_{ay} = \frac{d}{2} \qquad (13.13)$$

$$r_{bx} = \frac{d_e - d \cos \beta}{2 \cos \beta} \qquad (13.14)$$

$$r_{by} - R_{ri} d = -r_i \qquad (13.15)$$

The radii of curvature for the ball–outer-race contact are

$$r_{ax} = r_{ay} = \frac{d}{2} \qquad (13.16)$$

$$r_{bx} = -\frac{d_e + d \cos \beta}{2 \cos \beta} \qquad (13.17)$$

$$r_{by} = -R_{ro} d = -r_o \qquad (13.18)$$

In Equations (13.14) and (13.17), $\beta$ is used instead of $\beta_f$ because these equations are also valid when a load is applied to the contact. By setting $\beta = 0$, Equations (13.13) to (13.18) are equally valid for radial ball bearings. For thrust ball bearings $r_{bx} = \infty$, and the radii are defined as given in the preceding equations.

From the preceding radius-of-curvature expressions and Equations (8.4) and (8.5), for the *ball–inner-race* contact

$$R_{xi} = \frac{d(d_e - d \cos \beta)}{2d_e} \qquad (13.19)$$

$$R_{yi} = \frac{R_{ri} d}{2R_{ri} - 1} \qquad (13.20)$$

and for the *ball–outer-race* contact

$$R_{xo} = \frac{d(d_e + d \cos \beta)}{2d_e} \qquad (13.21)$$

$$R_{yo} = \frac{R_{ro} d}{2R_{ro} - 1} \qquad (13.22)$$

Substituting these equations into Equations (8.2) and (8.3) enables the curvature sum and difference to be obtained.

---

**EXAMPLE 13.2**

**Given:** A single-row, deep-groove ball bearing, shown in Figure 13.1, has a pitch diameter of 100 mm, an inner-race diameter of 80 mm, and a ball diameter of 19.9 mm. The outer-race radius is 10.5 mm, and the inner-race radius is 10.2 mm. The shoulder height angle is 30°.

**Find:**

(a) Diametral clearance
(b) Inner- and outer-race conformities as well as total conformity
(c) Free contact angle and endplay
(d) Shoulder heights for the inner and outer races
(e) Ellipticity parameter and the curvature sum and difference of the contacts

**Solution:**

(a) Using Equations (13.1) and (13.2) gives the diametral clearance as

$$c_d = d_o - d_i - 2d = 2d_e - 2d_i - 2d = 200 - 160 - 2(19.9) = 0.2 \text{ mm}$$

(b) Using Equation (13.3) gives the conformities of the outer and inner races as

$$R_{ro} = \frac{r_o}{d} = \frac{10.5}{19.9} = 0.528$$

$$R_{ri} = \frac{r_i}{d} = \frac{10.2}{19.9} = 0.513$$

From Equation (13.6) the total conformity is

$$B = R_{ro} + R_{ri} - 1 = 0.528 + 0.513 - 1 = 0.041$$

From Equation (13.5) the distance between the race curvature centers is

$$c_r = Bd = (0.041)(19.9) = 0.816 \text{ mm}$$

(c) From Equation (13.7) the free contact angle is

$$\cos \beta_f = 1 - \frac{c_d}{2c_r} = 1 - \frac{0.2}{2(0.816)} = 0.877$$

$$\therefore \qquad \beta_f = 28.7°$$

From Equation (13.10) the endplay is

$$c_e = 2c_r \sin \beta_f = 2(0.816) \sin 28.7° = 0.783 \text{ mm}$$

(d) From Equation (13.11) the shoulder heights are

$$s_{hi} = r_i(1 - \cos \theta_s) = 10.2(1 - \cos 30°) = 1.367 \text{ mm}$$
$$s_{ho} = r_o(1 - \cos \theta_s) = 10.5(1 - \cos 30°) = 1.407 \text{ mm}$$

From Equation (13.12) the minimum shoulder height for complete retention of the ball-race contact is

$$(s_{hi})_{min} = \frac{c_d r_i}{2c_r} = \frac{(0.2)(10.2)}{2(0.816)} = 1.25 \text{ mm}$$

$$(s_{ho})_{min} = \frac{c_d r_o}{2c_r} = \frac{(0.2)(10.5)}{2(0.816)} = 1.29 \text{ mm}$$

(e) *Ball–inner-race contact.* From Equations (13.3), (13.4), and (13.15)

$$r_{ax} = r_{ay} = \frac{d}{2} = \frac{19.9}{2} = 9.95 \text{ mm}$$

$$r_{bx} = \frac{d_e - d \cos \beta}{2 \cos \beta} = \frac{100 - 19.9 \cos 28.7°}{2 \cos 28.7°} = 47.1 \text{ mm}$$

$$r_{by} = -R_{ri}d = -r_i = -10.2 \text{ mm}$$

The curvature sums in the $x$ and $y$ directions are

$$\frac{1}{R_x} = \frac{1}{r_{ax}} + \frac{1}{r_{bx}} = \frac{1}{9.95} + \frac{1}{47.1} = 0.122/\text{mm}$$

$$\therefore \quad R_x = 8.215 \text{ mm}$$

$$\frac{1}{R_y} = \frac{1}{r_{ay}} + \frac{1}{r_{by}} = \frac{1}{9.95} - \frac{1}{10.2} = (2.46)(10^{-3})/\text{mm}$$

$$\therefore \quad R_y = 406.0 \text{ mm}$$

From Equations (8.2) and (8.3) the curvature sum and difference are

$$\frac{1}{R} = \frac{1}{R_x} + \frac{1}{R_y} = 0.122 + 0.00246 = 0.1245/\text{mm}$$

$$\therefore \quad R = 8.052 \text{ mm}$$

$$R_d = R\left(\frac{1}{R_x} - \frac{1}{R_y}\right) = 8.052(0.122 - 0.00246) = 0.9625$$

From Equations (8.6) and (8.11) the radius ratio and the ellipticity parameter are

$$\alpha_{ri} = \frac{R_y}{R_x} = \frac{406.0}{8.215} = 49.42$$

$$k_{ei} = \alpha_{ri}^{2/\pi} = (49.42)^{2/\pi} = 12.0$$

*Ball–outer-race contact.*

$$r_{ax} = r_{ay} = \frac{d}{2} = 9.95 \text{ mm}$$

$$r_{bx} = -\frac{d_e + d \cos \beta}{2 \cos \beta} = -\frac{100 + 19.9 \cos 28.7°}{2 \cos 28.7°}$$

$$r_{bx} = -66.95 \text{ mm}$$

$$r_{by} = -r_o = -10.5 \text{ mm}$$

The curvature sums in the $x$ and $y$ directions are

$$\frac{1}{R_x} = \frac{1}{r_{ax}} + \frac{1}{r_{bx}} = \frac{1}{9.95} - \frac{1}{66.95} = 0.0856/\text{mm}$$

$$\therefore \quad R_x = 11.69 \text{ mm}$$

$$\frac{1}{R_y} = \frac{1}{r_{ay}} + \frac{1}{r_{by}} = \frac{1}{9.95} - \frac{1}{10.5} = 0.0053/\text{mm}$$

$$\therefore \quad R_y = 190.0 \text{ mm}$$

The curvature sum and difference are

$$\frac{1}{R} = \frac{1}{R_x} + \frac{1}{R_y} = 0.0856 + 0.0053 = 0.0909/\text{mm}$$

$$\therefore \quad R = 11.00 \text{ mm}$$

$$R_d = R\left(\frac{1}{R_x} - \frac{1}{R_y}\right) = 11.00(0.0856 - 0.0053) = 0.8833$$

The radius ratio and the ellipticity parameter are

$$\alpha_{ro} = \frac{R_y}{R_x} = \frac{190.0}{11.69} = 16.25$$

$$k_{eo} = \alpha_r^{2/\pi} = (16.25)^{2/\pi} = 5.90$$

### 13.4.2   ROLLER BEARINGS

The equations developed for the pitch diameter $d_e$ and diametral clearance $c_d$ for ball bearings [Eqs. (13.1) and (13.2), respectively] are directly applicable for roller bearings.

#### Crowning

High stresses at the edges of the rollers in cylindrical roller bearings are usually prevented by machining a **crown** into the rollers, either fully or partially, as shown in Figure 13.7. (The crown curvature is greatly exaggerated for clarity.) Crowning the rollers also protects the bearing against the effects of slight misalignment. For cylindrical rollers $r_r/d \approx 10^2$. In contrast, for rollers in spherical roller bearings, as shown in Figure 13.7(a), $r_r/d \approx 4$. Observe in Figure 13.7 that the roller effective length $l_l$ is the length presumed to be in contact with the races under loading. Generally, the roller effective length can be written as

$$l_l = l_t - 2r_c$$

where $r_c$ is the roller corner radius or the grinding undercut, whichever is larger.

#### Race Conformity

Race conformity applies to roller bearings much as it applies to ball bearings. It is a measure of the geometric conformity of the race and the roller. Figure 13.8 shows a cross section of a spherical roller bearing. From this figure the race conformity can be written as

$$R_r = \frac{r}{2r_r}$$

(a)                                    (b)                                    (c)

**Figure 13.7**     (a) Spherical roller (fully crowned) and (b) cylindrical roller [from Hamrock and Anderson (1983).]; (c) section of toroidal roller bearing (CARB) [from SKF].

**Figure 13.8** Geometry of spherical roller bearing. [From Hamrock and Anderson (1983).]

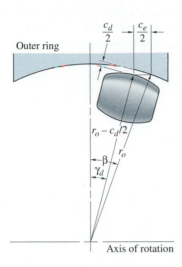

**Figure 13.9** Schematic diagram of spherical roller bearing, showing diametral play and endplay. [From Hamrock and Anderson (1983).]

In this equation, if $R_r$ and $r$ are subscripted with $i$ or $o$, the race conformity values are for the inner- or outer-race contacts, respectively.

### Free Endplay and Contact Angle

Cylindrical roller bearings have a contact angle of zero and may support thrust load only by virtue of axial flanges. Tapered-roller bearings must be subjected to a thrust load, or the inner and outer races (the cone and the cup) will not remain assembled; therefore, tapered-roller bearings do not exhibit free diametral play. Radial spherical roller bearings are, however, normally assembled with free diametral play and hence exhibit free endplay. The diametral play $c_d$ for a spherical roller bearing is the same as that obtained for ball bearings as expressed in Equation (13.2). Figure 13.9 shows diametral play, as well as endplay, for a spherical roller bearing. From this figure

$$r_o \cos\beta = \left(r_o - \frac{c_d}{2}\right)\cos\gamma_d$$

or

$$\beta = \cos^{-1}\left[\left(1 - \frac{c_d}{2r_o}\right)\cos\gamma_d\right]$$

Also, from Figure 13.9 the free endplay can be written as

$$c_e = 2r_o(\sin\beta - \sin\gamma_d) + c_d\sin\gamma_d$$

*Curvature Sum and Difference*

The same procedure used for ball bearings will be used for defining the curvature sum and difference for roller bearings. For spherical roller bearings (Fig. 13.8) the radii of curvature for the *roller–inner-race contact* can be written as

$$r_{ax} = \frac{d}{2}$$

$$r_{ay} = \frac{r_i}{2R_{ri}} = r_r$$

$$r_{bx} = \frac{d_e - d \cos \beta}{2 \cos \beta}$$

$$r_{by} = -2R_{ri} r_r = -r_i$$

The radii of curvature for the *roller–outer-race contact* can be written as

$$r_{ax} = \frac{d}{2}$$

$$r_{ay} = \frac{r_o}{2R_{ro}} = r_r$$

$$r_{bx} = -\frac{d_e + d \cos \beta}{2 \cos \beta}$$

$$r_{by} = -2R_{ro} r_r = -r_o$$

Once the radii of curvature for the respective contact conditions are known, the curvature sum and difference can be obtained directly from Equations (8.2) and (8.3). Furthermore, the radius-of-curvature expressions $R_x$ and $R_y$ for spherical roller bearings can be written for the *roller–inner-race contact* as

$$R_{xi} = \frac{d(d_e - d \cos \beta)}{2d_e} \tag{13.23}$$

$$R_{yi} = \frac{r_i r_r}{r_i - r_r} \tag{13.24}$$

and for the *roller–outer-race contact* as

$$R_{xo} = \frac{d(d_e + d \cos \beta)}{2d_e} \tag{13.25}$$

$$R_{yo} = \frac{r_o r_r}{r_o - r_r} \tag{13.26}$$

Substituting these equations into Equations (8.2) and (8.3) enables the curvature sum and difference to be obtained.

## 13.5  KINEMATICS

The relative motions of the separator, the balls or rollers, and the races of rolling-element bearings are important to understanding their performance. The relative velocities in a ball bearing are somewhat more complex than those in roller bearings, the latter being analogous to the specialized case of a zero- or fixed-contact-angle ball bearing. For that reason the

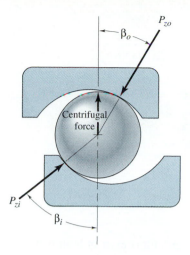

**Figure 13.10** Contact angles in ball bearing at appreciable speeds. [From Hamrock and Anderson (1983).]

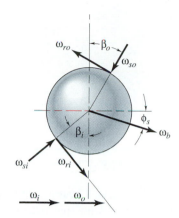

**Figure 13.11** Angular velocities of ball. [From Hamrock and Anderson (1983).]

ball bearing is used as an example here to develop approximate expressions for relative velocities. These expressions are useful for rapid but reasonably accurate calculation of elastohydrodynamic film thickness, which can be used together with surface roughnesses to calculate the lubrication life factor.

The precise calculation of relative velocities in a ball bearing where speed or centrifugal force effects, contact deformations, and elastohydrodynamic traction effects are considered requires a large computer to numerically solve the relevant equations. Such a treatment is beyond the scope of this text.

When a ball bearing operates at high speeds, the centrifugal force acting on the ball creates a divergency of the inner- and outer-race contact angles, as shown in Figure 13.10, in order to maintain force equilibrium on the ball. For the most general case of rolling and spinning at both inner- and outer-race contacts, the rolling and spinning velocities of the ball are as shown in Figure 13.11.

Jones (1964) developed the equations for ball and separator angular velocity for all combinations of inner- and outer-race rotation. Without introducing additional relationships to describe the elastohydrodynamic conditions at both ball-race contacts, however, the ball-spin axis orientation angle $\phi_s$ cannot be obtained. As mentioned, this requires a lengthy numerical solution except for the two extreme cases of outer- or inner-race control. Figure 13.12 illustrates these cases.

**Race control** assumes that pure rolling occurs at the controlling race, with all the ball spin occurring at the other race contact. The orientation of the ball rotation axis can then be easily determined from bearing geometry. Race control probably occurs only in dry bearings or dry-film-lubricated bearings, where Coulomb friction conditions exist in the ball-race contact ellipses. The spin-resisting moment will always be greater at one of the race contacts. Pure rolling will occur at the race contact with the higher-magnitude spin-resisting moment, usually the inner race at low speeds and the outer race at high speeds.

**Figure 13.12**     Ball-spin orientations for (a) outer-race control and (b) inner-race control. [From Hamrock and Anderson (1983).]

In oil-lubricated bearings in which elastohydrodynamic films exist in both ball-race contacts, rolling with spin occurs at both contacts. Therefore, precise ball motions can only be determined through a computer analysis. The situation can be approximated with a reasonable degree of accuracy, however, by assuming that the ball rolling axis is normal to the line drawn through the centers of the two ball-race contacts [Fig. 13.12(b)].

The angular velocity of the separator or ball set $\omega_c$ about the shaft axis can be shown to be (Anderson, 1970)

$$\omega_c = \frac{(v_i + v_o)/2}{d_e/2} = \frac{1}{2}\left[\omega_i\left(1 - \frac{d\cos\beta}{d_e}\right) + \omega_o\left(1 + \frac{d\cos\beta}{d_e}\right)\right] \tag{13.27}$$

where $v_i$ and $v_o$ are the linear velocities at the inner and outer contacts, respectively. The angular velocity of a ball about its own axis $\omega_b$, assuming no spin, is

$$\omega_b = \frac{v_i - v_o}{d_e/2} = \frac{d_e}{2d}\left[\omega_i\left(1 - \frac{d\cos\beta}{d_e}\right) - \omega_o\left(1 + \frac{d\cos\beta}{d_e}\right)\right] \tag{13.28}$$

It is convenient for calculating the velocities at the ball-race contacts, which are required for calculating elastohydrodynamic film thicknesses, to use a coordinate system that rotates at $\omega_c$. This system fixes the ball-race contacts relative to the observer. In the rotating coordinate system, the angular velocities of the inner and outer races become

$$\omega_{ri} = \omega_i - \omega_c = \frac{\omega_i - \omega_o}{2}\left(1 + \frac{d\cos\beta}{d_e}\right)$$

$$\omega_{ro} = \omega_o - \omega_c = \frac{\omega_o - \omega_i}{2}\left(1 - \frac{d\cos\beta}{d_e}\right)$$

The surface velocities entering the *ball–inner-race contact* for pure rolling are

$$u_{ai} = u_{bi} = \frac{\omega_{ri}(d_e - d\cos\beta)}{2} \tag{13.29}$$

$$u_{ai} = u_{bi} = \frac{d_e(\omega_i - \omega_o)}{4}\left(1 - \frac{d^2\cos^2\beta}{d_e^2}\right) \tag{13.30}$$

and those at the *ball–outer-race contact* are

$$u_{ao} = u_{bo} = \omega_{ro}\frac{d_e + d\cos\beta}{2} \tag{13.31}$$

$$u_{ao} = u_{bo} = \frac{d_e(\omega_o - \omega_i)}{4}\left(1 - \frac{d^2\cos^2\beta}{d_e^2}\right) \tag{13.32}$$

Thus,

$$|u_{ai}| = |u_{ao}|$$

For a cylindrical roller bearing $\beta = 0$, and Equations (13.27) to (13.32) become, if $d$ is roller diameter,

$$\omega_c = \frac{1}{2}\left[\omega_i\left(1 - \frac{d}{d_e}\right) + \omega_o\left(1 + \frac{d}{d_e}\right)\right]$$

$$\omega_b = \frac{d_e}{2d}\left[\omega_i\left(1 - \frac{d}{d_e}\right) + \omega_o\left(1 + \frac{d}{d_e}\right)\right]$$

$$u_{ai} = u_{bi} = \frac{d_e(\omega_i - \omega_o)}{4}\left(1 - \frac{d^2}{d_e^2}\right) \tag{13.33}$$

$$u_{ao} = u_{bo} = \frac{d_e(\omega_o - \omega_i)}{4}\left(1 - \frac{d^2}{d_e^2}\right)$$

Equations directly analogous to those for a ball bearing can be used for a tapered-roller bearing if $d$ is the average diameter of the tapered roller, $d_e$ is the diameter at which the geometric center of the rollers is located, and $\beta$ is the contact angle, as shown in Figure 13.13.

(a)

(b)

**Figure 13.13** Tapered-roller bearing. (a) Tapered-roller bearing with outer race removed to show rolling elements [from Timken, Inc.]; (b) simplified geometry for tapered-roller bearing [from Hamrock and Anderson (1983)].

**EXAMPLE 13.3**

**Given:** As shown in Figure 13.12, the motion of the balls can be guided by the inner-race contact or by the outer-race contact or by both. This is dependent on both the surface roughness and the surface geometry of angular-contact bearings. At very high speeds the centrifugal forces acting on the balls change both the contact angle of the outer race and the friction forces between the ball and the outer race. The outer-race contact angle is 20°, and the inner-race contact angle is 31°. The ball diameter is 14 mm, and the pitch diameter is 134 mm.

**Find:** The angular speed of the balls and the separator in an angular-contact ball bearing at an inner-race speed of 10,000 rpm when the balls are

(a)   Outer-race-guided
(b)   Inner-race-guided

**Solution:**

(a)   From Equation (13.27) the separator angular speed when the balls are outer-race-guided is

$$\omega_{co} = \frac{1}{2}\left[\omega_i\left(1 - \frac{d\cos\beta_o}{d_e}\right) + \omega_o\left(1 + \frac{d\cos\beta_o}{d_e}\right)\right]$$

$$= \frac{10,000\,(2\pi/60)}{2}\left(1 - \frac{14\cos 20°}{134}\right) = 472.2 \text{ rad/s} = 4509 \text{ rpm}$$

From Equation (13.28) the ball speed when the balls are outer-race-guided is

$$\omega_{bo} = \frac{d_e}{2d}\left[\omega_i\left(1 - \frac{d\cos\beta_o}{d_e}\right)\right]$$

$$= \frac{(134.0)(10,000)\,(2\pi/60)}{2(14.0)}\left(1 - \frac{14.0\cos 20°}{134.0}\right) = 4520 \text{ rad/s} = 43,160 \text{ rpm}$$

(b)   From Equation (13.27)

$$\omega_{ci} = \frac{1}{2}\left[\omega_i\left(1 - \frac{d\cos\beta_i}{d_e}\right)\right] = \frac{10,000(2\pi/60)}{2}\left(1 - \frac{14\cos 31°}{134}\right)$$

$$= 476.7 \text{ rad/s} = 4552 \text{ rpm}$$

From Equation (13.28)

$$\omega_{bi} = \frac{134.0}{2(14)}\left[(10,000)\frac{2\pi}{60}\left(1 - \frac{14\cos 31°}{134}\right)\right] = 4563 \text{ rad/s} = 43,570 \text{ rpm}$$

## 13.6    SEPARATORS

Ball and roller bearing **separators** (sometimes called cages or **retainers**) are bearing components that, although they never carry load, can exert a vital influence on bearing efficiency. In a bearing without a separator, the rolling elements contact each other during operation and experience severe sliding and friction. The primary functions of a separator are to maintain the proper distance between the rolling elements and to ensure proper load distribution and balance within the bearing. Another function of the separator is to maintain control of the rolling elements so as to produce the least possible friction through sliding contact. Furthermore, a separator is necessary for several types of bearing to prevent the rolling elements from falling out of the bearing during handling. Most separator troubles arise from improper mounting, misaligned bearings, or improper (inadequate or excessive) clearance in the rolling-element pocket.

The materials used for separators vary according to the type of bearing and the application. In ball bearings and some sizes of roller bearing, the most common type of separator is made from two strips of carbon steel that are pressed and riveted together. Called *ribbon separators,* they are the least expensive separators to manufacture and are entirely suitable for many applications; they are also lightweight and usually require little space.

The design and construction of angular-contact ball bearings allow the use of a one-piece separator. The simplicity and inherent strength of one-piece separators permit their fabrication from many desirable materials. Reinforced phenolic and bronze are the two most commonly used materials. Bronze separators offer strength and low-friction characteristics and can be operated at temperatures to 230°C (450°F). Machined, silver-plated, ferrous-alloy separators are used in many demanding applications. Because reinforced cotton-based phenolic separators combine the advantages of low weight, strength, and nongalling properties, they are used for such high-speed applications as gyroscope bearings. Lightness and strength are particularly desirable in high-speed bearings, since the stresses increase with speed but may be greatly minimized by reducing the separator weight. A limitation of phenolic separators, however, is that they have an allowable maximum temperature of about 135°C (275°F).

---

**Given:** A lightly loaded lubricated roller on a flat surface. The minimum oil film thickness is directly proportional to the roller radius. The cage pocket for a cylindrical roller bearing can be made with straight sides or with sides having a radius 5% larger than the roller radius.

**EXAMPLE 13.4**

**Find:** How much thicker will the oil film between the roller and the pockets be with a radius than with the straight pockets?

**Solution:** The geometry for a cage pocket with 5% larger radius than the roller is equivalent to that of a much larger roller on a flat. The equivalent radius $R$ giving the same oil film is

$$\frac{1}{R} = \frac{1}{r} - \frac{1}{1.05r} = \frac{0.05}{1.05r}$$

$$\therefore \qquad R = 21r$$

Since the oil film thickness is directly proportional to the equivalent radius of the contact, the film thickness will be 21 times as large as that for a straight cage pocket.

---

## 13.7 STATIC LOAD DISTRIBUTION

Since a simple analytical expression for the deformation in terms of load was defined in Section 8.3.1, it is possible to consider how the bearing load is distributed among the elements. Most rolling-element bearing applications involve steady-state rotation of the inner race or outer race or both. However, the rotational speeds are usually not so high as to cause ball or roller centrifugal forces or gyroscopic moments of significant magnitudes. In analyzing the loading distribution on the rolling elements, it is usually satisfactory to ignore these effects in most applications. In this section the load deflection relationships for ball and roller bearings are given, along with radial and thrust load distributions of statically loaded rolling elements.

### 13.7.1    LOAD DEFLECTION RELATIONSHIPS

For an elliptical conjunction, the load deflection relationship given in Equation (8.14) can be written as

$$w_z = K_{1.5}\delta^{3/2} \tag{13.34}$$

where
$$K_{1.5} = \pi k_e E' \left(\frac{2\mathcal{E}R}{9\mathcal{F}^3}\right)^{1/3} \tag{13.35}$$

Similarly, for a rectangular conjunction from Equation (8.18)

$$w_z = K_1\delta_m$$

where
$$K_1 = \frac{(\pi/2)lE'}{2\ln(4R_x/b^*) - 1} \tag{13.36}$$

and $l$ = length of rolling element.

In general, then,

$$w_z = K_j\delta_m^j \tag{13.37}$$

in which $j$ is 1.5 for ball bearings and 1.0 for roller bearings. The total normal approach between two races separated by a rolling element is the sum of the deformations under load between the rolling element and both races. Therefore,

$$\delta_m = \delta_{mo} + \delta_{mi} \tag{13.38}$$

where
$$\delta_{mo} = \left[\frac{w_z}{(K_j)_o}\right]^{1/j} \tag{13.39}$$

$$\delta_{mi} = \left[\frac{w_z}{(K_j)_i}\right]^{1/j} \tag{13.40}$$

Substituting Equations (13.38), (13.39), and (13.40) into Equation (13.37) gives

$$K_j = \frac{1}{\{[1/(K_j)_o]^{1/j} + [1/(K_j)_i]^{1/j}\}^j} \tag{13.41}$$

Recall that $(K_j)_o$ and $(K_j)_i$ are defined by Equations (13.35) and (13.36) for an elliptical and a rectangular conjunction, respectively. These equations show that $(K_j)_o$ and $(K_j)_i$ are functions of only the contact geometry and the material properties. The radial and thrust load analyses are presented in the following two sections and are directly applicable for radially loaded ball and roller bearings and thrust-loaded ball bearings.

## 13.7.2    RADIALLY LOADED BALL AND ROLLER BEARINGS

Figure 13.14 shows a radially loaded rolling-element bearing with radial clearance $c_d/2$. In the concentric arrangement [Fig. 13.14(a)] a uniform radial clearance between the rolling element and the races of $c_d/2$ is evident. The application of an arbitrarily small radial load to the shaft causes the inner race to move a distance $c_d/2$ before contact is made between a rolling element located on the load line and the inner and outer races. At any angle there will still be a radial clearance $c$ that, if $c_d$ is small relative to the radius of the groove, can be expressed with adequate accuracy by

$$c = (1 - \cos \psi) \frac{c_d}{2}$$

On the load line when $\psi = 0°$, the clearance is zero; but when $\psi = 90°$, the clearance retains its initial value of $c_d/2$.

The application of further load will elastically deform the balls and eliminate clearance around an arc $2\psi_l$. If the interference or total elastic compression on the load line is $\delta_{max}$,

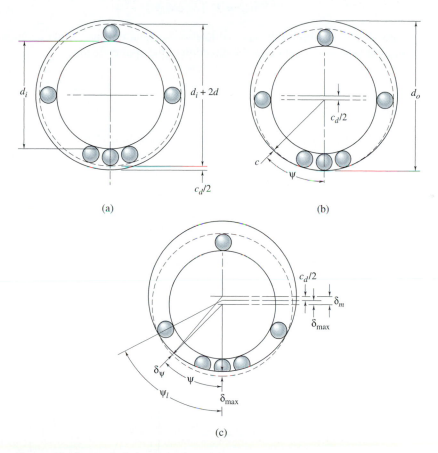

**Figure 13.14**    Radially loaded rolling-element bearing. (a) Concentric arrangement; (b) initial contact; (c) interference. [From Hamrock and Anderson (1983).]

the corresponding elastic compression of the ball $\delta_\psi$ along a radius at angle $\psi$ to the load line is given by

$$\delta_\psi = \delta_{max} \cos \psi - c = \left(\delta_{max} + \frac{c_d}{2}\right) \cos \psi - \frac{c_d}{2}$$

This assumes that the races are rigid. It is clear from Figure 13.14(c) that $\delta_{max} + c_d/2$ represents the total relative radial displacement of the inner and outer races. Hence,

$$\delta_\psi = \delta_m \cos \psi - \frac{c_d}{2} \tag{13.42}$$

The relationship between load and elastic compression along the radius at angle $\psi$ to the load vector is given by Equation (13.37) as

$$w_\psi = K_j \delta_\psi^j$$

Substituting Equation (13.42) into this equation gives

$$w_\psi = K_j \left(\delta_m \cos \psi - \frac{c_d}{2}\right)^j$$

For static equilibrium the applied load must equal the sum of the components of the rolling-element loads parallel to the direction of the applied load.

$$w_t = \sum w_\psi \cos \psi$$

Therefore,
$$w_t = K_j \sum \left(\delta_m \cos \psi - \frac{c_d}{2}\right)^j \cos \psi \tag{13.43}$$

The angular extent of the bearing arc $2\psi_l$ in which the rolling elements are loaded is obtained by setting the root expression in Equation (13.43) equal to zero and solving for $\psi$.

$$\psi_l = \cos^{-1} \frac{c_d}{2\delta_m}$$

The summation in Equation (13.43) applies only to the angular extent of the loaded region. This equation can be written for a roller bearing as

$$w_t = \left(\psi_l - \frac{c_d}{2\delta_m} \sin \psi_l\right) \frac{n K_1 \delta_m}{2\pi} \tag{13.44}$$

and similarly in integral form for a ball bearing as

$$w_t = \frac{n}{(w_z)_{max}} K_{1.5} \delta_m^{3/2} \int_0^{\psi_l} \left(\cos \psi - \frac{c_d}{2\delta}\right)^{3/2} \cos \psi \, d\psi$$

The integral in the equation can be reduced to a standard elliptic integral by the hypergeometric series and the beta function. If the integral is numerically evaluated directly, the following approximate expression is derived:

$$\int_0^{\psi_l} \left(\cos \psi - \frac{c_d}{2\delta}\right)^{3/2} \cos \psi \, d\psi = 2.491 \left\{ \left[ 1 + \left(\frac{c_d/2\delta_m - 1}{1.23}\right)^2 \right]^{1/2} - 1 \right\}$$

which fits the exact numerical solution to within 2% for a complete range of $c_d/(2\delta_m)$.

The load carried by the most heavily loaded rolling element is obtained by substituting $\psi = 0°$ in Equation (13.43) and dropping the summation sign.

$$(w_z)_{max} = K_j \delta_m^j \left(1 - \frac{c_d}{2\delta_m}\right)^j \tag{13.45}$$

Dividing this maximum load by the total radial load for a roller bearing [Eq. (13.44)] gives

$$w_z = \frac{[\psi_l - (c_d/2\delta_m)\sin\psi_l]\,n(w_z)_{max}/(2\pi)}{1 - c_d/(2\delta_m)} \tag{13.46}$$

and, similarly, for a ball bearing

$$w_z = \frac{n(w_z)_{max}}{Z_w} \tag{13.47}$$

where

$$Z_w = \frac{\pi(1 - c_d/2\delta_m)^{3/2}}{2.491\left\{\left[1 + \left(\frac{1 - c_d/2\delta_m}{1.23}\right)^2\right]^{1/2} - 1\right\}} \tag{13.48}$$

For roller bearings when the diametral clearance $c_d$ is zero, Equation (13.46) gives

$$w_z = \frac{n(w_z)_{max}}{4} \tag{13.49}$$

For ball bearings when the diametral clearance $c_d$ is zero, the value of $Z_w$ in Equation (13.47) becomes 4.37. This is the value derived by Stribeck (1901) for ball bearings of zero diametral clearance. The approach used by Stribeck was to evaluate the finite summation for various numbers of balls. He then derived the celebrated Stribeck equation for static load-carrying capacity by writing the more conservative value of 5 for the theoretical value of 4.37:

$$w_z = \frac{n(w_z)_{max}}{5} \tag{13.50}$$

In using Equation (13.50) it should be remembered that $Z_w$ was considered to be a constant and that the effects of clearance and applied load on load distribution were not taken into account. These effects were, however, considered in obtaining Equation (13.47).

### 13.7.3 THRUST-LOADED BALL BEARINGS

The **static thrust load-carrying capacity** of a ball bearing may be defined as the maximum thrust load that the bearing can endure before the contact ellipse approaches a race shoulder, as shown in Figure 13.15, or as the load at which the allowable mean compressive stress is reached, whichever is smaller. Both the limiting shoulder height and the mean compressive stress must be calculated to find the static thrust load-carrying capacity.

Each ball is subjected to an identical thrust component $W_t/n$, where $W_t$ is the total thrust load. The initial free contact angle prior to the application of a thrust load is

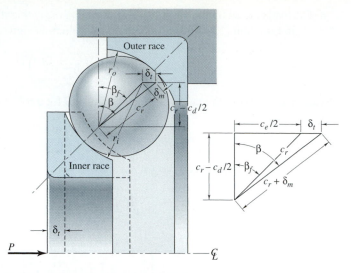

**Figure 13.15**   Contact ellipse in bearing race under load. [From Hamrock and Anderson (1983).]

**Figure 13.16**   Angular-contact ball bearing under thrust load. [From Hamrock and Anderson (1983).]

denoted by $\beta_f$. Under load, the normal ball thrust load $W_t$ acts at the contact angle $\beta$ and is written as

$$W = \frac{W_t}{n \sin \beta} \tag{13.51}$$

Figure 13.16 shows a cross section through an angular-contact bearing under a thrust load $W_t$. From this figure the contact angle after the thrust load has been applied can be written as

$$\beta = \cos^{-1} \left( \frac{c_r - c_d/2}{c_r + \delta_m} \right) \tag{13.52}$$

The initial free contact angle was given in Equation (13.7). Using that equation and rearranging terms in Equation (13.52) give, solely from geometry (Fig. 13.16),

$$\delta_m = c_r \left( \frac{\cos \beta_f}{\cos \beta} - 1 \right) \tag{13.53}$$

$$\delta_m = \delta_{mo} + \delta_{mi}$$

$$\delta_m = \left[ \frac{W}{(K_j)_o} \right]^{1/j} + \left[ \frac{W}{(K_j)_i} \right]^{1/j}$$

$$K_{1.5} = \frac{1}{\left\{ \left[ \frac{(4.5\mathcal{F}_o^3)^{1/2}}{\pi k_{eo} E_o' (R_o \mathcal{E}_o)^{1/2}} \right]^{2/3} + \left[ \frac{(4.5\mathcal{F}_i^3)^{1/2}}{\pi k_{ei} E_i' (R_i \mathcal{E}_i)^{1/2}} \right]^{2/3} \right\}^{3/2}} \tag{13.54}$$

$$W = K_{1.5} c_r^{3/2} \left( \frac{\cos \beta_f}{\cos \beta} - 1 \right)^{3/2} \tag{13.55}$$

where Equation (13.35) for $K_{1.5}$ is replaced by Equation (13.54) and $k$, $\mathcal{E}$, and $\mathcal{F}$ are given in Table 8.1.

From Equations (13.51) and (13.55)

$$\frac{W_t}{n \sin \beta} = W \tag{13.56}$$

$$\frac{W_t}{n K_{1.5} c_r^{3/2}} = \sin \beta \left( \frac{\cos \beta_f}{\cos \beta} - 1 \right)^{3/2} \tag{13.57}$$

This equation can be numerically solved by the Newton-Raphson method. The iterative equation to be satisfied is

$$\beta' - \beta = \frac{\dfrac{W_t}{n K_{1.5} c_r^{3/2}} - \sin \beta \left( \dfrac{\cos \beta_f}{\cos \beta} - 1 \right)^{3/2}}{\cos \beta \left( \dfrac{\cos \beta_f}{\cos \beta} - 1 \right)^{3/2} + \dfrac{3}{2} \cos \beta_f \tan^2 \beta \left( \dfrac{\cos \beta_f}{\cos \beta} - 1 \right)^{1/2}} \tag{13.58}$$

In this equation convergence is satisfied when $\beta' - \beta$ becomes essentially zero.

When a thrust load is applied, the shoulder height limits the axial deformation, which can occur before the pressure-contact ellipse reaches the shoulder. As long as the following inequality is satisfied, the contact ellipse will not exceed this limit:

$$\theta_s > \beta + \sin^{-1} \frac{D_y}{R_r d}$$

From Figure 13.5 and Equation (13.11) the angle $\theta_s$ used to define the shoulder height can be written as

$$\theta_s = \cos^{-1} \left( 1 - \frac{s_h}{R_r d} \right)$$

From Figure 13.16 the axial deflection $\delta_t$ corresponding to a thrust load can be written as

$$\delta_t = (c_r + \delta_m) \sin \beta - c_r \sin \beta_f \tag{13.59}$$

Substituting Equation (13.53) into Equation (13.59) gives

$$\delta_t = \frac{c_r \sin \left( \beta - \beta_f \right)}{\cos \beta} \tag{13.60}$$

Note that once $\beta$ has been determined from Equation (13.57) and $\beta_f$ from Equation (13.7), the relationship for $\delta_t$ can be easily evaluated.

### 13.7.4 PRELOADING

The use of angular-contact bearings as duplex pairs preloaded against each other is discussed in Section 13.3. As shown in Table 13.2, duplex bearing pairs are used in either back-to-back or face-to-face arrangements. Such bearings are usually preloaded against each other by providing what is called *stickout* in the manufacture of the bearing. Figure 13.17 illustrates this for a bearing pair used in a back-to-back arrangement. The magnitude of the stickout and the bearing design determine the level of preload on each bearing when the bearings

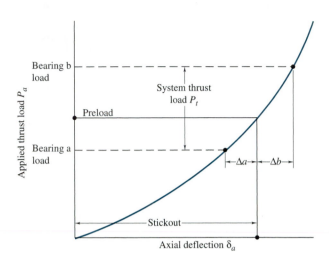

**Figure 13.17**   Angular-contact bearings in back-to-back arrangement, shown (a) individually as manufactured and (b) as mounted with preload. [From Hamrock and Anderson (1983).]

**Figure 13.18**   Thrust load–axial deflection curve for typical ball bearing. [From Hamrock and Anderson (1983).]

are clamped together as in Figure 13.17. The magnitude of preload and the load deflection characteristics for a given bearing pair can be calculated by using Equations (13.7), (13.34), (13.51), (13.54), and (13.57).

Figure 13.18 shows the relationship of initial preload, system load, and final load for bearings a and b. The load deflection curve follows the relationship $\delta_m = KP^{2/3}$. When a system thrust load $P_t$ is imposed on the bearing pairs, the magnitude of load on bearing b increases while that on bearing a decreases until the difference equals the system load. The physical situation demands that the change in each bearing deflection be the same ($\Delta a = \Delta b$ in Fig. 13.18). The increments in bearing load, however, are not the same. This is important because it always requires a system thrust load far greater than twice the preload before one bearing becomes unloaded. Prevention of bearing unloading, which can result in skidding and early failure, is an objective of preloading.

### 13.7.5   STATIC LOAD RATING

The **static load rating** for rolling-element bearings represents the loads at which, under defined operating conditions, the load carried by the most heavily loaded rolling element $w_{\max}$ equals the allowable rolling-element load $w_{\text{all}}$, which is the maximum load before plastic (rather than elastic) deformation occurs within the bearing.

The static load rating $C_0$ can be expressed as

$$C_0 = n_{s0} w_0 \tag{13.61}$$

where $n_{s0}$ = static safety factor

$w_0$ = equivalent static bearing load

Table 13.6 gives the static load rating for selected single-row, deep-groove ball bearings. This table gives the overall dimensions of the bearing $d_b$, $d_a$, and $b_w$ (see Fig. 13.1 for details); both static and dynamic load ratings; the allowable load limit; speed ratings for greases and oil; and the bearing designation. **Static thrust load-carrying capacity** is the load that gives small plastic deformation of the races.

The allowable load limit may be viewed as the maximum bearing load where fatigue failure will not occur. Thus, if the design load were equivalent to the allowable load limit, the safety factor would be 1 for infinite life. Also, the speed ratings in this table correspond to the maximum speed the bearing should experience before the bearing design must be altered to handle the high temperature developed in the bearing.

**Table 13.6**     Single-row, deep-groove ball bearings

| Principal Dimensions | | | Basic Load Ratings | | Allowable load limit | Speed Ratings | | Mass | Designation |
|---|---|---|---|---|---|---|---|---|---|
| | | | Dynamic | Static | | Grease | Oil | | |
| $d_b$ | $d_a$ | $b_w$ | $C$ | $C_0$ | $w_{all}$ | | | | |
| mm | | | N | | N | | | kg | |
| in | | | lb | | lb | rpm | | lbm | — |
| 15 | 32 | 8 | 5,590 | 2,850 | 120 | 22,000 | 28,000 | 0.025 | 16002 |
| 0.5906 | 1.2598 | 0.3150 | 1,260 | 641 | 27.0 | | | 0.055 | |
| | 32 | 8 | 5,590 | 2,850 | 120 | 22,000 | 28,000 | 0.030 | 6002 |
| | 1.2598 | 0.3150 | 1,260 | 641 | 27.0 | | | 0.066 | |
| | 35 | 11 | 7,800 | 3,750 | 160 | 19,000 | 24,000 | 0.045 | 6202 |
| | 1.3780 | 0.4331 | 1,750 | 843 | 36.0 | | | 0.099 | |
| | 35 | 13 | 11,400 | 5,400 | 228 | 17,000 | 20,000 | 0.082 | 6302 |
| | 1.3780 | 0.5118 | 2,560 | 1,210 | 51.3 | | | 0.18 | |
| 20 | 42 | 8 | 6,890 | 4,050 | 173 | 17,000 | 20,000 | 0.050 | 16004 |
| 0.7874 | 1.6535 | 0.3150 | 1,550 | 910 | 38.9 | | | 0.11 | |
| | 42 | 12 | 9,360 | 5,000 | 212 | 17,000 | 20,000 | 0.069 | 6004 |
| | 1.6535 | 0.4724 | 2,100 | 1,120 | 47.7 | | | 0.15 | |
| | 47 | 14 | 12,700 | 6,550 | 280 | 15,000 | 18,000 | 0.11 | 6204 |
| | 1.8504 | 0.5512 | 2,860 | 1,470 | 62.9 | | | 0.24 | |
| | 52 | 15 | 15,900 | 7,800 | 335 | 13,000 | 16,000 | 0.14 | 6304 |
| | 2.0472 | 0.5906 | 3,570 | 1,750 | 75.3 | | | 0.31 | |
| | 72 | 19 | 30,700 | 15,000 | 640 | 10,000 | 13,000 | 0.40 | 6404 |
| | 2.8346 | 0.7480 | 6,900 | 3,370 | 144 | | | 0.88 | |
| 25 | 47 | 12 | 11,200 | 6,550 | 275 | 15,000 | 18,000 | 0.080 | 6005 |
| 0.9843 | 1.8504 | 0.4724 | 5,520 | 1,470 | 61.8 | | | 0.18 | |
| | 52 | 15 | 14,000 | 7,800 | 335 | 12,000 | 15,000 | 0.13 | 6205 |
| | 2.0472 | 0.5906 | 3,150 | 1,750 | 75.3 | | | 0.29 | |
| | 62 | 17 | 22,500 | 11,600 | 490 | 11,000 | 14,000 | 0.23 | 6305 |
| | 2.4409 | 0.6693 | 5,060 | 2,610 | 110 | | | 0.51 | |
| | 80 | 21 | 35,800 | 19,300 | 815 | 9,000 | 11,000 | 0.53 | 6405 |
| | 3.1496 | 0.8268 | 8,050 | 4,340 | 183 | | | 1.17 | |

*(continued)*

**Table 13.6**     *(continued)*

| Principal Dimensions | | | Basic Load Ratings | | Allowable load limit | Speed Ratings | | | |
|---|---|---|---|---|---|---|---|---|---|
| | | | Dynamic | Static | | Grease | Oil | | |
| $d_b$ | $d_a$ | $b_w$ | $C$ | $C_0$ | $w_{all}$ | | | Mass | Designation |
| mm | | | N | | N | | | kg | |
| in | | | lb | | lb | rpm | | lbm | — |
| 30 | 55 | 15 | 13,300 | 8,300 | 355 | 12,000 | 15,000 | 0.12 | 6006 |
| 1.1811 | 2.1654 | 0.5118 | 2,990 | 1,870 | 79.8 | | | 0.26 | |
| | 62 | 16 | 19,500 | 11,200 | 475 | 10,000 | 13,000 | 0.20 | 6206 |
| | 2.4409 | 0.6299 | 4,380 | 2,520 | 107 | | | 0.44 | |
| | 72 | 19 | 28,100 | 16,000 | 670 | 9,000 | 11,000 | 0.35 | 6306 |
| | 2.8346 | 0.7480 | 6,320 | 3,600 | 151 | | | 0.77 | |
| | 90 | 23 | 43,600 | 23,600 | 1,000 | 8,500 | 10,000 | 0.74 | 6406 |
| | 3.5433 | 0.9055 | 9,800 | 5,310 | 225 | | | 1.63 | |
| 35 | 62 | 14 | 15,900 | 10,200 | 440 | 10,000 | 13,000 | 0.16 | 6007 |
| 1.3780 | 2.4409 | 0.5512 | 3,570 | 2,290 | 98.9 | | | 0.35 | |
| | 72 | 17 | 25,500 | 15,300 | 655 | 9,000 | 11,000 | 0.29 | 6207 |
| | 2.8346 | 0.6693 | 5,370 | 3,440 | 147 | | | 0.64 | |
| | 80 | 21 | 33,200 | 19,000 | 815 | 8,500 | 10,000 | 0.46 | 6307 |
| | 3.1496 | 0.8268 | 7,460 | 4,270 | 183 | | | 1.00 | |
| | 100 | 25 | 55,300 | 31,000 | 1,290 | 7,000 | 8,500 | 0.95 | 6407 |
| | 3.9370 | 0.9843 | 12,400 | 6,970 | 290 | | | 2.10 | |
| 40 | 68 | 15 | 16,800 | 11,600 | 490 | 9,500 | 12,000 | 0.19 | 6008 |
| 1.5748 | 2.6672 | 0.5906 | 3,780 | 2,610 | 110 | | | 0.42 | |
| | 80 | 18 | 30,700 | 19,000 | 800 | 8,500 | 10,000 | 0.37 | 6208 |
| | 3.1496 | 0.7087 | 6,900 | 4,270 | 147 | | | 0.82 | |
| | 90 | 23 | 41,000 | 24,000 | 1,020 | 7,500 | 9,000 | 0.63 | 6308 |
| | 3.5433 | 0.9055 | 9,220 | 5,400 | 229 | | | 1.40 | |
| | 110 | 27 | 63,700 | 36,500 | 1,530 | 6,700 | 8,000 | 1.25 | 6408 |
| | 4.3307 | 1.0630 | 14,300 | 8,210 | 344 | | | 2.75 | |
| 45 | 75 | 16 | 20,800 | 14,600 | 640 | 9,000 | 11,000 | 0.25 | 6009 |
| 1.7717 | 2.9528 | 0.6299 | 4,680 | 3,280 | 144 | | | 0.55 | |
| | 85 | 19 | 33,200 | 21,600 | 915 | 7,500 | 9,000 | 0.41 | 6209 |
| | 3.3465 | 0.7480 | 7,400 | 4,860 | 206 | | | 0.90 | |
| | 100 | 25 | 52,700 | 31,500 | 1,340 | 6,700 | 8,000 | 0.83 | 6309 |
| | 3.9370 | 0.9843 | 11,900 | 7,080 | 301 | | | 1.85 | |
| | 120 | 29 | 76,100 | 45,000 | 1,900 | 6,000 | 7,000 | 0.55 | 6409 |
| | 4.7244 | 1.1417 | 17,100 | 10,100 | 427 | | | 3.40 | |
| 50 | 80 | 16 | 21,600 | 16,000 | 710 | 8,500 | 10,000 | 0.26 | 6010 |
| 1.9685 | 3.1496 | 0.6299 | 4,860 | 3,600 | 160 | | | 0.57 | |
| | 90 | 20 | 35,100 | 23,200 | 1,600 | 7,000 | 8,500 | 0.46 | 6210 |
| | 3.5433 | 0.7874 | 7,890 | 5,220 | 220 | | | 1.00 | |
| | 110 | 27 | 61,800 | 38,000 | 980 | 6,300 | 7,500 | 1.05 | 6310 |
| | 4.3307 | 1.0630 | 13,900 | 8,540 | 360 | | | 2.30 | |
| | 130 | 31 | 87,100 | 52,000 | 2,200 | 5,300 | 6,300 | 1.90 | 6410 |
| | 5.1181 | 1.2205 | 19,600 | 11,700 | 495 | | | 4.20 | |

SOURCE: Adapted from SKF Catalog (1991).

**Table 13.7**    Single-row, cylindrical roller bearing

| Principal Dimensions | | | Basic Load Ratings | | Allowable load limit | Speed Ratings | | Mass | Designation |
|---|---|---|---|---|---|---|---|---|---|
| $d_b$ | $d_a$ | $b_w$ | Dynamic C | Static $C_0$ | $w_{all}$ | Grease | Oil | | |
| mm | | | N | | N | | | kg | |
| in | | | lb | | lb | rpm | | lbm | — |
| 15 | 35 | 11 | 12,500 | 10,200 | 1,220 | 18,000 | 22,000 | 0.047 | NU 202 EC |
| 0.5906 | 1.3780 | 0.4331 | 2,810 | 2,290 | 274 | | | 0.10 | |
| | 42 | 13 | 19,400 | 15,300 | 1,860 | 16,000 | 19,000 | 0.086 | NU 302 EC |
| | 1.6535 | 0.5118 | 4,360 | 3,440 | 418 | | | 0.19 | |
| 20 | 47 | 14 | 25,100 | 22,000 | 2,750 | 13,000 | 16,000 | 0.11 | NU 204 EC |
| 0.7874 | 1.8504 | 0.5512 | 5,640 | 4,950 | 618 | | | 0.24 | |
| | 52 | 15 | 30,800 | 26,000 | 3,250 | 12,000 | 15,000 | 0.15 | NU 304 EC |
| | 2.0472 | 0.5906 | 6,920 | 5,850 | 731 | | | 0.33 | |
| 25 | 52 | 15 | 28,600 | 27,000 | 3,350 | 11,000 | 14,000 | 0.13 | NU 205 EC |
| 0.9843 | 2.0472 | 0.5906 | 6,430 | 6,070 | 753 | | | 0.29 | |
| | 62 | 17 | 40,200 | 36,500 | 4,550 | 9,500 | 12,000 | 0.24 | NU 305 EC |
| | 2.4409 | 0.6693 | 9,040 | 8,210 | 1,020 | | | 0.53 | |
| 30 | 62 | 16 | 38,000 | 36,500 | 4,550 | 9,500 | 12,000 | 0.20 | NU 206 EC |
| 1.811 | 2.4409 | 0.6299 | 8,540 | 8,210 | 1,020 | | | 0.44 | |
| | 72 | 19 | 51,200 | 48,000 | 6,200 | 9,000 | 11,000 | 0.36 | NU 306 EC |
| | 2.8346 | 0.7480 | 11,500 | 10,800 | 1,390 | | | 0.79 | |
| 35 | 72 | 17 | 48,400 | 48,000 | 6,100 | 8,500 | 10,000 | 0.30 | NU 207 EC |
| 1.3780 | 2.8346 | 0.693 | 10,900 | 10,800 | 1,370 | | | 0.66 | |
| | 80 | 21 | 64,400 | 63,000 | 8,150 | 8,000 | 9,500 | 0.48 | NU 307 EC |
| | 3.1496 | 0.8268 | 14,500 | 14,200 | 1,830 | | | 1.05 | |
| 40 | 80 | 18 | 53,900 | 53,000 | 6,700 | 7,500 | 9,000 | 0.37 | NU 208 EC |
| 1.5748 | 3.1496 | 0.7087 | 12,100 | 11,900 | 1,510 | | | 0.82 | |
| | 90 | 23 | 80,900 | 78,000 | 10,200 | 6,700 | 8,000 | 0.65 | NU 308 EC |
| | 3.5433 | 0.9055 | 18,200 | 17,500 | 2,290 | | | 1.43 | |
| 45 | 85 | 19 | 60,500 | 64,000 | 8,150 | 6,700 | 8,000 | 0.43 | NU 209 EC |
| 1.7717 | 3.3465 | 0.7480 | 13,600 | 14,400 | 1,830 | | | 0.95 | |
| | 100 | 25 | 99,000 | 100,000 | 12,900 | 6,300 | 7,500 | 0.90 | NU 309 EC |
| | 3.9370 | 0.9843 | 22,300 | 22,500 | 2,900 | | | 2.00 | |
| 50 | 90 | 20 | 64,400 | 69,500 | 8,800 | 6,300 | 7,500 | 0.48 | NU 210 EC |
| 1.9685 | 3.5433 | 0.7874 | 14,500 | 15,600 | 1,980 | | | 1.05 | |
| | 110 | 27 | 110,000 | 112,000 | 15,000 | 5,000 | 6,000 | 1.15 | NU 310 EC |
| | 4.3307 | 1.0630 | 24,700 | 25,200 | 3,370 | | | 2.55 | |

SOURCE: Adapted from SKF Catalog (1991).

Table 13.7 gives similar information as Table 13.6 but for single-row, cylindrical roller bearings. In this text consideration is limited to these two types of bearing. Bearing catalogs are readily available for a wide variety of other bearing types.

### 13.7.6   EQUIVALENT STATIC LOAD

A load $P$ acting on a rolling-element bearing at an angle $\alpha_p$ (Fig. 13.19) is a combined load, since it contains both a radial and an axial component. The radial component is

$P_r = P \cos \alpha_p$

$P_a = P \sin \alpha_p$

**Figure 13.19**    Combined load acting on a radial deep-grove ball bearing.

$P_r = P \cos \alpha_p$ and the axial component is $P_a = P \sin \alpha_p$. The two components $P_r$ and $P_a$ combine to form the **equivalent static load** $P_0$, which can be expressed as

$$P_0 = X_0 P_r + Y_0 P_a \tag{13.62}$$

where $X_0$ = radial factor obtained from Table 13.8 and $Y_0$ = thrust factor obtained from Table 13.8.

In Table 13.8, $\beta$ is the contact angle.

**Table 13.8**    Radial factor $X_0$ and thrust factor $Y_0$ for statically stressed radial bearings

| Bearing type | | Single Row | | Double Row | |
|---|---|---|---|---|---|
| | | $X_0$ | $Y_0$ | $X_0$ | $Y_0$ |
| Radial deep-groove ball | | 0.6 | 0.5 | 0.6 | 0.5 |
| Radial angular-contact ball | $\beta = 20°$ | 0.5 | 0.42 | 1 | 0.84 |
| | $\beta = 25°$ | 0.5 | 0.38 | 1 | 0.76 |
| | $\beta = 30°$ | 0.5 | 0.33 | 1 | 0.66 |
| | $\beta = 35°$ | 0.5 | 0.29 | 1 | 0.58 |
| | $\beta = 40°$ | 0.5 | 0.26 | 1 | 0.52 |
| Radial self-aligning ball | | 0.5 | $0.22 \cot \beta$ | 1 | $0.44 \cot \beta$ |
| Radial spherical roller | | 0.5 | $0.22 \cot \beta$ | 1 | $0.44 \cot \beta$ |
| Radial tapered roller | | 0.5 | $0.22 \cot \beta$ | 1 | $0.44 \cot \beta$ |

**Given:** A single-row, angular-contact ball bearing loaded with a static load having an axial component of 1000 N and a radial component of 500 N. The static basic load rating is 1500 N.

**Find:** The contact angle between 20° and 40° that produces the largest static safety factor.

**Solution:** From Equation (13.62) for a radial angular-contact bearing (see Table 13.8)

$$P_0 = X_0 P_r + Y_0 P_a = (0.5)(500) + Y_0(1000)$$

$$\therefore \qquad P_0 = 250 + 1000 Y_0$$

The safety factor is the largest when $P_0$ is the smallest, which means that when $Y_0$ is the smallest. This implies that the contact angle should be 40°, thus producing $Y_0 = 0.26$.

$$\therefore \qquad P_0 = 250 + (0.26)(1000) = 510 \text{ N}$$

## 13.8 ELASTOHYDRODYNAMIC LUBRICATION

Nonconformal surfaces, such as those in rolling-element bearings, are lubricated elastohydrodynamically as first presented in Section 8.4.2. There it was shown that elastohydrodynamic lubrication is a form of fluid film lubrication where elastic deformation of the bearing surfaces becomes significant. It is usually associated with highly stressed machine components, such as rolling-element bearings. Historically, elastohydrodynamic lubrication may be viewed as one of the major developments in the field of lubrication in the 20th century. It not only revealed the existence of a previously unsuspected regime of lubrication in highly stressed nonconforming machine elements, but also brought order to the complete spectrum of lubrication regimes, ranging from boundary to hydrodynamic.

### 13.8.1 RELEVANT EQUATIONS

The relevant equations used in elastohydrodynamic lubrication are given here.

*Lubrication equation (Reynolds equation) first developed in Chapter 12:*

$$\frac{\partial}{\partial x}\left(\frac{\rho h^3}{\eta}\frac{\partial p}{\partial x}\right) + \frac{\partial}{\partial y}\left(\frac{\rho h^3}{\eta}\frac{\partial p}{\partial y}\right) = 12\tilde{u}\frac{\partial}{\partial x}(\rho h) \qquad (12.23)$$

where
$$\tilde{u} = \frac{u_a + u_b}{2}$$

*Viscosity variation:*

$$\eta = \eta_0 e^{\xi p} \qquad (13.63)$$

where $\eta_0$ is the coefficient of absolute or dynamic viscosity at atmospheric pressure and $\xi$ is the pressure-viscosity coefficient of the fluid.

*Density variation (for mineral oils):*

$$\rho = \rho_0 \left( 1 + \frac{0.6p}{1 + 1.7p} \right)$$

(13.64)

where $\rho_0$ is the density at atmospheric conditions and $p$ is in gigapascals.

*Elasticity equation:*

$$\delta = \frac{2}{E'} \iint\limits_{A} \frac{p(x, y)\, dx\, dy}{\sqrt{(x - x_1)^2 + (y - y_1)^2}}$$

(13.65)

where

$$E' = \frac{2}{\left(1 - v_a^2\right)/E_a + \left(1 - v_b^2\right)/E_b}$$

(13.66)

*Film thickness equation:*

$$h = h_0 + \frac{x^2}{2R_x} + \frac{y^2}{2R_y} + \delta(x, y)$$

(13.67)

where

$$\frac{1}{R_x} = \frac{1}{r_{ax}} + \frac{1}{r_{bx}}$$

(8.4)

$$\frac{1}{R_y} = \frac{1}{r_{ay}} + \frac{1}{r_{by}}$$

(8.5)

The radii ($r_{ax}$, $r_{bx}$, $r_{ay}$, and $r_{by}$) are expressed in Equations (13.13) to (13.18).

The elastohydrodynamic lubrication solution therefore requires calculating the pressure distribution within the conjunction, at the same time allowing for the effects that this pressure will have on the properties of the fluid and on the geometry of the elastic solids. The solution will also provide the shape of the lubricant film, particularly the minimum clearance between the solids. Hamrock et al. (2004) gives the complete elastohydrodynamic lubrication theory.

## 13.8.2   DIMENSIONLESS GROUPING

The variables resulting from the elastohydrodynamic lubrication theory are

$E'$ = effective elastic modulus, Pa
$P$ = normal applied load, N
$h$ = film thickness, m
$R_x$ = effective radius in $x$ (motion) direction, m
$R_y$ = effective radius in $y$ (transverse) direction, m
$\tilde{u}$ = mean surface velocity in $x$ direction, m/s
$\xi$ = pressure-viscosity coefficient of fluid, m²/N
$\eta_0$ = atmospheric viscosity, N · s/m²

From these variables the following five dimensionless groupings can be established:

*Dimensionless film thickness:*
$$H = \frac{h}{R_x} \qquad (13.68)$$

*Ellipticity parameter:*
$$k_e = \frac{D_y}{D_x} = \left(\frac{R_y}{R_x}\right)^{2/\pi} \qquad (8.24)$$

*Dimensionless load parameter:*
$$W = \frac{P}{E' R_x^2} \qquad (13.69)$$

*Dimensionless speed parameter:*
$$U = \frac{\eta_0 \tilde{u}}{E' R_x} \qquad (13.70)$$

*Dimensionless materials parameter:*
$$G = \xi E' \qquad (13.71)$$

The dimensionless minimum film thickness can thus be written as a function of the other four parameters

$$H = f(k_e, U, W, G)$$

The most important practical aspect of elastohydrodynamic lubrication theory is the determination of the minimum film thickness within a conjunction. That is, maintaining a fluid film thickness of adequate magnitude is extremely important to the operation of nonconformal machine elements such as rolling-element bearings. Specifically, elastohydrodynamic film thickness influences fatigue life, as discussed in Section 13.9.

### 13.8.3 MINIMUM-FILM-THICKNESS FORMULA

By using the numerical procedures outlined in Hamrock et al. (2004), the influence of the ellipticity parameter and of the dimensionless speed, load, and materials parameters on minimum film thickness has been investigated. The ellipticity parameter $k_e$ was varied from 1 (a ball-on-plate configuration) to 8 (a configuration approaching a rectangular contact). The dimensionless speed parameter $U$ was varied over a range of nearly two orders of magnitude; and the dimensionless load parameter $W$, over a range of one order of magnitude. Situations equivalent to using solid materials of bronze, steel, and silicon nitride and lubricants of paraffinic and naphthenic oils were considered in the investigation of the role of the dimensionless materials parameter $G$. Thirty-four cases were used in generating the minimum-film-thickness formula given here.

$$H_{\min} = 3.63 U^{0.68} G^{0.49} W^{-0.073} (1 - e^{-0.68 k_e}) \qquad (13.72)$$

In this equation the most dominant exponent occurs on the speed parameter, and the exponent on the load parameter is very small and negative. The materials parameter also carries a significant exponent, although the range of this variable in engineering situations is limited.

Similarly, from Hamrock et al. (2004) for rectangular contacts the dimensionless minimum-film-thickness formula is

$$H_{\min,r} = \frac{h_{\min}}{R_x} = 1.714(W')^{-0.128} U^{0.694} G^{0.568} \qquad (13.73)$$

where

$$W' = \frac{P'_z}{E'R_x} \qquad (13.74)$$

and $P'_z$ = normal applied load per unit width, N/m.

---

**EXAMPLE 13.6**

**Given:**  A deep-groove ball bearing has 12 balls radially loaded with 12,000 N. The groove and ball radii are matched to give an ellipticity ratio of 10. The lubricant viscosity is 0.075 N · s/m², and its pressure-viscosity coefficient is $1.6 \times 10^{-8}$ m²/N. The rotational speed is 1500 rpm, the mean surface velocity is 3 m/s, and the effective radius in the $x$ direction is 8 mm. The modulus of elasticity for the ball and races is 207 GPa, and Poisson's ratio is 0.3.

**Find:**  The minimum oil film thickness in the bearing.

**Solution:**  The Stribeck equation [Eq. (13.50)] gives the load acting on the most heavily loaded ball as

$$(w_z)_{\max} = \frac{5W_z}{n} = \frac{5(12,000)}{12} = 5000 \text{ N}$$

The effective elasticity modulus is

$$E' = \frac{E}{1 - v^2} = \frac{(207)(10^9)}{1 - (0.3)^2} = 227 \text{ GPa}$$

The dimensionless load, speed, and materials parameters from Equations (13.69), (13.70), and (13.71), respectively, can be expressed as

$$U = \frac{\eta_0 \tilde{u}}{E'R_x} = \frac{(0.075)(3)}{(227)(10^9)(8)(10^{-3})} = 1.236 \times 10^{-10}$$

$$W = \frac{P}{E'R_x^2} = \frac{5000}{(227)(10^9)(8)^2(10^{-6})} = 3.442 \times 10^{-4}$$

$$G = \xi E' = (1.6)(10^{-8})(227)(10^9) = 3632$$

The dimensionless minimum-film-thickness formula for elliptical contacts [Eq. (13.72)] produces

$$\begin{aligned}
H_{\min} = \frac{h_{\min}}{R_x} &= 3.63 U^{0.68} G^{0.49} W^{-0.073}(1 - e^{-0.68k_e}) \\
&= 3.63(1.236 \times 10^{-10})^{0.68}(3632)^{0.49}(3.442 \times 10^{-4})^{-0.073}(1 - e^{-0.68(10)}) \\
&= 6.596 \times 10^{-5}
\end{aligned}$$

The dimensional minimum film thickness is

$$h_{\min} = H_{\min} R_x = (6.596)(10^{-5})(8)(10^{-3}) = 0.53 \times 10^{-6} \text{ m} = 0.53 \ \mu\text{m}$$

Thus, the minimum oil film thickness in the deep-groove ball bearing considered in this example is 0.53 $\mu$m.

EXAMPLE 13.7

**Given:** A ball bearing is having problems with an oil film that is too thin. The geometries are such that the ellipticity parameter is 2. By decreasing the race radius of curvature in the axial direction, the balls fit more tightly with the races, and thus the ellipticity parameter increases from 2 to 10.

**Find:** How much thicker the oil film becomes in changing the ellipticity parameter from 2 to 10. Assume that all the other operating parameters remain constant.

**Solution:** From Equation (13.72)

$$\frac{(H_{min})_{k_e=10}}{(H_{min})_{k_e=2}} = \frac{1 - e^{-0.68(10)}}{1 - e^{-0.68(2)}} = 1.344$$

The film thickness will increase by 34.4% when the ellipticity parameter is increased from 2 to 10.

## 13.9 FATIGUE LIFE

### 13.9.1 CONTACT FATIGUE THEORY

Rolling fatigue is a material failure caused by the application of repeated stresses to a small volume of material. It is a unique failure type—essentially a process of seeking out the weakest point at which the first failure will occur. A typical spall is shown in Figure 13.20. On a microscopic scale there will be a wide dispersion in material strength or resistance to fatigue because of inhomogeneities in the material. Because bearing materials are complex alloys, they are not homogeneous or equally resistant to failure at all points. Therefore, the

(a)

(b)

**Figure 13.20** Typical fatigue spall. (a) Spall on tapered roller bearing; (b) detail of fatigue spall. [Courtesy of Timken, Inc.]

fatigue process can be expected to be one in which a group of seemingly identical specimens exhibit wide variations in failure time when stressed in the same way. For this reason it is necessary to treat the fatigue process statistically.

Predicting how long a particular bearing will run under a specific load requires the following essential pieces of information:

1.   An accurate, quantitative estimate of life dispersion or scatter

2.   The life at a given survival rate or reliability level

This translates to an expression for the *load-carrying capacity,* or the ability of the bearing to endure a given load for a stipulated number of stress cycles or revolutions. If a group of supposedly identical bearings is tested at a specific load and speed, the distribution in bearing life shown in Figure 13.21 will occur.

### 13.9.2    WEIBULL DISTRIBUTION

Weibull (1949) postulated that the fatigue lives of a homogeneous group of rolling-element bearings are dispersed according to

$$\ln \ln \frac{1}{\tilde{S}} = e_1 \ln \frac{\tilde{L}}{\tilde{A}} \tag{13.75}$$

where $\tilde{S}$ is the probability of survival, $\tilde{L}$ is the fatigue life, and $e_1$ and $\tilde{A}$ are constants. The Weibull distribution results from a statistical theory of strength based on probability theory, where the dependence of strength on volume is explained by the dispersion in material strength. This is the "weakest link" theory.

Figure 13.22 considers a volume being stressed that is broken up into $m$ similar volumes; the $\tilde{M}$'s represent the probability of failure and the $\tilde{S}$'s the probability of survival.

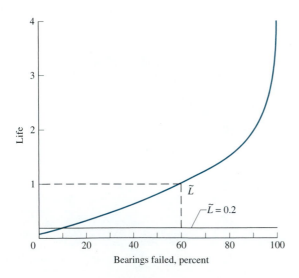

**Figure 13.21**    Distribution of bearing fatigue failures. [From Hamrock and Anderson (1983).]

**Figure 13.22**    Representation of $m$ similar stressed volumes.

For the entire volume

$$\tilde{S} = \tilde{S}_1 \tilde{S}_2 \tilde{S}_3 \cdots \tilde{S}_m$$

Then

$$1 - \tilde{M} = (1 - \tilde{M}_1)(1 - \tilde{M}_2)(1 - \tilde{M}_3) \cdots (1 - \tilde{M}_m)$$

$$= \prod_{i=1}^{m} (1 - \tilde{M}_i)$$

$$\tilde{S} = \prod_{i=1}^{m} (1 - \tilde{M}_i) \tag{13.76}$$

The probability of a crack starting in the $i$th volume is

$$\tilde{M}_i = f(x)\tilde{V}_i \tag{13.77}$$

where $f(x)$ is a function of the stress level, the number of stress cycles, and the depth into the material where the maximum stress occurs and $\tilde{V}_i$ is the elementary volume. Therefore, substituting Equation (13.77) into Equation (13.76) gives

$$\tilde{S} = \prod_{i=1}^{m} [1 - f(x)\tilde{V}_i]$$

$$\ln \tilde{S} = \sum_{i=1}^{m} \ln [1 - f(x)\tilde{V}_i] \tag{13.78}$$

Now, if $f(x)\tilde{V}_i \ll 1$, then $\ln[1 - f(x)\tilde{V}_i] = -f(x)\tilde{V}_i$ and $\ln \tilde{S} = -\sum_{i=1}^{m} f(x)\tilde{V}_i$. Letting $\tilde{V}_i \to 0$ then,

$$\sum_{i=1}^{\infty} f(x)\tilde{V}_i = \int f(x)\,d\tilde{V} = \bar{f}(x)\tilde{V} \tag{13.79}$$

where $\bar{f}(x)$ = volume-average value of $f(x)$.

Lundberg and Palmgren (1947) assumed that $f(x)$ could be expressed as a power function of shear stress $\tau_0$, number of stress cycles $\tilde{J}$, and depth of the maximum shear stress $z_0$:

$$f(x) = \frac{\tau_0^{c_1} \tilde{J}^{c_2}}{z_0^{c_3}} \tag{13.80}$$

They also chose the stressed volume as $\tilde{V} = D_y z_0 l_v$. Substituting Equations (13.79) and (13.80) into Equation (13.78) gives

$$\ln \tilde{S} = -\frac{\tau_0^{c_1} \tilde{J}^{c_2} D_y l_v}{z_0^{c_3 - 1}} \tag{13.81}$$

$$\ln \frac{1}{\tilde{S}} = \frac{\tau_0^{c_1} \tilde{J}^{c_2} D_y l_v}{z_0^{c_3 - 1}} \tag{13.82}$$

For a specific bearing and load (e.g., stress) $\tau_0$, $D_y$, $l_v$, and $z_0$ are all constant so that

$$\ln \frac{1}{\tilde{S}} \approx \tilde{J}^{c_2}$$

Designating $\tilde{J}$ as fatigue life $\tilde{L}$ in stress cycles gives

$$\ln \frac{1}{\tilde{S}} = \left(\frac{\tilde{L}}{\tilde{A}}\right)^{c_2}$$

$$\ln \ln \frac{1}{\tilde{S}} = c_2 \ln \frac{\tilde{L}}{\tilde{A}} \qquad (13.83)$$

This is the **Weibull distribution,** which relates the probability of survival and life. It has two principal functions. First, bearing fatigue lives plot as a straight line on Weibull coordinates (log-log versus log), so that the life at any reliability level can be determined. Of greatest interest are the $\tilde{L}_{10}$ life ($\tilde{S} = 0.9$) and the $\tilde{L}_{50}$ life ($\tilde{S} = 0.5$). Bearing load ratings are based on the $\tilde{L}_{10}$ life. Second, Equation (13.83) can be used to determine the $\tilde{L}_{10}$ life or to obtain a required life at any reliability level. The $\tilde{L}_{10}$ life is calculated from the load on the bearing and the bearing dynamic capacity, or by the load rating given in manufacturers' catalogs and engineering journals, by using the equation

$$\tilde{L} = \left(\frac{\bar{C}}{P}\right)^{m_k} \qquad (13.84)$$

where $\bar{C}$ = specific dynamic capacity or basic dynamic load rating, N

$\quad\quad\ P$ = equivalent bearing load, N

$\quad\quad\ m_k$ = load-life exponent; 3 for elliptical contacts and 10/3 for rectangular contacts

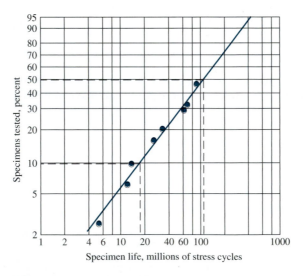

**Figure 13.23**    Typical Weibull plot of bearing fatigue failures. [From Hamrock and Anderson (1983).]

Note that $\tilde{L}$ in Equation (13.84) is bearing life expressed in millions of revolutions. The bearing life in hours is

$$(\tilde{L})_{\text{hour}} = \frac{10^6 \tilde{L}}{60 N_b} \qquad (13.85)$$

where $N_b$ = rotational speed, rpm. A typical Weibull plot is shown in Figure 13.23.

### 13.9.3 DYNAMIC LOAD RATING

The Lundberg-Palmgren theory, on which bearing ratings are based, is expressed by Equation (13.82). The exponents in this equation are determined experimentally from the dispersion of bearing lives and the dependence of life on load, geometry, and bearing size. As a standard of reference, all bearing load ratings are expressed in terms of the dynamic load rating, which, by definition, is the load that a bearing can carry for 1 million inner-race revolutions with a 90% chance of survival.

Factors on which specific dynamic capacity and bearing life depend are

1. Size of rolling element
2. Number of rolling elements per row
3. Number of rows of rolling elements
4. Conformity between rolling elements and races
5. Contact angle under load
6. Material properties
7. Lubricant properties
8. Operating temperature
9. Operating speed

Only factors 1 to 5 are incorporated in bearing dynamic load ratings developed from the Lundberg-Palmgren theory (1947, 1952).

Table 13.6 gives the dynamic load ratings for single-row, deep-groove ball bearings; and Table 13.7, those for single-row, cylindrical roller bearings. The dynamic load ratings are based on the material and manufacturing techniques used for producing standard production bearings. They apply to loads that are constant in both magnitude and direction; for radial bearings these loads are radial, and for thrust bearings they are axial loads that act centrally.

### 13.9.4 EQUIVALENT DYNAMIC LOAD

Equation (13.82), used for calculating the fatigue life, assumes a purely radial load in radial bearings and a purely axial load in thrust bearings, both loads of constant direction and magnitude. In many rolling-element bearing applications, the load does not satisfy these conditions, since the force acts on the bearing obliquely or changes in magnitude. In such cases, a constant radial or axial force must be determined for the rating life calculation, representing, with respect to the rating life, an equivalent stress.

**Table 13.9** Capacity formulas for rectangular and elliptical conjunctions for radial and angular bearings

| Bearing type | | $e$ | Single-Row Bearings | | | | Double-Row Bearings | | | |
|---|---|---|---|---|---|---|---|---|---|---|
| | | | $\dfrac{P_a}{P_r} \leq e$ | | $\dfrac{P_a}{P_r} > e$ | | $\dfrac{P_a}{P_r} \leq e$ | | $\dfrac{P_a}{P_r} > e$ | |
| | | | X | Y | X | Y | X | Y | X | Y |
| Deep-groove | $P_a/C_0 = 0.025$ | 0.22 | 1 | 0 | 0.56 | 2.0 | | | | |
| ball bearings | $P_a/C_0 = 0.04$ | 0.24 | 1 | 0 | 0.56 | 1.8 | | | | |
| | $P_a/C_0 = 0.07$ | 0.27 | 1 | 0 | 0.56 | 1.6 | | | | |
| | $P_a/C_0 = 0.13$ | 0.31 | 1 | 0 | 0.56 | 1.4 | | | | |
| | $P_a/C_0 = 0.25$ | 0.37 | 1 | 0 | 0.56 | 1.2 | | | | |
| | $P_a/C_0 = 0.50$ | 0.44 | 1 | 0 | 0.56 | 1 | | | | |
| Angular-contact | $\beta = 20°$ | 0.57 | 1 | 0 | 0.43 | 1 | 1 | 1.09 | 0.70 | 1.63 |
| ball bearings | $\beta = 25°$ | 0.68 | 1 | 0 | 0.41 | 0.87 | 1 | 0.92 | 0.67 | 1.41 |
| | $\beta = 30°$ | 0.80 | 1 | 0 | 0.39 | 0.76 | 1 | 0.78 | 0.63 | 1.24 |
| | $\beta = 35°$ | 0.95 | 1 | 0 | 0.37 | 0.66 | 1 | 0.66 | 0.60 | 1.07 |
| | $\beta = 40°$ | 1.14 | 1 | 0 | 0.35 | 0.57 | 1 | 0.55 | 0.57 | 0.93 |
| | $\beta = 45°$ | 1.33 | 1 | 0 | 0.33 | 0.50 | 1 | 0.47 | 0.54 | 0.81 |
| Self-aligning ball bearings | | $1.5 \tan \beta$ | | | | | 1 | $0.42 \cot \beta$ | 0.65 | $0.65 \cot \beta$ |
| Spherical roller bearings | | $1.5 \tan \beta$ | | | | | 1 | $0.45 \cot \beta$ | 0.67 | $0.67 \cot \beta$ |
| Tapered-roller bearings | | $1.5 \tan \beta$ | 1 | 0 | 0.40 | $0.4 \cot \beta$ | 1 | $0.42 \cot \beta$ | 0.67 | $0.67 \cot \beta$ |

**Table 13.10** Radial factor $X$ and thrust factor $Y$ for thrust bearings

| Bearing type | | $e$ | Single-Acting | | Double-Acting | | | |
|---|---|---|---|---|---|---|---|---|
| | | | $\dfrac{P_a}{P_r} > e$ | | $\dfrac{P_a}{P_r} \leq e$ | | $\dfrac{P_a}{P_r} > e$ | |
| | | | X | Y | X | Y | X | Y |
| Thrust ball | $\beta = 45°$ | 1.25 | 0.66 | 1 | 1.18 | 0.59 | 0.66 | 1 |
| | $\beta = 60°$ | 2.17 | 0.92 | 1 | 1.90 | 0.55 | 0.92 | 1 |
| | $\beta = 75°$ | 4.67 | 1.66 | 1 | 3.89 | 0.52 | 1.66 | 1 |
| Spherical roller thrust | | $1.5 \tan \beta$ | $\tan \beta$ | 1 | $1.5 \tan \beta$ | 0.67 | $\tan \beta$ | 1 |
| Tapered roller | | $1.5 \tan \beta$ | $\tan \beta$ | 1 | $1.5 \tan \beta$ | 0.67 | $\tan \beta$ | 1 |

The **equivalent load** is expressed as

$$P = XP_r + YP_a \tag{13.86}$$

where $X$ = radial factor

$Y$ = thrust factor

$P_r$ = radial load component, N

$P_a$ = axial load component, N

Table 13.9 shows values of $X$ and $Y$ for various radial bearing types at respective contact angles; also, $C_0$ is the static load rating given in Section 13.7.5. The $X$ and $Y$ for various thrust bearing types are given in Table 13.10.

## 13.9.5  LIFE ADJUSTMENT FACTORS

This section presents the factors affecting the fatigue life of bearings that were not taken into account in the Lundberg-Palmgren theory. It is assumed that the various environmental or bearing design factors are multiplicative in their effects on bearing life. The following equations result:

$$L_A = \bar{D}\bar{E}\bar{F}_l\bar{G}\bar{H}_m\tilde{L}_{10} \tag{13.87}$$

$$L_A = \bar{D}\bar{E}\bar{F}_l\bar{G}\bar{H}_m\left(\frac{\bar{C}}{\bar{P}}\right)^{m_k} \tag{13.88}$$

where  $\bar{D}$ = material factor
$\bar{E}$ = metallurgical processing factor
$\bar{F}_l$ = lubrication factor
$\bar{G}$ = speed effect factor
$\bar{H}_m$ = misalignment factor
$\bar{P}$ = bearing equivalent load, N
$m_k$ = load-life exponent; 3 for ball bearings or 10/3 for roller bearings
$\bar{C}$ = specific dynamic capacity, N

Factors $\bar{D}$, $\bar{E}$, and $\bar{F}_l$ are reviewed briefly here. Refer to Bamberger (1971) for a complete discussion of all five life adjustment factors.

### Material Factors

For over a century, AISI 52100 steel has been the predominant material for rolling-element bearings. In fact, the specific dynamic capacity as defined by the Anti-Friction Bearing Manufacturers Association (AFBMA) in 1949 is based on an air-melted 52100 steel, hardened to at least Rockwell C58. Since that time better control of air melting processes and the introduction of vacuum remelting processes have resulted in more homogeneous steels with fewer impurities. Such steels have extended rolling-element bearing fatigue lives to several times the catalog life; life extensions of 3 to 8 times are not uncommon. Other steels, such as AISI M–1 and AISI M–50, chosen for their higher-temperature capabilities and resistance to corrosion, also have shown greater resistance to fatigue pitting when vacuum melting techniques are employed. Case-hardened materials, such as AISI 4620, AISI 4118, and AISI 8620, used primarily for rolling-element bearings, have the advantage of a tough, ductile steel core with a hard, fatigue-resistant surface.

The recommended material factors $\bar{D}$ for various through-hardened alloys processed by air melting are shown in Table 13.11. Insufficient definitive life data were found for case-hardened materials to recommend values of $\bar{D}$ for them. Refer to the bearing manufacturer for the choice of a specific case-hardened material.

The processing variables considered in the development of the metallurgical processing factor $\bar{E}$ include melting practice (air and vacuum melting) and metalworking (thermomechanical working). Thermomechanical working of M-50 has also been shown to lengthen life, but in a practical sense it is costly and still not fully developed as a processing technique. Bamberger (1971) recommends an $\bar{E}$ of 3 for consumable-electrode vacuum melted materials.

**Table 13.11**        Material factors for
                       through-hardened
                       bearing materials

| Material | Material factor $\bar{D}$ |
|----------|---------------------------|
| 52100    | 2.0                       |
| M-1      | 0.6                       |
| M-2      | 0.6                       |
| M-10     | 2.0                       |
| M-50     | 2.0                       |
| T-1      | 0.6                       |
| Halmo    | 2.0                       |
| M-42     | 0.2                       |
| WB 49    | 0.6                       |
| 440C     | 0.6–0.8                   |

I  SOURCE: From Bamberger (1971).

### Lubrication Factor

Until approximately 1960 the role of the lubricant between surfaces in rolling contact was not fully appreciated; metal-to-metal contact was presumed to occur in all applications, with attendant required boundary lubrication. The development of elastohydrodynamic lubrication theory showed that lubricant films with thicknesses on the order of microinches and tenths of microinches occur in rolling contact. Since surface finishes are of the same order of magnitude as that of the lubricant film thicknesses, the significance of rolling-element

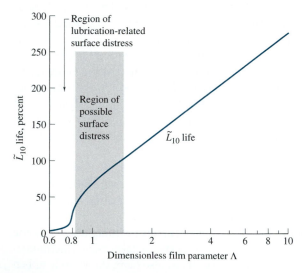

**Figure 13.24**    Group fatigue life $\tilde{L}_{10}$ as function of dimensionless film parameter. [From Tallian (1967).]

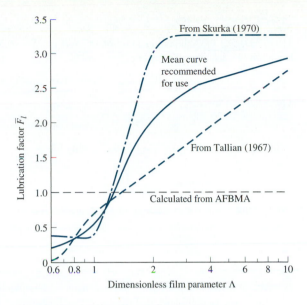

**Figure 13.25** Lubrication factor as function of dimensionless film parameter. [From Bamberger (1971).]

bearing surface roughnesses to bearing performance became apparent. Tallian (1967) was the first to report on the importance to bearing life of the ratio of elastohydrodynamic lubrication film thickness to surface roughness. Figure 13.24 shows calculated $\tilde{L}_{10}$ life as a function of the dimensionless film parameter $\Lambda$, which was introduced in Chapter 8, where

$$\Lambda = \frac{h_{\min}}{\left(R_{qa}^2 + R_{qb}^2\right)^{1/2}} \tag{13.89}$$

Figure 13.25 presents a curve of the recommended $\bar{F}_l$ as a function of $\Lambda$. A mean of the curves presented in Tallian (1967) for ball bearings and in Skurka (1970) for roller bearings is recommended. A formula for calculating the minimum film thickness $H_{\min}$ is given in Equation (13.72). Procedures for calculating $\tilde{D}$ and $\tilde{F}_l$ are at present less than definitive, reflecting the need for additional research, life data, and operating experience.

---

**EXAMPLE 13.8**

**Given:** A deep-groove 6304 ball bearing has a load rating of 15,900 N. It is radially loaded with 4000 N and axially loaded with 2000 N. The rotational speed is 3000 rpm.

**Find:** The number of hours the bearing will survive with 90% probability.

**Solution:** From Table 13.6 for a 6304 deep-groove ball bearing, the static load rating is 7800 N.

$$\therefore \quad \frac{P_a}{C_0} = \frac{2000}{7800} = 0.256$$

From Table 13.9 for $P_a/C_0 = 0.256$ for a deep-groove ball bearing, the constant $e$ is 0.37. Since $P_a/P_r = 0.5 > 0.37$, the radial and thrust factors are $X = 0.56$ and $Y = 1.2$, respectively. From Equation (13.86) the equivalent bearing load is

$$P = XP_r + YP_a = 0.56(4000) + 1.2(2000) = 4640 \text{ N}$$

The rotational speed is

$$N_b = 3000 \text{ rpm}$$

From Equations (13.84) and (13.85) the bearing life is

$$(\tilde{L}_{10})_{hour} = \frac{10^6}{60N_b} \left(\frac{\bar{C}}{P}\right)^3 = \frac{10^6}{(60)(3000)} \left(\frac{15,900}{4640}\right)^3 = 223.5 \text{ h}$$

The bearing has a 90% probability of surviving for 223.5 h.

---

**EXAMPLE 13.9**

**Given:** A machine manufacturer has dimensioned the roller bearings in a paper-making machine containing 400 bearings such that the $\tilde{L}_{10}$ for each bearing is 40 years.

**Find:** Use Figure 13.23 to calculate the $\tilde{L}_{50}$ life in years. Also, determine the mean time between failures in the steady state for an old paper-making machine where bearings are replaced whenever they fail.

**Solution:** From Figure 13.23 the $\tilde{L}_{50}$ life is about 110 million stress cycles while the $\tilde{L}_{10}$ life is 17 million stress cycles. The ratio is

$$\frac{\tilde{L}_{50}}{\tilde{L}_{10}} = \frac{110}{17} = 6.47$$

The $\tilde{L}_{50}$ life is thus 6.47 times the $\tilde{L}_{10}$ life, which in this example is 40 years.

$$\therefore \qquad \tilde{L}_{50} = (6.47)(40) = 258.8 \text{ yr}$$

As there are 400 independent bearings in the machine, the mean time between failures is

$$\frac{\tilde{L}_{50}}{400} = \frac{258.8}{400} = 0.647 \text{ yr} = 236 \text{ days}$$

---

**Case Study 13.1**  ROLLER BEARING DESIGN FOR LOCOMOTIVE WHEEL AXLE BOX

**Given:** Applications for roller bearings are truly everywhere; they can be found in washing machines, jet engines, process equipment of all kinds, lawnmowers, roller blades, etc. One of the most historically successful applications of rolling-element bearings is in vehicle axles, such as the railroad boxcar axle bearing shown in Figure 13.26. Such bearings are used for prolonged periods of maintenance-free operation and are commonly packed with a suitable

## Case Study (Continued)

(a)

(b)

**Figure 13.26** (a) Railroad boxcar wheel assembly and (b) disassembled axle, showing roller bearing components. [Courtesy of FAG Kugelfischer Georg Schafer Kga. Schweinfurt, Germany.]

bearing grease. This case study involves determining the film thickness in the bearing so that wear can be assessed. The bearings are highly polished, and a typical surface roughness of 0.05 $\mu$m is assumed.

The equations for elastohydrodynamic film thickness developed in Section 13.8.3 relate primarily to elliptical and rectangular conjunctions. The rectangular conjunctions correspond to line-contact problems, as would be found in such a cylindrical roller bearing. Figures 13.1 and 13.7 show a roller bearing describing the various dimensions used in this case study. Therefore, the minimum elastohydrodynamic film thicknesses are calculated on the inner and outer races of a cylindrical roller bearing with the following dimensions—for both the elliptical conjunction and the rectangular conjunction:

| | |
|---|---|
| Inner-race diameter $d_i$, mm | 64 |
| Outer-race diameter $d_o$, mm | 96 |
| Diameter of cylindrical rollers $d$, mm | 16 |
| Effective axial length of cylindrical rollers $l_l$, mm | 16 |
| Number of rollers in complete bearing $n$ | 9 |

A railroad bearing of this type might well experience the following static operating conditions for extended periods:

| | |
|---|---|
| Radial load $P$, N | 10,800 |
| Inner-race angular velocity $\omega_i$, rad/s | 52.4 |
| Outer-race angular velocity $\omega_o$, rad/s | 0 |
| Absolute viscosity at $p = 0$ and bearing effective operating temperature $\eta_0$, Pa $\cdot$ s | 0.1 |
| Viscosity-pressure coefficient $\xi$, m²/N | $2.2 \times 10^{-8}$ |
| Modulus of elasticity for both rollers and races $E$, Pa | $2.075 \times 10^{11}$ |
| Poisson's ratio for both rollers and races $\nu$ | 0.3 |

**Find:** The minimum film thickness for such a cylindrical roller bearing.

**Solution:** Since the diametral clearance $c_d$ is zero, from Equation (13.49) the most heavily loaded roller is

$$P_{z,\max} = \frac{4P_z}{n} = \frac{4(10,800 \text{ N})}{9} = 4800 \text{ N}$$

Therefore, the radial load per unit length on the most heavily loaded roller can be expressed as

$$(P_z')_{\max} = \frac{4800 \text{ N}}{0.016 \text{ m}} = 0.3 \text{ MN/m}$$

*(continued)*

## Case Study        (Continued)

From Figure 8.4 the radii of curvature are

$$r_{ax} = 0.008 \text{ m} \qquad r_{ay} = \infty$$
$$r_{bxi} = 0.032 \text{ m} \qquad r_{byi} = \infty$$
$$r_{bxo} = -0.048 \text{ m} \qquad r_{byo} = \infty$$

Then        $\dfrac{1}{R_{xi}} = \dfrac{1}{0.008} + \dfrac{1}{0.032} = \dfrac{5}{0.032}$

giving $R_{xi} = 0.0064$ m,

$$\frac{1}{R_{xo}} = \frac{1}{0.008} - \frac{1}{0.048} = \frac{5}{0.048}$$

giving $R_{xo} = 0.0096$ m, and

$$\frac{1}{R_{yi}} = \frac{1}{R_{yo}} = \frac{1}{\infty} + \frac{1}{\infty} = 0$$

giving $R_{yi} = R_{yo} = \infty$.

From the input information the effective modulus of elasticity can be written as

$$E' = \frac{2}{\left(1 - v_a^2\right) E_a + \left(1 - v_b^2\right) E_b} = 2.28 \times 10^{11} \text{ Pa}$$

For pure rolling, the mean surface velocity $\tilde{u}$ relative to the lubricated conjunctions for a cylindrical roller is

$$\tilde{u} = \frac{|\omega_i - \omega_o|\left(d_e^2 - d^2\right)}{4d_e}$$

where $d_e$ = pitch diameter
          $d$ = roller diameter
$$d_e = \frac{d_o + d_i}{2} = \frac{0.096 + 0.064}{2} = 0.08 \text{ m}$$

Hence,    $\tilde{u} = \dfrac{0.08^2 - 0.016^2}{4(0.08)} |52.4 - 0| = 1.006 \text{ m/s}$

The dimensionless speed, materials, and load parameters expressed in Equations (13.69) to (13.71) for the inner- and outer-race conjunctions thus become, respectively,

$$U_i = \frac{\eta_0 \tilde{u}}{E' R_{xi}} = \frac{(0.10)(1.006)}{(2.28)(10^{11})(0.0064)} = 6.895 \times 10^{-11}$$

$$G_o = G_i = \xi E' = 5016$$

$$W_i = \frac{(P_z)_{\max}}{E' R_{xi}^2} = \frac{4800}{(2.28)(10^{11})(0.0064)^2} = 5.140 \times 10^{-4}$$

$$U_o = \frac{\eta_0 \tilde{u}}{E' R_{xo}} = \frac{(0.10)(1.006)}{(2.28)(10^{11})(0.0096)} = 4.597 \times 10^{-11}$$

$$W_o = \frac{(P_z)_{\max}}{E' R_{xo}^2} = \frac{4800}{(2.28)(10^{11})(0.0096)^2} = 2.284 \times 10^{-4}$$

From Equation (13.72) the appropriate elastohydrodynamic film thickness equation for a fully flooded elliptical conjunction is

$$H_{\min} = \frac{\tilde{h}_{\min}}{R_x} = 3.63 U^{0.68} G^{0.49} W^{-0.073} (1 - e^{-0.68 k_e})$$

For a roller bearing $k_e = \infty$, and this equation reduces to

$$H_{\min} = 3.63 U^{0.68} G^{0.49} W^{-0.073}$$

The dimensionless film thickness for the *roller–inner-race conjunction* is

$$\tilde{H}_{\min} = \frac{h_{\min}}{R_{xi}} = (3.63)(1.231)(10^{-7})(65.04)(1.738)$$
$$= 50.5 \times 10^{-6}$$

and hence,

$$h_{\min} = (0.0064)(50.5)(10^{-6}) = 0.32 \times 10^{-6} \text{ m} = 0.32\ \mu\text{m}$$

The dimensionless film thickness for the *roller–outer-race conjunction* is

$$\tilde{H}_{\min} = \frac{h_{\min}}{R_{xo}} = (3.63)(9.343)(10^{-8})(65.04)(1.844)$$
$$= 40.7 \times 10^{-6}$$

and hence,

$$h_{\min} = (0.0096)(40.7)(10^{-6}) = 0.3905 \times 10^{-6} \text{ m}$$
$$= 0.39\ \mu\text{m}$$

It is clear from these calculations that the smaller minimum film thickness in the bearing occurs at the roller–inner-race conjunction, where the geometric conformity is less favorable. It was found that if the ratio of minimum film thickness to composite surface roughness is greater than 3,

## Case Study (Concluded)

an adequate elastohydrodynamic film is maintained. This result implies that a composite surface roughness of less than 0.1 $\mu$m is needed to ensure that an elastohydrodynamic film is maintained.

Now using, instead of the elliptical conjunction results, the rectangular conjunction results from Equation (13.73) gives

$$H_{min} = \frac{h_{min}}{R_x} = 1.714(W')^{-0.128}U^{0.694}G^{0.568}$$

where

$$W' = \frac{(P'_z)_{max}}{E'R_x}$$

For the *roller–inner-race conjunction*

$$W' = \frac{(P'_z)_{max}}{E'R_x} = \frac{(0.3)(10^6)}{(2.28)(10^{11})(0.0064)} = 2.056 \times 10^{-4}$$

$$H_{min} = 1.714(2.056 \times 10^{-4})^{-0.128}$$

$$\times (6.895 \times 10^{-11})^{0.694}(5016)^{0.568} = 56.98 \times 10^{-6}$$

$$h_{min} = 0.0064(56.98)(10^{-6}) = 0.3647 \times 10^{-6}\ \text{m}$$

$$= 0.3647\ \mu\text{m}$$

For the *roller–outer-race conjunction*

$$W' = \frac{(P'_z)_{max}}{E'R_x} = \frac{(0.3)(10^6)}{(2.28)(10^{11})(0.0096)} = 1.371 \times 10^{-4}$$

$$H_{min} = 1.714(1.371 \times 10^{-4})^{-0.128}$$

$$\times (4.597 \times 10^{-11})^{0.694}(5016)^{0.568} = 45.29 \times 10^{-6}$$

so that

$$h_{min} = 0.0096(45.29)(10^{-6}) = 0.4348 \times 10^{-6}\ \text{m}$$

$$= 0.4348\ \mu\text{m}$$

The film thickness obtained from the rectangular conjunction results produces approximately a 12% increase from the elliptical conjunction results. Since the original surface roughness is 0.05 $\mu$m, the ratio of film thickness to surface roughness is approximately 6.0. This ratio still ensures that the bearing is in a full film lubrication state and that adhesive wear is far less likely than fatigue wear.

## 13.10 SUMMARY

Rolling-element bearings are precise, yet simple, machine elements of great utility. This chapter drew together the current understanding of rolling-element bearings and attempted to present it in a concise manner. The detailed, precise calculations of bearing performance were only summarized, and appropriate references were given. The history of rolling-element bearings was briefly reviewed, and subsequent sections described the types of rolling-element bearing, their geometry, and their kinematics. Having defined ball bearing operation under unloaded and unlubricated conditions, the chapter then focused on static loading of rolling-element bearings. Most rolling-element bearing applications involve steady-state rotation of the inner race or outer race or both. However, the rotational speeds are usually not so high as to cause ball or roller centrifugal forces or gyroscopic moments of significant magnitudes; thus these were neglected. Radial, thrust, and preloaded bearings that are statically loaded were considered in the second major thrust of the chapter, which was on loaded but unlubricated rolling-element bearings.

The last major thrust of the chapter dealt with loaded and lubricated rolling-element bearings. Topics covered were fatigue life and dynamic analyses. The chapter concluded by applying the knowledge of this and previous chapters to roller and ball bearing applications. The use of the elastohydrodynamic lubrication film thickness was integrated with the rolling-element bearing theory developed in this chapter. It was found that the most critical conjunctions of both ball and roller bearings occurred between the rolling elements and the inner races.

## KEY WORDS

**angular-contact ball bearings**    bearings with a two-shouldered groove in one race and a single-shouldered groove in other.

**angular-contact bearings**    bearings that have clearance built into unloaded bearing, which allows operation under high thrust loads.

**ball bearings**    rolling-element bearings using spheres as rolling-element cage fitting or spacer to keep proper distance between balls in bearing track.

**conrad or deep-groove bearing**    non-separable bearing capable of operating at high speeds.

**crown**    curvature machined into rollers to eliminate high edge stresses.

**cylindrical roller bearings**    bearings using cylinders as rolling elements.

**deep-groove bearing**    ball bearing with race containing pronounced groove for rolling elements.

**duplex pairs**    sets of two angular-contact ball bearings that preload each other upon assembly to shaft.

**endplay**    maximum axial movement of inner race with respect to outer race under small forces.

**equivalent load**    resultant dynamic load when considering radial and axial components.

**equivalent static load**    resultant load when considering thrust and radial components.

**free contact angle**    angle made by line through points where ball contacts both races and plane perpendicular to bearing axis rotation under low loads.

**pitch diameter**    mean of inner- and outer-race contact diameters.

**race control**    condition where pure rolling occurs at controlling race.

**race depth**    same as **shoulder height.**

**races**    grooves within bearing rings for rolling elements to roll in.

**retainers**    same as **cage.**

**rolling-element bearings**    machinery elements where surfaces are nonconformal and motion is primarily rolling.

**self-aligning bearings**    bearings with one race having spherical shape to allow for large misalignment.

**separators**    same as **cages.**

**shoulder height**    depth of race groove.

**spherical roller bearings**    same as **self-aligning roller bearings.**

**split-ring bearing**    bearing that has one race made from two halves, allowing for accurate axial positioning of shafts.

**static load rating**    load at which, under defined operating conditions, load carried by most heavily loaded rolling element $W_{max}$ equals allowable rolling-element load $w_{all}$, or maximum load before plastic (rather than elastic) deformation occurs within bearing.

**static thrust load-carrying capacity**    maximum thrust load that bearing can endure before contact ellipse approaches race shoulder.

**thrust ball bearings**    bearings with race grooves arranged to support large axial or thrust forces.

**Weibull distribution**    relationship between survival and life, applied in this chapter to bearings.

## RECOMMENDED READINGS

Halling, J. (ed.) (1978) *Principles of Tribology,* Macmillan, New York.

Hamrock, B. J., and Dowson, D. (1981) *Ball Bearing Lubrication—The Elastohydrodynamics of Elliptical Contacts,* Wiley, New York.

Hamrock, B. J., Schmid, S. R., and Jacobson, B. O. (2004) *Fundamentals of Fluid Film Lubrication,* 2nd ed., Marcel Dekker, New York.

Harris, T. A. (1991) *Rolling Bearing Analysis,* 3rd ed., Wiley, New York.

Johnson, K. L. (1985) *Contact Mechanics,* Cambridge University Press, Cambridge, England.

Juvinall, R. C., and Marshek, K. M. (2003) *Fundamentals of Machine Component Design,* 3rd ed. Wiley, New York.

Mott, R. L. (1998) *Machine Elements in Mechanical Design,* 3rd ed., Prentice-Hall, Upper Saddle River, NJ.

Norton, R. L. (1996) *Machine Design,* Prentice-Hall, Englewood Cliffs, NJ.

Shigley, J. E., Mischke, C. R., and Budynas, R. G. (2003), *Mechanical Engineering Design,* 7th ed., McGraw-Hill, New York.

## REFERENCES

Anderson, W. J. (1970) Elastohydrodynamic Lubrication Theory as a Design Parameter for Rolling Element Bearings, *ASME Paper 70–DE–19.*

Bamberger, E. N. (1971) *Life Adjustment Factors for Ball and Roller Bearings—An Engineering Design Guide,* American Society for Mechanical Engineers, New York.

Hamrock, B. J. (1993) *Fundamentals of Fluid Film Lubrication,* McGraw-Hill, New York.

Hamrock, B. J., Schmid, S. R., and Jacobson, B. O. (2004) *Fundamentals of Fluid Film Lubrication,* 2nd ed., Marcel-Dekker, New York.

Hamrock, B. J., and Anderson, W. J. (1983) *Rolling-Element Bearings,* NASA Reference Publication 1105.

Hamrock, B. J., and Dowson, D. (1981) *Ball Bearing Lubrication—The Elastohydrodynamics of Elliptical Contacts,* Wiley-Interscience, New York.

Jones, A. B. (1964) The Mathematical Theory of Rolling Element Bearings, *Mechanical Design and Systems Handbook,* H. A. Rothbart (ed.), McGraw-Hill, New York, pp. 13-1 to 13-76.

Lundberg, G., and Palmgren, A. (1947) Dynamic Capacity of Rolling Bearings, *Acta Polytech. Mech. Eng. Sci.,* vol. 1, no. 3. (Also, Ingeniörs Vetenskaps Akademien Handlingar, no. 196.)

Lundberg, G., and Palmgren, A. (1952) Dynamic Capacity of Roller Bearing, *Acta Polytech. Mech. Eng. Sci.,* vol. 2, no. 4. (Also, Ingeniörs Vetenskaps Akademien Handlingar, no. 210.)

SKF Catalog (1991), SKF USA, Inc.

Skurka, J. C. (1970) Elastohydrodynamic Lubrication of Roller Bearings, *J. Lubr. Technol.,* vol. 92, no. 2, pp. 281–291.

Stribeck, R. (1901) Kugellager für beliebige Belastungen, *Z. Ver. Deutsch. Ing.,* vol. 45, pp. 73–125.

Tallian, T. E. (1967) On Competing Failure Modes in Rolling Contact, *Trans. ASLE,* vol. 10, pp. 418–439.

Weibull, W. (1949) A Statistical Representation of Fatigue Failures in Solids, *Transactions of the Royal Institute of Technology, Stockholm,* no. 27.

## PROBLEMS

### Section 13.3

**13.1**   In roller bearings the load is carried by rollers, and the contacts between rollers and races are rectangular. The bearing macrogeometry determines which types of load can be carried by the bearing: radial, axial, or moment loads. Find which type of load can be carried by the different types of bearing shown in Tables 13.4 and 13.5.

**13.2**   Too high a power loss can be a problem in extremely high-speed applications of angular-contact ball bearings. A large portion of that power loss comes from the churning of the oil when the rolling balls displace the oil and splash it around the bearing. It is therefore important not to collect too much oil in the rolling tracks. For a duplex arrangement with the lubricating oil passing through both bearings for cooling, find out if a back-to-back or face-to-face arrangement is preferred from a low-power-loss point of view.

### Section 13.4

★ **13.3**   A deep-groove ball bearing with its dimensions is shown in sketch $a$. All dimensions are in millimeters. Also, $n = 11$, $d = 21/32$ in $= 16.67$ mm. Find the maximum ball force and stress as well as the relative displacement of the race. The static radial load capacity of the bearing is 36,920 N. Assume the ball and race are made of steel. *Ans.* $(W_z)_{max} = 16,780$ N, $\delta = 0.1121$ mm.

---

| * Indicates problem of greater difficulty.

Sketch a, for Problem 13.3

**13.4** In determining how large the contact areas are in a rolling-element bearing under a certain load, the geometry of each bearing part must be known. For a deep-groove ball bearing, the width-to-length ratio of the contact area is given by the race conformity, as expressed in Equation (13.3). Calculate the race conformity for both the inner- and outer-race contacts when the ball diameter is 17 mm and the radii of curvature in the axial direction are 8.840 mm for the inner race and 9.180 mm for the outer race. Also calculate the free contact angle and the endplay when the diametral clearance is 136 $\mu$m. Ans. $R_{ri} = 0.52$, $R_{ro} = 0.54$, $\beta_f = 21.04°$, $c_e = 97.65$ $\mu$m.

**13.5** In a spherical roller bearing, the inner radius of the outer race (the radius to the spherical surface) is 175 mm. The roller diameter is 41 mm, and the crowning radius is 168 mm. The diametral play is 0.33 mm for the outer race. Calculate the race conformity, the free endplay, and the free contact angle when the angle between each roller center and the bearing center plane is 12°. Ans. $R_r = 0.5208$, $\beta_f = 12.25°$, $c_e = 1.562$ mm.

## Section 13.5

★ **13.6** For a cylindrical roller bearing, the separator has the form of a cylinder with outer diameter of 80 mm, inner diameter of 74 mm, and width of 16 mm. Fourteen 13 by 13-mm rectangular pockets with depth of 3 mm are evenly distributed around the circumference of the separator. In each pocket is a 12.5- by 12.5-mm cylindrical roller. The inner-race outer diameter is 65 mm, and the outer-race inner diameter is 90 mm. How fast can the bearing inner race rotate before the polyester separator fails? Use a safety factor of 10 and the lowest values in Table A.4 for polyester. Ans. $N = 25,790$ rpm.

**13.7** For the cylindrical roller bearing in Problem 13.6, calculate the sliding speed at the contact between the rollers and the separator if the inner race rotates at 10,000 rpm.

## Section 13.6

**13.8**   A separator for a cylindrical roller bearing is made of nylon 66 with an ultimate strength of 80 MPa at 20°C. At the maximum service temperature of 260°C the ultimate strength falls to 1 MPa. The cage material can withstand a bearing speed of 16,000 rpm at 50°C. What speed can it withstand at 150°C if the ultimate strength varies linearly with the temperature? *Ans.* $N = 11,660$ rpm

**13.9**   A deep-groove ball bearing has a nylon 66 snap cage that is snapped in between the balls after they have been mounted. If the arms separating the balls have a square 3-mm by 3-mm cross section and are 12 mm long, calculate how fast a 100-mm-diameter cage can rotate without overstressing the ball-separating arms. The maximum allowable bending stress is 20 MPa. *Ans.* $N = 14,910$ rpm.

## Section 13.8

★ **13.10**  Using the results from Problem 13.3, determine the minimum film thickness and the film parameter for the ball–inner-race contact and the ball–outer-race contact. The inner race rotates at 20,000 rpm, and the outer race is stationary. The absolute viscosity at $p = 0$ and bearing effective temperature is 0.04 Pa · s. Also, the viscosity-pressure coefficient is $2.3 \times 10^{-8}$ m²/N. The surface roughnesses are $R_q = 0.0625$ $\mu$m for the balls and $R_q = 0.175$ $\mu$m for the races. *Ans.* For the ball inner race, $h_{min} = 2.411$ $\mu$m, $\Lambda = 12.97$.

★ **13.11**  A 0.4-in-diameter ball is loaded against a plane surface with 25 lb. The other parameters are as follows:

| | |
|---|---:|
| RMS surface finish of ball | 2.5 $\mu$in |
| RMS surface finish of plane | 7.5 $\mu$in |
| Absolute viscosity at $p = 0$ and bearing effective temperature | $6 \times 10^{-6}$ lb · s/in² |
| Viscosity-pressure coefficient | $1.6 \times 10^{-4}$ in²/lb |
| Modulus of elasticity for ball and plane | $3 \times 10^{7}$ psi |
| Poisson's ratio for ball and plane | 0.3 |
| Film parameter | 3.0 |

Determine the following:

*a)*   Minimum film thickness. *Ans.* $h_{min} = 23.72$ $\mu$in.
*b)*   Mean speed necessary to achieve the above film thickness. *Ans.* $u = 508.5$ in/s.
*c)*   Contact dimensions. *Ans.* $D_y = 0.01221$ in.
*d)*   Maximum pressure. *Ans.* $\sigma_{max} = 320$ ksi.
*e)*   Maximum deformation. *Ans.* $\delta_{max} = 186$ $\mu$in.

★ **13.12**  A deep-groove ball bearing has steel races and silicon nitride balls. It is lubricated with a mineral oil that has a viscosity of 0.026 N · s/m² at the application temperature. The pressure-viscosity coefficient of the oil is $2 \times 10^{-8}$ m²/N. The ball diameter is 17 mm, and the radius of curvature in the axial direction is 8.84 mm for both races. The radii in the rolling direction are 30 mm for the inner race and 47 mm for the outer race. Calculate the minimum oil film thickness at the inner-race contact when the ball load is 20,000 N and the rolling speed is 3 m/s. What should the radius of curvature in the axial direction be for the outer race to have the same film thickness as that of the inner race? *Ans.* $h_{min} = 0.2328$ $\mu$m, $r_{by} = 10.65$ mm.

**13.13** For the bearing considered in Problem 13.12, the silicon nitride balls are changed to steel balls. Calculate how large the load can be and still maintain the same oil film thickness as obtained in Problem 13.12. *Ans. P = 25,880 N.*

**13.14** A steel roller, shown in sketch *b*, is used for rolling steel sheets that have an ultimate strength of 400 MPa. The 1-m-long roller has a diameter of 20 cm. What load per unit width will cause plastic deformation of the sheets? *Ans. w′ = 442 kN/m.*

Sketch *b*, for Problem 13.14

★ **13.15** An artificial hip (see sketch *c*) inserted in the body is lubricated by the fluid in the joint, known as the *synovial fluid*. The following conditions are typical for a total hip replacement:

| | |
|---|---|
| Equivalent radius $R_x = R_y$ | 1 m |
| Viscosity of synovial fluid $\eta_0$ | $2 \times 10^{-3}$ Pa·s |
| Pressure exponent of viscosity $\xi$ | $2.75 \times 10^{-8}$ m²/N |
| Mean entraining velocity $\tilde{u}$ | 0.075 m/s |
| Applied load $P_z$ | 4500 N |

Sketch *c*, for Problem 13.15 (Courtesy Zimmer, Inc.)

Metal-on-metal implants use a cobalt-chrome-molybdenum (Co-Cr-Mo) alloy with $E = 230$ GPa and $\nu = 0.30$, while metal-on-polymer implants use Co-Cr-Mo on polyethylene ($E = 0.78$ GPa, $\nu = 0.46$). Perform the following:

a) If the metal surfaces are polished to a surface roughness of 0.020 $\mu$m, estimate the film thickness and film parameter for metal-on-metal total hip replacements. *Ans.* $\Lambda = 0.785$.

b) If the polymer surfaces are machined to a 0.060-$\mu$m roughness, repeat part (a) for metal-on-polymer total hip replacements.

c) Ceramic-on-ceramic implants are made from $Al_2O_3$, with $E = 300$ GPa, $\nu = 0.26$, and $R_q = 0.005$ $\mu$m. Calculate the film parameter for ceramic-on-ceramic implants. *Ans.* $\Lambda = 2.98$.

d) Comment on the lubrication effectiveness for these options.

### Section 13.9

**13.16** A shaft with rolling bearings and loads is shown in sketch $d$. The bearings are mounted such that both the axial and radial forces caused by the external loads are absorbed. For bearing NU205EC the geometry is cylindrical with no flange on the inner race, so that only radial load can be absorbed. The full axial thrust load $P_a$ is thus taken up by the deep-groove ball bearing 6205. Determine the $\tilde{L}_{10}$ lives of each of the two bearings while assuming length dimensions are in meters. *Ans.* $(\tilde{L}_{10})_{NU205EC} = 71{,}500$ million revolutions, $(\tilde{L}_{10})_{6205} = 343$ million revolutions.

Sketch $d$, for Problem 13.16

**13.17** Given the shaft and bearings in Problem 13.16 but with the bearing positions shifted so the 6205 deep-groove ball bearing is in the left end of the shaft and bearing NU205EC in the middle, find the $\tilde{L}_{10}$ lives for the two bearings in their new positions. Which of the two positions would be preferred and why? *Ans.* $(\tilde{L}_{10})_{NU205EC} = 7098$ million revolutions, $(\tilde{L}_{10})_{6205} = 1091$ million revolutions.

★ **13.18** A bearing arrangement for a shaft in a gearbox is shown in sketch $e$. Bearing A is a cylindrical roller bearing (NU202EC) with $d = 15$ mm, $D = 35$ mm, $B = 11$ mm, and $C = 12{,}500$ N. Bearing B is a deep-groove ball bearing (6309) with $d = 45$ mm, $D = 100$ mm, $B = 25$ mm, $C = 52{,}700$ N, and $C_0 = 31{,}500$ N. Determine which bearing is likely to fail first and what the $\tilde{L}_{10}$ is for that bearing. Length dimensions are in millimeters. *Ans.* Bearing B fails at $\tilde{L}_{10} = 1556$ million revolutions.

Sketch e, for Problem 13.18

★ **13.19** The two bearings in Problem 13.18 have the same composite surface roughness of the inner-race, rolling-element contact, $(R_{qa}^2 + R_{qb}^2)^{0.5} = 0.10$ $\mu$m. For bearing NU202EC at A, the outer radius of the inner race is 9.65 mm, the roller diameter is 5.7 mm, and inner radius of the outer race is 15.35 mm. For bearing 6309 at B, the ball diameter is 17 mm, the outer radius of the inner race is 27.4 mm, and the inner radius of the outer race is 44.4 mm. The race conformity for the outer ring is 0.53 while for the inner ring it is 0.52. The lubricant used has the absolute viscosity $\eta_0 = 0.1$ N·s/m², and the pressure-viscosity coefficient $\xi = 2 \times 10^{-8}$ m²/N. The bearing rotational speed is 1500 rpm. The cylindrical roller bearing has 13 rollers, and the ball bearing has 10 balls. The ball bearing is made of AISI 440C ($E = 200$ GPa), and the roller bearing is made of AISI 52100 ($E = 207$ GPa); both materials have $\nu = 0.3$. Find the adjusted bearing lives $\tilde{L}_{10,\mathrm{adj}}$ for the two bearings. *Ans.* NU202EC: $\tilde{L}_{10,\mathrm{adj}} = 1.3\tilde{L}_{10}$, 6309: $\tilde{L}_{10,\mathrm{adj}} = 2.3\tilde{L}_{10}$.

★ **13.20** Given the shaft arrangement in Figure 11.3, with the applied forces indicated, bearing B is a cylindrical roller bearing of type NU304EC with $C = 30{,}800$ N, and bearing A is a deep-groove ball bearing of type 6404 ($C = 30{,}700$ N, $C_0 = 15{,}000$ N). Find the $\tilde{L}_{10}$ life for each bearing and judge if they are strong enough to be used at 1500 rpm for 20 years. *Ans.* $(\tilde{L}_{10})_{\mathrm{NU304EC}} = 1.887 \times 10^{11}$ r, $(\tilde{L}_{10})_{6404} = 2.343 \times 10^{11}$ r.

**13.21** Given the shaft arrangement in Problem 13.20 but with the belt drive at D changed to a worm gear giving the same radial force but also an additional axial force of 800 N directed from D to A, find the $\tilde{L}_{10}$ lives of the bearings and judge if they are strong enough to last 20 years at 1500 rpm. *Ans.* $(\tilde{L}_{10})_{\mathrm{NU304EC}} = 6.572 \times 10^9$ r, $(\tilde{L}_{10})_{6404} = 1.577 \times 10^{10}$ r.

**13.22** A flywheel has deep-groove ball bearings located close to the rotating disk. The disk mass is 50 kg, and it is mounted out of balance, so that the center of mass is situated at a radius 1 mm from the bearing centerline. The bearings are two 6305 deep-groove ball bearings with $C = 22{,}500$ N. Neglecting the shaft mass, find the nominal bearing $\tilde{L}_{10}$ in hours for each bearing at

a)   500 rpm. *Ans.* $\tilde{L} = 3.68 \times 10^5$ million revolutions (12.3 million h).
b)   2500 rpm. *Ans.* $\tilde{L} = 1510 \times 10^5$ million revolutions (10,000 h).
c)   12,500 rpm. *Ans.* $\tilde{L} = 0.142 \times 10^5$ million revolutions (11 min.).

**13.23** A ball bearing manufacturer wants to make sure that its bearings provide long service lives on different components. Extra polishing operation can be applied to

a)   The ring surfaces
b)   The ball surfaces
c)   All contact surfaces

The polishing decreases the roughness by one-half the original roughness, which for the races is 0.08 $\mu$m and for the balls is 0.04 $\mu$m. Find how much the $\Lambda$ value changes when $a$, $b$, or $c$ is polished.

★ **13.24** The shaft shown in sketch $f$ transfers power between the two pulleys. The tension on the slack side (right pulley) is 30% of that on the tight side. The shaft rotates at 900 rpm and is supported uniformly by a radial ball bearing at points 0 and B. Select a pair of radial ball bearings with 99% reliability and 30,000 h of life. All length dimensions are in millimeters.

Sketch $f$, for Problem 13.24

★ **13.25** In sketch $g$ the tension on the slack side of the left pulley is 20% of that on the tight side. The shaft rotates at 720 rpm. Select a pair of roller bearings to support the shaft for 99% reliability and a life of 24,000 h. All length dimensions are in millimeters.

Sketch $g$, for Problem 13.25

★ **13.26** In sketch *h* the shaft rotates at 1100 rpm and transfers power with light shock (increases load by 20%) from a 600-mm-diameter pulley to a 300-mm-diameter sprocket. Select a radial ball bearing for support at 0 and a roller bearing for support at A. The bearings should have a life of 24,000 h at 95% reliability. All length dimensions are in millimeters.

Sketch *h*, for Problem 13.26

★ **13.27** The power-transmitting system shown in sketch *i* consists of a helical gear, a bevel gear, and a shaft that rotates at 600 rpm and is supported by two roller bearings. The load on the bevel gear is $-0.5P\bar{i} - 0.41P\bar{j} + 0.44P\bar{k}$. The left bearing supports the axial load on the shaft. Select bearings for 36,000-h life at 98% reliability. All length dimensions are in millimeters.

Sketch *i*, for Problem 13.27

**13.28** A cylindrical roller bearing is used to support an 8-kN load for 1200 h at 600 rpm. Calculate the load-carrying capacity of this bearing for 95% reliability. What size bearing would you recommend? *Ans. $\bar{C} = 30,000$ N.*

**13.29** A deep-groove ball bearing is used to support a 6-kN load for 1000 h at 1200 rpm. For 99% reliability what should be the load-carrying capacity of the bearing? What size bearing should be chosen? *Ans. $\bar{C} = 50,900$ N.*

# GENERAL GEAR THEORY; SPUR GEARS

An assortment of gears
(Courtesy of Quality
Transmission
Components).

*Just stare at the machine. There is nothing wrong with that.*
*Just live with it for a while. Watch it the way you watch a*
*line when fishing and before long, as sure as you live,*
*you'll get a little nibble, a little fact asking in a timid, humble*
*way if you're interested in it. That's the way the world*
*keeps on happening. Be interested in it.*
　　　Robert Pirsig, Zen and the Art of Motorcycle Maintenance

## SYMBOLS

| | |
|---|---|
| $a$ | addendum, m |
| $A'$ | constant in Eq. (14.45) |
| $A, B, C$ | constants |
| $b$ | dedendum, m |
| $b^*$ | Hertzian half-width, m |
| $b_w$ | face width, m |
| $b_l$ | backlash, m |
| $C_e$ | mesh alignment correction factor |
| $C_H$ | hardness ratio factor |
| $C_{ma}$ | mesh alignment factor |
| $C_{mc}$ | lead correction factor |
| $C_{mf}$ | face load distribution factor |
| $C_{mt}$ | transverse load factor |
| $C_{pf}$ | pinion proportion factor |
| $C_{pm}$ | pinion proportion modifier |
| $C_r$ | contact ratio |
| $c$ | clearance, m |
| $c_d$ | center distance, m |
| $d$ | pitch diameter, m |
| $d_b$ | base diameter, m |
| $d_{bg}$ | base diameter of gear, m |
| $d_{bp}$ | base diameter of pinion, m |
| $d_g$ | pitch diameter of gear, m |
| $d_{og}$ | outside diameter of gear, m |
| $d_{op}$ | outside diameter of pinion, m |
| $d_p$ | pitch diameter of pinion, m |
| $d_r$ | root or fillet diameter, m |
| $E$ | modulus of elasticity, Pa |
| $E'$ | effective modulus of elasticity, Pa |
| $g_r$ | gear ratio |
| HB | Brinell hardness |
| $h$ | height, m |
| $h_k$ | working depth, m |
| $h_p$ | transmitted horsepower, hp |
| $h_t$ | total depth, m |
| $I$ | area moment of inertia, m$^4$ |
| $K_a$ | application factor |
| $K_b$ | rim thickness factor |
| $K_g$ | geometry factor for contact stress evaluation |
| $K_i$ | idler factor |
| $K_m$ | load distribution factor |
| $K_R$ | reliability factor |
| $K_s$ | size factor |
| $K_T$ | temperature factor |
| $K_v$ | dynamic factor |
| $L_{ab}$ | length of line of action, m |

| | |
|---|---|
| $L_{a^*b^*}$ | length of line from $a^*$ to $b^*$, m |
| $m$ | module, $d/N$, mm |
| $N$ | number of teeth |
| $N_a$ | gear rotational speed, rpm |
| $N_g$ | number of teeth in gear |
| $N_p$ | number of teeth in pinion |
| $N_v$ | equivalent number of teeth |
| $n_s$ | safety factor |
| $P$ | power, W |
| $P_z$ | normal load, N |
| $P_z'$ | normal load per face width, N/m |
| $p_a$ | axial pitch for helical gears, m |
| $p_b$ | base pitch, m |
| $p_c$ | circular pitch, $\pi d/N$, m |
| $p_{cn}$ | normal circular pitch, m |
| $p_d$ | diametral pitch, $N/d$, in$^{-1}$ |
| $p_{dn}$ | normal diametral pitch, in$^{-1}$ |
| $p_H$ | maximum Hertzian contact pressure, Pa |
| $p_p$ | pitch point |
| $Q_v$ | transmission accuracy level number |
| $R_q$ | rms surface finish of gear, m |
| $R_x$ | curvature sum in $x$ direction, m |
| $r$ | pitch radius, m |
| $r_b$ | base radius, m |
| $r_{bg}$ | base radius of gear, m |
| $r_{bp}$ | base radius of pinion, m |
| $r_g$ | pitch radius of gear, m |
| $r_o$ | outside radius, m |
| $r_{og}$ | outside radius of gear, m |
| $r_{op}$ | outside radius of pinion, m |
| $r_p$ | pitch radius of pinion, m |
| $S_c$ | contact stress number, Pa |
| $S_t$ | bending stress number, Pa |
| $T$ | torque, N · m |
| $t$ | tooth thickness, m |
| $t_h$ | circular tooth thickness measured on pitch circle, m |
| $t_{ha}$ | actual circular tooth thickness measured on pitch circle, m |
| $\tilde{u}$ | mean velocity, m/s |
| $v_t$ | pitch-line velocity, ft/min |
| $W$ | load, N |
| $W_r$ | radial load, N |
| $W_t$ | tangential load, N |
| $w'$ | load per unit width, N/m |
| $x$ | unknown distance, m |
| $Y$ | Lewis form factor |
| $Y_j$ | geometry factor |

| | | | |
|---|---|---|---|
| $Y_N$ | stress cycle factor for bending | $\phi_n$ | pressure angle in normal direction for helical gears, deg |
| $Z$ | angular velocity ratio | | |
| $Z_N$ | stress cycle factor for contact stress | $\psi$ | helix angle, deg |
| $\alpha$ | coefficient of thermal expansion, $(°C)^{-1}$ | $\Omega$ | angular velocity of shaft, rad/s |
| $\alpha_{ag}$ | angle of approach of gear, deg | $\omega$ | angular speed of gear, rad/s |
| $\alpha_{ap}$ | angle of approach of pinion, deg | | |
| $\alpha_{rg}$ | angle of recess of gear, deg | **SUBSCRIPTS** | |
| $\alpha_{rp}$ | angle of recess of pinion, deg | | |
| $\eta_0$ | atmospheric viscosity, Pa·s | $a$ | actual; axial |
| $\Lambda$ | film parameter | all | allowable |
| $\xi$ | pressure-viscosity coefficient, $m^2/N$ | $b$ | bending |
| $\sigma$ | stress, Pa | $c$ | contact |
| $\sigma_d$ | design stress, Pa | $g$ | gear |
| $\sigma_g$ | bending stress of gear, Pa | max | maximum |
| $\sigma_{max}$ | maximum normal stress, Pa | $o$ | outside |
| $\sigma_r$ | contact stress of gear, Pa | $p$ | pinion |
| $\phi$ | pressure angle, deg | $t$ | total |

# 14.1 INTRODUCTION

A **gear** may be thought of as a toothed wheel that when meshed with another smaller-in-diameter toothed wheel (the **pinion**) will transmit rotation from one shaft to another. The primary function of a gear is to transfer power from one shaft to another while maintaining a definite ratio between the velocities of the shaft rotations. The teeth of a driving gear mesh push on the driven gear teeth, exerting a force component perpendicular to the gear radius. Thus, a torque is transmitted, and because the gear is rotating, power is transferred. Gears are the most rugged and durable torque transmitters listed in Chapter 1 (Sec. 1.2). Their power transmission efficiency is as high as 98%. On the other hand, gears are usually more costly than other torque transmitters, such as chain drives and belt drives. Gears are highly standardized as to tooth shape and size. The American Gear Manufacturers Association (AGMA) publishes standards for gear design, manufacture, and assembly (AGMA, 1999).

This chapter attempts to follow the methods, recommendations, and terminology defined by the AGMA. However, there are occasions when the approach has been simplified. For example, AGMA follows a practice of assigning a different symbol to a variable depending on whether it is in U.S. Customary System or metric units. Also, some of the strength of materials terminology has been brought into agreement with the nomenclature consistent throughout this text.

# 14.2 TYPES OF GEAR

Gears can be divided into three major classes: parallel-axis gears, nonparallel but coplanar gears, and nonparallel and noncoplanar gears. This section describes gears in each of these classes. This chapter will emphasize spur gears, and helical, bevel and worm gears will be discussed in Chapter 15.

(a)                                      (b)

**Figure 14.1**    Spur gear drive. (a) Schematic illustration of meshing spur gears; (b) a collection of spur gears. [Courtesy of Boston Gear Works, Inc.]

### 14.2.1 PARALLEL-AXIS GEARS

Parallel-axis gears are the simplest and most popular type of gear. They connect parallel shafts and can transfer large amounts of power with high efficiency. Spur and helical gears are two of the primary gears in this classification.

#### Spur Gears

Figure 14.1 shows a **spur gear** drive with teeth on the outside of a cylinder and parallel to the cylinder axis. Spur gears are the simplest and the most common type of gear.

#### Helical Gears

Figure 14.2 shows a helical gear drive, with gear teeth cut on a spiral that wraps around a cylinder. Helical teeth enter the meshing zone progressively and, therefore, have a smoother action than spur gear teeth. Helical gears also tend to be quieter. Another positive feature of helical gears (relative to spur gears) is that the transmitted load is somewhat larger, thus implying that helical gear life will be longer for the same load. A smaller helical gear can transmit the same load as a larger spur gear.

    A disadvantage of helical gears (relative to spur gears) is that they produce an additional end thrust along the shaft axis that is not present with spur gears. This end thrust may require an additional component, such as a thrust collar, ball bearings, or tapered-roller bearings. Another disadvantage is that helical gears have slightly lower efficiency than that of equivalent spur gears. Efficiency depends on total normal tooth load, which is higher for spur gears. Although the total load-carrying capacity is larger for helical gears, the load is distributed normally and axially, whereas for a spur gear all load is distributed normally.

(a)                                                                          (b)

**Figure 14.2**    Helical gear drive. (a) Schematic illustration of meshing helical gears; (b) a collection of helical gears. [Courtesy of Boston Gear Works, Inc.]

## 14.2.2    NONPARALLEL, COPLANAR GEARS

This section discusses gears that have nonparallel axes (unlike spur and helical gears, which have parallel axes) and that are coplanar (like spur and helical gears). **Bevel,** straight, Zerol, and **spiral gears** are all in the nonparallel, coplanar class. The common feature of this class is the redirection of power around a corner, as might be required, for example, when a horizontally mounted engine is connected to the vertically mounted rotor shaft on a helicopter. Figure 14.3 shows a bevel gear drive with straight teeth and illustrates the gears in this classification. Observe that the axes of the gear drive are nonparallel but coplanar.

(a)                                                                          (b)

**Figure 14.3**    Bevel gear drive. (a) Schematic illustration of meshing bevel gears with straight teeth; (b) a collection of bevel gears. [Courtesy of Boston Gear Works, Inc.]

**Figure 14.4**     Worm gear drive. (a) Cylindrical teeth; (b) double enveloping; (c) a collection of worm gears. [Courtesy of Boston Gear Works, Inc.]

### 14.2.3     NONPARALLEL, NONCOPLANAR GEARS

Nonparallel, noncoplanar gears are more complex in both geometry and manufacturing than the two previous gear classifications. As a result, these gears are more expensive than the other gears discussed.

Figure 14.4, a **worm gear** drive with cylindrical teeth, illustrates this class of gear. Note that the axes are nonparallel and noncoplanar. These gears can provide considerably higher reduction ratios than those of coplanar or simple crossed-axis gear sets, but their load-carrying capacity is low, their contact pressure is extremely high, and their wear rate is high. Thus, worm gear drives are used only for light-load applications.

This brief discussion explains the various gear classes and is not meant to be all-inclusive. More detailed and inclusive discussion can be found in Drago (1992). The rest of this chapter focuses primarily on spur gears. Other types of gears will be addressed in Chapter 15.

---

**EXAMPLE 14.1**

**Given:**  A bicycle is normally driven by the pedals through a roller chain transmission to the back wheel. The power efficiency of such a chain drive is 95% if it is well lubricated and rather heavily loaded. To avoid getting oil on the trousers from the chain, a design change is considered. The chain drive is to be changed to a shaft with two sealed ball bearings inside a tube in the frame and a bevel gear drive at the pedals and another bevel gear drive at the back wheel. At the speed of 20 km/h the chain-driven bike requires 220 W to the rear wheel, and the same rear-wheel power is required for the new design.

**Find:**  How large does the input power have to be for the new drive to run 20 km/h if the newly designed shaft rotates at 1200 rpm and the bearing and seal friction torque in each bearing is 0.1 N · m? The power efficiency for the bevel gears is 0.98. Also, calculate the total power efficiency for the new drive when the bicycle is driven at 20 km/h.

**Solution:** The power required to drive the rear wheel at 20 km/h is 220 W. The input power needed to the bevel gear is

$$\frac{P}{0.98} = \frac{220}{0.98} = 224.5 \text{ W}$$

The power needed to overcome the friction in the bearings and seals is

$$T\omega = 2(0.1)\left(\frac{1200}{60}\right)(2\pi) = 25.13 \text{ W}$$

The output power from the bevel gear is $224.5 + 25.13 = 249.6$ W, and the input power needed from the pedals is

$$\frac{249.6}{0.98} = 254.7 \text{ W}$$

The power efficiency is

$$\frac{P_{\text{out}}}{P_{\text{in}}} = \frac{220}{254.7} = 0.864 \text{ or } 86.4\%$$

This efficiency is lower than that for the chain drive.

## 14.3 GEAR GEOMETRY

Figure 14.5 shows the basic spur gear geometry, and Figure 14.6 shows gear teeth with the nomenclature to be used. The geometry of gear teeth enables a normal to the tooth profiles at their point of contact to pass through a fixed point on the line of centers called the **pitch point.** From Figures 14.5 and 14.6 the addendum $a$ is the distance from the top land to the pitch circle, and the dedendum $b$ is the radial distance from the bottom land to the pitch circle. The clearance shown in Figure 14.6 is the amount by which the dedendum exceeds the addendum. A clearance is needed to prevent the end of one gear tooth from riding the bottom of the mating gear.

### 14.3.1 CENTER DISTANCE, CIRCULAR PITCH, AND DIAMETRAL PITCH

From Figure 14.5 the pitch circles of the gear and pinion make contact at the pitch point. The center distance between the axes of rotation of the rotating gears can be expressed as

$$c_d = \frac{d_p + d_g}{2} \tag{14.1}$$

where $d_p$ = pitch diameter of pinion, m
$d_g$ = pitch diameter of gear, m

The **circular pitch** $p_c$ is the distance measured on the pitch circle from one point on one tooth to a corresponding point on the adjacent tooth. Mathematically, it is

$$p_c = \text{tooth thickness} + \text{width of spacing} = \frac{\pi d}{N} = \frac{\pi d_p}{N_p} = \frac{\pi d_g}{N_g} \tag{14.2}$$

where $N_p$ = number of pinion teeth
$N_g$ = number of gear teeth

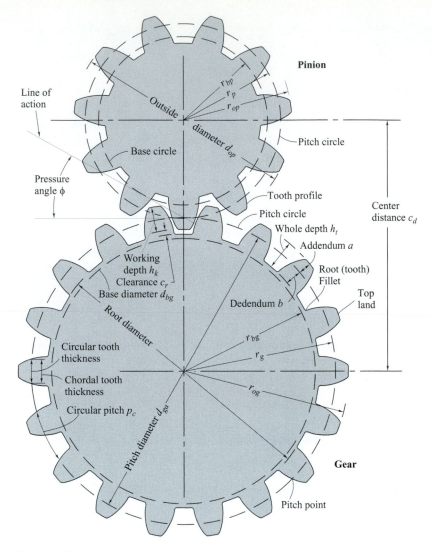

**Figure 14.5**    Basic spur gear geometry.

From Equation (14.2) the **gear ratio** is

$$g_r = \frac{d_g}{d_p} = \frac{p_c N_g}{\pi} \frac{\pi}{p_c N_p} = \frac{N_g}{N_p} \tag{14.3}$$

Substituting Equations (14.2) and (14.3) into Equation (14.1) gives

$$c_d = \frac{p_c}{2\pi}(N_p + N_g) = \frac{p_c N_p}{2\pi}\left(1 + \frac{N_g}{N_p}\right) = \frac{p_c N_p}{2\pi}(1 + g_r) \tag{14.4}$$

The **diametral pitch** $p_d$ is the number of teeth in the gear per pitch diameter. In the English system the unit of diametral pitch is reciprocal inches. The expression for diametral

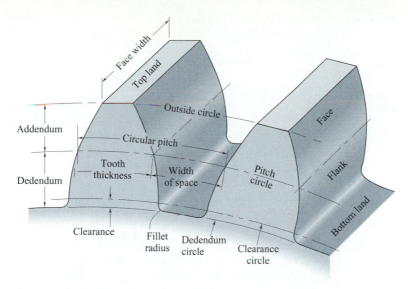

**Figure 14.6** Nomenclature of gear teeth.

pitch is

$$p_d = \frac{N_p}{d_p} = \frac{N_g}{d_g} \tag{14.5}$$

Figure 14.7 shows standard diametral pitches and compares them with tooth size for full-size gears. The smaller the tooth, the larger the diametral pitch. Table 14.1 shows preferred diametral pitches for four tooth classes. Coarse and medium-coarse gears are used for power transmission applications; fine and ultrafine gears are used for instrument and control systems. The pitch can also be selected from power transmission requirements, as shown in Figure 14.8.

**Figure 14.7** Standard diametral pitches compared with tooth size. Full size is assumed.

**Table 14.1**    Preferred diametral pitches for four tooth classes

| Class | Diametral pitch $p_d$, in$^{-1}$ |
|---|---|
| Coarse | $\frac{1}{2}$, 1, 2, 4, 6, 8, 10 |
| Medium-coarse | 12, 14, 16, 18 |
| Fine | 20, 24, 32, 48, 64 |
| | 72, 80, 96, 120, 128 |
| Ultrafine | 150, 180, 200 |

**Figure 14.8**    Transmitted power as a function of pinion speed for a number of diametral pitches. [From Mott (1999).]

Comparing Equations (14.2) and (14.5) yields the following:

$$p_c = \frac{\pi}{p_d}$$ (14.6)

Thus, the circular pitch is inversely proportional to the diametral pitch.

The **module** $m$ is the ratio of the pitch diameter to the number of teeth and is only expressed in the SI unit of millimeters.

$$m = \frac{d_g}{N_g} = \frac{d_p}{N_p}$$ (14.7)

From Equations (14.2) and (14.7)

$$p_c = \pi m$$ (14.8)

Also, from Equations (14.6) and (14.8)

$$m = \frac{1}{p_d}$$ (14.9)

---

**Given:** A 20-tooth pinion with diametral pitch of 5 per inch meshes with a 63-tooth gear. **EXAMPLE 14.2**

**Find:** The center distance, the circular pitch, and the gear ratio.

**Solution:** From Equation (14.5)

$$d_p = \frac{N_p}{p_d} = \frac{20}{5} = 4 \text{ in}$$

$$d_g = \frac{N_g}{p_d} = \frac{63}{5} = 12.6 \text{ in}$$

From Equation (14.1) the center distance is

$$c_d = \frac{d_p + d_g}{2} = \frac{4 + 12.6}{2} = 8.3 \text{ in}$$

From Equation (14.2) the circular pitch is

$$p_c = \frac{\pi d_p}{N_p} = \frac{\pi 4}{20} = 0.6283 \text{ in}$$

$$p_c = \frac{\pi d_g}{N_g} = \frac{\pi (12.6)}{63} = 0.6283 \text{ in}$$

From Equation (14.3) the gear ratio is

$$g_r = \frac{d_g}{d_p} = \frac{N_g}{N_p} = \frac{12.6}{4} = \frac{63}{20} = 3.150$$

---

## 14.3.2 ADDENDUM, DEDENDUM, AND CLEARANCE

Table 14.2 shows values of addendum, dedendum, and clearance for a spur gear with a 20° pressure angle and a full-depth involute. Both coarse-pitch and fine-pitch formulations are given. Note from Table 14.2 that the following is true:

$$c = b - a$$ (14.10)

**Table 14.2**    Formulas for addendum, dedendum and clearance (pressure angle, 20°; full-depth involute)

| Parameter | Symbol | Coarse pitch $(p_d < 20\text{ in}^{-1})$ | Fine pitch $(p_d \geq 20\text{ in}^{-1})$ | Metric module system |
|-----------|--------|-------------------|-----------------|----------------------|
| Addendum  | $a$    | $1/p_d$           | $1/p_d$         | 1.00 m               |
| Dedendum  | $b$    | $1.25/p_d$        | $1.200/p_d + 0.002$ | 1.25 m           |
| Clearance | $c$    | $0.25/p_d$        | $0.200/p_d + 0.002$ | 0.25 m           |

Once the addendum, dedendum, and clearance are known, a number of other parameters can be obtained. Some of these are

$$\text{Outside diameter:}\qquad d_o = d + 2a \qquad\qquad (14.11)$$

$$\text{Root diameter:}\qquad d_r = d - 2b \qquad\qquad (14.12)$$

$$\text{Total depth:}\qquad h_t = a + b \qquad\qquad (14.13)$$

$$\text{Working depth:}\qquad h_k = a + a = 2a \qquad\qquad (14.14)$$

These parameters can also be expressed in terms of diametral pitch and number of teeth rather than pitch diameter by using Equation (14.5). For example, the outside diameter given in Equation (14.11) can be expressed, by making use of the formulas given in Table 14.2 and Equation (14.5), as

$$d_o = \frac{N}{p_d} + 2\left(\frac{1}{p_d}\right) = \frac{N+2}{p_d} \qquad\qquad (14.15)$$

All formulas given in this section are equally applicable for the gear and the pinion.

### 14.3.3   LINE OF ACTION, PRESSURE ANGLE, AND GEAR INVOLUTE

The purpose of meshing gear teeth is to provide constant, instantaneous relative motion between the engaging gears. To achieve this tooth action, the common normal of the curves of the two meshing gear teeth must pass through a common point, called the pitch point, on the pitch circles.

Figure 14.9 shows the pitch and base circles for both the pinion and the gear. Also shown in this figure are the line of action, the straight line from $a$ to $b$, and the pressure angle $\phi$. The pressure angle is the angle between the line of centers and a line perpendicular to the line of action and going through the center of the gear or pinion. From Figure 14.9 observe that the radius of the base circle is

$$r_b = r \cos\phi$$

$$r_{bp} = r_p \cos\phi \qquad \text{and} \qquad r_{bg} = r_g \cos\phi \qquad\qquad (14.16)$$

The base pitch $p_b$ is the distance from a point on one tooth to the corresponding point on the adjacent tooth measured around the base circle. From Equation (14.2) the base pitch is

$$p_b = p_c \cos\phi = \frac{\pi d}{N}\cos\phi = \frac{\pi d_p}{N_p}\cos\phi = \frac{\pi d_g}{N_g}\cos\phi \qquad\qquad (14.17)$$

The shape of the gear tooth is obtained from the involute curve shown in Figure 14.10.

**Figure 14.9** Pitch and base circles for pinion and gear as well as line of action and pressure angle.

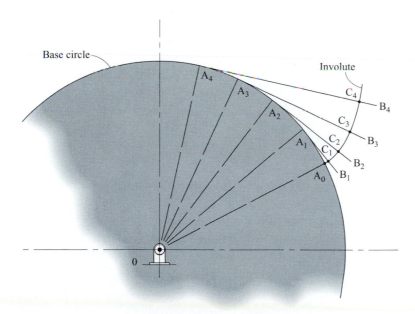

**Figure 14.10** Construction of involute curve.

PROCEDURE FOR CONSTRUCTING THE INVOLUTE CURVE

1. Divide the base circle into a number of equal distances, thus constructing $A_0$, $A_1$, $A_2$,...

2. Beginning at $A_1$, construct the straight line $A_1B_1$, perpendicular to $0A_1$, and likewise beginning at $A_2$ and $A_3$.

3. Along $A_1B_1$, lay off the distance $A_1A_0$, thus establishing $C_1$. Along $A_2B_2$, lay off twice $A_1A_0$, thus establishing $C_2$, etc.

4. Establish the involute curve by using points $A_0$, $C_1$, $C_2$, $C_3$,.... Gears made from the involute curve have at least one pair of teeth in contact with each other.

**EXAMPLE 14.3**

**Given:** A gear with 40 teeth, a pinion with 16 teeth, a diametral pitch of 2 per inch, and a pressure angle of 20°.

**Find:**

(a) Determine the circular pitch, the center distance, and the base diameter for the pinion and the gear.

(b) If the center distance is increased by 0.25 in, find the pitch diameters for the gear and pinion and the pressure angle.

**Solution:**

(a) From Equation (14.6) the circular pitch is $p_c = \pi/p_d = \pi/2$ in $= 1.571$ in. From Equation (14.5) the pitch diameters of the gear and pinion are

$$d_g = \frac{N_g}{p_d} = \frac{40}{2} = 20 \text{ in}$$

$$d_p = \frac{N_p}{p_d} = \frac{16}{2} = 8 \text{ in}$$

From Equation (14.1) the center distance is

$$c_d = \frac{d_p + d_g}{2} = \frac{8 + 20}{2} = 14 \text{ in}$$

From Equation (14.16) the radii of the base circles for the pinion and the gear are

$$r_{bp} = \frac{d_p}{2} \cos\phi = \frac{8}{2} \cos 20° = 3.759 \text{ in}$$

$$r_{bg} = \frac{d_g}{2} \cos\phi = \frac{20}{2} \cos 20° = 9.397 \text{ in}$$

(b) If the center distance is increased by 0.25 in,

$$c_d = 14.25 \text{ in}$$

This value corresponds to a 1.79% increase in the center distance. But

$$c_d = \frac{(d_p + d_g)}{2}$$

$$\therefore \quad d_p + d_g = 2(14.25 \text{ in}) = 28.5 \text{ in} \qquad \text{(a)}$$

From Equation (14.5)

$$\frac{N_p}{d_p} = \frac{N_g}{d_g}$$

$$\therefore \quad \frac{16}{d_p} = \frac{40}{d_g} \qquad \text{(b)}$$

Thus, there are two equations and two unknowns. From Equation (a)

$$d_p = 28.5 - d_g \qquad \text{(c)}$$

Substituting Equation (c) into Equation (b) gives

$$\frac{16}{28.5 - d_g} = \frac{40}{d_g}$$

$$16d_g = 40(28.5) - 40d_g$$

$$56d_g = 1140$$

$$d_g = 20.36 \text{ in} \qquad \text{(d)}$$

Substituting Equation (d) into Equation (a) gives

$$d_p = 28.50 - 20.36 = 8.143 \text{ in} \qquad \text{(e)}$$

From Equation (14.5) the diametral pitch is

$$p_d = \frac{N_p}{d_p} = \frac{16}{8.143} = 1.965/\text{in}$$

Also, $$p_d = \frac{N_g}{d_g} = \frac{40}{20.36} = 1.965/\text{in}$$

From Equation (14.6) the circular pitch is

$$p_c = \frac{\pi}{p_d} = \frac{\pi}{1.965} = 1.599 \text{ in}$$

Changing the center distance does not affect the base circle of the pinion or the gear. Thus, from Equation (14.16)

$$\phi = \cos^{-1} \frac{r_{bp}}{r_p} = \cos^{-1} \frac{3.759}{8.143/2} = \cos^{-1} 0.9232 = 22.60°$$

Also, $$\phi = \cos^{-1} \frac{r_{bg}}{r_g} = \cos^{-1} \frac{9.397}{20.36/2} = \cos^{-1} 0.9232 = 22.60°$$

Thus, a 1.79% increase in center distance results in a 2.60° increase in pressure angle.

## 14.4    GEAR RATIO

For meshing gears the rotational motion and the power must be transmitted from the driving gear to the driven gear with a smooth and uniform positive motion and with minor loss of power due to friction. The **fundamental law of gearing** states that the common normal to the tooth profile at the point of contact must always pass through a fixed point, called the pitch point, in order to maintain a constant velocity ratio of the two meshing gear teeth. The pitch point $p_p$ is shown in Figure 14.9. Since the velocity at this point must be the same for both the gear and the pinion,

$$r_p \omega_p = r_g \omega_g$$

Expressing this equation in diameter rather than in radius and making use of Equation (14.3) give the gear ratio as

$$g_r = \frac{d_g}{d_p} = \frac{\omega_p}{\omega_g} = \frac{N_g}{N_p} \qquad (14.18)$$

This equation assumes no slip between the meshing pinion and the gear.

---

**EXAMPLE 14.4**

**Given:** When two involute gears work properly, the contact point between the gear flanks moves along the straight line of action. Let $N_p = 20$, $N_g = 38$, and $\phi = 20°$. The diametral pitch is 8 per inch, and the pinion speed is 3600 rpm.

**Find:** The gear speed.

**Solution:** From Equation (14.5) the diametral pitch is

$$p_d = \frac{N_p}{d_p} \qquad \text{or} \qquad d_p = \frac{N_p}{p_d} = \frac{20}{8} = 2.5 \text{ in}$$

The speed at the contact point along the straight line of action is

$$\omega_p r_{bp} = \omega_p \left(\frac{d_p}{2}\right)(\cos 20°) = \left(\frac{3600}{60}\right)(2\pi)\left(\frac{2.5}{2}\right)(\cos 20°)$$

$$= 442.8 \text{ in/s}$$

---

## 14.5    CONTACT RATIO

To ensure smooth, continuous tooth action, as one pair of teeth ceases contact, a succeeding pair of teeth must already have come into contact. Figure 14.11 illustrates the important parameters in defining the contact ratio. The contact ratio is a ratio of the length of the line of action to the base pitch. Figure 14.12 shows more details of the length of the line of action. The **contact ratio** defines the average number of teeth in contact at any time. It is defined by

$$C_r = \frac{L_{ab}}{p_b} \qquad (14.19)$$

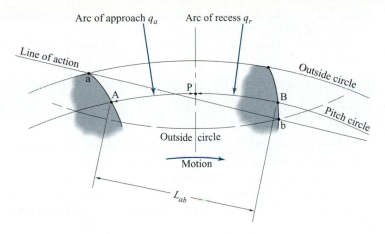

**Figure 14.11** Illustration of parameters important in defining contact.

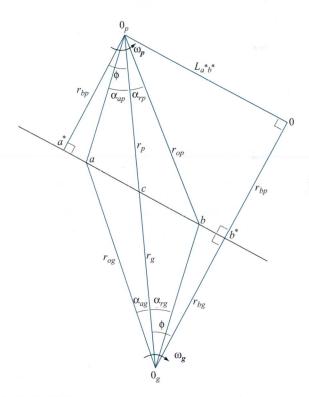

**Figure 14.12** Details of line of action, showing angles of approach and recess for both pinion and gear.

where $L_{ab}$ is the length of the line of action of the gear teeth, which can be calculated from the gear and pinion geometry as

$$L_{ab} = \sqrt{r_{op}^2 - r_{bp}^2} + \sqrt{r_{og}^2 - r_{bg}^2} - c_d \sin \phi \qquad \text{(14.20)}$$

By applying Equation (14.17), the contact ratio can be expressed as

$$C_r = \frac{L_{ab}}{p_c \cos \phi} = \frac{1}{p_c \cos \phi} \left( \sqrt{r_{op}^2 - r_{bp}^2} + \sqrt{r_{og}^2 - r_{bg}^2} \right) - \frac{c_d \tan \phi}{p_c} \qquad \text{(14.21)}$$

If the contact ratio is 1.0, then there will be a new pair of teeth coming in contact just as a pair is leaving contact, and there will be one tooth in contact during contact. This may seem acceptable, but leads to poor performance, since loads will be applied at tooth tips, and there will be additional vibration, noise, and backlash. It is good practice to maintain a contact ratio of 1.2 or greater. Most spur gear sets will have contact ratios between 1.4 and 2.0.

A contact ratio between 1 and 2 means that part of the time two pairs of teeth are in contact and during the remaining time one pair is in contact. A contact ratio between 2 and 3 means that part of the time two or three pairs of teeth are always in contact. This is a rare situation with spur gear sets.

The kinematics of gears is such that three velocities for the pinion and gear are important.

• Velocity of pinion and gear:

$$u_{bp} = \omega_p r_{bp} = \omega_p r_p \cos \phi \qquad \text{(14.22)}$$

$$u_{bg} = \omega_g r_{bg} = \omega_g r_g \cos \phi \qquad \text{(14.23)}$$

For no slip between meshing gears $\omega_p r_p = \omega_g r_g$.

• Sliding velocities in the contact

The sliding speed is the difference in speed perpendicular to the line of action for the two gears. At the pitch point the sliding speed is zero. The sliding speed increases on both sides of the pitch point (point $c$ in Figure 14.12) and reaches a maximum at the far ends of the line of action (points $a$ and $b$ in Figure 14.12). The sliding velocity at points $a$ and $b$ can be obtained from

$$u_{sa} = L_{ab*}\omega_g - L_{a*a}\omega_p \qquad \text{(14.24)}$$

$$u_{sb} = L_{a*b}\omega_p - L_{b*b}\omega_p \qquad \text{(14.25)}$$

The sliding velocity of the meshing gear set is the larger of the value $u_{sa}$ or $u_{sb}$. Values obtained for $u_{bp}$ and $u_{bg}$ in Equations (14.22) and (14.23) are typically 2 to 3 times larger than the values obtained for $u_{sa}$ and $u_{sb}$ in Equations (14.24) and (14.25).

• Lubrication velocities

For the pinion and gear the lubrication velocities are

$$u_p = \omega_p L_{a*c} = \omega_p r_p \sin \phi \qquad \text{(14.26)}$$

$$u_g = \omega_g L_{b*c} = \omega_g r_g \sin \phi \qquad \text{(14.27)}$$

EXAMPLE 14.5

**Given:** In a gearbox the pinion has 20 teeth, and the gear has 67 teeth. The diametral pitch is 10 per inch. The manufacturer needs to reduce the size of the gearbox but does not want to change the size of the gear teeth. The original gears are standard with an addendum of $1/p_d$ and a pressure angle of $20°$. The approach to be used is to decrease the size of the gearbox by stubbing (reducing the addendum on) one of the wheels.

**Find:** Which wheel should be stubbed, and what percentage smaller would the gearbox be due to the stubbing?

**Solution:** From Equation (14.6) the circular pitch is

$$p_c = \frac{\pi}{p_d} = \frac{\pi}{10} \text{ in}$$

From Equations (14.1) and (14.4) the center distance is

$$c_d = \frac{d_p + d_g}{2} = \frac{N_p + N_g}{2p_d} = \frac{20 + 67}{2(10)} = 4.35 \text{ in}$$

From Equation (14.5) the pitch circle radii for the pinion and gear are

$$r_p = \frac{N_p}{2p_d} = \frac{20}{2(10)} = 1 \text{ in}$$

$$r_g = \frac{N_g}{2p_d} = \frac{67}{2(10)} = 3.35 \text{ in}$$

From Equation (14.16) the base circle radii for the pinion and gear are

$$r_{bp} = r_p \cos\phi = 1 \cos 20° = 0.9397 \text{ in}$$

$$r_{bg} = r_g \cos\phi = 3.35 \cos 20° = 3.148 \text{ in}$$

The outside radii for the pinion and gear are

$$r_{op} = r_p + a$$

$$r_{og} = r_g + a$$

(a) **Standard gears:** For standard gears the addendum is

$$a = \frac{1}{p_d} = \frac{1}{10} = 0.1 \text{ in}$$

$$r_{op} = r_p + a = 1 + 0.1 = 1.1 \text{ in} \qquad \text{(a)}$$

$$r_{og} = r_g + a = 3.35 + 0.1 = 3.45 \text{ in} \qquad \text{(b)}$$

$$\therefore \quad r_{op} + r_{og} = 4.55 \text{ in}$$

(b) **Pinion-stubbed:** If the pinion is stubbed, the addendum radius of the pinion is

$$r_{op} = r_p + xa \qquad \text{(c)}$$

where $x =$ constant between 0 and 1. Note that if $x = 1$, a standard pinion exists; and if $x = 0$, the pinion is stubbed to the pitch circle. Either the pinion or gear is stubbed, but not both. The smallest contact ratio should be 1.1. Thus, using Equation (14.21), (b), and (c) gives

$$1.1 = \frac{10}{\pi \cos 20°} \left[ \sqrt{(1 + 0.1x)^2 - (0.9397)^2} + \sqrt{(3.45)^2 - (3.148)^2} \right] - \frac{4.35 \tan 20°}{\pi/10}$$

Solving for $x$ gives $x = 0.2164$. The outside radius for the pinion is

$$r_{op} = 1 + 0.1(0.2164) = 1.0216 \text{ in}$$

The sum of the outside radii is

$$r_{op} + r_{og} = 1.0216 + 3.45 = 4.472 \text{ in}$$

Thus, the percent reduction of the gearbox if the pinion is stubbed is

$$\frac{(4.55 - 4.472)(100)}{4.55} = 1.714\%$$

(c)   **Gear-stubbed:** If the gear is stubbed, the addendum radius of the gear is

$$r_{og} = r_g + ya \qquad\qquad (d)$$

where $y$ = constant between 0 and 1. Using Equation (14.21), (d), and (a) while setting the contact ratio to 1.1 gives

$$1.1 = \frac{10}{\pi \cos 20°} \left[ \sqrt{(1.1)^2 - (0.9397)^2} + \sqrt{(3.35 + 0.1y)^2 - (3.148)^2} \right] - \frac{4.35 \tan 20°}{\pi/10}$$

Solving for $y$ gives $y = 0.3370$. The outside radius for the gear is

$$r_{og} = 3.35 + 0.1(0.3370) = 3.384 \text{ in}$$

The sum of the outside radii is

$$r_{op} + r_{og} = 1.1 + 3.384 = 4.484 \text{ in}$$

Thus, the percent reduction of the gearbox if the gear is stubbed is

$$\frac{(4.55 - 4.484)(100)}{4.55} = 1.451\%$$

Therefore, the pinion should be stubbed because it will result in a larger (1.714%) reduction of the gearbox.

## 14.6    TOOTH THICKNESS, BACKLASH, AND INTERFERENCE

The circular tooth thickness $t_h$ is the arc length, measured on the pitch circle, from one side of the tooth to the other. Theoretically, the circular tooth thickness is one-half of the circular pitch.

$$t_h = \frac{p_c}{2} = \frac{\pi}{2 p_d} \qquad\qquad \text{(14.28)}$$

For perfect meshing of pinion and gear, the tooth thickness measured on the pitch circle must be exactly one-half of the circular pitch. Because of unavoidable inaccuracies it is

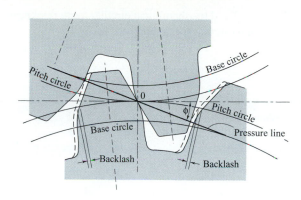

**Figure 14.13**    Illustration of backlash in gears.

necessary to cut the teeth slightly thinner to provide some clearance so that the gears will not bind but will mesh together smoothly. This clearance measured on the pitch circle is called **backlash.** Figure 14.13 illustrates backlash in meshing gears.

To obtain the amount of backlash desired, it is necessary to decrease the tooth thickness given in Equation (14.28). This decrease must be greater than the desired backlash owing to errors in manufacturing and assembly. Since the amount of the decrease in tooth thickness depends on machining accuracy, the allowance for a specified backlash varies according to manufacturing conditions.

In Example 14.6, fundamental equations are given for determining backlash in a single meshing gear and pinion. To determine backlash in gear trains (covered later in this chapter), it is necessary to sum the backlash of each mated gear pair. However, to obtain the total backlash for a series of meshes, it is necessary to take into account the gear ratio of each mesh relative to a chosen reference shaft in the gear train. This is beyond the scope of this text, but more details can be found in Mechalec (1962) or Townsend (1992).

Backlash might occur in two ways: (1) if the tooth thickness is less than that given in Equation (14.28) and (2) if the operating center distance is greater than that obtained when Equations (14.28) and (14.1) are valid. Table 14.3 gives recommended minimum backlash for coarse-pitch gears. All dimensions are given in inches. Thus, when the backlash $b_l$ from

**Table 14.3**    Recommended minimum backlash for coarse-pitched gears

| Diametral pitch $p_d$, in$^{-1}$ | Center Distance $c_d$, in | | | | |
|---|---|---|---|---|---|
| | 2 | 4 | 8 | 16 | 32 |
| 18 | 0.005 | 0.006 | — | — | — |
| 12 | 0.006 | 0.007 | 0.009 | — | — |
| 8 | 0.007 | 0.008 | 0.010 | 0.014 | — |
| 5 | — | 0.010 | 0.012 | 0.016 | — |
| 3 | — | 0.014 | 0.016 | 0.020 | 0.028 |
| 2 | — | — | 0.021 | 0.025 | 0.033 |
| 1.25 | — | — | — | 0.034 | 0.042 |

Table 14.3 and the theoretical circular tooth thickness $t_h$ from Equation (14.28) are known, the actual circular tooth thickness is

$$t_{ha} = t_h - b_l \qquad\qquad (14.29)$$

---

**EXAMPLE 14.6**

**Given:** The two situations given in Example 14.3.

**Find:** Determine the tooth thickness, the recommended backlash, and the contact ratio.

**Solution:**

(a) For the original center distance, from Equation (14.28) the tooth thickness is

$$t_h = \frac{p_c}{2} = \frac{1.571}{2} = 0.7855 \text{ in}$$

From Table 14.3 the interpolated recommended backlash for a diametral pitch of 2 per inch and a center distance of 14 in is

$$b_l = 0.024 \text{ in}$$

Thus, the actual tooth thickness, from Equation (14.29), is

$$t_{ha} = t_h - b_l = 0.7855 - 0.024 = 0.7615 \text{ in}$$

From Table 14.2 the addendum for both the pinion and gear is

$$a = \frac{1}{p_d} = \frac{1}{2} = 0.5 \text{ in}$$

The outside radii for the pinion and gear are

$$r_{op} = r_p + a = 4 + 0.5 = 4.5 \text{ in}$$
$$r_{og} = r_g + a = 10 + 0.5 = 10.5 \text{ in}$$

By making use of the above and Equation (14.28), the contact ratio can be expressed as

$$C_r = \frac{1}{1.571 \cos 20°} \left( \sqrt{4.5^2 - 3.759^2} + \sqrt{10.5^2 - 9.397^2} \right) - \frac{14 \tan 20°}{1.571} = 1.606$$

This implies that part of the time two pairs of teeth are in contact and during the remaining time one pair is in contact.

(b) When the center distance is increased by 0.25 in, from Equation (14.29) the tooth thickness is

$$t_h = \frac{p_c}{2} = \frac{1.599}{2} = 0.800$$

The actual tooth thickness is

$$t_{ha} = t_h - b_l = 0.800 - 0.024 = 0.776 \text{ in}$$

The outside radii for the pinion and gear are

$$r_{op} = r_p + a = 4.072 + 0.5 = 4.572 \text{ in}$$

$$r_{og} = r_g + a = 10.18 + 0.5 = 10.68 \text{ in}$$

$$C_r = \frac{1}{1.599 \cos 22.6°} \left( \sqrt{4.572^2 - 3.759^2} + \sqrt{10.68^2 - 9.397^2} \right) - \frac{14.25 \tan 22.6°}{1.599}$$

$$= 1.491$$

Thus, increasing the center distance increases the pressure angle and decreases the contact ratio.

## 14.7    GEAR TRAINS

**Gear trains** are machine elements consisting of multiple gears. They are used to obtain a desired angular velocity of an output shaft while the input shaft rotates at a different angular velocity. The angular velocity ratio between input and output gears is constant.

### 14.7.1    SINGLE GEAR MESH

Figure 14.14 shows *externally* meshing spur gears with fixed centers. Using Equation (14.18) and the sign convention that positive means the same direction and negative means the opposite direction gives the angular velocity ratio for the situation in Figure 14.14 as

$$Z_{21} = \frac{\omega_2}{\omega_1} = -\frac{N_1}{N_2} \tag{14.30}$$

Externally meshing gear teeth are the type of meshing gear discussed throughout this chapter. The gears rotate in opposite directions. The center distance for externally meshing spur

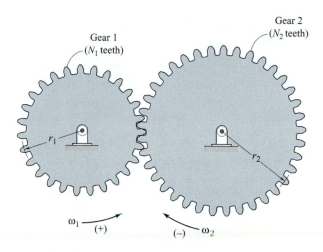

**Figure 14.14**    Externally meshing spur gears.

**Figure 14.15**    Internally meshing spur gears.

gears is

$$c_d = r_1 + r_2 \qquad (14.31)$$

Figure 14.15 shows *internally* meshing spur gears. Note that both gears are moving in the same direction and thus are positive. The angular velocity ratio is

$$Z_{21} = \frac{\omega_2}{\omega_1} = \frac{N_1}{N_2} \qquad (14.32)$$

The difference in sign in Equations (14.30) and (14.32) accounts for the fact that externally meshing gears rotate in opposite directions whereas internally meshing gears rotate in the same direction.

The length of the line of action of *externally* meshing gears is obtained from Equation (14.20). The changes necessary for determining the line of action for *internally* meshing gears are

$$r_{og} = r_g - a \qquad c_d = r_2 - r_1 = r_g - r_p$$

$$L_{ab} = \sqrt{r_{op}^2 - r_{bp}^2} - \sqrt{r_{og}^2 - r_{bg}^2} + c_d \sin \phi \qquad (14.33)$$

### 14.7.2    SIMPLE SPUR GEAR TRAINS

A simple gear train, as shown in Figure 14.16, has only one gear mounted on each shaft. If the $i$th gear ($i = 1,2,3, \ldots , n$) has $N_i$ teeth and rotates with angular velocity $\omega_i$ measured positive counterclockwise, the angular velocity ratio or *train value* $Z_{ji}$ of the $j$th gear relative to the $i$th gear is

$$Z_{ji} = \frac{\omega_j}{\omega_i} = \pm\frac{N_i}{N_j} \qquad (14.34)$$

**Figure 14.16**   Simple gear train.

A positive sign is used in the previous equation when $j$ is odd, and a negative sign is used when $j$ is even. The reason for this is that in Figure 14.16 all the gears are external, implying that the directions of rotation of adjacent shafts are opposite.

### 14.7.3   COMPOUND SPUR GEAR TRAINS

In a compound gear train (Fig. 14.17) at least one shaft carries two or more gears. In this type of train all gears are keyed to their respective shafts, so that the angular velocities of all gears are equal to that of the shaft on which they are mounted. The angular velocities of adjacent shafts are governed by the gear ratio of the associated mesh.

Let $\Omega_i$ represent the angular velocity of the $i$th shaft. Using Figure 14.17 while going from left to right and assuming that positive is counterclockwise and negative is clockwise give

$$\frac{\Omega_2}{\Omega_1} = -\frac{N_1}{N_2} \qquad \frac{\Omega_3}{\Omega_2} = -\frac{N_3}{N_4} \qquad \frac{\Omega_4}{\Omega_3} = -\frac{N_5}{N_6} \qquad \text{and} \qquad \frac{\Omega_5}{\Omega_4} = -\frac{N_7}{N_8}$$

Hence, the angular velocity ratio $Z_{51}$ of shaft 5 relative to shaft 1 is given by

$$Z_{51} = \frac{\Omega_5}{\Omega_1} \tag{14.35}$$

**Figure 14.17**   Compound gear train.

where

$$\frac{\Omega_5}{\Omega_1} = \frac{\Omega_5}{\Omega_4}\frac{\Omega_4}{\Omega_3}\frac{\Omega_3}{\Omega_2}\frac{\Omega_2}{\Omega_1} = \left(-\frac{N_7}{N_8}\right)\left(-\frac{N_5}{N_6}\right)\left(-\frac{N_3}{N_4}\right)\left(-\frac{N_1}{N_2}\right)$$

For the compound gear train shown in Figure 14.17

$$Z_{51} = \frac{N_1 N_3 N_5 N_7}{N_2 N_4 N_6 N_8} \qquad (14.36)$$

Note that

$$Z_{51} = \frac{\omega_8}{\omega_1} \qquad (14.37)$$

**EXAMPLE 14.7**

**Given:**  The gear train shown in Figure 14.18.

**Find:**  Determine the angular velocity ratio for the gear train. If the shaft carrying gear A rotates at 1750 rpm clockwise, determine the speed and direction of the shaft carrying gear E.

**Solution:**  From Figure 14.18 observe that

$$Z_{14} = \left(-\frac{N_B}{N_A}\right)\left(-\frac{N_D}{N_C}\right)\left(-\frac{N_E}{N_D}\right) = -\frac{N_B N_E}{N_A N_C} = -\frac{(70)(54)}{(20)(18)} = -10.5$$

Gear D is an *idler*, since it has no effect on the magnitude of the angular velocity ratio, but it does cause a direction reversal. Therefore, if gear D were not present, the directionality of gear E would be in the opposite direction, but the angular velocity ratio would have the same numerical value but opposite sign.

**Figure 14.18**    Gear train used in Example 14.7.

The output speed is

$$Z_{14} = \frac{\omega_A}{\omega_E} = -\frac{N_B N_E}{N_A N_C} = -10.5$$

$$\therefore \qquad \omega_E = \frac{\omega_A}{Z_{14}} = \frac{1750}{-10.5} = -166.7 \text{ rpm}$$

The negative sign implies that the direction is counterclockwise.

### 14.7.4 PLANETARY GEAR TRAINS

**Planetary gear trains** are very common and have surprising flexibility in their application, and they give high gear ratios in compact packages. A planetary or *epicyclic* gear train consists of a **sun gear,** a **ring gear,** a **planet carrier** or *arm,* and one or more **planet gears.** The planetary gear train shown in Figure 14.19(a) depicts an arm with three planets, which is a common arrangement. However, the kinematics of gear trains can often be conceptually simplified by recognizing that the system of Figure 14.19(b) gives the same angular velocities as in Figure 14.19(a), and can be used for analysis of gear trains. However, note that the additional planets are desirable to increase stability and reduce tooth and contact stresses as well as gear deformations.

Planetary gear trains are 2 degree-of-freedom trains; that is, two inputs are needed to obtain a predictable output. For example, the sun can be the input (driven by a motor or other power source), the ring can be bolted to a machine frame (so that its velocity is specified as zero), and the planet arm can be the output (driving other devices). In this case, the input and output have parallel axes, and this arrangement is popular for power transmission devices. Another mode of operation can be achieved by driving the sun gear, fixing the ring, and using a planet gear as an output; this results in a rotating shaft that itself rotates about the

(a)                                                                (b)

**Figure 14.19**    Illustration of planetary gear train. (a) With three planets; (b) with one planet (for analysis only).

input axis. This application of planetary gear trains is very common in mixers and agitators of all kinds.

Another use of a planetary gear train is in an automotive differential transmission. In this situation, the sun is an input (from the driveshaft), and the ring and armature are outputs. Normally, the outputs are frictionally coupled and are equal, but they are not rigidly constrained. Thus, one wheel can rotate independently of the other, a beneficial arrangement for slippery road surfaces, smooth ride, and low wheel wear.

Planetary gear trains can be analyzed through the application of the following equations:

$$\frac{\omega_{ring} - \omega_{arm}}{\omega_{sun} - \omega_{arm}} = -\frac{N_{sun}}{N_{ring}} \qquad (14.38)$$

$$\frac{\omega_{planet} - \omega_{arm}}{\omega_{sun} - \omega_{arm}} = -\frac{N_{sun}}{N_{planet}} \qquad (14.39)$$

$$N_{ring} = N_{sun} + 2N_{planet} \qquad (14.40)$$

The angular velocity ratio is more difficult to define, since any of these gears can be the first or last gear in the planetary gear train. The angular velocity ratio can be written as

$$Z_p = \frac{\omega_L - \omega_A}{\omega_F - \omega_A} \qquad (14.41)$$

where $Z_p$ = angular velocity ratio for a planetary gear train

$\omega_L$ = angular velocity for the last gear in the planetary gear train (output gear)

$\omega_A$ = angular velocity of the arm

$\omega_F$ = angular velocity for the first gear in the planetary gear train (driven gear)

Note that if the arm is the output or input gear, then the angular velocity ratio can be calculated from Equation (14.38).

---

**EXAMPLE 14.8**

**Given:** A planetary gear train is used as part of a transmission for a tractor. The sun gear is driven by the engine at 3000 rpm, the ring is bolted to the machine frame, and the armature is connected to the track drive system. The sun has 16 teeth, and there are three planets, each with 34 teeth.

**Find:** Determine the angular velocity of the armature, the planets, and the angular velocity ratio for the gear train.

**Solution:** The number of teeth in the ring needs to be obtained from Equation (14.40):

$$N_{ring} = N_{sun} + 2N_{planet} = 16 + 2(34) = 84 \text{ teeth}$$

Therefore, the angular velocity of the armature can be calculated from Equation (14.39) as

$$\frac{\omega_{ring} - \omega_{arm}}{\omega_{sun} - \omega_{arm}} = -\frac{N_{sun}}{N_{ring}}; \qquad \frac{0 - \omega_{arm}}{314 - \omega_{arm}} = -\frac{16}{84} \qquad \omega_{arm} = 50.24 \text{ rad/s} = 480 \text{ rpm}$$

Since this is a positive value, the armature rotates in the same direction as the sun gear. The angular velocity of the planet can now be calculated from Equation (14.40) as

$$\frac{\omega_{planet} - \omega_{arm}}{\omega_{sun} - \omega_{arm}} = -\frac{N_{sun}}{N_{planet}} \qquad \therefore \qquad \frac{\omega_{planet} - 50.24}{314 - 50.24} = -\frac{16}{34}$$

$$\omega_{planet} = -73.88 \text{ rad/s} = -705.5 \text{ rpm}$$

The negative sign indicates that the planet moves in the opposite direction to the sun. The angular velocity ratio is calculated from Equation (14.35) as

$$Z_p = \frac{\omega_{arm}}{\omega_{sun}} = \frac{480}{3000} = 0.16$$

## 14.8    GEAR MANUFACTURE AND QUALITY

### 14.8.1    QUALITY INDEX $Q_v$

Consider the shape of the spur gear shown in Figure 14.1 with the tooth profile in Figure 14.6. The manufacturing process used to create the gear has many far-reaching effects on the long-term gear performance. The importance of the involute tooth profile and the minimization of deviation from this profile are essential for the efficient operation of the gear. In addition, a smoother gear will run quieter and with a larger film parameter $\Lambda$ than a rougher gear. Finally, the strength of the gear teeth will be strongly influenced by the manufacturing process chosen for gear production. Clearly, the manufacturing processes used in producing the gear have far-reaching design impact. This section is intended to introduce some of the concepts of gear manufacture; additional information can be found in Kalpakjian and Schmid (2001).

The AGMA has defined a quality index $Q_v$ which is related to the pitch variation of a gear set. The manufacturing process used to produce the gear has the largest effect on the quality index, but also on the gear cost, as shown in Figure 14.20. From a design standpoint, selection of a quality index serves to constrain the available manufacturing methods for producing the gear. Table 14.4 lists typical quality index values for different applications and pitch-line velocities.

The quality index $Q_v$ for U.S. Customary System units is defined by

$$Q_v = 0.5048 \ln N_i - 1.144 \ln P_d - 2.852 \ln(V_{PA} \times 10^4) + 14.71 \qquad \text{(14.42)}$$

where $N_i$ = number of pinion or gear teeth
$P_d$ = normal diametral pitch, in$^{-1}$
$V_{PA}$ = absolute pitch variation, in (or $\mu$m)
$m$ = module, mm

It is important to use the proper units in Equation (14.42) to obtain meaningful values for the quality index. The quality index is an empirical variable to relate gear tolerance to performance. There is significant scatter in most gear performance evaluations, and the

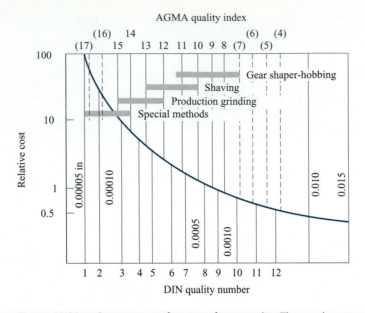

**Figure 14.20**    Gear cost as a function of gear quality. The numbers along the vertical lines indicate tolerances. [From Kalpakjian and Schmid (2001).]

**Table 14.4**    Quality index $Q_v$ for various applications

| Application | Quality index $Q_v$ |
|---|---|
| Cement mixer drum driver | 3–5 |
| Cement kiln | 5–6 |
| Steel mill drives | 5–6 |
| Corn pickers | 5–7 |
| Punch press | 5–7 |
| Mining conveyor | 5–7 |
| Clothes washing machine | 8–10 |
| Printing press | 9–11 |
| Automotive transmission | 10–11 |
| Marine propulsion drive | 10–12 |
| Aircraft engine drive | 10–13 |
| Gyroscope | 12–14 |

| Pitch Velocity | | |
|---|---|---|
| ft/min | m/s | Quality index $Q_v$ |
| 0–800 | 0–4 | 6–8 |
| 800–2000 | 4–10 | 8–10 |
| 2000–4000 | 10–20 | 10–12 |
| >4000 | >20 | 12–14 |

SOURCE: Mott (1999)

quality index should be recognized as an imperfect but valuable quantity. Indeed, the AGMA recommendation is to round the value of $Q_v$ to the next-lower integer.

## 14.8.2 GEAR MANUFACTURE

Metal gears are manufactured through a variety of processes, including casting, forging, extrusion, drawing, thread rolling, powder metallurgy, and blanking sheet metal. Nonmetallic gears can be made by injection molding, compression molding, or casting. These processes yield AGMA $Q_v$ numbers that are low, typically below 5. A better surface finish and more accurate profile can be achieved by machining processes. There are two basic methods of producing a gear profile by machining processes: *form cutting* and *generating*.

### *Form Cutting*

In form cutting, as shown in Figure 14.21, the cutting tool has the shape of the space between the gear teeth. The gear tooth shape is reproduced by cutting the gear blank around its periphery. The cutter travels axially along the length of the gear tooth at the appropriate depth to produce the gear tooth profile. After each tooth is cut, the cutter is withdrawn, the gear blank is rotated (indexed), and the cutter proceeds to cut another tooth. The process continues until all teeth are cut.

Broaching can also be used to produce gear teeth and is particularly applicable to internal teeth. The process is rapid and produces fine surface finish with high dimensional accuracy. However, because broaches are expensive and a separate broach is required for each gear size, this method is suitable mainly for high-quantity production.

Form cutting is a relatively simple process and can be used for cutting gear teeth with various profiles; however, it is a slow operation, and some types of machines require skilled

Form cutter

Gear blank

(a)   (b)   (c)

**Figure 14.21** Form cutting of gear teeth. (a) A form cutter. Notice that the tooth profile is defined by the cutter profile. (b) Schematic illustration of the form cutting process [from Kalpakjian and Schmid (2001)]. (c) Form cutting of teeth on a bevel gear. [Courtesy Schafer Gear Works, Inc.]

labor. Consequently, it is suitable only for low-quantity production. Machines with semiau-tomatic features can be used economically for form cutting on a limited-production basis.

### Generating

In gear generating, the tool may be a pinion-shaped cutter (Fig. 14.22), a rack-shaped straight cutter, or a *hob*. The pinion-shaped cutter can be considered as one of the gears in a conjugate pair, and the other as the gear blank. The cutter has an axis parallel to that of the gear blank and rotates slowly with the blank at the same pitch-circle velocity with an axial reciprocating motion. Cutting may take place at either the downstroke or upstroke of the machine. The process can be used for low-quantity as well as high-quantity production.

On a *rack shaper,* the generating tool is a segment of a rack which reciprocates parallel to the axis of the gear blank. Because it is not practical to have more than 6 to 12 teeth on a rack cutter, the cutter must be disengaged at suitable intervals and returned to the starting point, the gear blank, meanwhile remaining fixed.

A gear-cutting hob, shown in Figure 14.23, is basically a worm, or screw, that has been made into a gear-generating tool by machining a series of longitudinal slots, or gashes, into it to produce the cutting teeth. When hobbing a spur gear, the angle between the hob and gear blank axes is 90° minus the lead angle at the hob threads. All motions in hobbing are rotary, and the hob and gear blank rotate continuously as in two gears meshing until all teeth are cut.

### Gear-Finishing Processes

As produced by any of the processes described, the surface finish and dimensional accuracy of gear teeth may not be sufficiently accurate for certain applications. Several finishing processes are available for improving the surface quality of gears produced by form cutting or generating.

(a)

(b)

**Figure 14.22**    Production of gear teeth with a pinion-shaped cutter. (a) Schematic illustration of the process [from Kalpakjian and Schmid (2001)]; (b) photograph of the process with gear and cutter motions indicated [Courtesy Schafer Gear Works, Inc.]

Top view

Gear
blank

Hob

Gear
blank

(a)

Helical gear

Hob rotation

Hob

(b)

**Figure 14.23** Production of gears through the hobbing process. (a) A hob, along with a schematic illustration of the process [from Kalpakjian and Schmid (2001)]; (b) production of a worm gear through hobbing. [Courtesy Schafer Gear Works, Inc.]

In gear *shaving,* a cutter is made in the exact shape of the finished tooth profile and removes small amounts of metal from the gear teeth. Although the tools are expensive and special machines are necessary, shaving is rapid and is the most commonly used process for gear finishing. It produces gear teeth with improved surface finish and accuracy of tooth profile. Shaved gears may subsequently be heat-treated and then ground for improved hardness, wear resistance, and accurate tooth profile.

For highest dimensional accuracy, tooth spacing and form, and surface finish, gear teeth may subsequently be *ground, honed,* and *lapped.* There are several types of grinders for gears, with the single-index form grinder being the most commonly available. In form grinding, the grinding wheel is in the identical shape of the tooth spacing.

In honing, the tool is a plastic gear impregnated with fine abrasive particles. The process is faster than grinding and is used to improve surface finish. To further improve the surface finish, ground gear teeth are lapped using abrasive compounds with either a gear-shaped lapping tool (made of cast iron or bronze) or a pair of mating gears that are run together. Although production rates are lower and costs are higher, these finishing operations are particularly suitable for producing hardened gears of very high quality, long life, and quiet operation.

Note that *burnishing* is a common finishing process for gear teeth, Introduced in the 1960s, this is basically a surface plastic deformation process using a special hardened gear-shaped burnishing die that subjects the tooth surfaces to a surface-rolling action (gear rolling). Cold working of tooth surfaces improves the surface finish and induces surface compressive residual stresses on the gear teeth, thus improving their fatigue life. However, burnishing does not significantly improve gear tooth accuracy.

## 14.9    GEAR MATERIALS

### 14.9.1    ALLOWABLE STRESSES

The American Gear Manufacturers Association (AGMA) has promulgated a number of industry standards for the design of gears, and includes recommendations for the allowable loading of gear materials. The variety of materials used in gears is extensive and depends on the application of the gear. Recall that Chapter 3 focused on materials. This section addresses only a sampling of materials for gears. More information about different materials, different manufacturing processes, and different coatings applied to gear teeth can be found in AGMA (1999).

Section 14.10 explains that a gear tooth acts like a cantilevered beam in resisting the force exerted on it by the mating tooth. The point of highest tensile bending stress is at the root of the tooth, where the involute curve bends with the fillet. AGMA (1990) presents a set of allowable stress number for various steels as a function of Brinell hardnesses. The AGMA uses the term 'bending stress number' and 'contact stress number' instead of allowable stress as is the convention in this text. The terminology is intended to emphasize that the reported material properties are suitable only for gear applications. These allowable bending stress numbers are shown in Figure 14.24. The grades differ in degree of control

**(a)**

| Material | Grade | Allowable bending stress number | |
|---|---|---|---|
| | | **MPa** | **ksi** |
| Through-hardened steels | 1 | 0.703 HB + 113 | 0.0773 HB + 12.8 |
| | 2 | 0.533 HB + 88.3 | 0.102 HB + 16.4 |
| Nitriding through-hardened steels | 1 | 0.0823 HB + 12.15 | 0.568 HB + 83.8 |
| | 2 | 0.1086 HB + 15.89 | 0.749 HB + 110 |

**(b)**

| Material | Grade | Allowable bending stress number | |
|---|---|---|---|
| | | **MPa** | **ksi** |
| Niralloy | 1 | 0.594 HB + 87.76 | 0.0862 HB + 12.73 |
| | 2 | 0.784 HB + 114.81 | 0.1138 HB + 16.65 |
| 2.5% Chrome | 1 | 0.7255 HB + 63.89 | 0.1052 HB + 9.28 |
| | 2 | 0.7255 HB + 153.63 | 0.1052 HB + 22.28 |
| | 3 | 0.7255 HB + 201.91 | 0.1052 HB + 29.28 |

**Figure 14.24**  Effect of Brinell hardness on allowable bending stress number for steel gears. (a) Through-hardened steels; (b) flame or induction-hardened nitriding steels. Note that Brinell hardness refers to case hardness for these gears. [AGMA Standard 2001-C95, (1999).]

639

**Table 14.5**     Allowable bending and contact stresses for selected gear materials

| Material designation | Grade | Typical hardness[a] | Allowable Bending Stress $S_t$ | | Allowable Contact Stress $S_c$ | |
|---|---|---|---|---|---|---|
| | | | lb/in$^2$ | MPa | lb/in$^2$ | MPa |
| **Steel** | | | | | | |
| Through-hardened | 1 | — | See Fig. 14.24(a) | | See Fig. 14.25 | |
| | 2 | — | See Fig. 14.24(a) | | See Fig. 14.25 | |
| Carburized and hardened | 1 | 55-64 HRC | 55,000 | 380 | 180,000 | 1240 |
| | 2 | 58-64 HRC | 65,000[b] | 450[b] | 225,000 | 1550 |
| | 3 | 58-64 HRC | 75,000 | 515 | 275,000 | 1895 |
| Nitrided and | 1 | 83.5 HR15N | See Fig. 14.24(a) | | 150,000 | 1035 |
| through-hardened | 2 | — | See Fig. 14.24(a) | | 163,000 | 1125 |
| Nitralloy 135M and | 1 | 87.5 HR15N | See Fig. 14.24(b) | | 170,000 | 1170 |
| Nitralloy N, nitrided | 2 | 87.5 HR15N | See Fig. 14.24(b) | | 183,000 | 1260 |
| 2.5% Chrome, nitrided | 1 | 87.5 HR15N | See Fig. 14.24(b) | | 155,000 | 1070 |
| | 2 | 87.5 HR15N | See Fig. 14.24(b) | | 172,000 | 1185 |
| | 3 | 87.5 HR15N | See Fig. 14.24(b) | | 189,000 | 1305 |
| **Cast Iron** | | | | | | |
| ASTM A48 gray cast | Class 20 | — | 5,000 | 34.5 | 50,000–60,000 | 345–415 |
| iron, as-cast | Class 30 | 174 HB | 8,500 | 59 | 65,000–75,000 | 450–520 |
| | Class 40 | 201 HB | 13,000 | 90 | 75,000–85,000 | 520–585 |
| ASTM A536 ductile | 60-40-18 | 140 HB | 22,000–33,000 | 150–230 | 77,000–92,000 | 530–635 |
| (nodular) iron | 80-55-06 | 179 HB | 22,000–33,000 | 150–230 | 77,000–92,000 | 530–635 |
| | 100-70-03 | 229 HB | 27,000–40,000 | 185–275 | 92,000–112,000 | 635–770 |
| | 120-90-02 | 269 HB | 31,000–44,000 | 215–305 | 103,000–126,000 | 710–870 |
| **Bronze** | | | | | | |
| $S_{ut} > 40,000$ psi ($S_{ut} > 275$ GPa) | | | 5700 | 39.5 | 30,000 | 205 |
| $S_{ut} > 90,000$ psi ($S_{ut} > 620$ GPa) | | | 23,600 | 165 | 65,000 | 450 |

[a]Hardness refers to case hardness unless through-hardened.

[b]70,000 psi (485 MPa) may be used if bainite and microcracks are limited to grade 3 levels.

SOURCE: Adapted from AGMA 2001 (1999).

of the microstructure, alloy composition, cleanness, prior heat treatment, nondestructive testing performed, core hardness values, and other factors. Grade 2 materials are more tightly controlled than grade 1 materials. Allowable stresses for some other materials are given in Table 14.5.

A second, independent form of gear failure is caused by pitting and spalling of the tooth surfaces, usually near the pitch line, where the contact stresses are extremely high. The nonconformal contact of the meshing gear teeth produces high contact stresses and eventual failure due to fatigue caused by the cyclic variation of stresses. AGMA (1999) also presents allowable contact stresses for various materials. These allowable contact stresses are shown in Figure 14.25 for through-hardened steel and listed in Table 14.5 for some other materials.

**Figure 14.25** Effect of Brinell hardness on allowable contact stress number for two grades of through-hardened steel. [AGMA (1999).]

## 14.9.2 STRESS MODIFICATION FACTORS

AGMA recommends modifying the allowable bending stress number under certain circumstances. This results in an allowable bending stress given by

$$\sigma_{t,\text{all}} = \frac{S_t}{n_s} \frac{Y_N}{K_T K_R} \qquad (14.43)$$

where $S_t$ = allowable bending stress, MPa (or lb/in$^2$)
$\quad n_s$ = safety factor
$\quad Y_N$ = stress cycle factor for bending
$\quad K_T$ = temperature factor
$\quad K_R$ = reliability factor

The allowable contact stress must also be modified under certain conditions. AGMA gives the modified allowable contact stress as

$$\sigma_{c,\text{all}} = \frac{S_c}{n_s} \frac{Z_N C_H}{K_T K_R} \qquad (14.44)$$

where $S_c$ = allowable contact stress, MPa (or lb/in$^2$)
$\quad n_s$ = safety factor
$\quad Z_N$ = stress cycle factor for contact stress
$\quad K_T$ = temperature factor
$\quad K_R$ = reliability factor
$\quad C_H$ = hardness ratio factor

Note that the AGMA uses different variable labels depending on the unit system used by the designer, but this complication has been ignored here.

(a)

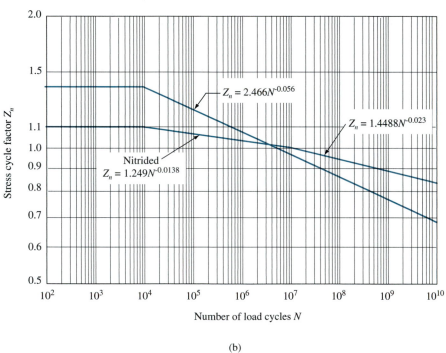

(b)

**Figure 14.26** Stress cycle factor. (a) Bending strength stress cycle factor $Y_N$; (b) pitting resistance stress cycle factor $Z_N$. [AGMA Standard 2001-C95, (1999).]

*Stress Cycle Factors $Y_N$ and $Z_N$*

The strengths given in Section 14.9.1 are based on $10^7$ load cycles. If the allowable stresses are to be calculated for a different life, then the stress cycle factors $Y_N$ for bending or $Z_N$ for contact stress are required. This is better defined for stress cycle factors less than $10^7$ load cycles. For longer lives, there is significant variation in the lives of gears, and a good correlation is difficult to obtain. For example, the life of gears with respect to pitting will be strongly affected by the quality of lubricant and the lubrication regime in which the gear operates.

Figure 14.26 shows the stress cycle correction factors for bending considerations.

*Temperature Factor $K_T$*

While the AGMA prescribes the use of a temperature factor, the only recommendation is that the temperature factor $K_T$ be taken as unity if the temperature does not exceed 120°C (250°F). For higher temperatures, a $K_T$ greater than 1 is needed to allow for the effect of temperature on oil film and material properties.

*Reliability Factor $K_R$*

The allowable stresses were based upon a statistical probability of 1% failure at $10^7$ cycles, or 99% reliability. For other reliabilities, the data in Table 14.6 should be used to obtain $K_R$.

*Hardness Ratio Factor $C_H$*

Calculation of contact stresses was discussed in Chapter 8, and further details of such calculations specific to gears are discussed in Section 14.9.2. A common result from such analyses is that the contact stresses are much larger than the yield stress of the gear material, indicating that plastic deformation takes place at the contact location. Gross plastic deformation does not occur, because the plastic zone is bounded by elastic regions that restrict metal flow, and the stress state is compressive and largely hydrostatic as a result.

Furthermore, since the pinion has fewer teeth than the gear, it will be subjected to larger contact stresses. There is some justification to using a harder pinion than gear, to obtain a uniform safety factor for contact stress and pitting resistance failures. However, a harder pinion results in higher stresses on the gear. Therefore, a correction factor $C_H$ is defined by AGMA, and it is a correction factor that is applied *on the gear only*.

**Table 14.6**  Reliability factor $K_R$

| Probability of survival, percent | Reliability factor[a] $K_R$ |
|---|---|
| 50 | 0.70[b] |
| 90 | 0.85[b] |
| 99 | 1.00 |
| 99.9 | 1.25 |
| 99.99 | 1.50 |

[a]Based on surface pitting. If tooth breakage is considered a greater hazard, a larger value may be required.
[b]At this value plastic flow may occur rather than pitting.
SOURCE: [From AGMA 2101-C95 (1999).]

The hardness ratio factor is defined by

$$C_H = 1.0 + A'(g_r - 1.0) \tag{14.45}$$

where $g_r$ is the gear ratio given by Equation (14.3) and $A'$ is defined by

$$A' = \begin{cases} 0 & \dfrac{\text{HB}_P}{\text{HB}_G} < 1.2 \\[2ex] (8.98 \times 10^{-3})\left(\dfrac{\text{HB}_P}{\text{HB}_G}\right) - 8.29 \times 10^{-3} & 1.2 \le \dfrac{\text{HB}_P}{\text{HB}_G} \le 1.7 \\[2ex] 0.00698 & \dfrac{\text{HB}_P}{\text{HB}_G} > 1.7 \end{cases} \tag{14.46}$$

where $\text{HB}_P$ = Brinell hardness of pinion
$\text{HB}_G$ = Brinell hardness of gear

A special case occurs when work-hardening of the gear teeth can occur. For hardened pinions (Rockwell C hardness of 48 or greater) and through-hardened steel gears (180 to 400 Brinell hardness), the correction factor is a function of surface finish, as shown in Figure 14.27. This can be calculated from

$$C_H = 1.0 + B(450 - HB_p) \tag{14.47}$$

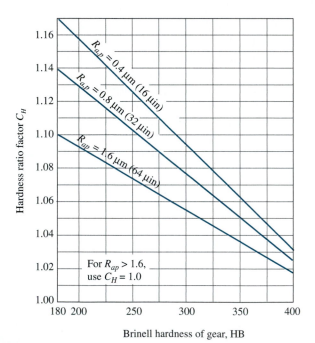

**Figure 14.27**    Hardness ratio factor $C_H$ for surface hardened pinions and through-hardened gears. [AGMA Standard 2001-C95, (1999).]

where $HB_p$ = Brinell hardness of pinion and $B$ is a function of the pinion surface roughness $R_{a,p}$, given by

$$B = \begin{cases} 0.00075e^{-0.448(R_{a,p})} & \text{where } R_{a,p} \text{ is in micrometers} \\ 0.00075e^{-0.0112(R_{a,p})} & \text{where } R_{a,p} \text{ is in microinches} \end{cases} \tag{14.48}$$

## 14.10 LOADS ACTING ON A GEAR TOOTH

Gears transfer power by the driving teeth exerting a load on the driven teeth while the reaction load acts back on the teeth of the driving gear. Figure 14.28 shows the loads acting on an individual gear tooth. The tangential $W_t$, radial $W_r$, and normal $W$ loads are shown along with the pressure angle and the pitch circle. The tangential load $W_t$, or transmitted force, in pounds force, can be obtained from Equation (4.40) as

$$W_t = \frac{126,050h_p}{dN_a} \tag{14.49}$$

where $h_p$ = transmitted horsepower, hp
$\quad\ d$ = pitch diameter, in
$\quad N_a$ = gear rotational speed, rpm

Equation (14.38) is valid only for English units. The comparable equation to Equation (14.38) in SI units can be obtained directly from Equation (4.41). The normal load $W$ and

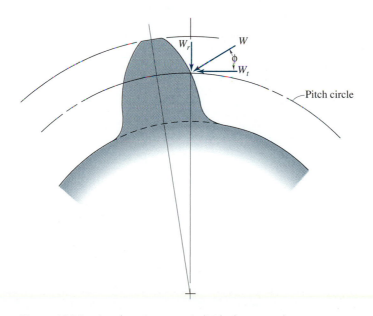

**Figure 14.28** Loads acting on an individual gear tooth.

the radial load $W_r$ can be computed from the following:

$$W = \frac{W_t}{\cos \phi}$$
(14.50)

$$W_r = W_t \tan \phi$$
(14.51)

The normal load is used in determining contact stress, contact dimensions, and elastohydrodynamic lubrication film thickness.

## 14.11  BENDING STRESSES IN GEAR TEETH

Gear failure can occur from tooth breakage, which results when the design stress due to bending is greater than the allowable stress. Figure 14.29(a) shows the actual tooth with the forces and dimensions used in determining bending tooth strength; Figure 14.29(b) shows a cantilevered beam simulating the forces and dimensions acting on the gear tooth. Note that the load components shown in Figure 14.28 have been moved to the tooth tip in Figure 14.29. Using the bending of a cantilevered beam to simulate the stresses acting on a gear tooth is only approximate at best, since the situations present in a long, thin uniform beam loaded by moments at the ends are quite different from those found in a gear tooth. Wilfred Lewis was the first to use this approach (Lewis, 1892), and thus the equation that results is often called the **Lewis equation.** This approach has had universal acceptance and will therefore be used.

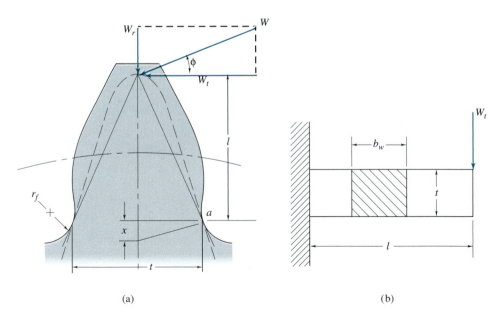

(a)                                        (b)

**Figure 14.29**  Loads and length dimensions used in determining tooth bending stress. (a) Tooth; (b) cantilevered beam.

From Equation (4.48) the bending stress can be written as

$$\sigma_t = \frac{Mc}{I} \qquad (14.52)$$

From similar triangles

$$\tan \alpha = \frac{x}{t/2} = \frac{t/2}{l}$$

$$l = \frac{t^2}{4x} \qquad (14.53)$$

For a rectangular section from Table 4.1 with base $b_w$ and height $t$, the area moment of inertia is

$$I = \frac{bh^3}{12} = \frac{b_w t^3}{12}$$

If $M = W_t l$ and $c = t/2$, by making use of Equation (14.53), Equation (4.48) becomes

$$\sigma_t = \frac{W_t l(t/2)}{b_w t^3/12} = \frac{6W_t l}{b_w t^2} \qquad (14.54)$$

where $b_w$ = face width of gear, m. Substituting Equation (14.53) into Equation (14.54) gives

$$\sigma_t = \frac{3W_t}{2b_w x} = \frac{3W_t p_d}{2b_w p_d x} = \frac{W_t p_d}{b_w Y} \qquad (14.55)$$

where $p_d$ = diametral pitch, in$^{-1}$, and

$$Y = \frac{2xp_d}{3} = \text{Lewis form factor} \qquad (14.56)$$

Equation (14.55) is known as the **Lewis equation,** and $Y$ is called the **Lewis form factor.** The Lewis equation considers only static loading and does not take the dynamics of meshing teeth into account.

Table 14.7 gives the Lewis form factors for various numbers of teeth while assuming a pressure angle of 20° and a full-depth involute. The Lewis form factor given in Equation (14.45) is dimensionless. It is also independent of tooth size and only a function of shape.

The Lewis equation [Eq. (14.55)] does not take the stress concentration $K_c$ that exists in the tooth fillet into account. Introducing a geometry factor $Y_j$, where $Y_j = Y/K_c$, changes the Lewis equation to

$$\sigma_t = \frac{W_t p_d}{b_w Y_j} \qquad (14.57)$$

Figure 14.30 gives values of geometry factors $Y_j$ for a spur gear with a pressure angle of 20° and a full-depth involute. The discontinuity at fewer than 18 teeth is due to the undercutting of pinions with fewer than 18 teeth. Equation (14.57) is known as the **modified Lewis equation.**

**Table 14.7**    Lewis form factor for various numbers of teeth (pressure angle, 20°; full-depth involute)

| Number of teeth | Lewis form factor | Number of teeth | Lewis form factor |
|---|---|---|---|
| 10 | 0.176 | 34 | 0.325 |
| 11 | 0.192 | 36 | 0.329 |
| 12 | 0.210 | 38 | 0.332 |
| 13 | 0.223 | 40 | 0.336 |
| 14 | 0.236 | 45 | 0.340 |
| 15 | 0.245 | 50 | 0.346 |
| 16 | 0.256 | 55 | 0.352 |
| 17 | 0.264 | 60 | 0.355 |
| 18 | 0.270 | 65 | 0.358 |
| 19 | 0.277 | 70 | 0.360 |
| 20 | 0.283 | 75 | 0.361 |
| 22 | 0.292 | 80 | 0.363 |
| 24 | 0.302 | 90 | 0.366 |
| 26 | 0.308 | 100 | 0.368 |
| 28 | 0.314 | 150 | 0.375 |
| 30 | 0.318 | 200 | 0.378 |
| 32 | 0.322 | 300 | 0.382 |

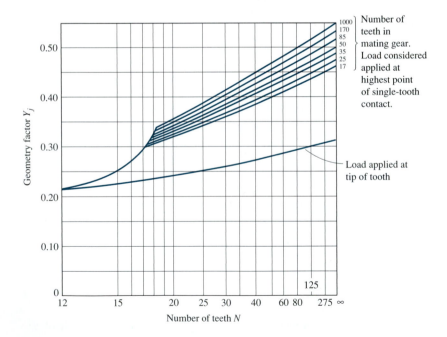

**Figure 14.30**    Spur gear geometry factors for pressure angle of 20° and full-depth involute profile. [AGMA Standard 1012-F90, (1999).]

Other modifications to Equation (14.55) are recommended in AGMA Standard No. 2001 (1999) for practical design to account for the variety of conditions that can be encountered in service. The AGMA bending stress equation is

$$\sigma_t = \frac{W_t \, p_d K_a K_s K_m K_v K_i K_b}{b_w Y_j} \tag{14.58}$$

where $W_t$ = transmitted load, N (or lb)
  $p_d$ = diametral pitch, in$^{-1}$
  $K_a$ = application or overload factor
  $K_s$ = size factor
  $K_m$ = load distribution factor
  $K_v$ = dynamic factor
  $K_i$ = idler factor
  $K_b$ = rim thickness factor

Each of these factors is discussed here.

### 14.11.1 APPLICATION FACTOR

The application factor $K_a$ considers the probability that load variations, vibrations, shock, speed changes, and other application-specific conditions may result in peak loads greater than $W_t$ being applied to the gear teeth during operation. Table 14.8 will be used, where the application factor is a function of both the driving power source and the driven machine. An application factor of 1.00 would be applied for a perfectly smooth-operating electric motor driving a perfectly smooth-operating generator through a gear speed reducer. Rougher conditions produce a value of $K_a$ greater than 1.00. Power source classifications with typical examples are

1.  **Uniform:** electric motor with constant-speed turbine
2.  **Light shock:** water turbine with variable-speed drive
3.  **Moderate shock:** multicylinder engine

Classification of driven machine roughness and examples are

1.  **Uniform:** continuous-duty generator
2.  **Light shock:** fans and low-speed centrifugal pumps, liquid agitators, variable-duty generators, uniformly loaded conveyors, rotary positive-displacement pumps

**Table 14.8** Application factor as function of driving power source and driven machine

| Power Source | Driven Machines | | | |
| | Uniform | Light Shock | Moderate Shock | Heavy Shock |
| | Application factor $K_a$ | | | |
| Uniform | 1.00 | 1.25 | 1.50 | 1.75 |
| Light shock | 1.20 | 1.40 | 1.75 | 2.25 |
| Moderate shock | 1.30 | 1.70 | 2.00 | 2.75 |

3.  **Moderate shock:** high-speed centrifugal pumps, reciprocating pumps and compressors, heavy-duty conveyors, machine tool drives, concrete mixers, textile machinery, meat grinders, saws

4.  **Heavy shock:** rock crushers, punch press drives, pulverizers, processing mills, tumbling barrels, wood chippers, vibrating screens, railroad car dumpers

Recognizing the classifications of the driving power source and the driven machine and using Table 14.8 enable the application factor to be obtained.

### 14.11.2  SIZE FACTOR

The AGMA indicates that the size factor $K_s$ is 1.00 for most gears. However, for gears with large teeth or large face widths, a value greater than 1.00 is recommended. Dudley (1984) recommends a size factor of 1.00 for diametral pitches of 5 in$^{-1}$ or greater or metric modules of 5 or smaller. For larger teeth the values of the size factors shown in Table 14.9 are recommended.

### 14.11.3  LOAD DISTRIBUTION FACTOR

The value of the load distribution factor $K_m$ is based on many variables in the design of gears, making it the most difficult factor to determine. It is defined as the ratio of the peak load intensity to the average or uniformly distributed load intensity. In other words, it is the ratio of peak to mean loading. AGMA suggests that the load distribution factor can be expressed as

$$K_m = f(C_{mt}, C_{mf}) \tag{14.59}$$

where $C_{mt}$ = transverse load distribution factor
$\phantom{where}$ $C_{mf}$ = face load distribution factor

The transverse load distribution factor $C_{mt}$ accounts for the nonuniform distribution of load among the gear teeth that share the load. It is mainly affected by the correctness of the tooth profiles. No standardized method has been established to evaluate $C_{mt}$, so AGMA recommends that this factor be set equal to unity, and therefore the load distribution factor is a function of the face load distribution factor only.

$\phantom{xxxx}$AGMA describes a simplified empirical approach for determining the load distribution factor, applicable to situations where

1.  The ratio of face width to pinion pitch diameter $b_w/d_p \leq 2$.

**Table 14.9**　Size factor as function of diametral pitch or module

| Diametral pitch $p_d$, in$^{-1}$ | Module $m$, mm | Size factor $K_s$ |
|---|---|---|
| ≥5 | ≤5 | 1.00 |
| 4 | 6 | 1.05 |
| 3 | 8 | 1.15 |
| 3 | 12 | 1.25 |
| 1.25 | 20 | 1.40 |

2. The gear elements are mounted between bearings. Overhung gears present additional complications and require a complex theoretical approach beyond the scope of this book. A discussion of this circumstance is contained in AGMA 2001-C95 (1999).

3. The face width is less than 1 m (40 in).

4. Contact is across the full face width of the narrowest member when loaded.

For situations that do not meet these conditions, a lengthy analytical approach discussed in AGMA 2001-C95 (1999) is applicable.

The load distribution factor can be expressed by

$$K_m = C_{mf} = 1.0 + C_{mc}(C_{pf}C_{pm} + C_{ma}C_e) \qquad (14.60)$$

where $C_{mc}$ = lead correction factor
$\quad C_{pf}$ = pinion proportion factor
$\quad C_{pm}$ = pinion proportion modifier
$\quad C_{ma}$ = mesh alignment factor
$\quad C_e$ = mesh alignment correction factor

These factors will be briefly described here.

### Lead Correction Factor $C_{mc}$

The lead correction factor $C_{mc}$ corrects the peak load when crowning or other forms of lead modification are applied.

$$C_{mc} = \begin{cases} 1.0 & \text{for uncrowned teeth} \\ 0.8 & \text{for crowned teeth} \end{cases} \qquad (14.61)$$

### Pinion Proportion Factor $C_{pf}$

The pinion proportion factor accounts for deflections under load and is shown in Figure 14.31. If the face width and pinion pitch diameter are in millimeters, $C_{pf}$ is given by

$$C_{pf} = \begin{cases} \dfrac{b_w}{10d_p} - 0.025 & b_w \leq 25 \text{ mm} \\[2mm] \dfrac{b_w}{10d_p} - 0.0375 + 0.000492b_w & 25 \text{ mm} < b_w \leq 432 \text{ mm} \\[2mm] \dfrac{b_w}{10d_p} - 0.1109 + 0.000815b_w - (3.53 \times 10^{-7})b_w^2 & 432 \text{ mm} < b_w \leq 1020 \text{ mm} \end{cases}$$
$$(14.62)$$

If the face width and pinion pitch diameter are in inches, $C_{pf}$ is given by

$$C_{pf} = \begin{cases} \dfrac{b_w}{10d_p} - 0.025 & b_w \leq 1 \text{ in} \\[2mm] \dfrac{b_w}{10d_p} - 0.0375 + 0.0125b_w & 1 \text{ in} < b_w \leq 17 \text{ in} \\[2mm] \dfrac{b_w}{10d_p} - 0.1109 + 0.02907b_w - (2.28 \times 10^{-4})b_w^2 & 17 \text{ in} < b_w \leq 40 \text{ in} \end{cases}$$
$$(14.63)$$

Also, if $b_w/d_p < 0.5$, then the value of $b_w/d_p = 0.5$ should be used in Equations (14.62) or (14.63).

**Figure 14.31**   Pinion proportion factor $C_{pf}$. [AGMA Standard 2001-C95, (1999).]

### *Pinion Proportion Modifier $C_{pm}$*

If a gear is mounted at midspan between two equally stiff bearings, any deflections that occur will not change the load distribution across the gear tooth face. However, if the gear is mounted away from midspan, deflections of the shaft will include a slope, which changes the stress distribution across the face width. The pinion proportion modifier $C_{pm}$ alters $C_{pf}$ based on the location of the gear relative to the bearing centerline.

$$C_{pm} = \begin{cases} 1.0 & S_1/S < 0.175 \\ 1.1 & S_1/S \geq 0.175 \end{cases} \tag{14.64}$$

where $S$ is the bearing span and $S_1$ is the offset of the pinion, as shown in Figure 14.32.

### *Mesh Alignment Factor $C_{ma}$*

The mesh alignment factor accounts for the misalignment of the pitch cylinder axes of rotation from all other causes other than elastic deformations. It can be obtained from Figure 14.33.

### *Mesh Alignment Correction Factor $C_e$*

When the manufacturing or assembly techniques improve the mesh alignment, the mesh alignment correction factor $C_e$ is used.

$$C_e = \begin{cases} 0.80 & \text{when gearing is adjusted at assembly} \\ 0.80 & \text{when compatability between gear teeth is improved by lapping} \\ 1.0 & \text{for all other conditions} \end{cases} \tag{14.65}$$

**Figure 14.32** Evaluation of $S$ and $S_1$. [AGMA Standard 2001-C95, (1999).]

## 14.11.4 IDLER FACTOR

In a gear train, an idler is usually subjected to more stress cycles per time and to larger alternating loads than nonidler gears. The AGMA approach is to reduce the allowable strength of a gear by 0.70 if that gear is an idler. While this approach works well, in this text the idler factor has been incorporated by using $K_i$ in Equation (14.58). For an idler, use $K_i = 1.42$. For a nonidler gear, use $K_i = 1.0$.

## 14.11.5 RIM FACTOR

Very large gears are more economically produced from rim-and-spoke designs instead of solid blanks. However, if the rim is thin, it is more likely that fracture will occur through

$$C_{ma} = A + Bb_w + Cb_w^2$$

If $b_w$ is in inches:

| Condition | A | B | C |
|---|---|---|---|
| Open gearing | 0.247 | 0.0167 | $-0.765 \times 10^{-4}$ |
| Commercial enclosed gears | 0.127 | 0.0158 | $-1.093 \times 10^{-4}$ |
| Precision enclosed gears | 0.0675 | 0.0128 | $-0.926 \times 10^{-4}$ |
| Extra-precision enclosed gears | 0.000380 | 0.0102 | $-0.822 \times 10^{-4}$ |

If $b_w$ is in millimeters:

| Condition | A | B | C |
|---|---|---|---|
| Open gearing | 0.247 | $6.57 \times 10^{-4}$ | $-1.186 \times 10^{-7}$ |
| Commercial enclosed gears | 0.127 | $6.22 \times 10^{-4}$ | $-1.69 \times 10^{-7}$ |
| Precision enclosed gears | 0.0675 | $5.04 \times 10^{-4}$ | $-1.44 \times 10^{-7}$ |
| Extra-precision enclosed gears | 0.000360 | $4.02 \times 10^{-4}$ | $-1.27 \times 10^{-7}$ |

**Figure 14.33** Mesh alignment factor. [AGMA Standard 2001-C95, (1999).]

the rim than through the tooth root. The **backup ratio** is defined as

$$m_B = \frac{t_R}{h_t}$$

(14.66)

If the backup ratio is greater than 1.2, which includes the extreme case of a solid (conventional) gear, use $K_b = 1$. Otherwise, the rim factor is defined by

$$K_b = -2m_B + 3.4$$

(14.67)

### 14.11.6    DYNAMIC FACTOR

Note that the American Gear Manufacturers Association changed the definition of the dynamic factor in 1995; before 1995, the factor had a value less than 1.0, and it appeared in the denominator of Equation (14.58). Equation (14.58) is consistent with the current AGMA standards, and therefore $K_v$ has a value greater than 1.0. Caution should be exercised when referring to technical literature that predates this definition change.

The dynamic factor $K_v$ accounts for the fact that the load is higher than the transmitted load, since there is some impact loading that is not accounted for in Equation (14.57). The value of $K_v$ depends on the accuracy of the tooth profile, the elastic properties of the tooth, and the speed with which the teeth come into contact.

Figure 14.34 gives the dynamic factors for pitch-line velocity and transmission accuracy level number. The pitch-line velocity $v_t$ used in Figure 14.34 is the linear velocity of a point

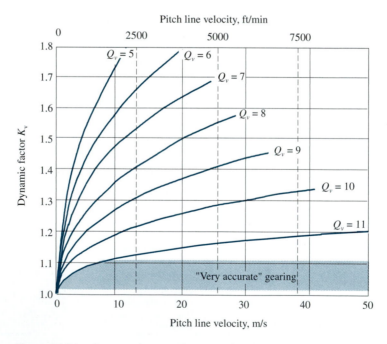

**Figure 14.34**    Dynamic factor as function of pitch-line velocity and transmission accuracy level number.

on the pitch circle of the gear, given by

$$v_t = \pi d_p N_{ap} = \pi d_g N_{ag} \tag{14.68}$$

where $v_t$ = pitch-line velocity, ft/min or m/s
$N_{ap}$ = rotational speed of pinion, rpm or rps
$N_{ag}$ = rotational speed of gear, rpm or rps
$d_p$ = pitch diameter of pinion, in or m
$d_g$ = pitch diameter of gear, in or m

The dynamic factor can be alternatively approximated for $5 \le Q_v \le 11$ by

$$K_v = \left( \frac{A + C\sqrt{v_t}}{A} \right)^B \tag{14.69}$$

where
$$A = 50 + 56(1.0 - B)$$

$$B = 0.25(12 - Q_v)^{0.667}$$

$$C = \begin{cases} 1 & \text{for } v_t \text{ in ft/min} \\ \sqrt{200} = 14.14 & \text{for } v_t \text{ in m/s} \end{cases}$$

The maximum recommended pitch line velocity for a given value of $Q_v$ is

$$v_{t,\max} = \frac{1}{C^2}[A + (Q_v - 3)]^2 \tag{14.70}$$

## 14.11.7 CONTACT STRESSES IN GEARS

In addition to considering the critical bending stress in a gear design, we need to check for failure caused by excessive contact stresses. These stresses can cause pitting, scoring, and scuffing of the surfaces. Pitting is the removal of small bits of metal from the gear surface due to fatigue, leaving small holes or pits. Pitting is caused by excessive surface stress due to high normal loads, a high local temperature due to high rubbing speeds, or inadequate lubricant.

Scoring is heavy scratch patterns extending from the tooth root to the tip. It appears as if a heavily loaded tooth pair has dragged particles between sliding teeth. It can be caused by lubricant failure, incompatible materials, and overload.

Scuffing is a surface destruction comprising plastic material flow plus superimposed gouges and scratches caused by loose metallic particles acting as an abrasive between teeth. Both scoring and scuffing are associated with welding or seizing and plastic deformation. The resulting contact stress in meshing gear teeth is high because of the nonconformal surfaces. This high stress results in pitting and spalling of the surfaces.

The contact stress is calculated by using a Hertzian contact stress analysis. The Hertzian contact stress of gear teeth is based on the analysis of two cylinders under a radial load. The radii of the cylinders are the radii of curvature of the involute tooth forms of the mating teeth at the point of contact. The load on the teeth is the normal load (see Figures 14.28 and 14.29) and is given in Equation (14.49).

The maximum Hertzian pressure in the contact can be written from Equation (8.19) as

$$p_H = E' \left( \frac{W'}{2\pi} \right)^{1/2} \tag{14.71}$$

where $E'$ = effective modulus of elasticity = $\dfrac{2}{(1 - v_a^2)/E_a + (1 - v_b^2)/E_b}$

$W'$ = dimensionless load = $\dfrac{w'}{E'R_x}$  (14.72)

$w'$ = load per unit width = $\dfrac{W}{b_w}$

$$\frac{1}{R_x} = \left(\frac{1}{r_p} + \frac{1}{r_g}\right)\frac{1}{\sin\phi} = \left(\frac{1}{d_p} + \frac{1}{d_g}\right)\frac{2}{\sin\phi} \qquad (14.73)$$

The parameter $R_x$ is the effective radius and for gears is a function of the pitch radius of the pinion and the gear as well as the pressure angle. Figure 14.9 should be used for a visualization of these parameters. Equation (14.73) thus states that 1 over the effective radius of the meshing spur gears is equal to 1 over the radius from $b$ to $p_p$ plus 1 over the radius from $a$ to $p_p$ in Figure 14.11.

The contact made between meshing spur gears is rectangular, having a Hertzian half-width $b^*$, and the length of the contact is the face width $b_w$. Equation (8.16) can be used to calculate the Hertzian half-width. Note that the maximum deformation in the center of the contact is given by Equation (8.18).

Just as modification factors [see Eq. (14.58)] were used to describe the bending stress acting on the meshing gears, there are also modification factors when one is dealing with contact stresses. Thus, the contact stress given in Equation (14.71) is modified to

$$\sigma_c = E'\left(\frac{W'K_a K_s K_m K_v}{2\pi}\right)^{1/2} = p_H(K_a K_s K_m K_v)^{1/2} \qquad (14.74)$$

The values for the application factor $K_a$, the size factor $K_s$, the load distribution factor $K_m$, and the dynamic factor $K_v$ are obtained in the same way as those for the bending stress in Equation (14.58).

## 14.12   ELASTOHYDRODYNAMIC FILM THICKNESS

The minimum-film-thickness formula for elastohydrodynamic lubrication (EHL) developed in Chapter 13 is equally applicable for gears or any other nonconformal machine element as long as the appropriate representations of the geometry, load, and speed are made. The best way to observe how EHL is applied is through the following example.

**EXAMPLE 14.9**

**Given:** A meshing spur gear with crowned teeth has pitch radii of 50 and 75 mm and a pressure angle of 20°. The angular velocity of the larger gear is 210 rad/s, the face width of the gear is 30 mm, and the transmitted load is 2250 N. The module is 5.0 mm, the AGMA quality number is 9 (commercially enclosed gears), and the power source and the driven machine are assumed to be uniform. The properties of the lubricant and the gear materials are $\eta_0 = 0.075\ \text{Pa}\cdot\text{s}$, $\xi = 2.2 \times 10^{-8}\ \text{m}^2/\text{N}$, $E = 207\ \text{GPa}$, and $v = 0.3$. The surface roughness of the gear is 0.3 $\mu$m.

**Find:** Calculate the minimum film thickness and the dimensionless film parameter of the meshing gear teeth. Also, determine the maximum Hertzian stress, the safety factor for contact stress analysis, the dimensions of the contact, and the maximum deformation in the contact.

**Solution:** From Equation (14.50) the effective radius of the meshing gears is

$$\frac{1}{R_x} = \left(\frac{1}{r_p} + \frac{1}{r_g}\right)\frac{1}{\sin\phi} = \left(\frac{1}{50} + \frac{1}{75}\right)\frac{1}{\sin 20°} = 0.0975/\text{mm}$$

$$\therefore \quad R_x = 10.26 \text{ mm} = 0.01026 \text{ m}$$

From Equation (14.3) the gear ratio is

$$g_r = \frac{r_g}{r_p} = \frac{75}{50} = 1.5$$

From Equation (14.18)

$$\omega_p = g_r\omega_g = 1.5(210) = 315 \text{ rad/s}$$

The mean velocity is

$$\tilde{u} = \frac{\omega_p r_p \sin\phi + \omega_g r_g \sin\phi}{2} = \frac{(315)(50)(\sin 20°) + (210)(75)(\sin 20°)}{2}$$

$$= 5387 \text{ mm/s} = 5.387 \text{ m/s}$$

Since no slip occurs between meshing gear teeth at the pitch point, the numerators of the two terms are equal.

The effective modulus of elasticity is

$$E' = \frac{2}{(1 - v_p^2)/E_p + (1 - v_g^2)/E_g} = \frac{E}{1 - v^2}$$

$$= \frac{(2.07)(10^{11})}{1 - 0.3^2} = 2.275 \times 10^{11} \text{ Pa}$$

The transmitted load $W_t$ was given as 2250 N. The normal load, from Equation (14.50), is

$$W = \frac{W_t}{\cos\phi} = \frac{2250}{\cos 20°} = 2394 \text{ N}$$

The normal load per face width is

$$P_z' = \frac{W}{b_w} = \frac{2394}{0.03} = 79,800 \text{ N/m}$$

The dimensionless parameters for load $W'$, speed $U$, and materials $G$ can be obtained from Equations (13.72), (13.70), and (13.71) for a rectangular contact.

$$W' = \frac{P_z'}{E'R_x} = \frac{(7.980)(10^4)}{(2.275)(10^{11})(1.026)(10^{-2})} = 3.419 \times 10^{-5}$$

$$U = \frac{\eta_0\tilde{u}}{E'R_x} = \frac{(0.075)(5.387)}{(2.275)(10^{11})(1.026)(10^{-2})} = 1.731 \times 10^{-10}$$

$$G = \xi E' = (2.2)(10^{-8})(2.275)(10^{11}) = 5005$$

Making use of Equation (13.73) gives the dimensionless minimum film thickness for a rectangular contact as

$$H_{\min} = \frac{h_{\min}}{R_x} = 1.714(W')^{-0.128}U^{0.694}G^{0.568}$$

$$= 135.3 \times 10^{-6}$$

The dimensional minimum film thickness is

$$h_{\min} = R_x H_{\min} = (1.026)(10^{-2})(1.353)(10^{-4})\,\text{m} = 1.388\,\mu\text{m}$$

From Equation (8.32) the dimensionless film parameter is

$$\Lambda = \frac{h_{\min}}{\left(R_{qa}^2 + R_{qb}^2\right)^{1/2}} = \frac{(1.388)(10^{-6})}{(10^{-6})(0.3^2 + 0.3^2)^{1/2}} = 3.272$$

Therefore, the meshing spur gears will enjoy the significant benefits of elastohydrodynamic lubrication. Having $\Lambda \approx 3$ is an optimum design, as discussed in Chapter 13 for rolling-element bearings.

From Equation (8.19) the maximum Hertzian stress in the contact of meshing gear teeth is

$$\sigma_c = p_H = E'\left(\frac{W'}{2\pi}\right)^{1/2} = 2.275(10^{11})\left(\frac{3.419 \times 10^{-5}}{2\pi}\right)^{1/2}$$

$$= 5.307 \times 10^8\,\text{Pa} = 530.7\,\text{MPa}$$

Incorporating the modification factors from Equation (14.74) gives

$$\sigma_c = (530.7)(10^6)(K_a K_s K_m K_v)^{1/2}$$

Both the power source and the driven machine have uniform operation, so that the application factor from Table 14.8 is 1.0. The size factor from Table 14.9 for $m = 5$ mm is $K_s = 1.0$. And $K_m$ will be calculated from Equation (14.60), but first note from Equation (14.61), $C_{mc} = 0.8$ since the teeth are crowned. From Equation (14.62), noting that $b_w = 30$ mm and using $b_w/d_p = 0.5$ [see note after Equation (14.62)],

$$C_{pf} = \frac{b_w}{10d_p} - 0.0375 + 0.000492b_w = \frac{0.5}{10} - 0.0375 + 0.000492(30) = 0.02726$$

From Equation (14.64), $C_{pm} = 1.1$ to allow flexibility in location of the mounting bearings. It should be recognized that the gear safety factor calculated below may be conservative as a result. From Figure 14.33, the mesh alignment factor $C_{ma}$ is

$$C_{ma} = A + Bb_w + Cb_w^2 = 0.127 + (5.04 \times 10^{-4})(30) + (-1.69 \times 10^{-7})(30)^2 = 0.1390$$

From Equation (14.65), $C_e = 1.0$, so that expensive manufacturing or complicated adjustments are not required. Therefore, from Equation (14.60),

$$K_m = 1.0 + C_{mc}(C_{pf}C_{pm} + C_{ma}C_e) = 1.0 + 0.8[(0.02726)(1.1) + (0.1390)(1.0)] = 1.135$$

The pitch-line velocity is calculated as

$$v_t = r_p \omega = \frac{0.1 \text{ m}}{2}(210 \text{ rad/s}) = 10.5 \text{ m/s}$$

Therefore, from Equation (14.69),

$$C = 14.14$$

$$B = 0.25(12 - Q_v)^{0.667} = 0.25(12 - 9)^{0.667} = 0.5202$$

$$A = 50 + 56(1.0 - B) = 50 + 56(1 - 0.5202) = 76.87$$

$$K_v = \left(\frac{A + C\sqrt{v_t}}{A}\right)^B = \left(\frac{76.87 + 14.14\sqrt{10.5}}{76.87}\right)^{0.5202} = 1.275$$

Therefore, from Equation (14.74), the corrected contact stress is

$$\sigma_c = P_H(K_a K_s K_m K_v)^{1/2} = (530.7 \text{ MPa})[(1.0)(1.0)(1.135)(1.275)]^{1/2} = 638.5 \text{ MPa}$$

From Figure 14.25 for a Brinell hardness of 400 and grade 2 steel, the allowable contact stress is

$$\sigma_{all} = 1.2 \text{ GPa}$$

The safety factor is

$$n_s = \frac{\sigma_{all}}{\sigma_c} = \frac{1.2}{0.6385} = 1.88$$

Thus, fatigue failure due to contact stress is not anticipated.

## 14.13 SUMMARY

The primary function of gears is to provide constancy of angular velocities or proportionality of power transmission. This chapter logically developed the important methodology needed to successfully design gears. Primary emphasis was on the methodology of spur gear design. Various types of gear were discussed. Gears are divided into three major classes: parallel-axis gears, nonparallel and coplanar gears, and nonparallel and noncoplanar gears. Although spur gears were the major focus, other gear types were explained.

The geometric factors as well as the kinematics, loads, and stresses acting in a spur gear were discussed. Gears transfer power by the driving teeth exerting a load on the driven teeth while the reacting loads act back on the teeth of the driving gear. The thrust load is the force that transmits power from the driving to the driven gear. It acts perpendicular to the axis of the shaft carrying the gear. This thrust load is used in establishing the bending stress in the tooth. The highest tensile bending stress occurs at the root of the tooth, where the involute curve blends with the fillet. The design bending stress is then compared with an allowable stress to establish whether failure will occur.

A second, independent form of gear failure results from surface pitting and spalling, usually near the pitch line, where the contact stresses are high. Thus, failure due to fatigue occurs. The Hertzian contact pressure with modification factors was used to establish the design stress, which was then compared with an allowable stress to determine if failure due to fatigue would occur. It was found that if an adequate protective elastohydrodynamic lubrication film exists, gear life is greatly extended.

## KEY WORDS

**backlash**    clearance measured on pitch circle.

**backup ratio**    ratio of tooth root to total depth.

**bevel gears**    nonparallel, coplanar gears with straight teeth.

**circular pitch**    distance from one tooth to corresponding point on adjacent tooth.

**contact ratio**    defines the average number of teeth in contact at any time.

**diametral pitch**    number of teeth per pitch diameter.

**fundamental law of gearing**    law stating that common normal to tooth profile must pass through fixed point, called the pitch point.

**gear**    toothed wheel that, when mated with another toothed wheel, transmits power between shafts.

**gear ratio**    ratio of number of teeth between two meshing gears.

**gear trains**    machine elements of multiple gears used to obtain desired velocity ratio.

**Lewis equation**    uses bending of a cantilevered beam to simulate the stresses acting on a gear tooth under static conditions.

**Lewis form factor**    geometrical considerations used in the Lewis equation.

**modified Lewis equation**    considers stress concentration of the tooth fillet which is not considered in the Lewis equation.

**module**    ratio of pitch diameter to number of teeth.

**pinion**    smaller of two meshed gears.

**pitch point**    point defined by normal to tooth profiles of meshing gears at point of contact on line of centers; see also **fundamental law of gearing.**

**planet carrier**    a wheel that carries the planet wheels in planetary gear trains.

**planet gear**    an epicyclic gear with a fixed annulus, a rotating sun wheel, a rotating planet carrier, and planet wheels rotating about their own spindles.

**planetary gear trains**    a system of planetary gears in which one or more wheel axes revolves about other fixed axes.

**ring gear**    a large–diameter gear in the form of an annulus.

**spiral gear**    a gear whose pitch curves are inclined to the pitch element at the spiral angle.

**spur gears**  parallel-axis gears with straight teeth.

**sun gear**  a gear wheel around which one or more planet wheels or planetary pinions rotate in mesh.

**worm gear**  a helical gear that meshes with a worm wheel in sliding contact.

# RECOMMENDED READINGS

Dudley, D. W. (1994) *Handbook of Practical Gear Design,* CRC Press, Boca Raton, FL.
Howes, M. A. (ed.) (1980) *Source Book on Gear Design Technology and Performance,* American Society for Metals, Metals Park, OH.
Juvinall, R. C., and Marshek, K. M. (2003) *Fundamentals of Machine Component Design,* 3rd ed., Wiley, New York.
Litvin, F. L. (1994) *Gear Geometry and Applied Theory,* Prentice-Hall, Englewood Cliffs, NJ.
Mott, R. L. (1999) *Machine Elements in Mechanical Design,* 3rd ed., Prentice-Hall, Upper Saddle River, NJ.
Norton, R. L. (1996) *Machine Design,* Prentice-Hall, Englewood Cliffs, NJ.
Shigley, J. E., Mischke, C. R., and Budynas, R. G. (2003) *Mechanical Engineering Design,* 7th ed., McGraw-Hill, New York.
Stokes, A. (1992) *Gear Handbook,* Butterworth Heinemann, Oxford, England.
Townsend, D. P. (1992) *Dudley's Gear Handbook,* 2nd ed., McGraw-Hill, New York.

# REFERENCES

AGMA (1989) *Geometry Factors for Determining the Pitting Resistance and Bending Strength of Spur, Helical, and Herringbone Gear Teeth,* ANSI/AGMA Standard 908-B89, American Gear Manufacturers Association, Alexandria, VA.
AGMA (1990) *Gear Nomenclature, Definitions of Terms with Symbols,* ANSI/AGMA Standard 1012-F90, American Gear Manufacturers Association, Alexandria, VA.
AGMA (1992) *Practice for Enclosed Cylindrical Wormgear Speed Reducers and Gearmotors,* ANSI/AGMA Standard 6034-B92, American Gear Manufacturing Association, Alexandria, VA.
AGMA (1997) *Rating the Pitting Resistance and Bending Strength of Generated Straight Bevel, Zerol Bevel and Spiral Bevel Gear Teeth,* ANSI/AGMA Standard 2003-B97, American Gear Manufacturers Association, Alexandria, VA.
AGMA (1999) *Fundamental Rating Factors and Calculation Methods for Involute Spur and Helical Gear Teeth,* ANSI/AGMA Standard 2101–C95, American Gear Manufacturers Association, Alexandria, VA.
Drago, R. J. (1992) Gear Types and Nomenclature, Chapter 2 in *Dudley's Gear Handbook,* D. P. Townsend (ed.), 2nd ed., McGraw-Hill, New York.
Dudley, D. W. (1984) *Handbook of Practical Gear Design,* McGraw-Hill, New York.
Kalpakjian, S., and Schmid, S. R. (2001) *Manufacturing Engineering and Technology,* Prentice-Hall, Upper Saddle River, NJ.

Lewis, W. (1892), "Investigation of the Strength of Gear Teeth," an Address to the Engineer's Club of Philadelphia, October 15, 1892.

Mechalec, G. W. (1962) *Gear Handbook,* McGraw-Hill, New York, chap. 9.

Mott, R. L. (1992) *Machine Elements in Mechanical Design,* 2nd ed., Merrill, New York.

Mott, R. L. (1999) *Machine Elements in Mechanical Design,* 3rd ed., Prentice-Hall, Upper Saddle River, NJ.

Townsend, D. P. (1992) *Dudley's Gear Handbook,* 2nd ed., McGraw-Hill, New York.

## PROBLEMS

### Section 14.2

**14.1**   *Hunting ratio* is a ratio of numbers of gear and pinion teeth which ensures that each tooth in the pinion will contact *every* tooth in the gear before it contacts any tooth a second time. (For example, 13 to 48 is a hunting ratio; 12 to 48 is not a hunting ratio.) Give three examples of hunting ratios. Explain the advantages of using hunting ratios in gear teeth.

### Section 14.3

**14.2**   A pinion has a pressure angle of 20°, a module of 3 mm, and 22 teeth. It is meshed with a gear having 32 teeth. The center distance between the shafts is 81 mm. Determine the gear ratio and the diametral pitch. *Ans.* $g_r = 1.45$, $p_d = 0.333$ mm$^{-1}$.

**14.3**   A gearbox has permanent shaft positions due to the bearing positions, but the gear ratio can be changed by changing the gear wheels. To achieve similar power transmission ability for different gear ratios, a manufacturer chooses to have the same module for two different gearboxes. One of the gearboxes has a pinion with 22 teeth, a gear with 68 teeth, and a center distance of 225 mm. How large is the gear module, and which gear ratios are possible for a pinion with 22 or more teeth for the same module? *Ans.* $m = 5$ mm.

★ **14.4**   The blank for a normal gear with 20 teeth was made too large during manufacturing. Therefore, the addendum for the gear teeth was moved outward, and the tooth thickness at the tops of the teeth was decreased. The module for the rack tool was 2 mm. Find out how large the maximum gear outside radius can be when the tool flank angle is 20° to avoid zero thickness at the gear teeth tops. *Ans.* 27 mm.

**14.5**   The inlet and outlet speeds for a reduction gear train are 2000 and 320 rpm, respectively. The pressure angle is designed to be 25°. The gears have a 5-mm module, and their center distance is 435 mm.

   *a)*   Determine the number of teeth for both gears; the pitch diameters; and the radii of the base, addendum, and dedendum circles. *Ans.* $N_p = 24$, $N_g = 150$, $d_p = 120$ mm, $d_g = 750$ mm.

   *b)*   Find the change in center distance if the pressure angle changes from 25° to 25.9°. *Ans.* $\Delta c_d = 3.27$ mm.

* Indicates problem of greater difficulty.

**14.6** The base circle radii of a gear pair are 100 and 260 mm, respectively. Their module is 8 mm, and the pressure angle is 20°. Mounting inaccuracy has caused a 5-mm change in center distance. The number of teeth in the pinion is 12. Calculate the change in the pitch radii, the center distance, and the pressure angle.

**14.7** A gear train has a 50.3-mm circular pitch and a 25° pressure angle, and the number of teeth in the pinion is 12. Design this gear train for the lowest volume occupied and the fewest teeth. Obtain the pitch diameters, the number of teeth, the speed ratio, the center distance, and the module of this gear train. The gearbox housing has a diameter of 620 mm.

### Section 14.4

★ **14.8** The pinion of the gear train for an endurance strength test machine (sketch a) has 5 kW of power and rotates clockwise at 1200 rpm. The gears have a 12.57-mm circular pitch and a 20° pressure angle.

a) Find the magnitude and direction of the output speed. *Ans. $\omega = 640$ rpm.*
b) Calculate the center distance between the input and output shafts. *Ans. $c_d = 330$ mm.*
c) Draw free-body diagrams of forces acting on gears 3 and 4.

Sketch a, for Problem 14.8

### Section 14.5

**14.9** A standard straight gear system consists of two gears with tooth numbers $N_g = 50$ and $N_p = 20$. Pressure angle $\phi = 20°$ and diametral pitch $p_d = 5.08$ per inch. Find the center distance $c_d$ and the contact ratio $C_r$.

**14.10** Given a straight gear with diametral pitch $p_d = 6.35$ per inch, pressure angle $\phi = 20°$, $N_g = 28$, and $N_p = 18$, find the contact ratio $C_r$. *Ans. $C_r = 1.582$.*

**14.11** Given an internal straight gear with diametral pitch $p_d = 5.08$ per inch and $N_p = 22$, find the contact ratio $C_r$ if center distance $c_d = 4.33$ in and pressure angle $\phi = 20°$. *Ans. $C_r = 1.933$.*

**14.12** A straight gear with diametral pitch $p_d = 6.35$ per inch and face width $b_w = 1.575$ in has center distance $c_d = 4.724$ in. The pinion has 25 teeth, the speed is 1000 rpm, and pressure angle $\phi = 20°$. Find the contact stress at the pitch point when the gear transmits 20 kW. Neglect friction forces. The gear steel has modulus of elasticity $E = 207$ GPa and a Poisson's ratio of 0.3.

★ **14.13** A spur gear has $N_g = 40$, $N_p = 20$, and $\phi = 20°$. The diametral pitch $p_d = 5.08$ per inch. Find the center distances between which the gear can work (i.e., from the backlash-free state to $C_r = 1$). *Ans.* Between 5.9055 and 6.042 in.

**14.14** Given a gear with the data $N_g = 45$, $N_p = 18$, $p_d = 6.35$ per inch, and $\phi = 20°$, find

  a)   Center distance $c_d$. *Ans.* $c_d = 4.961$ in.
  b)   Contact ratio $C_r$. *Ans.* $C_r = 1.633$.
  c)   Tooth thickness on the pitch circle for both pinion and gear. *Ans.* $t_h = 0.2474$ in.

★ **14.15** Given a gear with the center distance $c_d$, find out how much the pressure angle $\phi$ increases if the center distance increases by a small amount $\Delta c_d$. *Ans.* $\Delta\phi = \dfrac{\Delta c_d}{c_d \tan \phi}$.

## Section 14.6

**14.16** A straight gear is free of backlash at the theoretical center distance $c_d = a = 0.264$ m $= 10.394$ in. Pressure angle $\phi = 20°$, $N_g = 47$, $N_p = 19$, and $\alpha = 11 \times 10^{-6}$ (°C)$^{-1}$. Find the center distance needed to give a backlash of 0.008378 in (0.2128 mm). What gear temperature increase is then needed relative to the housing temperature to make the gear free from backlash? *Ans.* $\Delta t = 100.6°$C.

★ **14.17** A gear with a center distance of 8 in is going to be manufactured. Different diametral pitches can be chosen depending on the transmitted torque and speed and on the required running accuracy. Which diametral pitch should be chosen to minimize gear angular backlash if the gear ratio is 2? *Ans.* 12 per inch.

**14.18** A pinion having 25 teeth, a circular pitch of 0.2618 in, and rotating at 2000 rpm is to drive a gear rotating at 500 rpm. Determine the diametral pitch, number of teeth in the gear, and the center distance.

**14.19** What is the contact ratio for two coarse pitch spur gears having 20° pressure angles, addendums of 12.5 mm, and velocity ratio of 0.5? The pinion has 30 teeth.

## Section 14.7

**14.20** A drive line needs to be developed for a rear-wheel-drive car. The gear ratio in the gear from the car transmission to the rear wheels is 4.5:1. The rear wheel diameter is 550 mm. The motor can be used in the speed interval 1500 to 6000 rpm. At 6000 rpm the speed of the car should be 200 km/h in fifth gear. In first gear it should be possible to drive as slow as 10 km/h without slipping the clutch. Choose the gear ratios for the five gears in the gearbox.

**14.21** In the train drive shown in sketch $b$, gear 2 is the inlet of the gear train and transfers 2.1 kW of power at 2000 rpm. Each gear contact reduces the speed ratio by 5:3.

  a)   Find the number of teeth for gears 3 and 4. *Ans.* $N_4 = 50$, $N_4 = 120$.
  b)   Determine the rotational speed of each gear. *Ans.* $\omega_3 = \omega_4 = 1200$ rpm, $\omega_5 = 720$ rpm.
  c)   Calculate the torque of each shaft when, because of frictional losses, there is a 5% power loss in each matching pair. *Ans.* $T_{in} = 10.0$ N · m, $T_{out} = 25.13$ N · m.

Sketch *b*, for Problem 14.21

**14.22** In Problem 14.21 the module of all gears in the train drive is 8 mm, and the pressure angle is 25°. Draw a free-body diagram of the forces acting on each gear in the train.

**14.23** In the gear train shown in sketch *c*, the pinion rotates at 420 rpm and transfers 5.5 kW to the train. The circular pitch is 31.4 mm, and the pressure angle is 25°. Find the output speed, and draw a free-body diagram of the forces acting on each gear. *Ans.* $\omega_{out} = 168$ rpm.

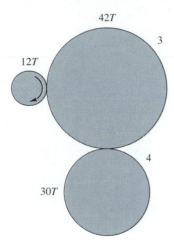

Sketch *c*, for Problem 14.23

**14.24** The input shaft of the planetary gear shown in sketch *d* has 10 kW of power and is rotating clockwise at 500 rpm. Gears 4 and 5 have a module of 4 mm. The pressure angle is 20°. The inlet and outlet shafts are coaxial. Find the following:

*a)* The output speed and its direction. *Ans.* $\omega = 13.89$ rpm.
*b)* The minimum inside diameter of the gearbox housing. *Ans.* $d_{min} = 408$ mm.
*c)* The contact loads acting on each gear. *Ans.* $W_{t5} = W_{t4} = 114$ kN.

Sketch d, for Problem 14.24

**14.25** An 18-tooth pinion rotates at 1420 rpm and transmits 52 kW of power to the gear train shown in sketch *e*. The power of the output shaft is 35% of the input power of the train. There is 3% power loss in every gear contact. The pressure angle for all the gears is 25°. Calculate the contact forces acting on each gear. All dimension units are in millimeters. *Ans.* Gear 3: $W_t = 6.25$ kN, $W_r = 2.92$ kN.

Sketch e, for Problem 14.25

## Section 14.8

**14.26** The third gear in the gearbox of a sports car is dimensioned to have a finite life in order to save weight. When the car is used in a race, the load spectrum is totally different from the spectrum produced by normal driving. Therefore, the third gear fails by bending stresses at the gear roots, and the gear teeth crack and fall off. For the load spectrum used to dimension the gearbox, bending stress failure at the gear tooth root has the same stress safety factor as surface pitting at the contact areas. The gears are made of grade 2 steel with Brinell hardness HB = 250. By changing the material to 400 Brinell hardness, the bending stress failures disappear. Is there then any risk of surface pitting? *Ans.* No.

★ **14.27** A speed of 520 rpm is used in a 2:1 gear reduction pair to transfer 3 kW of power for a useful life of 50 million turns with 90% reliability. A pressure angle of 20° is to be used along with a safety factor of 2. The loads are uniform with a regular mounting and hub-machined teeth. On the basis of contact stress, determine the proper material, module, and number of teeth for this pair, and calculate the pitch diameters and the face width.

## Section 14.9

★ **14.28** An endurance strength test machine (see sketch $f$) consists of a motor, a gearbox, two elastic shafts, bearings, and the test gears. The gear forces are created by twisting the shafts when the gears are mounted. At one location the shafts were twisted at an angle representing one gear tooth. Find the gear force in the test gears as a function of the parameters given in the sketch.

*Ans.* $P = \dfrac{G\pi^2 d^4 p_d}{16 N_p^2 l \cos \phi}$.

Sketch $f$, for Problem 14.28

★ **14.29** For a 2:1 gear reduction an ASTM A536 ductile cast iron pinion is rotating at 320 rpm and transfers 5.2 kW of power to a gear of the same material. The gears are standard full depth with a 25° pressure angle. Design this gear pair for a safety factor of 2 based on bending stress.

## Section 14.10

**14.30** To increase the power density of a machine, the rotational speeds are increased. Typical examples of this practice are high-speed electric motors and vacuum cleaners, where the speeds recently have been increased more than 10-fold by using modern frequency converters. If the output shaft from a high-speed motor must have a low speed, a reduction gear is used. The quality of the reduction gear determines how large the dynamic impact forces will be in the output shaft gear. Hobbed gears of quality level 5 are used for a gear rotating at 500 rpm at a pitch diameter of 50 mm. To increase the throughput power, the speed is increased to 5000 rpm. How much higher power can be transferred through the gear at that speed? How high would the power be with extremely accurate gearing?

**14.31** A 25-tooth pinion running at 1200 rpm and transferring 13 kW gives a 4:1 reduction gear pair. The gears are made of steel. The circular pitch is 15.7 mm, the face width is 49 mm, and the pressure angle is 20°. Determine the bending stress of the gears if the power source is uniform and the machine encounters moderate shock, the teeth are not crowned, and the gear is centered between supports. Use a quality factor of $Q_v = 8$ and assume the gear will be checked at assembly to ensure proper alignment. Consider the gears to be commercially enclosed gears.

## Section 14.11

**14.32** A standard gear commercial enclosed gearset ($Q_v = 6$) with crowned and pressure angle $\phi = 20°$ and module $m = 3$ mm transfers 24 kW at a pinion rotational speed of 1500 rpm. Face width $b_w = 15$ mm, $N_p = 24$, $N_g = 42$, and $E = 207$ GPa for the gear steel. Find the contact stress from the AGMA equations. Assume $K_a = 1.0$.

**14.33** A gearbox with spur gears has a fixed shaft center distance $c_d$. The contact pressure between the gears at the pitch point limits the power being transmitted through the gearbox. The rotational speed of the incoming shaft is fixed. The rotational speed of the outgoing shaft varies with the gear ratio. Find the gear ratio at which the maximum power can be transmitted through the gearbox.

**14.34** A standard commercial enclosed ($Q_v = 9$) gear with crowned teeth has $m = 3$ mm, $\phi = 20°$, $N_g = 40$, and $N_p = 20$. Find the AGMA contact stress if the force is split evenly between the two contacts. The transmitted power is 16 kW at 1500 rpm for the smaller wheel. Gear face width $b_w = 18$ mm, and modulus of elasticity $E = 210$ GPa. Assume $K_a = 1.25$.

**14.35** A spur gear has $m = 3$ mm, $\phi = 20°$, $N_g = 58$, and $N_p = 28$. The gear is loaded with 50 N·m of torque on the slow shaft. Gear face width $b_w = 20$ mm, and modulus of elasticity $E = 210$ GPa and Poisson's ratio of 0.3. Friction is neglected. Find the maximum contact pressure between the gears just as the small wheel comes into contact.

**14.36** An external spur gear has $p_d = 6.35$ per inch, $N_g = 52$, $N_p = 28$, $\phi = 20°$, pinion rotational speed of 1500 rpm, $b_w = 0.7874$ in, and $E = 30 \times 10^6$ psi and $V = 0.3$. Find the largest power the gear can transmit if the maximum allowable Hertzian contact pressure is $p_H = 142.1$ ksi. *Ans. $h_p = 72.7$ hp.*

**14.37** An 18-tooth pinion rotates at 1500 rpm, mates with a 72-tooth gear, and transfers 3.2 kW of power. The module is 4 mm and the pressure angle is 20°. If the gears are made of steel with a yield strength of 720 MPa, calculate the face width for a safety factor of 5 using the modified Lewis equation. Also calculate the maximum contact stress between the mating teeth. Assume $K_a = 1$. *Ans. $b_w = 3$ mm, $\sigma_c = 859$ MPa.*

**14.38** A 16-tooth pinion rotates at 720 rpm and transfers 5 kW to a 3:1 speed reduction gear pair. The circular pitch is 9.4 mm, the face width is 38 mm, and the pressure angle is 20°. Calculate the maximum contact stresses for this steel gear pair. *Ans. $\sigma_c = 676$ MPa.*

**14.39** Two commercial enclosed steel gears (grade 1 with Brinell hardness of 300 HB each, 20° pressure angle) are in contact. The pinion has 12 teeth with a diametral pitch of 8 teeth per in, and the gear has 48 teeth. The pinion speed is 4000 rpm, and 3100 ft·lb/s is transmitted. The gears are manufactured to a quality standard of $Q_v = 7$ and have a face width of 0.75 in. The electric power source is uniform, but the machine loading results in occasional light shock. Find the safety factor against failure due to excessive bending stresses.

**14.40** A spur gear set has 17 crowned teeth on the pinion and 51 crowned teeth on the gear. The pressure angle is 20°, the diametral pitch is 6 teeth per inch, and the face width is 2 in. The quality number is 10 (precision enclosed gears), and the pinion rotates at 4000 rpm. The material is a through-hardened steel (grade 1) with Brinell harness of 350 HB. (a) What are the center distance, gear ratio, and circular pitch? (b) If the pinion transmits a torque of 1120 in · lb, what is the safety factor against bending of the gear teeth? Use an application factor of 1.0. *Ans. $n_s = 3.56$.*

★ **14.41** A pair of crowned-tooth gears with pressure angles of 20° transmits 1 kW while the pinion rotates at 3450 rpm. The module is 1 mm, the pinion has 24 teeth, and the gear has 120 teeth. Both the pinion and the gear are grade 2 steel and heat-treated to 300 HB. For a safety factor of 3, determine the required face width. The input power is from an electric motor, and the drive is for a small machine tool with light shock. Assume $Q_v = 10$ and $K_m = 1.0$.

**14.42** Two gears (grade 2 through-hardened steel with Brinell hardness of 250 HB each, 20° pressure angle) are in contact. The pinion has 12 crowned teeth with a diametral pitch of 8 teeth per inch, and the gear has 48 crowned teeth. The pinion speed is 4000 rpm, and 3100 ft · lb/s is transmitted. The gears are manufactured to a quality standard of $Q_v = 7$ (commercial enclosed gears) and have a face width of 0.75 in. The electric power source is uniform, but the machine loading results in occasional light shock.

   a)   What are the allowable bending stress and contact stress for these gears? *Ans. $S_t = 41.9$ ksi, $S_c = 109.6$ ksi*

   b)   Assume that the gears have been hardened further so that the allowable bending stress number is 50 ksi. Find the safety factor against failure due to excessive bending stresses. *Ans. $n_s = 4.35$*

**14.43** Repeat Problem 14.42 for the following circumstances:

   a)   The gear will see a life of 100,000 cycles with a reliability of 99.99%. *Ans. $n_s = 3.6$*

   b)   The gear will see a life of 100,000 cycles with a reliability of 99.99%, and it is mounted off-center. The spacing between bearings is 5 in, and the center of the gear hub is 1.5 in from the center of one bearing. The teeth are crowned.

**14.44** The gear considered in Problem 14.42 was mistakingly produced from grade 1 steel and was only hardened to 200 HB. What are the allowable bending and contact stresses for this case? Will the gear fail?

**14.45** A vertical mixer uses an electric motor to drive the sun gear of a planetary gear train. The ring is attached to the machine frame, and the mixing attachment is attached to one of the planets. If the sun gear and the planet gears have 30 teeth, determine the angular velocity and articulating angular velocity of the mixing attachment if the sun gear is driven at 500 rpm. Repeat the problem if the sun gear has 20 teeth.

**14.46** A planetary gear train has $N_{sun} = 30$, $N_{planet} = 60$, a diametral pitch of 6 in$^{-1}$, and a 20° pressure angle and is to transmit 6 hp. What torque can be delivered by the armature if the sum is driven and the ring is stationary? *Ans. $T = 238$ in-lb*

★ **14.47** Design a suitable pitch and width for a pair of meshing gears that will transmit 7.5 hp. The center distance is around 10 in, and the angular velocity ratio is to be 3:4. The pinion rotates at 150 rpm. Assume the pressure angle is 20°, and the gears will be constructed from grade 1 through-hardened steel with a Brinell hardness of 300. The power source and the driven machine are subjected to light shock, and a life of 10 million cycles and 99.99% reliability is needed. Assume $K_m = K_i = K_b = 10$.

## Section 14.12

**14.48** An external gear has a module of 4 mm, 52 teeth in the gear, 28 teeth in the pinion, a pressure angle of 20°, pinion rotational speed of 1500 rpm, face width of 20 mm, and gear and pinion material having a modulus of elasticity of 206 GPa and a Poisson's ratio of 0.3. The Hertzian contact pressure is 0.980 GPa. The lubricant used has an absolute viscosity of 0.075 Pa · s and pressure viscosity coefficient of $2.2 \times 10^{-8}$ m²/N. The root-mean-square surface roughness for both the pinion and gear is 0.1 $\mu$m. Determine the following:

*a)*     Transmitted power. *Ans.* 54.85 kW
*b)*     Minimum film thickness. *Ans.* 2.203 $\mu$m
*c)*     The film parameter
*d)*     Is the film parameter adequate? Justify your answer. If it is not adequate, what changes would you suggest to improve the situation?

# HELICAL, BEVEL, AND WORM GEARS

A combined helical and worm gearset speed reducer. (Courtesy Boston Gear.)

*The main object of science is the freedom and happiness of man.*
*Thomas Jefferson*

## SYMBOLS

| | |
|---|---|
| $a$ | addendum, m |
| $b$ | dedendum, m |
| $b_w$ | face width, m |
| $b_{wp}$ | projected face width, m |
| $C_m$ | gear ratio correction factor |
| $C_r$ | contact ratio |
| $C_{rt}$ | total contact ratio |
| $C_{ra}$ | axial contact ratio |
| $C_s$ | materials factor |
| $c_d$ | center distance, m |
| $d$ | pitch diameter, m |
| $d_g$ | pitch diameter of gear, m |
| $d_o$ | outer diameter, m |
| $d_p$ | pitch diameter of pinion, m |
| $d_r$ | root diameter, m |
| $d_w$ | pitch diameter of worm, m |
| $E$ | modulus of elasticity, Pa |
| $E'$ | effective modulus of elasticity, Pa |
| $g_r$ | gear ratio |
| HB | Brinell hardness |
| $h_p$ | transmitted power, hp |
| $h_{pi}$ | input horsepower, hp |
| $I_b$ | geometry factor for contact stresses for bevel gears |
| $I_h$ | geometry factor for contact stresses for helical gears |
| $K_a$ | application factor |
| $K_b$ | rim factor |
| $K_i$ | idler factor |
| $K_m$ | load distribution factor |
| $K_s$ | size factor |
| $K_v$ | dynamic factor |
| $K_x$ | crowning factor |
| $L$ | lead of worm, m |
| $m$ | module, $d/N$, m |
| $N$ | number of teeth |
| $N_g$ | number of teeth in gear |
| $N_n$ | equivalent number of teeth |
| $N_p$ | number of teeth in pinion |
| $N_v$ | equivalent number of teeth |
| $N_w$ | number of teeth or starts in worm |
| $p_a$ | axial pitch for helical gears, m |
| $p_c$ | circular pitch, $\pi d/N$, m |
| $p_{cn}$ | normal circular pitch, m |
| $p_d$ | diametral pitch, $N/d$, in$^{-1}$ |
| $p_{dn}$ | normal diametral pitch, in$^{-1}$ |
| $p_H$ | maximum Hertzian contact pressure, Pa |
| $q$ | exponent in Equation (15.29) |
| $r$ | pitch radius, m |
| $r_g$ | pitch radius of gear, m |
| $r_p$ | pitch radius of pinion, m |
| $T$ | torque, N $\cdot$ m |
| $v_t$ | pitch-line velocity, m/s |
| $W$ | load, N |
| $W_a$ | axial or thrust load, N |
| $W_{ag}$ | thrust load on worm gear, N |
| $W_{aw}$ | thrust load on worm, N |
| $W_f$ | friction force, N |
| $W_r$ | radial load, N |
| $W_s$ | separation force, N |
| $W_t$ | tangential load, N |
| $W_{tg}$ | tangential load on worm gear, N |
| $W_{tw}$ | tangential load on worm, N |
| $Y_b$ | geometry factor in bending for bevel gears |
| $Y_h$ | geometry factor in bending for helical gears |
| $Z$ | angular velocity ratio |
| $\alpha_p$ | pitch cone angle, deg |
| $\lambda$ | lead angle, deg |
| $\mu$ | coefficient of friction |
| $\nu$ | Poisson's ratio |
| $\sigma$ | stress, Pa |
| $\sigma_c$ | contact stress of gear, Pa |
| $\sigma_t$ | bending stress of gear, Pa |
| $\phi$ | pressure angle, deg |
| $\phi_n$ | pressure angle in normal direction, deg |
| $\psi$ | helix angle, deg |
| $\Omega$ | angular velocity of shaft, rad/s |
| $\omega$ | angular speed of gear, rad/s |

## SUBSCRIPTS

| | |
|---|---|
| $a$ | actual; axial |
| all | allowable |
| $b$ | bending |
| $c$ | contact |
| $g$ | gear |
| max | maximum |
| $o$ | outside |
| $p$ | pinion |
| $t$ | total |

## 15.1 INTRODUCTION

Chapter 14 discussed the basic geometry of gears and presented the design methodology developed by the American Gear Manufacturers Association (AGMA). Similar design approaches have been developed for helical, bevel, and worm gears and are briefly summarized in this chapter.

The interested student, faced by a menu from which to select basic gear types, and then catalogs from which to select specific gears, may feel overwhelmed by the unconstrained nature of the design problem. Section 14.2 presented some fundamental considerations relating to the orientation of drive and driven shafts for a gear set. This often provides sufficient restrictions on the design such that a class, i.e., spur, worm, bevel, etc., can be selected and analysis can proceed to determine required gear dimensions. However, shaft orientation is often a design variable, and any gear type cannot be easily eliminated from consideration. Some of the advantages and disadvantages of gear types are given in Table 15.1. Further information to aid designers in gear selection is contained in Dudley (1994).

## 15.2 HELICAL GEARS

The material covered in Chapter 14 related primarily to spur gears. This section describes **helical gears.** Although helical gears were introduced in Section 14.2.1, this section describes them in greater detail. However, instead of providing as much detail as was given for spur gears, this section builds upon the information obtained from spur gears and describes only the differences that exist between spur gears and helical gears.

**Table 15.1**    Design considerations for gears

| Gear type | Advantages | Disadvantages |
|---|---|---|
| Spur | Inexpensive, simple to design, no thrust load is developed by the gearing, wide variety of manufacturing options. | Can generate significant noise, especially at high speeds, and are usually restricted to pitch-line speeds below 20 m/s (4000 ft/min). |
| Helical | Useful for high-speed and high-power applications, quiet at high speeds. Often used in lieu of spur gears for high-speed applications. | Generate a thrust load on a single face, more expensive than spur gears. |
| Bevel | High-efficiency (can be 98% or higher), can transfer power across nonintersecting shafts. Spiral bevel gears transmit loads evenly and are quieter than straight bevel. | Shaft alignment is critical, rolling element bearings are therefore often used with bevel gears. This limits power transfer for high-speed applications (where a journal bearing is preferable). Can be expensive. |
| Worm | Compact designs for large gear ratios. Efficiency can be 80% or higher. | Wear by abrasion is of greater concern than other gear types, can be expensive. Generate very high thrust loads. Worm cannot be driven by gear; worm must drive gear. |

**Figure 15.1**    Helical gear. (a) Front view; (b) side view.

Tooth strength is improved with helical gears (relative to spur gears) because of the elongated helical wrap-around tooth base support, as shown in Figure 14.2. For helical gears the contact ratio is higher owing to the axial tooth overlap. Helical gears, therefore, tend to have greater load-carrying capacity than spur gears of the same size. Spur gears, on the other hand, have a somewhat higher efficiency.

### 15.2.1    HELICAL GEAR RELATIONSHIPS

All the relationships governing spur gears apply to helical gears with some slight modifications to account for the axial twist of the teeth caused by the helix angle. The helix angle varies from the base of the tooth to the outside radius. The *helix angle $\psi$* is defined as the angle between an element of the pitch cylinder and the tangent to the helicoidal tooth at the intersection of the pitch cylinder and the tooth profile. Figure 15.1 defines the helix angle. Front and side views are shown.

### 15.2.2    PITCHES OF HELICAL GEARS

Helical gears have two related pitches: one in the plane of rotation and the other in a plane normal to the tooth. For spur gears the pitches were described only in terms of the plane of rotation, but there is an additional axial pitch for helical gears. Figure 15.2 shows

**Figure 15.2**    Pitches of helical gears.
(a) Circular; (b) axial.

the circular and axial pitches of helical gears, which are related by the **normal circular pitch**

$$p_{cn} = p_c \cos \psi \qquad (15.1)$$

where $\psi$ = helix angle, deg. The normal diametral pitch is

$$p_{dn} = \frac{p_d}{\cos \psi} \qquad (15.2)$$

The axial pitch of a helical gear is the distance between corresponding points on adjacent teeth measured parallel to the gear's axis [see Figure 15.2(b)]. The axial pitch is related to the circular pitch by the following expression:

$$p_a = p_c \cot \psi = \frac{p_{cn}}{\sin \psi} \qquad (15.3)$$

### 15.2.3  EQUIVALENT NUMBER OF TEETH AND PRESSURE ANGLE

The number of teeth for helical gears equivalent to that found for spur gears is given as

$$N_n = \frac{N}{\cos^3 \psi} \qquad (15.4)$$

For helical gears there is a normal pressure angle as well as the usual pressure angle in the plane of rotation, and they are related to the helix angle, or

$$\tan \phi = \frac{\tan \phi_n}{\cos \psi} \qquad (15.5)$$

### 15.2.4  HELICAL TOOTH PROPORTIONS

Helical gear tooth proportions follow the same standards as those for spur gears. Addendum, dedendum, whole depth, and clearance are exactly the same. This discussion will cover parallel-shaft helical gear meshes but not crossed-shaft helical gear meshes.

The pitch diameter is given by the same expression used for spur gears [Eq. (14.5)], but if the normal pitch is involved, it is a function of the helix angle.

$$d = \frac{N}{p_d} = \frac{N}{p_{dn} \cos \psi} \qquad (15.6)$$

where $d$ = pitch diameter, m (in)
$\quad\ N$ = number of teeth
$\quad\ p_d$ = diametral pitch, m$^{-1}$ (in$^{-1}$)
$\quad p_{dn}$ = normal diametral pitch, m$^{-1}$ (in$^{-1}$)
$\quad\ \psi$ = helix angle, deg

The center distance, using an analogy to Equation (14.5), is

$$c_d = \frac{d_p + d_g}{2 \cos \psi} = \frac{N_p + N_g}{2 p_{dn} \cos \psi} \qquad (15.7)$$

The contact ratio of helical gears is enhanced by the axial overlap of the teeth. The total contact ratio is the sum of the circumferential contact ratios when calculated in the same manner as done for spur gears [Eq. (14.21)] and added to the axial contact ratio $C_{ra}$, or

$$C_{rt} = C_r + C_{ra} \tag{15.8}$$

$$C_{ra} = \frac{b_w}{p_a} = \frac{b_w \tan \psi}{p_c} = \frac{b_w \sin \psi}{p_{cn}} \tag{15.9}$$

where $C_r$ = contact ratio obtained from Equation (14.21)
 $b_w$ = face width, m

### 15.2.5    LOADS AND STRESSES

The tangential load $W_t$ is the same for spur or helical gears. Recall that the tangential load is the force that transmits power from the driver to the driven gear. It acts perpendicular to the axis of the shaft carrying the gear. Thus, Equations (14.49) and (14.57) are equally valid for spur gears or helical gears.

The *axial,* or thrust, load in a helical gear is

$$W_a = W_t \tan \psi \tag{15.10}$$

Note that the axial load increases as the helix angle increases. Helix angles typically range from 15° to 45°.

The *radial* load is the force that acts toward the center of the gear (i.e., radially). The direction of the force is to push the two gears apart. By making use of Equation (14.51), the radial load is

$$W_r = W_t \tan \phi \tag{15.11}$$

The *normal* load used in the elastohydrodynamic lubrication evaluation is

$$W = \frac{W_t}{\cos \phi_n \cos \psi} \tag{15.12}$$

The bending and contact stresses for helical gears are the same as those given for spur gears in Equations (14.54) and (14.71) with the coefficients in those equations having the same meaning but different numerical values. This chapter explained the methodology for spur gears and gave a brief discussion of the difference in going from a spur gear to a helical gear. The same methodology can be applied to helical gears with the use of figures and tables available in AGMA (1990).

---

**EXAMPLE 15.1**

**Given:** An involute gear drives a high-speed centrifuge. The speed of the centrifuge is 18,000 rpm. It is driven by a 3000-rpm electric motor through a 6:1 speedup gearbox. The pinion has 21 teeth, and the gear has 126 teeth with a diametral pitch of 14 per inch. The width of the gears is 1.8 in, and the pressure angle is 20°.

**Find:**   The gear contact ratio for

 (a)   A spur gear
 (b)   A helical gear with helix angle of 30°

Also find the helix angle if the contact ratio is 3.

**Solution:**

(a) **Spur gear:** From Equation (14.6) the circular pitch is

$$p_c = \frac{\pi}{p_d} = \frac{\pi}{14} = 0.224 \text{ in}$$

From Equations (14.1) and (14.4) the center distance is

$$c_d = \frac{d_p + d_g}{2} = \frac{N_p + N_g}{2p_d} = \frac{21 + 126}{2(14)} = 5.25 \text{ in}$$

From Equation (14.5) the pitch circle radii for the pinion and gear are

$$r_p = \frac{N_p}{2p_d} = \frac{21}{2(14)} = 0.75 \text{ in}$$

$$r_g = \frac{N_g}{2p_d} = \frac{126}{2(14)} = 4.5 \text{ in}$$

The radii of the base circles for the pinion and gear, using Equation (14.16), are

$$r_{bp} = r_p \cos\phi = 0.75 \cos 20° = 0.7048 \text{ in}$$

$$r_{bg} = r_g \cos\phi = 4.5 \cos 20° = 4.2286 \text{ in}$$

The outside radii for the pinion and gear are, respectively,

$$r_{op} = r_p + a = r_p + \frac{1}{p_d} = 0.75 + \frac{1}{14} = 0.8214 \text{ in}$$

$$r_{og} = r_g + a = r_g + \frac{1}{p_d} = 4.5 + \frac{1}{14} = 4.5714 \text{ in}$$

From Equation (14.21) the contact ratio for the spur gear is

$$C_r = \frac{1}{0.2244 \cos 20°} \left[ \sqrt{(0.8214)^2 - (0.7048)^2} + \sqrt{(4.571)^2 - (4.229)^2} \right]$$

$$- \frac{5.250 \tan 20°}{0.2244} = 1.712$$

(b) **Helical gear:** The total contact ratio for a helical gear is given by Equations (15.8) and (15.9) as

$$C_{rt} = C_r + C_{ra} = 1.712 + \frac{1.8 \tan 30°}{0.2244} = 6.343$$

To get a contact ratio of 3, the helix angle has to be such that

$$C_{ra} = \frac{b_w \tan\psi}{p_c} = 3.00 - 1.712 = 1.288$$

$$\therefore \quad \tan\psi = \frac{(1.288)(0.2244)}{1.8}$$

$$\psi = 9.122°$$

Therefore, the helix angle should be 9.122° to get a contact ratio of 3.

### 15.2.6    AGMA Design Approach for Helical Gears

The design approaches for helical and spur gears are very similar. Indeed, AGMA (1999) uses the same equations for pitting resistance and bending strength for both spur and helical gears. Equation (14.58) still holds for bending stress in helical gears:

$$\sigma_t = \frac{W_t \, p_d \, K_a \, K_s \, K_m \, K_v \, K_i \, K_b}{b_w \, Y_h} \tag{15.13}$$

where   $W_t$ = transmitted load, N (lb)
$p_d$ = diametral pitch, m$^{-1}$ (in$^{-1}$)
$K_a$ = application factor, as given in Table 14.8
$K_s$ = size factor, as given in Table 14.9
$K_m$ = load distribution factor, as given by Equation (14.54)
$K_v$ = dynamic factor, as given by Equation (14.69) or Figure 14.34
$K_i$ = idler factor, as given in Section 14.11.4
$K_b$ = rim factor, as given in Section 14.11.5
$b_w$ = face width of gear, m (in)
$Y_h$ = geometry correction factor for bending

and the equation for contact stress is similar to Equation (14.73):

$$\sigma_c = p_H \left( \frac{K_a \, K_s \, K_m \, K_v}{I_h} \right)^{1/2} \tag{15.14}$$

where   $p_H$ = maximum Hertzian contact pressure given by Equation (14.71), Pa (psi)
$K_a$ = application factor, as given in Table 14.8
$K_s$ = size factor, as given in Table 14.9
$K_m$ = load distribution factor, as given by Equation (14.60)
$K_v$ = dynamic factor, as given by Equation (14.69) or Figure 14.34
$I_h$ = geometry correction factor for pitting resistance

All these variables maintain their definitions as from Chapter 14, except that $Y_h$ and $I_h$ are geometry factors for helical gears. The geometry factors can be calculated for any helical gear geometry by using the approach described by AGMA (1999). Table 15.2 gives some geometry factor values for common pressure and helix angles.

## 15.3    BEVEL GEARS

### 15.3.1    Types of Bevel Gears

As discussed in Section 14.2.1, bevel gears have nonparallel axes that lie in the same plane. Usually **bevel gears** are mounted perpendicular to each other, but almost any shaft angle can be accommodated. Bevel gear blanks are conical, but a number of tooth shapes can be produced. The simplest tooth form occurs on **straight bevel gears,** where the teeth have a profile as shown in Figure 14.3. Note that since the bevel gear is a conic section, the thickness of a gear tooth is not constant. Straight bevel gears are usually cut and may be finished by lapping or grinding.

**Table 15.2**  Geometry factors $Y_h$ and $I_h$ for helical gears loaded at tooth tip. (a) $\phi = 20°$

$\psi = 10°$

| Gear teeth | | 12 P | 12 G | 14 P | 14 G | 17 P | 17 G | 21 P | 21 G | 26 P | 26 G | 35 P | 35 G | 55 P | 55 G | 135 P | 135 G |
|---|---|---|---|---|---|---|---|---|---|---|---|---|---|---|---|---|---|
| 12 | $I_h$ | $U^a$ | | | | | | | | | | | | | | | |
| 14 | $I_h$ | $U^a$ | | 0.129 | | | | | | | | | | | | | |
| | $Y_h$ | $U^a$ | | 0.47 | 0.47 | | | | | | | | | | | | |
| 17 | $I_h$ | $U^a$ | | 0.144 | | 0.133 | | | | | | | | | | | |
| | $Y_h$ | $U^a$ | | 0.48 | 0.51 | 0.52 | 0.52 | | | | | | | | | | |
| 21 | $I_h$ | $U^a$ | | 0.159 | | 0.149 | | 0.136 | | | | | | | | | |
| | $Y_h$ | $U^a$ | | 0.48 | 0.55 | 0.52 | 0.55 | 0.56 | 0.56 | | | | | | | | |
| 26 | $I_h$ | $U^a$ | | 0.175 | | 0.165 | | 0.152 | | 0.139 | | | | | | | |
| | $Y_h$ | $U^a$ | | 0.49 | 0.58 | 0.53 | 0.58 | 0.57 | 0.59 | 0.60 | 0.60 | | | | | | |
| 35 | $I_h$ | $U^a$ | | $U^a$ | | 0.195 | | 0.186 | | 0.175 | | 0.162 | | | | | |
| | $Y_h$ | $U^a$ | | 0.50 | 0.61 | 0.54 | 0.62 | 0.57 | 0.63 | 0.61 | 0.64 | 0.64 | 0.64 | | | | |
| 55 | $I_h$ | $U^a$ | | 0.221 | | 0.215 | | 0.206 | | 0.195 | | 0.178 | | 0.148 | | | |
| | $Y_h$ | $U^a$ | | $U^a$ | | 0.51 | 0.65 | 0.55 | 0.66 | 0.58 | 0.67 | 0.62 | 0.68 | 0.65 | 0.69 | 0.70 | 0.70 |
| 135 | $I_h$ | $U^a$ | | 0.257 | | 0.255 | | 0.251 | | 0.246 | | 0.236 | | 0.215 | | 0.154 | |
| | $Y_h$ | $U^a$ | | 0.52 | 0.70 | 0.56 | 0.71 | 0.60 | 0.72 | 0.63 | 0.73 | 0.67 | 0.74 | 0.71 | 0.75 | 0.76 | 0.76 |

$\psi = 20°$

| Gear teeth | | 12 P | 12 G | 14 P | 14 G | 17 P | 17 G | 21 P | 21 G | 26 P | 26 G | 35 P | 35 G | 55 P | 55 G | 135 P | 135 G |
|---|---|---|---|---|---|---|---|---|---|---|---|---|---|---|---|---|---|
| 12 | $I_h$ | 0.128 | | | | | | | | | | | | | | | |
| | $Y_h$ | 0.47 | 0.47 | | | | | | | | | | | | | | |
| 14 | $I_h$ | 0.140 | | 0.131 | | | | | | | | | | | | | |
| | $Y_h$ | 0.47 | 0.50 | 0.50 | 0.50 | | | | | | | | | | | | |
| 17 | $I_h$ | 0.154 | | 0.145 | | 0.134 | | | | | | | | | | | |
| | $Y_h$ | 0.48 | 0.53 | 0.51 | 0.54 | 0.54 | 0.54 | | | | | | | | | | |
| 21 | $I_h$ | 0.169 | | 0.161 | | 0.150 | | 0.137 | | | | | | | | | |
| | $Y_h$ | 0.48 | 0.56 | 0.51 | 0.57 | 0.55 | 0.58 | 0.58 | 0.58 | | | | | | | | |
| 26 | $I_h$ | 0.184 | | 0.176 | | 0.166 | | 0.153 | | 0.140 | | | | | | | |
| | $Y_h$ | 0.49 | 0.59 | 0.52 | 0.60 | 0.55 | 0.60 | 0.59 | 0.61 | 0.62 | 0.62 | | | | | | |
| 35 | $I_h$ | 0.202 | | 0.196 | | 0.187 | | 0.176 | | 0.163 | | 0.144 | | | | | |
| | $Y_h$ | 0.49 | 0.62 | 0.53 | 0.63 | 0.56 | 0.64 | 0.60 | 0.64 | 0.62 | 0.65 | 0.66 | 0.66 | | | | |
| 55 | $I_h$ | 0.227 | | 0.222 | | 0.215 | | 0.206 | | 0.196 | | 0.178 | | 0.148 | | | |
| | $Y_h$ | 0.50 | 0.66 | 0.53 | 0.67 | 0.57 | 0.67 | 0.60 | 0.68 | 0.63 | 0.69 | 0.67 | 0.70 | 0.71 | 0.71 | | |
| 135 | $I_h$ | 0.258 | | 0.257 | | 0.255 | | 0.251 | | 0.246 | | 0.236 | | 0.214 | | 0.153 | |
| | $Y_h$ | 0.51 | 0.70 | 0.54 | 0.71 | 0.58 | 0.72 | 0.62 | 0.72 | 0.65 | 0.73 | 0.68 | 0.74 | 0.72 | 0.75 | 0.76 | 0.76 |

$\psi = 30°$

| Gear teeth | | 12 P | 12 G | 14 P | 14 G | 17 P | 17 G | 21 P | 21 G | 26 P | 26 G | 35 P | 35 G | 55 P | 55 G | 135 P | 135 G |
|---|---|---|---|---|---|---|---|---|---|---|---|---|---|---|---|---|---|
| 12 | $I_h$ | 0.130 | | | | | | | | | | | | | | | |
| | $Y_h$ | 0.46 | 0.46 | | | | | | | | | | | | | | |
| 14 | $I_h$ | 0.142 | | 0.133 | | | | | | | | | | | | | |
| | $Y_h$ | 0.47 | 0.49 | 0.49 | 0.49 | | | | | | | | | | | | |
| 17 | $I_h$ | 0.156 | | 0.147 | | 0.136 | | | | | | | | | | | |
| | $Y_h$ | 0.47 | 0.51 | 0.50 | 0.52 | 0.52 | 0.52 | | | | | | | | | | |
| 21 | $I_h$ | 0.171 | | 0.163 | | 0.151 | | 0.138 | | | | | | | | | |
| | $Y_h$ | 0.48 | 0.54 | 0.50 | 0.54 | 0.53 | 0.55 | 0.55 | 0.55 | | | | | | | | |
| 26 | $I_h$ | 0.186 | | 0.178 | | 0.167 | | 0.154 | | 0.141 | | | | | | | |
| | $Y_h$ | 0.48 | 0.56 | 0.51 | 0.56 | 0.53 | 0.57 | 0.56 | 0.57 | 0.58 | 0.58 | | | | | | |
| 35 | $I_h$ | 0.204 | | 0.198 | | 0.188 | | 0.176 | | 0.163 | | 0.144 | | | | | |
| | $Y_h$ | 0.49 | 0.58 | 0.51 | 0.59 | 0.54 | 0.59 | 0.56 | 0.60 | 0.58 | 0.60 | 0.61 | 0.61 | | | | |
| 55 | $I_h$ | 0.228 | | 0.223 | | 0.216 | | 0.207 | | 0.196 | | 0.178 | | 0.147 | | | |
| | $Y_h$ | 0.49 | 0.61 | 0.52 | 0.61 | 0.54 | 0.62 | 0.57 | 0.62 | 0.59 | 0.63 | 0.62 | 0.64 | 0.64 | 0.64 | | |
| 135 | $I_h$ | 0.259 | | 0.257 | | 0.255 | | 0.251 | | 0.245 | | 0.234 | | 0.212 | | 0.151 | |
| | $Y_h$ | 0.50 | 0.64 | 0.53 | 0.64 | 0.55 | 0.65 | 0.58 | 0.66 | 0.60 | 0.66 | 0.62 | 0.67 | 0.65 | 0.68 | 0.68 | 0.68 |

$^a$U indicates that this geometry would produce an undercut tooth form and should be avoided.

SOURCE: From AGMA (1999).

Two meshing straight bevel gears are depicted in Figure 15.3 with the appropriate terminology identified. The tooth profile closely approximates an involute tooth profile (see Sec. 14.3.3) for a spur gear with more teeth than the bevel gear; known as the *virtual number of teeth,* it is defined by

$$N_v = \frac{N}{\cos \alpha_p}$$

(15.15)

where $N_v$ = virtual number of teeth
        $N$ = actual number of teeth
        $\alpha_p$ = pitch cone angle, as shown in Figure 15.3

Straight bevel gears are usually limited to pitch-line velocities of 5 m/s (1000 ft/min) or less. Alignment of the bevel gear shaft is fairly critical, and for this reason bevel gears are usually supported by rolling element bearings instead of journal bearings. A complication for mounting bevel gears is that they must also be mounted at the correct distance from the cone center.

Straight bevel gears are commonly made with 14.5°, 17.5°, and 20° pressure angles, with 20° being the most popular. Straight bevel gears are the *least* costly form of bevel gears, and standard sizes are widely available.

**Figure 15.3**    Terminology for bevel gears.

(a)                                                                    (b)

**Figure 15.4**   Bevel gears with curved teeth. (a) Spiral bevel gears; (b) Zerol®. *Source:* Courtesy of Graessner GmBh.

**Zerol bevel gears** are a patented design that is similar to that for straight bevel gears, except that the tooth is curved in the lengthwise direction but with zero spiral angle [see Figure 15.4(b)]. Zerol gears can be generated on a rotary cutter and ground when necessary. Because of the tooth curvature, Zerol gears have a slight overlapping action, and are better suited than straight bevel gears for higher-speed applications. Zerol gears are produced in matched sets and are more expensive than straight bevel gears. Interchangeability (as for maintenance or replacement of gears) can be achieved by using master sets of gears to fit against production gears.

**Spiral bevel gears** are similar to Zerol bevel gears in that the teeth have a lengthwise curvature, but there is an appreciable spiral angle, as shown in Figure 15.4(a). Like Zerol gears, spiral bevel gears are produced in matched sets and are more expensive than straight bevel gears. Spiral gears have greater tooth overlap than Zerol gears and are therefore better suited for high-speed applications. However, spiral gears generate a significantly larger thrust force compared to Zerol gears, and this thrust force must be accommodated by the mounting bearings.

A gear commonly found in automotive and industrial drives is the **hypoid bevel gear**, which resembles a bevel gear in many respects. Hypoid gears are used on nonparallel, coplanar shafts, and the parts usually taper as bevel gears do. The main difference is that axes of the pinion and gear for hypoid gear sets do not intersect; the distance between the axes is called the *offset*. A spiral gear set can be considered as a hypoid gear set with zero offset.

Hypoid gears are different from spur, helical, and bevel gears in that their pitch diameters are not proportional to the number of teeth. The module, diametral pitch, pitch diameter, and pitch angle are used for the gear only, since the pitch for the pinion is smaller than that for the gear. Hypoid gears are produced in matched sets, and economic considerations similar to those for Zerol and spiral gears must be taken into account.

### 15.3.2    BEVEL GEAR GEOMETRY AND FORCES

Figure 15.3 shows two meshing bevel gears with important nomenclature. The projected cone distance is given by

$$\frac{r_p}{\sin \alpha_p} = \frac{r_g}{\sin \alpha_g} \tag{15.16}$$

The angular velocity ratio for bevel gears is defined as in Equation (14.30), but Equation (15.16) allows some simplification when the shaft angle is 90°:

$$Z_{21} = \frac{\omega_2}{\omega_1} = \frac{N_1}{N_2} = \tan \alpha_1 = \cot \alpha_2 \tag{15.17}$$

Note that the minus sign in Equation (14.30) has been omitted in Equation (15.17) because it is impossible to know the sign of the angular velocity without knowledge of the mounting geometry.

Figure 15.5 shows the resultant forces acting on a straight bevel gear. The force acting on a gear tooth has three components: the transmitted load $W_t$ (which also results in a transmitted torque), a radial force $W_r$, and an axial or thrust force $W_a$. Assuming these forces act at the midpoint of the gear tooth, statics yields for a straight bevel gear

$$W = \frac{W_t}{\cos \phi} = \frac{T}{r_{ave} \cos \phi} \tag{15.18}$$

$$W_a = W_t \tan \phi \sin \alpha \tag{15.19}$$

$$W_r = W_t \tan \phi \cos \alpha \tag{15.20}$$

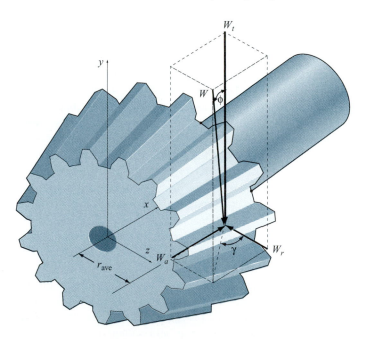

**Figure 15.5**    Forces acting on a bevel gear.

where $T$ = transmitted torque, N·m (lb·in)

$r_{ave}$ = pitch radius at midpoint of tooth, m (in)

$\phi$ = pressure angle, deg

$\alpha$ = pitch cone angle (see Figure 15.3), deg

For a spiral bevel gear, the thrust and radial forces are more complicated. For a right-handed spiral rotating clockwise when viewed from its large end, or for a left-handed spiral rotating counterclockwise from its large end, the radial and thrust forces acting on the tooth are

$$W_a = \frac{W_t}{\cos \psi} (\tan \phi \sin \alpha - \sin \psi \cos \alpha) \qquad (15.21)$$

$$W_r = \frac{W_t}{\cos \psi} (\tan \phi \cos \alpha + \sin \psi \sin \alpha) \qquad (15.22)$$

while for a right-handed spiral rotating counterclockwise or a left-handed spiral rotating clockwise,

$$W_a = \frac{W_t}{\cos \psi} (\tan \phi \sin \alpha + \sin \psi \cos \alpha) \qquad (15.23)$$

$$W_r = \frac{W_t}{\cos \psi} (\tan \phi \cos \alpha - \sin \psi \sin \alpha) \qquad (15.24)$$

where $\psi$ = helix angle, deg.

### 15.3.3 AGMA DESIGN APPROACH FOR BEVEL GEARS

The American Gear Manufacturers Association has developed a design methodology [AGMA (1997)] that is similar to that presented for spur gears in Chapter 14. As with the spur gear methodology, AGMA defines a bending or contact stress number, which is then compared to an allowable stress number for a candidate material. For the most part, allowable stresses for bevel and spur gears are similar for a given material, and the data presented in Section 14.9 can be used for illustrative purposes in this chapter. Note that extensive material properties are available in the technical literature and from AGMA, and these data are essential for real applications.

This section will provide a simplified approach for bevel gear analysis, using many of the correction factors defined in Chapter 14. More accurate correction factors can be obtained by using the full approach described in the AGMA standards [AGMA (1997)].

The equations for bending and contact stress numbers, using terminology consistent with this text when appropriate, are given as

$$\sigma_t = \begin{cases} \dfrac{2T_p}{b_w d_p} \dfrac{p_d K_a K_v K_s K_m}{K_x Y_b} & \text{English units} \\[3mm] \dfrac{2T_p}{b_w d_p} \dfrac{K_a K_v K_s K_m}{m_p K_x Y_b} & \text{SI units} \end{cases} \qquad (15.25)$$

$$\sigma_c = \begin{cases} \sqrt{\dfrac{2T_p E'}{\pi b_w d_p^2 I_b} K_a K_v K_m K_s K_x} & \text{English units} \\[4ex] \sqrt{\dfrac{2T_p E'}{\pi b_w d_p^2 I_b} K_a K_v K_m K_s K_x} & \text{SI units} \end{cases} \qquad (15.26)$$

where $T_p$ = pinion torque, $\text{N} \cdot \text{m}$ $(\text{lb} \cdot \text{in})$

$\quad b_w$ = face width, m (in)

$\quad d_p$ = pinion outer pitch diameter, m (in)

$\quad p_d$ = outer transverse diametral pitch, $\text{in}^{-1}$

$\quad m_p$ = metric outer transverse module, m

$\quad K_a$ = application factor, as given in Table 14.8

$\quad K_v$ = dynamic factor, as given in Figure 14.34

$\quad K_m$ = load distribution factor

$\quad K_s$ = size factor

$\quad K_x$ = crowning factor

$\quad Y_b$ = bending strength geometry factor

$\quad I_b$ = contact strength geometry factor

$\quad E'$ = effective elastic modulus, given by

$$E' = \frac{2}{\left(1 - v_a^2\right)/E_a + \left(1 - v_b^2\right)/E_b}$$

As indicated above, the application factor $K_a$ and dynamic factor $K_v$ can be obtained as shown in Chapter 14. The remaining factors will be briefly summarized here.

### *Load Distribution Factor $K_m$*

The load distribution factor can be calculated from

$$K_m = \begin{cases} K_{mb} + 0.0036 b_w^2 & \text{English units} \\ K_{mb} + 5.6 b_w^2 & \text{SI units} \end{cases} \qquad (15.27)$$

where

$$K_{mb} = \begin{cases} 1.00 & \text{for both gear and pinion straddle-mounted} \\ & \quad \text{(bearings on both sides of gear)} \\ 1.10 & \text{for only one member straddle-mounted} \\ 1.25 & \text{for neither member straddle-mounted} \end{cases}$$

### *Size Factor $K_s$*

The size factor depends on the face width for contact stress considerations, and on the outer transverse pitch or module for bending. Figure 15.6 gives the size factor for contact stress and bending circumstances.

### *Crowning Factor $K_x$*

The teeth of most bevel gears are crowned in the lengthwise direction to accommodate deflection of the mountings. Under light loads, this results in localized contact and higher

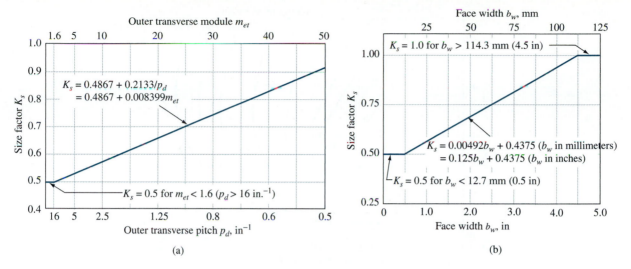

**Figure 15.6** Size factor for bevel gears. (a) Size factor for bending stress; (b) size factor for contact stress or pitting resistance. *From AGMA (1997).*

stresses. On the other hand, lack of crowning will result in localized contact and higher stresses at the end of the tooth. AGMA recommends that a crowning factor be included in *contact stress calculations* according to

$$K_x = \begin{cases} 1.5 & \text{for properly crowned teeth} \\ 2.0 & \text{(or larger) for noncrowned teeth} \end{cases} \qquad \textbf{(15.28)}$$

For *bending stress calculations,* AGMA recommends that $K_x$ be set equal to 1 for straight or Zerol bevel gears. For spiral bevel gears, $K_x$ is given by

$$K_x = 0.211 \left( \frac{r_c}{A_m} \right)^q + 0.789 \qquad \textbf{(15.29)}$$

where $r_c$ = cutter radius used for producing the gears, mm (in)
$\quad A_m$ = mean cone distance, mm (in)

and $q$ is given by

$$q = \frac{0.279}{\log(\sin \psi)} \qquad \textbf{(15.30)}$$

### Geometry Factors $Y_b$ and $I_b$

The geometry factors for bevel gears are different from those for spur or helical gears. The AGMA standard provides geometry factors for straight, spiral, and Zerol gears for a number of helix and pressure angles. A selection of these correction factors is provided in Figures 15.7 through 15.9.

(a)

(b)

**Figure 15.7** Geometry factors for straight bevel gears, with pressure angle $\phi = 20°$ and shaft angle $= 90°$. (a) Geometry factor for contact stress $I_b$; (b) geometry factor for bending $Y_b$. [*From AGMA (1997).*]

(a)

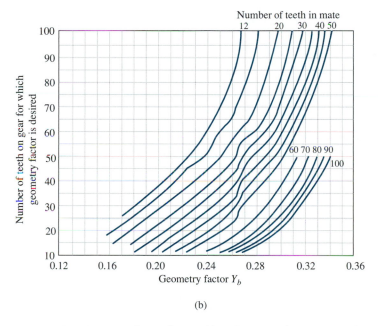

(b)

**Figure 15.8**  Geometry factors for spiral bevel gears, with pressure angle $\phi = 20°$, spiral angle $\psi = 25°$, and shaft angle $= 90°$. (a) Geometry factor for contact stress $I_b$; (b) geometry factor for bending $Y_b$. [*From AGMA (1997).*]

Number of gear teeth

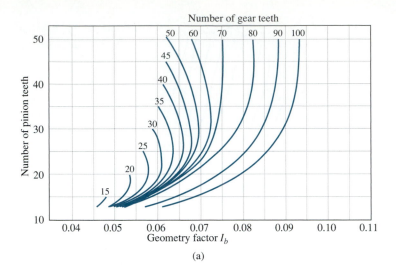

(a)

Number of teeth in mate

(b)

**Figure 15.9** Geometry factors for Zerol bevel gears, with pressure angle $\phi = 20°$, spiral angle $\psi = 25°$, and shaft angle $= 90°$. (a) Geometry factor for contact stress $I_b$; (b) geometry factor for bending $Y_b$. [*From AGMA (1997).*]

**EXAMPLE 15.2**

**Given:** Two spiral bevel gears mesh with a shaft angle of $90°$, and they have properly crowned teeth with a pressure angle of $\phi = 20°$ and a spiral angle of $\psi = 25°$. The face width is $b_w = 1.00$ in, the diametral pitch is $p_d = 5.60$ in$^{-1}$, the pinion has 14 teeth, and the gear has 40 teeth. They are produced from steel with a quality factor of $Q_v = 11$, an allowable bending stress of

28,200 psi, and an allowable contact stress of 200,000 psi. The pinion is overhung-mounted, and the gear is straddle-mounted. The pinion is driven at 1750 rpm with a torque of 1440 in · lb. The gears were produced with a cutter radius of 4.5 in, and the mean cone distance is 3.20 in. Assume the application factor is $K_a = 1.0$.

**Find:** What is the factor of safety based on the design of the pinion?

**Solution:** From Equation (14.5), the diameter of the pinion is

$$d_p = \frac{N_p}{p_d} = \frac{14}{5.60} = 2.50 \text{ in}$$

For steel, $E = 30 \times 10^6$ psi and $\nu = 0.3$, so

$$E' = \frac{2}{(1 - \nu_a^2)/E_a + (1 - \nu_b^2)/E_b} = \frac{E}{1 - \nu^2} = \frac{30 \times 10^6 \text{ psi}}{1 - 0.30^2} = 32.97 \times 10^6 \text{ psi}$$

The contact stress will be calculated from Equation (15.26), but the assorted factors must be obtained first. From Figure 15.8(a), $I_b = 0.095$. The pitch velocity is calculated as

$$v_t = \omega r = N_p(2\pi)\frac{d}{2} = (1750 \text{ rpm})(2\pi)\left(\frac{2.5 \text{ in}}{2}\right) = 13,740 \text{ in/min} = 1145 \text{ ft/min}$$

Therefore, using a quality factor of $Q_v = 11$, we can see from Figure 14.34 that $K_v = 1.08$. Since only one member is straddle-mounted, $K_{mb} = 1.10$, and from Equation (15.27),

$$K_m = K_{mb} + 0.0036b_w^2 = 1.10 + (0.0036)(1.0)^2 = 1.1036$$

From Figure 15.6,

$$K_s = 0.125b_w + 0.4375 = 0.5625$$

Note that to use these equations, $b_w$ needed to be expressed in inches. Since the teeth are properly crowned, Equation (15.28) suggests that $K_x = 1.5$. Therefore the contact stress is given by Equation (15.26) as

$$\sigma_c = \sqrt{\frac{T_p E'}{\pi b_w d_p^2 I_b} K_a K_v K_m K_s K_x}$$

$$= \sqrt{\frac{(1440 \text{ in} \cdot \text{lb})(32.97 \times 10^6 \text{ lb/in}^2)}{\pi(1.0 \text{ in})(2.5 \text{ in})^2(0.095)}(1.0)(1.08)(1.1036)(0.5625)(1.5)}$$

$$= 159,990 \text{ psi} = 160.0 \text{ ksi}$$

Therefore the safety factor against surface pitting failure is

$$n_s = \frac{200,000}{160,000} = 1.25$$

The bending stress is given by Equation (15.25). The factors used above can be utilized, except that the bending geometry factor is obtained from Figure 15.8(b) as $Y_b = 0.22$. Also, $K_x$ needs to be recalculated based on Equation (15.28). Since this is a spiral gear,

$$q = \frac{0.279}{\log \sin \psi} = \frac{0.279}{\log \sin 25°} = -0.7459$$

Therefore, from Equation (15.29),

$$K_x = 0.211 \left( \frac{4.5}{3.20} \right)^{-0.7459} + 0.789 = 0.9526$$

The bending stress is

$$\sigma_t = \frac{2T_p}{b_w d_p} \frac{p_d K_a K_v K_s K_m}{K_x Y_b}$$

$$= \frac{2(1440 \text{ in} \cdot \text{lb})}{(1.0 \text{ in})(2.50 \text{ in})} \frac{(5.60 \text{ in}^{-1})(1.0)(1.08)(0.5625)(1.1036)}{(1.0)(0.22)} = 19.66 \text{ ksi}$$

The safety factor against bending is therefore

$$n_s = \frac{28.20}{19.66} = 1.43$$

Therefore, surface pitting has the lower safety factor, with the value of 1.25.

## 15.4    WORM GEARS

**Worm gears** are extremely popular for situations where large speed reductions are needed. Worm gears are unique in that they cannot be back-driven; the worm can drive the worm gear, but the worm gear cannot drive the worm. This is a function of the thread geometry and is referred to as **self-locking,** a topic that is explored in detail in Section 16.3.3.

One of the other main differences between worm gears and other types of gears is that abrasive wear is the primary concern with worm gears, and contact and bending stresses are much less of a concern. This requires effective lubrication for the worm, and the thread geometry will be designed to avoid point contacts. Furthermore, *wear-in* effects can occur, where initial abrasive wear causes the worm gear thread profiles to conform to the worm teeth, which are usually constructed of a much harder material. Initial contact can be point contact, but wear-in leads to line contact as the teeth become better conforming with time. This wear-in results in more effective lubrication and efficient transfer of loads, so that the wear rates after wear-in are much lower.

Worm gears are produced with speed ratios of 1:1 up to 360:1, although most worm gears in operation vary from 3:1 to 100:1 speed ratios. Ratios greater than 30:1 use single threaded worms, while lower ratios will usually use worms with multiple threads, or **starts.** In general, the larger the center distance between the worm and the worm gear, the larger the number of starts. The majority of worm threads are right-handed.

### 15.4.1    TOOTH GEOMETRY

A worm will usually have two to three teeth in contact with the worm gear at any time, as shown in Figure 15.10. As the worm gear rotates, the teeth contact location shifts as shown in the figure. As a result, a worm and worm gear profile must be designed to allow for efficient transfer of force along line contacts. Worm tooth profiles are more complex than the involute tooth profiles described in Chapter 14, and the interested reader is referred to the American Gear Manufacturers Association standards (ANSI/AGMA 1993) for different forms of worm gear tooth profiles. For example, recessed worm designs can be used to increase the number of teeth in contact.

Line 2    Line 3

Line 1

Rotation of worm

Rotation of gear

Line 3

Line 2    Line 1

**Figure 15.10**    Illustration of worm contact with a worm gear, showing multiple teeth in contact. [*From AGMA (1997).*]

The pressure angle for worm gears is usually restricted to the values in Table 15.3. In general, higher pressure angles are used when high worm gear teeth are needed, but this can result in fewer teeth in contact. In addition the bearing loads can be higher with higher pressure angles, and the bending stresses and deflections in the worm must be considered.

AGMA recommends the minimum number of worm gear teeth as a function of pressure angle, as summarized in Table 15.3. The number of threads in the worm is determined according to

$$N_w = \frac{N_g}{Z} \tag{15.31}$$

where $N_w$ = number of teeth in worm
  $N_g$ = number of teeth in worm gear
  $Z$ = required speed ratio

The pitch of the worm is the distance in the axial plane from a point on one thread to the corresponding point on the next thread. Pitches can be selected for the worm gear just as they were selected for spur gears in Chapter 14. The worm pitch must equal the circular

**Table 15.3**    Suggested minimum number of worm gear teeth for customary designs

| Pressure angle $\phi$, deg | Minimum number of worm gear teeth |
|---|---|
| 14.5 | 40 |
| 17.5 | 27 |
| 20 | 21 |
| 22.5 | 17 |
| 25 | 14 |
| 27.5 | 12 |
| 30 | 10 |

SOURCE: ANSI/AGMA (1993).

pitch of the worm gear. The **lead** of the worm is defined as

$$L = N_w p \tag{15.32}$$

where $N_w$ is the number of threads or starts in the worm. The **lead angle** is given by

$$\lambda = \tan^{-1} \frac{L}{\pi d_{pw}} \tag{15.33}$$

AGMA recommends that the worm pitch diameter fall between the following ranges:

$$\frac{c_d^{0.875}}{3.0} \leq d_w \leq \frac{c_d^{0.875}}{1.6} \qquad \text{English units}$$

$$\frac{c_d^{0.875}}{2.0} \leq d_w \leq \frac{c_d^{0.875}}{1.07} \qquad \text{SI units} \tag{15.34}$$

where $c_d$ is the the center distance between the axes of the worm and worm gear. The worm gear pitch diameter is calculated as

$$d_g = 2c_d - d_w \tag{15.35}$$

The addendum and dedendum for worms are given by

$$a = \frac{p_b}{\pi} \tag{15.36}$$

and

$$b = \begin{cases} \dfrac{1.157 p_b}{\pi} & \text{for } p_b > 4.06 \text{ mm (0.160 in)} \\[2mm] \dfrac{1.200 p_b}{\pi} + 0.05 \text{ mm} & \text{for } p_b < 4.06 \text{ mm} \\[2mm] \dfrac{1.200 p_b}{\pi} + 0.002 \text{ in} & \text{for } p_b < 0.160 \text{ in} \end{cases} \tag{15.37}$$

The outside and root diameters of the worm are given by

$$d_o = d + 2a \tag{15.38}$$

$$d_r = d - 2b \tag{15.39}$$

---

**EXAMPLE 15.3**

**Given:** A double-threaded worm with a pitch diameter of 3 in meshes with a 25-tooth worm gear. The worm gear has a pitch diameter of 5 in.

**Find:** The center distance, the speed ratio, and the lead angle for the gear set.

**Solution:** From Equation (15.35), the center distance can be obtained as

$$d_g = 2c_d - d_{pw}$$

$$c_d = \frac{d_g + d_{pw}}{2} = \frac{5 \text{ in} + 3 \text{ in}}{2} = 4 \text{ in}$$

The speed ratio is obtained from Equation (15.31) as

$$Z = \frac{N_g}{N_w} = \frac{25}{2} = 12.5$$

The worm gear pitch is

$$p = \frac{\pi d_g}{N_g} = \frac{\pi(5 \text{ in})}{25} = 0.6283 \text{ in}$$

From Equation (15.32),

$$L = N_w p = 2(0.6283 \text{ in}) = 1.257 \text{ in}$$

Therefore, the lead angle is calculated from Equation (15.33) as

$$\lambda = \tan^{-1} \frac{L}{\pi d_{pw}} = \tan^{-1} \frac{1.257 \text{ in}}{\pi(3 \text{ in})} = 7.596°$$

### 15.4.2 FORCES ON WORM GEARS

Figure 15.11 shows the forces acting on a right-handed thread on a worm. The nomenclature of the forces is slightly more complicated than that for other gears, since the thrust force for the worm gear is equal to the tangential force for the worm, and vice versa. The tangential force for the worm gear and the thrust force for the worm are equal and are given by

$$W_{tg} = W_{aw} = \frac{2T_g}{d_m} \tag{15.40}$$

where $W_{tg}$ = worm gear tangential force, N (lb)
$\quad W_{aw}$ = worm axial or thrust force, N (lb)
$\quad T_g$ = worm gear torque, N · m (in · lb)
$\quad d_m$ = worm gear mean diameter, m (in)

The gear separation force is given by

$$W_s = \frac{W_{tg} \tan \phi_n}{\cos \lambda_n} \tag{15.41}$$

(a)  (b)

**Figure 15.11** Forces acting on a worm. (a) Side view, showing forces acting on worm and worm gear. (b) Three-dimensional view of worm, showing worm forces. The worm gear has been removed for clarity.

where $W_s$ = separation force, N (lb)

      $\phi_n$ = normal pressure angle, deg

      $\lambda$ = worm mean lead angle, deg

The worm gear thrust force and the worm tangential force are equal and given by

$$W_{ag} = W_{tw} = \frac{2T_w}{d_m} \tag{15.42a}$$

where $W_{ag}$ = worm gear axial or thrust force, N (lb)

      $W_{tw}$ = worm tangential force, N (lb)

      $T_w$ = worm torque, N · m (in · lb)

      $d_m$ = worm mean diameter, m (in)

It can be shown from statics that worm and wormgears thrust loads are related by the following expression:

$$W_{tw} = W_{tg} \left( \frac{\cos \phi_n \sin \lambda + \mu \cos \lambda}{\cos \phi_n \cos \lambda - \mu \sin \lambda} \right) \tag{15.42b}$$

### 15.4.3    AGMA Equations

A number of attempts have been made by various researchers and organizations to develop a power rating system for worm gears. This has been elusive since the long-term success of worm gears depends on many factors that are more difficult to adequately quantify than with other gear types. For example, lubrication is a critical concern, especially since the wear mechanism is abrasive wear instead of surface pitting, as for other gear types. The ability to remove wear particles from the lubricant, the operating temperature of the lubricant, the heat removed from the worm, the number of stops and starts, the frequency and severity of dynamic loadings, etc., are all important considerations that are somewhat addressed in current worm gear rating equations, but are not robustly understood. Regardless, the AGMA equations have been demonstrated to work well for most worm gear applications.

AGMA (1992) gives the following formula for the input power rating for a worm gear set:

$$h_{pi} = \begin{cases} \dfrac{N W_t d_g}{126{,}000 Z} + \dfrac{v_t W_f}{33{,}000} & \text{English units} \\[2ex] \dfrac{\omega W_t d_g}{2Z} + v_t W_f & \text{SI units} \end{cases} \tag{15.43}$$

where $h_{pi}$ = rated input power, W (hp)

      $N$ = rotational speed of the worm, rpm

      $d_g$ = gear diameter, m (in)

      $Z$ = gear ratio, $Z = N_g / N_w$

      $W_t$ = tangential load on worm gear tooth, N (lb)

      $W_f$ = friction force on worm gear tooth, N (lb)

      $\omega$ = amgular speed of worm, rad/s

      $v_t$ = sliding velocity at mean worm diameter, m/s (ft/min), given by

$$v_t = \begin{cases} \dfrac{\pi N d_{wm}}{12 \cos \lambda} & \text{English units} \\[2ex] \dfrac{\omega d_{wm}}{2 \cos \lambda} & \text{SI units} \end{cases} \qquad (15.44)$$

and where $d_{wm}$ is the mean worm diameter in millimeters (inches) and $\lambda$ is the lead angle given by Equation (15.33). Equation (15.43) contains two terms; the first term gives the output power at the gear, while the second gives the power loss at the tooth mesh. Recall that the worm will be constructed of a harder material than that of the worm gear, and the system is designed so that the worm gear conforms to the worm. Thus, the critical component for power ratings is the worm gear.

The tangential load for use in Equation (15.43) is prescribed by AGMA as

$$W_t = \begin{cases} C_s d_{gm}^{0.8} b_w C_m C_v & \text{English units} \\[2ex] \dfrac{C_s d_{gm}^{0.8} b_w C_m C_v}{75.948} & \text{SI units} \end{cases} \qquad (15.45)$$

where $C_s$ = materials factor, as given in Figure 15.12 and Table 15.4 for bronze worm gears, $N/m^2$ ($lb/in^2$)

$d_{gm}$ = mean gear diameter, mm (in)

$b_w$ = face width of gear, mm (in), not to exceed $0.67 d_{gm}$

$C_m$ = ratio correction factor, given by

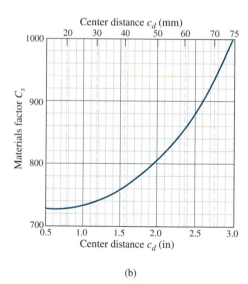

(a)                                                                (b)

**Figure 15.12**   Materials parameter $C_s$ for bronze worm gears and worm minimum surface hardness of 58 Rc. (a) Materials factor for center distances $c_d$ greater than 76 mm (3 in); (b) Materials factor for center distances $c_d$ less than 76 mm (3 in). When using the figure in (b), the value from part (a) should be checked, and the lower value used. See also Table 15.4. [From AGMA (1992)].

**Table 15.4**    Materials factor $C_s$ for bronze worm gears with worm having surface hardness of 58 Rc

| Manufacturing process | Pitch diameter | Units for Pitch Diameter | |
|---|---|---|---|
| | | in | mm |
| Sand casting | $d \leq 64$ mm (2.5 in) | 1000 | 1000 |
| | $d \geq 64$ mm | $1189.6365 - 476.5454 \log d$ | $1859.104 - 476.5454 \log d$ |
| Static chill cast or forged | $d \leq 200$ mm (8 in) | 1000 | 1000 |
| | $d > 200$ mm | $1411.6518 - 455.8259 \log d$ | $2052.012 - 455.8259 \log d$ |
| Centrifugally cast | $d \leq 625$ mm (25 in) | 1000 | 1000 |
| | $d > 625$ mm | $1251.2913 - 179.7503 \log d$ | $1503.811 - 179.7503 \log d$ |

| SOURCE: AGMA (1992).

$$C_m = \begin{cases} 0.0200 \left( -Z^2 + 40Z - 76 \right)^{0.5} + 0.46 & 3 \leq Z < 20 \\ 0.0107 \left( -Z^2 + 56Z + 5145 \right)^{0.5} & 20 \leq Z < 76 \\ 1.1483 - 0.00658Z & 76 \leq Z \end{cases} \quad \text{(15.46)}$$

$C_v$ = velocity correction factor, given by

$$C_v = \begin{cases} 0.659 \exp\left( -0.0011 v_t \right) & 0 < v_t \leq 700 \text{ ft/min} \\ 13.31 v_t^{-0.571} & 700 \text{ ft/min} < v_t \leq 3000 \text{ ft/min} \\ 65.52 v_t^{-0.774} & 3000 \text{ ft/min} < v_t \end{cases} \quad \text{(15.47)}$$

Note that the velocity in meters per second can be multiplied by 196.5 to obtain a velocity in feet per minute.

The friction force $W_f$ in Equation (15.41) is given by

$$W_f = \frac{\mu W_t}{\cos \lambda \cos \phi_n} \quad \text{(15.48)}$$

where $\mu$ = friction coefficient

$\lambda$ = lead angle at mean worm diameter, deg

$\phi_n$ = normal pressure angle at mean worm diameter, deg

For steel worms and bronze worm gears, AGMA offers the following expressions for the coefficient of friction:

$$\mu = \begin{cases} 0.150 & v_t = 0 \text{ ft/min} \\ 0.124 \exp\left( -0.074 v_t^{0.645} \right) & 0 < v_t \leq 10 \text{ ft/min} \\ 0.103 \exp\left( -0.110 v_t^{0.450} \right) + 0.012 & 10 \text{ ft/min} < v_t \end{cases} \quad \text{(15.49)}$$

The rated torque for a worm gear can be calculated from the tangential force $W_t$ by using Equation (15.40).

**EXAMPLE 15.4**

**Given:**   A worm with two starts and a face width of 1.0 in meshes with a 50-tooth worm gear with a pitch diameter of 8.0 in. The worm has a pitch diameter of 2.0 in and a pressure angle of 20°. The worm gear is sand cast from bronze, and the worm is heat-treated to a hardness of 65 Rc.

**Find:** Determine the rated input horsepower for the gear set if the worm is driven at 2000 rpm.

**Solution:** The rated horsepower is given by Equation (15.43), but a number of factors need to be determined before this equation can be used. Note that the gear ratio is

$$Z = \frac{N_g}{N_w} = \frac{50}{2} = 25$$

The pitch of the worm gear (which is also the pitch of the worm) is

$$p = \frac{\pi d_g}{N_g} = \frac{\pi (8.0 \text{ in})}{50} = 0.503 \text{ in}$$

The lead is given by Equation (15.32) as

$$L = N_w p = (2)(0.503 \text{ in}) = 1.006 \text{ in}$$

Therefore, the lead angle is calculated from Equation (15.33):

$$\lambda = \tan^{-1} \frac{L}{\pi d_{pw}} = \tan^{-1} \frac{1.006 \text{ in}}{\pi (2.0 \text{ in})} = 9.096°$$

Therefore, from Equation (15.44), the pitch-line velocity in feet per minute is calculated as

$$v_t = \frac{\pi N d_w}{12 \cos \lambda} = \frac{\pi (2000)(2.0)}{12 \cos 9.096°} = 1061 \text{ ft/min}$$

From Table 15.4 for a sand cast bronze worm gear,

$$C_s = 1189.6365 - 476.5454 \log d_g = 1189.6365 - 476.5454 \log 8 = 759.3$$

From Equation (15.46),

$$C_m = 0.0107 \left(-Z^2 + 56Z + 5145\right)^{0.5} = 0.0107[-(25)^2 + 56(25) + 5145]^{0.5} = 0.8233$$

Since $v_t = 1061$ ft/min, Equation (15.47) gives

$$C_v = 13.31(v_t)^{-0.571} = 0.2492$$

Therefore, the tangential load on the worm gear, $W_t$, is given by Equation (15.45) as

$$W_t = C_s d_g^{0.8} b_w C_m C_v = (759.3)(8)^{0.8}(1.0)(0.8233)(0.2492) = 822.2 \text{ lb}$$

The coefficient of friction is given by Equation (15.49) as

$$\mu = 0.103 \exp\left(-0.110 w_t^{0.45}\right) + 0.012 = 0.103 \exp[-0.110(1061)^{0.45}] + 0.012 = 0.0202$$

Therefore, the friction force is given by Equation (15.48) as

$$W_f = \frac{\mu W_t}{\cos \lambda \cos \phi} = \frac{(0.0202)(822.2 \text{ lb})}{\cos 9.096° \cos 20°} = 17.91 \text{ lb.}$$

So the rated horsepower, from Equation, (15.43), is

$$h_{pi} = \frac{N W_t d_g}{126{,}000 Z} + \frac{v_t W_f}{33{,}000}$$

$$= \frac{(2000)(822.2)(8)}{(126{,}000)(25)} + \frac{(1061)(17.91)}{33{,}000} = 4.752 \text{ hp}$$

Or, the rated input horsepower is roughly 4.75 hp.

---

**Case Study 15.1** | PORTION OF GEAR TRAIN FOR INDUSTRIAL FOOD MIXER

Figure 15.13 depicts a view of the gearing present in a common food mixer of the type found in restaurants and small bakeries. Different speeds can be achieved by engaging different gears with the drive, each gear with a prescribed number of teeth. The different speeds are required to perform various operations, such as kneading, mixing, and whipping. This problem involves the specification of a helical gear to effect mixing or kneading of dough.

**Given:** The motor is attached to a speed reducer whose output speed is 200 rpm and whose torque and speed curves are as shown in Figure 15.14. Use an allowable bending stress of 40 ksi and an allowable contact stress of 350 ksi.

**Figure 15.13** The gears used to transmit power from an electric motor to the agitators of a commercial mixer. (*Courtesy of Hobart.*)

## Case Study  (Continued)

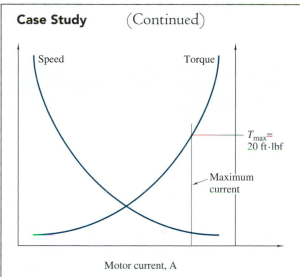

**Figure 15.14** Torque and speed of motor as a function of current for the industrial mixer used in Case Study 15.1.

**Find:** A helical gear mating pair, as shown in the figure, so that the output shaft speed is near 90 rpm. (This speed is then the input to a planet of a planetary gear train driving the agitator.)

**Solution:** The torque and speed characteristics are extremely important. If the mixer has been overloaded by using improper ingredients or by overfilling, a torque high enough to stall the motor may be developed. The highest torque that can be developed is that at the maximum current provided by the power supply, or 20 ft·lb. (The power supply circuitry is designed to limit the stall current through placement of resistive elements.) It is important that a stalled motor not lead to gear tooth breakage, since gears are a relatively expensive component and difficult to replace. Thus, the gears will be designed to survive the stall torque, even though ideally they should never be subjected to such high loading.

Given the required speed ratios, the number of teeth in the driven gears must be approximately 2.22 times the number of teeth in the driving gear. At this point an infinite number of gear tooth combinations will serve to fulfill the design requirements, since a face width has not been specified. However, because it is difficult to manufacture gears with few teeth, the driving gear will be arbitrarily

assigned 24 teeth; and the driven gear, 52 teeth. Note that this is a particular solution to a problem with many solutions.

This application is not so demanding as to require fine or ultrafine tooth classes; a coarse spacing may even be sufficient, but a medium-coarse spacing will be chosen as a balance between performance and cost. Using a diametral pitch of 12 per inch gives the gear pitch diameters as 2 and 4.32 in for the driving and driven gears, respectively. We will use standard values of $\phi = \psi = 20°$. The torque exerted onto the driveshaft causes a load of $W_t = T/r = T/1 = 240$ lb.

The gear and pinion must be designed based on both bending and contact stress based on Equations (15.13) and (15.14). Before these equations can be applied, the assorted factors need to be determined. From Table 14.8, we can reasonably assign a value of $K_a = 1.0$, since the power source is uniform, as is the machine operation. The size factor $K_s$, from Table 14.9, is clearly 1.00, since the diametral pitch is larger than 5. The load distribution factor is calculated from Equation 14.60, but first we note the following:

From Equation (14.55) using crowned teeth, $C_{mc} = 1.0$. From Equation (14.63), assuming $b_w \leq 1$ in,

$$C_{pf} = \frac{b_w}{10d_p} - 0.025$$

Note that if the face width needs to be larger than 1 in, a different relationship for $C_{pf}$ will be needed and this portion of the analysis will have to be repeated. From Equation (14.64), allowing flexibility for the design of bearing supports, we take $C_{pm} = 1.1$. From Figure (14.33),

$$C_{ma} = 0.127 + 0.0128b_w - 1.093 \times 10^{-4}b_w^2$$

And, finally, from Equation (14.65), $C_e = 1$, so that the gears don't need to be lapped or require complicated adjustment procedures. These can be substituted into Equation (14.60) to obtain the load distribution factor $K_m$ as a function of face width $b_w$:

$$K_m = 1.0 + (1.0)\left[\frac{b_w}{10(2\ \text{in})} - 0.025\right](1.1)$$
$$+ \left(0.127 + 0.0128b_w - 1.093 \times 10^{-4}b_w^2\right)(1.0)$$

*(continued)*

**Case Study**        (Concluded)

The pitch-line velocity is given by

$$v_t = \pi d_p N_{ap} = \pi(2\text{ in})(200\text{ rpm}) = 1256\text{ in/min}$$
$$= 105\text{ ft/min}$$

From Figure 14.34, all gears have $K_v \approx 1.0$ at this low speed, so we assign $K_v = 1$ and take $Q_v = 5$, although this may be modified at a later time with no design consequences, in order to reduce noise or for any other reason. Note that since the pinion is a nonidler gear, Section 14.11.4 suggests that we use $K_i = 1.0$. Similarly, this is not a large gear produced from a rim-and-spoke design, so we use $K_b = 1.0$. The geometry factor is obtained by interpolation from Table 15.2(a) as $Y_h = 0.51$.

Since $\sigma_t$ was defined as 40 ksi, Equation (15.13) gives

$$\sigma_t = \frac{W_t p_d K_a K_s K_m K_v K_i K_b}{b_w Y_h}$$

$$= \frac{(240)(12)(1.0)(1.0)\left[1 + (b_w/20 - 0.025)(1.1) + 0.127 + 0.0128b_w - 1.093 \times 10^{-4} b_w^2\right]}{b_w(0.51)}$$

This equation is solved numerically as $b_w = 0.15$ in, but since this is an awkward dimension, a face width of $b_w = 0.25$ in is selected. Note that the correction factors $C_{pf}$ and $K_m$ as calculated above are correct for this face width.

It is now important to determine a face width for acceptable contact stress. Equation (15.14) gives the contact stress, and many of the factors calculated above can be used directly. However, some additional values are needed before we solve Equation (15.14). First, from Table 15.2(a), $I_h$ is obtained as 0.187 by interpolation. From Equation (8.15), noting that for steel $E = 30 \times 10^6$ psi and $\nu = 0.30$,

$$E' = \frac{E}{1 - \nu^2} = \frac{30 \times 10^6 \text{ psi}}{1 - (0.30)^2} = 32.96 \times 10^6 \text{ psi}$$

From Equation (14.72),

$$\frac{1}{R_x} = \left(\frac{1}{r_p} + \frac{1}{r_g}\right)\frac{1}{\sin\phi} = \left(\frac{1}{2\text{ in}} + \frac{1}{4.32\text{ in}}\right)\frac{1}{\sin 20°}$$

or $R_x = 0.4676$ in. The applied load is $W = W_t/\cos\phi = 240/\cos 20° = 255$ lb. Therefore, from Equation (14.71),

$$p_H = E'\left(\frac{W}{2\pi b_w R_x E'}\right)^{1/2}$$

$$= (32.96 \times 10^6 \text{ psi})\left[\frac{255}{2\pi b_w(32.96 \times 10^6)(0.4676)}\right]^{1/2}$$

$$= 94,800(b_w)^{-1/2}$$

where $p_H$ is in pounds per square inch and $b_w$ is in inches. Therefore, we can apply Equation (15.14) to obtain the contact stress, which was restricted to a maximum value of 350 ksi:

$$\sigma_c = 350,000\text{ psi} = p_H\left(\frac{K_a K_s K_m K_v}{I_h}\right)$$

$$= [94,800(b_w)^{-1/2}\text{ psi}]\left[\frac{(1)(1)(1.1)(1)}{0.187}\right]^{1/2}$$

This is solved as $b_w = 0.137$ in. Therefore, a face width of 0.25 in is selected to complete the design.

This problem is typical of most design problems in that an infinite number of solutions exist. This case study merely demonstrates the design procedure used to obtain a reasonable answer, not the "right" answer.

## 15.5    SUMMARY

This chapter extended the discussion of gears beyond spur gears addressed in Chapter 14 to helical, bevel, and worm gears. All these gears have their advantages, with the main limitation being higher cost compared to spur gears.

Helical gears are similar to spur gears, but the tooth geometry is wound in a helix around the outside of the gear cylinder. Helical gears have the main advantage over spur gears of being quieter, especially at high speeds. Bevel gears are quiet and efficient, and they do not require parallel shafts. Worm gear sets consist of a worm, which is similar to a helical gear with a very small helix angle, and a worm gear that is similar to a spur gear. Worm gears are commonly used for compact designs that achieve large speed reductions, and they are unique among gears in that their dominant wear mechanism is abrasive wear.

The particular kinematics and dynamics of these gears were analyzed by using first principles, and design methodologies standardized by the American Gear Manufacturers Association were briefly outlined. These robust design approaches allow for the selection of gears and the analysis of specific gear geometries to ensure long service lives.

## KEY WORDS

**bevel gears**   nonparallel, coplanar gears with straight, hypoid, spiral, or Zerol teeth.

**circular pitch**   distance from one tooth to corresponding point on adjacent tooth.

**helical gears**   parallel-axis gears with teeth cut on helix that wraps around cylinder.

**gear**   toothed wheel that, when mated with another toothed wheel, transmits power between shafts.

**gear ratio**   ratio of number of teeth between two meshing gears.

**hypoid bevel gear**   a spiral gear set with zero offset.

**lead**   axial distance advanced in one revolution.

**lead angle**   relates the lead to pitch circumference.

**module**   ratio of pitch diameter to number of teeth.

**normal circular pitch**   in a plane normal to the tooth, the distance from one tooth to corresponding point on the adjacent tooth.

**normal pitch**   number of teeth per pitch diameter in a direction normal to tooth face.

**pinion**   smaller of two meshed gears.

**self-locking**   meshing gears friction is high enough to prevent movement when no load or torque is applied.

**spiral bevel gears**   nonparallel, coplanar gears with teeth cut on a spiral.

**straight bevel gears**   nonparallel, coplanar gears with straight teeth.

**starts**   number of threads.

**worm gear**   nonparallel, noncoplanar gear with typically high reduction ratio.

**Zerol bevel gears**   similar to straight bevel gears, but with curved tooth in the lengthwise direction and zero spiral angle.

## RECOMMENDED READINGS

Crosher, W. P. (2002) *Design and Application of the Worm Gear,* ASME, New York.
Dudley, D. W. (1994) *Handbook of Practical Gear Design,* CRC Press, Boca Raton, FL.

Howes, M. A., ed. (1980) *Source Book on Gear Design Technology and Performance,* American Society for Metals, Metals Park, OH.

Juvinall, R. C., and Marshek, K. M. (1991) *Fundamentals of Machine Component Design,* Wiley, New York.

Litvin, F. L. (1994) *Gear Geometry and Applied Theory,* Prentice-Hall, Englewood Cliffs, NJ.

Mechalec, G. W. (1962) *Gear Handbook,* McGraw-Hill, New York, chapter 9.

Mott, R. L. (1985) *Machine Elements in Mechanical Design,* Merrill Publishing Co., Columbus, OH.

Norton, R. L. (1996) *Machine Design,* Prentice-Hall, Englewood Cliffs, NJ.

Shigley, J. E., and Mischke, C. R. (1989) *Mechanical Engineering Design,* McGraw-Hill, New York.

Stokes, A. (1992) *Gear Handbook,* Butterworth Heinemann, Oxford, England.

Townsend, D. P. (1992) *Dudley's Gear Handbook,* 2nd ed., McGraw-Hill, New York.

# REFERENCES

AGMA (1989) *Geometry Factors for Determining the Pitting Resistance and Bending Strength of Spur, Helical, and Herringbone Gear Teeth,* ANSI/AGMA Standard 908-B89, American Gear Manufacturers Association, Alexandria, VA.

AGMA (1990) *Gear Nomenclature, Definitions of Terms with Symbols,* ANSI/AGMA Standard 1012-F90, American Gear Manufacturers Association, Alexandria, VA.

AGMA (1992) *Practice for Enclosed Cylindrical Wormgear Speed Reducers and Gearmotors,* ANSI/AGMA Standard 6034-B92, American Gear Manufacturing Association, Alexandria, VA.

AGMA (1993) *Design Manual for Cylindrical Wormgearing,* ANSI/AGMA Standard 6022-C93, American Gear Manufacturers Association, Alexandria, VA.

AGMA (1997) *Rating the Pitting Resistance and Bending Strength of Generated Straight Bevel, Zerol Bevel and Spiral Bevel Gear Teeth,* ANSI/AGMA Standard 2003-B97, American Gear Manufacturing Association, Alexandria, VA.

AGMA (1999) *Fundamental Rating Factors and Calculation Methods for Involute Spur and Helical Gear Teeth,* ANSI/AGMA Standard 2101-C95, American Gear Manufacturers Association, Alexandria, VA.

Dudley, D. W. (1994) *Handbook of Practical Gear Design,* McGraw-Hill, New York.

# PROBLEMS

## Section 15.2

**15.1**  For gear transmissions there is a direct proportionality between the local sliding speeds experienced by the contacting surfaces and the power loss. For a normal spur or helical gear, the mean sliding speed is typically 20% of the load transmission speed. The coefficient of friction in lubricated gears is typically between 0.05 and 0.1. This leads to a power loss of 1% to 2% and a power efficiency of 99% to 98%. For a hypoid gear in the rear axle of a car, the sliding speed can be 60% of the load transmission speed, and for a worm gear it can be 1200%. Estimate the power efficiencies for a hypoid gear and a worm gear. *Ans.* $e_{hypoid} = 94\%$ to 97%, $w_{worm} = 45.5\%$ to 62%.

**15.2**  A 16-tooth, commercial enclosed ($Q_v = 6$) right-handed helical pinion turns at 2000 rpm and transfers 8 kW to a 38-tooth gear. The helix angle is $30°$, the pressure angle is $25°$, and the normal module is 2 mm. For a safety factor of 3 for bending and 1 for contact stress, calculate the pitch diameters; the normal, axial, and tangential pitch; the normal pressure angle; and the face width. The material is 2.5% chrome steel (Grade 2), hardened to 315 HB. Assume $K_a = 1.0$.

**15.3** Two meshing commercial enclosed helical gears are made of steel (allowable bending stress = 60 ksi and allowable contact stress = 200 ksi) and are mounted on parallel shafts 10 in apart. The desired velocity ratio is 0.35, with a diametral pitch of 8.47 in$^{-1}$, and the gears have a 2-in face width, with a 30° helix angle and 30° pressure angle. What is the maximum horsepower that can be safely transmitted at a pinion speed of 1750 rpm? Use $Q_v = 9$.

**15.4** Two helical gears have shafts at 90° and helix angles of 45°. The speed ratio is 3:1, and the gears have a normal module of 5. Find the center distance if the pinion has 20 teeth. *Ans.* $c_d = 282.8$ mm.

**15.5** A 25-tooth pinion transmits 5 hp at 8000 rpm to a 100-tooth gear. Both gears have $\phi = \psi = 30°$ and a face width of 0.50 in. The pinion and gear are constructed of grade 2 through-hardened steel. Determine the minimum diameter and the required Brinell hardness based on bending stress.

**15.6** A 25-tooth helical gear with a 20° pressure angle in the plane of rotation has the helix angle $\psi = 25°$. What are the pressure angle in the normal plane $\phi_n$ and the equivalent number of teeth $N_n$? *Ans.* $N_n = 33.58$ teeth.

**15.7** Two meshing helical gears on parallel shafts have 25 teeth and 43 teeth, respectively, and have the helix angle $\psi = 30°$. The circular pitch is $p_c = 6.5$ mm and $\phi = 20°$. How wide do the gears have to be to obtain a total contact ratio of 2.5? *Ans.* $b_w = 6.68$ mm.

## Section 15.3

**15.8** A bevel gear with a 60° pitch angle and 40 teeth meshes with a pinion with a 30° pitch angle and 20 teeth. Calculate the virtual number of teeth for the gear and the pinion. *Ans.* $N_{vg} = 80$ teeth, $N_{vp} = 23.1$ teeth.

**15.9** Two 20° pressure angle straight bevel gears have a diametral pitch of 4 in$^{-1}$ at the outside radius, with a 2-in face width. The pinion has 20 teeth, and the gear has 40. The pinion is driven at 1000 rpm. Determine the maximum horsepower that can be transmitted by the gears. The allowable contact stress is 230 ksi and the allowable bending stress is 35 ksi. Both gears are made from steel, straddle mounted, properly crowned and have $Q_v = 11$.

**15.10** A 90° straight bevel gear set is needed for a 6:1 speed reduction. Determine the gear forces if the 20° pressure angle pinion has 20 teeth with a module of 3, and 750 W is transmitted at a pinion speed of 1000 rpm. *Ans.* $W_t = 238.7$ N, $W_a = 86.88$ N.

**15.11** Determine the safety factor of the gear set described in Example 15.2 based on analysis of the gear instead of the pinion.

**15.12** Two Zerol bevel gears with 90° shaft angles and mean cone distance of 80 mm have a face width of 20 mm, a module of 20 mm, properly crowned teeth and were produced with a cutter radius of 110 mm. The gears have $\phi = 20°$, $\psi = 25°$, $N_p = 15$, and $N_g = 40$; and both pinion and gear are straddle-mounted. If the allowable contact stress is 650 MPa and the allowable bending stress is 205 MPa, find the maximum power rating for the gear set if the pinion is driven at 1000 rpm. Assume $K_a = 1.0$ and $Q_v = 8$.

**15.13** A pair of bevel gears are mounted on shafts that are 90° apart, and are to transmit 10 hp at 300 rpm. The pinion has a 6-in outside pitch diameter, 2-in face width, and has $\phi = 20°$. For a velocity ratio of 0.5, draw a free-body diagram of the gears showing all forces. What is the torque produced about the gear axis?

**15.14** Determine the required bending and contact strength of the gear materials for Problem 15.13.

### Section 15.4

**15.15** A worm with two starts has a lead angle of 8° and meshes with a worm gear having 50 teeth and a module of 7. Find the center distance between the shafts. *Ans.* $c_d = 222.6$ mm.

★ **15.16** The efficiency $\eta$ of a gear set is the ratio of the output to input power. Obtain an expression for the efficiency of a worm gear set based on the AGMA approach. Using the gear set of Example 15.3, plot the efficiency of the gear set as a function of velocity. Repeat the analysis if the worm gear has (a) 40 teeth and (b) 50 teeth.

**15.17** A worm with two starts has a lead angle of 10°, a pressure angle of 35°, and a mean diameter of 75 mm. The worm is driven by a 2-kW motor at 1500 rpm. The worm gear has 40 teeth. Calculate the worm and worm gear tangential, separation, and thrust forces. *Ans.* $W_{tw} = 339.5$ N, $W_s = 1200$ N, $W_{aw} = 1688$ N.

**15.18** A triple-threaded worm has a lead angle of 12° and a pitch diameter of 2.281 in. Find the center distance when the worm is mated with a worm gear of 40 teeth.

**15.19** A single-threaded steel worm meshes with a 42-tooth forged bronze worm gear. If the pitch diameter of the worm gear is 3 in, the center distance is 12 in, the pressure angle is 20°, and the pinion turns at 860 rpm, find the worm gear diameter and maximum input horsepower if the facewidth is 2 in.

**15.20** A double-threaded worm gear has a worm pitch diameter of 50 mm, a pitch of 10 mm, and a speed reduction of 25:1. Find the lead angle, the worm gear diameter, and the center distance.

**15.21** For the worm gear in Problem 15.20, determine the forces acting on the worm and worm gear if the transmitted torque is 100 in · lb at 1000 rpm.

★ **15.22** Design a triple-threaded worm gear set to deliver 15 hp from a shaft rotating at 1500 rpm to another rotating at 75 rpm. Use $\phi = 25°$ and $\psi = 20°$, a forged bronze worm gear, and a maximum center distance of 15 in.

★ **15.23** A bakery carousel oven consists of six trays, 10 ft by 3 ft, mounted onto a wheel and rotated within a heated enclosure. It is desired to attain a rotational speed of the carousel of 2 rpm, while a 0.5-hp motor to be used to power the oven has an operating speed of 500 rpm. Design a worm gear set, including specification of worm and gear materials, numbers of teeth, face width, numbers of starts, and center distance. How large an enclosure would be needed for this gear set? Note that the motor can be mounted outside the oven, so no thermal effects need to be considered. Assume $C_m = 0.10$.

★ **15.24** Repeat Problem 15.23, using a helical planetary gear system in the speed reducer. Comment on the strengths and weaknesses of each design.

| * Indicates problem of greater difficulty.

# FASTENERS AND POWER SCREWS

A collection of threaded fasteners. (Courtesy of Clark Craft Fasteners)

*Engineers need to be continually reminded that nearly all engineering failures result from faulty judgments rather than faulty calculations.*
Eugene S. Ferguson, *Engineering and the Mind's Eye*

## SYMBOLS

| | |
|---|---|
| $A$ | cross-sectional area, m$^2$ |
| $A_g$ | gasket area per bolt, m$^2$ |
| $A_i$ | constant |
| $A_s$ | shear stress area, m$^2$ |
| $A_t$ | tensile stress area, m$^2$ |
| $a$ | distance from wall to applied load, m |
| $B_i$ | constant |
| $b$ | member width, m |
| $C_k$ | dimensionless stiffness parameter |
| $c$ | distance from neutral axis to outer fiber, m |
| $d$ | major diameter, largest possible diameter of thread, m |
| $d_c$ | crest diameter, largest actual diameter of thread, m |
| $d_i$ | smallest diameter of frustum cone, m |
| $d_p$ | pitch diameter, m |
| $d_r$ | root diameter, smallest diameter of thread, m |
| $d_s$ | distance traveled in axial direction, m |
| $d_w$ | washer diameter, m |
| $E$ | modulus of elasticity, Pa |
| $e$ | efficiency of raising load, % |
| $e_k$ | elongation, m |
| $h$ | maximum cantilever thickness, m |
| $h_e$ | weld leg length, m |
| $h_p$ | power, W |
| $h_t$ | largest possible thread height, m |
| $I$ | area moment of inertia, m$^4$ |
| $I_u$ | unit area moment of inertia, m$^3$ |
| $J$ | polar area moment of inertia, m$^4$ |
| $J_u$ | unit polar area moment of inertia, m$^3$ |
| $K_f$ | fatigue stress concentration factor |
| $k$ | stiffness, N/m |
| $k_f$ | surface finish factor |
| $k_m$ | miscellaneous factor |
| $k_r$ | reliability factor |
| $k_s$ | size factor |
| $L$ | length, m |
| $L_g$ | grip length, m |
| $L_s$ | shank length, m |
| $L_{se}$ | shank effective length, m |
| $L_t$ | threaded length, m |
| $L_{te}$ | effective threaded length, m |
| $L_w$ | weld length, m |
| $l$ | lead, m |
| $M$ | bending moment, N · m |
| $m$ | 1, 2, 3, for single-, double-, and triple-threaded screws, respectively |

| | |
|---|---|
| $N_a$ | rotational speed, rpm |
| $N_r$ | number of rivets in width of member |
| $n$ | number of threads per inch, in$^{-1}$ |
| $n_0$ | number of revolutions |
| $n_s$ | safety factor |
| $P$ | load; friction force; deflection force; stretching force; press force; shear force, N |
| $P_i$ | preload, N |
| $P_n$ | load acting on diagonal of parallelepiped, N |
| $P_0$ | load neglecting friction, N |
| $p$ | pitch, in |
| $p_0$ | minimum gasket seal pressure, Pa |
| $r$ | distance from centroid of weld group to farthest point in weld, m |
| $S_e$ | modified endurance limit, Pa |
| $S_e'$ | endurance limit, Pa |
| $S_p$ | proof strength, Pa |
| $S_y$ | yield strength, Pa |
| $S_u$ | ultimate tensile strength, Pa |
| $T$ | torque, N · m |
| $T_l$ | torque needed to lower load, N · m |
| $T_r$ | torque needed to raise load, N · m |
| $t_e$ | weld throat length, m |
| $t_m$ | thickness of thinnest member, m |
| $V$ | shear force, N |
| $W$ | load; mating force; tensile force, N |
| $w'$ | force per unit length of weld, N/m |
| $x, y$ | coordinates, m |
| $\bar{x}, \bar{y}$ | centroid, m |
| $y$ | maximum cantilever deflection, m |
| $Z_m$ | section modulus, m$^3$ |
| $\alpha$ | lead angle; wedge angle of cantilever tip; angle of inclination, deg |
| $\alpha_f$ | angle of frustum cone stress representation, deg |
| $\beta$ | thread angle, deg |
| $\gamma$ | cone angle, deg |
| $\epsilon$ | maximum allowable strain |
| $\theta_n$ | angle created by force $P_n$ for diagonal of parallelepiped shown in Figure 16.7(a), deg |
| $\mu$ | coefficient of friction between threads |
| $\mu_c$ | coefficient of friction of collar |
| $\sigma$ | normal stress, Pa |
| $\tau$ | shear stress, Pa |
| $\omega$ | rotational speed, rad/s |

## SUBSCRIPTS

| | |
|---|---|
| $a$ | alternating |
| all | allowable |

| | | | | |
|---|---|---|---|---|
| $b$ | bolt | | max | maximum |
| $c$ | collar | | min | minimum |
| cr | critical | | $n$ | normal, nominal |
| $D$ | dynamic | | $o$ | outer |
| $d$ | direct | | $p$ | proof |
| $g$ | gasket | | $s$ | shear; secant |
| $i$ | preload; inner | | $t$ | thread; torsional |
| $j$ | joint | | $x, y$ | coordinates |
| $m$ | mean | | | |

## 16.1 INTRODUCTION

The primary focus of this chapter is on fasteners. Since the complications of manufacturing intricate parts often require assembly of components, engineers are confronted with the task of fastening various members together. A number of fastener types can be used for this task, including threaded fasteners, riveted joints, welded joints, and adhesive joints. Each of these types is considered in this chapter. Various fastening methods are presented, and information is given about selecting a suitable type for a specific task. Interesting information about fastener materials indicates that the bolt, rivet, or weld material should be strong and tough whereas the members being fastened should ideally be soft and ductile. This chapter also considers power screws.

## 16.2 THREAD TERMINOLOGY, CLASSIFICATION, AND DESIGNATION

Each machine element has its unique terminology. Figure 16.1 describes the terminology and dimensions of threaded parts. Note in particular the differences between the various diameters (major, crest, pitch, and root), the largest possible thread height $h_t$, and the thread angle $\beta$. The **pitch** $p$ shown in Figure 16.1 is the distance from a point on one thread to the same point on an adjacent thread, and its unit is meters or inches. A parameter that can be used instead of pitch is **threads per inch** $n$. An important relationship between pitch and threads per inch is

$$p = \frac{1}{n} \tag{16.1}$$

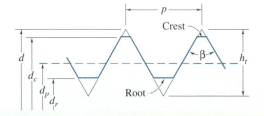

**Figure 16.1** Parameters used in defining terminology of thread profile.

**Figure 16.2**    (a) Single-, (b) double-, and (c) triple-threaded screws.

Another thread parameter is the **lead,** which is the distance that the screw would advance relative to the nut in 1 revolution. For a single-threaded screw $l = p$, for a double-threaded screw $l = 2p$, etc. Figure 16.2 shows the differences between single-, double-, and triple-threaded screws.

A number of different thread profiles can be used for a wide variety of applications. Figure 16.3 shows two types. The Acme thread profile is used for power screws and machine tool threads. The second profile, the unified (UN), is used extensively. The Acme profile has a thread angle of 29°, whereas the UN profile has a thread angle of 60°. The metric profile (M) is popular and is quite similar to the UN profile.

Figure 16.4 shows more details of the M and UN thread profiles than are shown in Figure 16.3. From this figure

$$h_t = \frac{0.5p}{\tan 30°} = 0.8660p \tag{16.2}$$

Once the pitch $p$ and the largest possible thread height $h_t$ are known, the various dimensions of the UN and M thread profiles can be obtained. This text focuses on the UN thread profile while using English units and the M profile for metric units. Information obtained from these thread profiles can be directly applied to other thread profiles.

The term **thread series** can be applied to threads of any size. The UN thread profile has eight thread series of constant pitch. Each thread series has the same number of threads

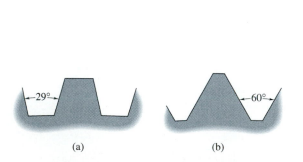

**Figure 16.3**    Thread profiles. (a) Acme; (b) UN.

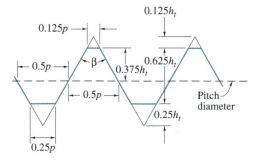

**Figure 16.4**    Details of M and UN thread profiles.

per inch. The helix angle that the thread makes around the fastener varies with the fastener diameter. However, the thread depth is constant, regardless of the fastener diameter, because the 60° thread angle remains constant. This situation can be clearly observed from Equation (16.2). The eight constant-pitch UN thread series are 4, 6, 8, 12, 16, 20, 28, and 32 threads per inch.

In addition to thread series, thread profiles are classified by **coarseness,** which refers to the quality and number of threads per inch produced on a common diameter of the fastener. These designations after UN have the following meanings:

1. C—coarse-pitch threads
2. F—fine-pitch threads
3. EF—extrafine-pitch threads

The coarseness designation is followed by the crest diameter in inches and the number of threads per inch. Thus, UNF $\frac{1}{2} \times 16$ means a UN thread profile with fine-pitch threads, a crest diameter of $\frac{1}{2}$ in, and 16 constant-pitch threads per inch.

For metric thread profiles the coarseness designation is usually considered only as coarse or fine, omitting the extrafine designation. Instead of using threads per inch, the metric thread series simply uses the pitch distance between two threads in millimeters. For example, MF $6 \times 1$ means an M thread profile with fine-pitch threads, a crest diameter of 6 mm, and a pitch distance of 1 mm.

The classifications given above are only applicable for individual threads and do not consider how the male and female parts of the fastener fit together. That is, is the fit loose and sloppy, or is it tight? In English units the letter A designates external threads and B designates internal threads. There are three fits: 1 (the loosest fit), 2 (a normal fit), and 3 (a tight fit). Thus, for a very dirty environment a 1 fit would be specified; a 3 fit requires an extra degree of precision and a very clean environment. A 2 fit is usually designated for normal applications, somewhere between the extremes of 1 and 3. Thus, the designation UNC $2 \times 8 - 1B$ means a UN thread profile with coarse-pitch threads, a crest diameter of 2 in, 8 constant-pitch threads per inch, and a normal fit, with the nut of the fastener being specified.

The fit designation has more options in metric units than in English units. There G and H denote internal threads, and e, f, g, and h denote external threads. For example, G and e define the loosest fit and the greatest clearance, and H and h define zero allowance, no deviation from the basic profile. Thus, a very dirty environment would require a G and e fit and an extra degree of precision, and a very clean environment would require an H and h fit.

The tolerance grade for metric threads is denoted by a number symbol. Seven tolerance grades have been established and are identified by numbers 3 to 9. Nine defines the loosest fit (the most generous tolerance) and 3 defines the tightest. Thus, MF $8 \times 2 - G6$ means an M thread profile with fine-pitch threads, a crest diameter of 8 mm, a pitch of 2 mm, and a normal fit of the internal threads.

Table 16.1 gives approximate equivalent fits of English and metric unit threads. This table does not show all possible equivalencies but does give an idea of how the two systems work.

**Table 16.1**    Inch and metric equivalent thread classifications

| Inch Series | | Metric Series | |
|---|---|---|---|
| **Bolts** | **Nuts** | **Bolts** | **Nuts** |
| 1A | 1B | 8g | 7H |
| 2A | 2B | 6g | 6H |
| 3A | 3B | 8h | 5H |

**EXAMPLE 16.1**

**Given:**   A screw thread is going to be chosen for a nuclear power plant pressure vessel lid. The outside diameter of the screws is limited to 160 mm, and the length of the threaded part screwed into the pressure vessel wall is 220 mm. For metric threads M160 three different pitches are preferred: 6, 4, and 3 mm. The manufacturing tolerance for the screw thread outer diameter is 0.6 mm, and the difference in thermal expansion of the pressure vessel wall relative to the screw is 0.2 mm because the screw heats more slowly than the wall during start-up. The manufacturing tolerance for the thread in the pressure vessel wall is 0.9 mm.

**Find:**   Which pitch should be chosen?

**Solution:**   The highest mean shear stress in the thread is equal to the force divided by $\pi$ times the root diameter of the screw (if there is no play between the screw and nut) times the length of the thread. If there is play between the screw and nut, a smaller part consisting of only the tops of the threads needs to be sheared off; and if a weaker material is used in the pressure vessel than in the screws, that material will break before the screw threads.

If the pitch is 3 mm, the height of the thread profile is

$$(0.625)\left(\frac{\sqrt{3}}{2}\right)(3) = 1.624 \text{ mm}$$

which is less than the combined thermal expansion and manufacturing error. For maximum safety the largest pitch of 6 mm should be chosen.

## 16.3   POWER SCREWS

**Power screws** are used to change angular motion into linear motion and usually to transmit power. More specifically, power screws are used

1.   To obtain greater mechanical advantage in order to lift weight, as in a screw type of jack for cars

2.   To exert large forces, as in a home compactor or a press

3.   To obtain precise positioning of an axial movement, as in a micrometer screw or the lead screw of a lathe

In each of these applications a torque is applied to the ends of the screws through a set of gears, thus creating a load on the device.

**Figure 16.5** Details of Acme thread profile. (All dimensions are in inches.)

Power screws use the Acme thread profile, first shown in Figure 16.3(a). Figure 16.5 shows more details of Acme threads. The thread angle is 29°, and the thread dimensions can be easily determined once the pitch is known.

Table 16.2 gives the crest diameter, the number of threads per inch, and the tensile and shear stress areas for Acme power screw threads. The tensile stress area is

$$A_t = \frac{\pi}{4}\left(\frac{d_r + d_p}{2}\right)^2 \qquad (16.3)$$

The shear stress area $A_s$ is also given in Table 16.2, but a simple formulation of it is not readily available. Note that $A_s$ represents the area in shear approximately at the pitch line of the threads for a 1.0-in length of engagement. The length of engagement is the axial length over which the threads are engaged with each other. Other lengths than 1.0 in would require that the shear stress area be modified by the ratio of the actual length to 1.0 in. Because of this, the modified shear stress area differs considerably from the tensile stress area. Table 16.2 and Figure 16.5 should be used jointly in evaluating Acme power screws. The pitch diameter of an Acme power screw thread is

$$d_p = d_c - 0.5p - 0.01 \qquad \text{or} \qquad d_p = d_c - 0.5p - 0.254 \qquad (16.4)$$

The equation to the left is valid only when inches are used for $d_c$ and $p$, while the equation to the right is valid only when millimeters are used for $d_c$ and $p$.

### 16.3.1 FORCES AND TORQUE

Figure 16.6 shows a load $W$ into which the supporting screw is threaded and which can be raised or lowered by rotating the screw. It is assumed that the load $W$ is prevented from turning when the screw rotates. Also shown in Figure 16.6 is the thread angle of an Acme power screw, which is 29°, and the lead angle. The lead angle $\alpha$ relates the lead to the pitch circumference through the equation

$$\alpha = \tan^{-1}\frac{l}{\pi d_p} \qquad (16.5)$$

where

$$l = \text{lead} = mp \qquad (16.6)$$

**Table 16.2**    Crest diameters, threads per inch, and stress areas for Acme threads

| Crest diameter $d_c$, in | Number of threads per inch $n$ | Tensile stress area $A_t$, in$^2$ | Shear stress area $A_s$, in$^2$ |
|---|---|---|---|
| $\frac{1}{4}$ | 16 | 0.02632 | 0.3355 |
| $\frac{5}{16}$ | 14 | 0.04438 | 0.4344 |
| $\frac{3}{8}$ | 12 | 0.06589 | 0.5276 |
| $\frac{7}{16}$ | 12 | 0.09720 | 0.6396 |
| $\frac{1}{2}$ | 10 | 0.1225 | 0.7278 |
| $\frac{5}{8}$ | 8 | 0.1955 | 0.9180 |
| $\frac{3}{4}$ | 6 | 0.2732 | 1.084 |
| $\frac{7}{8}$ | 6 | 0.4003 | 1.313 |
| 1 | 5 | 0.5175 | 1.493 |
| $1\frac{1}{8}$ | 5 | 0.6881 | 1.722 |
| $1\frac{1}{4}$ | 5 | 0.8831 | 1.952 |
| $1\frac{3}{8}$ | 4 | 1.030 | 2.110 |
| $1\frac{1}{2}$ | 4 | 1.266 | 2.341 |
| $1\frac{3}{4}$ | 4 | 1.811 | 2.803 |
| 2 | 4 | 2.454 | 3.262 |
| $2\frac{1}{4}$ | 3 | 2.982 | 3.610 |
| $2\frac{1}{2}$ | 3 | 3.802 | 4.075 |
| $2\frac{3}{4}$ | 3 | 4.711 | 4.538 |
| 3 | 2 | 5.181 | 4.757 |
| $3\frac{1}{2}$ | 2 | 7.338 | 5.700 |
| 4 | 2 | 9.985 | 6.640 |
| $4\frac{1}{2}$ | 2 | 12.972 | 7.577 |
| 5 | 2 | 16.351 | 8.511 |

where $m = 1$, single-threaded screw
$\quad m = 2$, double-threaded screw
$\quad m = 3$, triple-threaded screw
$\quad d_p$ = pitch diameter, m

Figure 16.6 also shows a thrust collar, which is covered later in this chapter.
The distance traveled in the axial direction can be expressed as

$$d_s = n_0 l = n_0 mp \tag{16.7}$$

where $n_0$ = number of revolutions. Thus, the power screw uses rotary motion to obtain a uniform axial linear motion.

### Raising the Load

To determine the force $P$ required to overcome a certain load $W$, it is necessary to observe the relationship of the load to the distance traveled. The total force on the threads is represented

**Figure 16.6** Dimensions and angles of power screw with collar.

by a single source $P_n$, shown in Figure 16.7(a), that is normal to the thread surface. From Figure 16.7 recall the representation of the lead angle $\alpha$ and the thread angle $\beta$. The angle $\theta_n$ is created by the force $P_n$, which is the diagonal of the parallelepiped. Side ABE0 is an axial section through the bolt and is shown in Figure 16.7(b). Note from Figure 16.7(a) and (b) that the sides DC, AB, and 0E of the parallelepiped are equal. Thus, the following must be true:

$$\sin \theta_n = \cos \theta_n \cos \alpha \tan \frac{\beta}{2}$$

$$\theta_n = \tan^{-1} \left( \cos \alpha \tan \frac{\beta}{2} \right)$$

(16.8)

In Figure 16.7(a) the projection of load $P_n$ into plane ABE0 is inclined at an angle $\beta/2$, where $\beta$ is the thread angle.

Side ACH0 lies in the plane tangent to the pitch point. The projection of $P_n$ into this plane is inclined at the lead angle $\alpha$ calculated at the screw pitch radius and is shown in Figure 16.7(c). This figure contains not only the normal components but also the friction forces due to the screw and nut as well as the friction forces due to the collar.

Summing the vertical forces shown in Figure 16.7(c) gives

$$P_n \cos \theta_n \cos \alpha = \mu P_n \sin \alpha + W$$

$$P_n = \frac{W}{\cos \theta_n \cos \alpha - \mu \sin \alpha}$$

(16.9)

All thread forces act at the thread pitch radius $d_p/2$. From Figure 16.6 the force of the collar acts at radius $r_c$, the midpoint of the collar surface. The torque required to raise the load is

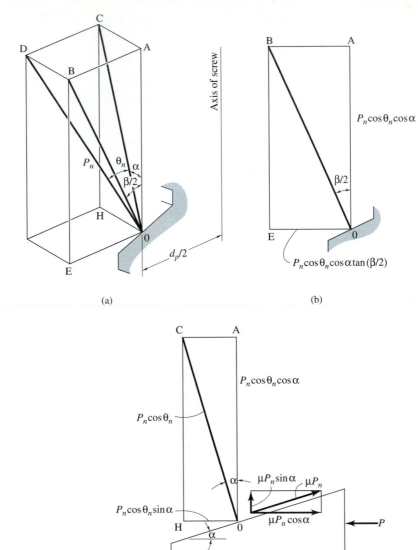

**Figure 16.7**    Forces acting in raising load of power screw. (a) Forces acting on parallelepiped; (b) forces acting on axial section; (c) forces acting on tangential plane.

obtained by multiplying the *horizontal* forces by the appropriate radii. Thus,

$$T_r = P_n \left(\frac{d_p}{2}\right)(\cos\theta_n \sin\alpha + \mu\cos\alpha) + r_c\mu_c W$$

Substituting Equation (16.9) into the above equation gives

$$T_r = W\left[\frac{(d_p/2)(\cos\theta_n \tan\alpha + \mu)}{\cos\theta_n - \mu\tan\alpha} + r_c\mu_c\right] \qquad (16.10)$$

Recall that $\mu$ is the coefficient of friction between the threads and $\mu_c$ is the coefficient of friction for the collar. Note that the torque equations [Eqs. (16.10) and (16.8)], although developed for power screws, are general, so that these equations can be applied equally well to other types of threaded fastener.

In typical power screw problems, the coefficients of friction $\mu$ and $\mu_c$ are given, and the thread and lead angles are known. Often, only approximate values of $\mu$ and $\mu_c$ can be obtained, leading to some uncertainty in the actual forces and torques required. Regardless, the angle $\theta_n$ can be determined from Equation (16.8). Once the pitch radius and the collar radius are known, the torque required to raise the load can be calculated.

If the collar consists of rolling-element bearings, the collar friction is extremely low and can be neglected ($\mu_c = 0$).

### Lowering the Load

Lowering the load differs from raising the load, given in Figure 16.7, only in that the friction force components [Fig. 16.7(c)] become negative. These changes result in the summation of the vertical forces being

$$P_n = \frac{W}{\cos\theta_n \cos\alpha + \mu\sin\alpha} \qquad (16.11)$$

The torque required to lower the load is obtained by multiplying the horizontal forces by the appropriate radii. Thus,

$$T_l = P_n \left(\frac{d_p}{2}\right)(\cos\theta_n \sin\alpha - \mu\cos\alpha) - r_c\mu_c W$$

Substituting Equation (16.11) into the above equation gives

$$T_l = -W\left[\frac{(d_p/2)(\mu - \cos\theta_n \tan\alpha)}{\cos\theta_n + \mu\tan\alpha} + r_c\mu_c\right] \qquad (16.12)$$

Using Equations (16.8) and (16.12) gives the torque required to lower the load.

### 16.3.2  POWER AND EFFICIENCY

Once the torque is known, the power transferred by the power screw can be obtained from Equations (4.40) and (4.42). From Equation (4.40) the power $h_p$ in horsepower can be expressed as

$$h_p = \frac{TN_a}{63,025}$$

where $T$ = torque, lb · in

$\quad$ $N_a$ = rotational speed, rpm

For metric units from Equation (4.42) the power $h_p$ in watts is

$$h_p = T\omega$$

where $T$ = torque, N · m

$\quad$ $\omega$ = rotational speed, rad/s

The efficiency of a screw mechanism, expressed in percent, is the ratio of work output to work input, or

$$e = \frac{100Wl}{2\pi T} \qquad (16.13)$$

where $l$ = lead, m

$\quad$ $W$ = load, N

Once the torque required to raise the load is known, the efficiency of the power screw mechanism can be calculated. Note from Equations (16.10) and (16.12) that if the coefficients of friction of the threads and the collar are equal to zero,

$$e = \frac{100Wl}{2\pi W(d_p/2)\tan\alpha} = \frac{(100)l}{2\pi (d_p/2)\tan\alpha}$$

Therefore, making use of Equation (16.5) shows that if $\mu$ and $\mu_c$ can be reduced to zero, the efficiency will be 100%.

---

**EXAMPLE 16.2**

**Given:**  A double-threaded Acme power screw has a load of 1000 lb, a crest diameter of 1 in, a collar diameter of 1.5 in, a thread coefficient of friction of 0.16, and a collar coefficient of friction of 0.12.

**Find:**  Determine the torque required to raise and lower the load. Also determine the efficiency for raising the load.

**Solution:**  From Table 16.2 for $d_c = 1$ in, the number of threads per inch $n = 5$, or $p = 0.2$ in. The lead $l = 2p = 0.4$ in. Also, Equation (16.4) gives

$$d_p = d_c - 0.5p - 0.01 = 1.0 - \frac{0.5}{5} - 0.01 = 0.89 \text{ in}$$

From Equation (16.5) the lead angle is

$$\alpha = \tan^{-1}\frac{l}{\pi d_p} = \tan^{-1}\frac{0.4}{\pi 0.89} = 8.142°$$

Recall that for an Acme screw the thread angle $\beta = 29°$. Making use of Equation (16.8) gives

$$\theta_n = \tan^{-1}\left(\cos\alpha\tan\frac{\beta}{2}\right) = \tan^{-1}(\cos 8.142° \tan 14.5°) = 14.36°$$

Note that $\theta_n$ and $\beta/2$ are nearly equal.

Using Equation (16.10) gives the torque required to *raise* the load as

$$T_r = W\left[\frac{(d_p/2)(\cos\theta_n\tan\alpha + \mu)}{\cos\theta_n - \mu\tan\alpha} + r_c\mu_c\right]$$

$$= 1000\left[\frac{(0.445)(\cos 14.36°\tan 8.142° + 0.16)}{\cos 14.36° - 0.16\tan 8.142°} + 0.75(0.12)\right]$$

$$= 1000(0.1405 + 0.0900) = 230.5 \text{ in} \cdot \text{lb}$$

Using Equation (16.12) gives the torque required to *lower* the load as

$$T_l = -W\left[\frac{(d_p/2)(\mu - \cos\theta_n\tan\alpha)}{\cos\theta_n + \mu\tan\alpha} + r_c\mu_c\right]$$

$$= -1000\left[\frac{(0.445)(0.16 - \cos 14.36°\tan 8.142°)}{\cos 14.36° + 0.16\tan 8.142°} + 0.75(0.12)\right]$$

$$= -1000(0.0096 + 0.0900) = -99.60 \text{ in} \cdot \text{lb}$$

Note that 90.4% of the torque is due to collar friction. As expected, less torque is required to lower the load than to raise the load. The minus sign indicates the opposite direction in lowering the load relative to raising the load.

Using Equation (16.13) gives the efficiency in raising the load as

$$e = \frac{100Wl}{2\pi T_r} = \frac{100(1000)(0.4)}{2\pi(230.5)} = 27.62\%$$

---

**EXAMPLE 16.3**

**Given:** The same conditions as in Example 16.2 except that a single-threaded power screw is used instead of a double-threaded one.

**Find:** Determine the torque required to raise and lower the load. Also determine the efficiency for raising the load.

**Solution:** The lead equals the pitch ($l = p = 0.2$ in). The lead angle is

$$\alpha = \tan^{-1}\frac{l}{\pi d_p} = \tan^{-1}\frac{0.2}{0.89\pi} = 4.091°$$

From Equation (16.8)

$$\theta_n = \tan^{-1}\left(\cos\alpha\tan\frac{\beta}{2}\right) = \tan^{-1}(\cos 4.091°\tan 14.5°) = 14.46°$$

Using Equation (16.10) gives the torque required to *raise* the load as

$$T_r = W\left[\frac{(d_p/2)(\cos\theta_n\tan\alpha + \mu)}{\cos\theta_n - \mu\tan\alpha} + r_c\mu_c\right]$$

$$= 1000\left[\frac{(0.445)(\cos 14.46°\tan 4.091° + 0.16)}{\cos 14.46° - 0.16\tan 4.091°} + 0.75(0.12)\right]$$

$$= 1000(0.1066 + 0.0900) = 196.6 \text{ in} \cdot \text{lb}$$

Note that to raise the load, the single-threaded screw requires lower torque than the double-threaded screw.

Using Equation (16.13) gives the efficiency in raising the load as

$$e = \frac{100Wl}{2\pi T_r} = \frac{100(1000)(0.2)}{2\pi(196.6)} = 16.19\%$$

The single-threaded screw is thus less efficient (by 11 percentage points) than the double-threaded screw given in Example 16.2.

Using Equation (16.12) gives the torque required to *lower* the load as

$$T_l = -W\left[\frac{(d_p/2)(\mu - \cos\theta_n \tan\alpha)}{\cos\theta_n + \mu\tan\alpha} + r_c\mu_c\right]$$

$$= -1000\left[\frac{(0.445)(0.16 - \cos 14.46° \tan 4.091°)}{\cos 14.46° + 0.16\tan 4.091°} + 0.75(0.12)\right]$$

$$= -1000(0.0412 + 0.0900) = -131.2 \text{ in}\cdot\text{lb}$$

### 16.3.3   SELF-LOCKING SCREWS

If the screw thread has a large lead angle, the friction force may not be able to stop the tendency of the load to slide down the plane, and gravity will cause the load to fall. Usually, for power screws the lead angle is small, and the friction force of the thread interaction is large enough to oppose the load and keep it from sliding down. Such a screw is called **self-locking,** a desirable characteristic to have in power screws.

If the collar uses a rolling-element bearing, the collar friction can be assumed to be zero. From Equation (16.12) the torque required to lower the load is negative if

$$\mu - \cos\theta_n \tan\alpha \geq 0$$

Thus, under static conditions self-locking occurs if

$$\mu > \cos\theta_n \tan\alpha = \frac{l\cos\theta_n}{\pi d_p} \tag{16.14}$$

Vibrations may cause the power screw to move.

**EXAMPLE 16.4**

**Given:**   The double-threaded screw of Example 16.2 and the single-threaded screw of Example 16.3.

**Find:**   Determine the thread coefficient of friction required to ensure that self-locking occurs. Assume that rolling-element bearings have been installed at the collar.

**Solution:**   For the double-threaded screw, self-locking occurs if

$$\mu > \cos\theta_n \tan\alpha = \cos 14.36° \tan 8.142° = 0.1386$$

For the single-threaded screw, self-locking occurs if

$$\mu > \cos \theta_n \tan \alpha = \cos 14.36° \tan 4.091° = 0.0693$$

Thus, the double-threaded screw requires twice the friction coefficient of the single-threaded screw.

---

**EXAMPLE 16.5**

**Given:** A triple-threaded Acme power screw has a crest diameter of 2 in and 4 threads per inch. A rolling-element bearing is used at the collar, so that the collar coefficient of friction is zero. The thread coefficient of friction is 0.15.

**Find:** Determine the load that can be raised by a torque of 400 in · lb. Also, check to see if self-locking occurs.

**Solution:** The pitch and the lead are, respectively,

$$p = \frac{1}{n} = \frac{1}{4} \text{ in} = 0.25 \text{ in}$$

$$l = mp = 3(0.25) = 0.75 \text{ in}$$

The pitch diameter is

$$d_p = d_c - 0.5p - 0.01 = 2 - 0.5(0.25) - 0.01 = 1.865 \text{ in}$$

From Equation (16.5) the lead angle is

$$\alpha = \tan^{-1} \frac{l}{\pi d_p} = \tan^{-1} \frac{0.75}{\pi(1.865)} = 7.295°$$

From Equation (16.8)

$$\theta_n = \tan^{-1} \left( \cos \alpha \tan \frac{\beta}{2} \right) = \tan^{-1} (\cos 7.295° \tan 14.5°) = 14.39°$$

Using Equation (16.10) gives the torque required to raise the load as

$$T_r = W \left[ \frac{(d_p/2)(\cos \theta_n \tan \alpha + \mu)}{\cos \theta_n - \mu \tan \alpha} \right]$$

$$400 = W \left[ \frac{(1.865/2)(\cos 14.39° \tan 7.295° + 0.15)}{\cos 14.39° - 0.15 \tan 7.295°} \right]$$

$$\therefore \qquad W = \frac{400}{0.2691} = 1486 \text{ lb}$$

From Equation (16.14) self-locking occurs if

$$\mu > \cos \theta_n \tan \alpha$$

$$0.15 > \cos 14.39° \tan 7.295° = 0.1240$$

Therefore, self-locking occurs.

## 16.4    THREADED FASTENERS

A fastener is a device used to connect or join two or more members. Many fastener types and variations are available for use. This chapter considers threaded, riveted, welded, and adhesive fasteners. The most common are threaded fasteners.

### 16.4.1    TYPES OF THREADED FASTENER

Figure 16.8 shows three types of threaded fastener: the bolt and nut, the cap screw, and the stud. Most threaded fasteners consist of a **bolt** passing through a hole in the members being joined and mating with a **nut,** as shown in Figure 16.8(a). Occasionally, the bolt mates with threads in one of the members rather than with a nut, as shown in Figure 16.8(b). This type is called a **cap screw.** A **stud** [Fig. 16.8(c)] is threaded on both ends and screwed, more or less permanently, into the threaded hole in one of the members being joined.

A selection of available bolt and screw types are shown in Table 16.3 along with some notes regarding their typical applications; head, washers and nut types are shown in Table 16.4. Most bolt heads (and nuts) are hexagonal, and this text considers only this type, although much of the mechanics remain applicable to other types. A more detailed coverage of bolted joints can be found in Bickford (1995).

### 16.4.2    LOAD ANALYSIS OF BOLTS AND NUTS

The bolt and nut shown in Figure 16.8(a) can be thought of as a spring system, as shown in Figure 16.9. The bolt is viewed as a spring in tension with stiffness $k_b$. The joint, with a number of members being joined, is viewed as a compressive spring with stiffness $k_j$.

Figure 16.10 shows the forces and deflections acting on the bolt and the joint. In Figure 16.10(a) the bolt and the joint are considered to be unassembled, and the load and the deflection are shown for each. For the bolt the force is tensile and the deflection is an extension of the bolt, whereas for the joint the deflection is a contraction. Figure 16.10(b) considers the assembled bolt, nut, and joint. The slopes of the load-deflection lines are the

(a)                          (b)                          (c)

**Figure 16.8**    Three types of threaded fastener. (a) Bolt and nut; (b) cap screw; (c) stud.

**Table 16.3**    Common types of bolts and screws

| Illustration | Type | Description | Application notes |
|---|---|---|---|
| | Hex-head bolt | An externally threaded fastener with a trimmed hex head, often with a washer face on the bearing side. | Used in a variety of general-purpose applications in different grades depending on the required loads and material being joined. |
| | Carriage bolt | A round-head bolt with a square neck under the head and a standard thread. | Used in slots where the square neck keeps the bolt from turning when being tightened. |
| | Elevator bolt or belt bolt | A bolt with a wide, countersunk flat head, a shallow conical bearing surface, an integrally formed square neck under the head, and a standard thread. | Used in belting and elevator applications where head clearances must be minimal. |
| | Serrated flange bolt | A hex bolt with integrated washer, but wider than standard washers and incorporating serrations on the bearing surface side. | Used in applications where loosening hazard exists, such as vibration applications. The serrations grip the surface so that greater torque is needed to loosen than tighten the bolt. |
| | Flat cap screw (slotted head shown) | A flat, countersunk screw with a flat top surface and conical bearing surface. | A common fastener for assembling joints where head clearance is critical. |
| | Buttonhead cap screw (socket head shown) | Dome-shaped head that is wider and has a lower profile than that of a flat cap screw. | Designed for light fastening applications where their appearance is desired. Not recommended for high-strength applications. |
| | Lag screw | A screw with spaced threads, a hex head, and a gimlet point. (Can also be made with a square head.) | Used to fasten metal to wood or with expansion fittings in masonry. |
| | Step bolt | A plain, circular, oval-head bolt with a square neck. The head diameter is about 3 times the bolt diameter. | Used to join resilient materials or sheet metal to supporting structures, or for joining wood since the large head will not pull through. |

Table 16.4    Common nuts and washers for use with threaded fasteners

| Illustration | Type | Description | Application notes |
|---|---|---|---|
| **Nuts** | Hex nut | A six-sided internally threaded fastener. Specific dimensions are prescribed in industry standards. | The most commonly used general-purpose nut. |
| | Nylon insert stop | A nut with a hex profile and an integral nylon insert. | The nylon insert exerts friction on the threads and prevents loosening due to vibration or corrosion. |
| | Cap nut | Similar to a hex nut with a dome top. | Used to cover exposed, dangerous bolt threads or for aesthetic reasons. |
| | Castle nut | A type of slotted nut. | Used for general-purpose fastening and locking. A cotter pin or wire can be inserted through the slots and the drilled shank of the fastener. |
| | Coupling nut | A six-sided double-chamfered nut. | Used to join two externally threaded parts of equal thread diameter and pitch. |
| | Hex jam nut | A six-sided, internally threaded fastener, thinner than a normal hex nut. | Used in combination with a hex nut to keep the nut from loosening. |
| | K-lock or keplock nut | A hex nut preassembled with a free-spinning external tool lock washer. When tightened, the teeth bite into the member to achieve locking. | A popular locknut because of ease of use and low cost. |
| | Wing nut | An internally threaded nut with integral pronounced flat tabs. | Used for applications where repetitive hand tightening is required. |
| | Serrated nut | A hex nut with integrated washer, but wider than standard washers and incorporating serrations on the bearing surface side. | Used in applications where loosening hazard exists, such as vibration applications. The serrations grip the surface so that greater torque is needed to loosen than tighten the bolt. |
| **Washers** | Flat washer | A circular disk with circular hole, produced in accordance with industry standards. Fender washers have larger surface area than that of conventional flat washer. | Designed for general-purpose mechanical and structural use. |

**Table 16.4** *(concluded)*

| Illustration | Type | Description | Application notes |
|---|---|---|---|
| | Belleville washer | A conical disk spring. | Used to maintain load in bolted connections. |
| | Split lock washer | A coiled, hardened, split circular washer with a slightly trapezoidal cross section. | Preferred for use with hardened bearing surfaces. Applies high bolt tension per torque, resists loosening caused by vibration and corrosion. |
| | Tooth lock washer | A hardened circular washer with twisted teeth or prongs. | Internal teeth are preferred from aesthstics since the teeth are hidden under the bolt head. External teeth give greater locking efficiency. Combination teeth are used for oversized or out-of-round holes or for electrical connections. |

**Figure 16.9**
Bolt-and-nut assembly modeled as bolt-and-joint spring.

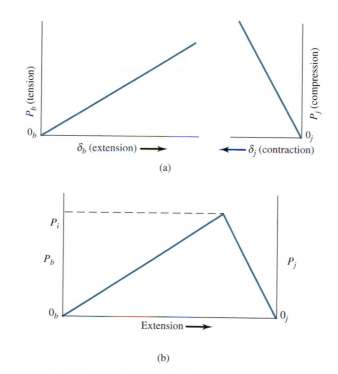

**Figure 16.10** Force versus deflection of bolt and member.
(a) Separated bolt and joint; (b) assembled bolt and joint.

**Figure 16.11**    Forces versus deflection of bolt and joint when external load is applied.

same as in Figure 16.10(a), but the slopes for the bolt and joint not only are opposite in sign but also have quite different values. The bolt and joint load-deflection lines intersect at a preload $P_i$.

Figure 16.11 shows the forces versus deflection for a bolt and a joint when an external load has been applied. The bolt is elongated by $e_k$, and the contraction of the joint is reduced, as shown in Figure 16.11. The force on the bolt becomes $P_i + k_b e_k$, and the force on the joint becomes $P_i - k_j e_k$.

From the equilibrium of the bolt

$$P + P_i - k_j e_k - P_i - k_b e_k = 0$$

$$\therefore \qquad e_k = \frac{P}{k_b + k_f} \tag{16.15}$$

Making use of Equation (16.15) gives the load on the bolt as

$$P_b = P_i + k_b e_k = P_i + \frac{P k_b}{k_b + k_j} = P_i + C_k P \tag{16.16}$$

where the dimensionless stiffness parameter is

$$C_k = \frac{k_b}{k_b + k_j} \tag{16.17}$$

From Equation (16.17) $C_k$ is less than 1.

The load on the joint is

$$P_j = P_i - k_j e_k = P_i - \frac{P k_j}{k_j + k_b} = P_i - (1 - C_k)P \tag{16.18}$$

The preceding analysis assumes that the external force is applied under the head of the bolt and under the head of the nut.

### 16.4.3    STIFFNESS PARAMETERS

Recall from Chapter 4 that the spring rate is the normal load divided by the elastic deflection for axial loading [Eq. (4.26)] and that the elastic deflection for axial loading is given in Equation (4.25). Making use of these equations gives

$$k = \frac{P}{PL/(AE)} = \frac{AE}{L} \tag{16.19}$$

where  $A$ = cross-sectional area, m$^2$
$\quad\quad L$ = length, m
$\quad\quad E$ = modulus of elasticity, Pa

### Bolt Stiffness

A bolt with thread is considered as a stepped shaft. The root diameter is used for the threaded section of the bolt, and the crest diameter is used for the unthreaded section, called the *shank*. The bolt is treated as a spring in series when one is considering the shank and the threaded section. The bolt can also have different diameters due to other specifications, so that the stiffness of the bolt can be expressed as

$$\frac{1}{k_b} = \frac{1}{k_{b1}} + \frac{1}{k_{b2}} + \frac{1}{k_{b3}} + \cdots \tag{16.20}$$

Figure 16.12(a) shows a bolt-and-nut assembly; and Figure 16.12(b), the stepped-shaft representation of the shank and the threaded section. In Figure 16.12(b) the effective length of the shaft and the threaded section includes additional lengths extending into the bolt head

(a)                                    (b)

**Figure 16.12**    Bolt and nut. (a) Assembled; (b) stepped-shaft representation of shank and threaded section.

and the nut. Making use of Equation (16.19) and Figure 16.12(b) gives

$$\frac{1}{k_b} = \frac{4}{\pi E}\left(\frac{L_{se}}{d_c^2} + \frac{L_{te}}{d_r^2}\right) = \frac{4}{\pi E}\left(\frac{L_s + 0.4d_c}{d_c^2} + \frac{L_t + 0.4d_r}{d_r^2}\right) \qquad (16.21)$$

where $d_c$ = crest diameter, m
$d_r$ = root diameter, m

Equation (16.21) is valid for a shank having a constant crest diameter. In some designs the shank has different diameters. This situation can be easily analyzed by using Equation (16.20) and the appropriate lengths and diameters. Note that $0.4d_c$ is added to the length only for the section closest to the bolt head. The other lengths in the shank should be the actual lengths.

Often, a threaded fastener dedicated to a particular application will be specifically fabricated, and any combination of diameter and thread length is possible. However, it is usually much more economical to use mass-produced threaded fasteners; these threads are generally of higher quality (as they may be rolled instead of cut), and they have a more repeatable strength. Usually, a bolt or cap screw will have as little of its length threaded as practicable to maximize bolt stiffness. However, for standardized threads the threaded length is given by

$$L_t = \begin{cases} 2d_c + 6 & L \le 125 \\ 2d_c + 12 & 125 < L \le 200 \\ 2d_c + 25 & L > 200 \end{cases} \qquad \text{(metric threads)} \qquad (16.22)$$

$$L_t = \begin{cases} 2d_c + 0.25 \text{ in} & L \le 6 \text{ in} \\ 2d_c + 0.50 \text{ in} & L > 6 \text{ in} \end{cases} \qquad \text{(inch series)} \qquad (16.23)$$

where $L_t$ = threaded length
$L$ = total bolt length
$d_c$ = crest diameter

### Joint Stiffness

Determining joint stiffness is much more complicated than determining bolt stiffness (which was covered in the preceding section). Thus, approximations must be employed. One of the most frequent of these approximations is that the stress induced in the joint is uniform throughout a region surrounding the bolt hole with zero stress outside this region. Two conical frusta symmetric about the joint midplane and each having a vertex angle of $2\alpha_f$ are often used to represent the stresses in the joint. Figure 16.13 shows the conical frustum stress representation of the joint in a bolt-and-nut assembly. Note that $d_w$ is the diameter of the washer through which the load is transferred. Shigley and Mischke (1989) arrived at the following expression for joint stiffness:

$$k_{ji} = \frac{\pi E_j d_c \tan \alpha_f}{\ln\left[(2L_i \tan \alpha_f + d_i - d_c)(d_i + d_c)/(2L_i \tan \alpha_f + d_i + d_c)(d_i - d_c)\right]} \qquad (16.24)$$

where $L_i$ = axial length of frustum cone, m
$d_i$ = frustum cone diameter, m

**Figure 16.13** Bolt-and-nut assembly with conical frustum stress representation of joint.

Table 16.5 summarizes some special case situations of member stiffness based on Equation (16.24).

In calculating stiffness, always use the smallest of the frustum cone diameters. For the member closest to the bolt head or to the nut, $d_i = d_w = 1.5 d_c$.

The resulting joint stiffness is

$$\frac{1}{k_j} = \frac{1}{k_{j1}} + \frac{1}{k_{j2}} + \frac{1}{k_{j3}} + \cdots \qquad (16.25)$$

Wileman et al. (1991) obtained an exponential expression for the joint stiffness that curve fits results from a finite-element analysis which uses a range of materials and geometry.

**Table 16.5** Member stiffness equations for common bolted joint configurations

| Description | Member stiffness $k_m$ |
|---|---|
| Single member, general case | $k_{ji} = \dfrac{\pi E_j d_c \tan \alpha_f}{\ln[(2L_i \tan \alpha_f + d_i - d_c)(d_i + d_c)/(2L_i \tan \alpha_f + d_i + d_c)(d_i - d_c)]}$ |
| Single member, $\alpha_f = 30°$ | $k_{ji} = \dfrac{1.813 E_j d_c}{\ln[(1.15 L_i + d_i - d_c)(d_i + d_c)/(1.15 L_i + d_i + d_c)(d_i - d_c)]}$ |
| Two members, same Young's modulus $E$, back-to-back frusta[a] | $k_j = \dfrac{\pi E_j d_c \tan \alpha_f}{2 \ln[(2L_i \tan \alpha_f + d_i - d_c)(d_i + d_c)/(2L_i \tan \alpha_f + d_i + d_c)(d_i - d_c)]}$ |
| Two members, same Young's modulus, back-to-back frusta, $\alpha = 30°$, $d_i = d_w = 1.5 d_c$[a] | $k_j = \dfrac{1.813 E_j d_c}{2 \ln[(2.885 L_i + 2.5 d_c)/(0.577 L_i + 2.5 d_c)]}$ |
| Two members, same material, Wileman method[a,b] | $k_j = E_i d_c A_i e^{B_i d_c/L_i}$ |

[a]Note that this is stiffness for the complete joint, not a member in the joint.
[b]See Table 16.6 for values of $A_i$ and $B_i$ for various materials.

**Table 16.6**    Constants used in joint stiffness formula [Eq. (16.26)]

| Material | Poisson's ratio $\nu$ | Modulus of Elasticity | | Numerical Constants | |
|---|---|---|---|---|---|
| | | $E$, GPa | M psi | $A_i$ | $B_i$ |
| Steel | 0.291 | 206.8 | 30 | 0.78715 | 0.62873 |
| Aluminum | 0.334 | 71.0 | 10 | 0.79670 | 0.63816 |
| Copper | 0.326 | 118.6 | 16 | 0.79568 | 0.63553 |
| Gray cast iron | 0.211 | 100.0 | 14.6 | 0.77871 | 0.61616 |

I SOURCE: Wileman et al. (1991).

The expression for joint stiffness is

$$k_j = E_i d_c A_i e^{B_i d_c/L} \tag{16.26}$$

where $A_i$, $B_i$ = numerical constants given in Table 16.6.

Note that $A_i$ and $B_i$ are dimensionless and that Equation (16.26) is equally applicable in English units.

Once the joint and bolt stiffnesses are known, the dimensionless stiffness parameter can be calculated from Equation (16.17).

---

**EXAMPLE 16.6**

**Given:**  A hexagonal bolt-and-nut assembly, shown in Figure 16.14(a), is used to join two members. The bolt and the nut are made of steel, and the frustum cone angle is 30°. The thread crest diameter is 14 mm, and the root diameter is 12 mm.

**Figure 16.14**    Hexagonal bolt-and-nut assembly used in Example 16.6.
(a) Assembly and dimensions; (b) dimension of frustum cone. (All dimensions are in millimeters.)

**Find:** Find the bolt and joint stiffnesses as well as the dimensionless stiffness parameter. Use the modulus of elasticity and Poisson's ratio from Table 16.6, and consider the cases where

(a) Both members are made of steel. Use both the Shigley and Mischke (1989) and Wileman et al. (1991) approaches.

(b) The 15-mm-thick member is made of steel, and the 10-mm-thick member is made of aluminum.

**Solution:** From Equation (16.21) the bolt stiffness is

$$\frac{1}{k_b} = \frac{4}{\pi(206.8)(10^9)}\left[\frac{(17)(10^{-3}) + 0.4(14)(10^{-3})}{(14)^2(10^{-6})} + \frac{8(10^{-3}) + (0.4)(12)(10^{-3})}{(12)^2(10^{-6})}\right]$$

$$k_b = 0.7954 \times 10^9 \text{ N/m} = 7.954 \times 10^8 \text{ N/m}$$

(a) *Steel joint.* Using the Shigley and Mischke (1989) method, we recognize that the joint consists of two back-to-back frusta with $\alpha = 30°$ and $d_i = 1.5d_c$, so from Table 16.5, the member stiffness is given by

$$k_j = \frac{1.813 E_j d_c}{2\ln\left[(2.885 L_i + 2.5d_c)/(0.577 L_i + 2.5d_c)\right]}$$

$$= \frac{1.813(2.068)(10^{11})(0.014)}{2\ln\left\{[2.885(25) + 2.5(14)]/[0.577(25) + 2.5(14)]\right\}} = 3.393 \times 10^9 \text{ N/m}$$

By using the Wileman method, Equation (16.26) yields

$$k_j = E_i d_c A_i e^{B_i d_c / L} = (2.068 \times 10^{11})(0.014)(0.78715)e^{(0.62873)(14)/(25)}$$

$$= 3.241 \times 10^9 \text{ N/m}$$

where the values of $A_i$ and $B_i$ have been taken from Table 16.6. Note that the Wileman et al. (1991) calculations are simple and quite close to the Shigley and Mischke (1989) calculations. The Wileman method will therefore be used for calculating joint stiffness. The dimensionless stiffness parameter [Eq. (16.17)] gives

$$C_k = \frac{k_b}{k_b + k_f} = \frac{0.7954}{0.7954 + 3.241} = 0.197$$

(b) *Steel-aluminum joint.* Using the Shigley and Mischke (1989) method, the approach is more complicated for this case. As shown in Figure 16.14, the members have to be considered as three frusta. The member stiffnesses are (using the simplified equation for $\alpha_f = 30°$ from Table 16.5)

$$k_{j1} = \frac{1.813 E_j d_c}{\ln\left[\dfrac{(1.15 L_i + d_i - d_c)(d_i + d_c)}{(1.15 L_i + d_i + d_c)(d_i - d_c)}\right]} = \frac{1.813(71 \times 10^9)(0.014)}{\ln\left\{\dfrac{[1.15(12.5) + 21 - 14](21 + 14)}{[1.15(12.5) + 21 + 14](21 - 14)}\right\}}$$

$$= 2.334 \times 10^9 \text{ N/m}$$

$$k_{j2} = \frac{1.813(71 \times 10^9)(0.014)}{\ln\left\{\dfrac{[1.15(2.5) + 32.55 - 14](32.55 + 14)}{[1.15(2.5) + 32.55 + 14](32.55 - 14)}\right\}} = 2.141 \times 10^{10} \text{ N/m}$$

$$k_{j3} = \frac{1.813(206.8 \times 10^9)(0.014)}{\ln\left\{\dfrac{[1.15(10) + 21 - 14](21 + 14)}{[1.15(10) + 21 + 14](21 - 14)}\right\}} = 2.620 \times 10^9 \text{ N/m}$$

Substituting the previous equations into Equation (16.25) gives

$$\frac{1}{k_j} = \frac{1}{2.334 \times 10^9 \text{ N/m}} + \frac{1}{2.141 \times 10^9 \text{ N/m}} + \frac{1}{2.620 \times 10^9 \text{ N/m}}$$

$$\therefore \quad k_j = 1.167 \times 10^9 \text{ N/m}$$

The dimensionless stiffness parameter [Eq. (16.17)] gives

$$C_k = \frac{k_b}{k_b + k_f} = \frac{0.7954}{0.7954 + 1.167} = 0.4053$$

### 16.4.4    STRENGTH

The **proof load** of a bolt is the maximum load that a bolt can withstand without acquiring a permanent set. The **proof strength** is the limiting value of the stress determined by using the proof load and the tensile stress area. Although proof strength and yield strength have something in common, the yield strength is usually higher because it is based on a 0.2% permanent deformation.

The proof strength $S_p$, as defined by the Society of Automotive Engineers (SAE), the American Society for Testing and Materials (ASTM), and International Organization for Standardization (ISO) specifications, defines bolt grades or classes that specify material, heat treatment, and minimum proof strength for the bolt or screw. Table 16.7 gives the strength

**Table 16.7**    Strength of steel bolts for various sizes in inches

| SAE grade | Head marking | Range of crest diameters, in | Ultimate tensile strength $S_u$, ksi | Yield strength $S_y$, ksi | Proof strength $S_p$, ksi |
|---|---|---|---|---|---|
| 1 | | $\frac{1}{4}-1\frac{1}{2}$ | 60 | 36 | 33 |
| 2 | | $\frac{1}{4}-\frac{3}{4}$ | 74 | 57 | 55 |
| | | $>\frac{3}{4}-1\frac{1}{2}$ | 60 | 36 | 33 |
| 4 | | $\frac{1}{4}-1\frac{1}{2}$ | 115 | 100 | 65 |
| 5 | | $\frac{1}{4}-1$ | 120 | 92 | 85 |
| | | $>1-1\frac{1}{2}$ | 105 | 81 | 74 |
| 5.2 | | $\frac{1}{4}-1$ | 120 | 92 | 85 |
| 7 | | $\frac{1}{4}-1\frac{1}{2}$ | 133 | 115 | 105 |
| 8 | | $\frac{1}{4}-1\frac{1}{2}$ | 150 | 130 | 120 |
| 8.2 | | $\frac{1}{4}-1$ | 150 | 130 | 120 |

**Table 16.8** Strength of steel bolts for various sizes in millimeters

| Metric grade | Head marking | Crest diameter $d_c$, mm | Ultimate tensile strength $S_u$, MPa | Yield strength $S_y$, MPa | Proof strength $S_p$, MPa |
|---|---|---|---|---|---|
| 4.6 | 4.6 | M5–M36 | 400 | 240 | 225 |
| 4.8 | 4.8 | M1.6–M16 | 420 | $340^a$ | 310 |
| 5.8 | 5.8 | M5–M24 | 520 | $415^a$ | 380 |
| 8.8 | 8.8 | M17–M36 | 830 | 660 | 600 |
| 9.8 | 9.8 | M1.6–M16 | 900 | $720^a$ | 650 |
| 10.9 | 10.9 | M6–M36 | 1040 | 940 | 830 |
| 12.9 | 12.9 | M1.6–M36 | 1220 | 1100 | 970 |

| $^a$Yield strengths approximate and not included in standard.

information for several SAE grades of bolt, and Table 16.8 gives similar information for metric bolts. SAE grade numbers range from 1 to 8, and metric grade numbers from 4.6 to 12.9, with higher numbers indicating greater strength.

Proof stress is comparable to the allowable stress, which is used throughout the text. Thus, the safety factor is equivalent to the proof stress divided by the design stress.

Tables 16.9 and 16.10 give the dimensions and the tensile stress areas for UN and M coarse and fine threads. Dimensions for common hexagonal bolts, cap screws, and nuts are given in Appendix C. The equation for the tensile stress areas for UN thread profiles given in Table 16.9 is

$$A_t = (0.7854)\left(d_c - \frac{0.9743}{n}\right)^2 \tag{16.27}$$

In Equation (16.27) the crest diameter is in inches, and $n$ is the number of threads per inch. The equation for the tensile stress areas for M thread profiles given in Table 16.10 is

$$A_t = (0.7854)(d_c - 0.9382p)^2 \tag{16.28}$$

In Equation (16.28) both the crest diameter $d_c$ and the pitch $p$ are in millimeters.

### 16.4.5 BOLT PRELOAD—STATIC LOADING

This section makes use of Figures 16.10 and 16.11 and the material covered in Section 16.4.2. Equation (16.16) can be written in terms of bolt stress as

$$\sigma_b = \frac{P_b}{A_t} = \frac{P_i}{A_t} + C_k\frac{P}{A_t} \tag{16.29}$$

**Table 16.9**     Dimensions and tensile stress areas for UN coarse and fine threads. Root diameter is calculated from Equation (16.2) and Figure 16.4

| Crest diameter $d_c$, in | Coarse Threads (UNC) | | | Fine Threads (UNF) | | |
|---|---|---|---|---|---|---|
| | Number of threads per inch $n$ | Root diameter $d_r$, in | Tensile stress area $A_t$, in$^2$ | Number of threads per inch $n$ | Root diameter $d_r$, in | Tensile stress area $A_t$, in$^2$ |
| 0.0600 | — | — | — | 80 | 0.04647 | 0.00180 |
| 0.0730 | 64 | 0.05609 | 0.00263 | 72 | 0.05796 | 0.00278 |
| 0.0860 | 56 | 0.06667 | 0.00370 | 64 | 0.06909 | 0.00394 |
| 0.0990 | 48 | 0.07645 | 0.00487 | 56 | 0.07967 | 0.00523 |
| 0.1120 | 40 | 0.08494 | 0.00604 | 48 | 0.08945 | 0.00661 |
| 0.1250 | 40 | 0.09794 | 0.00796 | 44 | 0.1004 | 0.00830 |
| 0.1380 | 32 | 0.1042 | 0.00909 | 40 | 0.1109 | 0.01015 |
| 0.1640 | 32 | 0.1302 | 0.0140 | 36 | 0.1339 | 0.01474 |
| 0.1900 | 24 | 0.1449 | 0.0175 | 32 | 0.1562 | 0.0200 |
| 0.2160 | 24 | 0.1709 | 0.0242 | 28 | 0.1773 | 0.0258 |
| 0.2500 | 20 | 0.1959 | 0.0318 | 28 | 0.2113 | 0.0364 |
| 0.3125 | 18 | 0.2523 | 0.0524 | 24 | 0.2674 | 0.0580 |
| 0.3750 | 16 | 0.3073 | 0.0775 | 24 | 0.3299 | 0.0878 |
| 0.4750 | 14 | 0.3962 | 0.1063 | 20 | 0.4194 | 0.1187 |
| 0.5000 | 13 | 0.4167 | 0.1419 | 20 | 0.4459 | 0.1599 |
| 0.5625 | 12 | 0.4723 | 0.182 | 18 | 0.5023 | 0.203 |
| 0.6250 | 11 | 0.5266 | 0.226 | 18 | 0.5648 | 0.256 |
| 0.7500 | 10 | 0.6417 | 0.334 | 16 | 0.6823 | 0.373 |
| 0.8750 | 9 | 0.7547 | 0.462 | 14 | 0.7977 | 0.509 |
| 1.000 | 8 | 0.8647 | 0.606 | 12 | 0.9098 | 0.663 |
| 1.125 | 7 | 0.9703 | 0.763 | 12 | 1.035 | 0.856 |
| 1.250 | 7 | 1.095 | 0.969 | 12 | 1.160 | 1.073 |
| 1.375 | 6 | 1.195 | 1.155 | 12 | 1.285 | 1.315 |
| 0.500 | 6 | 1.320 | 1.405 | 12 | 1.140 | 1.581 |
| 1.750 | 5 | 1.533 | 1.90 | — | — | — |
| 2.000 | 4.5 | 1.759 | 2.5 | — | — | — |

The limiting value of the bolt stress is the proof strength, which is given in Tables 16.7 and 16.8. Also, introducing the safety factor for bolt stress gives the proof strength as

$$S_p = \frac{P_i}{A_t} + \frac{P_{max} n_s C_k}{A_t}$$     (16.30)

where $A_t$ = tensile stress area given by Equation (16.27) or (16.28) and Table 16.9 or 16.10
       $P_i$ = preload, N

The safety factor is not applied to the preload stress. Equation (16.30) can be rewritten to give the bolt failure safety factor as

$$n_{sb} = \frac{A_t S_p - P_i}{P_{max,b} C_k}$$     (16.31)

where $P_{max,b}$ = maximum load applied to bolt, N.

**Table 16.10**   Dimensions and tensile stress areas for M coarse and fine threads. Root diameter is calculated from Equation (16.2) and Figure 16.4

| Crest diameter $d_c$, mm | Coarse Threads (MC) | | | Fine Threads (MF) | | |
|---|---|---|---|---|---|---|
| | Pitch $p$, mm | Root diameter $d_r$, mm | Tensile stress area $A_t$, mm² | Pitch $p$, mm | Root diameter $d_r$, mm | Tensile stress area $A_t$, mm² |
| 1 | 0.25 | 0.7294 | 0.460 | — | — | — |
| 1.6 | 0.35 | 1.221 | 1.27 | 0.20 | 1.383 | 1.57 |
| 2 | 0.4 | 1.567 | 2.07 | 0.25 | 1.729 | 2.45 |
| 2.5 | 0.45 | 2.013 | 3.39 | 0.35 | 2.121 | 3.70 |
| 3 | 0.5 | 2.459 | 5.03 | 0.35 | 2.621 | 5.61 |
| 4 | 0.7 | 3.242 | 8.78 | 0.5 | 3.459 | 9.79 |
| 5 | 0.8 | 4.134 | 14.2 | 0.5 | 4.459 | 16.1 |
| 6 | 1.0 | 4.917 | 20.1 | 0.75 | 5.188 | 22 |
| 8 | 1.25 | 6.647 | 36.6 | 1.0 | 6.917 | 39.2 |
| 10 | 1.5 | 8.376 | 58.0 | 1.25 | 8.647 | 61.2 |
| 12 | 1.75 | 10.11 | 84.3 | 1.25 | 10.65 | 92.1 |
| 16 | 2.0 | 13.83 | 157 | 1.5 | 14.38 | 167 |
| 20 | 2.5 | 17.29 | 245 | 1.5 | 18.38 | 272 |
| 24 | 3.0 | 20.75 | 353 | 2.0 | 21.83 | 384 |
| 30 | 3.5 | 26.21 | 561 | 2.0 | 27.83 | 621 |
| 36 | 4.0 | 31.67 | 817 | 3.0 | 32.75 | 865 |
| 42 | 4.5 | 37.13 | 1121 | — | — | — |
| 48 | 5.0 | 42.59 | 1473 | — | — | — |

Equation (16.31) suggests that the safety factor is maximized by having zero preload on the bolt. For statically loaded bolts where separation is not a concern, this is certainly true. However, a preload is often applied to bolted connections to make certain that members are tightly joined in order to minimize tolerances and provide tight fits. Further, many applications require members to avoid separation, as shown in Figure 16.15. Separation occurs when $P_j = 0$ in Equation (16.18). Thus, the safety factor guarding against separation is

$$n_{sj} = \frac{P_i}{P_{\max,j}(1 - C_k)} \tag{16.32}$$

where $P_{\max,j}$ = maximum load applied to joint, N.

The amount of preload that is in practice applied to bolts under static conditions is therefore a compromise between bolt overloading (where zero preload is most beneficial) and separation (where a large preload is desirable). The preload is given for reused and permanent connections as

$$P_i = \begin{cases} 0.75P_p & \text{for reused connection} \\ 0.90P_p & \text{for permanent connections} \end{cases} \tag{16.33}$$

where $P_p$ = proof load, $S_p A_t$. Equation (16.33) is applicable for static as well as dynamic conditions. Recall that the tensile stress area is obtained from either Table 16.9 or Table 16.10 depending on whether UN or metric threads are used. Also, the proof strength $S_p$

**Figure 16.15**    Separation of joint.

is obtained from Table 16.7 or 16.8 depending on whether thread size is expressed in inches or millimeters.

In design practice, preloads are rarely specified because preloads are extremely difficult to measure during assembly of bolted connections. Two alternatives are used:

1.  Specify a torque to be applied during tightening, which is then controlled by using a torque wrench.

2.  Define a number of rotations from a "snug state," such as a half-turn.

Both alternatives have difficulties associated with them: Specifying a torque does not accommodate changes in thread friction, which clearly affect the resulting preload; and a snug state is inherently subjective. Thus, although Equation (16.33) can be used in design problems and for determining a pretorque or deformation, it should be recognized that bolts in applications have a wide variance in preloads.

---

**EXAMPLE 16.7**

**Given:**  The results of Example 16.6 for the steel-aluminum joint.

**Find:**  Determine the maximum load for bolt-and-joint failure while assuming reused connections and a static safety factor of 2.5. Assume a 5.8 grade and coarse threads.

**Solution:**  From Table 16.8 for 5.8 grade, $S_p = 380$ MPa; and from Table 16.10 for a crest diameter of 14 mm and coarse threads, $A_t = (84.3 + 157)/2 = 120.7$ mm$^2$. For reused connections from Equation (16.33)

$$P_i = 0.75P_p = 0.75A_tS_p = (0.75)(120.7)(10^{-6})(380)(10^6) = 34,400 \text{ N}$$

From Equation (16.31) the maximum load that the bolt can carry while assuming a safety factor of 2.5 is

$$P_{\text{max},b} = \frac{A_tS_p - P_i}{n_{sb}C_k} = \frac{(120.7)(380) - 34,400}{(2.5)(0.4053)} = 11,320 \text{ N}$$

From Equation (16.32) the maximum load before separation occurs is

$$P_{\text{max},j} = \frac{P_i}{n_{sj}(1 - C_k)} = \frac{34,400}{(2.5)(1 - 0.4053)} = 23,140 \text{ N}$$

Thus, failure due to separation will occur before bolt failure.

## 16.4.6 BOLT PRELOAD—DYNAMIC LOADING

The preload effect is even greater for dynamically loaded joints than for statically loaded joints. Figure 16.16 attempts to describe how the forces versus deflection of the bolt and joint will vary as a function of time. While considering Figure 16.16, recall Figures 16.10 and 16.11, which consider static loading. From Figure 16.16 observe that the bolt load is in tension that varies with time from the preload $P_i$ to some maximum bolt load $P_{b,\text{max}}$. Simultaneously, the joint is in compression that varies with time from the preload $P_i$ to the minimum joint load $P_{j,\text{min}}$. When bolt load is a maximum, the joint load has a minimum value and vice versa. Also, because of the differences in slope of the load and deflection curves for the bolt and the joint, the amount of load variation for one cycle is much larger for the joint than for the bolt.

Figure 16.16 also shows the deflection variation with time for both the bolt and the joint. The quantitative amount is the same, but for the bolt the deflection is an increased elongation whereas for the joint the contraction is decreased.

Since failure due to cyclic loading is more apt to occur to the bolt, we only consider the bolt here. Equation (16.16) gives the alternating and mean loads acting on the bolt as,

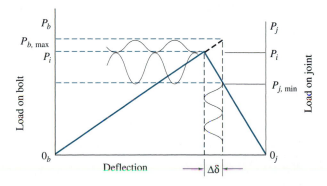

**Figure 16.16** Forces versus deflection of bolt and joint as function of time.

respectively,

$$P_{ba} = \frac{P_{b,\max} - P_{b,\min}}{2} = \frac{C_k(P_{\max} - P_{\min})}{2} = C_k P_a \tag{16.34}$$

$$P_{bm} = \frac{P_{b,\max} + P_{b,\min}}{2} = P_i + \frac{C_k(P_{\max} + P_{\min})}{2} = P_i + C_k P_m \tag{16.35}$$

The alternating and mean stresses for dynamic loading can be expressed as

$$\sigma_{Da} = \frac{C_k P_a n_s}{A_t} \tag{16.36}$$

$$\sigma_{Dm} = \frac{P_i + C_k P_m n_s}{A_t} \tag{16.37}$$

The safety factor is not applied to the preload.

The Goodman fatigue failure criterion given in Section 7.9.1 is used. Since the safety factor has been incorporated into the alternating and mean stresses [Eqs. (16.36) and (16.37)], the Goodman equation [Eq. (7.28) when $n_s = 1$] is

$$\frac{K_f \sigma_a}{S_e} + \frac{\sigma_m}{S_{ut}} = 1 \tag{16.38}$$

Substituting Equations (16.36) and (16.37) into Equation (16.38) gives

$$\frac{K_f C_k P_a n_s}{S_e A_t} + \frac{P_i + C_k P_m n_s}{A_t S_{ut}} = 1 \tag{16.39}$$

Equation (16.39) can also be expressed in terms of stresses as

$$\frac{n_s K_f C_k \sigma_a}{S_e} + \frac{\sigma_i + C_k \sigma_m n_s}{S_{ut}} = 1$$

Solving for the safety factor gives

$$n_s = \frac{S_{ut} - \sigma_i}{C_k[K_f \sigma_a(S_{ut}/S_e) + \sigma_m]} \tag{16.40}$$

The fatigue stress concentration factor $K_f$ for threaded elements can be obtained from Table 16.11. Cutting is the simplest method of producing the threads. However, rolling the threads instead of cutting them provides a smoother thread finish (fewer initial cracks). Rolling provides an unbroken flow of material grain in the thread region. A generous fillet between the head and the shank reduces the fatigue stress concentration, as shown in Table 16.11. It is usually safe to assume that the threads have been rolled, unless specific information is available.

**Table 16.11**    Fatigue stress concentration factors for threaded elements

| SAE grade | Metric grade | Rolled threads | Cut threads | Fillet |
|-----------|--------------|----------------|-------------|--------|
| 0–2       | 3.6–5.8      | 2.2            | 2.8         | 2.1    |
| 4–8       | 6.6–10.9     | 3.0            | 3.8         | 2.3    |

The smallest safety factor for fatigue loading is the ratio of ultimate stress to prestress (if the mean and alternating stresses are zero). Examining the values in Table 16.7 and the requirements of Equation (16.33) leads to the conclusion that the safety factor for bolt fatigue cannot be greater than 2.4 and is usually below 2.0.

The modified endurance limit given in Equation (7.16) will be used. The surface finish factor $k_f$ is assumed to have been incorporated into the fatigue stress concentration factor. Also, the size factor is 1 for axial loading. If no specific information is given about the reliability factor, it is recommended to assume a survival probability of 90%. The endurance limit is obtained from Eq. (7.7) for various types of loading.

The previous section discussed the importance of preload in statically loaded connections. With dynamic or cyclic loads a preload is especially important. The reasons can be seen in Equations (16.36) and (16.37) and by recalling some subtleties of fatigue analysis discussed in Chapter 7. As long as separation does not occur, the alternating stress experienced by the bolt is reduced by the dimensionless stiffness parameter $C_k$. The mean stress is increased by the preload. However, the failure theories in Sec. 7.9 (e.g., Gerber, Goodman, Soderberg, etc.) show that the alternating stress has a much greater effect on service life than the mean stress. That is, the endurance limit is usually a small fraction of the yield or ultimate strength. Thus, a large enough preload to prevent separation is critical. For the purposes of this text this preload can be obtained by using the relations for preload given by Equation (16.33).

---

**EXAMPLE 16.8**

**Given:** The hexagonal bolt-and-nut assembly with one steel and one aluminum member shown in Figure 16.14(a) and considered in Examples 16.6 and 16.7 has an external load of 11,000 N applied cyclically in a released-tension manner.

**Find:** Find the safety factor guarding against fatigue failure of the bolt. Assume a survival probability of 90%, rolled threads, and a 5.8 metric grade.

**Solution:** For released tension with a maximum load of 11,000 N

$$P_a = P_m = 5500 \text{ N}$$

The endurance limit for axial loading from Equation (7.7) and Table 16.8 for 5.8-metric-grade thread is

$$S'_e = 0.45 S_u = 0.45(520)(10^6) \text{ Pa} = 234 \text{ MPa}$$

From Table 16.11 for 5.8-metric-grade thread and rolled thread, the fatigue stress concentration factor $K_f = 2.2$. From Table 7.4 the reliability factor $k_r$ for 90% survival probability is 0.90. Thus, the modified endurance limit from Equation (7.16) is

$$S_e = k_f k_s k_r k_m S'_e = (1)(1)(0.9)(1)(234)(10^6) \text{ Pa} = 210.6 \text{ MPa}$$

From Table 16.10 for a crest diameter of 14 mm and coarse threads the tensile stress area is 120.7 mm². From Equation (16.40)

$$n_s = \frac{S_{ut} - \sigma_i}{\sigma_a C_k (K_f S_{ut}/S_e + 1)} = \frac{520(10^6) - 285(10^6)}{45.57(10^6)(0.4053)[2.2(520/210.6) + 1]} = 1.99$$

Bolt fatigue failure is not expected.

### 16.4.7   GASKETED JOINTS

Figure 16.17 shows a threaded fastener with an unconfined **gasket** joining two members. An unconfined gasket extends over the entire diameter (as shown in Fig. 16.17) of a joint, whereas a confined gasket extends over only a part of the joint diameter. A confined gasket may be considered as if the gasket were not present.

An unconfined gasket is made of a soft elastomeric material. Recall that the various members of the joint can be considered as individual springs with different spring rates depending on their materials. The modulus of elasticity is much less for the gasket than for the other joint members ($E_g \ll E_1, E_2, \ldots$). Since the modulus of elasticity is directly proportional to the stiffness,

$$k_g \ll k_1, k_2, \ldots \tag{16.41}$$

Furthermore,

$$\frac{1}{k_g} \gg \frac{1}{k_1}, \frac{1}{k_2}, \ldots$$

Making use of the above and Equation (16.25) gives for an unconfined gasket

$$\frac{1}{k_j} = \frac{1}{k_1} + \frac{1}{k_2} + \cdots + \frac{1}{k_g} \approx \frac{1}{k_g} \tag{16.42}$$

$$\therefore \qquad k_j = k_g$$

Thus, for unconfined gaskets the joint stiffness is equivalent to the gasket stiffness.

A gasketed joint must satisfy Equations (16.17) and (16.18) for preload. In addition, the preload must be large enough to achieve the minimum sealing pressure required for the gasket material. Thus,

$$P_i \geq A_g p_0 \tag{16.43}$$

**Figure 16.17**   Threaded fastener with unconfined gasket and two other members.

where $A_g$ = gasket area per bolt, m$^2$
  $p_0$ = minimum gasket seal pressure, Pa

Thus, for gasketed joints Equation (16.43) must be checked.

The material covered thus far is for axial loading. Threaded fasteners can also experience shear. Shear is considered next, since riveted joints in shear are treated the same as threaded fasteners in shear.

## 16.5    RIVETED FASTENERS

**Rivets** are in wide use as fasteners of members in buildings, bridges, boilers, tanks, ships, and (in general) for any framework. In design the rivet diameter is used rather than the hole diameter, even through the rivet diameter may expand on assembly and nearly fill the hole.

The reasons for choosing rivets over threaded fasteners are the following:

1.  Rivets will not shake loose.

2.  They are cheap, especially in assembly time.

3.  They are lightweight.

4.  They can be assembled from the blind side.

However, rivets cannot provide as strong an attachment as a threaded fastener of the same diameter. Also, rivets do not allow any disassembly of the joint.

Rivets must be spaced neither too close nor too far apart. The minimum rivet spacing, center to center, for structural steel work is usually taken as three rivet diameters. A somewhat closer spacing is used in boilers. Rivets should not be spaced too far apart, or else plate buckling will occur. The maximum spacing is usually taken as 16 times the thickness of the outside plate.

Riveted and threaded fasteners in shear are treated exactly alike in design and failure analysis. Failure due to a shear force is primarily caused by the four different modes of failure shown in Figure 16.18: bending of a member, shearing of the rivet, tensile failure of a member, and the bearing of the rivet on a member or the bearing of a member on the rivet. Each of these four modes is presented here:

1.  **Bending of a member.**    To avoid this failure, the following should be valid:

$$\sigma = \frac{PL_g}{2Z_m} < 0.6(S_y)_j \qquad\qquad (16.44)$$

where $L_g$ = length of grip, $L_s + L_t$ (see Fig. 16.12), m
  $Z_m$ = section modulus of weakest member, $I/c$, m$^3$
  $(S_y)_j$ = yield strength of weakest member, Pa

2.  **Shear of rivets.**    To avoid this failure, the following should be valid:

$$\tau = \frac{4P}{\pi d_c^2} < S_{sy} = 0.4S_y \qquad\qquad (16.45)$$

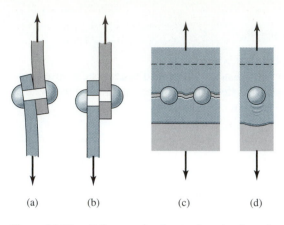

**Figure 16.18**    Failure modes due to shear loading of riveted fasteners. (a) Bending of member; (b) shear of rivet; (c) tensile failure of member; (d) bearing of member on rivet.

where $d_c$ = crest diameter, m

$S_{sy}$ = rivet yield strength in shear, Pa

In Equation (16.45) the diameter of the rivet rather than the diameter of the hole is used. Shear of rivets is a major failure mode considered in the design of threaded or riveted fasteners. This mode is also called the *direct shear* mode of failure.

3. **Tensile failure of member.**    To avoid this failure, the following should be valid:

$$\sigma = \frac{P}{(b - N_r d_c)t_m} < \left(S_y\right)_j \tag{16.46}$$

where $b$ = width of member, m

$N_r$ = number of rivets in width of member

$t_m$ = thickness of thinnest member, m

4. **Compressive bearing failure.**    To avoid this failure, the following should be valid:

$$\sigma = \frac{P}{d_c t_m} < 0.9\left(S_y\right)_j \tag{16.47}$$

These four failure modes relate to failure of an individual rivet or an individual member. Threaded and riveted fasteners are used in groups, and the torsional shear of the group must be considered as a failure mode. The result of the shear stress acting on the rivet is the vectorial sum of the direct and torsional shear stresses

$$\tau = \tau_d + \tau_t \tag{16.48}$$

where $\tau_d$ = shear stress of rivet, or direct shear, Pa

$\tau_t$ = shear stress of rivet due to torsional loading, Pa

When one is dealing with a group of threaded or riveted fasteners, the centroid of the group must be evaluated. Equilibrium of vertical and horizontal loads must be observed.

**Given:** The riveted member shown in Figure 16.19(a) is fastened to a beam, and an eccentric load is applied. Rivets A and C have $\frac{5}{8}$-in diameter, and rivets B and D have $\frac{7}{8}$-in diameter.

EXAMPLE 16.9

**Find:** Determine the following:

  (a)   Centroid of the rivet assembly
  (b)   Direct and resultant shear stresses acting on each rivet
  (c)   Safety factor guarding against shear of the rivet
  (d)   Safety factor guarding against bending of the member

Assume that the yield stress of the rivet is 85 ksi and that of the member is 54 ksi.

**Figure 16.19** Group of riveted fasteners used in Example 16.9. (a) Assembly of rivet group; (b) radii from centroid to center of rivets; (c) resulting triangles; (d) direct and torsional shear acting on each rivet; (e) side view of member. (All dimensions are in inches.)

**Solution:**

(a) The total area of the four rivets is

$$\sum A_i = \frac{\pi}{2}\left[\left(\frac{5}{8}\right)^2 + \left(\frac{7}{8}\right)^2\right] = 1.816 \text{ in}^2$$

In Figure 16.19(a) the $(x, y)$ coordinate is at the center of the C rivet, the rivet in the lower left corner. The centroid of the rivet group is

$$\bar{y} = \frac{\sum A_i \bar{y}_i}{\sum A_i} = \frac{0.3068(0+6) + 0.6013(0+6)}{1.816} = 3.000 \text{ in}$$

$$\bar{x} = \frac{\sum A_i \bar{x}_i}{\sum A_i} = \frac{0.3068(0+0) + 0.6013(7+7)}{1.816} = 4.635 \text{ in}$$

The eccentricity of the loading is

$$e_c = 8 + 7 - 4.635 = 10.36 \text{ in}$$

Figure 16.19(b) shows the distances from the center of the rivets to the centroid, and Figure 16.19(c) shows the resulting triangles. From these figures

$$r_A = r_C = \sqrt{3^2 + 4.635^2} = 5.521 \text{ in}$$

$$r_B = r_D = \sqrt{3^2 + 2.365^2} = 3.820 \text{ in}$$

(b) The direct shear stress acting on each of the rivets is

$$\tau_d = \frac{P}{\sum A} = \frac{1000 \text{ lb}}{1.816 \text{ in}^2} = 550.7 \text{ psi}$$

Figure 16.19(d) illustrates the direct and torsional shear stresses acting on each rivet. The torque due to eccentric loading is

$$T = Pe_c = (1000 \text{ lb})(10.36 \text{ in}) = 10{,}360 \text{ lb} \cdot \text{in}$$

$$\sum r_j^2 A_j = 2\left(r_C^2 A_C + r_D^2 A_D\right)$$

$$\sum r_j^2 A_j = 2\left[(5.525)^2 (0.3068) + (3.820)^2 (0.6013)\right] = 36.28 \text{ in}^4$$

The shear stress due to torsional loading is

$$\tau_{ti} = \frac{Tr_i}{\sum r_j^2 A_j} = \frac{10{,}360 r_i}{36.28} = 285.6 r_i$$

$$\tau_{tA} = \tau_{tC} = (285.6)(5.521) = 1577 \text{ psi}$$

$$\tau_{tB} = \tau_{tD} = (285.6)(3.820) = 1091 \text{ psi}$$

The $x$ and $y$ components of the torsional stresses are

$$\tau_{tAx} = \tau_{tCx} = \tau_{tA} \sin \alpha = 1577 \sin 32.91° = 856.9 \text{ psi}$$

$$\tau_{tAy} = \tau_{tCy} = \tau_{tA} \cos \alpha = 1577 \cos 32.91° = 1324 \text{ psi}$$

$$\tau_{tBx} = \tau_{tDx} = \tau_{tB} \sin \beta = 1091 \sin 51.75° = 856.8 \text{ psi}$$

$$\tau_{tBy} = \tau_{tDy} = \tau_{tB} \cos \beta = 1091 \cos 51.75° = 675.4 \text{ psi}$$

The resultant shear stress at the four rivets, from Figure 16.19(d), is

$$\tau_A = \tau_C = \sqrt{\left(\tau_{tAy} - \tau_d\right)^2 + \tau_{tAx}^2} = \sqrt{(1324 - 550.7)^2 + (856.8)^2} = 1154 \text{ psi}$$

$$\tau_B = \tau_D = \sqrt{\left(\tau_{tDy} + \tau_d\right)^2 + \tau_{tDx}^2} = \sqrt{(675.4 + 550.7)^2 + (856.8)^2} = 1496 \text{ psi}$$

The critical shear stress is at rivets B and D. Once the shear stress is known, the shear force can be obtained from

$$P_B = P_D = \tau_B A = (1496)(0.6013) = 899.5 \text{ lb}$$

From Equation (3.14) the allowable shear stress due to yielding is

$$\tau_{all} = 0.4 S_y = (0.4)(85)(10^3) \text{ psi} = 34.0 \text{ ksi}$$

(c)  The safety factor guarding against shear of the rivets is

$$n_s = \frac{\tau_{all}}{\tau_D} = \frac{34.0}{1.496} = 22.73$$

(d)  Figure 16.19(e) shows a side view of the member at a critical section where there are two rivets across the width. The distance from the neutral axis to the outer fiber is 5 in. The bending of the member, from Figure 16.19(a), is

$$M_B = (1000)(8) = 8000 \text{ lb} \cdot \text{in}$$

The area moment of inertia for the critical section shown in Figure 16.19(e) is

$$I = \tfrac{1}{12}\left(\tfrac{1}{2}\right)(10^3) - 2\left[\left(\tfrac{1}{12}\right)\left(\tfrac{1}{2}\right)\left(\tfrac{7}{8}\right)^3 + \left(\tfrac{1}{2}\right)\left(\tfrac{7}{8}\right)(3)^2\right]$$
$$= 41.67 - 2(0.0279 + 3.938) = 33.74 \text{ in}^4$$

The bending stress at rivet B is

$$\sigma_B = \frac{M_B c}{I} = \frac{(8000)(5)}{33.74} \text{ psi} = 1.186 \times 10^3 \text{ psi}$$

From Equation (3.15) the allowable stress due to bending is

$$\sigma_{all} = 0.6 S_y = (0.6)(54)(10^3) \text{ psi} = 32.4 \times 10^3 \text{ psi}$$

The safety factor guarding against bending of the member is

$$n_s = \frac{\sigma_{all}}{\sigma_B} = \frac{32.4}{1.186} = 27.3$$

Therefore, if failure occurred, it would occur first from shear of the rivets. However, because the safety factor guarding against shear of the rivets is 22.73, it is highly unlikely that failure will occur.

# 16.6   WELDED JOINTS

**A weld** is accomplished by forcing diffusion at the interface through the application of heat and/or pressure and sometimes with a filler metal. There are a number of welding methods, but this text focuses on one—shielded metal arc welding (generally called *arc welding*).

In this method the two members are placed in close proximity, both of them in contact with an electrical conductor. A low-voltage, high-current arc is struck with an electrode to complete the electric circuit at the joint. The distance between the electrode and the work is controlled such that the electric arc is sustained, creating temperatures high enough to melt the members at the joint. The electrode is gradually consumed in the process, supplying additional metal to the joint as it melts. A flux covering the electrode provides a shielding gas (hence, the term *shielded metal arc welding*) and a chalky covering (slag), both of which serve to protect the metal from oxidation. The slag is usually chipped off once the weld has cooled sufficiently.

Some advantages of welded joints over threaded fasteners are that they are inexpensive and there is no danger of the joint loosening. Some disadvantages of welded joints over threaded fasteners are that they produce residual stresses, they distort the shape of the member, metallurgical changes occur, and disassembly is a problem.

There are many configurations of welds and manufacturing processes for producing them. Figure 16.20 shows the standard symbols for welds, indicating that weld joints offer considerable design flexibility. Although welded joints are available in a wide variety of

| | | | Basic arc and gas weld symbols | | | | |
|---|---|---|---|---|---|---|---|
| Bead | Fillet | Plug or slot | Groove | | | | |
| | | | Square | V | Bevel | U | J |
| ⌒ | ◺ | ▽ | ‖ | ∨ | ⩔ | ⨆ | ⨆ |

| Basic resistance weld symbols | | | |
|---|---|---|---|
| Spot | Projection | Seam | Flash or upset |
| ✕ | ✕ | ✕✕✕ | | |

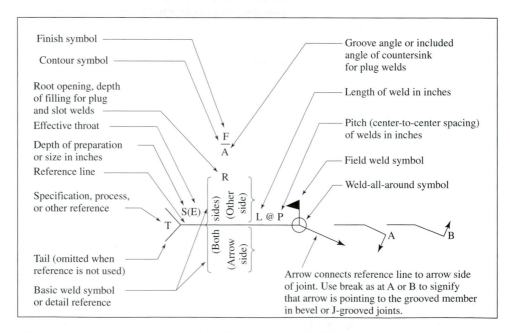

**Figure 16.20**   Basic weld symbols. [From Kalpakjian and Schmid (2001).]

**Figure 16.21** Fillet weld. (a) Cross section of weld showing throat and legs; (b) shear planes.

forms, only the fillet weld is considered here. This section examines one configuration with the expectation that, once it is understood, application to other configurations will be relatively simple.

A fillet weld, shown in Figure 16.21(a), is made with equal legs. The thinnest section is then at the **throat** of the weld, at 45° from the legs. The governing stress in fillet welds is shear on the throat of the weld, as shown in Figure 16.21(b). Observe from this figure that in the fillet weld aligned parallel to the load, the shear stress occurs along the throat of the fillet parallel to the load. In a fillet weld aligned transverse to the load, the shear stress occurs at 45° to the load, acting transverse to the axis of the fillet.

## 16.6.1 PARALLEL AND TRANSVERSE LOADING

Fillet welds fail by shearing at the minimum section, which is at the throat of the weld, shown in Figure 16.21(a). This is true whether the weld has parallel (on the side) or transverse [at the end, as shown in Fig. 16.21(b)] loading. The shear stress from these types of loading is

$$\tau = \frac{P}{t_e L_w} = \frac{P}{0.707 h_e L_w} = \frac{1.414P}{h_e L_w} \qquad \textbf{(16.49)}$$

where $t_e$ = weld throat length, $h_e \sin 45° = 0.707 h_e$, m
$\quad\ h_e$ = weld leg length, m
$\quad\ L_w$ = weld length, m

Thus, to avoid failure, the following should be valid:

$$\tau = \frac{P}{t_e L_w} < \left(S_{sy}\right)_{weld} \tag{16.50}$$

## 16.6.2    TORSIONAL LOADING

For torsional loading of a weld group, the resultant shear stress acting on the weld group is the vectorial sum of the direct and torsional shear stresses, as shown in Equation (16.48) for rivets. The direct (or transverse) shear stress in the weld is

$$\tau_d = \frac{V}{A} = \frac{\text{shear force}}{\text{total throat area}} \tag{16.51}$$

The torsional shear stress is

$$\tau_t = \frac{Tr}{J} \tag{16.52}$$

where $r$ = distance from centroid of weld group to farthest point in weld, m
$\quad\ T$ = torque applied to weld, N · m
$\quad\ J$ = polar area moment of inertia, m$^4$

The critical section for torsional loading is the throat section, as it is for parallel and transverse loading. The relationship between the unit polar moment of inertia and the polar moment of the fillet weld is

$$J = t_e J_u = 0.707 h_e J_u \tag{16.53}$$

where $J_u$ = unit polar area moment of inertia, m$^3$. Table 16.12 gives values of unit polar area moment of inertia for nine weld groups. Using this table simplifies obtaining torsional loading.

Thus, to avoid failure due to torsional loading, the following should be valid:

$$\tau = \tau_d + \tau_t < \left(S_{sy}\right)_{weld} \tag{16.54}$$

## 16.6.3    BENDING

In bending, the welded joint experiences a transverse shear stress as well as a normal stress. The direct (or transverse) shear stress is the same as that given in Equation (16.49). The moment $M$ produces a normal bending stress $\sigma$ in the welds. It is customary to assume that the stress acts normal to the throat area.

$$I = t_e I_u L_w = 0.707 h_e I_u L_w \tag{16.55}$$

where $I_u$ = unit area moment of inertia, m$^2$
$\quad\ L_w$ = length of weld, m

**Table 16.12**   Geometry of welds and parameters used when considering various types of loading

| Dimensons of weld | Bending | Torsion |
|---|---|---|

$A = d$

$I_u = d^2/6$

$M = Pa$

$J_u = d^3/12$

$T = Pa$

$c = d/2$

$A = 2d$

$I_u = d^2/3$

$J_u = \dfrac{d(3b^2 + d^2)}{6}$

$A = 2b$

$I_u = bd$

$J_u = \dfrac{b^3 + 3bd^2}{6}$

$A = b + d$

$\bar{x} = \dfrac{b^2}{2(b+d)}$

$\bar{y} = \dfrac{d^2}{2(b+d)}$

At top $I_u = \dfrac{4bd + d^2}{6}$

At bottom $I_u = \dfrac{d^2(4b + d)}{6(2b + d)}$

$J_u = \dfrac{(b+d)^4 - 6b^2d^2}{12(b+d)}$

$A = d + 2b$   $\bar{x} = \dfrac{b^2}{2(b+d)}$

$I_u = bd + d^2/6$

$J_u = \dfrac{(2b+d)^3}{12} - \dfrac{b^2(b+d)^2}{2(b+d)}$

*(continued)*

**Table 16.12**    (concluded)

| Dimensons of weld | Bending | Torsion |
|---|---|---|

$A = b + 2d$    $\bar{y} = \dfrac{d^2}{b + 2d}$

Weld — Weld

At top $I_u = \dfrac{2bd + d^2}{3}$

At bottom $I_u = \dfrac{d^2(2b + d)}{3(b + d)}$

Weld — Weld

$J_u = \dfrac{(b + 2d)^3}{12} - \dfrac{d^2(b + d)^2}{b + 2d}$

$A = 2b + 2d$

Weld all around

$I_u = bd + d^2/3$

Weld all around

$J_u = \dfrac{(b + d)^3}{6}$

$A = 2b + 2d$

Weld — Weld

$I_u = bd + d^2/3$

Weld — Weld

$J_u = \dfrac{b^3 + 3bd^2 + d^3}{6}$

$A = \pi b$

Weld all around

$I_u = \pi(d^2/4)$

Weld all around

$J_u = \pi(d^3/4)$

|  SOURCE: Mott (1998).

**Table 16.13** Minimum strength properties of electrode classes

| Electrode number | Ultimate tensile strength $S_u$, ksi | Yield strength $S_y$, ksi | Elongation $e_k$, % |
|---|---|---|---|
| E60XX | 62 | 50 | 17–25 |
| E70XX | 70 | 57 | 22 |
| E80XX | 80 | 67 | 19 |
| E90XX | 90 | 77 | 14–17 |
| E100XX | 100 | 87 | 13–16 |
| E120XX | 120 | 107 | 14 |

Table 16.12 gives values of unit area moment of inertia for nine weld groups. The force per unit length of the weld is

$$w' = \frac{Pa}{I_u} \tag{16.56}$$

where $a = $ distance from wall to applied load, m. The normal stress due to bending is

$$\sigma = \frac{Mc}{I} \tag{16.57}$$

where $c = $ distance from neutral axis to outer fiber, m. Once the shear stress and the normal stress are known, the principal shear stresses [from Eq. (2.18)] or the principal normal stresses [from Eq. (2.16)] can be determined. Once these principal stresses are obtained, the maximum-shear-stress theory (MSST) can be determined from Equation (6.6), or the distortion energy theory (DET) can be determined from Equation (6.7) to establish if the weld group will fail.

### 16.6.4 WELD STRENGTH

The electrodes used in arc welding are identified by the letter E followed by a four-digit number, such as E6018. The first two numbers indicate the strength of the deposited material in thousands of pounds per square inch. The last digit indicates variables in the welding technique, such as current supply. The next-to-last digit indicates the welding position as, for example, flat, vertical, or overhead.

Table 16.13 lists the minimum strength properties for some electrode classes. Once the yield strength of the weld and the type of loading are known from Equations (3.13) to (3.16), the allowable stress can be determined. Given the allowable and design stresses, the safety factor can be determined from Equation (1.1).

---

**EXAMPLE 16.10**

**Given:** A bracket is welded to a beam, as shown in Figure 16.22(a). Assume a steady loading of 20 kN and weld lengths $l_1 = 150$ mm and $l_2 = 100$ mm. Assume an electrode number of E60XX and a fillet weld.

**Find:** Determine the weld leg length for the eccentric loading shown in Figure 16.22(a) while considering torsion but no bending for a safety factor of 2.5.

**Figure 16.22**   Welded bracket used in Example 16.10. (a) Dimensions, load, and coordinates; (b) torsional shear stress components at points A and B. (All dimensions are in millimeters.)

**Solution:**   The sum of the weld areas is

$$\sum A_i = t_e(l_1 + l_2) = 0.7071 h_e(150 + 100) = 176.8 h_e = 250 t_e$$

The centroid of the weld group from Equations (4.5) and (4.6) is

$$\bar{x} = \frac{\sum A_i x_i}{\sum A_i} = \frac{0(150)t_e + 50(100)t_e}{250 t_e} = 20 \text{ mm}$$

$$\bar{y} = \frac{\sum A_i y_i}{\sum A_i} = \frac{75(150)t_e + 150(100)t_e}{250 t_e} = 105 \text{ mm}$$

The torque being applied is

$$T = Pr = (20,000)(0.3 - 0.02) \text{ N} \cdot \text{m} = 5.6 \text{ kN} \cdot \text{m}$$

From Table 16.12 for torsional consideration and the weld shown in Figure 16.22(a), the unit polar area moment of inertia is

$$J_u = \frac{(b + d)^4 - 6b^2 d^2}{12(b + d)} = \frac{(100 + 150)^4 - 6(100)^2(150)^2}{12(100 + 150)} = 852,083 \text{ mm}^3$$

The formulas given in Table 16.12 can be derived by using the parallel-axis theorem given in Equations (4.11) and (4.12). From Equation (4.11) for weld 1

$$I_{x'} = \frac{l_1^3}{12} + l_1\left(\bar{y} - \frac{l_1}{2}\right)^2 = \frac{(150)^3}{12} + 150(105 - 75)^2 = 416,250 \text{ mm}^3$$

From Equation (4.12) for weld 1

$$I_{y'} = l_1(\bar{x})^2 = 150(20)^2 = 60,000 \text{ mm}^2$$

The unit polar area moment for weld 1 is

$$(J_u)_1 = I_{x'} + I_{y'} = 476,250 \text{ mm}^3$$

For weld 2

$$I_{x'} = l_2(l_1 - \bar{y})^2 = 202{,}500 \text{ mm}^3$$

$$I_{y'} = \frac{l_2^3}{12} + l_2\left(\bar{x} - \frac{l_2}{2}\right)^2 = \frac{(100)^3}{12} + 100(50 - 20)^2 = 173{,}333 \text{ mm}^3$$

The unit polar area moment of inertia for weld 2 is

$$(J_u)_2 = I_{x'} + I_{y'} = 375{,}833 \text{ mm}^3$$

Thus, the unit polar area moment of inertia for welds 1 and 2 is

$$J_u = (J_u)_1 + (J_u)_2 = 852{,}083 \text{ mm}^3$$

This is the same result as that obtained by using the formula from Table 16.12.
The polar area moment of inertia is

$$J = t_e J_u = 0.7071 h_e J_u = 602{,}508 h_e$$

The polar area moment of intertia $J$ is expressed in millimeters raised to the fourth power.
From Equation (16.51) the direct (or transverse) shear at locations A and B shown in Figure 16.22 is

$$\tau_{dA} = \tau_{dB} = \frac{P}{A} = \frac{20{,}000}{176.8 h_e} = \frac{113.1}{h_e}$$

The weld leg length is in millimeters, and the shear stress is in megapascals.
From Figure 16.22(b) the torsional shear stress components at point A are

$$\tau_{tAx} = \frac{T(45)}{J} = \frac{(5.6)(10^6)(45)}{(0.6024)h_e} \text{ Pa} = \frac{418.3}{h_e} \quad \text{MPa}$$

$$\tau_{tAy} = \frac{T(80)}{J} = \left(\frac{80}{45}\right)\left(\frac{418.3 \times 10^6}{h_e}\right) \text{ Pa} = \frac{743.7}{h_e} \quad \text{MPa}$$

$$\tau_{Ax} = \tau_{tAx} = \frac{418.3}{h_e} \quad \text{MPa}$$

$$\tau_{Ay} = \tau_{dA} + \tau_{tAy} = \left(\frac{113.1}{h_e} + \frac{743.7}{h_e}\right)(10^6) \text{ Pa} = \frac{856.8}{h_e} \quad \text{MPa}$$

$$\therefore \quad \tau_A = \sqrt{\tau_{Ax}^2 + \tau_{Ay}^2} = \frac{1}{h_e}\sqrt{(418.3)^2 + (856.8)^2}(10^6) \text{ Pa} = \frac{953.5}{h_e} \quad \text{MPa}$$

Similarly, at point B,

$$\tau_{tBx} = \frac{T(105)}{J} = \frac{(5.6)(10^6)(105)}{(0.6024)h_e} \text{ Pa} = \frac{976.1}{h_e} \quad \text{MPa}$$

$$\tau_{tBy} = \frac{T(20)}{J} = \left(\frac{20}{105}\right)\left(\frac{976.1 \times 10^6}{h_e}\right) \text{ Pa} = \frac{185.9}{h_e} \quad \text{MPa}$$

$$\tau_{Bx} = \tau_{tBx} = \frac{976.1}{h_e} \quad \text{MPa}$$

$$\tau_{By} = \tau_{dB} - \tau_{tBy} = \left(\frac{185.9}{h_e} - \frac{113.1}{h_e}\right)(10^6)\ \text{Pa} = \frac{72.82}{h_e}\quad \text{MPa}$$

$$\therefore\quad \tau_B = \sqrt{\tau_{Bx}^2 + \tau_{By}^2} = \frac{1}{h_e}\sqrt{(976.1)^2 + (72.82)^2}(10^6)\ \text{Pa} = \frac{978.8}{h_e}\quad \text{MPa}$$

Because the shear stress is larger at point B, this point is selected for design.

From Table 16.13 for electrode E60XX, the yield strength is 50 ksi. From Equation (3.14) for shear, the allowable shear stress is

$$\tau_{all} = 0.40 S_y = 0.40(50)(10^3) = 20\ \text{ksi} = 138.0\ \text{MPa}$$

From Equation (1.1) the safety factor is

$$n_s = \frac{\tau_{all}}{\tau_B} = \frac{138.0}{978.8/h_e}$$

Given that the safety factor is 2.5,

$$\therefore\quad h_e = \frac{(978.8)(2.5)}{138.0} = 17.73\ \text{mm}$$

### 16.6.5    FATIGUE STRENGTH OF WELDS

When welded members are placed into an environment where cyclic loadings are experienced, the welds fail long before the welded members. Since the electrode material contains a large amount of alloying elements, it is relatively strong, and it is not clear why the weld strength is suspect. However, because a stress concentration is associated with every weld, the stresses are highest in the immediate vicinity of the weld. Further, cracks rarely propagate through the weld material, but rather fatigue failure arises by crack propagation through the *heat-affected zone* (HAZ) of the welded material. The HAZ is the area around the weld that recrystallizes as a wrought structure during the welding process. The HAZ material is weaker than the cold-worked substrate and can have micropores typical of castings.

Numerous machine design textbooks suggest that welded members should not be used at all in fatigue applications because of these shortcomings. However, since welded connections are routinely placed in situations where fatigue is possible, such admonitions provide no assistance to machine designers.

Recognizing the difficulties associated with determining the actual magnitude of the stress concentration associated with welds, Shigley and Mischke (1989) recommend the fatigue strength reduction factors shown in Table 16.14. These factors should be applied to the member's wrought material strength (regardless of the manufacturing history prior to welding) as well as to the weld material strength.

Note that some welding processes, notably laser and electron-beam welding, have much smaller heat-affected zones and are therefore less detrimental to fatigue applications than processes such as shielded metal arc welding. Also, it is common to perform shot peening (see Sec. 7.7.6) on highly stressed welds to improve both static and fatigue strength. For such situations, the recommendations in Table 16.14 are conservative.

**Table 16.14**      Fatigue strength reduction factors for welds

| Type of weld | Fatigue stress concentration factor $K_f$ |
|---|---|
| Reinforced butt weld | 1.2 |
| Tow of transverse fillet weld | 1.5 |
| End of parallel fillet weld | 2.7 |
| T-butt joint with sharp corners | 2.0 |

SOURCE: Shigley and Mischke (1989).

## 16.7   ADHESIVE BONDING

**Adhesive bonding** is the process of joining materials chemically through the formation of attractive interatomic or intermolecular forces. Adhesive bonding can be used to join a wide variety of materials (metal to metal, metal to ceramic, metal to polymer, etc.) for both structural and nonstructural uses. The application of adhesive technology in the aerospace industry is responsible for significant recent progress in the understanding of adhesive bonding and its use as a material joining method.

Some of the advantages of adhesive bonding over other fastening techniques covered earlier in this chapter are

1.   Uniform stress distribution with resultant increased life

2.   Reduced weight

3.   Fatigue resistance

4.   Ability to join thick or thin materials

5.   Ability to join dissimilar materials

6.   No stress concentration, since no perforation required

7.   Leakproof joints

8.   Vibration-damping and insulation properties

9.   Load spreading over large area

Some of the limitations of adhesive bonding are

1.   Possible need for surface preparation

2.   Temperature limitation of 500°F (260°C)

3.   Tendency to creep under sustained load

4.   Questionable long-term durability

Given these advantages and disadvantages it is understandable why adhesive bonding is so popular.

Four adhesive-bonded joints are shown in Figure 16.23. Figure 16.23(a) shows a lap joint with a large bond area, whereas Figure 16.23(b) shows a butt joint with a small bond area. The scarf joint shown in Figure 16.23(c) has a larger bond area than the butt joint, but

(a)

(b)

(c)

(d)

**Figure 16.23** Four methods of applying adhesive bonding. (a) Lap; (b) butt; (c) scarf; (d) double lap.

scarf joints are impractical for thin plates. The double lap joint shown in Figure 16.23(d) has the largest bonding area.

If the adhesion for a lap joint is uniform over the lapped surfaces, as shown in Figure 16.23(a), the average shear stress is

$$\tau_{avg} = \frac{P}{A} = \frac{P}{bL} \tag{16.58}$$

The shear stress would not be constant over the area and would be highest at the edges and lowest in the center. For a typical configuration we might expect that

$$\tau_{max} = 2\tau_{avg} = \frac{2P}{bL} \tag{16.59}$$

The stress distribution factor varies with aspect ratio $b/L$, but for an aspect ratio of 1 it is 2.

For a scarf joint adhesively bonded under axial load, as shown in Figure 16.24(a), while assuming a uniform stress distribution,

$$\sigma_x = \frac{P}{bt_m/\sin\theta} = \frac{P\sin\theta}{bt_m} \tag{16.60}$$

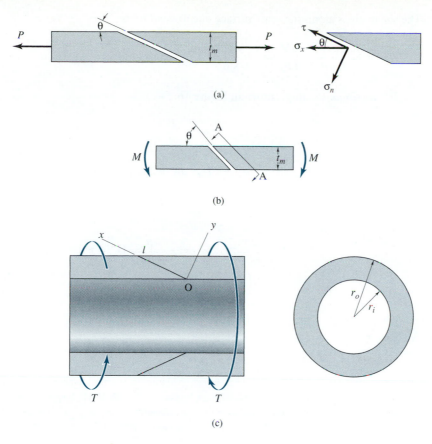

**Figure 16.24**    Scarf joint. (a) Axial loading; (b) bending; (c) torsion.

where $t_m$ = thickness of thinnest member, m. The normal and shear stress components can be written as, respectively,

$$\sigma_n = \sigma_x \sin\theta = \frac{P}{bt_m} \sin^2\theta \qquad (16.61)$$

$$\tau = \sigma_x \cos\theta = \frac{P}{bt_m} \sin\theta \cos\theta = \frac{P}{2bt_m} \sin 2\theta \qquad (16.62)$$

For *bending,* shown in Figure 16.24(b), the stress distribution along the scarf surface (A-A) is

$$\sigma_n = \frac{Mc}{I}$$

$$\therefore \qquad \sigma_n = \frac{M[t_m/(2\sin\theta)]}{(b/12)(t_m/\sin\theta)^3} = \frac{6M \sin^2\theta}{bt_m^2} \qquad (16.63)$$

The shear stress along the scarf surface due to *bending* is

$$\tau = \frac{\sigma_n}{\tan \theta} = \frac{6M}{bt_m^2} \sin^2 \theta \, \frac{\cos \theta}{\sin \theta} = \frac{6M}{bt_m^2} \sin \theta \cos \theta = \frac{3M}{bt_m^2} \sin 2\theta \qquad (16.64)$$

For *torsional* loading, shown in Figure 16.24(c),

$$\tau = \frac{Tr}{J}$$

where

$$J = \int r^2 \, dA = 2\pi \int_0^l r^3 \, dx$$

but

$$dx = \frac{dr}{\sin \theta}$$

$$J = \frac{2\pi}{\sin \theta} \int_{r_i}^{r_o} r^3 \, dr = \frac{\pi}{2 \sin \theta} \left( r_o^4 - r_i^4 \right)$$

$$\therefore \qquad \tau = \frac{2Tr \sin \theta}{\pi \left( r_o^4 - r_i^4 \right)} \qquad (16.65)$$

Also, $\sigma_n = 0$.

---

**EXAMPLE 16.11**

**Given:** A scarf joint in a hollow, round shaft transmits 150 kW at 1200 rpm. The shaft has an inner- to outer-radius ratio of 0.8, and the allowable shear stress is 30 MPa. The scarf joint has a 30° orientation with respect to the shaft axis. Use a safety factor of 2.0.

**Find:** The contact area.

**Solution:** From Equation (4.42) the torque in newton-meters is

$$T = \frac{h_p}{\omega}$$

where $h_p$ = power, W
$\omega$ = rotational speed, rad/s

Converting from revolutions per minute to radians per second requires that

$$\omega = \frac{2\pi N_a}{60}$$

$$\therefore \qquad \omega = \frac{2\pi (1200)}{60} = 125.7 \text{ rad/s}$$

Equation (4.42) becomes

$$T = \frac{(150)(10^3)}{125.7} = 1194 \text{ N} \cdot \text{m}$$

From Equation (16.65) when $r = r_o$

$$\tau_{max} = \frac{2Tr_o \sin\theta}{\pi\left(r_o^4 - r_i^4\right)} = \frac{2T\sin\theta}{\pi r_o^3\left[1 - (r_i/r_o)^4\right]} \leq \frac{\tau_{all}}{n_s}$$

$$(r_o)_{cr} = \left\{\frac{2Tn_s\sin\theta}{\pi\tau_{all}\left[1 - (r_i/r_o)^4\right]}\right\}^{1/3} = \left[\frac{2(1194)(2)\sin 30°}{\pi(30)(10^6)(1 - 0.8^4)}\right]^{1/3} = 0.035 \text{ m} = 3.5 \text{ cm}$$

The normal area is

$$A_n = \pi\left(r_o^2 - r_i^2\right) = \pi r_o^2(1 - 0.8^2) = \pi(3.5^2)(1 - 0.8^2) = 13.85 \text{ cm}^2$$

The scarf area is

$$A = \frac{A_n}{\sin\theta} = \frac{13.85}{\sin 30°} = 27.71 \text{ cm}^2$$

## 16.8 INTEGRATED SNAP FASTENERS

Although not a separate machinery element, **snap fasteners** are extremely popular when integrated into machine components because they greatly simplify assembly. Figure 16.25 shows common integrated snap fasteners. Integrated snap fasteners are commonly used with plastic housings or components, since they can be molded directly into the elements. This is one of the most powerful mechanisms for simplifying assemblies and reducing component and manufacturing costs.

Analyzing snap fasteners is complicated by the highly nonlinear elastic behavior of the polymers used. Although Renaud and Karr (1996) discuss in detail some of the difficulties and pitfalls of snap fastener design, here attention is restricted to the analysis of a cantilever

**Figure 16.25** Common examples of integrated fasteners. (a) Module with four cantilever lugs; (b) cover with two cantilever and two rigid lugs; (c) separable snap joints for chassis cover.

**Figure 16.26**    Cantilever snap joint.

snap joint, as depicted in Figure 16.26. For cantilever snap joints it is recommended that the following design rules be followed:

1.  $l/h \geq 5$. [LNP (1992) recommends a ratio of 10.] If this is not the case, deformations will not be restricted to the cantilever, and the beam stiffness can be much larger than expected.

2.  A secant modulus must be used because the polymers are usually highly nonlinear. The secant modulus is the slope of the straight line drawn from zero strain to $\epsilon$ on the stress–strain curve.

3.  $l/b \geq 5$. If the beam is wider, it approaches a plate configuration and will be much stiffer due to the Poisson's effect.

4.  Snaps should be located away from stress risers such as sharp corners and manufacturing complications such as mold gates or weld lines.

The maximum deflection for the cantilever shown is given by [Bayer (1996); LNP (1992)]

$$y = A\frac{\epsilon l^2}{h} \tag{16.66}$$

where $A$ = constant for given cantilever shape, as given in Figure 16.27
   $\epsilon$ = allowable strain for polymer
   $l$ = cantilever length
   $h$ = maximum cantilever thickness

For a rectangular cross section, the force required to obtain this deformation is

$$P = \frac{bh^2}{6}\frac{E_s\epsilon}{l} \tag{16.67}$$

where $b$ = cantilever width
   $E_s$ = secant modulus of elasticity
   $l$ = cantilever length

| | Shape of cross section | |
|---|---|---|
| | **A** | **B** |
| **Type of design** | **Rectangle** | **Trapezoid** |
| **1** Cross section constant over length | $y = 0.67 \dfrac{\varepsilon l^2}{h}$ | $y = \dfrac{a + b_{(1)}}{2a + b} \dfrac{\varepsilon l^2}{h}$ |
| **2** All dimensions in direction $y$ (e.g., $h$) decrease to one-half | $y = 1.09 \dfrac{\varepsilon l^2}{h}$ | $y = 1.64 \dfrac{a + b_{(1)}}{2a + b} \dfrac{\varepsilon l^2}{h}$ |
| **3** All dimensions in direction $z$ (e.g., $b$ and $a$) decrease to one-quarter | $y = 0.86 \dfrac{\varepsilon l^2}{h}$ | $y = 1.28 \dfrac{a + b_{(1)}}{2a + b} \dfrac{\varepsilon l^2}{h}$ |

(Permissible) deflection

Subscript (1) designates that these formulae apply when the tensile stress is in the small surface area b. If it occurs in the larger area a, however, then a and b must be interchanged.

**Figure 16.27** Permissible deflection of different snap fastener cantilever shapes.

During assembly, however, the mating force that must be overcome is directed along the cantilever axis and also includes frictional effects. The mating force is given by

$$W = P \frac{\mu + \tan \alpha}{1 - \mu \tan \alpha} \tag{16.68}$$

where $W$ = mating force, N
  $\mu$ = coefficient of friction
  $\alpha$ = wedge angle of cantilever tip, deg

Table 16.15 gives some typical friction coefficients for polymers used in integrated snap fasteners.

**Table 16.15**     Coefficients of friction for common snap fastener polymers

| | Coefficient of Friction | |
|---|---|---|
| Material | On steel | On self-mated polymer |
| Polytetrafluoroethylene PTFE (Teflon) | 0.12–0.22 | — |
| Polyethylene (rigid) | 0.20–0.25 | 0.40–0.50 |
| Polyethylene (flexible) | 0.55–0.60 | 0.66–0.72 |
| Polypropylene | 0.25–0.30 | 0.38–0.45 |
| Polymethylmethacrylate (PMMA) | 0.50–0.60 | 0.60–0.72 |
| Acrylonitrile-butadiene-styrene (ABS) | 0.50–0.65 | 0.60–0.78 |
| Polyvinylchloride (PVC) | 0.55–0.60 | 0.55–0.60 |
| Polystyrene | 0.40–0.50 | 0.48–0.60 |
| Polycarbonate | 0.45–0.55 | 0.54–0.66 |

I  SOURCE: *Bayer Snap-fit Joints for Plastics, A Design Guide*, Bayer Polymers Division, 1996.

---

**EXAMPLE 16.12**

**Given:**  A snap-fitting hook must be designed for a rectangular-cross-section cantilever with a constant decrease in thickness from $h$ at the root to $h/2$ at the end of the hook. The undercut dimension serves as the maximum deformation needed for assembly and is decided based on disassembly force requirements. The material is polycarbonate on polycarbonate, length $l = 19$ mm, width $b = 9.5$ mm, angle of inclination $\alpha = 30°$, undercut $y = 2.4$ mm, secant modulus of elasticity $E_s = 1815$ N/mm$^2$.

**Find:**

(a)  The thickness $h$ at which full deflection causes a strain of 2%
(b)  Deflection force $P$
(c)  Mating force $W$

**Solution:**

(a)  From Equation (16.66) and using $A = 1.09$ as shown in Figure 16.27,

$$y = 1.09\frac{\epsilon l^2}{h} \qquad h = 1.09\frac{\epsilon l^2}{y} = 1.09\frac{0.02(19 \text{ mm})^2}{2.4 \text{ mm}} = 3.28 \text{ mm}$$

(b)  From Equation (16.67)

$$P = \frac{bh^2}{6}\frac{E_s\epsilon}{l} = \frac{(9.5 \text{ mm})(3.28 \text{ mm})^2}{6}\frac{(1815 \text{ N/mm}^2)(0.02)}{19 \text{ mm}} = 32.5 \text{ N}$$

(c)  For polycarbonate from Table 16.12, an average coefficient of friction is 0.600. Using this value yields

$$\frac{\mu + \tan\alpha}{1 - \mu\tan\alpha} = 1.801$$

Thus, the mating force is given by

$$W = P\frac{\mu + \tan\alpha}{1 - \mu\tan\alpha} = (32.5 \text{ N})(1.801) = 58.54 \text{ N}$$

## Case Study 16.1          DESIGN OF BOLTS FOR A HYDRAULIC BALER

**B**alers are thought of mainly for their use in agriculture to tie hay into manageable shapes. However, balers are often used in industrial applications as well, usually to compress and bind solid materials or wastes into manageable bales before recycling or disposal. A baler uses a hydraulic ram to compress material fed through a hopper and to extrude it through an opening. By placing a binding material, usually wire or cable, over the opening, the material is tied into the baled shape.

The key to a baler is the hydraulic cylinder used to compress the material. This case study examines the design of a component of the cylinder and the bolts used to fasten the end plate onto the cylinder.

**Given:**  One of the more strenuous applications for industrial balers is the baling of scrap paper from a manufacturer of cardboard boxes. Such a machine will typically produce 2 completed bales per hour, with 10 to 12 compressions of scrap per bale, and the baler will be operated continuously. To achieve the required baling force, a cylinder with a 150-mm internal bore will be pressurized to a maximum of 7 MPa during operation, although usually it is too much lower than this value. The baler should have a design life of 30 years. The end cap and cylinder flange are shown in Figure 16.28, and each is 25 mm thick.

**Find:**  Design the bolts needed to affix the end caps to the cylinder of the hydraulic system for the baler. Specify the bolts needed so that they will not fail during the service life of the baler. Use a static safety factor of 3, and make sure that fatigue failure will not occur. Because of the importance of bolt tightness, use a fine thread. Restrict yourself to the standard (rolled) threads in Table 16.10.

**Solution:**  Obviously, the loading is cyclic, and the number of total loadings and unloadings is given by

$$N = (12 \text{ cycles/bale})(2 \text{ bales/h})(24 \text{ h/day})$$
$$\times (365 \text{ days/yr})(30 \text{ yr}) = 6.31 \times 10^6 \text{ cycles}$$

Thus, these bolts should be designed for high-cycle fatigue. Also, the O-ring seal is placed so that it does not

**Figure 16.28**   End cap and cylinder flange. *(Photo courtesy of Pacific Compactor Corporation.)*

*(continued)*

## Case Study    (Concluded)

affect the joint stiffness, although the effective pressurized diameter will be greater than the cylinder bore. As a worst-case analysis, the pressurized area will be set equal to the O-ring diameter, or 300 mm.

It is important to use a sufficient number of bolts so that warping in the end plate or cylinder flange will not lead to unloading of the O-ring gasket and leakage of the hydraulic fluid. At this point 12 bolts will be chosen, giving a bolt spacing of just over 85 mm. A denser bolt spacing can be used if an excessively large bolt diameter is required. The joint has the following stiffness from the Wileman method [Eq. (16.26)]:

$$k_j = E d_c A_i e^{B_i d_c / L} = E d_c (0.78715) e^{0.62873 d_c / 0.050}$$

The bolt stiffness cannot be determined until the threaded length is known. The necessary bolt length is the joint length plus the nut width plus a few millimeters to allow for assembly. Assuming the nut width is 15 mm, the total bolt length is approximately 70 mm, a value that can be modified, if necessary, when a bolt diameter has been selected. The total threaded length is then $2d_c + 6$. Thus, the shank length is $L_s = 0.064 - 2d_c$, and the threaded section under load is $L_t = 0.05 - L_s = 2d_c - 0.014$. The bolt stiffness is then obtained from Equation (16.21) as

$$\frac{1}{k_b} = \frac{4}{\pi E} \left( \frac{L_s + 0.4 d_c}{d_c^2} + \frac{L_t + 0.4 d_r}{d_r^2} \right)$$

$$= \frac{4}{\pi E} \left( \frac{0.064 - 1.6 d_c}{d_c^2} + \frac{2 d_c - 0.014 + 0.4 d_r}{d_r^2} \right)$$

The stiffness parameter is a complicated function of the root and crest diameters. The total load on the bolts is

$$P = (7 \text{ MPa}) \left( \frac{\pi}{4} \right) (0.150 \text{ m})^2 = 124 \text{ kN}$$

so that the load per bolt is roughly 10 kN. During extension the end cap sees the maximum pressure encountered by the cylinder. During retraction the pressure falls to a great extent. Assuming that the pressure during retraction falls to

zero, $P_a$ and $P_m$ are both 5 kN. The alternating and mean stresses acting on the bolt are given by Equations (16.36) and (16.37), respectively, and are functions of the root and crest diameters.

Since the safety factor against fatigue failure must be 3, Equation (16.41) will allow solution. The approach is as follows:

(a)   Select a grade for the bolts. This defines the ultimate yield and proof strengths.

(b)   Select a potential crest diameter and associated root diameter. Evaluate the joint and bolt stiffnesses, dimensionless stiffness parameters, alternating and mean stresses, and the safety factor.

(c)   If the safety factor is greater than 3 [per Eq. (16.31)], select a smaller bolt crest diameter. If the safety factor is less than 3, choose a larger crest diameter.

The results from this approach are as follows:

| Metric grade | Minimum required crest diameter, mm | Safety Factors | | |
|---|---|---|---|---|
| | | Bolt overload | Joint separation | Bolt fatigue |
| 4.6 | 20 | 3.17 | 6.8 | 1.64 |
| 4.8 | 20 | 4.37 | 9.4 | 1.25 |
| 5.8 | 16 | 3.51 | 6.97 | 1.27 |
| 8.8 | 12 | 3.42 | 5.93 | 1.30 |
| 9.8 | 12 | 3.70 | 6.43 | 1.31 |
| 10.9 | 10 | 3.45 | 5.36 | 1.20 |
| 12.9 | 10 | 4.03 | 6.26 | 1.21 |

An increase in bolt grade does not necessarily mean a reduction in bolt diameter, since a finite number of diameters are being considered. Also, the safety factor for bolt fatigue actually drops with an increase in bolt grade, a counterintuitive situation. Recall that the safety factor is due to the combination of applied loading and preloading; the proof strength is a higher fraction of the ultimate strength for higher-grade bolts; and a high-strength bolt has a higher preload [per Eq. (16.33)].

Any of these bolts would be suitable for use in the baler cylinder, and a final decision should be based on availability and cost.

## 16.9    SUMMARY

Members can be joined by using a number of different techniques. This chapter investigated threaded, riveted, welded, and adhesive joining of members. Power screws were also covered. A power screw is a device used to change angular motion into linear motion and to transmit power. A general analysis of the forces and torques required to raise and lower a load was presented. It was assumed that an Acme thread is specified for power screws. Collar friction was also considered. Power and efficiency of power screws, as well as when self-locking occurs, were discussed.

The types and classes of thread were given for threaded fasteners. It was found that fatigue strength can be increased by applying initial tension in the bolt. The lead and stiffness of the bolt and joint were analyzed. The most difficult of these analyses is to determine the joint stiffness. A needed assumption is that the stress induced in the joint is uniform throughout a region surrounding the bolt hole, with zero stress outside this region. Once the load is known, the design stresses imposed by the threaded fastener can be determined. The allowable stress for threaded fasteners was obtained by considering the proof stress and proof load. Given the design and allowable stresses, the safety factor can be determined. Both static and dynamic considerations were made, and gasketed joints were covered.

Riveted and threaded fasteners in shear are treated alike in design and failure analysis. Four modes of failure were presented: bending of a member, shear of a rivet, tensile failure of a member, and compressive bearing failure.

The design of welded fasteners was considered. Fillet welds were focused on, since they are the most frequently used type of weld. For axial loading when the fillet weld is aligned parallel to the load, the shear stress occurs along the throat of the fillet parallel to the load. The load-and-stress analysis occurs along the critical section, which is the plane created by the throat dimension extending the length of the weld. Failure could be predicted for various types of loading while the strength of the weld was also considered.

The last two fastener methods considered were adhesive bonding and snap fasteners. Adhesively bonded lap and scarf joints were briefly analyzed. In addition, integrated snap fasteners were examined.

## KEY WORDS

**adhesive bonding**    process of joining materials through interatomic or intermolecular attractive forces.

**bolt**    externally threaded fastener intended to be used with nut.

**cap screw**    externally threaded fastener intended to be used with internally threaded hole.

**coarseness**    quality and threads per inch of thread profiles on bolt, nut, or screw.

**gasket**    compliant member intended to provide seal.

**lead**     distance that screw would advance in 1 revolution.

**nut**     internally threaded mating member for bolt.

**pitch**     distance from point on one thread to same point on adjacent thread.

**power screws**     power transmission device using mechanical advantage of threads to apply large loads.

**proof load**     maximum load that bolt can withstand without acquiring permanent set.

**proof strength**     limiting value of stress determined by proof load and tensile stress area.

**rivets**     fasteners that function through mechanical interference, usually by upsetting one end of rivet extending outside of free hole.

**self-locking**     power screw where thread friction is high enough to prevent loads from lowering in absence of externally applied torque.

**snap fasteners**     integrated fasteners that operate through elastic deformation and recovery of part of structure after insertion into proper retainer.

**stud**     externally threaded member with threads on both ends in lieu of cap.

**thread series**     standardized thread profile, either UN (unified) or M (metric).

**threads per inch**     number of threads in 1 in, related to the pitch by $p = 1/n$.

**throat**     thinnest section of weld.

**weld**     junction formed through diffusion of two materials to be joined, combined with optional filler material.

## RECOMMENDED READINGS

*ASM Handbook,* vol. 6: *Welding, Brazing, and Soldering* (1993) American Society for Metals, Metals Park, OH.

Bickford, J. H. (1995) *An Introduction to the Design and Behavior of Bolted Joints,* 3rd ed., Marcel Dekker, New York.

Bickford, J. H., and Nassar, S. (1998) *Handbook of Bolts and Bolted Joints,* Marcel Dekker, New York.

Blodgett, O. W. (1976) *Design of Weldments,* James F. Lincoln Arc Welding Foundation, Cleveland, OH.

Blodgett, O. W. (1976) *Design of Welded Structures,* James F. Lincoln Arc Welding Foundation, Cleveland, OH.

*Fastener Standards,* 6th ed. (1988) Industrial Fastener Institute, Cleveland, OH.

Fisher, J. W., and Struik, J. H. A. (1974) *Guide to Design Criteria for Bolted and Riveted Joints,* Wiley, New York.

Juvinall, R. C., and Marshek, K. M. (2003) *Fundamentals of Machine Component Design,* 3rd ed., Wiley, New York.

Krutz, G. W., Schuelle, J. K., and Claar, P. W. (1994) *Machine Design for Mobile and Industrial Applications,* Society of Automotive Engineers, Warrendale, MI.

Mott, R. L. (1998) *Machine Elements in Mechanical Design,* 3rd ed., Prentice-Hall, Upper Saddle River, NJ.

Norton, R. L. (1996) *Machine Design,* Prentice-Hall, Englewood Cliffs, NJ.

Parmly, R. O. (1989) *Standard Handbook of Fastening and Joining,* 2nd ed., McGraw-Hill, New York.

Shigley, J. E., Mischke, C. R., and Budynas, R. G. (2003) *Mechanical Engineering Design,* 7th ed., McGraw-Hill, New York.

## REFERENCES

Bayer (1996) *Bayer Snap-fit Joints for Plastics, A Design Guide,* Bayer Polymers Division, Pittsburgh, PA.

Bickford, J. H. (1995) *An Introduction to the Design and Behavior of Bolted Joints,* 3rd ed., Marcel Dekker, New York.

Kalpakjian, S., and Schmid, S. R. (2001) *Manufacturing Engineering and Technology,* 4th ed., Prentice-Hall, Upper Saddle River, NJ.

LNP (1992) *Design and Engineering Guide,* LNP Engineering Plastics, Inc., Exton, PA.

Mott, R. L. (1998) *Machine Elements in Mechanical Design,* 2nd ed., Macmillan, New York.

Renaud, J. E., and Karr, P. (1996) "Development of an Integrated Plastic Snap Fastener Design Advisor," *Engineering Design and Automation,* vol. 2, pp. 291–304.

Shigley, J. E., and Mischke, C. R. (1989) *Mechanical Engineering Design,* 5th ed., McGraw-Hill, New York.

Wileman, J., Choudhury, M., and Green, I. (1991) "Computation of Member Stiffness in Bolted Connections," *J. Machine Design,* ASME, vol. 113, pp. 432–437.

## PROBLEMS

### Section 16.3

**16.1**  An Acme threaded power screw with a crest diameter of 1.125 in and single thread is used to raise a load of 25,000 lb. The collar mean diameter is 1.5 in. The coefficient of friction is 0.12 for both the thread and the collar. Determine the following:

   *a)*   Pitch diameter of the screw
   *b)*   Screw torque required to raise the load. *Ans.* $T_r = 4637$ in · lb.
   *c)*   Maximum thread coefficient of friction allowed to prevent the screw from self-locking if collar friction is eliminated

**16.2**  A car jack consists of a screw and a nut, so that the car is lifted by turning the screw. Calculate the torque needed to lift a load with a mass of 1000 kg if the lead of the thread $l = 9$ mm, its pitch diameter is 22 mm, and its thread angle is 30°. The coefficient of friction is 0.09 in the threads and 0 elsewhere. *Ans.* $T_r = 24.40$ N · m.

**16.3** A power screw gives the axial tool motions in a numerically controlled lathe. To get high accuracy in the motions, the heating and power loss in the screw have to be low. Determine the power efficiency of the screw if the coefficient of friction is 0.12, pitch diameter is 30 mm, lead is 6 mm, and thread angle is 25°. *Ans.* $\eta = 34.3\%$.

**16.4** Sketch $a$ shows a stretching device for steel wires used to stabilize the mast of a sailing boat. Both front and side views are shown, and all dimensions are in millimeters. A screw with square threads ($\beta = 0$), a lead and pitch of 4 mm, and an outer diameter of 20 mm is used. The screw can move axially but is prevented from rotating by flat guiding pins (side view in sketch $a$). Derive an expression for the tightening torque as a function of the stretching force $P$ when the coefficient of friction of all surface contacts is 0.20. Also, calculate the torque needed when the tightening force is 1000 N. *Ans.* For $P = 1000$ N, $T = 5.61$ N · m.

(a) Front view          (b) Side view

Sketch a, for Problem 16.4

**16.5** A flywheel of a motorcycle is fastened by a thread manufactured directly in the center of the flywheel, as shown in sketch $b$. The flywheel is mounted by applying a torque $T$. The cone angle is $\gamma$. Calculate the tensile force $W$ in the shaft between the contact line at $N$ and the thread as a function of $D_1$, $D_2$, $\gamma$, and $T$. The lead angle is $\alpha$ at the mean diameter $D_1$. The shaft is assumed to be rigid.

Sketch *b*, for Problem 16.5

**16.6**  To change its oil, a 20,000 lb truck is lifted a height of 5 ft by a screw jack. The power screw has Acme threads and a crest diameter of 5 in with 2 threads per inch, and the lead equals the pitch. Calculate how much energy has been used to lift and lower the truck if the only friction is in the threads, where the coefficient of friction is 0.10. *Ans.* $(hp)_r = 409$ kip-ft, $(hp)_l = 207$ kip-ft.

**16.7**  A single-threaded M32 × 3.5 power screw is used to raise a 12-kN load at a speed of 25 mm/s. The coefficients of friction are 0.08 for the thread and 0.12 for the collar. The collar mean diameter is 55 mm. Determine the power required. Also determine how much power is needed for lowering the load at 40 mm/s. *Ans.* $h_{pr} = 2748$ W, $h_{pl} = 3429$ W.

**16.8**  A double-threaded Acme power screw is used to raise a 1350-lb load. The outer diameter of the screw is 1.25 in, and the mean collar diameter is 2.0 in. The coefficients of friction are 0.13 for the thread and 0.16 for the collar. Determine the following:

    *a)*    Required torque for raising and lowering the load. *Ans.* $T_r = 408.1$ in · lb.
    *b)*    Geometric dimensions of the screw. *Ans.* $l = 0.4$ in, $d_p = 1.14$ in, $\alpha = 6.376°$, $\beta = 29°$.
    *c)*    Efficiency in raising the load. *Ans.* $e = 21.06\%$.
    *d)*    Load corresponding to the efficiency if the efficiency in raising the load is 18%. *Ans.* $W = 1154$ lb.

**16.9**  A 25-kN load is raised by two Acme threaded power screws with a minimum speed of 35 mm/s and a maximum power of 1750 W per screw. Because of space limitations the screw diameter should not be larger than 45 mm. The coefficient of friction for both the thread and the collar is 0.09. The collar mean diameter is 65 mm. Assuming that the loads are distributed evenly on both sides, select the size of the screw to be used and calculate its efficiency.

**16.10**  The lead screw of a small lathe is made from a $\frac{1}{2}$-in crest diameter Acme threaded shaft. The lead screw has to exert a force on the lathe carriage for a number of operations, and it is powered by a belt drive from the motor. If a force of 500 lb is desired, what is the torque required if the collar is twice the pitch diameter of the screw? Use $\mu = \mu_c = 0.25$. With what velocity does the lead screw move the crosshead if the lead screw is single-threaded and is driven at 500 rpm? *Ans.* $T = 92$ in · lb, $v = 0.833$ in/s.

### Section 16.4

**16.11** A screw with Acme thread can have more than one entrance to the thread per screw revolution. A single thread means that the pitch and the lead are equal, but for double and triple threads the lead is larger than the pitch. Determine the relationship between the number of threads per inch $n$, the pitch $p$, and the lead $l$.

**16.12** A section of a bolt circle on a large coupling is shown in sketch $c$. Each bolt is loaded by a repeated force $P = 6000$ lb. The members are steel, and all bolts have been carefully preloaded to $P_i = 25{,}000$ lb each. The bolt is to be an SAE grade 5, 0.75-in crest diameter with fine threads (so that $d_r = 0.674$ in), and the nut which fits on this bolt has a thickness of 0.50 in. The threads have been manufactured through rolling and use a survival probability of 90%.

    *a)*    If hardened steel washers 0.134 in thick are to be used under the bolt and nut, what length of bolts should be used? *Ans. L = 2.5 in.*

    *b)*    Find the stiffness of the bolt, the members, and the joint constant. *Ans. $k_b = 5.798 \times 10^6$ lb/in, $C_k = 0.193$.*

    *c)*    What is the factor of safety guarding against a fatigue failure? *Ans. $n_s = 4.06$*

Sketch c, for Problem 16.12          Sketch d, for Problem 16.13

**16.13** An M12, coarse-pitch, class-5.8 bolt with a hexagonal nut assembly is used to keep two machine parts together, as shown in sketch $d$. Determine the following:

    *a)*    Bolt stiffness and clamped member stiffness

    *b)*    Maximum external load that the assembly can support for a load safety factor of 2.5

    *c)*    Safety factor guarding against separation of the members

d) Safety factor guarding against fatigue if a repeated external load of 10 kN is applied to the assembly

* **16.14** Repeat Problem 16.13 if the 25-mm-thick member is made of steel.

**16.15** Find the total shear load on each of the three bolts for the connection shown in sketch *e*. Also, compute the shear stress and the bearing stress. Find the area moment of inertia for the 8-mm-thick plate on a section through the three bolt holes. *Ans.* $P_{center} = 4$ kN, $P_{top} = P_{bottom} = 37.7$ kN.

Sketch *e*, for Problem 16.15                    Sketch *f*, for Problem 16.16

**16.16** A coarse-pitch, SAE grade-5 bolt with a hexagonal nut assembly is used to keep two machine parts together, as shown in sketch *f*. The major diameter of the bolt is 0.5 in. The bolt and the bottom member are made of carbon steel. Assume that the connection is to be reused. Length dimension is in inches. Determine the following:

a) Length of the bolt
b) Stiffnesses of the bolt and the member
c) Safety factor guarding against separation of the members when the maximum external load is 5000 lb
d) Safety factor guarding against fatigue if the repeated maximum external load is 2500 lb in a released-tension loading cycle

**16.17** An electric-motor-driven press (sketch *g*) has the total press force $P = 5000$ lb. The screws are Acme type with $\beta = 29°$, $d_p = 3$ in, $p = l = 0.5$ in, and $\mu = 0.05$. The thrust bearings for the screws have $d_c = 5$ in and $\mu_c = 0.06$. The motor speed is 1720 rpm, the total speed ratio is 75 : 1, and the mechanical efficiency $e = 0.95$. Calculate

a) Press head speed
b) Power rating needed for the motor

| *Indicates problem of greater difficulty.

Sketch g, for Problem 16.17

**16.18** A valve for high-pressure air is shown in sketch $h$. The spindle has thread M12 with a pitch diameter of 10.9 mm, lead $l = 1.75$ mm, and a thread angle of $60°$. Derive the relationship between torque and axial thrust force, and calculate the axial force against the seating when the applied torque is 10 N · m during tightening. The coefficient of friction is 0.12. *Ans.* $T_r = (0.00104 \text{ m})P$.

Sketch h, for Problem 16.18

★ **16.19** Derive the expression for the power efficiency of a lead screw with a flat thread (thread angle $\beta = 0°$), and find the lead angle $\alpha$ that gives maximum efficiency in terms of the coefficient of friction. Also give results if $\mu = 0.1$. *Ans.* $e = (1 - \mu \tan \alpha)/(1 + \mu \cot \alpha)$.

★ **16.20** A car manufacturer has problems with the cylinder head studs in a new high-power motor. After a relatively short time the studs crack just under the nuts, the soft cylinder head gasket blows out, and the motor stops. To be able to analyze the problem, the car manufacturer experimentally measures the stiffnesses of the various components. The stiffness for all bolts together is 400 N/$\mu$m, the stiffness of the gasket is 600 N/$\mu$m, and the stiffness of the cylinder head that compresses the gasket is 10,000 N/$\mu$m. By comparing the life-stress relationships with those for rolling-element bearings, the car manufacturer estimates that the stress amplitude in the screws needs to be halved to get sufficient life. How can that be done?

★ **16.21** A pressure vessel of compressed air is used as an accumulator to make it possible to use a small compressor that works continuously. The stiffness parameter for the lid around each of the 10-mm bolt diameters is 900 MN/m. The shank length is 20 mm. Because the air consumption is uneven, the air pressure in the container varies between 0.2 and 0.8 MPa many times during a week. After 5 years of use, one of the bolts holding down the top lid of the pressure vessel cracks off. A redesign is then made, decreasing the stress variation amplitude by 25%, to increase the life of the bolts to at least 50 years. The stress variation amplitude is decreased by lengthening the bolts and using circular tubes with the same cross-sectional area as the solid circular cross section of the bolt to transfer the compressive force from the bolt head to the lid. Calculate how long the tubes should be. *Ans.* $l = 7.089$ mm.

★ **16.22** A loading hook of a crane is fastened to a block hanging in six steel wires. The hook and block are bolted together with four 10-mm-diameter screws prestressed to 20,000 N each. The shank length is 80 mm, and the thread length is 5 mm. The stiffness of the material around each screw is 1 GN/m. One of the screws of the crane cracks due to fatigue after a couple of years of use. Will it help to change the screws to 12-mm-diameter while other dimensions are unchanged if the stress variation needs to be decreased by at least 20%?

★ **16.23** Depending on the roughness of the contacting surfaces of a bolted joint, some plastic deformation takes place on the tops of the roughness peaks when the joint is loaded. The rougher the surfaces are, the more pressure in the bolted joint is lost by plastic deformation. For a roughness profile depth of 20 $\mu$m on each of the surfaces, a plastic deformation of 6.5 $\mu$m can be expected for the two surfaces in contact. For a bolt-and-nut assembly as shown in Figure 16.13, three sets of two surfaces are in contact. The stiffness of the two steel plates together is 700 MN/m when each is 40 mm thick. The bolt diameter is 16 mm with metric thread. The shank length is 70 mm. The bolt is prestressed to 25 kN before plastic deformation sets in. Calculate how much of the prestress is left after the asperities have deformed. *Ans.* $F_p = 19.8$ kN.

★ **16.24** An ISO M12 × 1.75 class-12.9 bolt is used to fasten three members, as shown in sketch $i$. The first member is made of cast iron, the second is low-carbon steel, and the third is aluminum. The static loading safety factor is 2.5. Dimensions are in millimeters. Determine

*a)* Total bolt length, threaded length, and threaded length in the joint. *Ans.* $L_{tot} = 80$ mm, $L_t = 30$ mm, $L_{tj} = 15$ mm.
*b)* Bolt-and-joint stiffness using a 30° cone. *Ans.* $k_b = 286.8$ MN/m, $k_j = 0.928$ GN/m.
*c)* Preload for permanent connections.
*d)* Maximum static load that the bolt can support.

Sketch *i*, for Problem 16.24

**★ 16.25** A pressurized cast iron cylinder shown in sketch *j* is used to hold pressurized gas at a static pressure of 8 MPa. The cylinder is joined to a low-carbon-steel cylinder head by bolted joints. The bolt to be used is metric grade 12.9 with a safety factor of 3. Dimensions are in millimeters. Determine the required number of bolts. Use grade-12.9 bolts, with M36 × 100-mm coarse threads. *Ans.* 47 bolts.

Sketch *j*, for Problem 16.25

Sketch *k*, for Problem 16.26

**16.26** In the bolted joint shown in sketch *k*, the first member is made of low-carbon steel, the second member is aluminum, and the third member is cast iron. Assuming that the members can be rearranged and the frustum cone angle is 45°, find the arrangement that can support the maximum load. Dimensions are in millimeters. *Ans.* Put the aluminum in the center.

**16.27** The cylinder shown in sketch *l* is pressurized up to 7 MPa and is connected to the cylinder head by 16 M24 × 3 metric-grade-8.8 bolts. The bolts are evenly spaced around the perimeters of the two circles with diameters of 1.2 and 1.5 m, respectively. The cylinder is made of cast iron, and its head is made of high-carbon steel. Assume that the force in each bolt is inversely related to its radial distance from the center of the cylinder head. Calculate the safety factor guarding against failure due to static loading.

Sketch *l*, for Problem 16.27

## Section 16.5

**16.28** A steel plate (sketch *m*) is riveted to a vertical pillar. The three rivets have a $\frac{5}{8}$-in diameter and carry the load and moment resulting from the external load of 1950 lb. All length dimensions are in inches. The yield strengths of the materials are $S_{y,\text{rivet}} = 85,000$ psi and $S_{y,\text{plate}} = 50,000$ psi. Calculate the safety factors for

a)   Shear of rivet when $S_{sy} = 0.57 S_y$
b)   Bearing of rivet
c)   Bearing of plate
d)   Bending of plate

State how failure should first occur.

Sketch *m*, for Problem 16.28

**16.29** The flange of a ship's propeller shaft is riveted in the radial direction against the hollow shaft. The outside diameter is 1 m, and there are 180 rivets around the circumference, each with a diameter of 25 mm. The rivets are made of AISI 1020 steel and placed in three rows. Calculate the maximum allowable propeller torque transmitted through the rivets for a safety factor of 3. *Ans.* $T = 1.738 \times 10^6$ N · m.

**16.30** A rectangular steel plate is connected with rivets to a steel beam, as shown in sketch *n*. Assume the steel to be low-carbon steel. The rivets have a yield strength of 600 MPa. A load of 24 kN is applied. For a safety factor of 3 calculate the diameter of the rivets. Length dimensions are in millimeters. *Ans.* $d = 14.8$ mm; use $d = 15$ mm.

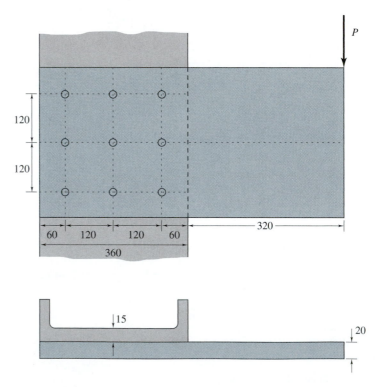

Sketch *n*, for Problem 16.30

**16.31** Repeat Problem 16.30 but with the plate and beam shown in sketch *o*.

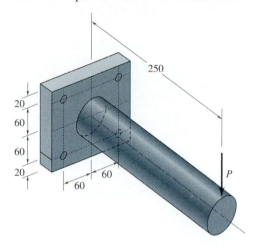

Sketch *o*, for Problem 16.31

## Section 16.6

**16.32** The steel plate shown in sketch *p* is welded against a wall. Length dimensions are in inches. The vertical load $W = 4000$ lb acts 6.8 in from the left weld. Both welds are made by AWS electrode number E8000. Calculate the weld size for a safety factor of $n_s = 3.0$. *Ans.* $h_e = 0.3125$ in.

Sketch *p*, for Problem 16.32

**16.33** Two medium-carbon steel (AISI 1040) plates are attached by parallel-loaded fillet welds, as shown in sketch $q$. E60 series welding rods are used. Each of the welds is 3 in long. What minimum leg length must be used if a load of 16.0 kN is to be applied with a safety factor of 3.5?

Sketch $q$, for Problem 16.33

**16.34** Determine the weld size required if only the top (AB) portion is welded in Problem 16.33.

**16.35** The universal joint on a car axle is welded to the 60-mm-outside-diameter tube and should be able to transfer 1500 N · m of torque from the gearbox to the rear axle. Calculate how large the weld leg should be to give a safety factor of 10 if the weld metal is of class E70.

**16.36** The steel bar shown in sketch $r$ is welded by an E60XX electrode to the wall. A 3.5-kN load is applied in the $y$ direction at the end of the bar. Calculate the safety factor against yielding. Also, would the safety factor change if the direction of load $P$ changed to the $z$ direction? Dimensions are in millimeters. *Ans.* $n_s = 5.53$.

Sketch $r$, for Problem 16.36

**16.37** The bar shown in sketch $s$ is welded to the wall by AWS electrodes. A 40-kN load is applied at the top of the bar. Dimensions are in millimeters. For a safety factor of 2.5 against yielding, determine the electrode number that must be used and the weld throat length, which should not exceed 10 mm.

Sketch s, for Problem 16.37

**16.38** AWS electrode number E100XX is used to weld a bar, shown in sketch $t$, to the wall. For a safety factor of 3 against yielding, find the maximum load that can be supported. What is the maximum stress in the weld, and where does it occur? Dimensions are in millimeters. *Ans.* $P_{max} = 46.8$ kN.

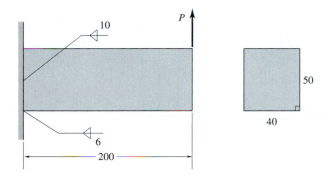

Sketch t, for Problem 16.38

**16.39** A cam is to be attached to an extruded channel by using arc welds. Three designs are proposed, as shown in sketch $u$. If E60 electrodes are used and the circular bar has a diameter of 2 in, and the square bars have side length of 2 in, find the factor of safety of the welded joint if $P = 200$ lb for each case. Use $a = 15$ in. You can assume that the extrusion fits tightly with the mating hole in the cam so that the loading is pure torsion. Design based on yielding of the weld.

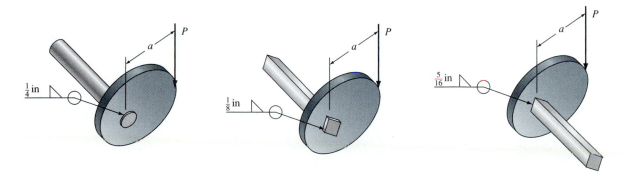

Sketch u, for Problem 16.39

### Section 16.7

**16.40** During the manufacturing of the fuselage of a commuter airplane, aluminum plates are glued together with lap joints. Because the elastic deformation for a single plate differs from the deformation for two plates glued together in a lap joint, the maximum shear stress in the glue is twice as high as the average shear stress. The shear strength of the glue is 20 MPa, the tensile strength of the aluminum plates is 95 MPa, and their thickness is 4.0 mm. Calculate the overlapping length needed to make the glue joint twice as strong as the aluminum plate.

**16.41** The ropes holding a children's swing are glued into two plastic tubes with an inner diameter of 10 mm and a length of 100 mm. The difference in elasticity between the rope and the rope plus the plastic tube gives a maximum shear stress 2.5 times as high as the mean shear stress in the glue. The glue is an epoxy type with an ultimate shear strength of 12 MPa. How heavy can the person on the swing be without overstressing the glue if the speed of the swing at its lowest point is 6.5 m/s and the distance from the center of gravity of the person to the fastening points of the ropes is 2 m? The safety factor is 10. *Ans.* Maximum mass is 97 kg.

★ **16.42** A fishing rod is made of carbon-fiber-reinforced plastic tube. To get optimum elastic properties along the length of the rod, and to therefore be able to make long and accurate casts, the concentrations of the fibers in the various parts of the rod have to be different. It is necessary to scarf-joint the rod parts. The tensile strength of the epoxy glue joint is 10 MPa, and its shear strength is 12.5 MPa. These strengths are independent of each other. Find the optimum scarf angle to make the rod as strong as possible in bending.

# SPRINGS

A collection of helical
compression springs.
(Courtesy of Danly Die)

*Entia non multiplicantor sunt prater necessitatum.*
(Do not complicate matters more than necessary.)
Galileo Galilei

## SYMBOLS

| | |
|---|---|
| $A$ | cross-sectional area, $m^2$ |
| $A_p$ | intercept, Pa |
| $a$ | length dimension, m |
| $b$ | width of leaf spring, m |
| $b_s$ | slope |
| $C$ | spring index, $D/d$ |
| $\bar{C}$ | intercept |
| $c$ | distance from neutral axis to outer fiber, m |
| $D$ | mean coil diameter, m |
| $D_i$ | coil inside diameter, m |
| $D_i'$ | coil inside diameter after loading, m |
| $D_o$ | coil outside diameter, m |
| $d$ | wire diameter, m |
| $E$ | modulus of elasticity, Pa |
| $f$ | bump height, m |
| $f_n$ | lowest natural frequency, Hz |
| $G$ | shear modulus of elasticity, Pa |
| $g$ | gravitational acceleration, 9.807 m/s² (32.174 ft/s²) |
| $g_a$ | gap, m |
| $h$ | height, m |
| $I$ | area moment of inertia, $m^4$ |
| $J$ | polar area moment of inertia, $m^4$ |
| $K_d$ | transverse shear factor, $(C + 0.5)/C$ |
| $K_i$ | spring stress concentration factor, $\dfrac{4C^2 - C - 1}{4C(C - 1)}$ |
| $K_w$ | Wahl curvature correction factor, $\dfrac{4C - 1}{4C - 4} + \dfrac{0.615}{C}$ |
| $K_1$ | defined in Equation (17.54) |
| $k$ | spring rate, N/m |
| $k_r$ | reliability factor |
| $k_t$ | spring rate considering torsional loading, N/m |
| $k_\theta$ | angular spring rate, N/m |
| $l$ | length, m |
| $l_w$ | length of wire in spring, m |
| $M$ | moment, N · m |
| $m$ | slope |
| $N$ | number of coils |
| $N_a$ | number of active coils |
| $N_a'$ | number of active coils after loading |
| $N_b$ | number of coils in body |
| $N_e$ | number of end coils |
| $N_t$ | total number of coils |
| $n$ | number of leaves |
| $n_s$ | safety factor |
| $P$ | force, N |

| | |
|---|---|
| $P_i$ | preload, N |
| $p$ | pitch, m; gas pressure, Pa |
| $R$ | radius used in applying torque; linear arm length for bracket, m |
| $R_d$ | diameter ratio, $D_o/D_i$ |
| $r$ | radius, m |
| $r_i$ | hook inner radius, m |
| $r_1, \ldots, r_4$ | hook radii, m |
| $S$ | strength, Pa |
| $S_{se}$ | modified endurance limit, Pa |
| $S_{se}'$ | endurance limit, Pa |
| $S_{sf}$ | modified shear fatigue strength, Pa |
| $S_{su}$ | shear ultimate strength, Pa |
| $S_{sy}$ | shear yield strength, Pa |
| $S_{ty}$ | tensile yield stress, Pa |
| $S_y$ | yield stress, Pa |
| $S_{ut}$ | tensile ultimate stress, Pa |
| $T$ | torque, N · m |
| $t$ | thickness, m |
| $U$ | stored elastic energy, N · m |
| $\Delta U$ | change in energy, N · m |
| $v$ | volume, $m^3$ |
| $x$ | Cartesian coordinate, m |
| $y$ | deflection from unloaded state, m |
| $\gamma_{\theta z}$ | shear strain due to torsional loading |
| $\Delta_v$ | loss coefficient |
| $d$ | deflection, m |
| $\delta_t$ | deflection due to torsional loading, m |
| $\zeta$ | cone angle |
| $\theta_{rad}$ | angular deflection in radians |
| $\theta_{rev}$ | angular deflection in revolutions |
| $\nu$ | Poisson's ratio |
| $\rho$ | density, kg/m³ |
| $\sigma$ | stress, Pa |
| $\theta$ | angle of twist, rad |
| $\tau$ | shear stress, Pa |

## SUBSCRIPTS

| | |
|---|---|
| $a$ | alternating |
| all | allowable |
| $b$ | body |
| $c$ | conical |
| $d$ | transverse (or direct) |
| $f$ | free (without load) |
| $h$ | hook |
| $i$ | installed; inside; preload |
| $l$ | loop |

| $m$ | mean | | $s$ | solid |
| max | maximum | | $t$ | torsional; total |
| min | minimum | | $u$ | ultimate |
| $o$ | operating; outside | | $0$ | initial |

## 17.1 INTRODUCTION

A **spring** is a flexible machine element used to exert a force or a torque and, at the same time, to store energy. Energy is stored in the solid that is bent, twisted, stretched, or compressed to form the spring. The energy is recoverable by the elastic return of the distorted material. Springs must have the ability to withstand large deflections elastically. The force can be a linear push or pull, or it can be radial. The torque can be used to cause rotation or to provide a counterbalance force for a machine element pivoting on a hinge. Springs frequently operate with high working stresses and with loads that are continuously varying.

Some more specific applications of springs are

1. To store and return energy, as in a gun recoil mechanism
2. To apply and maintain a definite force, as in relief valves and governors
3. To isolate vibrations, as in an automobile
4. To indicate and/or control load, as in a scale
5. To return or displace a component, as in a brake pedal or engine valve

## 17.2 SPRING MATERIALS

*Strength* is one of the most important characteristics to consider when one is selecting a spring material. Figure 3.19 shows modulus of elasticity $E$ against strength $S$. Strength means yield strength for metals and polymers, compressive crushing strength for ceramics, tear strength for elastomers, and tensile strength for composites and woods. As Ashby (1992) pointed out, the normalized strength $S/E$ is the parameter to be used in evaluating strength. If just evaluating strength, from Figure 3.19 we note that engineering polymers have values of $S/E$ in the range 0.01 to 0.1. The values for metals are 10 times lower. Even ceramics in compression are not as strong as engineering polymers, and in tension ceramics are far weaker. Composites and woods lie on the 0.01 contour, as good as the best metals. Because of their exceptionally low elastic modulus, elastomers have higher $S/E$, between 0.1 and 1.0, than does any other class of material.

The *loss coefficient* $\Delta_v$ is the second parameter that is important in selecting spring materials. The loss coefficient measures the fractional energy dissipated in a stress–strain cycle. Figure 17.1 shows the stress–strain variation for a complete cycle. The loss coefficient is

$$\Delta_v = \frac{\Delta U}{2U} \tag{17.1}$$

where $\Delta U$ = change of energy over 1 cycle, N · m
$U$ = stored elastic energy, N · m

**Figure 17.1**   Stress–strain curve for one complete cycle.

The loss coefficient is a dimensionless parameter. A material used for springs should have a *low* loss coefficient. Elastomers, as indicated by Ashby (1992), have the highest loss coefficients, and ceramics have the lowest with a 4 order-of-magnitude range between them. Ceramics are not a suitable spring material, since they are brittle in tension. High-carbon steels have just slightly higher loss coefficients than ceramics and are a more suitable spring material.

For the reasons stated above, most commercial spring materials come from the group of high-strength, low-loss-coefficient materials that includes high-carbon steel; cold-rolled, precipitation-hardened stainless steel; nonferrous alloys; and a few specialized nonmetallics such as laminated fiberglass. Table 17.1 gives values of typical properties for common spring materials (modulus of elasticity, shear modulus of elasticity, density, and maximum service temperature) and gives additional design information.

Springs are manufactured by either hot- or cold-working processes, depending on the size of the material, the spring index, and the properties desired. In general, prehardened wire *should not* be used if $D/d < 4$ or if $d > \frac{1}{4}$ in. Winding the spring induces residual stresses through bending, but these are normal to the direction of torsional working stresses in a coiled spring. When a spring is manufactured, it is quite frequently relieved after winding by a mild thermal treatment.

The ultimate strength of a spring material varies significantly with wire size, so that the ultimate strength cannot be specified unless the wire size is known. The material and its processing also have an effect on tensile strength. Results of extensive testing by Associated Spring Corp., Barnes Group Inc., show that a semilogarithmic plot of wire strength versus wire diameter is almost a straight line for some materials. The data for five materials can be fitted closely to the exponential form

$$S_{ut} = \frac{A_p}{d^m} \tag{17.2}$$

where $A_p$ = intercept of straight line
$\quad\quad m$ = slope

Table 17.2 gives values of $A_p$ and $m$ for five materials. Equation (17.2) is valid only for the limited range given in Table 17.2. Also, note that $A_p$ in thousand pounds per square inch requires $d$ to be in inches and that $A_p$ in megapascals requires $d$ to be in millimeters.

**Table 17.1** Typical properties of common spring materials

| Common name | Specification | Modulus of elasticity $E$, psi | Shear modulus of elasticity $G$, psi | Density $\rho$, lb/in³ | Maximum service temperature, °F | Principal characteristics |
|---|---|---|---|---|---|---|
| **High-carbon steels** | | | | | | |
| Music wire | ASTM A228 | $30 \times 10^6$ | $11.5 \times 10^6$ | 0.283 | 250 | High strength; excellent fatigue life |
| Hard-drawn | ASTM A227 | $20 \times 10^6$ | $11.5 \times 10^6$ | 0.283 | 250 | General-purpose use; poor fatigue life |
| **Stainless steels** | | | | | | |
| Martensitic | AISI 410, 420 | $29 \times 10^6$ | $11 \times 10^6$ | 0.280 | 500 | Unsatisfactory for subzero applications |
| Austenitic | AISI 301, 302 | $28 \times 10^6$ | $10 \times 10^6$ | 0.282 | 600 | Good strength at moderate temperatures; low stress relaxation |
| **Copper-based alloys** | | | | | | |
| Spring brass | ASTM B134 | $16 \times 10^6$ | $6 \times 10^6$ | 0.308 | 200 | Low cost; high conductivity; poor mechanical properties |
| Phosphor bronze | ASTM B159 | $15 \times 10^6$ | $6.3 \times 10^6$ | 0.320 | 200 | Ability to withstand repeated flexures; popular alloy |
| Beryllium copper | ASTM B197 | $19 \times 10^6$ | $6.5 \times 10^6$ | 0.297 | 400 | High elastic and fatigue strength; hardenable |
| **Nickel-based alloys** | | | | | | |
| Inconel 600 | — | $31 \times 10^6$ | $11 \times 10^6$ | 0.307 | 600 | Good strength; high corrosion resistance |
| Inconel X-750 | — | $31 \times 10^6$ | $11 \times 10^6$ | 0.298 | 1100 | Precipitation hardening; for high temperatures |
| Ni-Span C | — | $27 \times 10^6$ | $9.6 \times 10^6$ | 0.294 | 200 | Constant modulus over a wide temperature range |

SOURCE: Adapted from Relvas (1996).

**Table 17.2** Coefficients used in Equation (17.2) for five spring materials

| Material | Size Range in | Size Range mm | Exponent $m$ | Constant $A_p$ ksi | Constant $A_p$ MPa |
|---|---|---|---|---|---|
| Music wire[a] | 0.004–0.250 | 0.10–6.5 | 0.146 | 196 | 2170 |
| Oil-tempered wire[b] | 0.020–0.500 | 0.50–12 | 0.186 | 149 | 1880 |
| Hard-drawn wire[c] | 0.028–0.500 | 0.70–12 | 0.192 | 136 | 1750 |
| Chromium vanadium[d] | 0.032–0.437 | 0.80–12 | 0.167 | 169 | 2000 |
| Chromium silicone[e] | 0.063–0.375 | 1.6–10 | 0.112 | 202 | 2000 |

[a]Surface is smooth and free from defects and has a bright, lustrous finish.
[b]Surface has a slight heat-treating scale that must be removed before plating.
[c]Surface is smooth and bright with no visible marks.
[d]Aircraft-quality tempered wire; can also be obtained annealed.
[e]Tempered to Rockwell C49 but may also be obtained untempered.
SOURCE: *Engineering Guide to Spring Design*, Barnes Group, Inc., 1987.

In the design of springs the allowable stress is the torsional yield strength rather than the ultimate strength. Once the ultimate strength is known from Equation (17.2), the shear yield stress, which is the allowable shear stress, can be expressed as

$$S_{sy} = \tau_{all} = 0.40 S_{ut} \tag{17.3}$$

---

**EXAMPLE 17.1**

**Given:** A spring arrangement is being considered for opening the ashtray in a car. There can be either one spring behind the ashtray, or one weaker spring on each side of the ashtray, pushing it out after a snap mechanism is released. Except for the spring wire diameter, the spring force and the spring geometry are the same for the two options. The material chosen is music wire. For the one-spring option the wire diameter needed is 1 mm.

**Find:**

(a) Which option is cheaper to make if the cost of the springs is proportional to the weight of the spring material?
(b) Is there any risk that the two-spring option will suffer from fatigue if the one-spring option does not fatigue?

**Solution:**

(a) From Table 17.2 for music wire

$$m = 0.146 \quad \text{and} \quad A_p = 2170 \text{ MPa}$$

Substituting the above into Equation (17.2) gives

$$S_{ut} = \frac{2170}{1^{0.146}} = 2170 \text{ MPa}$$

This value is the allowable stress for option 1. From Table 5.1(a) for a concentrated load at the end of the beam ($a = x = l$), the deflection at the free end of the beam is

$$y = \frac{Pl^3}{3EI} \tag{a}$$

The spring has a circular cross section, so that the area moment of inertia is

$$I = \frac{\pi r^4}{4} \tag{b}$$

Substituting Equation (b) into Equation (a) gives

$$y = \frac{Pl^3 4}{3E\pi r^4} = \left(\frac{4l^3}{3\pi E}\right)\left(\frac{P}{r^4}\right) \tag{c}$$

The material used and the spring length are the same for options 1 and 2. Thus, if subscript 1 refers to option 1 and subscript 2 refers to option 2, from Equation (c)

$$\frac{P_1}{r_1^4} = \frac{P_1/2}{r_2^4}$$

$$\therefore \quad r_2 = \frac{r_1}{(2)^{0.25}} = 0.8409 r_1 \tag{d}$$

The cost will change proportionally to the weight, or

$$\frac{\pi r_1^2 l \rho}{2\pi r_2^2 l \rho} = \frac{r_1^2}{2r_2^2} = \frac{r_1^2}{2(0.8409 r_1)^2} = 0.7071 \tag{e}$$

Thus, the one-spring option costs only 0.7071 times the two-spring option and is therefore 29.3% cheaper.

(b) From Equation (4.48) the bending (design) stress for one spring (option 1) is

$$\sigma_1 = \frac{Mc}{I} = \frac{Plr_1}{\pi r_1^4/4} = \frac{4Pl}{\pi r_1^3} \tag{f}$$

For two springs (option 2) the design stress is

$$\sigma_2 = \frac{4(P/2)l}{\pi r_2^3} = \frac{2Pl}{\pi(0.8409)^3 r_1^3} = \frac{\sigma_1}{2(0.8409)^3} = 0.8409\sigma_1$$

Using Equation (17.2) gives the allowable stress for option 2 as

$$S_{ut} = \frac{2170}{(0.8409)^{0.146}} = 2226 \text{ MPa}$$

Thus, the allowable stress is higher for option 2 than for option 1, and the design stress is lower for option 2 than for option 1; so the safety factor for option 2 is much larger than that for option 1. Recall from Equation (1.1) that the safety factor is the allowable stress divided by the design stress. Therefore, if there is no risk of fatigue failure for option 1, there is no risk of fatigue failure for option 2.

## 17.3    HELICAL COMPRESSION SPRINGS

The helix is the spiral form of spring wound with constant coil diameter and uniform pitch. **Pitch** is the distance, measured parallel to the coil axis, from the center of one coil to the center of the adjacent coil. In the most common form of the **helical compression spring,** a round wire is wrapped into a cylindrical form with a constant pitch between adjacent coils. The best way to visualize this is to first begin with a straight wire of length $l$ and wire diameter $d$, to fit a bracket of linear arm length $R$ to each end, and then to subject the device to force $P$, as shown in Figure 17.2(a). The force $P$ applied at each end at a distance $R$ from the center of the wire produces a torsional moment

$$T = PR \tag{17.4}$$

In Figure 17.2(a) the wire is in equilibrium under the action of the forces $P$, and the length is described in the number of coils this straight wire is to form.

Figure 17.2(b) shows that the wire has been formed into a helix of $N$ coils each with mean coil radius $R$. The coiled wire is in equilibrium under the action of two equal and opposite forces $P$. The important difference between the parts of Figure 17.2 is that the straight wire in Figure 17.2(a) experiences only torsional shear whereas the helical coil spring in Figure 17.2(b) and (c) experiences transverse and torsional shear. Figure 17.2(b) illustrates transverse (sometimes called direct) shear; Figure 17.2(c) illustrates torsional

**Figure 17.2**    Helical coil. (a) Straight wire before coiling;
(b) coiled wire showing transverse (or direct) shear; (c) coiled wire
showing torsional shear.

shear. Torsional shear in a helical coil spring is the primary stress, but transverse (or direct) shear is important enough to consider.

## 17.3.1    TORSIONAL SHEAR STRESS

The stress in the straight wire shown in Figure 17.2(a) is shear caused by the torque given in Equation (17.4). The major stress in the helically coiled spring shown in Figure 17.2(c) is also torsional shear stress. From Equation (4.34), the maximum torsional shear stress assuming that the cross section of the wire is circular and using Table 4.1 is

$$\tau_{t,\max} = \frac{Tc}{J} = \frac{Td(32)}{2\pi d^4} = \frac{16PR}{\pi d^3} = \frac{8PD}{\pi d^3} \tag{17.5}$$

where $D$ = mean coil diameter, m
$\qquad d$ = wire diameter, m

The **spring index,** which is a measure of coil curvature, is

$$C = \frac{D}{d} \tag{17.6}$$

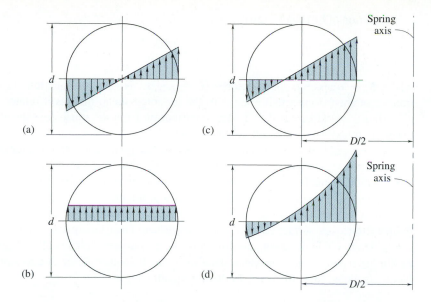

**Figure 17.3** Shear stresses acting on wire and coil. (a) Pure torsional loading; (b) transverse loading; (c) torsional and transverse loading with no curvature effects; (d) torsional and transverse loading with curvature effects.

For most springs the spring index is between 3 and 12. Figure 17.3, where the coil axis is always to the right, shows the stress distribution across the wire cross section. From Figure 17.3(a), for pure torsional stress, the shear stress is a maximum at the outer fiber of the wire and is zero at the center of the wire.

## 17.3.2 TRANSVERSE SHEAR STRESS

The maximum transverse (also called direct) shear stress shown in Figure 17.3(b) can be expressed for a solid circular cross section as

$$\tau_{d,\max} = \frac{P}{A} = \frac{4P}{\pi d^2} \tag{17.7}$$

In Figure 17.3(b) the maximum stress occurs at the midheight of the wire.

## 17.3.3 COMBINED TORSIONAL AND TRANSVERSE SHEAR STRESS

The maximum shear stress resulting from summing the torsional and transverse shear stresses, using Equations (17.5) and (17.7), is

$$\tau_{\max} = \tau_{t,\max} + \tau_{d,\max} = \frac{8PD}{\pi d^3} + \frac{4P}{\pi d^2} = \frac{8DP}{\pi d^3}\left(1 + \frac{d}{2D}\right) \tag{17.8}$$

Figure 17.3(c) shows that the maximum shear stress occurs at the midheight of the wire and at the coil inside diameter. Curvature effects are not considered in Equation (17.8) and are not shown in Figure 17.3(c).

Equation (17.8) can be rewritten as

$$\tau_{max} = \frac{8DK_dP}{\pi d^3}$$

(17.9)

where $K_d$ = transverse shear factor = $(C + 0.5)/C$. Note from Equation (17.8) that if the transverse shear were small relative to the torsional shear, then as it relates to Equation (17.10) $K_d$ would be equal to 1. Any contribution from the transverse shear term would make the transverse shear factor greater than 1.

Also, recall that the spring index $C$ is usually between 3 and 12. If that range is used, the range of transverse shear factor is 1.0417 to 1.1667. Thus, the contribution due to transverse shear is indeed small relative to that due to torsional shear. Equation (17.9) is used for static loading conditions and to check if buckling is a problem.

From Section 4.5.2 we found that for a curved member the stresses can be considerably higher at the inside surface than at the outside surface. Thus, incorporating curvature can play a significant role in the spring design. Considering curvature effects alters Equation (17.9) by simply replacing the transverse shear factor with another factor. A curvature correction factor attributed to A. M. Wahl results in the following:

$$\tau_{max} = \frac{8DK_wP}{\pi d^3}$$

(17.10)

where

$$K_w = \frac{4C - 1}{4C - 4} + \frac{0.615}{C}$$

(17.11)

The first fraction in Equation (17.11) accounts for the curvature effect, and the second fraction accounts for the transverse shear stress. Equations (17.10) and (17.11) should be used for cyclic loading.

Figure 17.3(d) shows the stress distribution when curvature effects and both torsional and transverse shear stresses are considered. The maximum stress occurs at the midheight of the wire and at the coil inside diameter. This location is where failure should first occur in the spring.

### 17.3.4   DEFLECTION

From Equation (4.27) the shear strain due to torsional loading is

$$\gamma_{\theta z} = \frac{r\theta}{l} = \frac{\text{deflection}}{\text{length}}$$

(17.12)

The deflection due to torsional loading is

$$\delta_t = r\theta = \frac{D}{2}\theta$$

(17.13)

Making use of Equation (4.31) gives

$$\delta_t = \left(\frac{D}{2}\right)\left(\frac{TL}{JG}\right)$$

(17.14)

Applying this equation to a helical coil spring, using Figure 17.2, and assuming that the wire has a circular cross section give

$$\delta_t = \frac{D}{2}\frac{(D/2)P(2\pi)(D/2)N_a}{G(\pi d^4/32)} = \frac{8PD^3 N_a}{Gd^4} = \frac{8PC^3 N_a}{Gd} \tag{17.15}$$

where $C$ = spring index, $D/d$
$N_a$ = number of active coils
$G$ = shear modulus of elasticity, Pa

The circumferential deflection due to torsional and transverse shear loading can be derived by using Castigliano's theorem (Sec. 5.6). The total strain energy from Table 5.2 is

$$U = \frac{T^2 l}{2GJ} + \frac{P^2 l}{2AG} \tag{17.16}$$

The first term on the right of the equals sign corresponds to torsional loading, and the second term corresponds to transverse shear for a circular cross section.

Applying Equation (17.16) to a helical coil spring of circular cross section gives the total strain energy as

$$U = \frac{(PD/2)^2(\pi DN_a)}{2G(\pi d^4/32)} + \frac{P^2(\pi DN_a)}{2G(\pi d^2/4)} = \frac{4P^2 D^3 N_a}{Gd^4} + \frac{2P^2 DN_a}{Gd^2}$$

Using Castigliano's theorem [Eq. (5.30)] gives

$$\delta = \frac{\partial U}{\partial P} = \frac{8PD^3 N_a}{Gd^4} + \frac{4PDN_a}{Gd^2} = \frac{8PD^3 N_a}{Gd^4}\left(1 + \frac{d^2}{2D^2}\right)$$

$$= \frac{8PC^3 N_a}{Gd}\left(1 + \frac{0.5}{C^2}\right) \tag{17.17}$$

Comparing Equation (17.15) with (17.17) shows that the second term in Equation (17.17) is the transverse shear term and that for a spring index in the normal range $3 \leq C \leq 12$ the deflection due to transverse shear is extremely small.

From Equations (17.15) and (17.17) the spring rate is

$$k_t = \frac{P}{\delta_t} = \frac{Gd}{8C^3 N_a} \tag{17.18}$$

$$k = \frac{P}{\delta} = \frac{Gd}{8C^3 N_a(1 + 0.5/C^2)} \tag{17.19}$$

Equation (17.18) accounts only for torsional shear; Equation (17.19) accounts for torsional as well as transverse shear loading.

The difference between spring index $C$ and **spring rate** $k$ is important. The spring index is dimensionless whereas spring rate is measured in newtons per meter. Also, a stiff spring has a small spring index and a large spring rate. (Recall that the spring index for most conventional springs varies between 3 and 12.) A spring with excessive deflection has a large spring index and a small spring rate.

### 17.3.5    END CONDITIONS AND SPRING LENGTH

Figure 17.4 shows four types of end commonly used in compression springs. The end coils produce an eccentric application of the load, increasing the stress on one side of the spring. Different end conditions have different amounts of eccentric loading, and compensation must be made in designing a spring. This design alteration is made by designating the number of active coils in a spring. A deduction is made from the total number of coils to account for the turns at the ends, which do not affect the deflection. It is difficult to identify just how much the deduction should be; however, an average number based on experimental results is used.

Figure 17.4(a) shows plain ends that have a noninterrupted helicoid; the spring rates for the ends are the same as if they were not cut from a longer coil. Figure 17.4(b) shows a plain end that has been ground. In Figure 17.4(c) a spring with plain ends that are squared (or closed) is obtained by deforming the ends to a 0° helix angle. Figure 17.4(d) shows squared and ground ends. A better load transfer into the spring is obtained by using ground ends.

Table 17.3 shows useful formulas for the pitch, length, and number of coils for compression springs for the four end conditions shown in Figure 17.4. Recall that the pitch is the distance measured parallel to the spring axis from the center of one coil to the center of an adjacent coil. In Table 17.3 the formulas differ for the various end conditions. The **solid length** is the length of the spring when all adjacent coils are in metal-to-metal contact. The **free length** is the length of the spring when no external forces are applied to it. Figure 17.5 shows the various lengths and forces applicable to helical compression springs.

Figure 17.6 shows the interdependent relationships between force, deflection, and spring length for four distinctively different positions: free, initial, operating, and solid.

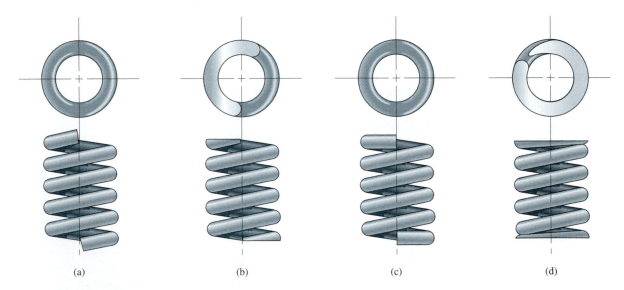

(a)    (b)    (c)    (d)

**Figure 17.4**    Four end types commonly used in compression springs. (a) Plain; (b) plain and ground; (c) squared; (d) squared and ground.

**Table 17.3** Useful formulas for compression springs with four end conditions

| Term | Plain | Plain and ground | Squared or closed | Squared and ground |
|---|---|---|---|---|
| | | **Type of Spring End** | | |
| Number of end coils $N_e$ | 0 | 1 | 2 | 2 |
| Total number of coils $N_t$ | $N_a$ | $N_a + 1$ | $N_a + 2$ | $N_a + 2$ |
| Free length $l_f$ | $pN_a + d$ | $p(N_a + 1)$ | $pN_a + 3d$ | $pN_a + 2d$ |
| Solid length $l_s$ | $d(N_t + 1)$ | $dN_t$ | $d(N_t + 1)$ | $dN_t$ |
| Pitch $p$ | $\dfrac{l_f - d}{N_a}$ | $\dfrac{l_f}{N_a + 1}$ | $\dfrac{l_f - 3d}{N_a}$ | $\dfrac{l_f - 2d}{N_a}$ |

The length is zero at the top right corner and moves to the left for positive lengths. Also, the free length $l_f$ is equal to the solid deflection plus the solid length, or

$$l_f = l_s + \delta_s \tag{17.20}$$

### 17.3.6 BUCKLING AND SURGE

Relatively long compression springs should be checked for buckling. Figure 17.7 shows critical buckling conditions for parallel and nonparallel ends. The critical deflection where buckling will first start to occur can be determined from this figure. If buckling is a problem, it can be prevented by placing the spring in a hole or over a rod. However, the coils rubbing on these guides will take away some of the spring force and thus reduce the load delivered at the spring ends.

A longitudinal vibration that should be avoided in spring design is a **surge,** or pulse of compression, passing through the coils to the ends, where it is reflected and returned. An

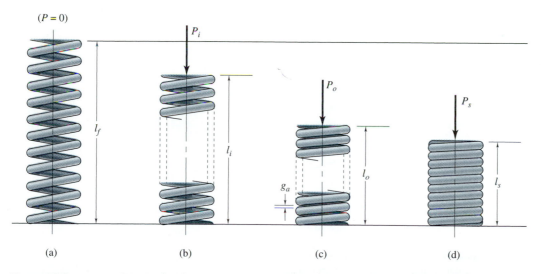

**Figure 17.5** Various lengths and forces applicable to helical compression springs. (a) Unloaded; (b) under initial load; (c) under operating load; (d) under solid load.

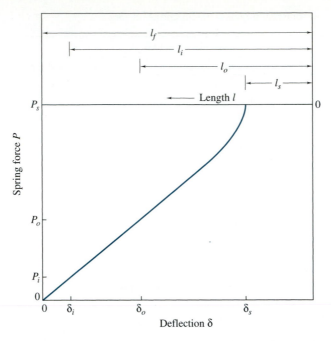

**Figure 17.6**   Graphical representation of deflection, force and length for four spring positions.

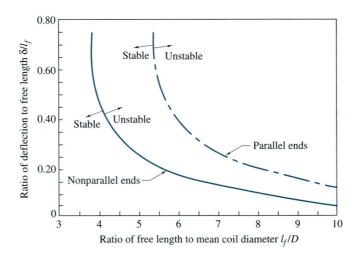

**Figure 17.7**   Critical buckling conditions for parallel and nonparallel ends of compression springs. [Barnes (1987).]

initial surge is sustained if the material frequency of the spring is close to the frequency of the repeated loading. The equation for the lowest natural frequency in cycles per second is

$$f_n = \frac{2}{\pi N_a} \frac{d}{D^2} \sqrt{\frac{G}{32\rho}} \qquad (17.21)$$

where $G$ = shear modulus of elasticity, Pa
$\rho$ = density, kg/m$^3$

Vibrations may also occur at whole multiples, such as 2, 3, and 4 times the lowest frequency. The spring design should avoid these frequencies.

---

**EXAMPLE 17.2**

**Given:** A helical compression spring with plain ends is made to have a spring rate of 100,000 N/m, a wire diameter of 10 mm, and a spring index of 5. The maximum allowable shear stress is 480 N/mm$^2$, and the shear modulus of elasticity is 80 GPa.

**Find:** The number of active coils, the maximum allowable static load, and the manufactured pitch so that the maximum load just compresses the spring to its solid length.

**Solution:** From Equation (17.19) the number of active coils is

$$N_a = \frac{Gd}{8C^3k\,(1 + 0.5/C^2)}$$

$$= \frac{(80)(10^9)\,(10)(10^{-3})}{8(5)^3(10^5)\,(1 + 0.5/25)} = 7.843 \text{ coils}$$

Therefore, let the number of active coils be 8. From Equation (17.9) the transverse shear factor is

$$K_d = \frac{C + 0.5}{C} = \frac{5 + 0.5}{5} = 1.10$$

For $\tau_{max} = \tau_{all} = 480$ N/mm$^2$, from Equation (17.9) the maximum force is

$$P_{max} = \frac{\pi d^3 \tau_{max}}{8DK_d} = \frac{\pi (10^{-2})^3 (480)}{(10^{-6})(8)(50)(10^{-3})(1.10)} = (3.427)(10^3) \text{ N} = 3.427 \text{ kN}$$

The maximum deflection just to bring the spring to solid length is

$$\delta_s = \delta_{max} = \frac{P_{max}}{k} = \frac{3427}{10^5} = (34.27)(10^{-3}) \text{ m} = 34.27 \text{ mm}$$

From Table 17.3 the solid length for plain ends is

$$l_s = d(N_a + 1) = (10)(10^{-3})(9) = (90)(10^{-3}) \text{ m} = 90 \text{ mm}$$

From Equation (17.20) the free length is

$$l_f = l_s + \delta_s = 90 + 34.27 = 124.27 \text{ mm}$$

From Table 17.3 the pitch is

$$p = \frac{l_f - d}{N_a} = \frac{124.27 - 10}{8} = 14.28 \text{ mm}$$

**EXAMPLE 17.3**

**Given:** A compression coil spring is made of music wire with squared and ground ends. The spring is to have a spring rate of 1250 N/m. The force corresponding to a solid length is 60 N. The spring index is fixed at 10. Static loading conditions are assumed.

**Find:** Find the wire diameter and the mean coil diameter for the limit when the spring is compressed solid. Give the free and solid lengths, and indicate whether buckling is a problem. Also give a design recommendation.

**Solution:** The tensile ultimate strength can be determined from Equation (17.2) and Table 17.2 as

$$S_{ut} = \frac{A_p}{d^m} = \frac{(2.170)(10^9)}{d^{0.146}}$$

where $d$ is in millimeters. If $d$ were expressed in meters rather than in millimeters, the above equation would become

$$S_{ut} = \frac{(2.170)(10^9)}{(1000)^{0.146}d^{0.146}} = \frac{(791.5)(10^6)}{d^{0.146}}$$

where $d$ is in meters. From Equation (17.3) the shear yield stress is

$$S_{sy} = 0.40S_{ut} = \frac{(316.6)(10^6)}{d^{0.146}} \tag{a}$$

For static loading and given that the spring index is 10, the transverse shear factor from Equation (17.9) is

$$K_d = 1 + \frac{0.5}{C} = 1.05$$

From Equation (17.9) the maximum design shear stress is

$$\tau_{max} = \frac{8CK_dP}{\pi d^2} = \frac{8(10)(1.05)(60)}{\pi d^2} = \frac{1604}{d^2} \tag{b}$$

Equating Equations (a) and (b) gives

$$\frac{(316.6)(10^6)}{d^{0.146}} = \frac{1604}{d^2} \tag{c}$$

$$\therefore \quad d = 1.393 \text{ mm}$$

The mean coil diameter is

$$D = Cd = 10(1.393) = 13.93 \text{ mm}$$

Since the spring rate is 1250 N/m and the force corresponding to a solid length is 60 N,

$$\delta_s = \frac{P_s}{k} = \frac{60}{1250} = 0.0480 \text{ m} = 48 \text{ mm}$$

From Table 17.1 the shear modulus of elasticity for music wire is

$$G = 11.5 \times 10^6 \text{ psi} = 0.7929 \times 10^{11} \text{ Pa}$$

From Equation (17.18) the number of active coils is

$$N_a = \frac{Gd}{8k_tC^3} = \frac{(0.7929)(10^{11})(1.393)(10^{-3})}{8(1250)(10^3)} = 11.05 \text{ coils}$$

From Table 17.3 for squared and ground ends, the total number of coils is

$$N_t = N_a + 2 = 13.05 \text{ coils}$$

The solid length is

$$l_s = dN_t = (1.393)(13.05) = 18.18 \text{ mm}$$

The free length from Equation (17.20) is

$$l_f = l_s + \delta_s = 18.18 + 48 = 66.18 \text{ mm}$$

The pitch is

$$p = \frac{l_f - 2d}{N_a} = \frac{66.18 - 2.786}{11.05} = 5.737 \text{ mm}$$

Making use of Figure 17.7 when

$$\frac{l_f}{D} = \frac{66.18}{13.93} = 4.751 \quad \text{and} \quad \frac{\delta_s}{l_f} = \frac{48.00}{66.18} = 0.7253$$

and assuming parallel ends show that stability should not be a problem.

Recognize that in Equation (c) the allowable and maximum designs are assumed to have a safety factor of 1. If instead the safety factor were 2, the wire diameter would be

$$\frac{(316.6)(10^6)}{d^{0.146}} = \frac{3208}{d^2}$$

$$\therefore \qquad d = 2.024 \text{ mm} \qquad \text{and} \qquad D = Cd = 20.24 \text{ mm}$$

From Equation (17.18) the number of active coils is

$$N_a = \frac{Gd}{8k_t C^3} = 16.00 \text{ coils}$$

$$N_t = N_a + 2 = 18.00 \text{ coils}$$

The solid and free lengths are

$$l_s = dN_t = (2.024)(18.00) = 36.43 \text{ mm}$$

$$l_f = l_s + \delta_s = 36.43 + 48 = 84.43 \text{ mm}$$

The pitch is

$$p = \frac{l_f - 2d}{N_a} = \frac{84.43 - 4.048}{16} = 5.024 \text{ mm}$$

Also,

$$\frac{l_f}{D} = \frac{84.43}{20.24} = 4.171$$

$$\frac{\delta_s}{l_f} = \frac{48.00}{84.43} = 0.5685$$

There is thus no problem with stability for parallel ends.

### 17.3.7    CYCLIC LOADING

Because spring loading is most often continuously fluctuating, allowance must be made in designing for fatigue and stress concentration. The material developed in Chapter 7 is applied in this section. Helical springs *are never* used as both compression and extension springs. Furthermore, springs are assembled with a preload in addition to the working stress, thus preventing the stress from being zero. The cyclic variation is that on the nonzero mean considered in Section 7.3. The worst condition would occur if there were no preload (i.e., when $\tau_{min} = 0$).

For cyclic loading the Wahl curvature correction factor given in Equation (17.11) should be used instead of the transverse shear factor defined in Equation (17.9), which is used for static conditions. The Wahl curvature correction factor can be viewed as a fatigue stress concentration factor. However, a major difference for springs in contrast to other machine elements is that for springs the Wahl curvature correction factor is applied to both the mean stress and the stress amplitude. The reason is that in its true sense the Wahl curvature correction factor is not a fatigue stress concentration factor (as considered in Sec. 7.7.1), but is a way of calculating the shear stress inside the coil.

When Equations (7.1) and (7.2) are applied to force rather than to stress, the alternating and mean forces can be expressed as

$$P_a = \frac{P_{max} - P_{min}}{2} \tag{17.22}$$

$$P_m = \frac{P_{max} + P_{min}}{2} \tag{17.23}$$

The alternating and mean stresses, from Equation (17.10), are

$$\tau_a = \frac{8DK_wP_a}{\pi d^3} \tag{17.24}$$

$$\tau_m = \frac{8DK_wP_m}{\pi d^3} \tag{17.25}$$

For springs the safety factor guarding against torsional endurance limit fatigue is

$$n_s = \frac{S_{se}}{\tau_a} \tag{17.26}$$

Against torsional yielding it is

$$n_s = \frac{S_{sy}}{\tau_a + \tau_m} \tag{17.27}$$

And against torsional fatigue strength it is

$$n_s = \frac{S_{sf}}{\tau_a} \tag{17.28}$$

where $S_{sf}$ = modified shear fatigue strength, Pa. Equations (7.11) to (7.14) should be used to determine the torsional fatigue strength.

The best data for the torsional endurance limits of spring steels are those of Zimmerli (1957). The surprising fact about these data is that size, material, and tensile strength have no effect on the endurance limits of spring steels with wire diameters under $\frac{3}{8}$ in (10 mm). Zimmerli's (1957) results are

$$S'_{se} = \begin{cases} 45.0 \text{ ksi (310 MPa)} & \text{for unpeened springs} \\ 67.5 \text{ ksi (465 MPa)} & \text{for peened springs} \end{cases} \quad (17.29)$$

These results are valid for all materials in Table 17.2. Shot peening is working the surface material to cause compressive residual stresses that toughen the surface. The endurance limit given in Equation (17.29) is corrected for all factors given in Equation (7.16) except the reliability factor.

The modulus of rupture, or shear ultimate strength $S_{su}$, for spring steels can be expressed as

$$S_{su} = 0.60S_u \quad (17.30)$$

---

**EXAMPLE 17.4**

**Given:** A helical compression spring has 14 active coils, a free length of 1.25 in, and an outside diameter of $\frac{7}{16}$ in. The ends of the spring are squared and ground, and the end plates are constrained to be parallel. The material is music wire with presetting. The wire diameter is fixed at 0.042 in.

**Find:**

(a) For *static* conditions compute the spring rate, the solid length, and the stress when the spring is compressed to the solid length. Will static yielding occur while the spring is brought to its solid length?

(b) For *dynamic* conditions with $P_{min} = 0.9$ lb and $P_{max} = 2.9$ lb, will the spring experience torsional endurance limit fatigue, torsional yielding, or torsional fatigue failure? Assume a survival probability of 90%, unpeened coils, and fatigue failure based on $50 \times 10^3$ cycles.

**Solution:**

(a) The spring index from Equation (17.6) is

$$C = \frac{D}{d} = \frac{D_o - d}{d} = \frac{\frac{7}{16} - 0.042}{0.042} = 9.417$$

From Equation (17.9) the transverse shear factor is

$$K_d = 1 + \frac{0.5}{C} = 1 + \frac{0.5}{9.417} = 1.053$$

From Equation (17.11) the Wahl curvature correction factor is

$$K_w = \frac{4C - 1}{4C - 4} + \frac{0.615}{C} = \frac{4(9.417) - 1}{4(9.417) - 4} + \frac{0.615}{9.417} = 1.154$$

From Table 17.3 for squared and ground ends

$$N_t = N_a + 2 = 14 + 2 = 16 \text{ coils}$$

The solid length is

$$l_s = dN_t = (0.042)(16) = 0.6720 \text{ in}$$

The pitch is

$$p = \frac{l_f - 2d}{N_a} = \frac{1.25 - 0.084}{14} = 0.0833 \text{ in}$$

Also, from Equation (17.20) the deflection to solid length is

$$\delta_s = l_f - l_s = 1.25 - 0.6720 = 0.5780 \text{ in}$$

From Equation (17.2) the ultimate strength in tension is

$$S_{ut} = \frac{A_p}{d^m}$$

From Table 17.2 for music wire

$$S_{ut} = \frac{(196)(10^3)}{(0.042)^{0.146}} \text{ psi} = 311.4 \times 10^3 \text{ psi}$$

From Equation (17.3) the torsional yield stress, or the allowable shear stress for static loading, is

$$\tau_{\text{all}} = S_{sy} = 0.40 S_{ut} = 0.40(311.4)(10^3) = 124.6 \times 10^3 \text{ psi}$$

From Equation (17.15) the force required to compress the coils to a solid length while using the shear modulus of elasticity for music wire in Table 17.1 of $11.5 \times 10^6$ psi is

$$P_s = \frac{Gd\delta_s}{8C^3 N_a} = \frac{(11.5)(10^6)(0.042)(0.5780)}{8(9.417)^3(14)} = 2.985 \text{ lb}$$

From Equation (17.9) the maximum design stress is

$$\tau_{\text{max}} = \frac{8DK_d P_s}{\pi d^3} = \frac{8CK_d P_s}{\pi d^2} = \frac{8(9.417)(1.053)(2.985)}{\pi(0.042)^2} = 42,729 \text{ psi}$$

The safety factor guarding against static yielding is

$$n_s = \frac{\tau_{\text{all}}}{\tau_{\text{max}}} = \frac{124,600}{42,729} = 2.916$$

Thus, failure should not occur due to static yielding.
    Checking for buckling gives

$$\frac{\delta_s}{l_f} = \frac{0.5780}{1.25} = 0.4624$$

$$\frac{l_f}{D} = \frac{1.25}{\frac{7}{16} - 0.042} = 3.161$$

From Figure 17.7 buckling is not a problem.

(b)    The alternating and mean forces are, respectively,

$$P_a = \frac{P_{\text{max}} - P_{\text{min}}}{2} = \frac{2.9 - 0.9}{2} = 1.0 \text{ lb}$$

$$P_m = \frac{P_{\text{max}} + P_{\text{min}}}{2} = \frac{2.9 + 0.9}{2} = 1.9 \text{ lb}$$

From Equation (17.10) the alternating and mean stresses are

$$\tau_a = \frac{8CK_w P_a}{\pi d^2} = \frac{8(9.417)(1.154)(1.0)}{\pi(0.042)^2} = 15.69 \times 10^3 \text{ psi}$$

$$\tau_m = \frac{8CK_w P_m}{\pi d^2} = \tau_a\left(\frac{P_m}{P_a}\right) = 29.81 \times 10^3 \text{ psi}$$

The safety factor guarding against torsional yielding is

$$n_s = \frac{S_{sy}}{\tau_a + \tau_m} = \frac{124.6}{15.69 + 29.81} = 2.738$$

Therefore, failure should not occur from torsional yielding.

From Table 7.4 and Equation (17.29) the modified endurance limit is

$$S_{se} = k_r S'_{se} = (0.9)(45)(10^3) \text{ psi} = 40.5 \times 10^3 \text{ psi}$$

The safety factor guarding against torsional endurance limit fatigue is

$$n_s = \frac{S_{se}}{\tau_a} = \frac{(40.5)(10^3)}{(15.69)(10^3)} = 2.581$$

Therefore, failure should not occur from torsional endurance limit fatigue.

From Equation (17.30) the modulus of rupture for spring steel is

$$S_{su} = 0.60 S_{ut} = 0.60(311.4)(10^3) \text{ psi} = 186.8 \times 10^3 \text{ psi}$$

From Equation (7.12) the slope used to calculate the fatigue strength is

$$b_s = -\frac{1}{3} \log \frac{0.72 S_{su}}{S_{se}} = -\frac{1}{3} \log \frac{(0.72)(186.8)(10^3)}{(40.5)(10^3)} = -0.1738$$

From Equation (7.13) the intercept used to calculate the fatigue strength is

$$\bar{C} = \log \frac{(0.72 S_{su})^2}{S_{se}} = \log \frac{(0.72)^2(186.8)^2(10^3)^2}{(40.5)(10^3)} = 5.650$$

From Equation (7.14) the fatigue strength is

$$S_{sf} = 10^{\bar{C}}(N_t)^{b_s} = 10^{5.650}[(50)(10^3)]^{-0.1738} = 68.12 \times 10^3 \text{ psi}$$

From Equation (17.28) the safety factor guarding against torsional fatigue strength failure is

$$n_s = \frac{S_{sf}}{\tau_a} = \frac{68.12}{15.69} = 4.342$$

Therefore, failure should not occur under either static or dynamic conditions.

## 17.4 HELICAL EXTENSION SPRINGS

In applying the load in **helical extension springs,** the shape of the hooks, or end turns, must be designed so that the stress concentration effects caused by the presence of sharp bends are decreased as much as possible. In Figure 17.8(a) and (b) the end of the extension spring is formed by merely bending up a half-loop. If the radius of the bend is small, the stress concentration at the cross section, at B in Figure 17.8(b), will be large.

**Figure 17.8**    Ends for extension springs. (a) Conventional design; (b) side view of Figure 17.8(a); (c) improved design over Figure 17.8(a); (d) side view of Figure 17.8(c).

**Figure 17.9**    Dimensions of helical extension spring.

The most obvious method for avoiding these severe stress concentrations is to make the mean radius of the hook $r_2$ larger. Figure 17.8(c) shows another method of achieving this goal. Here, the hook radius is smaller, which increases the stress concentration. The stress is greatly reduced, however, because the coil diameter is reduced. There is a lower stress because of the shorter moment arm. The largest stress occurs at cross section B, shown in Figure 17.8(d), a side view of Figure 17.8(c).

Figure 17.9 shows some of the important dimensions of a helical extension spring. All the coils in the body are assumed to be active. The total number of coils is

$$N_t = N_a + 1 \qquad\qquad\qquad \textbf{(17.31)}$$

The length of the body is

$$l_b = dN_t \qquad (17.32)$$

The free length is measured from the inside of the end loops, or hooks, and is

$$l_f = l_b + l_h + l_l \qquad (17.33)$$

For close-wound extension springs, the force–deflection curve is such that some initial force is required before any deflection occurs, and after the initial force the force–deflection curve is linear. Thus, the force is

$$P = P_i + \frac{\delta G d^4}{8 N_a D^3} \qquad (17.34)$$

where $P_i$ = preload, N. The spring rate is

$$k = \frac{P - P_i}{\delta} = \frac{d^4 G}{8 N_a D^3} = \frac{dG}{8 N_a C^3} \qquad (17.35)$$

The shear stresses for static and dynamic conditions are given by Equations (17.9) and (17.10), respectively. The preload stress $\tau_i$ is obtained by using Equation (17.5).

The preferred range of 3 to 12 for the spring index is equally valid for extension and compression springs. The initial tension wound into the spring, giving a preload, is described by

$$P_i = \frac{\pi \tau_i d^3}{8D} = \frac{\pi \tau_i d^2}{8C} \qquad (17.36)$$

Recommended values of $\tau_i$ depend on the spring index and are given in Figure 17.10. Springs should be designed for midway in the preferred spring index range in Figure 17.10.

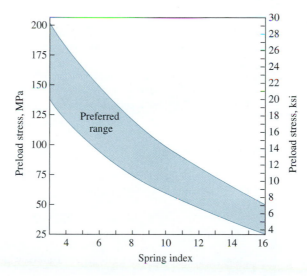

**Figure 17.10** Preferred range of preload stress for various spring indexes. [*Adapted from Almen and Laszlo (1936).*]

The critical stresses in the hook occur at sections A and B, as shown in Figure 17.8. At section A the stress is due to bending and transverse shear, and at section B the stress is due to torsion. The bending moment and transverse shear stresses acting at section A can be expressed as

$$\sigma_A = \left(\frac{Mc}{I}\right)\left(\frac{r_1}{r_3}\right) + \frac{P_A}{A} = \left(\frac{32 P_A r_1}{\pi d^3}\right)\left(\frac{r_1}{r_3}\right) + \frac{4 P_A}{\pi d^2} \qquad (17.37)$$

The shear stress acting at section B is

$$\tau_B = \frac{8 P_B C}{\pi d^2}\left(\frac{r_2}{r_4}\right) \qquad (17.38)$$

Radii $r_1, r_2, r_3,$ and $r_4$ are given in Figure 17.8. Recommended design practice is that $r_4 > 2d$. Values of $\sigma_A$ and $\tau_B$ given in Equations (17.37) and (17.38) are the design stresses. These stresses are compared with the allowable stresses, taking into account the type of loading to determine if failure will occur. The allowable shear stress to be used with Equation (17.38) is the shear yield stress given in Equation (17.3). The allowable stress to be used with Equation (17.37) is the yield strength in tension, given as

$$S_{ty} = 0.60 S_{ut} \qquad (17.39)$$

---

**EXAMPLE 17.5**

**Given:** A helical extension spring similar to that shown in Figure 17.8(a) and (b) is made of hard-drawn wire with a mean coil diameter of 10 mm, a wire diameter of 2 mm, and 120 active coils. The hook radius is 6 mm ($r_1 = 6$ mm), and the bend radius is 3 mm ($r_2 = 3$ mm). The preload is 30 N, and the free length is 264 mm.

**Find:** Determine the following:

(a)  Tensile and torsional yield strength of the wire
(b)  Initial torsional shear stress of the wire
(c)  Spring rate
(d)  Force required to cause the normal stress in the hook to reach the tensile yield strength
(e)  Force required to cause the torsional stress in the hook to reach the yield stress
(f)  Distance between hook ends if the smaller of the two forces found in parts $d$ and $e$ is applied

**Solution:**

(a)  From Table 17.2 for hard-drawn wire $m = 0.192$ and $A_p = 1750$ MPa. From Equation (17.2) the ultimate strength in tension is

$$S_{ut} = \frac{A_p}{d^m} = \frac{(1750)(10^6)}{(2)^{0.192}}\ \text{Pa} = 1.532\ \text{GPa}$$

From Equation (17.3) the torsional yield stress, which is the allowable shear stress, is

$$S_{sy} = 0.40 S_{ut} = 0.40(1.532)(10^9)\ \text{Pa} = 0.6128\ \text{GPa}$$

From Equation (17.39) the tensile yield strength is

$$S_{ty} = 0.60 S_{ut} = 0.60(1.532)(10^9)\ \mathrm{Pa} = 0.9192\ \mathrm{GPa}$$

(b) For static loading the transverse shear factor given in Equation (17.9) is

$$K_d = 1 + \frac{0.5}{C} = 1 + \frac{0.5}{5} = 1.1$$

Using Equation (17.36) gives

$$P_i = \frac{\pi \tau_i d^2}{8C} = \frac{\pi \tau_i (2)^2 (10^{-6})}{8(5)} = 30\ \mathrm{N}$$

$$\tau_i = 95.50\ \mathrm{MPa}$$

(c) From Table 17.1 for hard-drawn wire $G = 11.5 \times 10^6$ psi $= 79.29 \times 10^9$ Pa. From Equation (17.35) the spring rate is

$$k = \frac{dG}{8 N_a C^3} = \frac{(2)(10^{-3})(79.29)(10^9)}{8(120)(5)^3} = 1322\ \mathrm{N/m}$$

(d) The critical bending stress in the hook, from Equation (17.37) and the yield stress in tension given above, is

$$\sigma_A = \left( \frac{32 P_A r_1}{\pi d^3} \right) \left( \frac{r_1}{r_3} \right) + \frac{4 P_A}{\pi d^2} = S_{ty}$$

$$\frac{32 P_A (10^{-3})(6)^2}{\pi (2)^3 (10^{-9})(6-1)} + \frac{4 P_A}{\pi (2)^2 (10^{-6})} = (0.9192)(10^9)$$

$$P_A(9.167 + 0.3183)(10^6) = (0.9192)(10^9)$$

The transverse shear stress is only 3.5% of the stress due to both bending and transverse shear. Solving for $P_A$ gives the critical load where failure will first start to occur by normal stress in the hook.

$$P_A = 96.91\ \mathrm{N}$$

(e) The critical torsional shear stress in the hook can be determined from Equation (17.38), or

$$\tau_B = \frac{8 P_B C}{\pi d^2} \left( \frac{r_2}{r_4} \right) = S_{sy}$$

$$\therefore \qquad \frac{8 P_B (5)}{\pi (2)^2 (10^{-6})} \left( \frac{3}{2} \right) = (0.6128)(10^9)$$

$$P_B = 128.3\ \mathrm{N}$$

The smaller load is $P_A = 96.91$ N, which indicates that failure will first occur by normal stress in the hook.

(f) Using $P_A$ in Equation (17.35) gives the deflection as

$$\delta = \frac{P_A - P_i}{k} = \frac{96.91 - 30}{1322} = 0.05061\ \mathrm{m} = 50.61\ \mathrm{mm}$$

The distance between hook ends is

$$l_f + \delta = 264 + 50.61 = 314.6\ \mathrm{mm}$$

**EXAMPLE 17.6**

**Given:** The ends of a helical extension spring are manufactured according to the improved design shown in Figure 17.8(c). The conical parts have five coils at each end with a winding diameter changing according to

$$D = D_o\left(1 - \frac{\zeta}{15\pi}\right)$$

where $\zeta$ is a constant. The cylindrical part of the spring has an outside diameter $D_o$ and 20 coils.

**Find:** How much softer does the spring become with the conical parts at the ends?

**Solution:** From Equation (17.4) the torque in the spring is

$$T = PR = \frac{PD}{2}$$

From Equation (4.31) the change of angular deflection for a small length is

$$d\theta = \frac{T}{GJ}dl$$

The axial compression of the conical part of the spring is

$$d\delta_c = R\,d\theta = \frac{RT}{GJ}dl = \frac{PR^2}{GJ}R\,d\zeta = \frac{PR^3}{GJ}d\zeta$$

The total deflection of the two conical ends is

$$\delta_c = 2\int_0^{10\pi}\frac{PR^3}{GJ}d\zeta = \frac{2P}{GJ}\int_0^{10\pi}\left(\frac{D_o}{2}\right)^3\left(1 - \frac{\zeta}{15\pi}\right)^3 d\zeta$$

$$= \frac{PD_o^3(-15\pi)}{4GJ(4)}\left[\left(1 - \frac{\zeta}{15\pi}\right)^4\right]_0^{10\pi} = \frac{75\pi}{81}\frac{PD_o^3}{GJ}$$

From Equation (17.14) the deflection in the main part of the spring is

$$\delta = \frac{D_o}{2}\frac{Tl}{GJ} = \frac{D_o}{2}\frac{PD_o\pi D_o(20)}{2GJ} = \frac{5\pi PD_o^3}{GJ}$$

$$\frac{\delta_t}{\delta} = \frac{\delta + \delta_c}{\delta} = \frac{(5\pi + 75\pi/81)[PD_o^3/(GJ)]}{5\pi[PD_o^3/(GJ)]} = 1.185$$

Therefore the spring becomes 18.52% softer by including the conical ends.

## 17.5   HELICAL TORSION SPRINGS

The previous section shows that the helical coil spring can be loaded either in compression or in tension. This section considers a torsional load being applied. Figure 17.11 shows a typical example of a **helical torsion spring.** The coil ends can have a great variety of shapes to suit the various applications. The coils are usually close-wound as an extension spring is, as shown in Figure 17.11, but differ from extension springs in that torsion springs do not have any initial tension. Also, note from Figure 17.11 that the ends are shaped to transmit torque rather than force, as is the case for compression and tension springs. The torque is

**Figure 17.11** Helical torsion spring.

applied about the axis of the helix. The torque acts as a bending moment on each section of the wire. The primary stress in a torsional spring is bending. The bending moment $M = Pa$ produces a normal stress in the wire. (In contrast, in a compression or extension spring, the load produces a torsional stress in the wire.) This equality means that the residual stresses built in during winding are in the same direction, but of opposite sign, as the working stresses that occur during use. These residual stresses are useful in making the spring stronger by opposing the working stress, provided that the load is always applied to cause the spring to wind up. Because the residual stress opposes the working stress, torsional springs are designed to operate at stress levels that equal or even exceed the yield strength of the wire.

The maximum bending stress occurs at the inner fiber of the coil, and for wire of circular cross section, it is given as

$$\sigma = \frac{K_i M c}{I} = \frac{32 K_i M}{\pi d^3} \tag{17.40}$$

where

$$K_i = \frac{4C^2 - C - 1}{4C(C - 1)} \tag{17.41}$$

The angular deflection in radians is

$$\theta_{rad} = \frac{M l_w}{EI} \tag{17.42}$$

where $l_w = \pi D N_a = $ length of wire in spring, m.

The angular deflection in revolutions (r) is

$$\theta_{rev} = \frac{M \pi D N_a}{E(\pi d^4/64)} \left( \frac{1 \text{ r}}{2\pi \text{ rad}} \right) = \frac{32 M D N_a}{\pi E d^4} = \frac{10.186 M D N_a}{E d^4} \tag{17.43}$$

The angular spring rate is

$$k_\theta = \frac{M}{\theta_{rev}} = \frac{E d^4}{10.186 D N_a} \tag{17.44}$$

The number of active coils is

$$N_a = N_b + N_e \tag{17.45}$$

where $N_b = $ number of coils in body
$N_e = $ number of end coils $= (l_1 + l_2)/(3\pi D)$
$l_1, l_2 = $ length of ends, m

Torsion springs are frequently used over a round bar. When a load is applied to a torsion spring, the spring winds up, causing a decrease in the inside diameter. For design purposes the inside diameter of the spring must never become equal to the diameter of the bar; otherwise, a spring failure will occur. The inside diameter of a loaded torsion spring is

$$D_i' = \frac{N_a D_i}{N_a'} \tag{17.46}$$

where $N_a$ = number of active coils at no load
$D_i$ = coil inside diameter at no load, m
$N_a'$ = number of active coils when loaded
$D_i'$ = coil inside diameter when loaded, m

---

**EXAMPLE 17.7**

**Given:** A torsion spring similar to that shown in Figure 17.11 is made of 0.055-in-diameter music wire and has six coils in the body of the spring and straight ends. The distance $a$ in Figure 17.11 is 2 in. The outside diameter of the coil is 0.654 in.

**Find:**

(a) If the maximum stress is set equal to the yield strength of the wire, what is the corresponding moment?
(b) With load applied, what is the angular deflection in revolutions?
(c) When the spring is loaded, what is the resulting inside diameter?
(d) What size bar should be placed inside the coils?

**Solution:**

(a) The mean coil diameter is

$$D = D_o - d = 0.654 - 0.055 = 0.599 \text{ in}$$

The coil inside diameter without a load is

$$D_i = D - d = 0.599 - 0.055 = 0.544 \text{ in}$$

The spring index is

$$C = \frac{D}{d} = \frac{0.599}{0.055} = 10.89$$

From Equation (17.41)

$$K_i = \frac{4C^2 - C - 1}{4C(C-1)} = \frac{4(10.89)^2 - 10.89 - 1}{4(10.89)(10.89-1)} = 1.074$$

From Equation (17.2) and Table 17.2 for music wire

$$S_{ut} = \frac{A_p}{d^m} = \frac{196(10^3)}{(0.055)^{0.146}} \text{ psi} = 299.3 \times 10^3 \text{ psi}$$

$$S_y = 0.6S_{ut} = 179.6 \times 10^3 \text{ psi}$$

After we equate the bending stress to the yield strength, Equation (17.40) gives

$$M = \frac{\pi d^3 S_y}{32 K_i} = \frac{\pi (0.055)^3 (179.6)(10^3)}{32(1.074)} = 2.731 \text{ lb} \cdot \text{in}$$

(b)   The number of active coils is

$$N_a = N_b + N_e = 6 + \frac{4}{3\pi(0.599)} = 6.709$$

From Table 17.1 the modulus of elasticity for music wire is $30 \times 10^6$ psi. From Equation (17.44) the angular spring rate is

$$k_\theta = \frac{E d^4}{10.18 D N_a} = \frac{(30)(10^6)(0.055)^4}{10.18(0.599)(6.709)} = 6.710 \text{ lb} \cdot \text{in/r}$$

The angular deflection in revolutions is

$$\theta_{\text{rev}} = \frac{M}{k_\theta} = \frac{2.731}{6.710} = 0.4070 \text{ r}$$

(c)   The number of coils when loaded is

$$N_a' = N_a + \theta_{\text{rev}} = 6.709 + 0.4070 = 7.116$$

From Equation (17.46) the inside diameter after loading is

$$D_i' = \frac{N_a}{N_a'} D_i = \frac{6.709}{7.116}(0.544) = 0.5129 \text{ in}$$

(d)   Thus, if the bar inside the coils has $\frac{1}{2}$-in diameter, the spring should not fail.

---

**Given:**   A hand-squeeze grip training device with two 100-mm-long handles is connected to a helical torsion spring with 4.375 coils. The handles and coils are made of music wire. The spring rate at the end of the spring should be 2000 N/m.

**EXAMPLE 17.8**

**Find:**   The spring dimensions for a safety factor of 2 against breakage when the handles are moved together so that the helical torsion spring has 4.5 coils.

**Solution:**   From Equation (17.40) the maximum bending stress is

$$\sigma = \frac{K_i M c}{I} = \frac{K_i M r}{I} \tag{a}$$

Using Equation (17.42) gives the moment needed to deform the spring to 4.5 coils as

$$M = \frac{E I \theta_{\text{rad}}}{l_w} = \frac{E I (4.5 - 4.375) 2\pi}{l_w} = \frac{0.25 \pi E I}{l_w} \tag{b}$$

Substituting Equation (b) into Equation (a) gives

$$\sigma = \frac{\pi}{8} \frac{K_i d E}{l_w}$$

But $l_w = \pi D N_a$,

$$\sigma = \frac{\pi K_i d E}{8\pi D N_a} = \frac{K_i E}{8 C N_a}$$

For music wire $E = (2.07)(10^{11})$ Pa and $N_a = 4.375$, the stress in gigapascals is

$$\therefore \qquad \sigma = \frac{K_i}{C} \frac{(2.07)(10^{11})}{8(4.375)} \text{ Pa} = 5.914 \frac{K_i}{C} \qquad \qquad \text{(c)}$$

To get the spring rate of 2000 N/m at the ends of the handles (see Figure 17.11), the angular spring rate is

$$k_\theta = (2000 \text{ N/m})(0.1 \text{ m})(0.1 \text{ m})/\text{rad} = 20 \text{ N} \cdot \text{m/rad}$$

From Equation (17.44)

$$d^3 = \frac{10.18 C N_a k_\theta}{E} = C \frac{(10.18)(4.375)(20)}{(2.07)(10^{11})} = 4.303 \times 10^{-9} C \qquad \qquad \text{(d)}$$

As a first try, guess a value of $C = 6$. From Equation (17.41)

$$K_i = \frac{4C^2 - C - 1}{4C(C - 1)} = \frac{4(6)^2 - 6 - 1}{4(6)(5)} = 1.142$$

From Equation $(d)$

$$d^3 = 25.82 \times 10^{-9}$$

$$d = 2.956 \times 10^{-3} \text{ m} = 2.956 \text{ mm}$$

$$\therefore \qquad D = 6(2.956)(10^{-3}) = 17.73 \times 10^{-3} \text{ m} = 17.73 \text{ mm}$$

From Equation $(c)$ the maximum design stress is

$$\sigma = (5.914)\frac{1.142}{6} = 1.126 \text{ GPa} \qquad \qquad \text{(e)}$$

From Equation (17.2) the allowable stress is

$$\sigma_{\text{all}} = S_{ut} = \frac{A_p}{d^m} = \frac{(2170)(10^6)}{(2.956)^{0.146}} \text{ Pa} = 1.852 \text{ GPa}$$

The safety factor is

$$n_s = \frac{\sigma_{\text{all}}}{\sigma} = \frac{1.852}{1.126} = 1.645$$

Since the safety factor is too low, we try $C = 7.5$. From Equation (17.41)

$$K_i = \frac{4C^2 - C - 1}{4C(C - 1)} = \frac{4(7.5)^2 - 7.5 - 1}{4(7.5)(6.5)} = 1.110$$

From Equation ($d$)

$$d^3 = 4.303 \times 10^{-9}(7.5) = 32.27 \times 10^{-9}$$

$$d = 3.184 \times 10^{-3} \text{ m} = 3.184 \text{ mm}$$

$$\therefore \quad D = 7.5(3.184)(10^{-3}) = 23.88 \times 10^{-3} \text{ m} = 23.88 \text{ mm}$$

From Equation ($c$) the maximum design stress is

$$\sigma = (5.914)\frac{1.110}{7.5} = 0.8753 \text{ GPa}$$

From Equation (17.2) the allowable stress is

$$\sigma_{\text{all}} = S_{ut} = \frac{A_p}{d^m} = \frac{(2170)(10^6)}{(3.184)^{0.146}} \text{ Pa} = 1.832 \text{ GPa}$$

$$\therefore \quad n_s = \frac{\sigma_{\text{all}}}{\sigma} = \frac{1.832}{0.8753} = 2.09$$

The design is adequate.

## 17.6   LEAF SPRINGS

Multiple-**leaf springs** are in wide use, especially in the automobile and railway industries. An exact analysis of this type of spring is complicated. A multiple-leaf spring can be considered as a simple cantilever type, as shown in Figure 17.12(b). It can also be considered as a triangular plate, as shown in Figure 17.12(a). The triangular plate is cut into $n$ strips of width $b$ and stacked in the graduated manner shown in Figure 17.12(b).

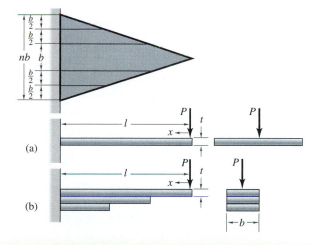

**Figure 17.12**   Leaf spring. (a) Triangular plate, cantilever spring; (b) equivalent multiple-leaf spring.

Before we analyze a multiple-leaf spring, first we consider a single-leaf, cantilevered spring with constant rectangular cross section. For bending of a straight beam from Equation (4.48)

$$\sigma = \frac{Mc}{I} \tag{17.47}$$

For a rectangular cross section of base $b$ and height $t$, Equation (17.47) becomes

$$\sigma = \frac{6M}{bt^2} = \frac{6Px}{bt^2} \tag{17.48}$$

The moment is just $Px$. The maximum moment occurs at $x = l$ and at the outer fiber of the cross section, or

$$\sigma_{max} = \frac{6Pl}{bt^2} \tag{17.49}$$

From Equation (17.48) the stress is a function of $x$ along the beam and is constant. To accomplish this, either $t$ is held constant and $b$ is allowed to vary, or conversely, so that the stress is constant for any $x$,

$$\frac{b(x)}{x} = \frac{6P}{t^2\sigma} = \text{constant} \tag{17.50}$$

Equation (17.50) is linear, giving the triangular shape shown in Figure 17.12(a) and a constant stress for any $x$.

The triangular-plate spring and the equivalent multiple-leaf spring have identical stresses and deflection characteristics with two exceptions:

1.   Interleaf friction provides damping in the multiple-leaf spring.
2.   The multiple-leaf spring can carry a full load in only one direction.

The deflection and spring rate for the ideal leaf spring are

$$\delta = \frac{6Pl^3}{Enbt^3} \tag{17.51}$$

$$k = \frac{P}{\delta} = \frac{Enbt^3}{6l^3} \tag{17.52}$$

**EXAMPLE 17.9**

**Given:**   A 35-in-long cantilever spring is composed of eight graduated leaves. The leaves are 7/4 in wide. A load of 500 lb at the end of the spring causes a deflection of 3 in. The spring is made of steel with modulus of elasticity of $3 \times 10^7$ psi.

**Find:**   Determine the thickness of the leaves and the maximum bending stress.

**Solution:**   From Equation (17.51)

$$\delta = \frac{6Pl^3}{Enbt^3} \qquad \text{or} \qquad t^3 = \frac{6Pl^3}{\delta Enb}$$

$$t^3 = \frac{6(500)(35)^3}{3(3)(10^7)(8)(1.75)} = 0.1021 \text{ in}^3$$

$$t = 0.4674 \text{ in}$$

From Equation (17.49) the maximum bending stress is

$$\sigma_{max} = \frac{6Pl}{nbt^2} = \frac{6(500)(35)}{(8)(1.75)(0.4674)^2} = 34{,}337 \text{ psi}$$

---

**Given:** A leaf spring for a locomotive wheel set is made of spring steel with a thickness of 20 mm and an allowable bending stress of 1050 MPa. The spring has a modulus of elasticity of 207 GPa, is 1.6 m long from tip to tip, and carries a weight of 12,500 N at the middle of each leaf spring.

**EXAMPLE 17.10**

**Find:**

(a) The width of the spring for a safety factor of 3
(b) How high the locomotive has to be lifted during overhauls to unload the springs

**Solution:**

(a) From Equation (1.1) the design stress is

$$\sigma_{max} = \frac{\sigma_{all}}{3} = \frac{(1.05)(10^9)}{3} = 0.350 \text{ GPa} \tag{a}$$

From Equation (17.49) the maximum bending stress is

$$\sigma_{max} = \frac{6Pl}{bt^2}$$

Because $P$ is force applied at the tip, the load is 6250 N and the length is 0.8 m in the locomotive spring.

$$\therefore \quad \sigma_{max} = \frac{6Pl}{bt^2} = \frac{6(6.25)(10^3)(9.807)(0.8)}{b(0.020)^2} = \frac{(0.7355)(10^9)}{b}$$

Making use of Equation ($a$) gives

$$b = 2.101 \text{ m}$$

Splitting into 10 leaves gives the width of the leaf spring as

$$\frac{b}{10} = 0.2101 \text{ m} = 210.1 \text{ mm}$$

(b) Using Equation (17.52) gives the amount the locomotive has to be lifted to unload the spring (deflection) as

$$\delta = \frac{P}{k} = \frac{6Pl^3}{Ebt^3} = \frac{6(6250)(9.807)(0.8)^3}{(2.07)(10^{11})(2.101)(0.020)^3} = 0.05412 \text{ m} = 54.1 \text{ mm}$$

## 17.7   BELLEVILLE SPRINGS

**Belleville springs** are named after their inventor, J. F. Belleville, who patented their design in 1867. Shaped like a coned disk, these springs are especially useful where large forces are desired for small spring deflections. In fact, because many lock washers used with bolts follow this principle to obtain a bolt preload, these springs are often referred to as Belleville washers. Typical applications include clutch plate supports, gun recoil mechanisms, and a wide variety of bolted connections.

Figure 17.13 shows the cross section of a Belleville spring. Two of the critical parameters affecting a Belleville spring are the diameter ratio $R_d = D_o/D_i$ and the height-to-thickness ratio $h/t$. From Figure 17.14 the behavior of a Belleville spring is highly nonlinear and varies considerably with a change in $h/t$. For low $h/t$ values the spring acts almost linearly, whereas large $h/t$ values lead to highly nonlinear behavior. A phenomenon occurring at large $h/t$ is *snap-through buckling*. In snap-through the spring deflection is unstable once the maximum force is applied; the spring quickly deforms, or snaps, to the next stable position.

The force–deflection curves for Belleville springs are given by

$$ P = \frac{4E\delta}{K_1 D_o^2 (1 - v^2)} \left[ (h - \delta) \left( h - \frac{\delta}{2} \right) t + t^3 \right] \qquad \textbf{(17.53)} $$

where $E$ is the elastic modulus, $\delta$ is the deflection from the unloaded state, $D_o$ is the coil outside diameter, $v$ is Poisson's ratio for the material, $h$ is the spring height, and $t$ is the

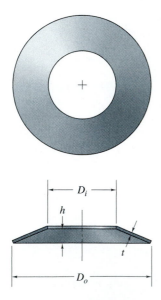

**Figure 17.13**   Typical Belleville spring.

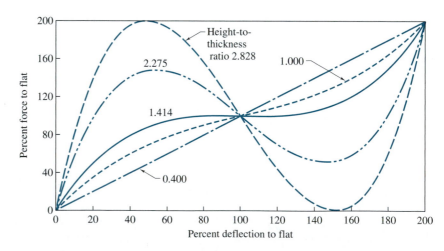

**Figure 17.14**   Force-deflection response of Belleville spring. [Norton (1996)]

(a)                 (b)

**Figure 17.15** Stacking of Belleville springs. (a) in parallel; (b) in series.

spring thickness. The factor $K_1$ is given by

$$K_1 = \frac{6}{\pi \ln R_d} \frac{(R_d - 1)^2}{R_d^2}$$ (17.54)

where $R_d$ is the diameter ratio, given by

$$R_d = \frac{D_o}{D_i}$$ (17.55)

The force required to totally flatten a Belleville spring is given by

$$P_{\text{flat}} = \frac{4Eht^3}{K_1 D_o^2 (1 - v^2)}$$ (17.56)

The forces associated with a Belleville spring can be multiplied by stacking them in parallel, as shown in Figure 17.15(a). The deflection associated with a given force can be increased by stacking the springs in series, as shown in Figure 17.15(b). Because such a configuration may be susceptible to buckling, a central support is necessary.

---

**Case Study 17.1** | DESIGN OF COMPRESSION SPRING FOR DICKERMAN FEED UNIT

**Given:** Figure 17.16 shows a Dickerman feed unit, used to automatically feed material into mechanical presses for progressive die work. During a stroke of a mechanical press, a cam attached to the press ram or punch holder lowers, pushing the follower shown to the right and causing one or more springs to compress. When the ram rises, the compressed spring forces the gripping unit to the left, and since the gripping unit mechanically interferes with notched strip or sheet stock, it forces new material into the press.

Dickerman feeds are among the oldest and least expensive feeding devices in wide use. They can be mounted to feed material in any direction, on both mechanical and

hydraulic presses, and the feed length can be adjusted by changing the low-cost, flat cams. Grippers can be blades (as shown) or cylinder assemblies.

This case study involves designing a helical compression spring for a Dickerman feed unit. It is desired to use a spring that will provide at least 20 lb and have a 6-in stroke. The pin diameter is 0.5 in. Also, the Dickerman unit should be compactly designed; the chamber for the spring will be not more than 8 in long.

**Find:** All necessary dimensions of the helical compression spring.

*(continued)*

**Case Study**   (Continued)

**Figure 17.16**   Dickerman feed unit. [*Tool and Manufacturing Engineering Handbook* (1984)]

**Solution:**   Not much information is given about the spring. Noting that the Dickerman drive can be used for different applications, it is necessary that 20 lb be provided whether the deflection is 6 in, 4 in, or a very small distance. Thus, the spring will be chosen so that at a length of 8 in, it is loaded with 20 lb of compressive force. Another requirement on the spring is that it must have a solid length not greater than 4 in, or else it will bottom out with every stroke of the press. Fortunately, buckling of the spring does not need to be considered because of the guide rod. However, the spring should not be very much longer than the guide rod, since this complicates the Dickerman unit motion.

Economics is not a critical issue here. As long as the spring is selected with standard dimensions and made from spring steel, the cost difference between candidate springs will be low, especially when spread over the total number of parts that will be produced with the Dickerman feed. It is much more essential that a highly reliable spring be chosen because an hour of downtime (required to replace a spring and set up the Dickerman drive) costs far more than a spring. Thus, the ends will be squared and ground for, as discussed in Section 17.3.5, better load transfer into the spring. The material used will be music wire, an outstanding spring material.

No maximum force requirement is given; indeed it appears that the maximum force is inconsequential. However, if the compressed spring contains too large a force, it could buckle the press stock, or else dynamic effects could cause too much material to feed with each stroke.

An inconsequential effect is the reduction of press capacity; loads on the order of tens of pounds are not high enough to affect press capacity.

Since mechanical presses used for progressive die work are operated unattended for long periods and typically will operate at 100 strokes per minute, the Dickerman feed unit must be designed for fatigue. The important equations are Equations (17.22) to (17.28). The solution method involves a variation of the unknown independent parameters (that is, $d$, $D$, and $N_a$). All other quantities can be calculated from these parameters. However, the problem is actually much simpler, for the following reasons:

1.  The more active coils in the spring, the lower the spring stiffness and hence the lower the maximum force. In addition to reducing the stress levels encountered by the spring, a large number of coils makes a more smoothly operating drive for the reasons discussed above. In terms of this case study the number of active coils is given by

$$N_a = \frac{4}{d} - 2$$

because the ends are squared and ground.

2.  The spring diameter is not specified. However, the guide rod in the center of the spring has 0.5-in diameter, so that the minimum spring diameter is $0.5 + d$. Normally, the spring diameter would be treated as another parameter to be varied, but in this problem

## Case Study    (Continued)

it has been set at 0.75 in. The reasons for this are twofold: First, this dimension ensures adequate clearance and ease of assembly onto the guide rod. Second, this standard size will be more readily available, so that suppliers are more likely to provide high-quality springs quickly.

3. A Goodman failure criterion is used, so that the safety factor is given by

$$\frac{\tau_a}{S_{se}} + \frac{\tau_m}{S_{su}} = \frac{1}{n_s} \quad \text{or} \quad n_s = \frac{S_{se}S_{su}}{\tau_a S_{su} + \tau_m S_{se}}$$

The solution approach is as follows:

1. Given $D$, choose a value for the wire diameter $d$. For this case the wire diameter varies over the useful range 0.02 to 0.2 in.

2. Calculate $C = D/d$.

3. Calculate the Wahl and tranverse shear stress factors, given by, respectively,

$$K_w = \frac{4C - 1}{4C - 4} + \frac{0.615}{C}$$

$$K_d = \frac{C + 0.5}{C}$$

4. Obtain the number of active coils $N_a$ from

$$N_a = \frac{4}{d} - 2$$

5. The wire properties are

$$S_{ut} = \frac{A_p}{d^m} \qquad S'_{se} = 45.0 \text{ ksi} \qquad S_{su} = 0.60S_u$$

6. By using the results from steps 1 to 5, calculate the following:

$$k_t = \frac{P}{\delta_t} = \frac{Gd}{8C^3 N_a}$$

$$P_a = \frac{P_{max} - P_{min}}{2}$$

$$P_m = \frac{P_{max} + P_{min}}{2}$$

$$\tau_a = \frac{8DK_w P_a}{\pi d^3}$$

$$\tau_m = \frac{8DK_w P_m}{\pi d^3}$$

$$n_s = \frac{S_{se}S_{su}}{\tau_a S_{su} + \tau_m S_{se}}$$

(16.24)

7. Output the maximum force, the safety factor, and all other desired quantities. Figure 17.17 shows maximum force and the safety factor as a function of wire diameter.

**Figure 17.17**    Performance of the spring in case study.

From the charts the safety factor is highest for a spring using a wire diameter of approximately 0.08 in. Thus, a wire diameter of 0.081 in (2 mm) is chosen. The resulting spring will have 47.4 active coils and result in a maximum force (when compressed to a length of 4 in) of 24.31 lb, with a free length of approximately 11.9 in. The safety factor for this case is 1.23.

The fact that the safety factor is small for extremely thin wires should not be surprising, since the stresses are inversely proportional to the cube of wire diameter. However, it is not immediately obvious that the safety factor would eventually decrease for increasing wire diameters.

*(continued)*

**Case Study**    (Concluded)

The ultimate strength of the wire decreases rapidly with increasing wire diameter, so that there is in fact an optimum wire diameter.

This analysis should be repeated for additional spring diameters to obtain a true optimum solution but is not done here because of space limitations. In this circumstance it was quite fortunate that the forces were reasonable at the maximum safety factor location. If this were not the case, we would have to consider another spring diameter, another spring material, or an entirely different class of spring.

## 17.8   SUMMARY

When a machine element needs flexibility or deflection, some form of spring is usually selected. Springs are used to exert forces or torques in a mechanism or to absorb the energy of suddenly applied forces. Springs frequently operate with high stresses and continuously varying forces.

This chapter provided information about spring design without considering the endless types of spring available to engineers. Different spring materials were presented. Spring strength is a material parameter that is obviously important in spring design. The loss coefficient, which measures the fractional energy dissipated in a stress–strain cycle, is an equally important parameter.

This chapter emphasized helical compression springs. Both torsional shear stress and transverse shear stress were considered. Transverse shear is small relative to torsional shear. The maximum stress occurs at the midheight of the wire at the coil inside diameter.

The spring index is dimensionless and is the mean coil diameter divided by the wire diameter. The spring rate is the force divided by the deflection, thus having the unit of newtons per meter. The difference between these terms is important. Different conditions cause different amounts of eccentric loading that must be compensated for in the design of a compression spring. Relatively long compression springs must be checked for buckling. Also, a surge, or longitudinal vibration, should be avoided in spring design. Avoiding the natural frequency is thus recommended. Spring loading is most often continuously fluctuating, so this chapter considered the design allowance that must be made for fatigue and stress concentrations.

In helical extension springs it was found that the hooks need to be shaped so that the stress concentration effects are decreased as much as possible. Two critical locations within a hook were analyzed. In one, the designer must consider normal stress caused by the bending moment and the transverse shear stresses; and in the other, shear stress must be considered. Both should be checked.

A spring can be designed for axial loading, either compressive or tensile, or it can be designed for torsional loading, as discovered when helical torsion springs were considered. The ends of torsion springs are shaped to transmit a torque rather than a force as for compression and tensile springs. The torque is applied about the axis of the helix and acts as a bending moment on each section of the wire.

The leaf spring, used extensively in the automobile and railway industries, was also considered. An approximate analysis considered a triangular-plate, cantilever spring and an equivalent multiple-leaf spring. Finally, Belleville springs were considered, and their advantages explained.

# KEY WORDS

**Belleville spring**  coned disk spring.

**free length**  length of spring when no forces are applied to it.

**helical compression spring**  most common spring, wherein round wire is wrapped into cylindrical form with constant pitch between adjacent coils and is loaded in compression.

**helical extension spring**  spring wherein round wire is wrapped into cylindrical form with constant pitch between adjacent coils and is loaded in tension.

**helical torsion spring**  helical coil spring loaded in torsion.

**leaf spring**  spring based on cantilever action.

**pitch**  distance, measured parallel to coil axis, from center of one coil to center of adjacent coil.

**solid length**  length of spring when all adjacent coils are in metal-to-metal contact.

**spring**  flexible machine element used to exert force or torque and, at same time, to store energy.

**spring index**  ratio of coil to wire diameter, measure of coil curvature.

**spring rate**  ratio of applied force to spring deflection.

**surge**  stress pulse that propagates along spring.

# RECOMMENDED READINGS

Carlson, H. (1978) *Spring Designers Handbook,* Marcel Dekker, New York.

Juvinall, R. C., and Marshek, K. M. (2003) *Fundamentals of Machine Component Design,* 3rd ed., Wiley, New York.

Krutz, G. W., Schuelle, J. K., and Claar, P. W. (1994) *Machine Design for Mobile and Industrial Applications,* Society of Automotive Engineers, Warrendale, PA.

Mott, R. L. (1998) *Machine Elements in Mechanical Design,* 3rd ed., Prentice-Hall, Upper Saddle River, NJ.

Norton, R. L. (1996) *Machine Design,* Prentice-Hall, Englewood Cliffs, NJ.

Rothbart, H. A., ed. (1996) *Mechanical Design Handbook,* McGraw-Hill, New York.

Shigley, J. E., Mischke, C. R., and Budynas, R. G. (2003) *Mechanical Engineering Design,* 7th ed., McGraw-Hill, New York.

SAE (1982) *Manual on Design and Application of Leaf Springs,* Society of Automotive Engineers, Warrendale, PA.

SAE (1996) *Spring Design Manual,* 2nd ed., Society of Automotive Engineers, Warrendale, PA.

Shigley, J. E., and Mischke, C. R. (1989) *Mechanical Engineering Design,* 5th ed., McGraw-Hill, New York.

## REFERENCES

Almen, J. O., and Laszlo, A. (1936) "The Uniform Section Dise Springs," *Trans. ASME,* vol. 58, no. 4, pp. 305–314.

Ashby, M. F. (1992) *Materials Selection in Mechanical Design,* Pergamon Press, Oxford.

Design Handbook (1987) *Engineering Guide to Spring Design,* Associated Spring Corp., Barnes Group Inc., Bristol, CT.

Norton, R. L. (1996) *Machine Design,* Prentice-Hall, Englewood Cliffs, NJ.

Relvas, A. A. (1996) Springs, chapter 28 in *Mechanical Design Handbook,* H. A. Rothbart, ed., McGraw-Hill, New York.

SME (1984) *Tool and Manufacturing Engineers Handbook,* vol. 2, Society of Manufacturing Engineers, Dearborn, MI, p. 58.

Zimmerli, F. P. ( 1957) "Human Failures in Spring Applications," *The Mainspring,* no. 17, Associated Spring Corp., Barnes Group, Inc., Bristol, CT.

## PROBLEMS

### Section 17.3

**17.1**  Sketch *a* shows a guide wire, used to deliver catheters in the human body during angioplasty operations. The compression spring at the end is used as a flexible support for the nose as it opens the artery and prevents its puncture by the guide wire. The spring has an outside diameter of 0.0260 in, a wire diameter of 0.003 in, a free length of 0.110 in and 30 total coils, and has squared and ground ends. The wire is a stainless steel ($E = 29$ Mpsi, $G = 11$ Mpsi), but has a strength so that $A_p = 100$ ksi, $m = 0.0474$. Determine

*a)*   The solid length and spring rate. *Ans.* $l_s = 0.090$ in, $k = 0.324$ lb/in

*b)*   The force needed to compress the spring to its solid length. *Ans.* $P = 0.00648$ lb.

*c)*   The safety factor against static overload if the spring is compressed to its solid length. Use the Wahl factor. *Ans.* $n_s = 3.14$.

Sketch *a,* for Problem 17.1

**17.2**  A helical compression spring with square and ground ends has an outside diameter of 0.560 in, 19 total coils, a wire diameter of 0.085 in (music wire), and a free length of 4.22 in. Calculate the spring index, the pitch, the solid length, and the shear stress in the wire when the spring is compressed to its solid length.

**17.3**  An overflow valve, shown in sketch *b*, has a piston diameter of 15 mm and a slit length of 5 mm. The spring has mean coil diameter $D = 10$ mm and wire diameter $d = 2$ mm. The valve should open at 1 bar pressure and be totally open at 3 bar pressure when the spring is fully compressed. Calculate the number of active coils, the free length, and the pitch of the spring. The shear modulus for the spring material $G = 80$ GPa. The spring ends are squared and ground. Determine the maximum shear stress for this geometry. *Ans.* $N_a = 22.6, l_f = 56.7$ mm.

Sketch *b*, for Problem 17.3

**17.4** A vehicle has individual wheel suspension in the form of helical springs. The free length of the spring $l_f = 360$ mm, and the solid length $l_s = 160$ mm at a compressive force of 5000 N. The shear modulus $G = 80$ GPa. Use $D/d = 9$ and calculate the shear stress for pure torsion of the spring wire. The spring ends are squared and ground. Find $N_a$, $p$, $d$, $D$, and $\tau_{max}$. *Ans.* $N_a = 8.43$, $p = 39.1$ mm, $d = 15.3$ mm.

**★ 17.5** Two equally long cylindrical helical compression springs are placed one inside the other (see sketch *c*) and loaded in compression. How should the springs be dimensioned to get the same shear stresses in both springs?

Sketch *c*, for Problem 17.5

| * Indicates problem of greater difficulty.

* **17.6** A mechanism is used to press as hard as possible against a moving horizontal surface, shown in sketch $d$. The mechanism consists of a stiff central beam and two flexible bending springs made of circular rods with length $l$, diameter $d$, modulus of elasticity $E$, and allowable stress $\sigma_{all}$. Wheels are mounted on these rods and can roll over a bump with height $f$. Calculate the diameter of the springs so that the prestress of the wheels against the moving surface is as high as possible without plastically deforming the springs when the bump is rolled over. The deflection of a spring is shown in sketch $e$.

Sketch $d$, for Problem 17.6

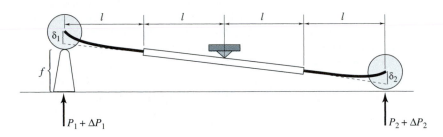

Sketch $e$, for Problem 17.6

**17.7** A helical compression spring is used as a catapult. Calculate the maximum speed of a body weighing 10 kg being thrown by the catapult, given that $\tau_{max} = 500$ MPa, $D = 50$ mm, $d = 8$ mm, $N_a = 20$, and $G = 80$ GPa. *Ans.* Using a Wahl factor, $v = 4.0$ m/s.

* **17.8** A spring is preloaded with force $P_i$ and is then exposed to a force increase whereby the shear stress increases to a certain value $\tau$. Choose the mean coil diameter $D$ to maximize the energy absorption caused by the force increase. *Ans.* $D = \dfrac{\pi}{8\sqrt{3}}\dfrac{\tau d^3}{P_i}$.

**17.9** A compression spring made of music wire is used for static loading. Wire diameter $d = 1.4$ mm, coil outside diameter $D_o = 12.1$ mm, and there are 8 active coils. Also, assume that the spring ends are squared and ground. Find the following:

*a)* Spring rate and solid length. *Ans.* $k = 3850$ N/m, $l_s = 14$ mm.

*b)* Greatest load that can be applied without causing a permanent set in excess of 2%. *Ans.* $P_{max} = 78.1$ N.

*c)* The spring free length if the load determined in part $b$ causes the spring to compress to a solid state. *Ans.* 34.3 mm.

*d)* Whether buckling is a problem. If it is, recommend what you would change in the redesign.

**17.10** A helical compression spring will be used in a pressure relief valve. When the valve is closed, the spring length is 2.0 in, and the spring force is to be 1.50 lb. The spring uses a wire diameter of 0.0577 in and a spring diameter of 0.375 in, and it has a solid length of 1.25 in. Use hard-drawn ASTM A227 wire, squared and ground ends, and ignore buckling since the spring is in a cage. Find the force needed to compress the spring to its solid length, the number of active coils, the spring's free length, the spring rate, and the safety factor against torsional yielding.

**17.11** Two compression springs will be used in a stamping press to open a die after it has been pushed closed by the action of the press ram. The springs will stay in a retainer, which means buckling is not an issue and also means that the springs will be allowed to be preloaded. Also, this retainer geometry requires that $D = 1$ in. It is desired to have a total deflection of 7 in, and each spring should exert 250 lb on the die when the spring is compressed and 150 lb when extended. Squared and ground ends will be used to ensure good performance. If 25 active coils are used in each spring, what is the required wire diameter and spring solid length? Use $G = 11.6$ Mpsi.

**17.12** Design a helical compression spring made of music wire with squared and parallel ends. Assume a spring index $C = 12$, a spring rate $k = 300$ N/m, and a force to solid length of 60 N. Find the wire diameter and the mean coil diameter while assuming a safety factor $n_s = 1.5$. Also, check whether buckling is a problem. Assume steady loading. *Ans. $d = 1.90$ mm, $D = 22.8$ mm.*

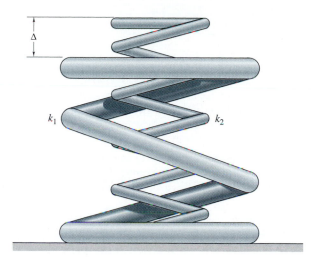

Sketch $f$, for Problem 17.13

**★ 17.13** Two helical springs with spring rates of $k_1$ and $k_2$ are mounted one inside the other as shown in sketch $f$. The difference in unstressed length is $\Delta$, and the second spring is longer. A loose clamping device is mounted at the top of the springs so that they are deformed and become equal in length. Calculate the forces in the two springs. An external force $P$ is the load applied thereafter to the spring. In a diagram show how the spring forces vary with $P$. *Ans.*

$$P_1 = P_0 + \frac{k_1 P}{k_1 + k_2}, \quad P_2 = P_0 - \frac{k_2 P}{k_1 + k_2}.$$

**★ 17.14** Using Equations (17.5) and (17.18), describe how the spring diameter can be chosen to give as low a maximum shear stress as possible.

**17.15** Consider a helical compression spring with plain ends made of hard-drawn wire with a spring index of 12 and a stiffness of 0.3 kN/m. The applied load on the spring is 60 N. For a safety factor of 1.5 guarding against yielding, find the spring diameter, the wire diameter, the number of coils, and the free and solid lengths of the spring. Does this spring have buckling and/or dynamic instability problems? *Ans. $d = 2.18$ mm, $D = 26.16$ mm, $N_a = 41.5$.*

**17.16** An 18-mm-mean-diameter helical compression spring has 22 coils, has 2-mm wire diameter, and is made of chromium vanadium. Determine the following:

   *a)*   The maximum load-carrying capacity for a safety factor of 1.5 guarding against yielding. *Ans. $P = 78.5$ N.*
   *b)*   The maximum deflection of the spring. *Ans. $\delta_t = 0.0635$ m.*
   *c)*   The free length for squared and ground ends. *Ans. $l_f = 0.111$ m.*

**17.17** A 60-mm-mean-diameter helical compression spring with plain ends is made of hard-drawn steel and has a wire diameter of 2.5 mm. The shear strength of the wire material is 750 MPa, and the spring has 20 coils. The free length is 500 mm. Find the following:

   *a)*   The required load needed to compress the spring to its solid length. *Ans. $P = 40.11$ N.*
   *b)*   By applying the load found in part *a* and then unloading it, whether the spring will return to its free length.

## Section 17.4

**17.18** A desk lamp has four helical extension springs to make it possible to position it over different parts of the desk. Each spring is preloaded to 15 N, so that no deflection takes place for forces below 15 N. Above 15-N force the spring rate is 100 N/m. The mean coil diameter is 10 mm, and the wire diameter is 1 mm. Calculate the torsion needed on the wire during manufacturing to get the correct preload, and calculate how many coils are needed to get the correct spring rate. *Ans. $T = 0.075$ N · m, $N_a = 99.125$.*

**★ 17.19** A spring balance for weighing fish needs to be dimensioned. The weighing mechanism is a sharp hook hanging in a helical extension spring. To make it easy to read the weight of the fish in the range from 0 to 10 kg, the length of the scale should be 100 mm. The spring material is music wire.

**★ 17.20** A muscle-training device consists of two handles with three parallel 500-mm-long springs in between. The springs are tightly wound but without prestress. When the springs are fully extended to 1600 mm, the force in each spring is 100 N. The springs are made of music wire. Dimension the springs.

**17.21** A 45-mm-diameter extension spring [similar to that shown in Figure 17.8(a) and (b)] has 102 coils and is made of 4-mm-diameter music wire. The stress due to preload is equivalent to 10% of the yield shear strength. The hook radius is 5 mm, and the bend radius is 2.5 mm. Determine the following:

   *a)*   The solid length of the spring and the spring stiffness. *Ans. $k = 273$ N/m, $l_s = 0.412$ m.*
   *b)*   The preload and the load that causes failure. *Ans. $P_i = 39.6$ N.*

**17.22** The extension spring shown in sketch *g* is used in a cyclic motion in turning on and off a power switch. The spring has a 15-mm outer diameter and is made of a 1.5-mm-diameter wire of hard-drawn steel. The spring has no preload. In a full stroke of the spring the force varies between 25 and 33 N. Determine the following:

a) The number of coils, the free length, the maximum and minimum lengths during cyclic loading, and the spring rate. *Ans.* $N_a = 38.0$, $l_f = 73.5$ mm.

b) For infinite life with 99% reliability the safety factors guarding against static and fatigue failure. *Ans.* $n_{s,\text{static}} = 1.66$.

Sketch g, for Problem 17.22

**17.23** Calculate the safety factor guarding against fatigue failure if the spring given in Problem 17.22 is designed for 50,000 strokes of motion with 50% reliability.

**17.24** The extension spring shown in sketch h is used in a car braking system. The spring is made of 2-mm-diameter wire of hard-drawn steel, has a mean coil diameter of 10 mm, and has a free length of 106 mm. The spring has 18 active coils and a deflection under braking force of 11 mm. Determine the following:

a) The safety factor of the spring

b) The torsional and normal stresses at the hook

Sketch h, for Problem 17.24

## Section 17.5

**17.25** A torsion rod is used as a vehicle spring. The torque on the rod is created by a force $P = 1500$ N acting on radius $R = 200$ mm. Maximum torsion angle is $45°$. Calculate the diameter and length of the rod if the maximum shear stress is 500 MPa.

★ **17.26** A torsional spring, shown in sketch i, consists of a steel cylinder with a rubber ring glued to it. The dimensions of the ring are $D = 45$ mm, $d = 15$ mm, and $h = 20$ mm. Calculate the torque as a function of the angular deflection. The shear modulus of elasticity for rubber is 150 N/cm$^2$. *Ans. T = (23.9 N · m/rad)$\phi$.*

Sketch *i*, for Problem 17.26

★ **17.27** To keep a sauna door shut, helical torsion springs are mounted at each of the two door hinges. The friction torque in each hinge is 0.2 N · m. It should be possible to open the door 180° without plastically deforming the spring. Dimension the springs so that a 10-kg door 700 mm wide will shut itself in 2 s. The length of the wire ends can be neglected. The air drag on the door is also neglected. The springs are manufactured from music wire and have a diameter of 2.5 mm.

★ **17.28** A helical torsion spring used in a door handle is made of hard-drawn steel wire with a diameter of 3 mm. The spring has 10 active coils and a coil diameter of 21 mm. After 5 years of use the spring cracks due to fatigue. To get longer fatigue life, the stresses should be decreased by 5%, but the only available space is that used by the present spring, which has an outside diameter of 24 mm, an inside diameter of 18 mm, and a length of 30 mm. The spring rate for the new spring should be the same as for the old one. Dimension a new spring with a rectangular cross section to satisfy the new stress requirement.

**17.29** A mouse trap uses two antisymmetric torsion springs. The wire has a diameter of 2 mm, and the inside diameter of the spring in the unset position is 12 mm and the spring has 10 turns in the unset position. About 50 N is needed to set the trap (not to start setting, but rather the largest load needed to deflect the spring to the set position). (a) Find the number of coils and diameter of the spring prior to assembly into the mouse trap, that is, in unstressed state. (b) Find the maximum stress in the spring wire when the trap is set.

**17.30** The oven door on a kitchen stove is kept shut by a helical torsion spring. The spring torque is 1 N · m when the door is shut. When the oven door is fully open, it must stay open by gravitational force. The door height is 450 mm and it weighs 4 kg. Dimension the helical torsion spring, using music wire as the spring material and a wire diameter of 4.5 mm. Is it also possible to use 3-mm-diameter wire? *Ans. For D/d = 10, $N_a$ = 5.89, $l_w$ = 0.83 m.*

**17.31** A helical torsion spring, shown in sketch $j$, is made from hard-drawn steel with a wire diameter of 2.2 mm and 8.5 turns. Dimensions are in millimeters.

  *a)*   Using a safety factor of 2, find the maximum force and the corresponding angular displacement.
  *b)*   What will the coil inside diameter be when the maximum load is applied?
  *c)*   For 100,000 loading cycles calculate the maximum moment and the corresponding angular displacement for a safety factor of 2.5 guarding against fatigue failure.

Sketch $j$, for Problem 17.31

**17.32** The helical torsion spring shown in sketch $k$ has a coil outside diameter of 22 mm, 8.25 turns, and a wire diameter of 2 mm. The spring material is hard-drawn steel. What would the applied moment be if the maximum stress equaled the yield limit? Calculate the inside diameter of the loaded spring and the corresponding angular displacement. Dimensions are in millimeters. *Ans.* $M = 0.668$ N · m, $D'_i = 17.29$ mm.

Sketch $k$, for Problem 17.32

**17.33** A helical torsion spring made of music wire has a coil diameter of 17.5 mm and a wire diameter of 1.5 mm while supporting a 0.15-N · m moment with a 20% fluctuation. The maximum number of turns is 12, and the load is 22 mm from the center of the spring. For infinite life with 99% reliability, find the safety factors guarding against static and fatigue failure. Also, determine the inside diameter of the spring when load has been applied.

## Section 17.6

**17.34** A fishing rod is made like an ideal leaf spring with rectangular cross sections. It is made of carbon-fiber-reinforced plastic with a 150 GPa modulus of elasticity. The thickness is constant at 8 mm and the length is 2 m. Find how large the cross section must be at the handle to carry a 0.3-kg fish by the hook without bending the top of the rod more than 200 mm. Neglect the weight of the lure. Also, calculate the bending stress.

**17.35** A trampoline is made like a leaf spring with variable width so that the maximum bending stress in each section of the trampoline is constant. The material is glass-fiber-reinforced plastic with a modulus of elasticity of 28 GPa and a bending strength of 300 MPa. Calculate the spring rate at the tip of the trampoline and the corresponding safety factor if a swimmer weighing 80 kg jumps onto the trampoline from a height of 2 m. The active length of the trampoline is 3 m, its width is 1.2 m, and its thickness is 38 mm.

**17.36** The leaf spring of a truck should be able to accommodate 55-mm deflections (up and down) of the wheels from an equilibrium position when the truck is driven on a rough road. The static load in the middle of the leaf spring is 50,000 N. Assume an allowable stress of 1050 MPa, a safety factor of 3, leaf half-length of 0.8 m, a leaf thickness of 0.02 m, and that there are 10 leaf layers. Determine the width of the spring. The modulus of elasticity for the spring is 207 GPa. *Ans. b = 0.1714 m.*

## Section 17.7

**17.37** A Belleville spring is formed from cold-rolled steel (AISI 1040) with $t = 1$ mm, $h = 2.5$ mm, $D_o/D_i = 2.0$, and $D_i = 7$ mm. Calculate the force needed to flatten the spring to $h_f = 1.5$ mm. What force is needed to fully flatten the spring? Explain your answer.

**17.38** A Belleville spring is to be used to control the pretension applied to a bolt by a nut. Thus, it is desired that the inner diameter be set equal to the crest diameter and that $R_d = 1.5$ so that the Belleville spring will have the same basic size as the bolt head. If the head height is to be $h = 3$ mm, what thickness is needed for a metric grade-4.6 bolt with a 12-mm crest diameter if the connection is to be reused? Is this a feasible design?

# Brakes and Clutches

A long-shoe, expanding,
internal rim clutch.
(Courtesy of Bucyrus Erie)

*Nothing has such power to broaden the mind as
the ability to investigate systematically and truly
all that comes under thy observation in life.*
        Marcus Aurelius, Roman Emperor

## SYMBOLS

| | |
|---|---|
| $A$ | area, m²; constant |
| $a$ | distance, m |
| $b$ | cone or face width, m |
| $C$ | cost |
| $c$ | constant |
| $D$ | largest diameter of cone, m |
| $d$ | smallest diameter of cone, m |
| $d_1, d_2,$ $d_3, d_4$ | distances used for short-shoe brakes, m |
| $d_5, d_6,$ $d_7$ | distances used for long-shoe rim brakes, m |
| $d_8, d_9,$ $d_{10}$ | distances used for band brakes, m |
| $F$ | friction force, N |
| $F_1$ | pin reaction force for band brakes, N |
| $F_2$ | actuating force for band brakes, N |
| $h_p$ | work or energy conversion rate, W |
| $M$ | moment, N·m |
| $N$ | number of sets of disks |
| $n_s$ | safety factor |
| $P$ | normal force, N |
| $p$ | contact pressure, Pa |
| $p_0$ | uniform pressure, Pa |
| $R$ | reaction force, N |
| $r$ | radius, m |
| $T$ | torque, N·m |
| $\bar{T}$ | dimensionless torque, $T/(2\mu Pr_o)$ |
| $t_m$ | temperature, °C |
| $u$ | sliding velocity, m/s |
| $W$ | actuating force, N |
| $\alpha$ | half-cone angle of cone clutch, deg |
| $\beta$ | radius ratio, $r_i/r_o$ |
| $\beta_o$ | optimum radius ratio |
| $\delta$ | wear depth, m |
| $\theta$ | circumferential coordinate, deg |
| $\theta_a$ | angle where $p = p_{max}$, deg |
| $\theta_1$ | location where shoe begins, deg |
| $\theta_2$ | location where shoe ends, deg |
| $\mu$ | coefficient of friction |
| $\phi$ | wrap angle, deg |
| $\omega$ | angular velocity, rad/s |

## SUBSCRIPTS

| | |
|---|---|
| $B$ | braking |
| $d$ | deenergizing |
| $F$ | friction force |
| $H$ | horizontal |
| $i$ | inner |
| $m$ | mean |
| max | maximum |
| $o$ | outer |
| $P$ | normal force |
| $p$ | uniform pressure |
| $s$ | self-energizing |
| $V$ | vertical |
| $w$ | uniform wear |
| $x, y$ | coordinates |

## 18.1   INTRODUCTION

Brakes and clutches are examples of machine elements that use friction in a useful way. Clutches are required when shafts must be frequently connected and disconnected. The function of a clutch is twofold: first, to provide a gradual increase in the angular velocity of the driven shaft, so that its speed can be brought up to the speed of the driving shaft without shock; second, when the two shafts are rotating at the same angular velocity, to act as a coupling without slip or loss of speed in the driving shaft. A **brake** is a device used to bring a moving system to rest, to slow its speed, or to control its speed to a certain value under varying conditions. The function of the brake is to turn mechanical energy into heat. The design of brakes and clutches is subjected to uncertainties in the value of the coefficient of friction that must necessarily be used. Materials covered in Sections 3.5 and 3.7 should be recalled for use in this chapter.

Figure 18.1 shows five types of brake and clutch that are covered in this chapter. These include a **rim type** that has internal expanding and external contracting shoes, a **band**

**Figure 18.1** Five types of brake and clutch. (a) Internal, expanding rim type; (b) external contracting rim type; (c) band brake; (d) thrust disk; (e) cone disk.

**brake,** a **thrust disk,** and a **cone disk.** This figure also shows the actuating forces being applied to each brake or clutch.

Brakes and clutches are all similar, but different from other machinery elements, in that they are tribological systems where friction is intended to be high. Therefore, much effort has been directed toward identifying and developing materials that result in high coefficients of friction and low wear so that a reasonable service life can occur. In previous years brake and clutch materials were asbestos-fiber-containing composites, but the wear particles associated with these materials resulted in excessive health hazards to maintenance personnel. Modern brakes and clutches use "semimetallic" materials (i.e., metals produced using powder metallurgy techniques) in the tribological interface, even though longer life could be obtained by using the older asbestos-based linings. This substitution is a good example of multidisciplinary design, in that a consideration totally outside of mechanical engineering has eliminated a class of materials from consideration.

Typical brake and clutch design also involves selecting components of sufficient size and capacity to attain reasonable service life. Many of the problems are solid mechanics–oriented; the associated theory is covered in Chapters 4 to 7 and not repeated here. Because this chapter is mainly concerned with the performance of brake and clutch systems, the focus here is on the actuating forces and resultant torques. Component size will not be emphasized.

A critical consideration in the design of brake and clutch components is temperature. Whenever brakes or clutches are activated, one high-friction material slides over another with a large normal force. The large amounts of energy created must be dissipated into heat. If the components become too hot, their performance and life can be severely reduced. For example, when brakes are overused, the surfaces can become glazed or "glassy" and will have a marked reduction in friction. The glazed surface must be removed either by replacing the component or by sanding the damaged surface. Further, overheating brake drums and rotors can cause heat checking (small cracks caused by excessively high local temperatures), warping, and even cracking of the component. Obviously, such circumstances should be avoided, but they can only be discovered through periodic inspection. Therefore, because brakes and clutches have unavoidable wear associated with them, regular maintenance of brake and clutch systems is essential, and their service lives are often much lower than those of other machinery elements.

## 18.2    THRUST DISK CLUTCHES

A thrust disk has its axis of rotation perpendicular to the contacting surfaces, as shown in Figure 18.1(d). Figure 18.2 shows the various radii of the thrust disk **clutch.** The design procedure is to obtain the axial force $P$ necessary to produce a certain torque $T$ and the resulting contact pressure $p$ and wear depth $\delta$. For some elemental area

$$dA = (r\,d\theta)\,dr \tag{18.1}$$

The normal force and the torque can be expressed as

$$dP = p\,dA = pr\,d\theta\,dr \tag{18.2}$$

$$T = \int r\,dF = \int \mu r\,dP = \int\int \mu pr^2\,dr\,d\theta \tag{18.3}$$

Two methods of analysis will be used. Only a single set of disks will be analyzed, but the torque for a single set of disks is multiplied by $N$ to get the torque for $N$ sets of disks.

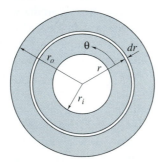

**Figure 18.2**    Thrust disk clutch surface with various radii.

## 18.2.1    UNIFORM PRESSURE MODEL

For new, accurately flat, and aligned disks the pressure will be uniform, or $p = p_0$. Substituting this into Equations (18.2) and (18.3) gives

$$P_p = \pi p_o\left(r_o^2 - r_i^2\right) \tag{18.4}$$

$$T_p = \frac{2\pi \mu p_0}{3}\left(r_o^3 - r_i^3\right) = \frac{2\mu P_p\left(r_o^3 - r_i^3\right)}{3\left(r_o^2 - r_i^2\right)} \tag{18.5}$$

Thus, expressions for the normal load and torque can be determined from the uniform pressure, the geometry ($r_o$ and $r_i$), and the coefficient of friction $\mu$.

## 18.2.2    UNIFORM WEAR MODEL

If the mating surfaces of the clutch are sufficiently rigid, it can be assumed that uniform wear will occur. This assumption generally holds true after some initial running in. After initial wear has taken place, the disks have worn down to the point where uniform wear becomes possible. Thus, for the disk shown in Figure 18.2 the wear is constant over the surface area $r_i \leq r \leq r_o$ and around the circumference of the disk.

   If the rate of wear is assumed to be proportional to the work or energy, the conversion rate is

$$h_p = Fu = \mu Pu = \mu pAu \tag{18.6}$$

where  $F$ = friction force, N
   $\mu$ = coefficient of friction
   $P$ = normal force, N
   $u$ = velocity, m/s
   $A$ = area, m$^2$

A new clutch wears most rapidly at the outer radius, where the velocity $u$ is the greatest. Where the surfaces wear the most, the pressure decreases the most. Thus, multiplying pressure by velocity will produce a constant work or energy conversion, implying that the wear should be uniform at any radius. Then, $\mu pAu$ remains constant; and if $\mu A$ is constant, $p$ is inversely proportional to $u$, and hence for any radius $r$

$$p = \frac{c}{r} \tag{18.7}$$

Substituting Equation (18.7) into Equation (18.2) gives

$$P_w = 2\pi c(r_o - r_i) \tag{18.8}$$

Since $p = p_{max}$ at $r = r_i$, it follows from Equation (18.7) that

$$c = p_{max}r_i \tag{18.9}$$

Substituting Equation (18.9) into Equation (18.8) gives

$$P_w = 2\pi p_{max}r_i(r_o - r_i) \tag{18.10}$$

Substituting Equation (18.7) into Equation (18.3) gives

$$T_w = \mu c \int \int r \, dr \, d\theta = \frac{2\pi \mu c}{2}(r_o^2 - r_i^2)$$

(18.11)

Substituting Equation (18.9) into Equation (18.11) gives

$$T_w = \pi \mu r_i \, p_{max}(r_o^2 - r_i^2)$$

(18.12)

Substituting Equation (18.10) into Equation (18.12) gives

$$T_w = F_w r_m = \frac{\mu P_w(r_o + r_i)}{2}$$

(18.13)

By coincidence, Equation (18.13) gives the same result as if the torque were obtained by multiplying the mean radius $r_m = (r_o + r_i)/2$ by the friction force $F$.

Equations (18.5) and (18.13) can be expressed as the dimensionless torque for uniform pressure $\bar{T}_p$ and for uniform wear $\bar{T}_w$ by the equations

$$\bar{T}_w = \frac{T_w}{2\mu P_w r_o} = \frac{1 + \beta}{4}$$

(18.14)

$$\bar{T}_p = \frac{T_p}{2\mu P_p r_o} = \frac{1 - \beta^3}{3(1 - \beta^2)}$$

(18.15)

where
$$\beta = \text{radius ratio} = \frac{r_i}{r_o}$$

(18.16)

Figure 18.3 shows the effect of the radius ratio on dimensionless torque for the uniform pressure and uniform wear models. The largest difference between these models occurs at a radius ratio of 0, and the smallest difference occurs at a radius ratio of 1. Also, for the same dimensionless torque the uniform wear model requires a larger radius ratio than does the

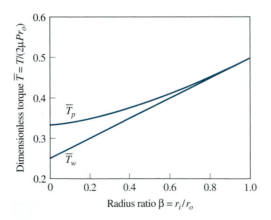

**Figure 18.3**    Effect of radius ratio on dimensionless torque for uniform pressure and uniform wear models.

**Table 18.1**    Representative properties of contacting materials operating dry

| Friction material[a] | Coefficient of friction $\mu$ | Maximum Contact Pressure[b] $p_{max}$ | | Maximum Bulk Temperature $t_{m,\,max}$ | |
|---|---|---|---|---|---|
| | | psi | kPa | °F | °C |
| Molded | 0.25–0.45 | 150–300 | 1030–2070 | 400–500 | 204–260 |
| Woven | 0.25–0.45 | 50–100 | 345–690 | 400–500 | 204–260 |
| Sintered metal | 0.15–0.45 | 150–300 | 1030–2070 | 400–1250 | 204–677 |
| Cork | 0.30–0.50 | 8–14 | 55–95 | 180 | 82 |
| Wood | 0.20–0.30 | 50–90 | 345–620 | 200 | 93 |
| Cast iron; hard steel | 0.15–0.25 | 100–250 | 690–1720 | 500 | 260 |

[a]When rubbing against smooth cast iron or steel.

[b]Use of lower values will give longer life.

uniform pressure model. This larger radius ratio implies that a larger area is needed for the uniform wear model. Thus, the uniform wear model may be viewed as the safer approach. This safety, along with its simplicity, makes it often the preferred equation to be used.

Table 18.1 gives the coefficient of friction for several materials rubbing against smooth cast iron or steel under dry conditions. It also gives the maximum contact pressure and the maximum bulk temperature for these materials. Table 18.2 gives the coefficient of friction for several materials, including those in Table 18.1, rubbing against smooth cast iron or steel in oil. As would be expected, the coefficients of friction are much smaller in oil than under dry conditions.

Equations (18.4) and (18.5) for the uniform pressure model and Equations (18.10) and (18.13) for the uniform wear model, which are applicable for thrust disk clutches, are also applicable for thrust disk brakes provided that the disk shape is similar to that shown in Figure 18.2. A detailed analysis of disk brakes gives equations that result in slightly larger torques than those resulting from the clutch equations. This text assumes that the brake and clutch equations are identical.

**Table 18.2**    Coefficient of friction for contacting materials operating in oil

| Friction material[a] | Coefficient of friction $\mu$ |
|---|---|
| Molded | 0.06–0.09 |
| Woven | 0.08–0.10 |
| Sintered metal | 0.05–0.08 |
| Paper | 0.10–0.14 |
| Graphitic | 0.12 (avg.) |
| Polymeric | 0.11 (avg.) |
| Cork | 0.15–0.25 |
| Wood | 0.12–0.16 |
| Cast iron; hard steel | 0.03–0.16 |

[a]When rubbing against smooth cast iron or steel.

**EXAMPLE 18.1**

**Given:** A single set of thrust disk clutches is to be designed for use in an engine with a maximum torque of 150 N · m. A woven material will contact steel in a dry environment. A safety factor of 1.5 is assumed to account for slippage at full engine torque. The outside diameter should be as small as possible.

**Find:** Determine the appropriate values for $r_o$, $r_i$, and $P$.

**Solution:** For woven material in contact with steel in a dry environment, Table 18.1 gives the coefficient of friction as $\mu = 0.35$ and the maximum contact pressure as $p_{max} = 345$ kPa $= 0.345$ MPa. The average coefficient of friction has been used, but the smallest pressure.
Making use of the above and Equation (18.12) gives

$$r_i \left( r_o^2 - r_i^2 \right) = \frac{n_s T_w}{\pi \mu p_{max}} = \frac{(1.5)(150)}{\pi (0.35)(0.345)(10^6)} = 0.5931 \times 10^{-3} \text{ m}^3$$

Solving for the outside radius $r_o$ gives

$$r_o = \sqrt{\frac{(5.931)(10^{-4})}{r_i} + r_i^2} \qquad \text{(a)}$$

The minimum outside radius is obtained by taking the derivative of the outside radius with respect to the inside radius and setting it equal to zero, to give

$$\frac{dr_o}{dr_i} = \frac{0.5}{\sqrt{(5.931)(10^{-4})/r_i + r_i^2}} \left( -\frac{5.931 \times 10^{-4}}{r_i^2} + 2r_i \right) = 0$$

$$r_i^3 = 2.966 \times 10^{-4} = 0.2966 \times 10^{-3} \text{ m}^3$$

$$r_i = 0.06669 \text{ m} = 66.69 \text{ mm} \qquad \text{(b)}$$

Substituting Equation (b) into Equation (a) gives

$$r_o = \sqrt{\frac{(5.931)(10^{-4})}{(666.9)(10^{-4})} + (6.669)^2 (10^{-4})} = 0.1155 \text{ m} = 115.5 \text{ mm}$$

The radius ratio is

$$\beta = \frac{r_i}{r_o} = \frac{66.69}{115.5} = 0.5774$$

The radius ratio required to maximize the torque capacity is the same as the radius ratio required to minimize the outside radius for a given torque capacity. Thus, the radius ratio for maximizing the torque capacity or for minimizing the outside radius is

$$\beta_o = \sqrt{\tfrac{1}{3}} = 0.5774$$

From Equation (18.13) the maximum normal force that can be applied to the clutch is

$$P = \frac{2 n_s T_w}{\mu (r_o + r_i)} = \frac{2(1.5)(150)}{(0.35)(0.1155 + 0.06669)} = 7057 \text{ N}$$

## 18.3 CONE CLUTCHES

The cone clutch uses wedging action to increase the normal force on the lining, thus increasing the friction force and the torque. Figure 18.4 shows a conical surface with forces acting on an element. The area of the element and the normal force on the element are

$$dA = (r\,d\theta)\frac{dr}{\sin\alpha} \tag{18.17}$$

$$dP = p\,dA \tag{18.18}$$

where $\alpha$ = half-cone angle, deg. The actuating force is the thrust component $dW$ of the normal force $dP$, or

$$dW = dP\sin\alpha = p\,dA\sin\alpha = pr\,dr\,d\theta$$

Using Equation (18.2) gives the actuating force as

$$W = \int\int pr\,dr\,d\theta = 2\pi\int_{d/2}^{D/2} pr\,dr \tag{18.19}$$

Similarly, Equation (18.3) gives the torque as

$$T = \int \mu r\,dP = \frac{2\pi}{\sin\alpha}\int_{d/2}^{D/2} \mu pr^2\,dr \tag{18.20}$$

**Figure 18.4** Forces acting on elements of cone clutch.

### 18.3.1    UNIFORM PRESSURE MODEL

As was discovered in Section 18.2.1, the pressure for a thrust disk clutch is assumed to be uniform over the surfaces, or $p = p_0$. Substituting this into Equation (18.19) gives the actuating force as

$$W = \frac{\pi p_o}{4}(D^2 - d^2) \tag{18.21}$$

Similarly, the torque is

$$T = \frac{2\pi p_o \mu}{3 \sin \alpha}\left(\frac{1}{8}\right)(D^3 - d^3) = \frac{\pi p_o \mu}{12 \sin \alpha}(D^3 - d^3) \tag{18.22}$$

Making use of Equation (18.21) enables Equation (18.22) to be rewritten as

$$T = \frac{\mu W(D^3 - d^3)}{3(\sin \alpha)(D^2 - d^2)} \tag{18.23}$$

### 18.3.2    UNIFORM WEAR MODEL

Substituting Equation (18.7) into Equation (18.19) gives the actuating force as

$$W = 2\pi c \int_{d/2}^{D/2} dr = \pi c(D - d) \tag{18.24}$$

Similarly, substituting Equation (18.7) into Equation (18.20) gives the torque as

$$T = \frac{2\pi \mu c}{\sin \alpha} \int_{d/2}^{D/2} r\, dr = \frac{\pi \mu c}{4 \sin \alpha}(D^2 - d^2) \tag{18.25}$$

Making use of Equation (18.24) enables Equation (18.25) to be rewritten as

$$T = \frac{\mu W}{4 \sin \alpha}(D + d) \tag{18.26}$$

---

**EXAMPLE 18.2**

**Given:**  A cone clutch similar to that shown in Figure 18.4 has the following dimensions: $D = 330$ mm, $d = 306$ mm, and $b = 60$ mm. The coefficient of friction is assumed to be 0.26, and the torque transmitted is 200 N · m.

**Find:**  Determine the actuating force and the contact pressure by using the uniform pressure and uniform wear models.

**Solution:**  **Uniform wear:** From Figure 18.4 the half-cone angle of the cone clutch is

$$\tan \alpha = \frac{D - d}{2b} = \frac{165 - 153}{60} = \frac{12}{60} = 0.200$$

$$\alpha = 11.31°$$

The pressure is a maximum when $r = d/2$. Thus, making use of Equations (18.25) and (18.7) gives

$$T = \frac{\pi \mu d p_{max}}{8 \sin \alpha}(D^2 - d^2)$$

$$p_{max} = \frac{8T \sin \alpha}{\pi \mu d(D^2 - d^2)} = \frac{8(200) \sin 11.31°}{\pi(0.26)(0.306)(0.330^2 - 0.306^2)} \text{ Pa} = 82.25 \text{ kPa}$$

From Equation (18.26) the actuating force can be written as

$$W = \frac{4T \sin \alpha}{\mu(D + d)} = \frac{4(200) \sin 11.31°}{(0.26)(0.330 + 0.306)} = 948.8 \text{ N}$$

**Uniform pressure:** From Equation (18.23) the actuating force can be expressed as

$$W = \frac{3T(\sin \alpha)(D^2 - d^2)}{\mu(D^3 - d^3)} = \frac{3(200)(\sin 11.31°)(0.330^2 - 0.306^2)}{(0.26)(0.330^3 - 0.306^3)} = 948.4 \text{ N}$$

From Equation (18.21) the uniform pressure, which is also the maximum pressure, is

$$p_{max} = p_o = \frac{4W}{\pi(D^2 - d^2)} = \frac{4(948.4)}{\pi(0.330^2 - 0.306^2)} \text{ Pa} = 79.11 \text{ kPa}$$

## 18.4 BLOCK, OR SHORT-SHOE, BRAKES

A block, or short-shoe, brake can be guided to move radially against a cylindrical drum, as shown in Figure 18.5. A normal force $P$ develops a friction force $F = \mu P$ on the drum, where $\mu$ is the coefficient of friction. The actuating force $W$ is also shown in Figure 18.5 along with critical hinge pin dimensions $d_1$, $d_2$, $d_3$, and $d_4$. The normal force $P$ and the friction force $\mu P$ (Fig. 18.5) are the forces acting on the brake. For block, or short-shoe,

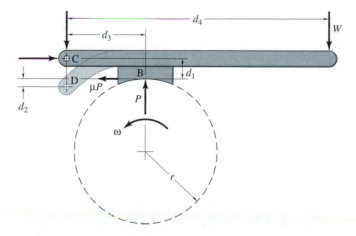

**Figure 18.5** Block, or short-shoe brake, with two configurations.

brakes a constant pressure is assumed over the pad surface. As long as the pad is short relative to the circumference of the drum, this assumption is relatively accurate.

A brake is considered to be **self-energizing** if the friction moment assists the actuating moment in applying the brake. This implies that the signs of the friction and actuating moments are the same. **Deenergizing** effects occur if the friction moment counteracts the actuating moment in applying the brake. Figure 18.5 can be used to illustrate self-energizing and deenergizing effects.

Summing the moments about the hinge at C (see Fig. 18.5) and setting the sum equal to zero give

$$d_4 W + \mu P d_1 - d_3 P = 0$$

Since the signs of the friction and actuating moments are the same, the brake hinged at C is self-energizing. Solving for the normal force gives

$$P = \frac{d_4 W}{d_3 - \mu d_1} \tag{18.27}$$

The braking torque at C (see Fig. 18.5) is

$$T = Fr = \mu r P = \frac{\mu r d_4 W}{d_3 - \mu d_1} \tag{18.28}$$

where $r$ = radius of drum, m.

Summing the moments about the hinge at D (see Fig. 18.5) and setting the sum equal to zero give

$$-W d_4 + \mu P d_2 + d_3 P = 0$$

Since the signs of the friction and actuating moments are opposite, the brake hinged at D is deenergizing. Solving for the normal force gives

$$P = \frac{W d_4}{d_3 + \mu d_2} \tag{18.29}$$

The torque of the brake hinged at D in Figure 18.5 is

$$T = \frac{\mu d_4 r W}{d_3 + \mu d_2} \tag{18.30}$$

A brake is considered to be self-locking if the actuating force ($W$ in Fig. 18.5) equals zero. Self-locking brakes are not desired, since they seize or grab, thus operating unsatisfactorily or even dangerously.

---

**EXAMPLE 18.3**

**Given:**   A 14-in-radius brake drum contacts a single short shoe, as shown in Figure 18.6, and sustains 2000 in · lb of torque at 500 rpm. Assume that the coefficient of friction for the drum and shoe combination is 0.3.

**Figure 18.6** Short-shoe brake used in Example 18.3.

**Find:** Determine the following:

(a) Normal force acting on the shoe
(b) Required actuating force $W$ when the drum has clockwise rotation
(c) Required actuating force $W$ when the drum has counterclockwise rotation
(d) Required change in the 1.5-in dimension (Fig. 18.6) for self-locking to occur if the other dimensions do not change

**Solution:**

(a) The torque of the brake is

$$T = rF = r\mu P$$

$$P = \frac{T}{\mu r} = \frac{2000}{(0.3)(14)} = 476.2 \text{ lb}$$

$$\mu P = (0.3)(476.2) = 142.9 \text{ lb}$$

(b) For clockwise rotation, summing the moments about the hinge pin and setting the sum equal to zero give

$$(1.5)(142.9) + 36W - 14(476.2) = 0 \qquad \text{(a)}$$

$$\therefore \qquad W = 179.2 \text{ lb}$$

Since the signs of the friction and actuating moments are the same, the brake is self-energizing.

(c) For counterclockwise rotation, summing the moments about the hinge pin and setting the sum equal to zero give

$$(1.5)(142.9) - 36W + 14(476.2) = 0$$

$$\therefore \qquad W = 191.1 \text{ lb}$$

Since the signs of the friction and actuating moments are not the same, the brake is deenergizing.

(d)    If in Equation (a) $W = 0$ and 1.5 is set to $x$,

$$x = \frac{(14)(476.2)}{142.9} = 46.65 \text{ in}$$

Therefore, self-locking will occur if in Figure 18.6 the distance of 1.5 in is changed to 46.65 in. Since self-locking is *not* a desirable effect in a brake and 1.5 in is quite a distance from 46.65 in, we would not expect the brake to have a self-locking effect.

## 18.5    LONG-SHOE, INTERNAL, EXPANDING RIM BRAKES

Figure 18.7 shows a long-shoe, internal, expanding rim brake with two pads. The hinge pin for the right shoe is at A in Figure 18.7. The *heel* of the pad is the region closest to the hinge pin, and the *toe* is the region closest to the actuating force $W$. A major difference between a short shoe (Fig. 18.5) and a long shoe (Fig. 18.7) is that the pressure can be considered constant for a short shoe but not for a long shoe. In a long shoe no or little pressure is applied at the heel, and the pressure increases as the toe is approached. This sort of pressure variation suggests that the pressure may vary sinusoidally. Thus, a relationship of the contact pressure $p$ in terms of the maximum pressure $p_{max}$ may be written as

$$p = p_{max}\left(\frac{\sin\theta}{\sin\theta_a}\right) \tag{18.31}$$

**Figure 18.7**    Long-shoe, internal, expanding rim brake with two shoes.

**Figure 18.8** Forces and dimensions of long-shoe, internal, expanding rim brake.

where $\theta_a$ = angle where $p = p_{max}$. Observe from Equation (18.31) that $p = p_{max}$ when $\theta = 90°$ and that for any angular extent of the shoe less than 90° (in Fig. 18.7, $\theta_2 < 90°$) $p = p_{max}$ when, for example, $\theta_a = \theta_2$.

Observe also in Figure 18.7 that the distance $d_6$ is perpendicular to the actuating force $W$. Figure 18.8 shows the forces and critical dimensions of a long-shoe, internal, expanding rim brake. In Figure 18.8 the $\theta$ coordinate begins with a line drawn from the center of the drum and the center of the hinge pin. Also, the shoe lining does not begin at $\theta = 0°$ but at some $\theta_1$ and extends until $\theta_2$. At any angle $\theta$ of the lining the differential normal force $dP$ is

$$dP = pbr\, d\theta \qquad (18.32)$$

where $b$ = face width, m. The face width is the distance perpendicular to the paper. Substituting Equation (18.31) into Equation (18.32) gives

$$dP = \frac{p_{max}br\sin\theta\, d\theta}{\sin\theta_a} \qquad (18.33)$$

From Equation (18.33) the normal force moment with moment arm $d_7\sin\theta$ is

$$M_P = \int d_7\sin\theta\, dP = \frac{d_7 br p_{max}}{\sin\theta_a}\int_{\theta_1}^{\theta_2}\sin^2\theta\, d\theta$$

$$= \frac{br d_7 p_{max}}{4\sin\theta_1}\left[2(\theta_2 - \theta_1)\frac{\pi}{180°} - \sin 2\theta_2 + \sin 2\theta_1\right] \qquad (18.34)$$

where $\theta_1$ and $\theta_2$ are in degrees. From Equation (18.33) the friction force moment with moment arm $r - d_7 \cos \theta$ is

$$M_F = \int (r - d_7 \cos \theta) \mu \, dP = \frac{\mu p_{\max} br}{\sin \theta_1} \int_{\theta_1}^{\theta_2} (r - d_7 \cos \theta) \sin \theta \, d\theta$$

$$= \frac{\mu p_{\max} br}{\sin \theta_a} \left[ -r(\cos \theta_2 - \cos \theta_1) - \frac{d_7}{2} \left( \sin^2 \theta_2 - \sin^2 \theta_1 \right) \right] \qquad (18.35)$$

### 18.5.1   SELF-ENERGIZING SHOE

Summing the moments about the hinge pin and setting the sum equal to zero give

$$-Wd_6 - M_F + M_P = 0 \qquad (18.36)$$

Since the actuating and friction moments have the same sign in Equation (18.36), the shoe shown in Figure 18.8 is self-energizing. It can also be concluded just from Figure 18.8 that the shoe is self-energizing because $W_x$ and $\mu \, dP \sin \theta$ are in the same direction. Solving for the actuating force in Equation (18.36) gives

$$W = \frac{M_P - M_F}{d_6} \qquad (18.37)$$

From Equation (18.33) the braking torque is

$$T = \int r\mu \, dP = \frac{\mu p_{\max} br^2}{\sin \theta_a} \int_{\theta_1}^{\theta_2} \sin \theta \, d\theta$$

$$T = \frac{\mu p_{\max} br^2 (\cos \theta_1 - \cos \theta_2)}{\sin \theta_a} \qquad (18.38)$$

Figure 18.8 shows the reaction forces as well as the friction force and the normal force. Summing the forces in the $x$ direction and setting the sum equal to zero give

$$R_{xs} + W_x - \int \cos \theta \, dP + \int \mu \sin \theta \, dP = 0 \qquad (18.39)$$

Substituting Equation (18.33) into Equation (18.39) gives the reaction force in the $x$ direction for a self-energizing shoe as

$$R_{xs} = \frac{p_{\max,s} br}{\sin \theta_a} \int_{\theta_1}^{\theta_2} \sin \theta \cos \theta \, d\theta - \frac{\mu p_{\max,s} br}{\sin \theta_a} \int_{\theta_1}^{\theta_2} \sin^2 \theta \, d\theta - W_x$$

$$= -W_x + \frac{p_{\max,s} br}{4 \sin \theta_a} \left\{ 2 \left( \sin^2 \theta_2 - \sin^2 \theta_1 \right) - \mu \left[ 2(\theta_2 - \theta_1) \frac{\pi}{180°} - \sin 2\theta_2 + \sin 2\theta_1 \right] \right\}$$

$$\qquad (18.40)$$

where $\theta_1$ and $\theta_2$ are in degrees. From Figure 18.8, summing the forces in the $y$ direction and setting the sum equal to zero give

$$R_{ys} + W_y - \int \mu \, dP \cos \theta - \int dP \sin \theta = 0 \qquad (18.41)$$

$$R_{ys} = -W_y + \frac{p_{max,s} br}{4 \sin \theta_a} \left[ 2(\theta_2 - \theta_1) \frac{\pi}{180°} - \sin 2\theta_2 + \sin 2\theta_1 + 2\mu \left( \sin^2 \theta_2 - \sin^2 \theta_1 \right) \right] \qquad (18.42)$$

In Equations (18.39) to (18.42) the reference system has its origin at the center of the drum. The positive $x$ axis is taken to be through the hinge pin. The positive $y$ axis is in the direction of the shoe.

## 18.5.2 DEENERGIZING SHOE

If in Figure 18.8 the direction of rotation is changed from clockwise to counterclockwise, the friction forces change direction. Thus, summing the moments about the hinge pin and setting the sum equal to zero give

$$-Wd_6 + M_F + M_P = 0 \qquad (18.43)$$

The only difference between Equation (18.36) and Equation (18.43) is the sign of the friction moment. Solving for the actuating force in Equation (18.43) gives

$$W = \frac{M_P + M_F}{d_6} \qquad (18.44)$$

The evaluations of the normal moment in Equation (18.34) and the friction moment in Equation (18.35) are the same whether the shoe is self-energizing or deenergizing with the exception of the maximum pressure.

For a deenergizing shoe the only changes from the equations derived for the self-energizing shoe are that in Equations (18.40) and (18.42) a sign change occurs for terms containing the coefficient of friction $\mu$, resulting in the following:

$$R_{xd} = -W_x + \frac{p_{max,d} br}{4 \sin \theta_a} \left\{ 2 \left( \sin^2 \theta_2 - \sin^2 \theta_1 \right) + \mu \left[ 2(\theta_2 - \theta_1) \frac{\pi}{180°} - \sin 2\theta_2 + \sin 2\theta_1 \right] \right\} \qquad (18.45)$$

$$R_{yd} = -W_y + \frac{p_{max,d} br}{4 \sin \theta_a} \left[ 2(\theta_2 - \theta_1) \frac{\pi}{180°} - \sin 2\theta_2 + \sin 2\theta_1 - 2\mu \left( \sin^2 \theta_2 - \sin^2 \theta_1 \right) \right] \qquad (18.46)$$

The maximum contact pressure used in evaluating a self-energizing shoe is taken from Table 18.1. The maximum contact pressure used in evaluating a deenergizing shoe is less than that for the self-energizing shoe, since the actuating force is the same for both types of shoe.

**EXAMPLE 18.4**

**Given:** Figure 18.9 shows four long shoes in an internal, expanding rim brake. The brake drum has a 400-mm diameter. Each hinge pin (A and B) supports a pair of shoes. The actuating mechanism is to be arranged to produce the same actuating force $W$ on each shoe. The shoe face width is 75 mm. The material of the shoe and drum produces a coefficient of friction of 0.24 and a maximum contact pressure of 1 MPa. Dimensions are the following: $d = 50$ mm, $b = 165$ mm, and $a = 150$ mm.

**Find:** Determine the following:

(a) Which shoes are self-energizing and which are deenergizing?
(b) What are the actuating forces and total torques for the four shoes?
(c) What are the hinge pin reactions as well as the resultant reaction?

**Solution:**

(a) With the drum rotating in the clockwise direction (Fig. 18.9) the top right and bottom left shoes have their actuating and friction moments acting in the same direction. Thus, they are self-energizing shoes. The top-left and bottom-right shoes have their actuating and friction moments acting in opposite directions. Thus, they are deenergizing shoes.

(b) The dimensions given in Figure 18.9 correspond to the dimensions given in Figures 18.7 and 18.8 as $d_5 = d = 50$ mm, $d_6 = b = 165$ mm, and $d_7 = a = 150$ mm. Also, since $\theta_2 < 90°$, then $\theta_2 = \theta_a$. Because the hinge pins in Figure 18.9 are at A and B, $\theta_1 = 10°$ and $\theta_2 = \theta_a = 75°$.

**Figure 18.9**    Four-long-shoe, internal expanding rim brake used in Example 18.4.

**Self-energizing shoes:** Making use of the previous and Equation (18.34) gives the normal force moment as

$$M_{Ps} = \frac{br d_7 p_{max,s}}{4 \sin \theta_a} \left[ 2(\theta_2 - \theta_1) \frac{\pi}{180°} - \sin 2\theta_2 + \sin 2\theta_1 \right]$$

$$= \frac{(0.075)(0.2)(0.15)(10^6)}{4 \sin 75°} \left[ 2(75 - 10) \frac{\pi}{180°} - \sin 150° + \sin 20° \right]$$

$$= 1229 \text{ N} \cdot \text{m}$$

From Equation (18.35) the friction moment is

$$M_{Fs} = \frac{\mu p_{max,s} br}{\sin \theta_a} \left[ -r(\cos \theta_2 - \cos \theta_1) - \frac{d_7}{2} \left( \sin^2 \theta_2 - \sin^2 \theta_1 \right) \right]$$

$$= \frac{(0.24)(10^6)(0.075)(0.20)}{\sin 75°}$$

$$\times [-(0.20)(\cos 75° - \cos 10°) - 0.075(\sin^2 75° - \sin^2 10°)]$$

$$= 288.8 \text{ N} \cdot \text{m}$$

From Equation (18.37) the actuating force for both the self-energizing and deenergizing shoes is

$$W_s = W_d = \frac{M_P - M_F}{d_6} = \frac{1229 - 288.8}{0.165} = 5698 \text{ N}$$

From Equation (18.38) the braking torque for each self-energizing shoe is

$$T_s = \frac{\mu p_{max,s} br^2 (\cos \theta_1 - \cos \theta_2)}{\sin \theta_a}$$

$$= \frac{(0.24)(10^6)(0.075)(0.2)^2 (\cos 10° - \cos 75°)}{\sin 75°} = 541.2 \text{ N} \cdot \text{m}$$

**Deenergizing shoes:** The only change in the calculation of the normal and friction moments for the deenergizing shoes is the maximum pressure.

$$M_{Pd} = \frac{M_{Ps} p_{max,d}}{p_{max,s}} = \frac{1229 p_{max,d}}{10^6}$$

$$M_{Fd} = \frac{M_{Fs} p_{max,d}}{p_{max,s}} = \frac{288.8 p_{max,d}}{10^6}$$

From Equation (18.44) the actuating load for the deenergizing shoes is

$$W_d = \frac{M_{Pd} + M_{Fd}}{d_6} = \frac{1229 + 288.8}{10^6} \frac{p_{max,d}}{0.165} = W_s = 5698 \text{ N}$$

$$\therefore \quad p_{max,d} = \frac{(5698)(0.165)(10^6)}{1229 + 288.8} \text{ Pa} = 0.6194 \text{ MPa}$$

The braking torque for the deenergizing shoes is

$$T_d = T_s \left( \frac{p_{max,d}}{p_{max,s}} \right) = 541.2 \left( \frac{0.6194}{1.000} \right) = 335.2 \text{ N} \cdot \text{m}$$

The total braking torque of the four shoes, two of which are self-energizing and two of which are deenergizing, is

$$T = 2(T_s + T_d) = 2(541.2 + 335.2) = 1753 \text{ N} \cdot \text{m}$$

(c) **Self-energizing shoes:** From Equation (18.40)

$$R_{xs} = -W_y + \frac{p_{max,s}br}{4\sin\theta_a}\left\{2\left(\sin^2\theta_2 - \sin^2\theta_1\right) - \mu\left[2(\theta_2 - \theta_1)\frac{\pi}{180°} - \sin 2\theta_2 + \sin 2\theta_1\right]\right\}$$

$$= -5698 + \frac{(10^6)(0.075)(0.2)}{4\sin 75°}$$

$$\times \left\{2(\sin^2 75° - \sin^2 10°) - 0.24\left[2(75° - 10°)\frac{\pi}{180°} - \sin 150° + \sin 20°\right]\right\}$$

$$= -654.6 \text{ N}$$

From Equation (18.42)

$$R_{ys} = -W_y + \frac{p_{max,s}br}{4\sin\theta_a}\left[2(\theta_2 - \theta_1)\frac{\pi}{180°} - \sin 2\theta_2 + \sin 2\theta_1 + 2\mu\left(\sin^2\theta_2 - \sin^2\theta_1\right)\right]$$

$$= -0 + \frac{(10^6)(0.075)(0.2)}{4\sin 75°}$$

$$\times \left[2(75° - 10°)\frac{\pi}{180°} - \sin 150° + \sin 20° + 2(0.24)(\sin^2 75° - \sin^2 10°)\right]$$

$$= 9878 \text{ N}$$

**Deenergizing shoes:** From Equation (18.45)

$$R_{xd} = -5698 + \frac{(0.6194)(10^6)(0.075)(0.2)}{4\sin 75°}$$

$$\times \left\{2(\sin^2 75° - \sin^2 10°) + 0.24\left[2(75° - 10°)\frac{\pi}{180°} - \sin 150° + \sin 20°\right]\right\}$$

$$= 5560 - 5698 = -137.5 \text{ N}$$

From Equation (18.46)

$$R_{yd} = -0 + \frac{(0.6194)(10^6)(0.075)(0.2)}{4\sin 75°}$$

$$\times \left[2(75° - 10°)\frac{\pi}{180°} - \sin 150° + \sin 20° - 2(0.24)(\sin^2 75° - \sin^2 10°)\right]$$

$$= 4034 \text{ N}$$

The resultant forces of the reactions in the hinge pin in the horizontal and vertical directions are

$$R_H = -654.6 - 137.5 = -792.1 \text{ N}$$

$$R_V = 9878 - 4034 = 5844 \text{ N}$$

The resultant force at the hinge pin is

$$R = \sqrt{R_H^2 + R_V^2} = \sqrt{(-792.1)^2 + (5844)^2} = 5897 \text{ N}$$

## 18.6 LONG-SHOE, EXTERNAL, CONTRACTING RIM BRAKES

Figure 18.10 shows the forces and dimensions of a long-shoe, external, contracting rim brake. In Figure 18.8 the brake is internal to the drum, whereas in Figure 18.10 the brake is external to the drum. The symbols used in these figures are similar. The equations developed in Section 18.5 for internal shoe brakes are exactly the same as those for external shoe brakes as long as one properly identifies whether the brake is self-energizing or deenergizing.

The *internal* brake shoe in Figure 18.8 was found to be *self-energizing* for clockwise rotation, since in the moment summation [(Eq. (18.36)] the actuating and friction moments have the same sign. The *external* brake shoe in Figure 18.10 is *deenergizing* for clockwise rotation. Summing the moments and equating the sum to zero give

$$Wd_6 - M_F - M_P = 0 \qquad\qquad (18.47)$$

The actuating and friction moments have opposite signs, and thus the external brake shoe shown in Figure 18.10 is deenergizing.

If in Figures 18.8 and 18.10 the direction of rotation were changed from clockwise to counterclockwise, the friction moments in Equations (18.36) and (18.47) would have

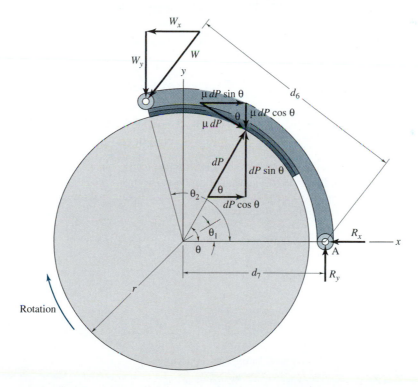

**Figure 18.10**  Forces and dimensions of long-shoe, external, contracting rim brake.

opposite signs. Therefore, the internal brake shoe would be deenergizing, and the external brake shoe would be self-energizing.

---

**EXAMPLE 18.5**

**Given:** An external, long-shoe rim brake is to be cost-optimized. Three lining geometries are being considered, covering the entire 90° of the shoe, covering only 45° of the central portion of the shoe, or covering only 22.5° of the central portion of the shoe. The braking torque must be the same for all three geometries, and the cost of changing any of the brake linings is one-half of the cost of changing a 22.5° lining. The cost of the lining material is proportional to the wrap angle. The wear rate is proportional to the pressure. The input parameters for the 90° lining are $d_7 = 100$ mm, $r = 80$ mm, $b = 25$ mm, $\theta_1 = 0°$, $\theta_2 = 90°$, $\mu = 0.27$, and $T = 125$ N · m.

**Find:** Which of the wrap angles (90°, 45°, or 22.5°) would be the most economical?

**Solution:** The braking torque is given by Equation (18.38) and is the same for all three geometries. For the 90° wrap angle ($\theta_1 = 0°$, $\theta_2 = 90°$, and $\theta_a = 90°$)

$$(p_{max})_{90°} = \frac{T \sin\theta_a}{\mu b r^2(\cos\theta_1 - \cos\theta_2)} = \frac{125 \sin 90°}{(0.27)(0.025)(0.08)^2(\cos 0° - \cos 90°)}$$

$$= 2.894 \times 10^6 \text{ Pa} = 2.894 \text{ MPa}$$

For the 45° wrap angle ($\theta_1 = 22.5°$, $\theta_2 = 67.5°$, and $\theta_a = 67.5°$)

$$(p_{max})_{45°} = \frac{125 \sin 67.5°}{(0.27)(0.025)(0.08)^2(\cos 22.5° - \cos 67.5°)}$$

$$= 4.940 \times 10^6 \text{ Pa} = 4.940 \text{ MPa}$$

For the 22.5° wrap angle ($\theta_1 = 33.75°$, $\theta_2 = 56.25°$, and $\theta_a = 56.25°$)

$$(p_{max})_{22.5°} = \frac{125 \sin 56.25°}{(0.27)(0.025)(0.08)^2(\cos 33.75° - \cos 56.25°)}$$

$$= 8.720 \times 10^6 \text{ Pa} = 8.720 \text{ MPa}$$

The cost of changing the lining is $C$. The lining costs are $2C$ for a 22.5° lining, $4C$ for a 45° lining, and $8C$ for a 90° lining. The wear rate is proportional to the pressure, or the time it takes for the shoe to wear out is inversely proportional to the pressure. Thus, the times it takes for the shoe to wear out for the three geometries are

$$t_{90°} = \frac{A}{(p_{max})_{90°}} = \frac{A}{(2.894)(10^6)} = 3.455 \times 10^{-7} A$$

where $A = $ constant independent of geometries, and

$$t_{45°} = \frac{A}{(p_{max})_{45°}} = \frac{A}{(4.940)(10^6)} = 2.024 \times 10^{-7} A$$

$$t_{22.65°} = \frac{A}{(p_{max})_{22.5°}} = \frac{A}{(8.720)(10^6)} = 1.147 \times 10^{-7} A$$

The costs per unit time for the three geometries are for a *90° wrap angle:*

$$\frac{8C + C}{(3.455)(10^{-7})A} = 26.05 \times 10^6 \, C/A$$

*45° wrap angle:*

$$\frac{4C + C}{(2.024)(10^{-7})A} = 24.70 \times 10^6 \, C/A$$

*22.5° wrap angle:*

$$\frac{2C + C}{(1.147)(10^{-7})A} = 26.16 \times 10^6 \, C/A$$

The 45° wrap angle shoe gives the lowest cost, 5.6% lower than the 22.5° wrap angle shoe and 5.2% lower than the 90° wrap angle shoe.

## 18.7 SYMMETRICALLY LOADED PIVOT-SHOE BRAKES

Figure 18.11 shows a symmetrically loaded pivot-shoe brake. The major difference between the internal and external rim brakes shown in Figures 18.8 and 18.10, respectively, and the symmetrically loaded pivot-shoe brake shown in Figure 18.11 is the pressure distribution around the shoe. Recall from Equation (18.31) for the internal rim brake that the maximum pressure was at $\theta = 90°$ and the pressure distribution from the heel to the top of the brake was sinusoidal. For the symmetrically loaded pivot-shoe brake (Fig. 18.11) the maximum pressure is at $\theta = 0°$, which suggests the pressure variation to be

$$p = \frac{p_{max} \cos \theta}{\cos \theta_a} = p_{max} \cos \theta \qquad (18.48)$$

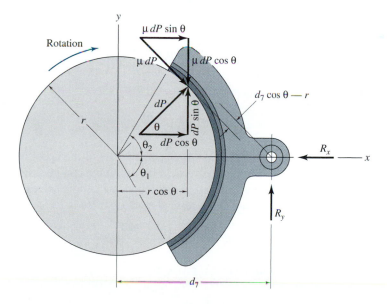

**Figure 18.11** Symmetrically loaded pivot-shoe brake.

For any $\theta$ from the pivot a differential normal force $dP$ acts with a magnitude of

$$dP = pbr \, d\theta = p_{\max} br \, \cos\theta \, d\theta \tag{18.49}$$

The design of a symmetrically loaded pivot-shoe brake is such that the distance $d_7$, which is measured from the center of the drum to the pivot, is chosen such that the resulting friction moment acting on the brake shoe is zero. From Figure 18.11 the friction moment, when set equal to zero, is

$$M_F = 2 \int_0^{\theta_2} \mu \, dP \, (d_7 \cos\theta - r) = 0 \tag{18.50}$$

Substituting Equation (18.49) into Equation (18.50) gives

$$2\mu p_{\max} br \int_0^{\theta_2} (d_7 \cos^2\theta - r\cos\theta) = 0$$

This reduces to

$$d_7 = \frac{4r \sin\theta_2}{2\theta_2(\pi/180°) + \sin 2\theta_2} \tag{18.51}$$

This value of $d_7$ produces a friction moment equal to zero ($M_F = 0$).

The braking torque is

$$T = 2 \int_0^{\theta_2} r\mu \, dP = 2\mu r^2 b p_{\max} \int_0^{\theta_2} \cos\theta \, d\theta = 2\mu r^2 b p_{\max} \sin\theta_2 \tag{18.52}$$

Note from Figure 18.11 that for any $x$ the horizontal *friction force* components of the upper half of the shoe are equal and opposite in direction to the horizontal friction force components of the lower half of the shoe. For a fixed $x$ the horizontal *normal* components of both halves of the shoe are equal and in the same direction, so that the horizontal reaction force is

$$R_x = 2 \int_0^{\theta_2} dP \cos\theta = \frac{p_{\max} br}{2}\left[2\theta_2\left(\frac{\pi}{180°}\right) + \sin 2\theta_2\right] \tag{18.53}$$

Making use of Equation (18.51) gives

$$R_x = \frac{2br^2 p_{\max} \sin\theta_2}{d_7} \tag{18.54}$$

For a fixed $y$ the vertical *friction force* components of the upper half of the shoe are equal and in the same direction as the vertical friction force components of the lower half of the shoe. For a fixed $y$ the vertical *normal* components of both halves of the shoe are equal and opposite in direction, so that the vertical reaction force is

$$R_y = 2 \int_0^{\theta_2} \mu \, dP \cos\theta = \frac{\mu p_{\max} br}{2}\left[2\theta_2\left(\frac{\pi}{180°}\right) + \sin 2\theta_2\right] \tag{18.55}$$

Making use of Equation (18.53) gives

$$R_y = \frac{2\mu br^2 p_{\max} \sin\theta_2}{d_7} = \mu R_x$$

EXAMPLE 18.6

**Given:** A symmetrically loaded, pivot-shoe brake has the distance $d_7$ shown in Figure 18.11 optimized for a 180° wrap angle. When the brake lining is worn out, it is replaced with a 90° wrap angle lining symmetrically positioned in the shoe. The actuating force is 11,000 N, the coefficient of friction is 0.31, the brake drum radius is 100 mm, and the brake width is 45 mm.

**Find:** Calculate the pressure distribution in the brake shoe and the braking torque.

**Solution:** The distance $d_7$ can be expressed from Equation (18.51) for the 180° wrap angle as

$$(d_7)_{180°} = \frac{4r \sin \theta_2}{2\theta_2(\pi/180°) + \sin 2\theta_2}$$

$$= \frac{4(0.1) \sin 90°}{2(90)(\pi/180°) + \sin 180°} = 0.1273 \text{ m}$$

For the 90° wrap angle, symmetrically loaded, pivot-shoe brake, the pressure distribution will be unsymmetric. The maximum pressure, which will occur at $\theta_2$, needs to be determined from the shoe equilibrium.

$$\therefore \quad \int_{-\pi/4}^{\pi/4} pr \, d\theta b(d_7 \sin \theta) - \int_{-\pi/4}^{\pi/4} \mu pr \, d\theta \, b(d_7 \cos \theta - r) = 0 \tag{a}$$

If the wear rate is proportional to the pressure and the pressure maximum is at $\theta = \theta_0$, the pressure distribution is

$$p = p_{max} \cos(\theta - \theta_0) \tag{b}$$

Substituting Equation (b) into Equation (a) gives

$$\int_{-\pi/4}^{\pi/4} d_7 \cos(\theta - \theta_0) \sin \theta \, d\theta - \int_{-\pi/4}^{\pi/4} \mu \cos(\theta - \theta_0)(d_7 \cos \theta - r) d\theta = 0$$

But

$$\cos(\theta - \theta_0) = \cos \theta \cos \theta_0 + \sin \theta \sin \theta_0$$

$$\therefore \quad d_7 \int_{-\pi/4}^{\pi/4} (\cos \theta_0 \cos \theta \sin \theta + \sin \theta_0 \sin^2 \theta) \, d\theta$$

$$= \mu \int_{-\pi/4}^{\pi/4} (\cos \theta_0 \cos \theta + \sin \theta_0 \sin \theta)(d_7 \cos \theta - r) \, d\theta$$

Integrating gives

$$\frac{d_7 \cos \theta_0}{2} \left[ \sin^2 \frac{\pi}{4} - \sin^2 \left(-\frac{\pi}{4}\right) \right] + d_7 \sin \theta_0 \left( \frac{\pi}{4} - \frac{1}{2} \right) = \mu d_7 \cos \theta_0 \left( \frac{\pi}{4} + \frac{1}{2} \right)$$

$$+ \mu d_7 \sin \theta_0 \left[ \frac{\sin^2(\pi/4)}{2} - \frac{\sin^2(-\pi/4)}{2} \right] - \mu r \cos \theta_0 \sqrt{2} + \mu r \sin \theta_0 \left( \frac{1}{\sqrt{2}} - \frac{1}{\sqrt{2}} \right)$$

This reduces to

$$d_7 \sin \theta_0 \left( \frac{\pi}{4} - \frac{1}{2} \right) = \mu \cos \theta_0 \left[ d_7 \left( \frac{\pi}{4} + \frac{1}{2} \right) - r\sqrt{2} \right]$$

$$\tan \theta_0 = \frac{\mu \left[ d_7 \left( \pi/4 + \frac{1}{2} \right) - r\sqrt{2} \right]}{d_7 \left( \pi/4 - \frac{1}{2} \right)} = \frac{0.31 \left[ 0.1273 \left( \pi/4 + \frac{1}{2} \right) - (0.100)\sqrt{2} \right]}{0.1273 \left( \pi/4 - \frac{1}{2} \right)} = 0.1895$$

$$\therefore \quad \theta_0 = 10.73°$$

The actuating force is

$$W = \int_{-\pi/4}^{\pi/4} pr\, d\theta\, b\cos\theta + \int_{-\pi/4}^{\pi/4} \mu pr\, d\theta\, b\sin\theta \qquad (c)$$

Substituting Equation (b) into Equation (c) gives

$$W = rbp_{max}\left[\int_{-\pi/4}^{\pi/4}\cos(\theta-\theta_0)\cos\theta\, d\theta + \int_{-\pi/4}^{\pi/4}\mu\cos(\theta-\theta_0)\sin\theta\, d\theta\right] \qquad (d)$$

But

$$\int_{-\pi/4}^{\pi/4}\cos(\theta-\theta_0)\cos\theta\, d\theta = \int_{-\pi/4}^{\pi/4}(\cos\theta\cos\theta_0 + \sin\theta\sin\theta_0)\cos\theta\, d\theta = \left(\frac{\pi}{4}+\frac{1}{2}\right)\cos\theta_0 \qquad (e)$$

and

$$\int_{-\pi/4}^{\pi/4}\cos(\theta-\theta_0)\sin\theta\, d\theta = \left(\frac{\pi}{4}-\frac{1}{2}\right)\sin\theta_0 \qquad (f)$$

Substituting Equations (e) and (f) into Equation (d) while solving for the maximum pressure gives

$$p_{max} = \frac{W}{rb\left[\left(\pi/4+\frac{1}{2}\right)\cos\theta_0 + \mu\left(\pi/4-\frac{1}{2}\right)\sin\theta_0\right]}$$

$$= \frac{11,000}{(0.100)(0.045)\left[\left(\pi/4+\frac{1}{2}\right)\cos 10.73° + 0.31\left(\pi/4-\frac{1}{2}\right)\sin 10.73°\right]}$$

$$= 1.911 \times 10^6\,\text{Pa} = 1.911\,\text{MPa}$$

From Equation (b) the pressure distribution can be expressed as

$$p = (1.911)(10^6)\cos(\theta - 10.73°)$$

The braking torque is

$$T = \mu p_{max} b r^2 \int_{-\pi/4}^{\pi/4}\cos(\theta-\theta_0)\, d\theta$$

$$= \mu p_{max} b r^2\left[\sin\left(\frac{\pi}{4}-\theta_0\right) + \sin\left(\frac{\pi}{4}+\theta_0\right)\right]$$

$$= (0.31)(1.911)(10^6)(0.045)(0.1)^2(\sin 34.27° + \sin 55.73°)$$

$$= 370.4\,\text{N}\cdot\text{m}$$

## 18.8   BAND BRAKES

Figure 18.12 shows a band brake, which consists of a band wrapped partly around a drum. The brake is actuated by pulling the band tighter against the drum, as shown in Figure 18.12(a). The band is assumed to be in contact with the drum over the entire wrap angle $\phi$ in Figure 18.12(a). The pin reaction force is given as $F_1$ and the actuating force as $F_2$. In Figure 18.12(a) the heel of the brake is near $F_1$, and the toe is near $F_2$. Since some friction will exist between the band and the drum, the actuating force will be less than the pin reaction force, or $F_2 < F_1$.

**Figure 18.12**    Band brake. (a) Forces acting on band; (b) forces acting on element.

Figure 18.12(b) shows the forces acting on an element of the band. The forces are the normal force $P$ and the friction force $F$. Summing the forces in the *vertical* (radial) direction while using Figure 18.12(b) gives

$$(F + dF) \sin \frac{d\theta}{2} + F \sin \frac{d\theta}{2} - dP = 0$$

$$dP = 2F \sin \frac{d\theta}{2} + dF \sin \frac{d\theta}{2}$$

Since $dF \ll F$,

$$dP = 2F \sin \frac{d\theta}{2}$$

Since $d\theta/2$ is small, then $\sin d\theta/2 \approx d\theta/2$. Therefore,

$$dP = F\, d\theta \qquad\qquad \text{(18.56)}$$

Summing the forces in the *horizontal* (tangential) direction while using Figure 18.12(b) gives

$$(F + dF) \cos \frac{d\theta}{2} - F \cos \frac{d\theta}{2} - \mu\, dP = 0$$

$$dF \cos \frac{d\theta}{2} - \mu\, dP = 0$$

Since $d\theta/2$ is small, then $\cos(d\theta/2) \approx 1$. Therefore,

$$dF - \mu\, dP = 0 \qquad\qquad \text{(18.57)}$$

Substituting Equation (18.56) into Equation (18.57) gives

$$dF - \mu F d\theta = 0$$

$$\int_{F_2}^{F_1} \frac{dF}{F} = \mu \int_0^\phi d\theta$$

Integrating gives

$$\ln \frac{F_1}{F_2} = \frac{\mu \phi \pi}{180°}$$

$$\frac{F_1}{F_2} = e^{\mu \phi \pi / 180°} \tag{18.58}$$

where $\phi$ = wrap angle, deg.

The torque applied to the drum is

$$T = r(F_1 - F_2) \tag{18.59}$$

The differential normal force $dP$ acting on the element in Figure 18.12(b), with width $b$ (coming out of the paper) and length $r\, d\theta$, is

$$dP = pbr\, d\theta \tag{18.60}$$

where $p$ = contact pressure, Pa. Substituting Equation (18.60) into Equation (18.56) gives

$$p = \frac{F}{br} \tag{18.61}$$

The pressure is proportional to the tension in the band. The maximum pressure occurs at the heel, or near the pin reaction force, and has the value

$$p_{max} = \frac{F_1}{br} \tag{18.62}$$

---

**EXAMPLE 18.7**

**Given:**  The band brake shown in Figure 18.13 has $r = 4$ in, $b = 1$ in, $d_9 = 9$ in, $d_8 = 2$ in, $d_{10} = 0.5$ in, $\phi = 270°$, $\mu = 0.2$, and $p_{max} = 75$ psi.

**Find:**  Determine the following:

  (a)  Brake torque
  (b)  Actuating force
  (c)  Value of $d_{10}$ when the brake force locks

**Solution:**  From Equation (18.62) the pin reaction force is

$$F_1 = p_{max} br = (75)(1)(4) = 300 \text{ lb}$$

From Equation (18.58) the actuating force is

$$F_2 = F_1 e^{-\mu \phi \pi / 180°} = 300 e^{-0.2(270)\pi / 180} = 116.9 \text{ lb}$$

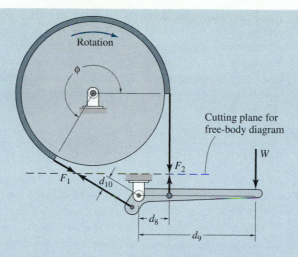

**Figure 18.13**    Band brake used in Example 18.7.

From Equation (18.59) the braking torque is

$$T = r(F_1 - F_2) = 4(300 - 116.9) = 732.4 \text{ lb} \cdot \text{in}$$

Summing the moments about the hinge pin and setting the sum equal to zero give

$$-d_9 W + d_8 F_2 - d_{10} F_1 = 0$$

Solving for the actuating force $W$ gives

$$W = \frac{d_8 F_2 - d_{10} F_1}{d_9} = \frac{2(116.9) - (0.5)(300)}{9} = 9.311 \text{ lb}$$

If $W = 0$, the brake will self-lock.

$$d_8 F_2 - d_{10} F_1 = 0$$

$$d_{10} = \frac{d_8 F_2}{F_1} = \frac{2(116.9)}{300} = 0.7793 \text{ in}$$

The brake will self-lock if $d_{10} \geq 0.7793$ in.

## 18.9    SLIP CLUTCHES

A clutch will often be used as a torque-limiting device, usually to prevent machinery damage in a malfunction or undesired event. A **slip clutch** consists of two surfaces held together by a constant force so that they slip when a preset level of torque is applied to them. Slip clutches come in a wide variety of sizes but are very compact. They are designed to be actuated only rarely, and thus the friction elements do not need to be sized for wear. Also, slip clutches are almost always contacting disks, mainly because it is imperative to prevent the possibility of a self-energizing shoe (which would compromise the torque-limiting control).

Slip clutches are mainly used to protect machinery elements and are not relied on for accident prevention. After slip clutches spend long time periods in contact, the friction surfaces can stick or weld together, requiring a larger torque to initiate slippage. This torque is usually not large enough to break a gear, for example, but can be enough of an increase to result in a serious injury. Regardless, a slip clutch is an effective torque limiter and can be used in lieu of shear pins or keyways with the advantage that no maintenance is required after the excessive torque condition has been corrected.

## 18.10    TEMPERATURE CONSIDERATIONS

As mentioned in Section 18.1, thermal effects are important in braking and clutch systems. If temperatures become too high, damage to components could result, which could compromise the useful life or performance of brake and clutch systems.

Predicting temperatures of brake and clutch systems is extremely difficult in practice because they are operated under widely varying conditions. The first law of thermodynamics requires that

$$Q_{\text{friction}} = Q_{\text{conduction}} + Q_{\text{convection}} + Q_{\text{storage}}$$

where $Q_{\text{friction}}$ = energy input into brake or clutch system from friction between sliding elements

$Q_{\text{conduction}}$ = heat lost from conduction through machinery elements

$Q_{\text{convection}}$ = heat lost from convection to surrounding environment

$Q_{\text{storage}}$ = energy stored in brake and clutch components, resulting in temperature increase

If conduction and convection are negligible, the temperature rise in the brake or clutch material is given by

$$\Delta t_m = \frac{Q_{\text{friction}}}{C_p m_a}$$

where $C_p$ = specific heat of material, J/(kg · °C)

$m_a$ = mass, kg

This equation is useful for determining the instantaneous temperature rise in a brake or clutch pad, since the frictional energy is dissipated directly on the contacting surfaces and does not have time to be conducted or conveyed. Brake pads and clutches usually have an area in contact that then moves out of contact and can cool. The maximum operating temperature is a complicated function of heat input and cooling rates.

The main difficulty in predicting brake system temperatures is that the heat conducted and the heat convected depend on the machine ambient temperature and the brake or clutch geometry and vary widely. In previous circumstances in this text a worst-case analysis would be performed, which in this case quickly reduces to circumstances where brakes and clutches become obviously overheated. This result is not incorrect: most brake and clutch systems are overheated when abused and can sustain serious damage as a result. The alternative is to use such massive brake and clutch systems as to make the economic

**Table 18.3** Product of contact pressure and sliding velocity for brakes and clutches

| Operating condition | *pu* | |
|---|---|---|
| | (kPa)(m/s) | (psi)(ft/min) |
| Continuous: poor heat dissipation | 1050 | 30,000 |
| Occasional: poor heat dissipation | 2100 | 60,000 |
| Continuous: good heat dissipation as in oil bath | 3000 | 85,000 |

I SOURCE: From Juvinall and Marshek (1991).

burden unbearable to responsible users. It is far more reasonable to use brake systems that require periodic maintenance and can be damaged through abuse than to incur the economic costs of surviving worst-case analyses. This differs from previous circumstances, where a worst-case analysis still resulted in a reasonable product.

Some clutches are intended to be used with a fluid (wet clutches) to aid in cooling the clutch. Similarly, some pads or shoes will include grooves so that air or fluid can be better entrained, resulting in increased heat convection from the clutch or brake. Predicting brake temperatures is a complex problem and requires numerical, usually finite element, methods.

Obviously, the proper size of brake components is extremely difficult to determine with certainty. For the purposes of this text the values given by Juvinall and Marshek (1991) for the product of brake shoe or pad contact pressure and sliding velocity *pu* can be used to estimate component sizes (Table 18.3). Most manufacturers rely heavily on experimental verification of designs; the application of these numbers in the absence of experimental verification requires extreme caution.

---

**Case Study 18.1** | SELECTION OF BRAKE FOR MOBILE HYDRAULIC CRANE

**Given:** Mobile hydraulic cranes are extremely common and valuable construction and material-handling vehicles. This case study illustrates the selection of a brake for one of the crane's functions. A crane can perform a number of different functions, such as hoisting (by reeling in the hoist line), elevating or lowering the boom (by reeling in the gantry line), and rotating. Each of these functions requires its own control system (i.e., brakes and clutches). This case study is restricted to a tractable problem—choosing a brake for the hoist line.

Crane capacities are given in terms of the maximum load they can lift while hoisting, which occurs when the boom is at maximum elevation. At lower elevations the crane will tip over at much lower loads, so the crane capacity is the true maximum load which will be experienced by the crane. This problem will deal with clutch selection for a 30-ton (60,000-lb) crane, which has a 12-in-diameter drum for the cable. The drum diameter has been chosen

to give reasonable wire rope service life, a topic further explored in Chapter 19.

**Find:** Select a brake for the hoist line.

**Solution:** A large crane achieves its full capacity only if multiple-part hoist lines are used. That is, the hoist line is run through multiple sheaves so that each part supports a fraction of the total load. A 30-ton crane may or may not use multiple-part hoist lines. A worst-case analysis suggests that the brake must withstand the entire hoist line force.

It is imperative that the crane be able to hold a load that has been hoisted. Also, dynamic loads are possible. For example, in dam construction wooden forms are placed, concrete is poured, and the forms are then removed by attaching the crane's hoist line and prying the forms

*(continued)*

**Case Study**　　　(Continued)

off the now-cured concrete. If the hoist line has insufficient tension, the form will drop a certain distance until the wire rope is taut, thereby causing a dynamic load to be applied to the hoist line. For this reason an application factor of 1.5 will be used. In most cases this allows for large safety factors, since the hoist line is rarely supporting the crane's capacity because of tipping as discussed.

The type of brake to be used is a long-shoe, internal, expanding rim type. A brake used for a swing mechanism on a similar crane is shown in the chapter's opening illustration. The brake used for this application is similar and is sketched in Figure 18.14. The reasons for using such a brake are multiple and include the following:

1. Space is really not an issue; the crane superstructure has sufficient room for any kind of brake discussed in this chapter.

2. To maximize maintenance-free service life, drum or disk brakes are the most obvious candidates. This statement is not meant to imply that long lives are not possible from band brakes or the like, merely that drum or disk brakes have long lives with easily replaceable shoes and pads.

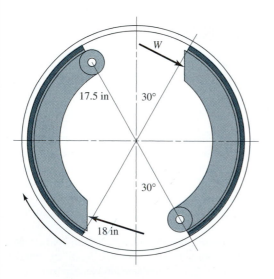

**Figure 18.14**　　Hoist line brake for mobile hydraulic crane. The cross section of brake with relevant dimensions is given.

3. Self-actuating behavior is desirable in this case, so a drum brake with an internal shoe is chosen.

4. A long-shoe brake efficiently uses the braking material, since little pressure occurs near the hinge pin.

Although a crane manufacturer will occasionally manufacture its own brake components, usually design work is performed in collaboration with a component supplier. The sizes of components are selected based on previous design experience for crane applications and on stress analyses using relations such as the Archard wear law. Such a design synthesis problem has been discussed in previous case studies.

In this design a mechanical spring normally would apply the brakes, and the actuating cylinder would force the shoes away from the drum. Thus, in actuating the brakes, an operator really is disengaging the hydraulic cylinder. The reason for this approach is that it is imperative that loads not drop in the event of a power failure or if all the crane's fuel is consumed. The brake is a fail-safe system in this respect. Further, through proper linkage design both shoes can be self-energizing, limiting the required force multiplication in the hydraulic actuation system.

The design problem becomes one of specifying the actuating force required to obtain the desired braking torque on the cable drum. This actuating force will then be achieved through proper hydraulic multiplication of the operator's exertions on the hoist control. The required torque is given by

$$T = 1.5F\left(\frac{D}{2}\right) = 1.5(60,000)(6) = 540,000 \text{ lb} \cdot \text{in}$$

Because both shoes are self-energizing, each shoe must develop one-half the required torque. Using Equation (18.38) gives

$$270,000 = \frac{\mu p_{max} b r^2 (\cos \theta_1 - \cos \theta_2)}{\sin \theta_a}$$

From the defined geometry the maximum pressure occurs at $\theta_a = 90°$ where $\theta_1 = 30°$ and $\theta_2 = 150°$. Similar cranes have brake shoes approximately 8 in wide, so $b = 8$ in. The drum radius has been specified as 18 in, although in a general design problem this dimension could be altered to ensure proper performance. Table 18.1 gives the coefficients of friction and the maximum temperatures and

**Case Study** (Concluded)

pressures for a wide variety of friction materials. For applications such as this mobile crane sintered powder metal brake components are usually used because they allow the largest pressures and are most forgiving in terms of local heating. Therefore, recognizing that the brake lining is a sintered metal, $\mu = 0.25$ will be used. The coefficient of friction can be smaller than 0.25, but for the brake components used in craning this is a reasonable lower bound. Equation (18.38) then yields, upon rearranging,

$$p_{max} = \frac{270{,}000}{(0.25)(8)(18)^2(\cos 30° - \cos 150°)} = 240.6 \text{ psi}$$

The maximum pressure is also a reasonable value, as can be seen from Table 18.1. The actuating force is given by

Equation (18.37). Thus,

$$M_P = \frac{br\,d_7\,p_{max}}{4}\left[2(\theta_2 - \theta_1)\frac{\pi}{180°} - \sin 2\theta_2 + \sin 2\theta_1\right]$$

$$= 87{,}500 \text{ lb} \cdot \text{in}$$

$$M_F = \mu p_{max} br\left[-r(\cos\theta_2 - \cos\theta_1) - \frac{d_7}{2}\left(\sin^2\theta_2 - \sin^2\theta_1\right)\right]$$

$$= 270{,}000 \text{ lb} \cdot \text{in}$$

$$W = \frac{M_P - M_F}{d_6} = 20{,}700 \text{ lb}$$

This large load obviously requires force multiplication; in addition to hydraulic actuation systems linkages are used to obtain the required forces. Also, the design value is far in excess of the requirements for all but the rarest applications.

## 18.11  SUMMARY

This chapter focused on two machine elements, clutches and brakes, that are associated with rotation and have the common function of storing and/or transferring rotating energy. In analyzing the performance of clutches and brakes, the actuating force, the torque transmitted, and the reaction force at the hinge pin were the major focuses of this chapter. The torque transmitted is related to the actuating force, the coefficient of friction, and the geometry of the clutch or brake. This is a problem in statics, where different geometries were studied separately.

For clutches two theories were studied: the uniform pressure model and the uniform wear model. It was found that for the same dimensionless torque, the uniform wear model requires a larger radius ratio than does the uniform pressure model for the same maximum pressure. This larger radius ratio implies that a larger area is needed for the uniform wear model. Thus, the uniform wear model was viewed as a safer approach.

## Key Words

**band brake**   brake that uses contact pressure of flexible band against outer surface of drum.

**brake**   device used to bring moving system to rest through dissipation of energy to heat by friction.

**clutch**   power transfer device that allows coupling and decoupling of shafts.

**cone disk**   brake or clutch that uses shoes pressed against convergent surface of cone.

**deenergizing**   brake or clutch shoe where frictional moment hinders engagement.

**rim type**   brake or clutch that uses internal shoes which expand onto inner surface of drum.

**self-energizing**   brake or clutch shoe where frictional moment assists engagement.

**slip clutch**   clutch where maximum transferred torque is limited.

**thrust disk**   brake or clutch that uses flat shoes pushed against rotating disk.

---

## RECOMMENDED READINGS

Army (1976) *Analysis and Design of Automotive Brake Systems,* U.S. Department of the Army
   Manual DARCOM–P706–358, Alexandria, VA.

Baker, A. K. (1986) *Vehicle Braking,* Pentech Press, London.

Crouse, W. H., and Anglin, D. L. (1983) *Automotive Brakes, Suspension and Steering,* 6th ed.,
   McGraw-Hill, New York.

Juvinall, R. C., and Marshek, K. M. (2003) *Fundamentals of Machine Component Design,* 3rd ed.,
   Wiley, New York.

Krutz, G. W., Schuelle, J. K, and Claar, P. W. (1994) *Machine Design for Mobile and Industrial
   Applications,* Society of Automotive Engineers, Warrendale, PA.

Monroe, T. (1977) *Clutch and Flywheel Handbook,* H. P. Books, Tucson, AZ.

Mott, R. L. (1998) *Machine Elements in Mechanical Design,* 3rd ed., Prentice Hall, Upper Saddle
   River, NJ.

Norton, R. L. (2000) *Machine Design,* 2nd ed., Prentice Hall, Englewood Cliffs, NJ.

Orthwein, W. C. (1986) *Clutches and Brakes: Design and Selection,* Marcel Dekker, New York.

Shigley, J. E., Mischke, C. R., and Budynas, R. G. (2003), *Mechanical Engineering Design,* 7th ed.,
   McGraw-Hill, New York.

---

## REFERENCES

Juvinall, R. C., and Marshek, K. M. (1991) *Fundamentals of Machine Component Design,* Wiley,
   New York.

---

## PROBLEMS

### Section 18.2

**18.1**   The disk brake shown in sketch *a* has brake pads in the form of circular sections with inner
   radius $r$, outer radius $2r$, and section angle $\pi/4$. Calculate the brake torque when the pads
   are applied with normal force $P$. The brake is worn in so that $pu$ is constant, where $p$ is
   the contact pressure and $u$ is the sliding velocity. The coefficient of friction is $\mu$. *Ans.* $T = 3\mu rP$.

Sketch a, for Problem 18.1

**18.2** A plate clutch has a single pair of mating friction surfaces with 300-mm outer diameter by 225-mm inner diameter. The mean value of the coefficient of friction is 0.25, and the actuating force is 5 kN. Find the maximum pressure and torque capacity when the clutch is new, as well as after a sufficiently long run-in period. *Ans.* $T_p = 165$ N · m, $p_{max,p} = 0.162$ MPa, $T_w = 164$ N · m.

★ **18.3** An automotive clutch with a single friction surface is to be designed with a maximum torque of 140 N · m. The materials are chosen such that $\mu = 0.35$ and $p_{max} = 0.35$ MPa. Use safety factor $n_s = 1.3$ with respect to slippage at full engine torque and as small an outside diameter as possible. Determine appropriate values of $r_o$, $r_i$, and $P$ by using both the uniform pressure and the uniform wear models. *Ans.* $P_p = 8743$ N, $P_w = 6160$ N.

**18.4** The brakes used to stop and turn a tank are built like a multiple-disk clutch with three loose disks connected through splines to the drive shaft and four flat rings connected to the frame of the tank. The brake has an outer contact diameter of 600 mm, an inner contact diameter of 300 mm, and six surface contacts. The wear of the disks is proportional to the contact pressure multiplied by the sliding distance. The coefficient of friction of the brake is 0.12, and the friction between the caterpillar and the ground is 0.16, which gives a braking torque of 12,800 N · m needed to block one caterpillar track so that it slides along the ground. Calculate the force needed to press the brake disks together to block one caterpillar track. Also calculate the force when the brake is new. *Ans.* $P_w = 79$ kN, $P_p = 76.2$ kN.

Sketch b, for Problem 18.5

| * Indicates problem of greater difficulty.

**★ 18.5**  A disk brake used in a printing machine is designed as shown in sketch $b$. The brake pad is mounted on an arm that can swivel around point 0. Calculate the braking torque when the force $P = 5000$ N. The friction pad is a circular sector with the inner radius equal to one-half the outer radius. Also, $a = 150$ mm, $b = 50$ mm, $D = 300$ mm, and $\mu = 0.25$. The wear of the brake lining is proportional to the pressure and the sliding distance.

**★ 18.6**  A disk brake for a flywheel is designed as shown in sketch $c$. The hydraulic pistons actuating the brake need to be placed at a radius $r_p$ so that the brake pads wear evenly over the entire contact surface. Calculate the actuating force $P$ and the radius so that the flywheel can be stopped within 4 s when it rotates at 1000 rpm and has a kinetic energy of $5 \times 10^5$ N · m. The input parameters are $\mu = 0.3$, $\alpha = 30°$, $r_o = 120$ mm, $r_i = 60$ mm. *Ans. $P = 44.3$ kN, $r_p = 0.089$ m.*

Sketch $c$, for Problem 18.6

**18.7**  Three pairs of thrust disk clutches are mounted on a shaft. Each has a pair of frictional surfaces. The hardened-steel clutches are identical, with an inside diameter of 100 mm and an outside diameter of 245 mm. What is the torque capacity of these clutches based on a) uniform wear and b) uniform pressure? *Ans. $T_w = 2839$ N · m, $T_p = 5167$ N · m.*

**18.8**  A pair of disk clutches has an inside diameter of 250 mm and an outside diameter of 420 mm. A normal force of 18.5 kN is applied, and the coefficient of friction of the contacting surfaces is 0.215. Using the uniform wear and uniform pressure assumptions, determine the maximum pressure acting on the clutches. Which of these assumptions would produce results closer to reality? *Ans. $p_{\max, p} = 206.8$ kPa, $p_{\max, w} = 277$ kPa.*

**★ 18.9**  A disk clutch is made of cast iron and has a maximum torque of 210 N · m. Because of space limitations the outside diameter must be minimized. Using the uniform wear assumption and a safety factor of 1.3, determine

*a)*   The inner and outer radii of the clutch. *Ans. $r_i = 0.068$ m, $r_o = 0.118$ m.*
*b)*   The maximum actuating force needed. *Ans. $P_w = 11.3$ kN.*

**18.10** A disk clutch produces a torque of 125 N · m and a maximum pressure of 315 kPa. The coefficient of friction of the contacting surfaces is 0.28. Assume a safety factor of 1.8 for maximum pressure, and design the smallest disk clutch for the above constraints. What should the normal force be? *Ans. $P_w = 4.39$ kN.*

## Section 18.3

**18.11** A leather-face cone clutch is to transmit 1200 lb · in of torque. The half-cone angle $\alpha = 10°$, the maximum diameter of the friction surface is 12 in, and face width $b = 2$ in. For coefficient of friction $\mu = 0.25$, find the normal force $P$ and the maximum contact pressure $p$ by using both the uniform pressure and the uniform wear models. *Ans. $P_p = 139$ lb, $p_{max,p} = 10.5$ psi, $P_w = 140$ lb, $p_{max,w} = 10.9$ psi.*

**18.12** The synchronization clutch for the second gear of a car has a major cone diameter of 50 mm and a minor diameter of 40 mm. When the stick shift is moved to second gear, the synchronized clutch is engaged with an axial force of 100 N, and the moment of inertia of 0.005 kg · m² is accelerated 200 rad/s² in 1 s to make it possible to engage the gear. The coefficient of friction of the cone clutch is 0.09. Determine the smallest cone clutch width that still gives large enough torque. Assume the clutches are worn in. *Ans. $b = 24.2$ mm.*

**18.13** A safety brake for an elevator is a self-locking cone clutch. The minor diameter is 120 mm, the width is 60 mm, and the major diameter is 130 mm. The force applying the brake comes from a prestressed spring. Calculate the spring force needed if the 2-ton elevator must stop from a speed of 3 m/s in a maximum distance of 3 m while the cone clutch rotates 5 r/m of elevator motion. The coefficient of friction in the cone clutch is 0.26. *Ans. $P = 3678$ N.*

**★ 18.14** A cone clutch is used in a car automatic gearbox to fix the planet wheel carriers to the gearbox housing when the gear is reversing. The car weighs 1300 kg with 53% loading on the front wheels. The gear ratio from the driven front wheels to the reversing clutch is 16.3:1 (i.e., the torque on the clutch is 16.3 times lower than the torque on the wheels if all friction losses are neglected). The car wheel diameter is 550 mm, the cone clutch major diameter is 85 mm and the minor diameter is 80 mm, and the coefficients of friction are 0.3 in the clutch and 1.0 between the wheel and the ground. Dimension the width of the cone clutch so that it is not self-locking. Calculate the axial force needed when the clutch is worn in. *Ans. $b < 0.0083$ m, $P = 2647$ N.*

**18.15** A cone clutch has a major diameter of 328 mm and a minor diameter of 310 mm, is 50 mm wide, and transfers 250 N · m of torque. The coefficient of friction is 0.31. Using the assumptions of uniform pressure and uniform wear, determine the actuating force and the contact pressure. *Ans. $P_p = 895$ N, $p_{max,p} = 99.3$ kPa.*

**★ 18.16** The coefficient of friction of a cone clutch is 0.25. It can support a maximum pressure of 410 kPa while transferring a maximum torque of 280 N · m. The width of the clutch is 65 mm. Minimize the major diameter of the clutch. Determine the clutch dimensions and the actuating force.

## Section 18.4

**18.17** A block brake is used to stop and hold a rope used to transport skiers from a valley to the top of a mountain. The distance between cars used to transport the skiers is 100 m, the length of the rope from the valley to the top of the mountain is 4 km, and the altitude difference is 1.4 km. The rope is driven by a V-groove wheel with a diameter of 2 m. The rope is stopped and held with a block brake mounted on the shaft of the V-groove wheel, shown in sketch *d*. Neglect all friction in the different parts of the ropeway except the friction in the driving sheave, and assume that the slope of the mountain is constant. Dimension the brake for 20 passengers with each passenger's equipment weighing 100 kg, and assume that all ropeway cars descending from the top are empty of people. The direction of rotation for the drive motor is shown in the figure. Calculate the braking force $W$ needed to hold the ropeway still if all passengers are on their way up. Do the same calculation if all passengers continue on down to the valley with the ropeway. The coefficient of friction in the brake is 0.23. *Ans.* $P_{up} = 2.388$ MN, $P_{down} = 857$ kN.

Sketch d, for Problem 18.17

**18.18** The motion of an elevator is controlled by an electric motor and a block brake. On one side the rotating shaft of the electric motor is connected to the gearbox driving the elevator, and on the other side it is connected to the block brake. The motor has two magnetic poles and can be run on either 60- or 50-Hz electricity (3600 or 3000 rpm). When the elevator motor is driven by 50-Hz electricity, the braking distance needed to stop is 52 cm when going down and 31 cm when going up with maximum load in the elevator. To use it with 60-Hz electricity and still be able to stop exactly at the different floor levels without changing the electric switch positions, the brake force at the motor should be changed. How should it be changed for going up and for going down? The brake geometry is like that shown in Figure 18.5 with $d_1 = 0.030$ m, $d_3 = 0.100$ m, $d_4 = 0.400$ m, $r = 0.120$ m, and $\mu = 0.20$. Only the inertia of the elevator must be considered, not the rotating parts. Make the brake self-energizing when the elevator is going down.

**18.19** An anchor winding is driven by an oil hydraulic motor with a short-shoe brake to stop the anchor machinery from rotating and letting out too much anchor chain when the wind moves the ship. The maximum force transmitted from the anchor through the chain is 1.1 MN at a radius of 2 m. Figure 18.5 describes the type of block brake used, which is self-energizing. Calculate the brake force $W$ needed when the the brake dimensions are $d_1 = 0.9$ m, $d_3 = 1.0$ m, $d_4 = 6$ m, $r = 3$ m, and $\mu = 0.31$. Also calculate the contact force between the brake shoe and the drum. *Ans.* $W = 284$ kN, $P = 2.36$ MN.

**18.20** The hand brake shown in sketch *e* has an average pressure of 600 kPa across the shoe and is 50 mm wide. The wheel runs at 150 rpm, and the coefficient of friction is 0.25. Dimensions are in millimeters. Determine the following:

a)    If $x = 150$ mm, what should the actuating force be? *Ans.* $P_n = 2.97$ kN.

Sketch e, for Problem 18.20

b)    What value of $x$ causes self-locking? *Ans.* $x = 2.0$ m.
c)    What torque is transferred? *Ans.* $T = 303$ N · m.
d)    If the direction of rotation is reversed, how will the answers to parts *a* to *c* change?

**18.21** The short-shoe brake shown in sketch $f$ has an average pressure of 1 MPa and a coefficient of friction of 0.32. The shoe is 250 mm long and 45 mm wide. The drum rotates at 310 rpm and has a diameter of 550 mm. Dimensions are in millimeters.

a)    Obtain the value of $x$ for the self-locking condition. *Ans.* $x = 1.81$ m.
b)    Calculate the actuating force if $x = 275$ mm. *Ans.* $P = 3982$ N.
c)    Calculate the braking torque. *Ans.* $T = 3.44$ kN · m.
d)    Calculate the reaction at point A.

Sketch $f$, for Problem 18.21

## Section 18.5

**18.22** Sketch $g$ shows a long-shoe, internal expanding shoe brake. The inside rim diameter is 280 mm, and the dimension $d_7$ is 90 mm. The shoes have a face width of 30 mm.

a)    Find the braking torque and the maximum pressure for each shoe if the actuating force is 1000 N, the drum rotation is counterclockwise, and the coefficient of friction is 0.30. *Ans.* $T = 218.7$ N · m.
b)    Find the braking torque if the drum rotation is clockwise.

Sketch *g*, for Problem 18.22.

**18.23** The brake on the rear wheel of a car is the long-shoe internal type. The brake dimensions according to Figure 18.7 are $\theta_1 = 10°$, $\theta_2 = 120°$, $r = 95$ mm, $d_7 = 73$ mm, $d_6 = 120$ mm, and $d_5 = 30$ mm. The brake shoe lining is 38 mm wide, and the maximum allowable contact pressure is 5 MPa. Calculate the braking torque and the fraction of the torque produced from each brake shoe when the brake force is 5000 N. Also calculate the safety factor against contact pressure that is too high. The coefficient of friction is 0.29. *Ans.* $T_{\text{tot}} = 612 \text{ N} \cdot \text{m}$, $T_{se} = 416 \text{ N} \cdot \text{m}$.

**18.24** Sketch *h* shows a 450-mm-inner-diameter brake drum with four internally expanding shoes. Each of the hinge pins A and B supports a pair of shoes. The actuating mechanism is a linkage which produces the same actuating force $W$ on each shoe as shown. The lining material is

Sketch *h*, for Problem 18.24.

to be sintered metal. The face width of each shoe is 70 mm. Using the midrange values of friction coefficient and contact pressure,

*a)*     Identify the self-energizing and deenergizing shoes and label them.
*b)*     Determine the maximum pressure on the deenergizing shoe.
*c)*     Determine the maximum force $W$ and the braking torque $T$ at this force.

★ **18.25** The maximum volume of the long-shoe internal brake on a car is given as $10^{-3}$ m$^3$. The brake should have two equal shoes, one self-energizing and one deenergizing, so that the brake can fit on both the right and the left sides of the car. Calculate the brake width and radius for maximum braking power if the space available inside the wheel is 400 mm in diameter and 100 mm wide. The brake lining material has a maximum allowed contact pressure of 4 MPa and a coefficient of friction of 0.18. Also calculate the maximum braking torque. *Ans. $T_{max} = 730$ N · m.*

**18.26** A long-shoe brake in a car is designed to give as high a braking torque as possible for a given force on the brake pedal. The ratio between the actuating force and the pedal force is given by the hydraulic area ratio between the actuating cylinder and the cylinder under the pedal. The brake shoe angles are $\theta_1 = 10°$, $\theta_2 = 170°$, and $\theta_a = 90°$. The maximum brake shoe pressure is 5 MPa, the brake shoe width is 40 mm, and the drum radius is 100 mm. Find the distance $d_7$ that gives the maximum braking power for a coefficient of friction of 0.2 at any pedal force. What braking torque would result if the coefficient of friction were 0.25?

### Section 18.6

**18.27** An external drum brake assembly (see sketch $i$) has a normal force $P = 200$ lb acting on the lever. Dimensions are in inches. Assume that coefficient of friction $\mu = 0.25$ and maximum contact pressure $p = 100$ psi. Determine the following from long-shoe calculations:

*a)*     Free-body diagram with the directionality of the forces acting on each component.
*b)*     Which shoe is self-energizing and which is deenergizing.
*c)*     Total braking torque. *Ans. $T = 20.45$ kip · in.*
*d)*     Pad width as obtained from the self-energizing shoe (deenergizing shoe width equals self-energizing shoe width). *Ans. $b = 1.26$ in.*
*e)*     Pressure acting on the deenergizing shoe.

Sketch $i$, for Problem 18.27

**18.28** A long-shoe external brake as shown in Figure 18.10 has a pivot point such that $d_7 = 4r$, $d_6 = 2r$, $\theta_1 = 5°$, and $\theta_2 = 45°$. Find the coefficient of friction needed to make the brake self-lock if the rotation is in the direction shown in Figure 18.10. If the shaft rotates in the opposite direction, calculate the drum radius needed to get a braking torque of 180 N · m for the actuating force of 10,000 N. *Ans.* $r = 43.32$ mm.

**18.29** An external, long brake shoe is mounted on an elastic arm. When a load is applied, the arm and the brake lining bend and redistribute the pressure. Instead of the normal sine pressure distribution, the pressure becomes constant along the length of the lining. For a given actuating force, calculate how the brake torque changes when the pressure distribution changes from sinusoidal to a constant pressure. Also assume that $d_7 = 110$ mm, $r = 90$ mm, $b = 40$ mm, $\theta_1 = 20°$, $\theta_2 = 160°$, $d_6 = 220$ mm, $\mu = 0.25$, and $W = 12$ kN.

★ **18.30** A long-shoe external brake has two identical shoes coupled in series so that the peripheral force from the first shoe is directly transferred to the second shoe. No radial force is transmitted between the shoes. Each of the two shoes covers 90° of the circumference, and the brake linings cover the central 70° of each shoe, leaving 10° at each end without lining, as shown in sketch $j$. The actuating force is applied tangentially to the brake drum at the end of the loose shoe, 180° from the fixed hinge point of the other shoe. Calculate the braking torques for both rotational directions when $d_7 = 150$ mm, $r = 125$ mm, $b = 50$ mm, $W = 14,000$ N, and $\mu = 0.2$. Also show a free-body diagram of these forces acting on the two shoes. *Ans.* $T = 717$ N · m. If reversed, $T = 1197$ N · m.

Sketch $j$, for Problem 18.30

**18.31** A special type of brake is used in a car factory to hold the steel panels during drilling operations so that the forces from the drill bits cannot move the panels. The brake is shown in sketch $k$. Calculate the braking force $P_B$ on the steel panel when it moves to the right with the speed $u_b$

between drilling operations. The actuating force is $P_M$. The brake lining is thin relative to the other dimensions. *Ans.* $P_b = \dfrac{9}{7}\mu P_M$.

Sketch k, for Problem 18.31

**18.32** Redo Problem 18.20 for long-shoe assumptions. The average contact pressure occurs at 40°. Determine the maximum contact pressure and its location. Assume that the distance $x$ is 150 mm. What is the braking torque? Also, repeat this problem while reversing the direction of rotation. Discuss the changes in the results.

## Section 18.7

**18.33** A symmetrically loaded, pivot-shoe brake has a wrap angle of 180° and the optimum distance $d_7$, giving a symmetric pressure distribution. The coefficient of friction of the brake lining is 0.30. A redesign is considered that will increase the breaking torque without increasing the actuating force. The wrap angle is decreased to 80° (+40° to −40°), and the $d_7$ distance is decreased to still give a symmetric pressure distribution. How much does the brake torque change? *Ans.* 15% decrease.

## Section 18.8

**18.34** The band brake shown in sketch *l* is activated by a compressed-air cylinder with diameter $d_c$. The brake cylinder is driven by air pressure $p = 0.7$ MPa. Calculate the maximum possible brake torque if the coefficient of friction between the band and the drum is 0.25. The mass force on the brake arm is neglected, $d_c = 50$ mm, $r = 200$ mm, $l_1 = 500$ mm, $l_2 = 200$ mm, and $l_3 = 500$ mm. *Ans.* $T = 983$ N · m.

Sketch *l*, for Problem 18.34

**18.35** The band brake shown in sketch *m* has wrap angle $\phi = 225°$ and cylinder radius $r = 80$ mm. Calculate the brake torque when the lever is loaded by 100 N and coefficient of friction $\mu = 0.3$. How long is the braking time from 1200 rpm if the rotor mass moment of inertia is 2.5 kg · m²? *Ans.* $T = 54$ N · m, $t = 5.825s$.

Sketch *m*, for Problem 18.35

**18.36** The band brake shown in sketch *n* has wrap angle $\phi = 215°$ and cylinder radius $r = 60$ mm. Calculate the brake torque when coefficient of friction $\mu = 0.25$. How long is the braking time from 1500 rpm if the rotor moment of inertia is $J = 2$ kg · m²?

Sketch n, for Problem 18.36

**18.37** A brake (see sketch $o$) consists of a drum with a brake shoe pressing against it. Drum radius $r = 80$ mm. Calculate the brake torque when $P = 7000$ N, $\mu = 0.35$, and brake pad width $b = 40$ mm. The wear is proportional to the contact pressure times the sliding distance.

Sketch o, for Problem 18.37

★ **18.38** For the band brake shown in sketch $p$, the following conditions are given: $d = 350$ mm, $p_{max} = 1.2$ MPa, $\mu = 0.25$, and $b = 50$ mm. All dimensions are in millimeters. Determine the following:

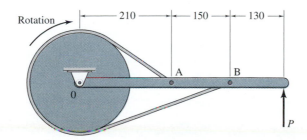

Sketch p, for Problem 18.38

- a)     The braking torque
- b)     The actuating force
- c)     The forces acting at hinge 0

**18.39** The band brake shown in sketch $q$ is 40 mm wide and can take a maximum pressure of 1.1 MPa. All dimensions are in millimeters. The coefficient of friction is 0.3. Determine the following:

- a)     The maximum allowable actuating force.
- b)     The braking torque. *Ans. T* = 307 Nm.
- c)     The reaction supports at $0_1$ and $0_2$.
- d)     Whether it is possible to change the distance $0_1 A$ in order to have self-locking. Assume point A can be anywhere on line $C0_1 A$.

Sketch *q*, for Problem 18.39

# FLEXIBLE MACHINE ELEMENTS

An assortment of chains and belts and their associated sprockets and pulleys. (Courtesy of Sterling Instrument Co.)

*Scientists study the world as it is, engineers create the world that never has been.*

Theodore von Karman

## SYMBOLS

| | |
|---|---|
| $A$ | $\dfrac{L}{p_t} - \dfrac{N_1 + N_2}{2}$ |
| $A_m$ | cross-sectional area of metal strand in rope, m$^2$ |
| $a$ | link plate thickness, m |
| $a_1$ | service factor |
| $a_2$ | multiple-strand factor in rolling chain |
| $B$ | $(N_2 - N_1)/(2\pi)$ |
| $c$ | distance from neutral axis to outer fiber, m |
| $c_d$ | center distance, m |
| $D$ | sheave or pulley diameter, m |
| $d$ | diameter, m |
| $d_w$ | wire diameter, m |
| $E$ | modulus of elasticity, Pa |
| $F$ | friction force, N |
| $F_a$ | force due to acceleration, N |
| $F_c$ | centrifugal force, N |
| $F_f$ | fatigue allowable force, N |
| $F_i$ | initial tensile force, N |
| $F_r$ | rope weight, N |
| $F_t$ | total friction force, N |
| $F_w$ | deadweight, N |
| $F_1$ | driver friction force, N |
| $F_2$ | driven friction force, N |
| $f_1$ | overload service factor |
| $f_2$ | extra overload factor |
| $G$ | gantry line tension, N |
| $g$ | gravitational acceleration, 9.807 m/s$^2$ = 32.2 ft/s$^2$ |
| $g_r$ | velocity ratio |
| $H_B$ | Brinell hardness |
| $h_p$ | power transmitted, W |
| $h_{pr}$ | rated power transmitted, W |
| $h_t$ | V-belt height, m |
| $I$ | area moment of inertia, m$^4$ |
| $L$ | belt length, m |
| $l_p$ | hoist rope length, m |
| $M$ | bending moment, N · m |

| | |
|---|---|
| $m$ | mass, kg |
| $m'$ | mass per unit length, kg/m |
| $N$ | number of teeth in sprocket |
| $N_a$ | speed of revolution, rpm |
| $n_s$ | safety factor |
| $p$ | bearing pressure, Pa |
| $p_{all}$ | allowable bearing pressure, Pa |
| $p_t$ | pitch, m |
| $r$ | sheave or pulley radius, m |
| $\Delta r$ | chordal rise, m |
| $r_c$ | chordal radius, m |
| $S_u$ | ultimate strength, Pa |
| $T$ | torque, N · m |
| $t$ | thickness, m |
| $u$ | belt velocity, m/s |
| $w_t$ | belt width, m |
| $w_z$ | weight, N |
| $w_z'$ | weight per unit length, N/m |
| $\alpha$ | angle used to describe loss in arc of contact, deg |
| $\beta$ | sheave or pulley angle, deg |
| $\zeta$ | mast angle, deg |
| $\theta$ | boom angle, deg |
| $\theta_r$ | angle of rotation to give chordal rise, deg |
| $\mu$ | coefficient of friction |
| $\sigma$ | normal stress, Pa |
| $\sigma_{all}$ | allowable normal stress, Pa |
| $\sigma_b$ | bending stress, Pa |
| $\sigma_t$ | tensile normal stress, Pa |
| $\phi$ | wrap angle; gantry line angle, deg |
| $\psi$ | angle from horizontal; gantry line system angle, deg |
| $\omega$ | angular velocity, s$^{-1}$ |

## SUBSCRIPTS

| | |
|---|---|
| 1 | driver pulley or sheave |
| 2 | driven pulley or sheave |

## 19.1    INTRODUCTION

The machine elements considered in this chapter, like the machine elements considered in Chapter 18, use friction as a useful agent to produce a high and uniform force and to transmit power. A belt, rope, or chain provides a convenient means for transferring power from one shaft to another. Belts are frequently necessary to reduce the higher rotative speeds of electric motors to the lower values required by mechanical equipment. Chains are effective in transferring power between parallel shafts. The design of belts and ropes is subject to uncertainties in the value of the coefficient of friction that must be used.

## 19.2    FLAT BELTS

**Flat belts** find considerable use in applications requiring small pulley diameters, high belt surface speeds, low noise levels, low weight, or low inertia. They cannot be used where absolute synchronization between pulleys must be maintained because they rely on friction for their proper functioning. All flat belts are subject to creep because relative motion occurs between the pulley surface and the adjacent belt surface that is under load deformation from the combined tension and flexural stresses. Flat belts must be kept under tension to function, and they therefore require tensioning devices.

### 19.2.1    BELT LENGTH

Figure 19.1 shows the dimensions, angles of contact, and center distance of an open flat belt. The term *open* is used to distinguish it from the geometry of a crossed belt, where the belt forms a figure eight. Note that distance $0_2D$ is equal to $r_1 = D_1/2$, distance BD is equal to $r_2 - r_1 = (D_2 - D_1)/2$, and distance AD is the center distance $c_d$. Also, triangle ABD is a right triangle so that

$$AB^2 + BD^2 = AD^2$$

$$\therefore \qquad AB^2 + (r_2 - r_1)^2 = c_d^2$$

$$AB = \sqrt{c_d^2 - (r_2 - r_1)^2} \qquad\qquad (19.1)$$

The length of the open flat belt can be expressed as

$$L = 2AB + r_1\phi_1\frac{\pi}{180} + r_2\phi_2\frac{\pi}{180} \qquad\qquad (19.2)$$

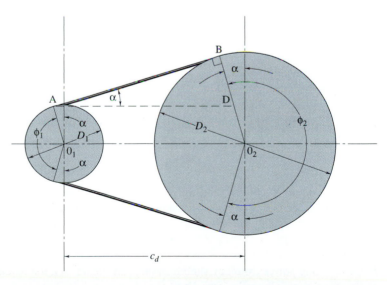

**Figure 19.1**    Dimensions, angles of contact, and center distance of open flat belt.

where $\phi$ = wrap angle, deg. The wrap angles can be expressed as

$$\phi_1 = 180° - 2\alpha \quad \text{and} \quad \phi_2 = 180° + 2\alpha \tag{19.3}$$

Also from right triangle ABD in Figure 19.1 the angle used to describe the loss in arc of contact is

$$\sin \alpha = \frac{D_2 - D_1}{2c_d} \quad \text{or} \quad \alpha = \sin^{-1} \frac{D_2 - D_1}{2c_d} \tag{19.4}$$

The angle $\alpha$ is in degrees and is created because the pulleys do not have a 1-to-1 ratio. By making use of Equations (19.3) and (19.4) the belt length in Equation (19.2) can be expressed as

$$L = \sqrt{(2c_d)^2 - (D_2 - D_1)^2} + \frac{D_1 \pi}{360} \left( 180° - 2 \sin^{-1} \frac{D_2 - D_1}{2c_d} \right)$$

$$+ \frac{D_2 \pi}{360} \left( 180° + 2 \sin^{-1} \frac{D_2 - D_1}{2c_d} \right)$$

$$= \sqrt{(2c_d)^2 - (D_2 - D_1)^2} + \frac{\pi}{2}(D_1 + D_2) + \frac{\pi(D_2 - D_1)}{180} \sin^{-1} \frac{D_2 - D_1}{2c_d} \tag{19.5}$$

### 19.2.2    BELT FORCES

The basic equations developed for band brakes are also applicable here. From Section 18.8 the following torque and friction equations can be written:

$$T = \frac{(F_1 - F_2)D_1}{2} \tag{18.59}$$

$$\frac{F_1}{F_2} = e^{\mu \phi \pi / 180°} \tag{18.58}$$

where $\phi$ = wrap angle, deg
$\mu$ = coefficient of friction
$F_1$ = tight-side or driver friction force, N
$F_2$ = slack-side or driven friction force, N

In obtaining the preceding equations it is assumed that the coefficient of friction on the belt is uniform throughout the entire wrap angle and that centrifugal forces on the belt can be neglected.

The required initial belt tension (or tensile force) $F_i$ depends on the elastic characteristics of the belt but can be approximated by

$$F_i = \frac{F_1 + F_2}{2} \tag{19.6}$$

In transmitting power from one shaft to another by means of a flat belt and pulleys, the belt must have the initial tensile force $F_i$ given in Equation (19.6). Also, from Equation (18.59) note that when power is being transmitted, the tensile force $F_1$ in the tight side exceeds the tension in the slack side $F_2$.

The power transmitted, in horsepower, is

$$h_p = (F_1 - F_2)u \qquad (19.7)$$

where $u$ = belt velocity, m/s. The power transmitted, in English units of horsepower (hp), is

$$h_p = \frac{(F_1 - F_2)u}{33,000} \qquad (19.8)$$

where $F_1$ and $F_2$ are in pounds force and $u$ is in feet per minute.

The centrifugal force can be expressed as

$$F_c = m'u^2 = \frac{w_z'}{g}u^2 \qquad (19.9)$$

where $m'$ = mass per unit length, kg/m
$\quad u$ = belt velocity, m/s
$\quad w_z'$ = weight per unit length, N/m

When the centrifugal force is considered, Equation (18.58) becomes

$$\frac{F_1 - F_c}{F_2 - F_c} = e^{\mu\phi\pi/180°} \qquad (19.10)$$

Equation (18.59) is applicable whether or not centrifugal forces are considered.

---

**EXAMPLE 19.1**

**Given:** A flat belt is 6 in wide and $\frac{1}{3}$ in thick and transmits 15 hp. The center distance is 8 ft. The driving pulley has 6-in diameter and rotates at 2000 rpm such that the loose side of the belt is on top. The driven pulley has 18-in diameter. The belt material weighs 0.035 lb/in$^3$.

**Find:** Determine the following:

(a) If $\mu = 0.30$, determine $F_1$ and $F_2$.
(b) If $\mu$ is reduced to 0.20 because of oil getting on part of the pulley, what are $F_1$ and $F_2$? Would the belt slip?
(c) What is the belt length?

**Solution:**

(a) The belt velocity is

$$u = \frac{\pi D_1 N_1}{12} = \frac{\pi(6)(2000)}{12} = 3142 \text{ ft/min}$$

From Equation (19.8)

$$F_1 - F_2 = \frac{(33,000)h_p}{u} = \frac{(33,000)(15)}{3142} = 157.6 \text{ lb} \qquad \text{(a)}$$

The weight per volume was given as

$$\frac{w_z}{Lw_i t} = 0.035 \text{ lb/in}^3$$

$$\therefore \quad \frac{w_z}{L} = 0.035 w_i t = (0.035)(6)\left(\tfrac{1}{3}\right) = 0.070 \text{ lb/in} = 0.84 \text{ lb/ft}$$

The centrifugal force acting on the belt is

$$F_c = \frac{w_z}{L}\frac{u^2}{g} = \frac{(0.84)(3142)^2}{(32.2)(60)^2} = 71.54 \text{ lb}$$

From Equation (19.4)

$$\alpha = \sin^{-1}\frac{D_2 - D_1}{2c_d} = \sin^{-1}\frac{18 - 6}{(2)(8)(12)} = 3.583° \qquad \text{(b)}$$

The wrap angle is

$$\phi = 180° - 2(\alpha) = 180° - 2(3.583°) = 172.8° \qquad \text{(c)}$$

Making use of Equation (19.10) gives

$$\frac{F_1 - 71.54}{F_2 - 71.54} = e^{(0.3)(172.8)\pi/180} = 2.472$$

$$F_1 - 71.54 = 2.472 F_2 - 176.8$$

$$\therefore \qquad F_1 = 2.472 F_2 - 105.3 \qquad \text{(d)}$$

Substituting Equation (d) into Equation (a) gives

$$2.472 F_2 - F_2 = 105.3 + 157.6$$

$$F_2 = \frac{262.9}{1.472} = 178.6 \text{ lb}$$

$$\therefore \qquad F_1 = 157.6 + 178.6 = 336.2 \text{ lb}$$

From Equation (19.6) the initial belt tension is

$$F_i = \frac{F_1 + F_2}{2} = \frac{336.2 + 178.6}{2} = 257.4 \text{ lb}$$

(b)  If $\mu = 0.20$ instead of 0.30,

$$\frac{F_1 - 71.54}{F_2 - 71.54} = e^{(0.2)(172.8)\pi/180} = 1.828$$

$$F_1 = 1.828 F_2 - (71.54)(1.828) + 71.54$$

$$= 1.828 F_2 - 59.23 \qquad \text{(e)}$$

Substituting Equation (e) into Equation (19.6) gives

$$\frac{1.828 F_2 - 59.23 + F_2}{2} = 257.4$$

$$\therefore \qquad F_2 = 203.0 \text{ lb}$$

Substituting this into Equation (e) gives

$$F_1 = 1.828(203.0) - 59.23 = 311.8 \text{ lb}$$

From Equation (19.8) the power transmitted, in English units, is

$$h_p = \frac{(F_1 - F_2)u}{33,000} = \frac{(311.8 - 203.0)(3142)}{33,000} = 10.36 \text{ hp}$$

Since this is less than 15 hp, the belt will slip.

(c) From Equation (19.5) while making use of Equations (b) and (c)

$$L = 2\sqrt{8^2(12)^2 - (9-3)^2} + \frac{3\pi}{180°}(172.8°) + \frac{9\pi}{180°}(187.2°)$$

$$= 191.6 + 9.048 + 29.41 = 230.1 \text{ in}$$

## 19.2.3 SLIP

To eliminate slip, the initial belt tension needs to be retained. But as the belt stretches over time, some of the initial tension will be lost. One solution might be to have excessive initial tension, but this would put large loads on the pulley and shaft and also shorten the belt life. Some better approaches are the following:

1. Develop ways of adjusting tension during operation.
2. Increase the wrap angle.
3. Change the belt material to increase the coefficient of friction.
4. Use a larger belt section.

Figure 19.2 illustrates a way of maintaining belt tension. The slack side of the belt is on the top, so that the sag of the belt acts to increase its wrap angle.

## 19.3 SYNCHRONOUS BELTS

**Synchronous belts,** or **timing belts,** are basically flat belts with a series of evenly spaced teeth on the inside circumference, thereby combining the advantages of flat belts with the excellent traction of gears and chains (covered later in Sec. 19.6). A synchronous, or timing, belt is shown in Figure 19.3.

Because synchronous belts do not slip or creep, unlike flat belts, the required belt tension is low, producing very small bearing loads. Synchronous belts will not stretch and require no lubrication. Speed is transmitted uniformly because there is no chordal rise and fall of the pitch line as in rolling chains (Sec. 19.6). The equations developed for flat-belt length and torque in Section 19.2 are equally valid for synchronous belts.

**Figure 19.2**    Weighted idler used to maintain desired belt tension.

**Figure 19.3** Synchronous, or timing, belt.

---

**EXAMPLE 19.2**

**Given:** A timing belt is used to transfer power from a high-speed turbine to a grinding wheel. The timing belt is 750 mm long and weighs 180 g. The maximum allowable force in the belt is 2000 N, and the speed of the turbine is the same as the speed of the grinding wheel, 5000 rpm.

**Find:** Calculate the optimum pulley pitch diameter for maximum power transfer.

**Solution:** If the total maximum force in the belt is $F_1$ and the centrifugal force is $F_c$, the maximum power transmitted is

$$h_p = u(F_1 - F_c)$$

Letting $F_1$ be equal to the maximum allowable force and making use of Equation (19.9) for the centrifugal force give

$$h_p = u\left(2000 - \frac{0.180}{0.750}u^2\right) = u(2000 - 0.24u^2)$$

The optimum power transmitted occurs when

$$\frac{\partial h_p}{\partial u} = 2000 - (0.24)(3)u^2 = 0$$

$$\therefore \quad u = 52.70 \text{ m/s}$$

The pulley diameter that produces the maximum power transfer is

$$\frac{\omega D}{2} = u$$

$$D = \frac{2u}{\omega} = \frac{2(52.70)}{(5000)(2\pi/60)} = 0.2013 \text{ m} = 201.3 \text{ mm}$$

Thus, the optimum pulley diameter for maximum power transfer is 201.3 mm.

## 19.4 V-BELTS

**V-belts** are used with electric motors to drive a great number of components, such as blowers, compressors, or machine tools. One or more V-belts are used to drive accessories on automotive and most other internal combustion engines. V-belts are made to standard lengths and with standard cross-sectional sizes, the details of which can be found in catalogs. The grooved pulleys that V-belts run in are called **sheaves.** They are usually made of cast iron, pressed steel, or die-cast metal. Figure 19.4 shows a V-belt in a sheave groove.

A number of different types of V-belt are available. The various types will not be discussed here. Instead, information about the light-duty type with three different sections $2L$, $3L$, and $4L$ will be considered. Once the design of the light-duty type of V-belt is known, other types of V-belt can be easily designed by using company catalogs.

V-belts find frequent application where synchronization between shafts is not important. V-belts are easily installed and removed, quiet in operation (but not quite as quiet as flat belts), and low in maintenance, and they provide shock absorption between the driver and driven shafts. V-belt drives normally operate best at belt velocities between 1500 and 6500 ft/min. Optimum velocity (peak capacity) is 4500 ft/min.

V-belts can operate satisfactorily at velocity ratios

$$g_r = \frac{N_{a1}}{N_{a2}} = \frac{r_2}{r_1} \qquad (19.11)$$

to approximately 7 to 1. V-belts typically operate at 90% to 98% efficiency, lower than that found for flat belts.

V-belts have a fiberglass-reinforced neoprene core and a fabric-impregnated neoprene jacket that protects the interior and provides a wear-resistant surface for the belt. Because

**Figure 19.4** V-belt in sheave groove.

their interior tension cords are stretch-resistant, V-belts (unlike flat belts) *do not* require frequent adjustment of initial tension.

A major advantage of V-belts over flat belts is their wedging action, which increases the normal force from $dN$ for flat belts to $(dN/2)/\sin\beta$ (as shown in Fig. 19.4) for V-belts, where $\beta$ is the sheave angle. Because the V-belt has a trapezoidal cross section, the belt rides on the side of the groove and a wedging action increases the traction. Pressure and friction forces act on the side of the belt. The force equations developed in Section 19.2 for flat belts are equally applicable for V-belts if the coefficient of friction $\mu$ is replaced with $\mu/\sin\beta$. Also, the belt length equation for flat belts [Eq. (19.5)] is equally valid for V-belts. The only difference when one is using Equation (19.5) for V-belts is that the pitch radius or pitch diameter is used, whereas for flat belts the outside radius or outside diameter of the pulley is used.

### 19.4.1   INPUT NORMAL POWER RATING

It is essential that the possible maximum load conditions be considered in designing a V-belt. The normal power rating of the driver is

$$(h_{pr})_{\text{input}} = h_p(f_1 + f_2) \tag{19.12}$$

where $f_1$ = overload service factors for various types of driven unit, given in Table 19.1
$f_2$ = extra overload factor of 0.4 if any of the following conditions prevail:
  (1) Frequent starting and stopping
  (2) High starting motor torque
  (3) Gas engine

Thus, a properly designed belt drive should use the power rating from Equation (19.12).

### 19.4.2   DRIVE SIZE

The design of a V-belt drive should use the largest possible pulleys. Small pulleys are less efficient and greatly reduce belt life because of slip and extreme flexing of the belt. Where small pulleys must be used and speed is high, select the lightest belt (2L). Table 19.2 shows minimum pitch diameters in inches, and Table 19.3 shows recommended pulley diameters in inches for electric motors. The criteria used in arriving at these critical sizes are bearing loading and shaft loading and not belt flexing life. Tables 19.2 and 19.3 give belt and pulley sizes for a specific belt drive.

### 19.4.3   ARC CORRECTION FACTOR

The rated power tables to be considered in Section 19.4.4 are based on a 1-to-1 ratio with 180° of belt wrap around each pulley. When pulleys of different diameters are used, there is a loss in traction effort on the smaller pulley. The angle $\alpha$ used to describe the loss in arc of contact is expressed in Equation (19.4) and shown in Figure 19.1. Recall that $\alpha$ is expressed in degrees rather than radians. Table 19.4 shows the arc correction factors for various values of $\alpha$.

**Table 19.1**  Overload service factors $f_1$ for various types of driven unit

| Driven unit | Overload factor |
|---|---|
| **Agitators** | |
| Liquid | 1.2 |
| Semisolid | 1.4 |
| **Compressor** | |
| Centrifugal | 1.2 |
| Reciprocating | 1.4 |
| **Conveyors and elevators** | |
| Package and oven | 1.2 |
| Belt | 1.4 |
| **Fans and blowers** | |
| Centrifugal, calculating | 1.2 |
| Exhausters | 1.4 |
| **Food machinery** | |
| Slicers | 1.2 |
| Grinders and mixers | 1.4 |
| **Generators** | |
| Farm lighting and exciters | 1.2 |
| **Heating and ventilating** | |
| Fans and oil burners | 1.2 |
| Stokers | 1.4 |
| **Laundry machinery** | |
| Dryers and ironers | 1.2 |
| Washers | 1.4 |
| **Machine tools** | |
| Home workshop and woodworking | 1.4 |
| **Pumps** | |
| Centrifugal | 1.2 |
| Reciprocating | 1.4 |
| **Refrigeration** | |
| Centrifugal | 1.2 |
| Reciprocating | 1.4 |
| **Worm gear speed reducers, input side** | 1.0 |

**Table 19.2**  Recommended minimum pitch diameters of pulley for three belt sizes

| Belt type | Size of belt, in | Minimum Pitch Diameter, in | |
|---|---|---|---|
| | | Recommended | Absolute |
| 2L | $\frac{1}{4} \times \frac{1}{8}$ | 1.0 | 1.0 |
| 3L | $\frac{3}{8} \times \frac{7}{32}$ | 1.5 | 1.5 |
| 4L | $\frac{1}{2} \times \frac{5}{16}$ | 2.5 | 1.8 |

**Table 19.3**    Recommended pulley diameters in inches for three
electric motor sizes

| Motor horsepower, hp | Motor Speed, rpm | | | | |
|---|---|---|---|---|---|
| | 575 | 695 | 870 | 1160 | 1750 |
| | Recommended pulley diameter, in | | | | |
| 0.50 | 2.50 | 2.50 | 2.50 | — | — |
| 0.75 | 3.00 | 2.50 | 2.50 | 2.50 | — |
| 1.00 | 3.00 | 3.00 | 2.50 | 2.50 | 2.25 |

### 19.4.4    DESIGN POWER RATING AND CENTER DISTANCE

Table 19.5 gives the rated power for three light-duty belt types. The three types considered are 2L with $w_t = \frac{1}{4}$ in and $h_t = \frac{1}{8}$ in, 3L with $w_t = \frac{3}{8}$ in and $h_t = \frac{1}{4}$ in, and 4L with $w_t = \frac{1}{2}$ in and $h_t = \frac{9}{32}$ in. (Figure 19.4 clarifies the width $w_t$ and height $h_t$ of the belt.) From these tables for a designated speed of the faster shaft, given in the first column, and the driver pulley's effective outside diameter, given in the top row of the table, the power rating can be obtained.

Table 19.6 shows the center distances of the 3L and 4L V-belt types. All dimensions are in inches. These tables give the center distance for various sizes of the driven and driver pulleys. The belt velocity expressed in feet per minute is

$$u = \frac{\pi D_1 N_{a1}}{12}$$

(19.13)

**Table 19.4**    Arc correction factor for various
angles of loss in arc of contact

| Loss in arc of contact, deg | Correction factor |
|---|---|
| 0 | 1.00 |
| 5 | 0.99 |
| 10 | 0.98 |
| 15 | 0.96 |
| 20 | 0.95 |
| 25 | 0.93 |
| 30 | 0.92 |
| 35 | 0.89 |
| 40 | 0.89 |
| 45 | 0.87 |
| 50 | 0.86 |
| 55 | 0.84 |
| 60 | 0.83 |
| 65 | 0.81 |
| 70 | 0.79 |
| 75 | 0.76 |
| 80 | 0.74 |
| 85 | 0.71 |
| 90 | 0.69 |

**Table 19.5** Power ratings for light-duty V-belts. (a) $2L$ section with $w_t = \frac{1}{4}$ in and $h_t = \frac{1}{8}$ in; (b) $3L$ section with $w_t = \frac{3}{8}$ in and $h_t = \frac{1}{4}$ in; (c) $4L$ section with $w_t = \frac{1}{2}$ in and $h_t = \frac{9}{32}$ in

| Speed of faster shaft, rpm | Pulley Effective Outside Diameter, in | | | | | |
|---|---|---|---|---|---|---|
| | 1.00 | 2.00 | 3.00 | 4.00 | 5.00 | 6.00 |
| | Rated horsepower, hp | | | | | |
| 500 | 0.04 | 0.05 | 0.06 | 0.08 | 0.10 | 0.12 |
| 1000 | 0.05 | 0.08 | 0.12 | 0.16 | 0.19 | 0.21 |
| 1500 | 0.06 | 0.12 | 0.17 | 0.22 | 0.25 | 0.30 |
| 2000 | 0.08 | 0.15 | 0.23 | 0.28 | 0.32 | 0.39 |
| 2500 | 0.10 | 0.18 | 0.27 | 0.34 | 0.39 | 0.44 |
| 3000 | 0.12 | 0.21 | 0.31 | 0.40 | 0.31 | 0.47 |
| 3500 | 0.14 | 0.24 | 0.35 | 0.44 | 0.44 | |
| 4000 | 0.16 | 0.28 | 0.38 | 0.46 | | |
| 4500 | 0.18 | 0.31 | 0.42 | | | |
| 5000 | 0.20 | 0.35 | | | | |

(a)

| Speed of faster shaft, rpm | Pulley Effective Outside Diameter, in | | | | | | |
|---|---|---|---|---|---|---|---|
| | 1.50 | 1.75 | 2.00 | 2.25 | 2.50 | 2.7 | 3.00 |
| | Rated horsepower, hp | | | | | | |
| 500 | 0.04 | 0.07 | 0.09 | 0.12 | 0.14 | 0.17 | 0.19 |
| 1000 | 0.07 | 0.12 | 0.16 | 0.21 | 0.25 | 0.30 | 0.34 |
| 1500 | 0.09 | 0.15 | 0.22 | 0.29 | 0.35 | 0.41 | 0.47 |
| 2000 | 0.10 | 0.19 | 0.27 | 0.35 | 0.43 | 0.51 | 0.59 |
| 2500 | 0.11 | 0.21 | 0.31 | 0.41 | 0.51 | 0.60 | 0.69 |
| 3000 | 0.11 | 0.23 | 0.35 | 0.45 | 0.57 | 0.68 | 0.78 |
| 3500 | 0.11 | 0.25 | 0.38 | 0.50 | 0.62 | 0.74 | 0.84 |
| 4000 | 0.11 | 0.26 | 0.40 | 0.54 | 0.66 | 0.78 | 0.88 |
| 4500 | 0.10 | 0.25 | 0.42 | 0.56 | 0.68 | 0.80 | 0.90 |
| 5000 | 0.09 | 0.26 | 0.42 | 0.57 | 0.69 | 0.80 | 0.89 |

(b)

| Speed of faster shaft, rpm | Pulley Effective Outside Diameter, in | | | | | | | | |
|---|---|---|---|---|---|---|---|---|---|
| | 2.00 | 2.25 | 2.50 | 2.75 | 3.00 | 3.25 | 3.50 | 3.75 | 4.00 |
| | Rated horsepower, hp | | | | | | | | |
| 500 | 0.08 | 0.14 | 0.19 | 0.24 | 0.29 | 0.34 | 0.39 | 0.44 | 0.49 |
| 1000 | 0.11 | 0.21 | 0.31 | 0.41 | 0.50 | 0.60 | 0.69 | 0.78 | 0.87 |
| 1500 | 0.12 | 0.26 | 0.40 | 0.54 | 0.67 | 0.81 | 0.94 | 1.07 | 1.20 |
| 2000 | 0.11 | 0.30 | 0.47 | 0.65 | 0.82 | 0.99 | 1.15 | 1.31 | 1.47 |
| 2500 | 0.09 | 0.31 | 0.53 | 0.73 | 0.94 | 1.13 | 1.32 | 1.51 | 1.69 |
| 3000 | 0.06 | 0.31 | 0.56 | 0.79 | 1.02 | 1.24 | 1.45 | 1.65 | 1.84 |
| 3500 | 0.02 | 0.30 | 0.57 | 0.83 | 1.07 | 1.31 | 1.53 | 1.73 | 1.92 |
| 4000 | — | 0.27 | 0.56 | 0.83 | 1.09 | 1.33 | 1.55 | 1.75 | 1.92 |
| 4500 | — | 0.22 | 0.54 | 0.81 | 1.07 | 1.30 | 1.51 | 1.69 | 1.84 |
| 5000 | — | 0.15 | 0.47 | 0.75 | 1.01 | 1.23 | 1.41 | 1.66 | 1.65 |

(c)

**Table 19.6** Center distances for various pitch diameters of driver and driven pulleys (a) 3*L* type of V-belt; (b) 4*L* type of V-belt

| Pulley Combination | | Nominal Center Distance | | | |
|---|---|---|---|---|---|
| | | Short Center | | Medium Center | |
| Driver pitch diameter, in | Driven pitch diameter, in | Belt number | Center distance, in | Belt number | Center distance, in |
| 2.0 | 2.0 | 3L200 | 6.4 | 3L250 | 9.4 |
| 3.0 | 3.0 | 3L250 | 7.4 | 3L310 | 10.4 |
| 2.0 | 2.5 | 3L210 | 6.6 | 3L270 | 9.6 |
| 2.0 | 3.0 | 3L220 | 6.7 | 3L280 | 9.7 |
| 3.0 | 4.5 | 3L290 | 8.2 | 3L350 | 11.2 |
| 2.0 | 3.5 | 3L240 | 7.3 | 3L300 | 10.3 |
| 2.0 | 4.0 | 3L250 | 7.2 | 3L310 | 10.3 |
| 2.25 | 4.5 | 3L270 | 7.7 | 3L330 | 10.7 |
| 2.5 | 5.0 | 3L290 | 8.1 | 3L350 | 11.1 |
| 3.0 | 6.0 | 3L330 | 8.9 | 3L390 | 11.9 |
| 2.0 | 5.0 | 3L250 | 8.0 | 3L340 | 11.0 |
| 2.0 | 6.0 | 3L310 | 8.6 | 3L370 | 11.6 |
| 3.0 | 9.0 | 3L410 | 10.3 | 3L470 | 13.4 |
| 2.0 | 7.0 | 3L340 | 9.2 | 3L400 | 12.2 |
| 2.0 | 9.0 | 3L390 | 9.9 | 3L450 | 13.0 |
| 2.0 | 10.0 | 3L420 | 10.4 | 3L480 | 13.6 |
| 1.5 | 9.0 | 3L390 | 10.1 | 3L450 | 13.3 |

(a)

| Pulley Combination | | Nominal Center Distance | | | | | |
|---|---|---|---|---|---|---|---|
| | | Minimum Center | | Short Center | | Medium Center | |
| Driver pitch diameter, in | Driven pitch diameter, in | Belt number | Center distance, in | Belt number | Center distance, in | Belt number | Center distance, in |
| 2.5 | 2.5 | 4L170 | 4.0 | 4L150 | 8.0 | 4L330 | 12.0 |
| 3.0 | 3.0 | 4L200 | 4.8 | 4L280 | 8.8 | 4L360 | 12.8 |
| 3.0 | 4.5 | 4L240 | 5.5 | 4L320 | 9.6 | 4L400 | 13.6 |
| 4.0 | 6.0 | 4L300 | 6.5 | 4L380 | 10.6 | 4L460 | 14.5 |
| 3.0 | 6.0 | 4L280 | 6.2 | 4L360 | 10.3 | 4L440 | 14.3 |
| 3.5 | 7.0 | 4L320 | 7.0 | 4L400 | 11.1 | 4L480 | 15.1 |
| 3.0 | 7.5 | 4L320 | 6.8 | 4L400 | 11.0 | 4L480 | 15.0 |
| 4.0 | 10.0 | 4L410 | 8.5 | 4L490 | 12.5 | 4L570 | 16.7 |
| 3.0 | 9.0 | 4L360 | 7.5 | 4L440 | 11.7 | 4L520 | 15.8 |
| 3.5 | 10.5 | 4L420 | 8.8 | 4L500 | 13.0 | 4L580 | 17.1 |
| 4.0 | 12.0 | 4L470 | 9.6 | 4L550 | 13.8 | 4L630 | 18.0 |
| 3.0 | 10.5 | 4L410 | 8.6 | 4L490 | 12.9 | 4L570 | 17.0 |
| 4.0 | 14.0 | 4L530 | 10.5 | 4L610 | 15.0 | 4L690 | 19.1 |
| 3.0 | 12.0 | 4L450 | 9.1 | 4L530 | 13.4 | 4L610 | 17.6 |
| 3.5 | 14.0 | 4L520 | 10.4 | 4L600 | 14.8 | 4L680 | 19.0 |
| 2.0 | 9.0 | 4L350 | 7.5 | 4L430 | 11.8 | 4L510 | 15.9 |
| 4.0 | 18.0 | 4L650 | 12.8 | 4L730 | 17.3 | 4L810 | 21.6 |
| 2.4 | 12.0 | 4L440 | 8.9 | 4L520 | 13.3 | 4L600 | 17.5 |
| 2.5 | 18.0 | 4L480 | 9.3 | 4L560 | 14.3 | 4L640 | 18.5 |
| 2.8 | 14.0 | 4L510 | 10.3 | 4L590 | 14.7 | 4L670 | 19.0 |
| 2.0 | 11.0 | 4L400 | 8.0 | 4L480 | 12.5 | 4L560 | 16.7 |
| 2.0 | 12.0 | 4L430 | 8.5 | 4L510 | 13.0 | 4L590 | 17.3 |
| 2.5 | 15.0 | 4L530 | 10.3 | 4L610 | 14.9 | 4L690 | 19.2 |
| 3.0 | 13.0 | 4L630 | 12.2 | 4L710 | 16.8 | 4L790 | 21.2 |
| 2.0 | 14.0 | 4L490 | 9.5 | 4L570 | 14.1 | 4L650 | 18.4 |

(b)

EXAMPLE 19.3

**Given:** A $\frac{1}{3}$-hp, capacitor, alternating-current motor is driving a reciprocating air compressor at 720 rpm. The motor speed is 1750 rpm on approximately 12-in centers. The maximum pitch diameter of the compressor pulley is to be 8 in. Design the belt drive with a light-duty V-belt.

**Find:** Give the size of the pulleys, the belt dimensions, and the rated horsepower.

**Solution:** From Table 19.1 for a reciprocating compressor $f_1 = 1.4$. Since the motor has a high starting torque, $f_2 = 0.4$. The given power was $\frac{1}{3}$ hp. The normal power rating of the drive, from Equation (19.12), is

$$(h_{pr})_{input} = h_p(f_1 + f_2) = \frac{1}{3}(1.4 + 0.4) = 0.600 \text{ hp}$$

The velocity ratio input is

$$g_r = \frac{\text{speed of motor}}{\text{speed of compressor}} = \frac{1750}{720} = 2.43$$

Given that the maximum allowable pitch diameter of the driven pulley is 8 in, assume that the driven pitch diameter at the limit is $D_2 = 8$ in. Then the driver diameter is

$$D_1 = \frac{D_2}{g_r} = \frac{8.0}{2.43} = 3.292 \text{ in}$$

A pitch diameter of 3.292 in corresponds to an outside diameter of 3.5 in. From Table 19.5(c) the best belt type to handle the requirements is $4L$. Specifically, for a pulley outside diameter of 3.5 and speed of 1750 rpm, the design rated power is 1.05 hp.

Table 19.6(b) does not contain the identical situation in this example—a speed ratio of 2.4 and pulley pitch diameters of 8 and 3.292 in. However, from this table one could estimate that a center distance of 12 in should work. With the above values the angle used to describe the loss in arc contact and the belt length can be expressed from Equations (19.4) and (19.5) as

$$\alpha = \sin^{-1}\frac{D_2 - D_1}{2c_d} = \sin^{-1}\frac{8 - 3.292}{24} = 11.31°$$

$$L = \sqrt{(2c_d)^2 - (D_2 - D_1)^2} + \frac{\pi}{2}(D_1 + D_2) + \frac{\pi(D_2 - D_1)\alpha}{180}$$

$$= \sqrt{(2)^2(12)^2 - (8 - 3.292)^2} + \frac{\pi}{2}(8 + 3.292) + \frac{\pi(8 - 3.292)(11.31°)}{180}$$

$$= 23.53 + 17.74 + 0.929 = 42.20 \text{ in}$$

Recall that the design power of 1.05 hp is for a 1-to-1 speed rating. From Table 19.4 for $2\alpha = 22.62°$ the arc correction factor is 0.94. The corrected design power rating is $0.94(1.05) = 0.99$ hp.

The belt velocity can be determined from Equation (19.13) or

$$u = \frac{\pi}{12}D_1 N_{a1} = \frac{\pi}{12}(3.292)(1750) = 1508 \text{ ft/min}$$

Summarizing the $4L$ V-belt design, we have

$$D_1 = 3.292 \text{ in} \qquad D_2 = 8 \text{ in} \qquad c_d = 12 \text{ in} \qquad L = 42 \text{ in} \qquad g_r = 2.4$$

$$N_{a1} = 1750 \qquad u = 1508 \text{ ft/min} \qquad (h_p)_{input} = 0.6 \text{ hp} \qquad (h_p)_{design} = 1 \text{ hp}$$

## 19.5    WIRE ROPES

**Wire ropes** are used instead of flat belts or V-belts when power must be transmitted over long center distances as in hoists, elevators, and ski lifts. Figure 19.5 shows a cross section of a wire rope. The center portion (dark section) is the **core** of the rope and is often made of hemp (a tall Asiatic herb). The purposes of the core are to lubricate and thus prevent excessive wire wear and to elastically support the strands. Some other core materials are polypropylene wire and steel wire.

**Figure 19.5**    Cross section of wire rope.

The **strands** are groupings of wires placed around the core. In Figure 19.5 there are six strands, and each strand consists of 19 wires. Wire ropes are typically designated as, for example, "$1\frac{1}{8}$, $6 \times 19$ hauling rope." The $1\frac{1}{8}$ gives the wire rope diameter in inches, designated by the symbol $d$. The 6 designates the number of strands and 19 designates the number of wires in a strand. The wire diameter is designated by the symbol $d_w$. The term *hauling rope* designates the application in which the wire rope is to be used.

Figure 19.6 shows two lays of wire rope. The regular **lay** [Fig. 19.6(b)] has the wire twisted in one direction to form strands and the strands twisted in the opposite direction to form the rope. Visible wires are approximately parallel to the rope axis. The major advantage of the regular lay is that the rope *does not kink or untwist* and is easy to handle. The Lang lay [Fig. 19.6(a)] has the wires in the strands and the strands in the rope twisted in the same direction. This type of lay has *greater resistance to abrasive wear and bending fatigue* than the regular lay. Lang-lay ropes are, however, more susceptible to handling abuses, pinching in undersized grooves, and crushing when improperly wound on drums. Also, the twisting moment acting in the strand tends to unwind the strand, causing excessive rope rotation. Lang-lay ropes should therefore always be secured at the ends to prevent the rope from unlaying.

Although steel is most popular, wire rope is made of many kinds of metal, such as copper, bronze, stainless steel, and wrought iron. Table 19.7 lists some of the various ropes

(a)                                                (b)

**Figure 19.6**    Two lays of wire rope. (a) Lang; (b) regular.

**Table 19.7**   Wire rope data

| Rope | Weight per height, lb/ft | Minimum sheave diameter, in | Rope diameter $d$, in | Material | Size of outer wires | Modulus of elasticity,[a] psi | Strength,[b] psi |
|---|---|---|---|---|---|---|---|
| 6 × 7 Haulage | $1.50d^2$ | $42d$ | $\frac{1}{4}$–$1\frac{1}{2}$ | Monitor steel | $d/9$ | $14 \times 10^6$ | $100 \times 10^3$ |
| | | | | Plow steel | $d/9$ | $14 \times 10^6$ | $88 \times 10^3$ |
| | | | | Mild plow steel | $d/9$ | $14 \times 10^6$ | $76 \times 10^3$ |
| 6 × 19 Standard hoisting | $1.60d^2$ | $26d$–$34d$ | $\frac{1}{4}$–$2\frac{3}{4}$ | Monitor steel | $d/13$–$d/16$ | $12 \times 10^6$ | $106 \times 10^3$ |
| | | | | Plow steel | $d/13$–$d/16$ | $12 \times 10^6$ | $93 \times 10^3$ |
| | | | | Mild plow steel | $d/13$–$d/16$ | $12 \times 10^6$ | $80 \times 10^3$ |
| 6 × 37 Special flexible | $1.55d^2$ | $18d$ | $\frac{1}{4}$–$3\frac{1}{2}$ | Monitor steel | $d/22$ | $11 \times 10^6$ | $100 \times 10^3$ |
| | | | | Plow steel | $d/22$ | $11 \times 10^6$ | $88 \times 10^3$ |
| 8 × 19 Extra flexible | $1.45d^2$ | $21d$–$26d$ | $\frac{1}{4}$–$1\frac{1}{2}$ | Monitor steel | $d/15$–$d/19$ | $10 \times 10^6$ | $92 \times 10^3$ |
| | | | | Plow steel | $d/15$–$d/19$ | $10 \times 10^6$ | $80 \times 10^3$ |
| 7 × 7 Aircraft | $1.70d^2$ | — | $\frac{1}{16}$–$\frac{3}{8}$ | Corrosion-resistant steel | — | — | $124 \times 10^3$ |
| | | | | Carbon steel | — | — | $124 \times 10^3$ |
| 7 × 9 Aircraft | $1.75d^2$ | — | $\frac{1}{8}$–$1\frac{3}{8}$ | Corrosion-resistant steel | — | — | $135 \times 10^3$ |
| | | | | Carbon steel | — | — | $143 \times 10^3$ |
| 19-Wire aircraft | $2.15d^2$ | — | $\frac{1}{32}$–$\frac{5}{16}$ | Corrosion-resistant steel | — | — | $165 \times 10^3$ |
| | | | | Carbon steel | — | — | $165 \times 10^3$ |

[a] The modulus of elasticity is only approximate; it is affected by the loads on the rope and, in general, increases with the life of the rope.

[b] The strength is based on the nominal area of the rope. The figures given are only approximate and are based on 1-in rope sizes and $\frac{1}{4}$-in aircraft cable sizes.

SOURCE: From Shigley and Mitchell (1983).

that are available, together with their characteristics and properties. The cross-sectional area of the metal strand in standard hoisting and haulage ropes is $A_m \approx 0.38d^2$.

## 19.5.1   TENSILE STRESS

The total force acting on the rope is

$$F_t = F_w + F_r + F_a \tag{19.14}$$

where $F_w$ = deadweight being supported, N
$F_r$ = rope weight, N
$F_a$ = force due to acceleration, N

The tensile stress is

$$\sigma_t = \frac{F_t}{A_m} \tag{19.15}$$

where $A_m$ = cross-sectional area of metal strand in standard hoisting and haulage ropes. The allowable stress is obtained from Table 19.7.

From Equation (1.1) the safety factor is

$$n_s = \frac{\sigma_{\text{all}}}{\sigma_t} \tag{19.16}$$

**Table 19.8**    Minimum safety factors for a variety
of wire rope applications

| Application | Safety factor[a] $n_s$ |
|---|---|
| Track cables | 3.2 |
| Guys | 3.5 |
| **Mine shafts, ft** | |
| Up to 500 | 8.0 |
| 500 to 1000 | 7.5 |
| 1000–2000 | 7.0 |
| 2000–3000 | 6.0 |
| Over 3000 | 5.0 |
| Hoisting | 5.0 |
| Haulage | 6.0 |
| Cranes and derricks | 6.0 |
| Electric hoists | 7.0 |
| Hand elevators | 5.0 |
| Private elevators | 7.5 |
| Hand dumbwaiters | 4.5 |
| Grain elevators | 7.5 |
| **Passenger elevators, ft/min** | |
| 50 | 7.60 |
| 300 | 9.20 |
| 800 | 11.25 |
| 1200 | 11.80 |
| 1500 | 11.90 |
| **Freight elevators, ft/min** | |
| 50 | 6.65 |
| 300 | 8.20 |
| 800 | 10.00 |
| 1200 | 10.50 |
| 1500 | 10.55 |
| **Powered dumbwaiters, ft/min** | |
| 50 | 4.8 |
| 300 | 6.6 |
| 800 | 8.0 |
| 500–1000 | 7.5 |

[a]Use of these factors does not preclude a fatigue failure.
SOURCE: Shigley and Mitchell (1983).

Table 19.8 gives minimum safety factors for a variety of wire rope applications. The safety factor obtained from Equation (19.16) should be larger than the safety factor obtained from Table 19.8.

## 19.5.2    BENDING STRESS

From Section 4.5 [Eqs. (4.48) and (4.50)] the bending moment of the wires in a rope passing over a pulley is

$$M = \frac{EI}{r} \qquad \text{and} \qquad M = \frac{\sigma I}{c} \qquad\qquad (19.17)$$

where $E$ = modulus of elasticity of wire, Pa. The symbols are the same as those used in Chapter 4. Equating the two equations and solving for the bending stress give

$$\sigma_b = \frac{Ec}{r}$$

where $r$ = radius of curvature that rope will experience, m
   $c$ = distance from neutral axis to outer fiber of wire, m

The radius of curvature that the rope experiences is similar to the pulley radius $D/2$, and $c$ is similar to $d_w/2$.

$$\therefore \qquad \sigma_b = \frac{Ed_w}{D} \qquad\qquad (19.18)$$

where $d_w$ = wire diameter, m
   $D$ = pulley diameter, m

Note from Equation (19.18) that the bigger $D/d_w$ is, the smaller the bending stress. Suggested minimum pulley diameters in Table 19.7 are based on a $D/d_w$ ratio of 400. If possible, the pulleys should be designed for a larger ratio. If the ratio $D/d_w$ is less than 200, heavy loads will often cause a permanent set of the rope. Thus, for a safe design assume that $D/d_w \geq 400$.

These design rules are hardly ever followed, for a number of reasons. First, consider a 2-in-diameter wire rope, such as would typically be used in the crane of Case Study 18.1. According to the rules just stated, for safety the pulleys on the crane would have to be 80 in, or over 6 ft, in diameter. Further, since the wire rope is wound on a drum and failure can occur in the rope adjacent to the drum, such a large diameter would also be needed for the drum. Because the required motor torque is the product of the hoist rope tension and the drum radius, a huge motor would be required for lifting relatively light loads.

The design rules just stated are recommendations for applications where the wire rope should attain infinite life. The economic consequences of infinite-life wire rope are usually too large to bear, and thus smaller pulleys are prescribed. To prevent failures that result in property damage or personal injury, the wire ropes are periodically examined for damage. Since a broken wire will generally be easily detected with a cotton cloth dragged over the rope's surface, rope life requirements are often expressed in terms of the number of broken wires allowed per length of wire rope. For example, the American Society of Mechanical Engineers [ASME (1994)] requires inspections of crane wire ropes every 6 months; and if any section has more than six broken wires within one lay of the rope, or three in any strand within a lay, the entire wire rope must be replaced. The same standard calls for pulley-to-wire diameter ratios of 12:1, which obviously results in wire rope with finite service life. Figures 19.7 and 19.8 relate the decrease in strength and service life associated with smaller pulley diameters.

### 19.5.3 BEARING PRESSURE

The rope stretches and rubs against the pulley, causing wear of both the rope and pulley. The amount of wear depends on the pressure on the rope in the pulley groove, or

$$p = \frac{2F_t}{dD} \qquad\qquad (19.19)$$

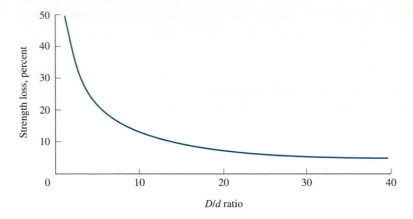

**Figure 19.7**    Percent strength loss in wire rope for different $D/d$ ratios.

where $d$ = rope diameter, m. The pressure obtained from Equation (19.19) should be less than the maximum pressure obtained from Table 19.9 for various pulley materials and types of rope.

### 19.5.4    FATIGUE

For the rope to have a long life, the total force $F_t$ must be less than the fatigue allowable force ($F_t \leq F_f$), where

$$F_f = \frac{S_u dD}{2000} \qquad\qquad \textbf{(19.20)}$$

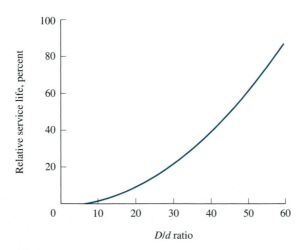

**Figure 19.8**    Service life for different $D/d$ ratios.

**Table 19.9** Maximum allowable bearing pressures for various sheave materials and types of rope

| | Material | | | | |
|---|---|---|---|---|---|
| | Wood[a] | Cast Iron[b] | Cast Steel[c] | Chilled Cast Iron[d] | Manganese Steel[e] |
| **Rope** | Allowable bearing pressure $p_{all}$, psi | | | | |
| **Regular lay** | | | | | |
| 6 × 7 | 150 | 300 | 550 | 650 | 1470 |
| 6 × 19 | 250 | 480 | 900 | 1100 | 2400 |
| 6 × 37 | 300 | 585 | 1075 | 1325 | 3000 |
| 8 × 19 | 350 | 680 | 1260 | 1550 | 3500 |
| **Lang lay** | | | | | |
| 6 × 7 | 165 | 350 | 600 | 715 | 1650 |
| 6 × 19 | 275 | 550 | 1000 | 1210 | 2750 |
| 6 × 37 | 330 | 660 | 1180 | 1450 | 3300 |

[a]On end grain of beech, hickory, or gum.
[b]For $H_B$ (min.) = 125.
[c]30–40 carbon; $H_B$ (min.) = 160.
[d]Use only with uniform surface hardness.
[e]For high speeds with balanced sheaves having ground surfaces.
SOURCE: Shigley and Mitchell (1983).

The ultimate strength given by Equation (19.20) for three materials is

| | | |
|---|---|---|
| Improved plow steel (monitor) | $240 \leq S_u \leq 280 \times 10^3$ psi | |
| Plow steel | $210 \leq S_u \leq 240 \times 10^3$ psi | **(19.21)** |
| Mild plow steel | $180 \leq S_u \leq 210 \times 10^3$ psi | |

**Given:** A hand elevator is to travel a height of 90 ft. The maximum load to be hoisted is 3000 lb at a velocity not exceeding 2 ft/s and an acceleration of 4 ft/s². Use 1-in plow steel, 6 × 19 standard hoisting ropes. | **EXAMPLE 19.4**

**Find:** Determine the safety factor while considering

(a) Tensile stress
(b) Bending stress
(c) Bearing pressure
(d) Fatigue

**Solution:** From Table 19.7 for 6 × 19 standard hoisting wire rope, assuming the use of only one rope,

$$F_r = 1.60d^2h_2 = 1.60(1)^2(90) = 144.0 \text{ lb/rope}$$

$$F_w = \text{deadweight} = W_{max} = 3000 \text{ lb}$$

The force due to acceleration is

$$F_a = ma = \frac{W}{g}a = \frac{(3000 + 144)(4)}{32.3} = 390.6 \text{ lb}$$

The total force on the rope is

$$F_t = F_w + F_r + F_a = 3000 + 144 + 390.6 = 3535 \text{ lb}$$

(a)   **Tensile stress:**

$$\sigma_t = \frac{F_t}{A_m} = \frac{3535}{0.38} = 9302 \text{ psi}$$

From Table 19.7 for $6 \times 19$ standard hoisting wire rope made of plow steel,

$$\sigma_{all} = 93,000 \text{ psi}$$

The safety factor is

$$n_s = \frac{\sigma_{all}}{\sigma_t} = \frac{93,000}{9301} = 10.0$$

From Table 19.8 the recommended safety factor for hand elevators is 5.0. Thus, one rope is sufficient as far as the tensile stress is concerned.

(b)   **Bending stress:** From Table 19.7 the minimum pulley diameter for $6 \times 19$ standard hoisting wire rope is $26d$ to $34d$. Choose $D = 34d = 34$ in. Also from the same table the wire diameter should be between $d/13$ and $d/16$. Choose $d_w = d/16 = \frac{1}{16}$ in.

$$\therefore \qquad \frac{D}{d_w} = \frac{34}{\frac{1}{16}} = 544$$

Permanent set should be avoided, since $D/d_w \geq 400$. From Table 19.7 the modulus of elasticity is $12 \times 10^6$ psi. The bending stress is

$$\sigma_b = E\frac{d_w}{D} = \frac{12(10^6)}{544} = 22.06 \times 10^3 \text{ psi}$$

The safety factor due to bending is

$$n_s = \frac{\sigma_{all}}{\sigma_b} = \frac{93,000}{22,060} = 4.22$$

This safety factor is less than the 5.0 recommended, and increasing the number of ropes will not alter the results. Changing the material from plow steel to monitor steel would produce a safety factor of 4.81, closer to 5. However, to use a lower than recommended safety factor, a larger pulley is needed.

(c)   **Bearing pressure:** From Equation (19.19)

$$p = \frac{2F_t}{dD} = \frac{2(3535)}{(1)(34)} = 207.9 \text{ psi}$$

From Table 19.9 for a $6 \times 19$ Lang lay for a cast steel pulley, $p_{all} = 1000$ psi. The safety factor is

$$n_s = \frac{p_{all}}{p} = \frac{1000}{207.9} = 4.81$$

(d) **Fatigue:** For monitor steel the ultimate strength is $280 \times 10^3$ psi. The allowable fatigue force from Equation (19.20) is

$$F_f = \frac{S_u dD}{2000} = \frac{(280)(10^3)(1)(34)}{2000} = 4760 \text{ lb}$$

The safety factor is

$$n_s = \frac{F_f}{F_t} = \frac{4760}{3535} = 1.35$$

Thus, fatigue failure is the most likely failure to occur, since it produced the smallest safety factor. Using four ropes instead of one would produce a safety factor greater than 5.

# 19.6  ROLLING CHAINS

**Rolling chains** are used to transmit power through sprockets rotating in the same plane. The machine element that it most resembles is a timing belt (shown in Fig. 19.3). The major advantage of using a rolling chain over a belt is that rolling chains do not slip. Large center distances can be dealt with more easily with rolling chains with fewer elements and in less space than with gears. Rolling chains also have high efficiency. No initial tension is necessary, and shaft loads are therefore smaller than with belt drives. The only maintenance required after careful alignment of the elements is lubrication, and with proper lubrication long life can be ensured.

## 19.6.1  OPERATION OF ROLLING CHAINS

Figure 19.9 shows the various parts of a rolling chain with pins, bushings, rollers, and link plates. The rollers turn on bushings that are press-fit to the inner link plates. The pins are prevented from rotating in the outer plates by the press-fit assembly. Table 19.10 gives dimensions for standard sizes. The size range given in Table 19.10 is large, so that chains can be used for both large and small amounts of power transmission. A large reduction in

**Figure 19.9**  Various parts of rolling chain.

**Table 19.10**     Standard sizes and strengths of rolling chains

| Chain number | Pitch $p_t$, in | Roller Diameter, in | Roller Width, in | Pin diameter $d$, in | Link plate thickness $a$, in | Average ultimate strength $S_u$, lb | Weight per foot, lb |
|---|---|---|---|---|---|---|---|
| [a]25 | 1/4 | 0.130 | 1/8 | 0.0905 | 0.030 | 875 | 0.084 |
| [a]35 | 3/8 | 0.200 | 3/16 | 0.141 | 0.050 | 2,100 | 0.21 |
| [a]41 | 1/2 | 0.306 | 1/4 | 0.141 | 0.050 | 2,000 | 0.28 |
| 40 | 1/2 | 5/16 | 5/16 | 0.156 | 0.060 | 3,700 | 0.41 |
| 50 | 5/8 | 2/5 | 3/8 | 0.200 | 0.080 | 6,100 | 0.68 |
| 60 | 3/4 | 15/32 | 1/2 | 0.234 | 0.094 | 8,500 | 1.00 |
| 80 | 1 | 5/8 | 5/8 | 0.312 | 0.125 | 14,500 | 1.69 |
| 100 | $1\frac{1}{4}$ | 3/4 | 3/4 | 0.375 | 0.156 | 24,000 | 2.49 |
| 120 | $1\frac{1}{2}$ | 7/8 | 1 | 0.437 | 0.187 | 34,000 | 3.67 |
| 140 | $1\frac{3}{4}$ | 1 | 1 | 0.500 | 0.219 | 46,000 | 4.93 |
| 160 | 2 | $1\frac{1}{8}$ | $1\frac{1}{4}$ | 0.562 | 0.250 | 58,000 | 6.43 |
| 180 | $2\frac{1}{4}$ | $1\frac{13}{32}$ | $1\frac{13}{32}$ | 0.687 | 0.281 | 76,000 | 8.70 |
| 200 | $2\frac{1}{2}$ | $1\frac{9}{16}$ | $1\frac{1}{2}$ | 0.781 | 0.312 | 95,000 | 10.51 |
| 240 | 3 | $1\frac{7}{8}$ | $1\frac{7}{8}$ | 0.937 | 0.375 | 130,000 | 16.90 |

|  [a]Without rollers.

speed can be obtained with rolling chains if desired. The tolerances for a chain drive are larger than for gears, and the installation of a chain is relatively easy.

The minimum wrap angle of the chain on the smaller sprocket is 120°. A smaller wrap angle can be used on idler sprockets, which are used to adjust the chain slack where the center distance is not adjustable. Horizontal drives (the line connecting the axes of sprockets is parallel to the ground) are recommended; vertical drives (the line connecting the axes of sprockets is perpendicular to the ground) are less desirable. Vertical drives, if used, should be used with idlers to prevent the chain from sagging and to avoid disengagement from the lower sprocket.

### 19.6.2   KINEMATICS

The velocity ratio, comparable to the gear ratio given in Equation (14.18), is

$$g_r = \frac{d_2}{d_1} = \frac{\omega_1}{\omega_2} = \frac{N_2}{N_1}$$

(19.22)

where $N$ = number of teeth in sprocket
$\omega$ = angular velocity, rad/s
$d$ = diameter, m

For a one-step transmission it is recommended that $g_r < 7$. Values of $g_r$ between 7 and 10 can be used at low speed ($< 650$ ft/min).

**Figure 19.10** Chordal rise in rolling chains.

### 19.6.3 Chordal Rise

An important factor affecting the smoothness of rolling chain drive operation, especially at high speeds, is chordal rise, shown in Figure 19.10. From right triangle 0CA

$$r_c = r \cos \theta_r \qquad (19.23)$$

The chordal rise while using Equation (19.23) gives

$$\Delta r = r - r_c = r(1 - \cos \theta_r) = r\left(1 - \cos \frac{180}{N}\right) \qquad (19.24)$$

where $N$ = number of teeth in sprocket. Note also from triangle 0CA

$$\sin \theta_r = \frac{p_t/2}{r} \qquad \text{or} \qquad p_t = 2r \sin \theta_r = D \sin \theta_r \qquad (19.25)$$

### 19.6.4 Chain Length

The number of links is

$$\frac{L}{p_t} = \frac{2c_d}{p_t} + \frac{N_1 + N_2}{2} + \frac{(N_2 - N_1)^2}{4\pi^2(c_d/p_t)} \qquad (19.26)$$

where $c_d$ = center distance between sprockets, m. It is normally recommended that $c_d/p_t$ lie between 30 and 50 pitches. If the center distance per pitch is not given, the designer is free to fix $c_d/p_t$ and to calculate $L/p_t$ from Equation (19.26). The nearest larger (preferably even) integer $L/p_t$ should be chosen. With $L/p_t$ as an integer, the center distance per pitch becomes

$$\frac{c_d}{p_t} = A + \sqrt{A^2 - \frac{B^2}{2}} \qquad (19.27)$$

where
$$A = \frac{1}{4}\left(\frac{L}{p_t} - \frac{N_1 + N_2}{2}\right) \tag{19.28}$$

$$B = \frac{N_2 - N_1}{2\pi} \tag{19.29}$$

The value of $c_d/p_t$ obtained from Equation (19.27) should be decreased by about 1% to provide slack in the nondriving chain strand.

The chain velocity in feet per minute is

$$u_1 = \frac{\pi N_{a1} D_1}{12} \tag{19.30}$$

$$u_1 = \frac{N_{a1} p_t N_1}{12} \tag{19.31}$$

where $N_{a1}$ = speed of member 1, rpm.

### 19.6.5   POWER RATING

The power that can be transmitted is expressed as

$$h_{pr} = h_p \frac{a_2}{a_1} \tag{19.32}$$

where $h_p$ = power that can be transmitted for a single strand, obtained from Table 19.11
$\quad a_1$ = service factor obtained from Table 19.12
$\quad a_2$ = multiple-strand factor obtained from Table 19.13

Rolling chains are almost always the $\frac{1}{4}$-pitch, no. 25, single-strand type, and Table 19.11 gives the transmitted power for this type. The four types of lubrication given in Table 19.11 are

Type I—Manual lubrication, oil applied periodically with brush or spout can

Type II—Drip lubrication, oil applied between link plate edges from a drop lubricator

Type III—Oil bath or oil slinger, oil level maintained in casing at predetermined height

Type IV—Oil stream, oil supplied by circulating pump inside chain loop or lower span

### 19.6.6   SELECTION OF SPROCKET SIZE AND CENTER DISTANCE

To ensure smooth performance and long life, the sprockets should consist of at least 17 teeth and no more than 67 teeth. For special situations where the speed is low or where space limitation is a factor, sprockets with fewer than 17 teeth can be used. Experience has shown that the velocity ratio given in Equation (19.22) should not exceed 7.

The small sprocket should have a chain wrap angle of 120° to have satisfactory operation and performance. The chain wrap angle of the smaller sprocket of a two-sprocket drive with speed ratio less than 3.5 will always be 120° or more. Chain wrap angle increases as the center distance is increased. For normal applications the center distance is between 30 and 50 pitches (30 and 50 $p_t$) of the chain.

**Table 19.11**   Transmitted power of single-strand, no. 25 rolling chain

| Number of teeth in small sprocket | Small Sprocket Speed, rpm | | | | | | | | | | | | | | | | | | | |
|---|---|---|---|---|---|---|---|---|---|---|---|---|---|---|---|---|---|---|---|---|
| | 100 | 500 | 900 | 1200 | 1800 | 2500 | 3000 | 3500 | 4000 | 4500 | 5000 | 5500 | 6000 | 6500 | 7000 | 7500 | 8000 | 8500 | 9000 | 10000 |
| 11 | 0.054 | 0.23 | 0.39 | 0.50 | 0.73 | 0.98 | 1.15 | 1.32 | 1.42 | 1.19 | 1.01 | 0.88 | 0.77 | 0.68 | 0.61 | 0.55 | 0.50 | 0.46 | 0.42 | 0.36 |
| 12 | 0.059 | 0.25 | 0.43 | 0.55 | 0.80 | 1.07 | 1.26 | 1.45 | 1.62 | 1.36 | 1.16 | 1.00 | 0.88 | 0.78 | 0.70 | 0.63 | 0.57 | 0.52 | 0.48 | 0.41 |
| 13 | 0.064 | 0.27 | 0.47 | 0.60 | 0.87 | 1.17 | 1.38 | 1.58 | 1.78 | 1.53 | 1.30 | 1.13 | 0.99 | 0.88 | 0.79 | 0.71 | 0.64 | 0.59 | 1.54 | 0.46 |
| 14 | 0.070 | 0.30 | 0.50 | 0.65 | 0.94 | 1.27 | 1.49 | 1.71 | 1.93 | 1.71 | 1.46 | 1.26 | 1.11 | 0.94 | 0.88 | 0.79 | 0.72 | 0.66 | 0.60 | 0.51 |
| 15 | 0.075 | 0.32 | 0.54 | 0.70 | 1.01 | 1.36 | 1.61 | 1.85 | 2.06 | 1.89 | 1.62 | 1.40 | 1.23 | 1.09 | 0.96 | 0.88 | 0.80 | 0.73 | 0.67 | 0.57 |
| 16 | 0.081 | 0.34 | 0.58 | 0.75 | 1.09 | 1.46 | 1.72 | 1.98 | 2.23 | 2.08 | 1.78 | 1.54 | 1.35 | 1.20 | 1.07 | 0.97 | 0.88 | 0.80 | 0.74 | 0.63 |
| 17 | 0.086 | 0.37 | 0.62 | 0.81 | 1.16 | 1.56 | 1.84 | 2.11 | 2.38 | 2.28 | 1.95 | 1.69 | 1.48 | 1.31 | 1.18 | 1.06 | 0.96 | 0.88 | 0.81 | 0.69 |
| 18 | 0.097 | 0.39 | 0.66 | 0.86 | 1.23 | 1.66 | 1.95 | 2.25 | 2.53 | 2.49 | 2.12 | 1.84 | 1.52 | 1.43 | 1.28 | 1.16 | 1.05 | 0.96 | 0.88 | 0.75 |
| 19 | 0.097 | 0.41 | 0.70 | 0.91 | 1.31 | 1.76 | 2.07 | 2.38 | 2.69 | 2.70 | 2.30 | 2.00 | 1.75 | 1.55 | 1.39 | 1.25 | 1.14 | 1.04 | 0.96 | 0.81 |
| 20 | 0.103 | 0.44 | 0.74 | 0.96 | 1.38 | 1.86 | 2.19 | 2.52 | 2.84 | 2.91 | 2.49 | 2.16 | 1.89 | 1.68 | 1.50 | 1.35 | 1.23 | 1.12 | 1.03 | 0.88 |
| 21 | 0.108 | 0.46 | 0.78 | 1.01 | 1.46 | 1.96 | 2.31 | 2.65 | 2.99 | 3.13 | 2.68 | 2.32 | 2.04 | 1.80 | 1.61 | 1.46 | 1.32 | 1.21 | 1.11 | 0.95 |
| 22 | 0.114 | 0.48 | 0.82 | 1.06 | 1.53 | 2.06 | 2.43 | 2.79 | 3.15 | 3.36 | 2.87 | 2.49 | 2.18 | 1.93 | 1.73 | 1.56 | 1.42 | 1.29 | 1.19 | 1.01 |
| 23 | 0.119 | 0.51 | 0.86 | 1.12 | 1.61 | 2.16 | 2.55 | 2.93 | 3.30 | 3.59 | 3.07 | 2.66 | 2.33 | 2.07 | 1.85 | 1.67 | 1.51 | 1.38 | 1.27 | 1.06 |
| 24 | 0.125 | 0.53 | 0.90 | 1.17 | 1.69 | 2.26 | 2.67 | 3.07 | 3.46 | 3.83 | 3.27 | 2.83 | 2.48 | 2.20 | 1.97 | 1.78 | 1.61 | 1.47 | 1.35 | 1.16 |
| 25 | 0.131 | 0.56 | 0.94 | 1.22 | 1.76 | 2.37 | 2.79 | 3.20 | 3.61 | 4.07 | 3.48 | 3.01 | 2.64 | 2.34 | 2.10 | 1.89 | 1.72 | 1.52 | 1.44 | 1.23 |
| 28 | 0.148 | 0.63 | 1.07 | 1.38 | 1.99 | 2.67 | 3.15 | 3.62 | 4.28 | 4.54 | 4.12 | 3.57 | 3.13 | 2.78 | 2.49 | 2.24 | 2.04 | 1.86 | 1.71 | 1.46 |
| 30 | 0.159 | 0.68 | 1.15 | 1.49 | 2.14 | 2.88 | 3.39 | 3.90 | 4.40 | 4.89 | 4.57 | 3.96 | 3.47 | 3.08 | 2.76 | 2.49 | 2.26 | 2.06 | 1.89 | 1.62 |
| 32 | 0.170 | 0.73 | 1.23 | 1.60 | 2.30 | 3.09 | 3.64 | 4.18 | 4.71 | 5.24 | 5.03 | 4.36 | 3.83 | 3.39 | 3.04 | 2.74 | 2.49 | 2.27 | 2.06 | 1.78 |
| 35 | 0.188 | 0.80 | 1.36 | 2.76 | 2.53 | 3.40 | 4.31 | 4.62 | 5.19 | 5.78 | 5.76 | 4.99 | 4.38 | 3.88 | 3.48 | 3.13 | 2.85 | 2.60 | 2.38 | 2.04 |
| 40 | 0.217 | 0.92 | 1.57 | 2.03 | 2.93 | 3.93 | 4.63 | 5.32 | 6.00 | 6.67 | 7.04 | 6.10 | 5.35 | 4.75 | 4.25 | 3.83 | 3.48 | 3.17 | 2.91 | 2.49 |
| 45 | 0.246 | 1.05 | 1.78 | 2.31 | 3.32 | 4.46 | 5.26 | 6.04 | 6.81 | 7.58 | 8.33 | 7.28 | 6.39 | 5.66 | 5.07 | 4.57 | 4.15 | 3.79 | 3.48 | 2.97 |
| 50 | 0.276 | 1.18 | 1.99 | 2.58 | 3.72 | 5.00 | 5.89 | 6.77 | 7.64 | 8.49 | 9.33 | 8.52 | 7.48 | 6.63 | 5.93 | 5.35 | 4.96 | 4.44 | 4.07 | 3.48 |
| 55 | 0.306 | 1.30 | 2.21 | 2.96 | 4.12 | 5.54 | 6.53 | 7.51 | 8.46 | 9.41 | 10.3 | 9.83 | 8.63 | 7.65 | 6.85 | 6.17 | 5.60 | 5.12 | 4.76 | 4.01 |
| 60 | 0.336 | 1.43 | 2.43 | 3.15 | 4.53 | 6.09 | 7.18 | 8.25 | 9.30 | 10.3 | 11.3 | 11.2 | 9.83 | 8.72 | 7.80 | 7.03 | 6.38 | 5.83 | 5.35 | 4.57 |

Lubrication zones indicated within the table: Type I, Type II, Type III, Type II lubrication, Type IV lubrication.

**Table 19.12**    Service factors for rolling chains

| Type of driven load | Type of Input Power | | |
|---|---|---|---|
| | Internal combustion engine with hydraulic drive | Electric motor or turbine | Internal combustion engine with mechanical drive |
| Smooth | 1.0 | 1.0 | 1.2 |
| Moderate shock | 1.2 | 1.3 | 1.4 |
| Heavy shock | 1.4 | 1.5 | 1.7 |

**Table 19.13**    Multiple-strand factors for rolling chains

| Number of strands | Multiple-strand factor $a_2$ |
|---|---|
| 2 | 1.7 |
| 3 | 2.5 |
| 4 | 3.3 |

**EXAMPLE 19.5**

**Given:** A four-strand, no. 25 rolling chain transmits power from a 21-tooth driving sprocket that turns at 1200 rpm. The speed ratio is 4:1.

**Find:** Determine the following:

(a) Power that can be transmitted for this drive
(b) Tension in the chain
(c) Safety factor of the chain based on minimum tensile strength
(d) Chain length if center distance is about 10 in

**Solution:**

(a) From Table 19.11 for a smaller sprocket of 21 teeth and a speed of 1200 rpm, the power that can be transmitted is 1.01 hp and type II lubrication is required. From Table 19.13 for four strands $a_2 = 3.3$. Nothing is mentioned about the type of input power or drive load, so assume that $a_1 = 1$. From Equation (19.32) the required power is

$$h_{pr} = h_p a_2 / a_1 = (1.01)(3.3)/(1) = 3.333 h_p$$

From Table 19.10 for no. 25 chain, the pitch is 0.25 in.

(b) The velocity in feet per minute can be obtained from Equation (19.31) as

$$u_1 = \frac{N_{a1} p_t N_1}{12} = \frac{(1200)(0.25)(21)}{12} = 525.0 \text{ ft/min}$$

Making use of Equation (4.37) gives

$$P_1 = \frac{33,000 h_{pr}}{u_1} = \frac{(33,000)(3.333)}{525} = 209.5 \text{ lb}$$

(c) From Table 19.10 the ultimate strength for a single strand is 875 lb. For four strands the ultimate strength is $4(875) = 3500$ lb. The safety factor is

$$n_s = \frac{P_{\text{all}}}{P_1} = \frac{3500}{209.5} = 16.71$$

(d) The number of teeth on the larger sprocket given a speed ratio of 4 is $4(21) = 84$ teeth. However,

$$\frac{c_d}{p_t} = \frac{10}{\frac{1}{4}} = 40$$

which is between the 30 and 50 pitches recommended. From Equation (19.26)

$$\frac{L}{p_t} = 2\left(\frac{c_d}{p_t}\right) + \frac{N_1 + N_2}{2} + \frac{(N_2 - N_1)^2}{4\pi^2(c_d/p_t)}$$

$$= 80 + \frac{21 + 84}{2} + \frac{(84 - 21)^2}{4\pi^2(40)}$$

$$= 80 + 52.5 + 2.513 = 135.0$$

Therefore, $L = 135$ pitches.

## Case Study 19.1 | DESIGN OF GANTRY FOR DRAGLINE

**Given:** A dragline, often used for mining or dredging operations, uses a large bucket to remove material and transport it elsewhere, usually into a trailer or a train box-car. The gantry line system is the portion of the dragline that supports the boom and fixes the boom angle, as shown in Figure 19.11. The boom weighs 10,200 lb, is 100 ft long, and can be considered to have its center of gravity at its geometric center. The dragline tips if the moment from the hoist rope and the boom exceeds 960 kip · ft. Figure 19.11 shows the dimensions of the dragline. The gantry line lengths are selected so that the 20-ft mast is vertical at a boom angle of $\theta = 30°$.

**Find:** Design a gantry line system for the dragline.

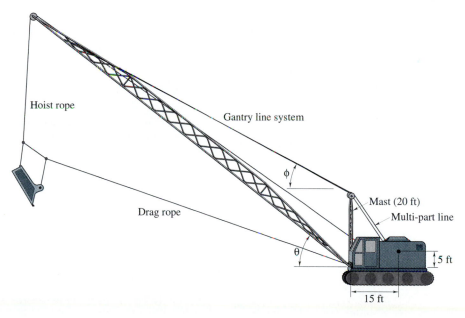

Hoist rope

Gantry line system

$\phi$

Mast (20 ft)

Multi-part line

Drag rope

$\theta$

5 ft

15 ft

**Figure 19.11** Typical dragline. *(continued)*

## Case Study    (Continued)

**Solution:** Since draglines have booms, many people confuse them with cranes. There are significant differences, however. Dragline load is limited by the size of the bucket, whereas a crane can attempt to lift extremely large loads. Further, a dragline will operate at a set boom angle for extended periods of time and will rarely, if ever, operate at very high or very low boom angles. Therefore, the range of boom angles that must be achieved is somewhat limited.

Just as with cranes, however, the gantry line system serves to eliminate bending forces on the boom. Thus, the gantry line must attach at the tip of the boom, known as the *boom point*. Attaching the gantry line on the crane's superstructure near the boom pivot will translate to extremely large forces because of static equilibrium. A mast is used to offset the gantry line system and obtain reasonably low stresses in the gantry line.

A number of clever design features have been incorporated into draglines in the past (Fig. 19.11). The lines closest to the crane have multiple pulleys (and hence are called a multiple-part line) to share the load and reduce the stress on the cable. Above the multiple-part line is a single line attached by proper couplings to the boom. Because this single line is never wound over a pulley, its size is determined by the rated strength of the cable.

The gantry line length can be obtained from geometry by considering the case when the boom is at a 30° elevation. The boom point is then 86.6 ft from the boom pivot and elevated 50 ft above the boom pivot. Since the mast is vertical and 20 ft long, the gantry rope length is given by

$$l_p = \sqrt{(86.6)^2 + (50 - 20)^2} = 91.65 \text{ ft}$$

The angle of the gantry line system from horizontal $\phi$ is then 19.1°. At any other boom angle $\theta$ the gantry line angle is given by

$$\phi = 19.1° + \theta - 30° = \theta - 10.9°$$

The gantry line load can be expressed as a function of $\theta$ by taking the moment equilibrium of the boom about the boom pivot:

$$L(100)\cos\theta + W(50)\cos\theta - (G\cos\phi)(100\sin\theta)$$
$$+ (G\sin\phi)(100\cos\theta) = 0$$

But the dragline tips over when the load and boom moments combined exceed 960 kip·ft. Thus, the gantry line tension is given by

$$G = \frac{960}{100(\sin\theta\cos\phi - \cos\theta\sin\phi)} = \frac{9.6}{\sin(\theta - \phi)}$$
$$= 51 \text{ kips}$$

The gantry line system must support 51 kips. Two lines, each to support 25.5 kips, are selected.

Because the mast rotates as the boom angle changes, the load supported by the gantry lines will change with the boom angle. From moment equilibrium of the mast, the total load on the gantry lines is

$$T_g = \frac{T_p \cos 19.1°}{\sin(\zeta - \phi)}$$

where $\zeta$ is the angle of the mast from horizontal and $\psi$ is the angle of the gantry line system from horizontal as shown above. These angles can be shown to be functions of $\theta$:

$$\zeta = \theta + 60°$$

$$\psi = \tan^{-1}\frac{20\sin\zeta - 5}{20\cos\zeta + 15}$$

The gantry line tension is highest at a boom angle of 0°. At this angle the multi-part line system must support a tension of 86.7 kips. Although draglines are not operated at such low boom angles because the lifting capacity is severely limited, this result places an upper bound on the load in the gantry line system.

In Figure 19.11 a number of equalizer pulleys (a multiple-part line) are used so that the load in the gantry line is reduced. The load in the gantry line depends on the number of pulleys or parts, each of which supports an equal share of the load. Thus, for $n$ parts to the line, the gantry line system tension is

$$T_g = \frac{86.7 \text{ kips}}{n}$$

The earlier discussion (Sec. 19.5.2) regarding pulley sizing is relevant. If the pulley and drum are too large, the motor capacity will be excessive. Per ASME requirements, pulley and drum diameters of $12d$ are to be chosen. These diameters reduce gantry line strength by approximately 12% (Fig. 19.7). Table 19.14 summarizes the results. From this summary, a number of reasonable alternatives exist. A five-part gantry line is selected as a good compromise between cost and safety.

## Case Study (Concluded)

**Table 19.14** Summary of results of Case Study 19.1

| Number of parts in gantry line | Minimum diameter of 6 × 37 line with 100-ksi breaking strength, in | Diameter selected (next-largest 0.2-in increment), in | Actual safety factor |
|---|---|---|---|
| 4 | 0.90 | 1.0 | 7.4 |
| 5 | 0.80 | 1.0 | 9.2 |
| 6 | 0.73 | 0.75 | 6.24 |
| 7 | 0.68 | 0.75 | 7.27 |
| 8 | 0.63 | 0.75 | 10.7 |

## 19.7 SUMMARY

Belts and ropes are machine elements (such as brakes and clutches) that use friction as a useful agent, in contrast to other machine elements in which friction is to be kept as low as possible. Belts, ropes, and chains provide a convenient means for transferring power from one shaft to another. This chapter discussed both flat and V-belts. All flat belts are subject to creep due to the relative motion between the pulley surface and the adjacent belt surface that is under load deformation from the combined tension and flexural stresses. Also, flat belts must be kept under tension to function and therefore require tensioning devices. One major advantage of V-belts over flat belts is that the wedging action of V-belts increases the normal force from $dN$ to $dN/\sin \beta$, where $\beta$ is the sheave angle.

Wire ropes are used instead of flat or V-belts when power must be transmitted over very long center distances. The major advantage of using rolling chains over flat or V-belts is that rolling chains do not slip. Another advantage of rolling chains over belt drives is that no initial tension is necessary and thus the shaft loads are smaller. The required length, power rating, and modes of failure were considered for belts, ropes, and rolling chains.

## KEY WORDS

**core** center of wire rope, mainly intended to support outer strands.

**flat belts** power transmission device that consists of loop of rectangular cross section placed under tension between pulleys.

**lay** type of twist in wire rope (regular or Lang); distance for strand to revolve around wire rope.

**rolling chains** power transmission device using rollers and links to form continuous loop, used with sprockets.

**sheaves** grooved pulleys in which V-belts run.

**strands** grouping of wire used to construct wire rope.

**synchronous belt**    flat belt with series of evenly spaced teeth on inside circumference, intended to eliminate slip and creep.

**timing belt**    same as **synchronous belt.**

**V-belt**    power transmission device with trapezoidal cross section placed under tension between grooved sheaves.

**wire rope**    wound collection of strands.

## RECOMMENDED READINGS

American Chain Association, *Chains for Power Transmission and Material Handling,* Marcel Dekker, New York.

Dickie, D. E. (1985) *Rigging Manual,* Construction Safety Association of Ontario, Toronto, Canada.

Juvinall, R. C., and Marshek, K. M. (2003) *Fundamentals of Machine Component Design,* 3rd ed. Wiley, New York.

Krutz, G. W., Schuelle, J. K., and Claar, P. W. (1994) *Machine Design for Mobile and Industrial Applications,* Society of Automotive Engineers, Warrendale, PA.

Mott, R. L. (1998) *Machine Elements in Mechanical Design,* 3rd ed. Merrill Publishing Co., Columbus, OH.

Rossnagel, W. E., Higgins, L. R., and MacDonald, J. A. (1988) *Handbook of Rigging,* 4th ed., McGraw-Hill, New York.

Shapiro, H. I., et al. (1991) *Cranes and Derricks,* 2nd ed., McGraw-Hill, New York.

Shigley, J. E., and Mischke, C. R., and Budynas, R. G. (2003) *Mechanical Engineering Design,* 7th ed., McGraw-Hill, New York.

*Wire Rope Users Manual* (1972) Armco Steel Corp, Granbury, TX.

## REFERENCES

ASME (1994) B30.5 *Mobile and Locomotive Cranes,* American Society of Mechanical Engineers, New York.

Shigley, J. E., and Mitchell, L. D. (1983) *Mechanical Engineering Design,* 4th ed., McGraw-Hill, New York.

## PROBLEMS

### Section 19.2

**19.1**   An open flat belt 10 in wide and 0.13 in thick connects a 16-in-diameter pulley with a 36-in-diameter pulley over a center distance of 15 ft. The belt speed is 3600 ft/min. The allowable preload per unit width of the belt is 100 lb/in, and the weight per volume is 0.042 lb/in$^3$. The coefficient of friction between the belt material and the pulley is 0.8. Find the length of the belt, the maximum forces acting on the belt before failure is experienced, and the maximum power that can be transmitted. *Ans.* $L = 441$ in, $h_{p,\max} = 169$ hp.

**19.2**   The driving pulley and the driven pulley in an open-flat-belt transmission each have a diameter of 160 mm. Calculate the preload needed in the belt if a power of 7 kW should be transmitted at 1000 rpm by using only one-half the wrap angle on each pulley. The coefficient of friction is 0.25, the density of the belt material is 1500 kg/m$^3$, and the allowable belt preload stress is 5 MPa. Also calculate the cross-sectional area of the belt. *Ans. $F_i = 2213$ N.*

Sketch *a*, for Problem 19.3

**19.3**   A belt transmission is driven by a motor hinged at point 0 in sketch *a*. The motor mass is 50 kg and its speed is 1400 rpm. Calculate the maximum transferable power for the transmission. Check, also, that the largest belt stress is lower than $\sigma_{all}$ when the bending stresses are included. Belt dimensions are 100 by 4 mm$^2$, the coefficient of friction is 0.32, the modulus of elasticity is 100 MPa, the belt density is 1200 kg/m$^3$, and the allowable stress is 20 MPa. *Ans. $h_p = 10.29$ kW.*

Sketch *b*, for Problem 19.4

★ **19.4**   The tension in a flat belt is given by the motor weight, as shown in sketch *b*. The mass is 80 kg and is assumed to be concentrated at the motor shaft position. The motor speed is 1405 rpm, and the pulley diameter is 400 mm. Calculate the belt width when the allowable belt stress is

| *Indicates problem of greater difficulty.

6.00 MPa, the coefficient of friction is 0.5, the belt thickness is 5 mm, the modulus of elasticity is 150 MPa, and the density is 1200 kg/m$^3$. *Ans. $w_t = 33.84$ mm.*

**19.5**    Calculate the maximum possible power transmitted when a full wrap angle is used as shown in sketch $c$. The belt speed is 5 m/s, and the coefficient of friction is 0.19. *Ans. $h_p = 400$ W.*

210°

Mass, 5 kg

180°

Driver

Sketch c, for Problem 19.5

★ **19.6**    Equation (19.10) gives the belt forces, where $F_1$ is the largest belt force. What is the maximum power the belt can transmit for a given value of $F_1$?

**19.7**    In a flat-belt drive with parallel belts, the motor and pulley are pivoted at point 0, as shown in sketch $d$. Calculate the maximum possible power transmitted at 1200 rpm. The motor and pulley together have a 50-kg mass at the center of the shaft. The centrifugal forces on the belt have to be included in the analysis. The coefficient of friction is 0.3, belt area is 300 mm$^2$, and density is 1500 kg/m$^3$. *Ans. $h_p = 11.3$ kW.*

$R = 100$ mm

$R$

$L_2 = 1.5R$

$\omega$

$L_1 = 0.8R$

0

Sketch d, for Problem 19.7

★ **19.8**    A flat-belt drive, shown in sketch $e$, has the top left pulley driven by a motor and is driving the other two pulleys. Calculate the wrap angles for the two driven pulleys when the full wrap angle is used on the driving pulley and the power from pulley 2 divided by the power from pulley 3 is equal to 2. The coefficient of friction is 0.3 and $l = 4r$. *Ans. $\alpha_3 = 37.2°$.*

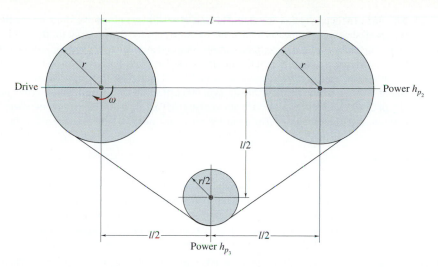

Sketch e, for Problem 19.8

★ **19.9** A flat belt 6 mm thick and 60 mm wide is used in a belt transmission with a speed ratio of 1. The radius of the pulleys is 100 mm, their angular speed is 50 rad/s, the center distance is 1000 mm, the belt modulus of elasticity is 200 MPa, and the density is 1 kg/dm³. The belt is pretensioned by increasing the center distance. Calculate this increase in center distance if the transmission should use only one-half the wrap angle to transmit 1 kW. Centrifugal effects should be considered. The coefficient of friction is 0.3. *Ans.* $\Delta a = 8$ mm.

**19.10** A 120-mm-wide and 5-mm-thick flat belt transfers power from a 250-mm-diameter driving pulley to a 700-mm-diameter driven pulley. The belt has a mass per length of 2.1 kg/m and a coefficient of friction of 0.3. The input power of the belt drive is 60 kW at 1000 rpm. The center distance is 3.5 m. Determine

*a)* The maximum tensile stress in the belt. *Ans.* $\sigma = 13.45$ MPa.
*b)* The loads for each pulley on the axis.

**19.11** A flat-belt drive is used to transfer 100 kW of power. The diameters of the driving and driven pulleys are 300 and 850 mm, respectively, and the center distance is 2 m. The belt has a width of 200 mm, thickness of 10 mm, speed of 20 m/s, and coefficient of friction of 0.35. For a safety factor of 2.0 for static loading, determine the following:

*a)* The belt tensions, contact angles, and length for an open configuration. *Ans.* $L = 5.84$ m.
*b)* The loads for each pulley on the axis. *Ans.* $F_x = 10.7$ kN, $F_y = 687$ N.

## Section 19.3

**19.12** A synchronous belt transmission is used as a timing belt for an overhead camshaft on a car motor. The belt is elastically prestressed at standstill to make sure it does not slip at high speeds due to centrifugal forces. The belt is 1100 mm long and weighs 200 g. The elastic spring constant for 1 m of belt material is $10^5$ N/m. The belt prestress elongates it 2 mm. It needs to elongate 4 mm more to start slipping. Calculate the maximum allowable motor speed if the pulley on the motor shaft has a diameter of 100 mm. *Ans.* $\omega = 1095$ rad/s.

**19.13** A timing belt for power transfer should be used at a velocity ratio of one-third, so that the outgoing speed should be 3 times as high as the incoming speed. The material for the belt reinforcement can be chosen from glass fiber, carbon fiber, and steel wire. These materials give different belt densities as well as different tensile strengths. The glass-fiber-reinforced belt has a density of 1400 kg/m³ and an allowable stress of 300 MPa. The carbon-fiber-reinforced belt has a density of 1300 kg/m³ and an allowable stress of 600 MPa. The steel-wire-reinforced belt has a density of 2100 kg/m³ and an allowable stress of 400 MPa. Calculate the maximum power for each belt type if the belt speed is 30 m/s, the belt cross section is 50 mm², and the safety factor is 10.

### Section 19.4

**19.14** A 25-hp, 1750-rpm electric motor drives a machine through a multiple-V-belt drive. The driver sheave has 3.7-in diameter, and the wrap angle of the driver is $165°$. The weight per length is 0.012 lb/in. Maximum belt preload is 150 lb, and the coefficient of friction of the belt material acting on the sheave is 0.2. How many belts should be used? Assume a sheave angle of $18°$. *Ans.* Three are needed.

★ **19.15** A V-belt drive has $r_1 = 200$ mm, $r_2 = 100$ mm, $2\beta = 36°$, and $c_d = 700$ mm. The speed of the smaller sheave is 1200 rpm. The cross-sectional area of the belt is 160 mm², and its density $\rho = 1500$ kg/m³. How large is the maximum possible power transmitted by six belts if each belt is prestressed to 200 N? The coefficient of friction $\mu = 0.30$. If 12 kW is transmitted, how large a part of the periphery of the smaller wheel is then active? *Ans.* $\phi = 63.4°$.

★ **19.16** A combined V-belt and flat-belt drive, shown in sketch $f$, has a speed ratio of 4. Three V-belts drive the sheave in three grooves but connect to the larger sheave as flat belts do on the cylindrical surface. The center distance is 6 times the smaller sheave radius, which is 80 mm with a speed of 1500 rpm. Determine which sheave can transmit the largest power without slip. Determine the ratio between the powers transmitted through the two sheaves. The angle $2\beta$ is $36°$, and the coefficient of friction is 0.30. *Ans.* $h_{p,\text{small}} = 2.64 h_{p,\text{large}}$.

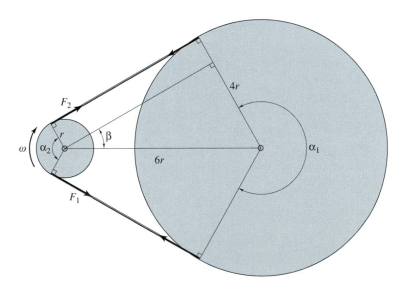

Sketch $f$, for Problem 19.16

★ **19.17** A V-belt drive for an industrial fan transmits 4 kW at a motor speed of 3000 rpm through four parallel V-belts of the $4L$ type. The sheave effective pitch diameter is 4 in, and the speed ratio is 1. The fan is running continuously. Calculate what the manufacturers presumed the coefficient of friction to be to get the rated power for the belt drive. Neglect power variations, and assume that all friction forces between the belt and the sheaves are tangent to the sheaves. The density of the V-belt is assumed to be 1000 kg/m³, and the prestress gives $F_1 + F_2 = $ constant independent of speed. Use the power ratings at 3000 and 5000 rpm for the calculations. *Ans.* $\mu = 0.12$.

**19.18** The input power to shaft A, shown in sketch $g$, is transferred to shaft B through a pair of mating spur gears and then to shaft C through a $2L$-section V-belt drive. The sheaves on shafts B and C have 76- and 200-mm diameters. For the maximum power the belt can transfer, determine the following:

*a)* The input and output torques of the system. *Ans.* $T_B = 0.829$ N · m, $T_C = 2.18$ N · m.
*b)* The belt length for an approximate center distance of 550 mm. *Ans.* $L = 1.32$ m.

Recall that 1 hp = 746 W.

Sketch $g$, for Problem 19.18

**19.19** An automobile fan is driven by an engine through a V-belt drive. The engine's sheave has a 350-mm diameter and is running at 880 rpm. The fan's sheave has a 220-mm diameter. The net air drag load of 2.3 N is applied tangentially at 330-mm radial distance from the fan axis. Consider 20% loss in other parts of this belt drive, and select the size of the V-belt.

## Section 19.5

**19.20** A hoist uses a 2-in, $6 \times 19$ monitor steel wire rope with Lang lay on cast steel sheaves. The rope is used to haul loads up to 8000 lb a distance of 480 ft.

*a)* Using a maximum acceleration of 2 ft/s², determine the tensile and bending stresses in the rope and their corresponding safety factors.

b)    Determine the bearing pressure in the rope and the safety factor when it interacts with the sheave. Also, determine the stretch of the rope.

c)    Determine the safety factor due to fatigue, and anticipate the number of bends until failure.

**19.21**  A steel rope for a crane has a cross-sectional area of 31 mm$^2$ and an ultimate strength of 1500 N/mm$^2$. The rope is 12 m long and is dimensioned to carry a maximum load of 1000 kg. If that load is allowed to free-fall 1 m before the rope is tightened, how large will the stress be in the rope? The modulus of elasticity is 68 GPa. *Ans.* $\sigma = 2236$ MPa.

**19.22**  Two 2-cm, $6 \times 19$ plow-steel wire ropes are used to haul mining material to a depth of 150 m at a speed of 8 m/s and an acceleration of 1 m/s$^2$. The pulley diameter is 80 cm. Using a proper safety factor, determine the maximum hauling load for these ropes. Are two ropes enough for infinite life? *Ans.* $F_w = 4000$ lb.

**19.23**  Using six 16-mm, $6 \times 19$ wire ropes, an elevator is to lift a 2500-kg weight to a height of 81 m at a speed of 4 m/s and an acceleration of 1 m/s$^2$. The 726-mm-diameter pulley has a strength of 6.41 MPa. Determine the safety factors for tension, bending, bearing pressure, and fatigue failure. Also, calculate the maximum elongation of the ropes. *Ans.* $n_{s,\text{tension}} = 11.1, n_{s,\text{fatigue}} = 1.80$.

★ **19.24**  Sketch *h* shows a wire rope drive used on an elevator to transport dishes between floors in a restaurant. The maximum mass being transported is 300 kg and hangs by a steel wire. The line drum is braked during the downward motion by a constant moment of 800 N · m. The rotating parts have a mass moment of inertia of 13 kg · m$^2$. Calculate the maximum force in the wire, and find when it first appears. *Ans.* $F_{\text{max}} = 3960$ N.

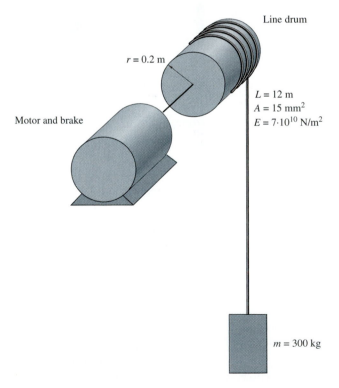

Sketch *h*, for Problem 19.24

★ **19.25** End and front views of a hoisting machine are shown in sketch $i$. Calculate the lifting height per drum revolution. Calculate the maximum force in the wire when the drum is instantly started with angular speed $\omega$ to lift a mass $m$. The free length of the wire is $L$, and its cross-sectional area is $A$. The modulus of elasticity of the wire is $E$.

Sketch $i$, for Problem 19.25

**19.26** A hauling unit with 25-mm, $8 \times 19$ wire rope made of fiber-core plow steel is used to raise a 45-kN load to a height of 150 m at a speed of 5 m/s and an acceleration of 2 m/s². Design for a minimum pulley diameter while determining the safety factors for tensile strength, bending stress, bearing pressure, and fatigue endurance.

**19.27** A building elevator operates at a speed of 5 m/s and acceleration of 1.2 m/s² and is designed for 10-kN deadweight, five 70-kg passengers, and 150-kg overload. The building has twelve 4-m stories. Using the proper safety factor, design the wire rope required for the elevator. A maximum of four ropes may be used.

## Section 19.6

Since the text only provides information about chain size 25, assume chain size 25 for problems 19.28 through 19.31.

**19.28** A three-strand, ANSI 50-roller chain is used to transfer power from a 20-tooth driving sprocket that rotates at 500 rpm to a 65-tooth driven sprocket. The input power is from an internal combustion engine, and the chain experiences medium shock. Determine the following:

a) The normal power. *Ans.* $h_p = 10.5$ kW per chain.
b) The length of the chain for an approximate center distance of 550 mm. *Ans.* $L = 1810$ mm.
c) The chordal speed variation. *Ans.* $\Delta V / V = 1.23\%$.

**19.29** To transfer 100 kW of power at 400 rpm, two strands of roller chain are needed. The load characteristics are heavy shock, poor lubrication, average temperature, and 16 h of service per day. The driving sprocket has 13 teeth, and the driven sprocket has 42 teeth. Determine the kind, length, and size of the chain for a center distance close to 50 pitches.

**19.30** A two-strand AISI 80-roller chain system is used to transmit the power of an electric motor rotating at 500 rpm. The driver sprocket has 12 teeth, and the driven sprocket has 60 teeth. The operating condition is specified by precision mounting and light shock. Calculate the power rating, the chordal speed variation, and the chain length for the maximum center distance. *Ans.* $h_{pr} = 6.15$ kW per chain, $L = 3.505$ m.

**19.31** A roller chain is used to transfer 9 kW of power from a 20-tooth driving sprocket running at 600 rpm to a driven sprocket with a speed ratio of 4:1. Design the chain drive for heavy shocks, 20 h of service per day, and poor lubrication. Also, determine the required chain length for a center distance of approximately 70 pitches.

# ELEMENTS OF MICROELECTROMECHANICAL SYSTEMS (MEMS)

A spider mite approaches a gear train. [Courtesy Sandia National Laboratory]

*There is plenty of room at the bottom.*
Richard Feynman

## SYMBOLS

| | |
|---|---|
| $a$ | radius or width, m |
| $A$ | cross-sectional area, $m^2$ |
| $b$ | width of rectangular plate, m |
| $d$ | piezoelectric coefficient |
| $E$ | Young's modulus, Pa |
| $F$ | electrostatic force, N |
| $h$ | gap, m |
| $K$ | electromechanical conversion factor |
| $l$ | length, m |
| $p$ | pressure, Pa |
| $Q$ | volumetric flow rate, $m^3/s$ |
| $r$ | radius |
| $t$ | plate thickness, m |
| $V$ | voltage, V |
| $\alpha$ | constant in Equation (20.7) |
| $\beta$ | constant in Equation (20.8) |
| $\delta$ | deflection, m |
| $\epsilon$ | permittivity, pF/m, or strain, m/m |
| $\epsilon_r$ | relative permittivity |
| $\nu$ | Poisson's ratio |
| $\sigma$ | stress, Pa |

## SUBSCRIPTS

| | |
|---|---|
| max | maximum |
| $r$ | radial direction |
| $x, y$ | Cartesian coordinates, m |
| $\theta$ | circumferential or hoop direction |

## 20.1    INTRODUCTION

A relatively new area of machine design relates to *microelectromechanical systems* or micromechanical devices. The term **microelectromechanical system** (or **MEMS**) refers to a machine that is mechanical in nature, has microscopic length scales, and is fully integrated with an electrical system in the form of an integrated circuit that controls the machine. **Microelectromechanical devices,** often referred to as **MEMS devices,** do not use circuit integration.

MEMS and MEMS devices promise to revolutionize society, and their potential applications include medical devices, sensors of all types, printers, magnetic storage devices, and communication systems. Already a multibillion-dollar industry, MEMS are commonly applied as sensors, in computer hard drives and printers, and in military applications. The field of MEMS is relatively new, but already has developed into a large field of study.

MEMS devices have been constructed from polycrystalline silicon (polysilicon) and single-crystal silicon because the technologies for integrated-circuit manufacture are well developed and exploited for these devices, and other new processes have been developed that are compatible with the existing processing steps. The use of anisotropic etching techniques allows the fabrication of devices with well-defined walls and high aspect ratios, and for this reason, some single-crystal silicon MEMS devices have been fabricated.

One of the recognized difficulties associated with silicon for MEMS devices is the high adhesion encountered at small length scales and the associated rapid wear. Most commercial devices are designed to avoid friction by, for example, using flexing springs instead of bushings; however, this complicates designs and makes some MEMS devices unfeasible. Therefore, significant research is being conducted to identify materials and lubricants that provide reasonable life and performance. Silicon carbide, diamond, and metals such as aluminum, tungsten, and nickel have been investigated as potential MEMS materials. Lubricants have also been investigated; it is known that surrounding the MEMS device in a silicone oil, for example, practically eliminates adhesive wear (see Sec. 8.9.1), but it also limits the performance of the device. Self-assembling layers of polymers are being investigated as well as novel and new materials with self-lubricating characteristics; however, the tribology of MEMS devices remains the main technological barrier to their widespread use.

An example of a commercial MEMS-based product is the digital pixel technology (DPT) device illustrated in Figure 20.1; this device uses an array of digital micromirror

**Figure 20.1**    The Texas Instruments digital pixel technology (DPT) device. (a) Exploded view of a single digital micromirror device (DMD). (b) View of two adjacent DMD pixels. (c) Images of DMD arrays, with some mirrors removed for clarity. Each mirror measures approximately 17 $\mu$m (670 $\mu$in) on a side. (d) A typical DPT device, used for digital projection systems, high-definition televisions, and other image display systems. The device shown contains 1,310,720 micromirrors and measures less than 2 in per side. (Source: Texas Instruments Corp.)

devices (DMDs) to project a digital image, such as in computer-driven projection systems. The aluminum mirrors can be tilted so that light is directed into or away from the optics that focus light onto a screen, so that each mirror can represent a pixel of an image's resolution. The mirror allows light or dark pixels to be projected, but levels of gray can also be accommodated. Since the switching time is about 15 $\mu$s, which is much faster than the human eye can respond, the mirror will switch between the on and off states to reflect the proper light dose to the optics.

An array of such mirrors represents a gray-scale screen; using three mirrors (one each for red, green, and blue light) for each pixel results in a color image with millions of discrete colors. Digital pixel technology is widely applied for digital projection systems, high-definition televisions, and other optical applications. However, to produce the device shown in Figure 20.1 requires the manufacture of full three-dimensional, multipart assemblies.

Small devices such as these cannot be designed through intuition based on macromachine elements discussed earlier in this text. Some of the main reasons are the following:

1.  Adhesive forces are much larger than internal stresses or other body forces. This can result in high friction and wear in improperly designed components.

2.  Material properties are very different from macro-scale bulk properties. For example, material strength is much higher than for larger structures; this can be more than 2 orders of magnitude higher strength.

3.  Lubrication is extremely difficult for MEMS devices, since the size of traditional lubricant molecules is often larger than the clearances required.

4.  Time scales of MEMS devices will be much shorter than those in conventional devices. Therefore, phenomenon such as heating or cooling can be performed very quickly, suggesting phenonema for machinery that cannot perform adequately at larger length scales.

## 20.2   INTRODUCTION TO MEMS MANUFACTURE

The manufacture of MEMS devices is a broad topic, worthy of a textbook in its own right. This section is intended to introduce the manufacturing processes associated with production of MEMS and micromechanical devices and closely follows the discussion of Kalpakjian and Schmid (2003); the interested reader is referred to Kalpakjian and Schmid (2003) or, Madou (1997) for further details regarding MEMS manufacture.

### 20.2.1   LITHOGRAPHY

**Lithography** is the process by which the geometric patterns that define the devices are transferred from a reticle or mask to the surface. A reticle is a glass or quartz plate with a pattern of the chip deposited onto it with a chromium film. The reticle image can be the same size as the desired structure on the chip, but it is often an enlarged image (usually 5X to 20X larger, although 10X magnification is most common). Enlarged images are then focused onto a wafer through a lens system, and this process is referred to as reduction lithography.

A silicon wafer will be coated with a photoresist, which is sensitive to ultraviolet (UV) light. The wafer is then baked to remove the solvent from and harden the photoresist. The wafer is aligned under the desired reticle and exposed to UV radiation. Upon development

**Figure 20.2**    Pattern transfer by lithography. Note that the mask in step 3 can be a positive or negative image of the pattern. [From Kalpakjian and Schmid (2003).]

and removal of the exposed photoresist, a duplicate of the reticle pattern will appear in the photoresist layer. As can be seen in Figure 20.2, the reticle can be a negative or a positive image of the desired pattern. A positive reticle uses the UV radiation to break down the chains in the organic film, so that these are preferentially removed by the developer. Positive masking is more common than negative masking because with negative masking the photoresist can swell and distort, making it unsuitable for small geometries.

The lithography process may be repeated as many as 25 times in the fabrication of the most advanced integrated circuits and MEMS.

## 20.2.2  ETCHING

**Etching** is the process by which entire films, or particular sections of films or the substrate, are removed; it plays an important role in the fabrication sequence. There are two basic types of etching: wet etching and dry etching.

Wet etching involves immersing the wafers in a liquid, usually acidic, solution. The main drawback to most wet etching operations is that they are isotropic; that is, they etch in all directions of the workpiece at the same rate, as illustrated in Figure 20.3. This results in undercuts beneath the mask material and limits the resolution of geometric features in the substrate.

Microelectronic devices and MEMS require accurate machining of structures, and this is done through masking. However, masking is a challenge with isotropic etchants. The strong acids etch aggressively and produce rounded cavities. Furthermore, the etch rate is very sensitive to agitation, and, therefore, lateral and vertical features are difficult to control. The size of features in MEMS determines their performance, and for this reason there is a strong desire to produce well-defined, extremely small structures. Such small features cannot be attained through isotropic etching because of the poor definition that results from undercutting of masks.

**Figure 20.3**     Etching directionality. (a) Isotropic etching: Etch proceeds vertically and horizontally at approximately the same rate, with significant mask undercut. (b) Orientation-dependent etching (ODE): Etch proceeds vertically, terminating on 111 crystal planes with little mask undercut. (c) Vertical etching: Etch proceeds vertically with little mask undercut. (Source: Courtesy K. R. Williams, Agilent Laboratories.)

Anisotropic etching takes place when etching is strongly dependent on compositional or structural variations in the material. There are two basic kinds of anisotropic etching: orientation-dependent etching (ODE) and vertical etching, although most vertical etching is done with dry plasmas and is discussed below. Orientation-dependent etching commonly occurs in a single crystal when etching takes place at different rates in different directions, as shown in Figure 20.3(b). When performed properly, these etchants produce geometric shapes with walls defined by the crystallographic planes that resist the etchants.

Modern integrated circuits and many MEMS are etched with dry etching, which involves the use of a plasma or discharge in areas of high electric and magnetic fields; any gases that are present are dissociated to form ions, photons, electrons, or highly reactive molecules. In contrast to the wet process, dry etching can have a high degree of directionality, resulting in highly anisotropic etch profiles [Fig. 20.3(c)]. Also, the dry process requires only small amounts of the reactant gases, whereas the solutions used in the wet process have to be refreshed periodically.

There are several specialized dry etching techniques, as discussed by Kalpakjian and Schmid (2003) and Madou (1997).

### 20.2.3     MICROMACHINING OF MEMS DEVICES

#### *Bulk Micromachining*

Until the early 1980s, **bulk micromachining** was the most common form of machining at micrometer scales. This process uses orientation-dependent etches on single-crystal silicon. The approach depends on etching down into a surface, stopping on certain crystal faces, doped regions, and etchable films to form a desired structure. As an example of the process, consider the fabrication of the silicon cantilever shown in Figure 20.4. By using the masking techniques described previously, a rectangular patch of the n-type silicon substrate is changed to p-type silicon through boron doping. Etchants such as potassium hydroxide will not be able to etch heavily boron-doped silicon, hence this patch will not be etched.

A mask is then produced, such as with silicon nitride on silicon. When etched with potassium hydroxide, the undoped silicon will be removed rapidly, while the mask and the doped patch essentially will be unaffected. Etching progresses until the (111) planes are

(a)

Diffused layer
(e.g., p-type Si)

Substrate
(e.g., n-type Si)

(b)

Nonetching mask
(e.g., silicon nitride)

(c)

Freestanding
cantilever

(111) planes

**Figure 20.4**    Schematic illustration of bulk micromachining. (a) Diffuse dopant in desired pattern. (b) Deposit and pattern masking film. (c) Orientation-dependent etch, leaving behind a freestanding structure. (Source: K. R. Williams, Agilent Laboratories.)

exposed in the n-type silicon substrate, and they undercut the patch, leaving a suspended cantilever as shown.

## Surface Micromachining

The basic steps in **surface micromachining** are illustrated in Figure 20.5 for silicon devices. A spacer or sacrificial layer is deposited onto a silicon substrate coated with a thin dielectric layer (isolation or buffer layer). Phosphosilicate glass deposited by chemical vapor deposition is the most common material for a spacer layer because it etches very rapidly in hydrofluoric acid. Figure 20.5(b) shows the spacer layer after application of masking and etching. At this point, a structural thin film is deposited onto the spacer layer; the film can be polysilicon, metal, metal alloy, or dielectric [Fig. 20.5(c)]. The structural film is then patterned, usually through dry etching to maintain vertical walls and tight dimensional

**Figure 20.5**    Schematic illustration of the steps in surface micromachining. (a) deposition of a phosphosilicate glass (PSG) spacer layer; (b) etching of spacer layer; (c) deposition of polysilicon; (d) etching of polysilicon; (e) selective wet etching of PSG, leaving the silicon substrate and deposited polysilicon unaffected. From Kalpakjian and Schmid [2003].

tolerances. Finally, wet etching of the sacrificial layer leaves a freestanding, three-dimensional structure [Fig. 20.5(e)]. Note that the wafer must be annealed to remove the residual stresses in the deposited metal before it is patterned. If this is not done, the structural film will severely warp once the spacer layer is removed.

Surface micromachining is a very widespread approach for the production of MEMS. Applications include accelerometers, pressure sensors, micropumps, micromotors, actuators, and microscopic locking mechanisms. Often, these devices require very large vertical walls which cannot be directly manufactured because the high vertical structure is difficult to deposit. This is overcome by machining large flat structures horizontally and then rotating or folding them into an upright position.

---

**EXAMPLE 20.1**

**Given:** A micromirror system is needed to reflect light that is parallel to a surface into a photosensor, as shown in Figure 20.6. It will use comb drives, as discussed in Section 20.4.1, and will then fold the material into a deployed position. To do so, special hinges such as shown in Figure 20.6 are integrated into the design.

**Figure 20.6**    (a) SEM image of a deployed micromirror. (b) Detail of the micromirror hinge. (Source: Sandia National Laboratories.)

**Figure 20.7**    Schematic illustration of the steps required to manufacture a hinge. (a) Deposition of a phosphosilicate glass (PSG) spacer layer and polysilicon layer. (b) Deposition of a second spacer layer. (c) Selective etching of the PSG. (d) Deposition of polysilicon to form a staple for the hinge. (e) After selective wet etching of the PSG, the hinge can rotate. [From Kalpakjian and Schmid (2003).]

**Find:** Describe the manufacturing steps necessary to produce the micromirror.

**Solution:** The manufacturing steps required to produce the hinges are illustrated in Figure 20.7. Hinges such as these have very high friction. If mirrors, as shown, are manually and carefully manipulated with probe needles, they will remain in position. Often, such mirrors will be combined with linear actuators to precisely control their deployment.

### 20.2.4    LIGA

**LIGA** is a German acronym for the combined process of X-ray lithography, electrodeposition, and molding (X-ray Lithographie, Galvanoformung und Abformung); a schematic illustration of this process is given in Figure 20.8.

The LIGA process involves the following steps:

1.  A very thick (up to hundreds of micrometers) resist layer of polymethylmethacrylate (PMMA) is deposited onto a primary substrate.
2.  The PMMA is exposed to columnated X-rays and developed.

**Figure 20.8**    The LIGA (lithography, electrodeposition, and molding) technique. (a) Primary production of a metal final product or mold insert. (b) Use of the primary part for secondary operations, or replication. (Source: IMM Institute für Mikrotechnik.)

3.  Metal is electrodeposited onto the primary substrate.

4.  The PMMA is removed or stripped, resulting in a freestanding metal structure.

5.  Plastic injection molding takes place.

Depending on the application, the final product from a LIGA process may be

1.  A freestanding metal structure, resulting from the electrodeposition process

2.  A plastic injection-molded structure

3.  An investment-cast metal part that is produced by using the injection-molded structure as a blank

4.  A slip-cast ceramic part, produced with the injection molded parts as the molds

The processing steps used to make freestanding metal structures are extremely time-consuming and expensive. The main advantage of LIGA is that these structures serve as molds for the rapid replication of submicrometer features through molding operations.

### 20.2.5    SOLID FREE-FORM FABRICATION OF DEVICES

Solid free-form fabrication is another term for rapid prototyping. This method is unique in that complex three-dimensional structures are produced through additive manufacturing, as opposed to material removal. Stereolithography is the most common form of solid free-form fabrication applied to MEMS, and it involves curing of a liquid thermosetting polymer by using a photoinitiator and a highly focused light source. Conventional stereolithography uses layers between 75 and 500 $\mu$m thick, with a laser dot focused to a 0.25-mm diameter. In microstereolithography, the process uses the same approach, but the laser is more highly focused (to a diameter as small as 1 $\mu$m) and layer thicknesses are around 10 $\mu$m. This technique has a number of cost advantages, but the MEMS devices are difficult to integrate with the controlling circuitry because stereolithography produces nonconducting polymer structures.

## 20.3    SPRINGS

Traditional macro-scale machine elements are often mounted on shafts or bolted onto beams or columns. This is not possible with MEMS because of the unavailability of long-lived bearings. In fact, much of the art of MEMS design is the substitution of other elements for bearings, with the goal of eliminating sliding between surfaces. (Recall that at small length scales, adhesive forces are large compared to applied forces, so even surfaces under nominal external loads will still wear quickly when they slide over each other.)

A common approach with MEMS devices is to mount elements onto elastic spring supports which deflect sufficiently to allow for proper operation. The springs are usually cantilevers or beams that are easily fabricated through micromachining. Commonly encountered geometries are summarized here, using the basic theory of Chapters 4 and 5. Further solutions can be found in Young (1989) or Timoshenko and Goodier (1970).

## 20.3.1   BEAMS

Cantilevered and anchored beams are common mounts for MEMS devices. They are commonly used with accelerometers, where a suspended mass deflects the beam when the system is subjected to acceleration. Position-sensing devices such as the probes in atomic force microscopes are affixed to the end of cantilevers. Often, devices are suspended at the end of cantilevers merely to lift them off the surface to eliminate high stiction (adhesion) forces.

The common situations for MEMS beam supports are summarized in Table 20.1. The simply supported situation shown is a good approximation for beams that are anchored to a surface. Note that the complete moment or deflection distributions are not normally needed for MEMS devices; but if needed, they can be calculated by using the methods of Chapter 4.

**Table 20.1**   Summary of important beam situations for MEMS devices

| Situation | Maximum deflection |
|---|---|
| Simply supported or anchored beam | $\delta_{max} = \dfrac{Pl^3}{48EI}$ |
| Cantilever | $\delta_{max} = \dfrac{Pl^3}{3EI}$ |
| Fixed-fixed beam | $\delta_{max} = \dfrac{Pl^3}{192EI}$ |
| Cantilever loaded in torsion | $\theta_{max} = \dfrac{Tl}{KG}$ $\quad K = ab^3\left[\dfrac{16}{3} - 3.36\dfrac{b}{a}\left(1 - \dfrac{b^4}{12a^4}\right)\right]$ $\quad$ for $a \geq b$ |

SOURCE: From Hsu (2002) and Young (1989).

**Given:**  A diamond-tipped stainless steel cantilever probe for an atomic force microscope is shown in Figure 20.9. The key dimensions are

$$l = 372 \ \mu m \qquad w_t = 93 \ \mu m \qquad t = 7.25 \ \mu m$$

Use $E = 190$ GPa and $\nu = 0.3$ for stainless steel.

**EXAMPLE 20.2**

**Find:**  Determine the normal and torsional stiffness of the cantilever.

**Solution:**

1. **Normal stiffness.**  For this cantilever, note that

$$I = \tfrac{1}{12}bh^3 = \tfrac{1}{12}(93 \times 10^{-6} \ m)(7.25 \times 10^{-6} \ m)^3 = 2.953 \times 10^{-21} \ m^4$$

Therefore, from Table 20.1,

$$\delta_{\text{max}} = -\frac{Pl^3}{3EI}$$

$$\therefore \qquad k = \frac{P}{\delta} = \frac{3EI}{l^3} = \frac{3(190 \ \text{GPa})(2.953 \times 10^{-21} \ m^4)}{(372 \times 10^{-6} \ m)^3} = 32.70 \ \text{N/m}$$

2. **Torsional stiffness.**  Note from Table 20.1 for the cantilever loaded in torsion that we need to use $a = 46.5 \ \mu m$ and $b = 3.625 \ \mu m$. Therefore,

$$K = ab^3\left[\frac{16}{3} - 3.36\frac{b}{a}\left(1 - \frac{b^4}{12a^4}\right)\right]$$

$$= (46.5 \times 10^{-6} \ m)(3.625 \times 10^{-6} \ m)^3\left[\frac{16}{3} - (3.36)\frac{3.625}{46.5}\left(1 - \frac{(3.625)^4}{12(46.5)^4}\right)\right]$$

$$= 1.123 \times 10^{-20} \ m^4$$

(a)                                    (b)

**Figure 20.9**  Scanning electron microscope images of a diamond-tipped cantilever probe used in atomic force microscopy. (a) Side view with detail of diamond; (b) bottom view of entire cantilever.

The shear modulus is obtained as

$$G = \frac{E}{2(1+v)} = \frac{190 \text{ GPa}}{2(1+0.3)} = 73.08 \text{ GPa}$$

Therefore, the stiffness is calculated from Table 20.1 as

$$k = \frac{T}{\theta} = \frac{KG}{l} = \frac{(1.123 \times 10^{-20} \text{ m}^4)(73.08 \text{ GPa})}{372 \times 10^{-6} \text{ m}} = 2.206 \times 10^{-6} \text{ N} \cdot \text{m/rad}$$

In practice, analytical evaluation of cantilever stiffness is limited in accuracy by the estimates of the cantilever dimensions. For example, the length of the cantilever can be seen from Figure 20.9(b) as being somewhat subjective, since it is not clear what role the adhesive will play in the cantilever deformation. Further, the error associated with length measurements of these scales can be relatively high.

### 20.3.2    PLATES

Plates are often used to support devices, but in such cases they are relatively thick and rigid, with low stresses and strains under normal loads. A different but common situation occurs when a plate is thin enough to deflect under an applied pressure, so that the plate can serve as a pressure sensor. Consider a circular plate of radius $a$ subjected to a pressure $p$. The important configuration occurs where the plate is supported around its edges to define a diaphragm. The maximum deflection occurs at the center of the plate and is given by

$$\delta_{\text{max}} = -\frac{3p(1-v^2)a^4}{16Et^3} \tag{20.1}$$

where $v$ = Poisson's ratio
$a$ = plate radius
$E$ = Young's modulus, Pa (psi)
$t$ = plate thickness, m (in)

The maximum radial and hoop stresses occur at the edge of the plate and are given by

$$\sigma_{r,\text{max}} = \frac{3a^2 p}{4t^2} \tag{20.2}$$

$$\sigma_{\theta,\text{max}} = \frac{3va^2 p}{4t^2} \tag{20.3}$$

At the center of the plate, these stresses are equal and are given by

$$\sigma_r = \sigma_\theta = \frac{3va^2 p}{8t^2} \tag{20.4}$$

Square plates are commonly used for pressure sensors, because they can be easily machined from a silicon substrate. For a square plate with edges fixed, thickness $t$, side length $a$, and applied pressure $p$, the maximum deflection occurs at the center of the plate and is given by

$$\delta_{\text{max}} = -\frac{0.0138pa^4}{Et^3} \tag{20.5}$$

**Table 20.2**    Coefficients $\alpha$ and $\beta$ for analysis of rectangular plate pressure sensor

| a/b | 1.0 | 1.2 | 1.4 | 1.6 | 1.8 | 2.0 | $\infty$ |
|---|---|---|---|---|---|---|---|
| $\alpha$ | 0.0138 | 0.0188 | 0.0226 | 0.0251 | 0.0267 | 0.0277 | 0.0284 |
| $\beta$ | 0.3078 | 0.3834 | 0.4356 | 0.4680 | 0.4872 | 0.4974 | 0.5000 |

| SOURCE: From Hsu (2002).

The maximum stress occurs at the center of each edge and is given by

$$\sigma_{\max} = \frac{0.308\,pa^2}{t^2} \tag{20.6}$$

For the general case of a rectangular plate with sides $a$ and $b$ and thickness $t$, the maximum deflection occurs at the center of the plate and is given by

$$\delta_{\max} = -\alpha \frac{pb^4}{Et^3} \tag{20.7}$$

The maximum stress occurs at the center of the longer edge and is given by

$$\sigma_{\max} = \beta \frac{pb^2}{t^2} \tag{20.8}$$

where the coefficients $\alpha$ and $\beta$ are given in Table 20.2.

## 20.4   ACTUATORS

Actuators used in macromachines are usually power screws, pneumatic cylinders, cams, or other common arrangements. These machine elements are difficult to manufacture for MEMS devices and besides are very inefficient at the small length scales involved. Similarly, the types of actuators used in MEMS are not suitable for macromachinery applications.

### 20.4.1   ELECTROSTATIC ACTUATORS

Consider the two parallel plates shown in Figure 20.10(a). According to electrostatics, the attractive force between the two plates when a potential voltage $V$ is applied is given by

$$F_y = \frac{\epsilon \epsilon_r V^2 A}{2h^2} \tag{20.9}$$

where   $F_y$ = attractive force (in the $y$ direction), N (lb)
      $\epsilon$ = permittivity of air, 8.85 pF/m
      $\epsilon_r$ = relative permittivity of gap material
      $V$ = applied voltage difference between plates
      $A$ = cross-sectional area of plates
      $h$ = gap clearance, m (in)

Consider now the case where the two plates are at the same potential, and a third plate at an elevated potential is partially inserted between them, as shown in Figure 20.10(b). The

**Figure 20.10**    Illustration of electrostatic actuation. (a) Attractive forces between charged plates; (b) forces resulting from eccentric charged plate between two other plates; (c) schematic illustration of a comb drive.

attractive stresses in the $y$ direction will cancel, but an attractive force in the $x$ direction is generated, given by

$$F_x = \frac{\epsilon \epsilon_r V^2 a}{h}$$

(20.10)

where $a$ is the width of the plate.

This is the basic principle behind a **comb drive,** an example of which is schematically illustrated in Figure 20.10(c) and shown in Figure 20.11. Note that there are many interstitial plates, since the attractive force predicted by Equation (20.10) is very small; the use of many plates amplifies this force and gives such drives the appearance of a comb, for which they are named.

A micromotor can be constructed using the same principle of electrostatic attraction. Figure 20.12(a) shows the schematic of a rotary micromotor, while Figure 20.12(b) shows a micromotor produced through the LIGA process. The stators shown are stationary, while the rotors are mounted on a bearing so that they can rotate. Note in the sketch that there 12 stators and 8 rotors, spaced so that there are always sufficient rotors eccentric to the stator. Some of the stators can be brought to a different voltage, while others (those aligned with the rotor) are placed at the same potential as the rotor. This ensures that a tangential force can always be generated according to Equation (20.2), and the rotor will be able to spin.

**Figure 20.11**    A comb drive. Note the springs in the center provide a restoring force to return the electrostatic comb teeth to their original position. (From Sandia National Laboratories.)

### 20.4.2    PIEZOELECTRIC ACTUATORS

A material exhibiting the **piezoelectric effect** generates a voltage potential when subjected to stress or strain. The behavior is reversible; application of a voltage will cause a strain in a piezoelectric material. Many materials exhibit a piezoelectric effect, including quartz, tourmaline, sodium potassium tartrate, barium titanate, and lead zirconate titanate (PZT). The most common use of the piezoelectric effect is for the generation of high voltages through application of large compressive stresses for impact detonation.

Piezoelectric materials are exploited for a number of MEMS devices, but are rarely used in macro-scale machines because the deflection of the materials is low, even for high voltages. However, a piezoelectric material can be used as an actuator, where it serves as a structural support; or it can be a sensor to measure strains or stresses.

The efficiency of a piezoelectric device is determined by the electromechanical conversion factor $K$, defined as

$$K = \sqrt{\frac{\text{output of mechanical energy}}{\text{input of electrical energy}}} = \sqrt{\frac{\text{output of electrical energy}}{\text{input of mechanical energy}}} \qquad (20.11)$$

(a)                                                              (b)

**Figure 20.12**    (a) Schematic illustration of a rotary electrostatic motor, sometimes called a *slide motor*; (b) scanning electron microscope image of a rotary micromotor. (Courtesy R. Kassing, University of Kassel.)

The electromechanical conversion factor is summarized for assorted materials in Table 20.3. The voltage produced by a stress is given by

$$V = \frac{\epsilon l}{d} = \frac{\sigma l}{dE} \qquad (20.12)$$

where $V$ = applied voltage, V
$\quad l$ = length of piezoelectric material in the direction of strain, m (in)
$\quad d$ = material constant, summarized in Table 20.3, m/V (in/V)
$\quad E$ = Young's modulus, Pa (psi)

**Table 20.3**    Properties of common piezoelectric materials

| Material | Coefficient $d$, $10^{-12}$ m/V | Electromechanical conversion factor $K$ |
|---|---|---|
| Quartz (single-crystal $SiO_2$) | 2.3 | 0.1 |
| Barium titanate ($BaTiO_3$) | 100–190 | 0.49 |
| Lead zirconate titanate, PZT ($PbTi_{1-x}Zr_xO_3$) | 480 | 0.72 |
| $PbZrTiO_6$ | 250 | |
| $PbNb_2O_6$ | 80 | |
| Rochelle salt ($NaKC_4H_4O_6\text{-}4H_2O$) | 350 | 0.78 |
| Polyvinylidene fluoride, PVDF | 18 | |

I  SOURCE: From Hsu (2002).

### 20.4.3  GEARS AND GEAR TRAINS

Spur gears have been widely applied in MEMS to achieve complicated motions. The kine-matic analysis of gears and gear trains is identical to that in Chapter 14, but note that no bending stress or contact stress equations analogous to the AGMA equations exist for gears in MEMS, nor have common MEMS materials had allowable stress numbers prescribed to date. Further, the profiles of gears are rarely as sophisticated as involute (they are usually linear), and usually have comparatively high backlash. Some gear trains are manually as-sembled, but they can also be produced through a number of surface micromachining steps.

## 20.5  FLUID FLOW DEVICES

For many MEMS and micromechanical devices, it is important to be able to pump fluid through small microchannels for various reasons, including sensing of chemicals, generation of ink droplets, or delivery of drugs for bio-MEMS devices. Conventional pumps are not suited for microcavity pumping; the pressure drop in a fluid flowing in a channel is given by [White (1994)]

$$\Delta p = \frac{8\eta l\, Q}{\pi a^4} \tag{20.13}$$

where $\Delta p$ = pressure loss over tube length $l$, Pa
$\eta$ = viscosity of fluid, N · s/m$^2$
$Q$ = volumetric flow rate of fluid, m$^3$/s
$a$ = radius of circular channel, m
$l$ = channel length, m

Note that the power required to overcome this pressure loss is the product of the pressure and the velocity of the fluid. Since the pressure loss is inversely proportional to the channel size to the fourth power, the pressure loss is very high for microchannels, and the use of a volumetric pressure to force fluid flow through channels is not a viable option.

Fortunately a large number of micropumps have been demonstrated for various ap-plications. Figure 20.13 illustrates a piezoelectric pump, which consists of a thin-walled flexible tube. The outside of the tube is coated with a piezoelectric material (see Sec. 20.4.2) and aluminum interdigital transducers (AITs). By applying a radio-frequency voltage to the AIT, the tube wall can be made to deform in a wave, transporting the fluid through surface forces instead of volumetric pressure.

Thermal printers are another option for fluid flow, and they have been used successfully in ink-jet printers. These printers operate by ejecting nano- or picoliters ($10^{-12}$ L) of ink from a nozzle toward paper. Ink-jet printers use a variety of designs, but silicon machin-ing technology is most applicable to high-resolution printers. It should be realized that a resolution of 1200 dpi (dots per inch) requires a nozzle spacing of approximately 20 $\mu$m.

The mode of operation, a thermal ink-jet printer is shown in Figure 20.14. When an ink droplet is to be generated and expelled, a tantalum resistor below a nozzle is heated. This heats a thin film of ink so that a bubble forms within 5 $\mu$s, which expands rapidly with internal pressures reaching 1.4 MPa (200 psi), and as a result forcing fluid rapidly out of the nozzle. Within 24 $\mu$s, the tail of the ink-jet droplet separates because of surface tension,

**Figure 20.13**    Capillary tube for microflow. (a) Schematic illustration of tube construction; (b) induced traveling wave and fluid flow. [From Hsu (2002).]

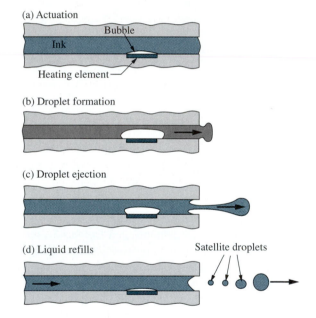

**Figure 20.14**    (a) Sequence of operation of a thermal ink-jet printer. (b) Resistive heating element is turned on, rapidly vaporizing ink and forming a bubble. (b) Within 5 $\mu$s, the bubble has expanded and displaced liquid ink from the nozzle. (c) Surface tension breaks the ink stream into a bubble, which is discharged at high velocity. The heating element is turned off at this time, so that the bubble collapses as heat is transferred to the surrounding ink. (d) Within 24 $\mu$s, an ink droplet (and undesirable satellite droplets) is ejected, and surface tension of the ink draws more liquid from the reservoir. [From Gad-el-hak (2002).]

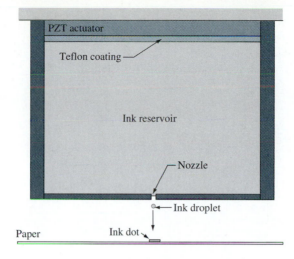

**Figure 20.15**    Schematic illustration of a piezoelectric-driven ink-jet printer head. [From Hsu (2002).]

the heat source is removed (turned off), and the bubble collapses inside the nozzle. Within 50 $\mu$s, sufficient ink has been drawn into the nozzle from a reservoir to form the desired meniscus for the next droplet.

Another type of printing mechanism uses a piezoelectric actuator and is shown schematically in Figure 20.15. A piezoelectric layer (usually PZT) is supported rigidly, with a Teflon boundary layer between the PZT and the liquid ink. By applying a voltage to the PZT, it can expand, displacing a volume of fluid in the reservoir. The liquid is forced to flow out of the small orifice and, because the expansion is rapid, is ejected as a ballistic particle.

**Given:**   An ink-jet printer head, as shown in Figure 20.15, is to be used to produce 11.8 dots per millimeter (300 dpi) with a 1-$\mu$m film thickness on paper. The PZT thickness is 15 $\mu$m, and the inkwell has a diameter of 2 mm and is fully replenished after each droplet is ejected.

**EXAMPLE 20.3**

**Find:**   Calculate the required orifice diameter, the droplet volume, and the required voltage.

**Solution:**   The required droplet volume can be calculated by considering the ink volume on paper. The ink dot has a diameter of 1/11.8 mm = 0.0847 mm = 84.7 $\mu$m. Since the thickness of the ink dot is 1 $\mu$m, the volume of the droplet is

$$\text{Volume} = \frac{\pi}{4}D^2 t = \frac{\pi}{4}(84.7 \times 10^{-6}\ \text{m})^2(1 \times 10^{-6}\ \text{m}) = 5.634 \times 10^{-15}\ \text{m}^3$$

Note that this could also be expressed as 5.634 pL. Assuming the droplet takes the shape of a sphere with the same radius as the orifice, we can equate the volume to the orifice radius as

$$\frac{4\pi r^3}{3} = 5.634 \times 10^{-15}\ \text{m}^3$$

$$\therefore \quad r = 11.04 \times 10^{-6}\ \text{m} = 11.04\ \mu\text{m}$$

so that the orifice needs a diameter of 22.08 $\mu$m. To generate the required volume in the inkwell, the PZT needs to expand by an amount $\delta$ given by

$$\delta \frac{\pi D_{\text{inkwell}}^2}{4} = 5.634 \times 10^{-15} \text{ m}^3$$

$$\therefore \qquad \delta = \frac{4(5.634 \times 10^{-15} \text{ m}^3)}{\pi (0.002 \text{ m})^2} = 1.793 \times 10^{-9} \text{ m}$$

Since the PZT thickness is 15 $\mu$m, the required strain in the PZT is

$$\epsilon = \frac{1.793 \times 10^{-9} \text{ m}}{15 \ \mu\text{m}} = 1.196 \times 10^{-4}$$

From Table 20.3, we obtain for PZT that $d = 480 \times 10^{-12}$ m/V. Therefore, from Equation (20.12),

$$V = \frac{\epsilon l}{d} = \frac{(1.196 \times 10^{-4})(15 \ \mu\text{m})}{480 \times 10^{-12} \text{ m/V}} = 3.736 \text{ V}$$

## 20.6 SELECTED SENSORS

Perhaps the most important application of MEMS and MEMS devices is for sensors of all types. Because of the small length scales involved, these sensors can give extremely fast response times and can be embedded into small volumes. A wide variety of sensors are available, incorporating the elements described in this chapter as well as specialized detectors for the particular quantity being measured.

### 20.6.1 CHEMICAL SENSORS

Chemical sensors are used to measure specific compounds, often gases in small concentrations. Miniature devices are desirable for these situations, since diffusion or chemical reaction-based phenomena will be more rapidly noticed for a small test volume. Chemical sensors are usually of the following types:

1. **Chemiresistor sensors.** In these sensors, organic polymers are embedded with metallic particles and tuned so that diffusion of certain gases into the polymer changes the resistance of the system. For example, the polymer phthalocyanide is used with copper to sense ammonia ($NH_3$) and nitrogen dioxide ($NO_2$) gases.

2. **Chemicapacitor sensors.** Special polymers can be placed between metal plates and then act as the dielectric in a capacitor. Exposure to certain gas species then changes the dielectric properties of the polymer, which is noted as a change in capacitance between the metal plates. An example is the use of polyphenylacetylene to sense gases such as carbon monoxide and carbon dioxide.

3. **Chemimechanical sensors.** Many polymers change shape when subjected to particular chemical species. Perhaps the most common circumstance of chemimechanical effect is the warpage of polymers when exposed to water vapor.

4. **Metal oxide gas sensors.** Similar to chemiresistor sensors, metal oxide sensors are commonly applied in automotive applications to monitor the combustion process and allow control of combustion to minimize production of pollutants. These sensors involve coating a sensor material, sometimes with the presence of a catalyst, onto a

**Table 20.4**      Common metal oxide sensors

| Metal coating | Catalyst | Detected gas |
|---|---|---|
| $BaTiO_3/CuO$ | $La_2O_3$, $CaCO_3$ | $CO_2$ |
| $SNO_2$ | $Pt + Sb$ | CO |
| $SNO_2$ | Pt | Alcohols |
| $SNO_2$ | $Sb_2O_3$ | $H_2$, $O_2$, $H_2S$ |
| $SNO_2$ | CuO | $H_2S$ |
| $WO_3$ | Pt | $NH_3$ |
| $Fe_2O_3$ | Ti-doped Au | CO |
| $Ga_2O_3$ | Au | CO |
| $MoO_3$ | — | $NO_2$, CO |
| $In_2O_3$ | — | $O_3$ |

| SOURCE: From Kovacs (1998) and Hsu (2002).

semiconductor and then measuring the resistance change when exposed to gases. Some of the common sensor materials are summarized in Table 20.4.

## 20.6.2   ACCELEROMETERS

Acceleration-measuring devices can be very simple, involving a suspended mass on a spring (cantilever). Automotive accelerometers are more elaborate, involving spring-mass systems with an integrated measurement system to measure deflection. For example, the accelerometer shown in Figure 20.16(b) and schematically in Figure 20.16(a) consists of a

(a)                                                                 (b)

**Figure 20.16**   (a) Schematic illustration of accellerometer; (b) photograph of Analog Devices ADXL-50 accelerometer with a surface micromachined capacitive sensor (center), on-chip excitation, self-test and signal conditioning circuitry. The entire chip measures 0.500 by 0.625 mm. [From Core (1993).]

suspended mass supported by a set of cantilevers anchored to the surface. The mass needs to be suspended above the surface to avoid sticking due to high adhesion. In the sensor shown in Figure 20.16(b), the position of the mass is monitored by measuring the change in capacitance between the support cantilevers and the stationary electrodes.

Automotive accelerometers will incorporate numerous suspended masses in order to measure accelerations in different directions. The sensor shown in Figure 20.16(b) is used for deployment of air bags, and the MEMS device measures $3 \times 3$ mm.

## 20.7   SUMMARY

Micro-Electro-Mechanical Systems (MEMS) and devices are a relatively new form of machinery, and use very different manufacturing technologies and machine elements from conventional, macro, machines. A Micro-Electro-Mechanical system is a mechanical system that is fully integrated with control circuitry, while a MEMS device is a mechanical device without the associated electrical system. The manufacturing processes are direct applications or extensions of the processes used in microcircuit manufacture, and include the important steps of lithography, etching, coating application and surface or bulk micromachining. Some unique manufacturing processes such as LIGA and solid freeform fabrication.

Machine element design is a challenge for MEMS applications, because of the manufacturing challenges, but also because of tribological considerations. Adhesive forces and wear rates are very high for small devices, and effective lubrication is extremely difficult to achieve. Common machine elements at the macro scale such as journal and rolling element bearings cannot be applied directly to MEMS and MEMS devices, and instead design solutions such as the use of spring supports or moving electrostatic drives have to be pursued.

The MEMS elements described in this chapter include structural components such as beams and plates, electrostatic and piezoelectric actuators, gears and gear trains, fluid flow devices and a number of sensors.

## KEY WORDS

**bulk micromachining**   machining of substrate material.

**comb drive**   an electrostatic actuator that uses a series of plates and stationary electrodes.

**etching**   controlled removal of material layers.

**LIGA**   a German acronym for X-ray lithography, electrodeposition, and molding.

**lithography**   the process of transferring a pattern to a surface.

**MEMS**   see **microelectromechanical system.**

**microelectromechanical device**   a machine with microscopic length scales that incorporates mechanical elements but does not involve an integrated electronic system.

**microelectromechanical system**   a machine with microscopic length scales that incorporates mechanical elements and fully integrated electronics.

**piezoelectric effect**   a material property wherein a voltage is generated as a result of stress or strain, and where a stress or strain is developed as a result of an applied voltage.

**surface micromachining**   machining of deposited layers, followed by removal of sacrificial layers.

# RECOMMENDED READINGS

Bhushan, B. (ed.) (1998) *Tribology Issues and Opportunities in MEMS.* Kluwer, Dordrecht.

Elwenspoek, M., and Jansen, H. (1998) *Silicon Micromachining,* Cambridge University Press.

Elwenspoek, M., and Wiegerink, R. (2001) *Mechanical Microsensors,* Springer-Verlag, Berlin.

Gad-el-Hak, M. (ed.) (2002) *The MEMS Handbook,* CRC Press, Boca Raton, FL.

Histand, M. B., and Alciatore, D. C. (1999) *Introduction to Mechatronics and Measurement Systems.* McGraw-Hill, New York.

Hsu, T-R. (2002) *MEMS and Microsystems Design and Manufacture.* McGraw-Hill, New York.

Kalpakjian, S., and Schmid, S. R. (2003) *Manufacturing Processes for Engineering Materials,* 4th ed., Prentice-Hall, Upper Saddle River, NJ.

Madou, M. J. (1997) *Fundamentals of Microfabrication,* CRC Press, Boca Raton, FL.

Maluf, N. (2000) *An Introduction to Microelectromechanical Systems Engineering,* Artech House, Boston, MA.

# REFERENCES

Core, T. A., Tsang, W. K., and Sherman, S. J., "Fabrication Technology for an Integrated Surface-Micromachined Sensor," *Solid State Technology,* v. 36, 1993, 00. 39–47.

Hsu, T-R. (2002) *MEMS and Microsystems Design and Manufacture.* McGraw-Hill, New York.

Kovacs, G. T. A., (1998) Micromachined Transducers Sourcebook, McGraw-Hill, New York.

Kalpakjian, S., and Schmid, S. R. (2003) *Manufacturing Processes for Engineering Materials,* 4th ed., Prentice-Hall, Upper Saddle River, NJ.

Madou, M. J. (1997) *Fundamentals of Microfabrication,* CRC Press, Boca Raton, FL.

Timoshenko, S. P., and Goodier, J. N. (1970) *Theory of Elasticity,* 3rd ed., McGraw-Hill, New York.

Young, W. C. (1989) *Roark's Formulas for Stress and Strain,* 6th ed., McGraw-Hill, New York.

White, F. M. (1994) *Fluid Mechanics,* 3rd ed., McGraw-Hill, New York.

# PROBLEMS

## Section 20.3

**20.1** A circular diaphragm is to be designed to produce a micro pressure sensor. The diaphragm has a diameter of 1 mm and is rigidly fixed to a silicon support around its entire periphery. The silicon diaphragm material ($S_y = 7000$ MPa, $E = 190{,}000$ MPa, $v = 0.25$) must withstand a pressure of 25 MPa. Calculate the minimum diaphragm thickness to avoid yielding. What is the maximum deflection of the diaphragm?

**20.2** Consider a rectangular plate with the same cross-sectional area as that of the circular diaphragm in Problem 20.1 and with 100 $\mu$m thickness and the same material and pressure range. Calculate the maximum stress and deformation in the rectangular diaphragm for

a)    $a/b = 1.0$
b)    $a/b = 2.0$

**20.3** Sketch $a$ shows typical designs for atomic force microscope probes produced from silicon. Estimate the normal stiffness for these designs and suggest methods for their manufacture. Use $E = 190$ GPa

Contact AFM cantilevers
Length = 450 $\mu$m
Width = 40 $\mu$m
Thickness = 3 $\mu$m

450 $\mu$m

40 $\mu$m

Tapping mode AFM cantilevers
Length = 125 $\mu$m
Width = 30 $\mu$m
Thickness = 5 $\mu$m

125 $\mu$m

30 $\mu$m

Sketch $a$ for Prob. 20.3

**20.4** Consider the beam in Example 20.2. Note that the thickness dimension is difficult to obtain accurately. Estimate the change in stiffness if the error in the thickness dimension measurement can be as high as 15%.

## Section 20.4

**20.5** Two square plates separated by a 2-$\mu$m air gap have a side dimension of 1 mm. Find the normal separating force between the plates if they have a 5 volt potential. What is the separating force if the gap is 0.2 $\mu$m?

**20.6** A comb drive needs to be designed to apply a force of 5 $\mu$N, where the gap between the comb tooth and the stationary electrodes is 2 $\mu$m and the width of the teeth and electrodes is 5 $\mu$m. How many combs are needed if the maximum voltage available is 150 V?

**20.7** Design a beam to support the comb drive considered in Problem 20.6 so that the maximum deflection of the comb drive is 10 $\mu$m. Use $E = 190$ GPa.

**20.8** A thin piezoelectric crystal film of PZT is used to measure the deflection of a cantilevered beam-type accelerometer shown in sketch $b$. Find the voltage output by the PZT at the maximum design acceleration of 100 m/s$^2$. Assume that the stiffness of PZT is much less than that of the aluminum beam.

PZT film

10 μm

10 μm

1 mm

50 μm

10-mg mass

Sketch *b* for Prob. 20.8

**20.9** Repeat Problem 20.8 if the piezoelectric film is made of PVDF.

## Section 20.5

**20.10** An ink-jet printing mechanism uses a 20 μm thick PZT piezoelectric pumping mechanism as shown in Figure 20.15. The printer needs to have a resolution of 600 dpi, and the ink film needs to be 0.5 μm thick. The ink reservoir has a diameter of 1 mm. Calculate the voltage required.

**20.11** Repeat Problem 20.10 for an ink-jet printer with a resolution of 1200 dpi.

## Section 20.6

**20.12** Consider the accelerometer shown in Figure 20.16(a). The four beams that support the central mass are all 20 μm wide, 50 μm deep, and 500 μm long. The combs on the mass overlap the stationary plates by 100 μm, and the gap at static equilibrium is 10 μm. If the suspended mass is 10 mg, obtain an expression for $(C_1 - C_2)/(C_1 + C_2)$. Use $E = 190$ GPa.

# Appendix A

# MATERIAL PROPERTIES

**Table A.1**    Properties of ferrous metals

| Material | Density, kg/m$^3$ | Modulus of elasticity, GPa (psi × 10$^6$) | Yield strength, MPa (ksi) | Ultimate strength, MPa (ksi) | Ductility, %EL in 2 in | Poisson's ratio | Thermal conductivity, W/(m·°C) | Coefficient of thermal expansion, (°C)$^{-1}$ × 10$^{-6}$ |
|---|---|---|---|---|---|---|---|---|
| Iron | 7870 | 207 (30) | 130 (19) | 260 (38) | 45 | 0.29 | 80 | 11.8 |
| Gray cast iron | 7150 | Variable | — | 125 (18) | — | Variable | 46 | 10.8 |
| Nodular cast iron | 7120 | 165 (24) | 275 (40) | 415 (60) | 18 | 0.28 | 33 | 11.8 |
| Malleable cast iron | 7120 | 172 (25) | 220 (32) | 345 (50) | 10 | 0.26 | 51 | 11.9 |
| Low-carbon steel (AISI 1020) | 7860 | 207 (30) | 295 (43) | 395 (57) | 37 | 0.30 | 52 | 11.7 |
| Medium-carbon steel (AISI 1040) | 7850 | 207 (30) | 350 (51) | 520 (75) | 30 | 0.30 | 52 | 11.3 |
| High-carbon steel (AISI 1080) | 7840 | 207 (30) | 380 (55) | 615 (89) | 25 | 0.30 | 48 | 11.0 |
| Stainless steels | | | | | | | | |
| Ferritic, type 446 | 7500 | 200 (29) | 345 (50) | 552 (80) | 20 | 0.30 | 21 | 10.4 |
| Austenitic, type 316 | 8000 | 193 (28) | 207 (30) | 552 (80) | 60 | 0.30 | 16 | 16.0 |
| Martensitic, type 410 | 7800 | 200 (29) | 275 (40) | 483 (70) | 30 | 0.30 | 25 | 9.9 |

**Table A.2**    Properties of nonferrous metals

| Material | Density, kg/m³ | Modulus of elasticity, GPa (psi × 10⁶) | Yield strength, MPa (ksi) | Ultimate strength, MPa (ksi) | Ductility, %EL in 2 in | Poisson's ratio | Thermal conductivity, W/(m · °C) | Coefficient of thermal expansion, (°C)⁻¹ × 10⁻⁶ |
|---|---|---|---|---|---|---|---|---|
| Aluminum (>99.5%) | 2,710 | 69 (10) | 17  (2.5) | 55   (8) | 25 | 0.33 | 231 | 23.6 |
| Aluminum alloy 2014 | 2,800 | 72 (10.5) | 97 (14) | 186 (27) | 18 | 0.33 | 192 | 22.5 |
| Copper (99.95%) | 8,940 | 110 (16) | 69 (10) | 220 (32) | 45 | 0.35 | 398 | 16.5 |
| Brass (70Cu–30 Zn) | 8,530 | 110 (16) | 75 (11) | 303 (44) | 68 | 0.35 | 120 | 20.0 |
| Bronze (92Cu–8Sn) | 8,800 | 110 (16) | 152 (22) | 380 (55) | 70 | 0.35 | 62 | 18.2 |
| Magnesium (>99%) | 1,740 | 45  (6.5) | 41  (6) | 165 (24) | 14 | 0.29 | 122 | 27.0 |
| Molybdenum (>99%) | 10,220 | 324 (47) | 565 (82) | 655 (95) | 35 | — | 142 | 4.9 |
| Nickel (>99%) | 8,900 | 207 (30) | 138 (20) | 483 (70) | 40 | 0.31 | 80 | 13.3 |
| Silver (>99%) | 10,490 | 76 (11) | 55  (8) | 125 (18) | 48 | 0.37 | 418 | 19.0 |
| Titanium (>99%) | 4,510 | 107 (15.5) | 240 (35) | 330 (48) | 30 | 0.34 | 17 | 9.0 |

**Table A.3**    Properties of ceramics

| Material | Density, kg/m³ | Modulus of elasticity, GPa (psi × 10⁶) | Poisson's ratio | Approximate hardness (Knoop) | Fracture strength, MPa (ksi) | Thermal conductivity, W/(m · °C) | Coefficient of thermal expansion, (°C)⁻¹ × 10⁻⁶ |
|---|---|---|---|---|---|---|---|
| Alumina ($Al_2O_3$) | 3970 | 393 (57) | 0.27 | 2100 | 275–550 (40–80) | 30 | 8.8[a] |
| Magnesia (MgO) | 3580 | 207 (30) | 0.36 | 370 | 105[b] (15) | 48 | 13.5[a] |
| Spinel ($MgAl_2O_4$) | 3550 | 284 (36) | — | 1600 | 83–220[b] (12–32) | 15.0[a] | 7.6[a] |
| Zirconia[c] ($ZrO_2$) | 5560 | 152 (22) | 0.32 | 1200 | 138–240[b] (20–35) | 2.0 | 10.0[a] |
| Fused silica ($SiO_2$) | 2200 | 75 (11) | 0.16 | 500 | 110 (16) | 1.3 | 0.5[a] |
| Soda-lime glass | 2500 | 69 (10) | 0.23 | 550 | 69 (10) | 1.7 | 9.0[d] |
| Borosilicate glass | 2230 | 62 (9) | 0.20 | — | 69 (10) | 1.4 | 3.3[d] |
| Silicon carbide (SiC) | 3220 | 414 (60) | 0.19 | 2500 | 450–520[b] (65–75) | 90 | 4.7 |
| Silicon nitride ($Si_3N_4$) | 3440 | 304 (44) | 0.24 | 2200 | 414–580[b] (60–80) | 16–33[a] | 3.6[a] |
| Titanium carbide (TiC) | 4920 | 462 (67) | — | 2600 | 275–450[b] (40–65) | 17.2 | 7.4 |

[a] Mean value taken over the temperature range 0–1000°C.

[b] Sintered and containing approximately 5% porosity.

[c] Stabilized with calcium oxide.

[d] Mean value taken over the temperature range 0–300°C.

**Table A.4**   Properties of polymers

| Polymers | State | Density, kg/m³ | Modulus of elasticity, GPa (ksi) | Ultimate strength, MPa (ksi) | Elongation at break, % | Glass transition temperature, °C | Maximum service temperature, °C | Thermal conductivity, W/(m·°C) | Coefficient of thermal expansion, (°C)$^{-1}$ × 10$^{-6}$ |
|---|---|---|---|---|---|---|---|---|---|
| **Thermoplastics** | | | | | | | | | |
| Polyethylene | High density, 70–80% crystalline | 952.1–965.0 | 1.07–1.09 (155–158) | 22–31 (3.2–4.5) | 10–1200 | −90 | 130–137 | 0.48 | 60–110 |
| | Low density, 40–50% crystalline | 917.1–932.1 | 0.17–0.28 (25–41) | 8.3–31.0 (1.2–4.5) | 100–650 | −110 | 98–115 | 0.33 | 100–220 |
| Polytetrafluoroethylene | 50–70% Crystalline | 2140–2200 | 0.40–0.55 (58–80) | 14–34 (2.0–5.0) | 200–400 | −90 | 327 | 0.25 | 70–120 |
| Polyvinyl chloride | Highly amorphous | 1300–1580 | 2.4–4.1 (350–600) | 41–52 (6.0–7.5) | 40–80 | 75–105 | 212 | 0.18 | 50–100 |
| Polypropylene | 50–60% Crystalline | 900–910 | 1.14–1.55 (165–225) | 31–41 (4.5–6.0) | 100–600 | −20 | 168–175 | 0.12 | 80–100 |
| Polystyrene | Amorphous | 1040–1050 | 2.28–3.28 (330–475) | 36–52 (5.2–7.5) | 1.2–2.5 | 74–105 | — | 0.13 | 50–83 |
| Polymethyl methacrylate | Amorphous | 1170–1200 | 2.24–3.24 (325–470) | 48–76 (7–11) | 2–10 | 85–105 | — | 0.21 | 50–90 |
| Nylon 6,6 polyhexamethylene adipamide) | 30–40% Crystalline | 1130–1150 | 1.58–3.79 (230–550) | 76–94 (11–13.7) | 15–300 | 57 | 255–265 | 0.24 | 80 |
| Polyethylene terephthalate (PET, a polyester) | 0–30% Crystalline | 1290–1400 | 2.76–4.14 (400–600) | 48–72 (7.0–10.5) | 30–300 | 73–80 | 245–265 | 0.14 | 65 |
| Polycarbonate (polybisphenol, a carbonate) | Amorphous | 1200 | 2.38 (345) | 65.5 (9.5) | 110 | 150 | — | 0.20 | 68 |
| **Thermosets** | | | | | | | | | |
| Epoxy | Complex network, amorphous | 1110–1140 | 2.41 (350) | 28–90 (4.0–13.0) | 3–6 | — | — | 0.19 | 45–65 |
| Phenolic | Complex network, amorphous | 1240–1320 | 2.76–4.83 (400–700) | 34–62 (5–9) | 1.5–2.0 | — | — | 0.15 | 68 |
| Polyester | Complex network, amorphous | 1040–1460 | 2.04–4.41 (300–640) | 41–90 (6–13) | <2 | — | — | 0.19 | 55–100 |

**Table A.5**    Properties of natural rubbers

| Name/repeat unit | Density, kg/m³ | Ultimate strength, MPa (ksi) | Maximum elongation, % | Modulus, at 100% elongation, MPa (psi) | Minimum service temperature, °C (°F) | Maximum service temperature, °C (°F) | Abrasion resistance | Tear resistance | Oxidation resistance |
|---|---|---|---|---|---|---|---|---|---|
| Natural polyisoprene (natural rubber, NR) | 920–1037 | 24–32 (3.5–4.6) | 500–760 | 3.3–5.9 (480–850) | −60 (−75) | 120 (250) | Excellent | Excellent | Good |
| Styrene-butadiene (SBR, GRS) | 940 | 12–21 (1.8–3.0) | 450–500 | 2.1–10.3 (300–1500) | −60 (−75) | 120 (250) | Excellent | Fair | Good |
| Acrylonitrile-butadiene (nitrile, Buna A, NBR) | 980 | 7–24 (1.0–3.5) | 400–600 | 3.4 (490) | −50 (−60) | 150 (300) | Excellent | Good | Fair–good |
| Chloroprene (neoprene, CR) | 1230–1250 | 3.5–24 (0.5–3.5) | 100–800 | 0.7–2.0 (100–3000) | −50 (−60) | 105 (225) | Excellent | Good | Very good |
| Pulybutadiene (BR) | 910 | 14–17 (2.0–2.5) | 450 | 2.1–10.3 (300–1500) | −100 (−150) | 90 (200) | Excellent | Good | Good |
| Polyurethane | 1020–1250 | 5.5–55 (0.8–8.0) | 250–800 | 0.17–34.5 (25–5000) | −55 (−65) | 120 (250) | Excellent | Outstanding | Excellent |
| Polydimethylsiloxane (silicone) | 1100–1600 | 10 (1.5) | 100–800 | — | −115 (−175) | 315 (600) | Poor | Fair | Excellent |

SOURCE: Adapted from *Materials Engineering*, a Penton publication.

# Appendix B

# STRESS–STRAIN RELATIONSHIPS

## SYMBOLS

$C$    elastic material coefficients, Pa
$E$    modulus of elasticity, Pa
$G$    shear modulus, Pa
$K$    bulk modulus, Pa
$x, y, z$    Cartesian coordinate system, m
$x', y', z'$    rotated Cartesian coordinate system, m
$\gamma$    shear strain
$\delta$    deformation, m
$\epsilon$    normal strain

$\lambda$    Lame's constant, Pa
$\nu$    Poisson's ratio
$\sigma$    normal stress, Pa
$\tau$    shear stress, Pa

## SUBSCRIPTS

$x, y, z$    Cartesian coordinates
$x', y', z'$    rotated Cartesian coordinates
$1, 2, 3$    principal axes

## B.1    LAWS OF STRESS TRANSFORMATION

Let a new orthogonal coordinate system $0'x'y'z'$ be placed having origin P, having the $z'$ axis coincident with the normal $n$ to the plane, and having the $x'$ and $y'$ axes (which must parallel the plane with normal $n$) coincident with the desired shear stress directions. Such a coordinate system is shown in Figure B.1, where the plane ABC is imagined to pass through point P, since the tetrahedron is very small. Figure B.1 also shows the three mutually perpendicular stress components $\sigma_{z'}$, $\tau_{z'x'}$, and $\tau_{z'y'}$ acting on the inclined plane. The laws of stress transformation give these stresses in terms of $\sigma_x, \sigma_y, \sigma_z, \tau_{xy}, \tau_{yz}$, and $\tau_{zx}$. The previously established rules for shear stress subscripts are equally applicable here. For example, $\tau_{z'x'}$ is the shear stress directed in the $x'$ direction and acting in a plane through P whose normal is directed in the $z'$ direction.

Because the normal direction coincides with the $z'$ direction,

$$p_{nz} = \sigma_z \cos(z', z) + \tau_{xz} \cos(z', x) + \tau_{yz} \cos(z', y)$$
$$p_{ny} = \tau_{zx} \cos(z', z) + \sigma_x \cos(z', x) + \tau_{yx} \cos(z', y) \tag{B.1}$$
$$p_{nx} = \tau_{zy} \cos(z', z) + \tau_{xy} \cos(z', x) + \sigma_y \cos(z', y)$$

where $\cos(z', x) =$ cosine of angle between $z'$ and $x$. The stress $\sigma_{z'}$ must equal the sum of the projections of $p_{nz}, p_{ny}$, and $p_{nx}$ onto the $z'$ axis

$$\sigma_{z'} = p_{nz} \cos(z', z) + p_{nx} \cos(z', x) + p_{ny} \cos(z', y) \tag{B.2}$$

Similarly,

$$\tau_{z'x'} = p_{nz} \cos(x', z) + p_{nx} \cos(x', x) + p_{ny} \cos(x', y) \tag{B.3}$$

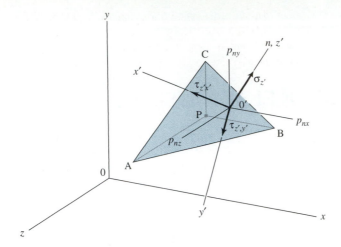

**Figure B.1**   Small tetrahedral element cut from body at point P.
The three perpendicular stress components $\sigma_{z'}$, $\tau_{z'x'}$, and $\tau_{z'y'}$
acting on the inclined plane are shown.

By substituting Equation (B.1) into Equations (B.2) and (B.3) while also making use of
the fact developed earlier that the shear stresses are symmetric (for example, $\tau_{x'y'} = \tau_{y'x'}$),
Equations (B.4) and (B.5) can be developed. Equations (B.6) to (B.9) can be derived in
an entirely similar way by using two more tetrahedrons with inclined planes having outer
normals parallel to the $x'$ and $y'$ axes. The result of all the above is

$$\sigma_{z'} = \sigma_z \cos^2(z', z) + \sigma_x \cos^2(z', x) + \sigma_y \cos^2(z', y) + 2\tau_{zx} \cos(z', z) \cos(z', x)$$
$$+ 2\tau_{xy} \cos(z', x) \cos(z', y) + 2\tau_{yz} \cos(z', y) \cos(z', z) \tag{B.4}$$

$$\sigma_{x'} = \sigma_x \cos^2(x', x) + \sigma_y \cos^2(x', y) + \sigma_z \cos^2(x', z) + 2\tau_{xy} \cos(x', x) \cos(x', y)$$
$$+ 2\tau_{xy} \cos(x', y) \cos(x', z) + 2\tau_{zx} \cos(x', z) \cos(x', x) \tag{B.5}$$

$$\sigma_{y'} = \sigma_y \cos^2(y', y) + \sigma_z \cos^2(y', z) + \sigma_x \cos^2(y', x) + 2\tau_{yz} \cos(y', y) \cos(y', z)$$
$$+ 2\tau_{zx} \cos(y', z) \cos(y', x) + 2\tau_{xy} \cos(y', x) \cos(y', y) \tag{B.6}$$

$$\tau_{z'x'} = \sigma_z \cos(z', z) \cos(x', z) + \sigma_x \cos(z', x) \cos(x', x) + \sigma_y \cos(z', y) \cos(x', y)$$
$$+ \tau_{zx}[\cos(z', z) \cos(x', x) + \cos(z', x) \cos(x', z)]$$
$$+ \tau_{xy}[\cos(z', x) \cos(x', y) + \cos(z', y) \cos(x', x)]$$
$$+ \tau_{yz}[\cos(z', y) \cos(x', z) + \cos(z', z) \cos(x', y)] \tag{B.7}$$

$$\tau_{x'y'} = \sigma_x \cos(x', x) \cos(y', x) + \sigma_y \cos(x', y) \cos(y', y) + \sigma_z \cos(x', z) \cos(y', z)$$
$$+ \tau_{xy}[\cos(x', x) \cos(y', y) + \cos(x', y) \cos(y', x)]$$
$$+ \tau_{yz}[\cos(x', y) \cos(y', z) + \cos(x', z) \cos(y', y)]$$
$$+ \tau_{zx}[\cos(x', z) \cos(y', x) + \cos(x', x) \cos(y', z)] \tag{B.8}$$

$$\tau_{y'z'} = \sigma_y \cos(y', y) \cos(z', y) + \sigma_z \cos(y', z) \cos(z', z) + \sigma_x \cos(y', x) \cos(z', x)$$
$$+ \tau_{yz}[\cos(y', y) \cos(z', z) + \cos(y', z) \cos(z', y)]$$
$$+ \tau_{zx}[\cos(y', z) \cos(z', x) + \cos(y', x) \cos(z', z)]$$
$$+ \tau_{xy}[\cos(y', x) \cos(z', y) + \cos(y', y) \cos(z', x)] \tag{B.9}$$

It is important to recognize the meaning of these equations. For example, Equation (B.7) can be used to find the shear stress acting in the $x'$ direction on a surface having an outer normal directed in the $z'$ direction if the six Cartesian stress components $\sigma_x, \sigma_y, \sigma_z, \tau_{xy}, \tau_{yz}$, and $\tau_{zx}$ at the point and the orientation of the coordinate system $0'x'y'z'$ are known.

## B.2   LAWS OF STRAIN TRANSFORMATION

In the study of stresses we found that three normal stresses $\sigma_x, \sigma_y$, and $\sigma_z$ and three shear stresses $\tau_{xy}, \tau_{yz}$, and $\tau_{zx}$ act on planes parallel to the Cartesian planes. Similarly, three normal strains $\epsilon_x, \epsilon_y$, and $\epsilon_z$ and three shear strains $\gamma_{xy}, \gamma_{yz}$, and $\gamma_{zx}$ characterize the behavior of line segments originally parallel to the Cartesian axes. From Durelli et al. (1958) the laws of strain transformation can be written directly from the laws of stress transformation given in Equations (B.4) to (B.9) if the following replacements are made to these equations:

$$\sigma_x \leftarrow \epsilon_x, \qquad \sigma_y \leftarrow \epsilon_y, \qquad \sigma_z \leftarrow \epsilon_z, \qquad 2\tau_{xy} \leftarrow \gamma_{xy}, \qquad 2\tau_{yz} \leftarrow \gamma_{yz}, \qquad 2\tau_{zx} \leftarrow \gamma_{zx}$$

$$\text{(B.10)}$$

## B.3   HOOKE'S LAW GENERALIZED

What is the relationship between stresses and strains when the stress system is not in simple tension or compression, as in Equation (3.22)? The stress system can be defined by the six components $\sigma_x, \sigma_y, \sigma_z, \tau_{xy}, \tau_{yz}$, and $\tau_{zx}$; and the strain system, by the six components $\epsilon_x, \epsilon_y, \epsilon_z, \gamma_{xy}, \gamma_{yz}$, and $\gamma_{zx}$. Thus, a generalization of Hooke's law is to make each stress component a linear function of the strain components, or

$$\sigma_x = C_{11}\epsilon_x + C_{12}\epsilon_y + C_{13}\epsilon_z + C_{14}\gamma_{xy} + C_{15}\gamma_{yz} + C_{16}\gamma_{zx}$$
$$\sigma_y = C_{21}\epsilon_x + C_{22}\epsilon_y + C_{23}\epsilon_z + C_{24}\gamma_{xy} + C_{25}\gamma_{yz} + C_{26}\gamma_{zx}$$
$$\vdots \qquad\qquad\qquad\qquad\qquad\qquad\qquad\qquad \text{(B.11)}$$
$$\tau_{zx} = C_{61}\epsilon_x + C_{62}\epsilon_y + C_{63}\epsilon_z + C_{64}\gamma_{xy} + C_{65}\gamma_{yz} + C_{66}\gamma_{zx}$$

where $C_{11}, C_{12}, \ldots, C_{66}$ are elastic material coefficients independent of stress or strain. Equation (B.11) is valid for many hard materials over a strain range of practical interest in designing machine elements.

The assumption of an isotropic material implies that the stress and strain components referred to a coordinate system $0x'y'z'$ of any arbitrary orientation must be related by the same elastic material coefficients $C_{11}, C_{12}, \ldots, C_{66}$. Thus, the six Equations (B.11) imply that the subscripts change, $x \rightarrow x', y \rightarrow y'$, and $z \rightarrow z'$, while the constants $C_{11}, C_{12}, \ldots, C_{66}$ remain the same. Thus,

$$\sigma_{x'} = C_{11}\epsilon_{x'} + C_{12}\epsilon_{y'} + C_{13}\epsilon_{z'} + C_{14}\gamma_{x'y'} + C_{15}\gamma_{y'z'} + C_{16}\gamma_{z'x'}$$
$$\sigma_{y'} = C_{21}\epsilon_{x'} + C_{22}\epsilon_{y'} + C_{23}\epsilon_{z'} + C_{24}\gamma_{x'y'} + C_{25}\gamma_{y'z'} + C_{26}\gamma_{z'x'}$$
$$\vdots \qquad\qquad\qquad\qquad\qquad\qquad\qquad\qquad \text{(B.12)}$$
$$\tau_{z'x'} = C_{61}\epsilon_{x'} + C_{62}\epsilon_{y'} + C_{63}\epsilon_{z'} + C_{64}\gamma_{x'y'} + C_{65}\gamma_{y'z'} + C_{66}\gamma_{z'x'}$$

Recall that $0x'y'z'$ is an arbitrary coordinate system.

**Figure B.2**    Coordinate system $0x'y'z'$ obtained by rotating $0z$ by 180°.

First, obtaining a new coordinate system $0x'y'z'$ by rotating $0z$ through a 180° angle results in Figure B.2. Recall that

$$(x, y) = \text{angle between } x \text{ and } y$$

From Figure B.2 and the above notation

$$
\begin{array}{lll}
(x', x) = 180° & (x', y) = 90° & (x', z) = 90° \\
(y', y) = 180° & (y', z) = 90° & (y', x) = 90° \\
(z', z) = 0° & (z', x) = 90° & (z', y) = 90°
\end{array}
\tag{B.13}
$$

Making use of Equation (B.13) reduces Equations (B.4) to (B.9) to

$$\sigma_{z'} = \sigma_z, \qquad \sigma_{x'} = \sigma_x, \qquad \sigma_{y'} = \sigma_y, \qquad \tau_{x'y'} = \tau_{xy}, \qquad \tau_{y'z'} = -\tau_{yz}, \qquad \tau_{z'x'} = -\tau_{zx} \tag{B.14}$$

Similarly, for the strains

$$\epsilon_{z'} = \epsilon_z, \qquad \epsilon_{x'} = \epsilon_x, \qquad \epsilon_{y'} = \epsilon_y, \qquad \gamma_{x'y'} = \gamma_{xy}, \qquad \gamma_{y'z'} = -\gamma_{yz}, \qquad \gamma_{z'x'} = -\gamma_{zx} \tag{B.15}$$

Substituting Equations (B.14) and (B.15) into Equation (B.12) gives

$$
\begin{aligned}
\sigma_x &= C_{11}\epsilon_x + C_{12}\epsilon_y + C_{13}\epsilon_z + C_{14}\gamma_{xy} - C_{15}\gamma_{yz} - C_{16}\gamma_{zx} \\
\sigma_y &= C_{21}\epsilon_x + C_{22}\epsilon_y + C_{23}\epsilon_z + C_{24}\gamma_{xy} - C_{25}\gamma_{yz} - C_{26}\gamma_{zx} \\
&\;\;\vdots \\
\tau_{zx} &= -C_{61}\epsilon_x - C_{62}\epsilon_y - C_{63}\epsilon_z - C_{64}\gamma_{xy} + C_{65}\gamma_{yz} + C_{66}\gamma_{zx}
\end{aligned}
\tag{B.16}
$$

Comparing Equations (B.11) and (B.16) shows that

$$C_{15} = -C_{15} \qquad C_{16} = -C_{16}$$

implying that
$$C_{15} = C_{16} = C_{25} = C_{26} = C_{35} = C_{36} = C_{45} = C_{46} = 0$$
$$C_{51} = C_{52} = C_{53} = C_{54} = C_{61} = C_{62} = C_{63} = C_{64} = 0$$
(B.17)

Substituting Equation (B.17) into Equation (B.11) gives

$$\sigma_x = C_{11}\epsilon_x + C_{12}\epsilon_y + C_{13}\epsilon_z + C_{14}\gamma_{xy}$$
$$\sigma_y = C_{21}\epsilon_x + C_{22}\epsilon_y + C_{23}\epsilon_z + C_{24}\gamma_{xy}$$
$$\sigma_z = C_{31}\epsilon_x + C_{32}\epsilon_y + C_{33}\epsilon_z + C_{34}\gamma_{xy}$$
$$\tau_{xy} = C_{41}\epsilon_x + C_{42}\epsilon_y + C_{43}\epsilon_z + C_{44}\gamma_{xy}$$
$$\tau_{yz} = C_{55}\gamma_{yz} + C_{56}\gamma_{zx}$$
$$\tau_{zx} = C_{65}\gamma_{yz} + C_{66}\gamma_{zx}$$
(B.18)

Thus, using the coordinate system of Figure B.2 with the isotropic assumption reduces the 36 elastic material constants of Equation (B.11) to the 20 expressed in Equation (B.18).

Second, obtaining a new coordinate system by rotating $0x$ through a 180° angle results in Figure B.3, giving

$$(x', x) = 0° \qquad (x', y) = 90° \qquad (z', z) = 90°$$
$$(y', y) = 180° \qquad (y', x) = 90° \qquad (y', z) = 90°$$
$$(z', z) = 180° \qquad (z', x) = 90° \qquad (z', y) = 90°$$
(B.19)

Making use of Equations (B.19) reduces Equations (B.4) to (B.9) to

$$\sigma_{z'} = \sigma_z \qquad \sigma_{x'} = \sigma_x \qquad \sigma_{y'} = \sigma_y$$
$$\tau_{z'x'} = -\tau_{zx} \qquad \tau_{x'y'} = -\tau_{xy} \qquad \tau_{y'z'} = \tau_{yz}$$
$$\epsilon_{x'} = \epsilon_x \qquad \epsilon_{y'} = \epsilon_y \qquad \epsilon_{z'} = \epsilon_z$$
$$\gamma_{x'y'} = -\gamma_{xy} \qquad \gamma_{y'z'} = \gamma_{yz} \qquad \gamma_{z'x'} = -\gamma_{zx}$$
(B.20)

The assumption of an isotropic material requires that the new coordinate system $0x'y'z'$ shown in Figure B.3 be related to the same elastic material constants $C_{11}, C_{12}, \ldots, C_{66}$

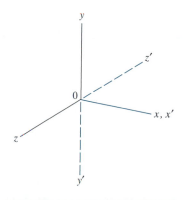

**Figure B.3** Coordinate system $0x'y'z'$ obtained by rotating $0x$ by 180°.

expressed in Equation (B.18), implying that

$$
\begin{aligned}
\sigma_{x'} &= C_{11}\epsilon_{x'} + C_{12}\epsilon_{y'} + C_{13}\epsilon_{z'} + C_{14}\gamma_{x'y'} \\
\sigma_{y'} &= C_{21}\epsilon_{x'} + C_{22}\epsilon_{y'} + C_{23}\epsilon_{z'} + C_{24}\gamma_{x'y'} \\
\sigma_{z'} &= C_{31}\epsilon_{x'} + C_{32}\epsilon_{y'} + C_{33}\epsilon_{z'} + C_{34}\gamma_{x'y'} \\
\tau_{x'y'} &= C_{41}\epsilon_{x'} + C_{42}\epsilon_{y'} + C_{43}\epsilon_{z'} + C_{44}\gamma_{x'y'} \\
\tau_{y'z'} &= C_{55}\gamma_{y'z'} + C_{56}\gamma_{z'x'} \\
\tau_{z'x'} &= C_{65}\gamma_{y'z'} + C_{66}\gamma_{z'x'}
\end{aligned}
\tag{B.21}
$$

Substituting Equation (B.20) into Equation (B.21) gives

$$
\begin{aligned}
\sigma_{x} &= C_{11}\epsilon_{x} + C_{12}\epsilon_{y} + C_{13}\epsilon_{z} - C_{14}\gamma_{xy} \\
\sigma_{y} &= C_{21}\epsilon_{x} + C_{22}\epsilon_{y} + C_{23}\epsilon_{z} - C_{24}\gamma_{xy} \\
\sigma_{z} &= C_{31}\epsilon_{x} + C_{32}\epsilon_{y} + C_{33}\epsilon_{z} - C_{34}\gamma_{xy} \\
-\tau_{xy} &= C_{41}\epsilon_{x} + C_{42}\epsilon_{y} + C_{43}\epsilon_{z} - C_{44}\gamma_{xy} \\
\tau_{yz} &= C_{55}\gamma_{yz} - C_{56}\gamma_{zx} \\
-\tau_{zx} &= C_{65}\gamma_{yz} - C_{66}\gamma_{zx}
\end{aligned}
\tag{B.22}
$$

Equations (B.18) and (B.22) will agree only if

$$
C_{14} = C_{24} = C_{34} = C_{41} = C_{42} = C_{43} = C_{56} = C_{65} = 0
\tag{B.23}
$$

Substituting Equation (B.13) into Equation (B.8) gives

$$
\begin{aligned}
\sigma_{x} &= C_{11}\epsilon_{x} + C_{12}\epsilon_{y} + C_{13}\epsilon_{z} \\
\sigma_{y} &= C_{21}\epsilon_{x} + C_{22}\epsilon_{y} + C_{23}\epsilon_{z} \\
\sigma_{z} &= C_{31}\epsilon_{x} + C_{32}\epsilon_{y} + C_{33}\epsilon_{z} \\
\tau_{xy} &= C_{44}\gamma_{xy} \\
\tau_{yz} &= C_{55}\gamma_{yz} \\
\tau_{zx} &= C_{66}\gamma_{zx}
\end{aligned}
\tag{B.24}
$$

Thus, using the coordinate system of Figure B.3 with the isotropic assumption reduces the 20 elastic material constants of Equation (B.18) to the 12 expressed in Equation (B.24). Third, rotating $0x$ through a $90°$ angle results in Figure B.4, implying that

$$
\begin{aligned}
(x', x) &= 0° & (x', y) &= 90° & (x', z) &= 90° \\
(y', y) &= 90° & (y', x) &= 90° & (y', z) &= 0° \\
(z', z) &= 90° & (z', x) &= 90° & (z', y) &= 180°
\end{aligned}
\tag{B.25}
$$

Making use of Equation (B.25) reduces Equations (B.4) to (B.9) and Equation (2.37) to

$$
\begin{aligned}
\sigma_{z'} &= \sigma_{y} & \sigma_{x'} &= \sigma_{x} & \sigma_{y'} &= \sigma_{z} \\
\tau_{z'x'} &= -\tau_{xy} & \tau_{x'y'} &= \tau_{zx} & \tau_{y'z'} &= -\tau_{yz} \\
\epsilon_{z'} &= \epsilon_{y} & \epsilon_{x'} &= \epsilon_{x} & \epsilon_{y'} &= \epsilon_{z} \\
\gamma_{z'x'} &= -\gamma_{xy} & \gamma_{x'y'} &= \gamma_{zx} & \gamma_{y'z'} &= -\gamma_{yz}
\end{aligned}
\tag{B.26}
$$

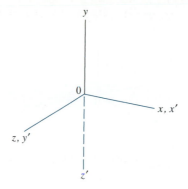

**Figure B.4**   Coordinate system $0x'y'z'$ obtained by rotating $0x$ by $90°$.

The assumption of an isotropic material requires that the new coordinate system $0x'y'z'$ shown in Figure B.4 be related to the same elastic material constants expressed in Equation (B.24), implying that

$$\sigma_{x'} = C_{11}\epsilon_{x'} + C_{12}\epsilon_{y'} + C_{13}\epsilon_{z'}$$
$$\sigma_{y'} = C_{21}\epsilon_{x'} + C_{22}\epsilon_{y'} + C_{23}\epsilon_{z'}$$
$$\sigma_{z'} = C_{31}\epsilon_{x'} + C_{32}\epsilon_{y'} + C_{33}\epsilon_{z'}$$
$$\tau_{x'y'} = C_{44}\gamma_{x'y'}$$
$$\tau_{y'z'} = C_{55}\gamma_{y'z'}$$
$$\tau_{z'x'} = C_{66}\gamma_{z'x'}$$

(B.27)

Substituting Equation (B.26) into Equation (B.27) gives

$$\sigma_x = C_{11}\epsilon_x + C_{12}\epsilon_y + C_{13}\epsilon_z$$
$$\sigma_y = C_{21}\epsilon_x + C_{22}\epsilon_y + C_{23}\epsilon_z$$
$$\sigma_z = C_{31}\epsilon_x + C_{32}\epsilon_y + C_{33}\epsilon_z$$
$$\tau_{xy} = C_{44}\gamma_{xy}$$
$$\tau_{yz} = C_{55}\gamma_{yz}$$
$$\tau_{zx} = C_{66}\gamma_{zx}$$

(B.28)

Equations (B.24) and (B.28) will agree only if

$$C_{12} = C_{13} \qquad C_{21} = C_{31} \qquad C_{23} = C_{32} \qquad C_{22} = C_{33} \qquad C_{44} = C_{66} \qquad \text{(B.29)}$$

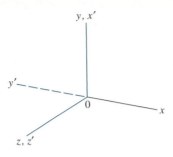

**Figure B.5**   Coordinate system $0x'y'z'$ obtained by rotating $0z$ by $90°$.

Substituting Equation (B.29) into Equation (B.28) gives

$$\sigma_x = C_{11}\epsilon_x + C_{12}(\epsilon_y + \epsilon_z)$$
$$\sigma_y = C_{21}\epsilon_x + C_{22}\epsilon_y + C_{23}\epsilon_z$$
$$\sigma_z = C_{21}\epsilon_x + C_{22}\epsilon_z + C_{23}\epsilon_y \qquad\qquad \text{(B.30)}$$
$$\tau_{xy} = C_{44}\gamma_{xy}$$
$$\tau_{yz} = C_{55}\gamma_{yz}$$
$$\tau_{zx} = C_{44}\gamma_{zx}$$

Thus, using the coordinate system of Figure B.4 with the isotropic assumption reduces the 12 elastic material constants of Equation (B.24) to the 7 expressed in Equation (B.30).

Fourth, rotating $0z$ through a $90°$ angle results in Figure B.5, giving

$$
\begin{array}{lll}
(x', x) = 90° & (x', y) = 0° & (x', z) = 90° \\
(y', y) = 90° & (y', x) = 180° & (y', z) = 90° \\
(z', z) = 0° & (z', x) = 90° & (z', y) = 90°
\end{array}
\qquad \text{(B.31)}
$$

Making use of Equation (B.31) reduces Equation (B.4) to (B.9) and Equation (2.37) to

$$
\begin{array}{lll}
\sigma_{z'} = \sigma_z & \sigma_{x'} = \sigma_y & \sigma_{y'} = \sigma_x \\
\tau_{z'x'} = \tau_{yz} & \tau_{x'y'} = -\tau_{xy} & \tau_{y'z'} = -\tau_{zx} \\
\epsilon_{z'} = \epsilon_z & \epsilon_{x'} = \epsilon_y & \epsilon_{y'} = \epsilon_x \\
\gamma_{z'x'} = \gamma_{yz} & \gamma_{x'y'} = -\gamma_{xy} & \gamma_{y'z'} = -\gamma_{zx}
\end{array}
\qquad \text{(B.32)}
$$

The assumption of an isotropic material requires that Equation (B.30) be written as

$$\sigma_{x'} = C_{11}\epsilon_{x'} + C_{12}(\epsilon_{y'} + \epsilon_{z'})$$
$$\sigma_{y'} = C_{21}\epsilon_{x'} + C_{22}\epsilon_{y'} + C_{23}\epsilon_{z'}$$
$$\sigma_{z'} = C_{21}\epsilon_{x'} + C_{22}\epsilon_{z'} + C_{23}\epsilon_{y'} \qquad\qquad \text{(B.33)}$$
$$\tau_{x'y'} = C_{44}\gamma_{x'y'}$$
$$\tau_{y'z'} = C_{55}\gamma_{y'z'}$$
$$\tau_{z'x'} = C_{44}\gamma_{z'x'}$$

Substituting Equation (B.32) into Equation (B.33) gives

$$\sigma_x = C_{11}\epsilon_y + C_{12}(\epsilon_x + \epsilon_z)$$
$$\sigma_y = C_{21}\epsilon_y + C_{22}\epsilon_x + C_{23}\epsilon_z$$
$$\sigma_z = C_{21}\epsilon_y + C_{22}\epsilon_z + C_{23}\epsilon_x$$
$$\tau_{xy} = C_{44}\gamma_{xy}$$
$$\tau_{yz} = C_{44}\gamma_{yz}$$
$$\tau_{zx} = C_{55}\gamma_{zx}$$

(B.34)

Equations (B.30) and (B.34) will agree only if

$$C_{22} = C_{11} \qquad C_{21} = C_{12} \qquad C_{23} = C_{12} \qquad C_{55} = C_{44}$$

(B.35)

Substituting Equation (B.37) into Equation (B.34) gives

$$\sigma_x = C_{11}\epsilon_x + C_{12}(\epsilon_y + \epsilon_z)$$
$$\sigma_y = C_{11}\epsilon_y + C_{12}(\epsilon_x + \epsilon_z)$$
$$\sigma_z = C_{11}\epsilon_z + C_{12}(\epsilon_x + \epsilon_y)$$
$$\tau_{xy} = C_{44}\gamma_{xy}$$
$$\tau_{yz} = C_{44}\gamma_{yz}$$
$$\tau_{zx} = C_{44}\gamma_{zx}$$

(B.36)

Thus, using the coordinate system of Figure B.5 with the isotropic assumption reduces the seven elastic material constants of Equation (B.30) to the three expressed in Equation (B.36).

Fifth and finally, rotating $0z$ through a $45°$ angle results in Figure B.6, implying that

$$(x', x) = 45° \qquad (x', y) = 45° \qquad (x', z) = 90°$$
$$(y', y) = 45° \qquad (y', x) = 135° \qquad (y', z) = 90°$$
$$(z', z) = 0° \qquad (z', x) = 90° \qquad (z', y) = 90°$$

(B.37)

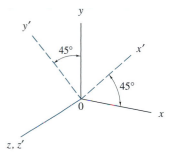

**Figure B.6** Coordinate system $0x'y'z'$ obtained by rotating $0z$ by $45°$.

Making use of Equation (B.37) reduces Equations (B.4) to (B.9) and Equation (2.37) to

$$
\begin{aligned}
\sigma_{z'} &= \sigma_z & \sigma_{x'} &= \tfrac{1}{2}(\sigma_x + \sigma_y) + \tau_{xy} & \sigma_{y'} &= \tfrac{1}{2}(\sigma_x + \sigma_y) - \tau_{xy} \\
\tau_{z'x'} &= \tfrac{1}{\sqrt{2}}(\tau_{zx} + \tau_{yz}) & \tau_{x'y'} &= \tfrac{1}{2}(\sigma_y - \sigma_x) & \tau_{y'z'} &= \tfrac{1}{\sqrt{2}}(\tau_{yz} - \tau_{zx}) \\
\epsilon_{z'} &= \epsilon_z & \epsilon_{x'} &= \tfrac{1}{2}(\epsilon_x + \epsilon_y + \gamma_{xy}) & \epsilon_{y'} &= \tfrac{1}{2}(\epsilon_x + \epsilon_y - \gamma_{xy}) \\
\gamma_{z'x'} &= \tfrac{1}{\sqrt{2}}(\gamma_{zx} + \gamma_{yz}) & \gamma_{x'y'} &= \tfrac{1}{2}(\epsilon_y - \epsilon_x) & \gamma_{y'z'} &= \tfrac{1}{\sqrt{2}}(\gamma_{yz} - \gamma_{zx})
\end{aligned}
\tag{B.38}
$$

The assumption of an isotropic material requires that Equation (B.37) be written as

$$
\begin{aligned}
\sigma_{x'} &= C_{11}\epsilon_{x'} + C_{12}(\epsilon_{y'} + \epsilon_{z'}) \\
\sigma_{y'} &= C_{11}\epsilon_{y'} + C_{12}(\epsilon_{x'} + \epsilon_{z'}) \\
\sigma_{z'} &= C_{11}\epsilon_{z'} + C_{12}(\epsilon_{x'} + \epsilon_{y'}) \\
\tau_{x'y'} &= C_{44}\gamma_{x'y'} \\
\tau_{y'z'} &= C_{44}\gamma_{y'z'} \\
\tau_{z'x'} &= C_{44}\gamma_{z'x'}
\end{aligned}
\tag{B.39}
$$

Substituting Equation (B.38) into the first of Equation (B.39) gives

$$
\frac{1}{2}(\sigma_x + \sigma_y) + \tau_{xy} = C_{11}\left(\frac{\gamma_{xy}}{2} + \frac{\epsilon_x + \epsilon_y}{2}\right) + C_{12}\left(\frac{\epsilon_x + \epsilon_y}{2} - \frac{\gamma_{xy}}{2} + \epsilon_z\right)
$$

and substituting the expressions for $\sigma_x$ and $\sigma_y$ in Equation (B.36) into the above equation gives

$$
\tau_{xy} = (C_{11} - C_{12})\frac{\gamma_{xy}}{2}
\tag{B.40}
$$

Comparing Equation (B.40) with the $\tau_{xy}$ expression in Equation (B.36) gives

$$
C_{11} = C_{12} + 2C_{44}
\tag{B.41}
$$

Letting $C_{12} = \lambda$ and $C_{44} = G$ and making use of Equation (B.41) gives Equation (B.36) as

$$
\begin{aligned}
\sigma_x &= (\lambda + 2G)\epsilon_x + \lambda(\epsilon_y + \epsilon_z) \\
\sigma_y &= (\lambda + 2G)\epsilon_y + \lambda(\epsilon_x + \epsilon_z) \\
\sigma_z &= (\lambda + 2G)\epsilon_z + \lambda(\epsilon_x + \epsilon_y) \\
\tau_{xy} &= G\gamma_{xy} \\
\tau_{yz} &= G\gamma_{yz} \\
\tau_{zx} &= G\gamma_{zx}
\end{aligned}
\tag{B.42}
$$

Thus, using Figure B.6 with the isotropic assumption reduces the number of elastic material constants from the 36 expressed in Equation (B.11) to the 2 expressed in Equation (B.42).

Equation (B.42) can be solved for strains to give

$$\epsilon_x = \frac{(\lambda + G)\sigma_x}{G(3\lambda + 2G)} - \frac{\lambda(\sigma_y + \sigma_z)}{2G(3\lambda + 2G)}$$

$$\epsilon_y = \frac{(\lambda + G)\sigma_y}{G(3\lambda + 2G)} - \frac{\lambda(\sigma_x + \sigma_z)}{2G(3\lambda + 2G)}$$

$$\epsilon_z = \frac{(\lambda + G)\sigma_z}{G(3\lambda + 2G)} - \frac{\lambda(\sigma_x + \sigma_y)}{2G(3\lambda + 2G)}$$

$$\gamma_{xy} = \frac{\tau_{xy}}{G}$$                            (B.43)

$$\gamma_{yz} = \frac{\tau_{yz}}{G}$$

$$\gamma_{zx} = \frac{\tau_{zx}}{G}$$

For isotropic but *nonhomogeneous* materials, the constants $\lambda$ and $G$ are functions of the space coordinates $x$, $y$, and $z$ and vary from point to point. For isotropic and *homogeneous* materials, these constants are not functions of the space coordinates and do not vary from point to point. They depend only on the particular material.

If the $0x$, $0y$, and $0z$ axes are chosen along the principal axes of stress, $\tau_{xy} = \tau_{yz} = \tau_{zx} = 0$. From Equation (B.43) it follows that $\gamma_{xy} = \gamma_{yz} = \gamma_{zx} = 0$. Thus, for isotropic materials the principal axes of stress and strain coincide.

## B.4   PHYSICAL SIGNIFICANCE OF ELASTIC MATERIAL CONSTANTS

The first elastic material constant is shear modulus, or modulus of rigidity, $G$ in pascals. The second elastic material constant $\lambda$, known as Lamé's constant, is of no particular physical significance.

In a uniaxial stress state where $\sigma_y = \sigma_z = \tau_{xy} = \tau_{yz} = \tau_{zx} = 0$ and $\sigma_x$ is the applied uniaxial stress, Equation (B.43) gives

$$\epsilon_x = \frac{(\lambda + G)\sigma_x}{G(3\lambda + 2G)}$$                     (B.44)

$$\epsilon_y = \epsilon_x = -\frac{\lambda\sigma_x}{2G(3\lambda + 2G)}$$                     (B.45)

Comparing Equation (B.44) with Equation (3.23) gives

$$\lambda = \frac{G(E - 2G)}{3G - E}$$ 

(B.46)

where $E$ = modulus of elasticity, the third elastic material constant, covered in Section 3.5.2. Relating the transverse strain in Equation (B.45) to Poisson's ratio in Equation (3.4) gives

$$\nu = \frac{\lambda}{2(\lambda + G)}$$ 

(B.47)

or, by solving for $\lambda$,

$$\lambda = \frac{2G\nu}{1 - 2\nu}$$ 

(B.48)

Thus, Equations (B.46) and (B.48) express $\lambda$ in terms of two known elastic material constants. Equating Equations (B.46) and (B.48) gives

$$\nu = \frac{E - 2G}{2G}$$ 

(B.49)

where Poisson's ratio $\nu$ is the fourth elastic material constant. Equation (B.49) is equivalent to Equation (3.6). The range of $\lambda$ is between 0 and 0.5. At $\nu = 0$ no transverse deformation, but rather longitudinal deformation, occurs. At $\nu = 0.5$ the material exhibits constant volume. The volume increase longitudinally is the same as the shrinkage in the transverse direction.

Besides the four elastic material constants $G$, $\lambda$, $E$, and $\nu$, a fifth is provided by bulk modulus $K$, the ratio of applied hydrostatic pressure to observed volume shrinkage per unit volume, which is in pascals (pounds per square inch). The bulk modulus is obtained from hydrostatic compression or when

$$\sigma_x = \sigma_y = \sigma_z = -p \quad \text{for } p > 0$$
$$\tau_{xy} = \tau_{yz} = \tau_{zx} = 0$$

Substituting the above into Equation (B.43) gives

$$\epsilon_x = \epsilon_y = \epsilon_z = -\frac{p}{3\lambda + 2G}$$ 

(B.50)

But the total strain is

$$\epsilon = \epsilon_x + \epsilon_y + \epsilon_z \rightarrow p = -\frac{(3\lambda + 2G)\epsilon}{3} = -K\epsilon$$ 

(B.51)

**Table B.1**    Relationships between elastic material constants for isotropic materials

| Constants involved | Lamé's constant $\lambda$ | Shear modulus $G$ | Modulus of elasticity $E$ | Poisson's ratio $\nu$ | Bulk modulus $K$ |
|---|---|---|---|---|---|
| $\lambda$  $G$ | — | — | $E = \dfrac{G(3\lambda + 2G)}{\lambda + G}$ | $\nu = \dfrac{\lambda}{2(\lambda + G)}$ | $K = \dfrac{3\lambda + 2G}{3}$ |
| $\lambda$  $E$ | — | $G = \dfrac{A^\dagger + E - 3\lambda}{4}$ | — | $\nu = \dfrac{A - (E + \lambda)}{4\lambda}$ | $K = \dfrac{A + 3\lambda + E}{6}$ |
| $\lambda$  $\nu$ | — | $G = \dfrac{\lambda(1 - 2\nu)}{2\nu}$ | $E = \dfrac{\lambda(1 + \nu)(1 - 2\nu)}{\nu}$ | — | $K = \dfrac{\lambda(1 + \nu)}{3\nu}$ |
| $\lambda$  $K$ | — | $G = \dfrac{3(K - \lambda)}{2}$ | $E = \dfrac{9K(K - \lambda)}{3K - \lambda}$ | $\nu = \dfrac{\lambda}{3K - \lambda}$ | — |
| $G$  $E$ | $\lambda = \dfrac{G(2G - E)}{E - 3G}$ | — | — | $\nu = \dfrac{E - 2G}{2G}$ | $K = \dfrac{GE}{3(3G - E)}$ |
| $G$  $\nu$ | $\lambda = \dfrac{2G\nu}{1 - 2\nu}$ | — | $E = 2G(1 + \nu)$ | — | $K = \dfrac{2G(1 + \nu)}{3(1 - 2\nu)}$ |
| $G$  $K$ | $\lambda = \dfrac{3K - 2G}{3}$ | — | $E = \dfrac{9KG}{3K + G}$ | $\nu = \dfrac{3K - 2G}{2(3K + G)}$ | — |
| $E$  $\nu$ | $\lambda = \dfrac{\nu E}{(1 + \nu)(1 - 2\nu)}$ | $G = \dfrac{E}{2(1 + \nu)}$ | — | — | $K = \dfrac{E}{3(1 - 2\nu)}$ |
| $E$  $K$ | $\lambda = \dfrac{3K(3K - E)}{9K - E}$ | $G = \dfrac{3EK}{9K - E}$ | — | $\nu = \dfrac{3K - E}{6K}$ | — |
| $\nu$  $K$ | $\lambda = \dfrac{3K\nu}{1 + \nu}$ | $G = \dfrac{3K(1 - 2\nu)}{2(1 + \nu)}$ | $E = 3K(1 - 2\nu)$ | — | — |

$\dagger A = \sqrt{(E + \lambda)^2 + 8\lambda^2}$

where
$$K = \frac{3\lambda + 2G}{3} \tag{B.52}$$

It is possible to express any of the elastic material constants ($G$, $\lambda$, $E$, $\nu$, and $K$) if two of these five constants are given. These relationships are expressed in Table B.1.

## B.5  STRESS–STRAIN EQUATIONS IN TERMS OF MODULUS OF ELASTICITY AND POISSON'S RATIO

From Table B.1, $\lambda$ and $G$ can be expressed in terms of $E$ and $\nu$ as

$$\lambda = \frac{\nu E}{(1 + \nu)(1 - 2\nu)} \quad \text{and} \quad G = \frac{E}{2(1 + \nu)} \tag{B.53}$$

Substituting Equation (B.53) into Equation (B.43) gives

$$\epsilon_x = \frac{1}{E}[\sigma_x - \nu(\sigma_y + \sigma_z)]$$

$$\epsilon_y = \frac{1}{E}[\sigma_y - \nu(\sigma_z + \sigma_x)]$$

$$\epsilon_z = \frac{1}{E}[\sigma_z - \nu(\sigma_x + \sigma_y)]$$

$$\gamma_{xy} = \frac{2(1+\nu)}{E}\tau_{xy}$$

(B.54)

$$\gamma_{yz} = \frac{2(1+\nu)}{E}\tau_{yz}$$

$$\gamma_{zx} = \frac{2(1+\nu)}{E}\tau_{zx}$$

Similarly, the stress components can be expressed in terms of strains as

$$\sigma_x = \frac{E}{(1+\nu)(1-2\nu)}[(1-\nu)\epsilon_x + \nu(\epsilon_y + \epsilon_z)]$$

$$\sigma_y = \frac{E}{(1+\nu)(1-2\nu)}[(1-\nu)\epsilon_y + \nu(\epsilon_x + \epsilon_z)]$$

$$\sigma_z = \frac{E}{(1+\nu)(1-2\nu)}[(1-\nu)\epsilon_z + \nu(\epsilon_x + \epsilon_y)]$$

(B.55)

$$\tau_{xy} = \frac{E}{2(1+\nu)}\gamma_{xy}$$

$$\tau_{yz} = \frac{E}{2(1+\nu)}\gamma_{yz}$$

$$\tau_{zx} = \frac{E}{2(1+\nu)}\gamma_{zx}$$

Although the stress and strain equations given in Equations (B.54) and (B.55), respectively, are expressed in terms of $E$ and $\nu$, by using Table B.1 these equations can be rewritten in terms of any two of the five elastic material constants $(G, \lambda, E, \nu,$ and $K)$. The reason for using $E$ and $\nu$ is that data for a particular material can be readily obtained for these constants, as demonstrated in Tables 3.2 and 3.3. For rubber $(\nu \to 0.5)$ it is more accurate to use the bulk modulus $K$ together with the shear modulus $G$, since $1 - 2\nu$ appears in the denominator.

For the special case in which the $x$, $y$, and $z$ axes are coincidental with the principal axes 1, 2, and 3 (the *triaxial stress state*), Equations (B.54) and (B.55) are simplified by virtue of all shear stresses and shear strains being equal to zero:

$$\epsilon_1 = \frac{1}{E}[\sigma_1 - v(\sigma_2 + \sigma_3)]$$

$$\epsilon_2 = \frac{1}{E}[\sigma_2 - v(\sigma_1 + \sigma_3)] \tag{B.56}$$

$$\epsilon_3 = \frac{1}{E}[\sigma_3 - v(\sigma_2 + \sigma_1)]$$

$$\sigma_1 = \frac{E}{(1+v)(1-2v)}[(1-v)\epsilon_1 + v(\epsilon_2 + \epsilon_3)]$$

$$\sigma_2 = \frac{E}{(1+v)(1-2v)}[(1-v)\epsilon_2 + v(\epsilon_1 + \epsilon_3)] \tag{B.57}$$

$$\sigma_3 = \frac{E}{(1+v)(1-2v)}[(1-v)\epsilon_3 + v(\epsilon_1 + \epsilon_2)]$$

For the commonly encountered *biaxial stress state,* one of the principal stresses (say, $\sigma_3$) is zero and Equation (B.56) becomes

$$\epsilon_1 = \frac{\sigma_1 - v\sigma_2}{E}$$

$$\epsilon_2 = \frac{\sigma_2 - v\sigma_1}{E} \tag{B.58}$$

$$\epsilon_3 = -\frac{v(\sigma_1 + \sigma_2)}{E}$$

For $\sigma_3 = 0$ the third of Equation (B.57) gives

$$\epsilon_3 = -\frac{v(\epsilon_1 + \epsilon_2)}{1 - v} \tag{B.59}$$

Substituting Equation (B.59) into Equation (B.57) gives

$$\sigma_1 = \frac{E(\epsilon_1 + v\epsilon_2)}{1 - v^2}$$

$$\sigma_2 = \frac{E(\epsilon_2 + v\epsilon_1)}{1 - v^2} \tag{B.60}$$

$$\sigma_3 = 0$$

For the *uniaxial stress state,* Equations (B.58) and (B.60) must reduce to

$$\epsilon_1 = \frac{\sigma_1}{E} \qquad \epsilon_2 = \epsilon_3 = -\frac{v\sigma_1}{E} \tag{B.61}$$

$$\sigma_1 = E\epsilon_1 \qquad \sigma_2 = \sigma_3 = 0 \tag{B.62}$$

These expressions for the uniaxial, biaxial, and triaxial stress states can be expressed in tabular form as shown in Table B.2. Recall that we chose to express the stress and strain in terms of the modulus of elasticity and Poisson's ratio. Furthermore, we are considering the special case in which the $x$, $y$, and $z$ axes are coincidental with the principal axes 1, 2, and

**Table B.2**     Principal stresses and strains in terms of modulus of elasticity and Poisson's ratio for uniaxial, biaxial, and triaxial stress states (It is assumed that the x, y, and z axes are coincident with the principal axes 1, 2, and 3, thus implying that the shear stresses and strains are equal to zero.)

| Type of stress | Principal strains | Principal stresses |
|---|---|---|
| Uniaxial | $\epsilon_1 = \dfrac{\sigma_1}{E}$ | $\sigma_1 = E\epsilon_1$ |
| | $\epsilon_2 = -\nu\epsilon_1$ | $\sigma_2 = 0$ |
| | $\epsilon_3 = -\nu\epsilon_1$ | $\sigma_3 = 0$ |
| Biaxial | $\epsilon_1 = \dfrac{\sigma_1}{E} - \dfrac{\nu\sigma_2}{E}$ | $\sigma_1 = \dfrac{E(\epsilon_1 + \nu\epsilon_2)}{1 - \nu^2}$ |
| | $\epsilon_2 = \dfrac{\sigma_2}{E} - \dfrac{\nu\sigma_1}{E}$ | $\sigma_2 = \dfrac{E(\epsilon_2 + \nu\epsilon_1)}{1 - \nu^2}$ |
| | $\epsilon_3 = -\dfrac{\nu\sigma_1}{E} - \dfrac{\nu\sigma_2}{E}$ | $\sigma_3 = 0$ |
| Triaxial | $\epsilon_1 = \dfrac{\sigma_1}{E} - \dfrac{\nu\sigma_2}{E} - \dfrac{\nu\sigma_3}{E}$ | $\sigma_1 = \dfrac{E\epsilon_1(1 - \nu) + \nu E(\epsilon_2 + \epsilon_3)}{1 - \nu - 2\nu^2}$ |
| | $\epsilon_2 = \dfrac{\sigma_2}{E} - \dfrac{\nu\sigma_1}{E} - \dfrac{\nu\sigma_3}{E}$ | $\sigma_2 = \dfrac{E\epsilon_2(1 - \nu) + \nu E(\epsilon_1 + \epsilon_3)}{1 - \nu - 2\nu^2}$ |
| | $\epsilon_3 = \dfrac{\sigma_3}{E} - \dfrac{\nu\sigma_1}{E} - \dfrac{\nu\sigma_2}{E}$ | $\sigma_3 = \dfrac{E\epsilon_3(1 - \nu) + \nu E(\epsilon_1 + \epsilon_2)}{1 - \nu - 2\nu^2}$ |

3, thus implying that the shear stresses and strains are equal to zero. For this case the shear stress and strain are equal to zero.

**EXAMPLE B.1**

**Given:**   Equations (B.42) and (B.43).

**Find:**   Determine the modulus of elasticity $E$ as a function of $G$ and $\nu$, where $\nu$ is given by Hooke's law for uniaxial tension.

$$\epsilon_x = \frac{\sigma_x}{E}$$

$$\epsilon_y = -\frac{\nu\sigma_x}{E}$$

$$\epsilon_z = -\frac{\nu\sigma_x}{E}$$

**Solution:**   Equation (B.43) gives

$$\epsilon_x = \frac{(\lambda + G)\sigma_x}{G(3\lambda + 2G)} - \frac{\lambda(\sigma_y + \sigma_z)}{2G(3\lambda + 2G)} = \frac{\sigma_x}{E} - \frac{\nu(\sigma_y + \sigma_z)}{E}$$

$$\epsilon_y = \frac{(\lambda + G)\sigma_y}{G(3\lambda + 2G)} - \frac{\lambda(\sigma_x + \sigma_z)}{2G(3\lambda + 2G)} = \frac{\sigma_y}{E} - \frac{\nu(\sigma_x + \sigma_z)}{E}$$

$$\epsilon_z = \frac{(\lambda + G)\sigma_z}{G(3\lambda + 2G)} - \frac{\lambda(\sigma_x + \sigma_y)}{2G(3\lambda + 2G)} = \frac{\sigma_z}{E} - \frac{\nu(\sigma_x + \sigma_y)}{E}$$

$$\left.\begin{array}{c} \dfrac{\lambda + G}{G(3\lambda + 2G)} = \dfrac{1}{E} \\[3mm] \dfrac{\lambda}{2G(3\lambda + 2G)} = \dfrac{\nu}{E} \end{array}\right\} \quad \dfrac{\lambda}{2(\lambda + G)} = \nu$$

$$(\lambda + G)E = 3\lambda G + 2G^2$$

$$G(E - 2G) = \lambda(3G - E) \qquad \lambda = \dfrac{G(E - 2G)}{3G - E}$$

$$\nu = \dfrac{G(E - 2G)}{2(3G - E)[G(E - 2G)/(3G - E) + G]} = \dfrac{E - 2G}{2G} = \dfrac{E}{2G} - 1$$

$$E = (\nu + 1)2G \qquad \text{and} \qquad G = \dfrac{E}{2(1 + \nu)}$$

The modulus of elasticity $E = 2G(\nu + 1)$, the shear modulus $G = E/[2(1 + \nu)]$, and Poisson's ratio $\nu = E/(2G) - 1$.

## B.6    SUMMARY

Hooke's law was generalized in this appendix. That is, the linear relationship between stress and strain in the elastic range was generalized for the six components of stress: three normal stresses and three shear stresses. By using the laws of stress and strain transformation, the 36 elastic material constants were reduced to 2 through five different coordinate orientations. It was also found that, for the isotropic and homogeneous materials assumed throughout this chapter, these elastic material constants are not functions of the space coordinates and do not vary from point to point. They depend only on the particular material.

Relationships were presented between the four elastic material constants (the shear modulus or modulus of rigidity $G$, the modulus of elasticity $E$, the bulk modulus $K$, and Poisson's ratio $\nu$) and Lamé's constant $\lambda$, which has no particular physical significance. It was shown how any two of the five constants can express the other constants. For the special case in which the $x$, $y$, and $z$ axes are coincidental with the principal axes, simplified equations were expressed where all shear stresses and shear strains were zero. For these situations uniaxial, biaxial, and triaxial principal stresses could be expressed. These equations serve as the foundation for the main text of this book.

---

## REFERENCE

Durelli, A. J., Phillips, E. A., and Tsao, C. H. (1958) *Introduction to the Theoretical and Experimental Analysis of Stress and Strain,* McGraw-Hill, New York.

# Appendix C

## STRESS INTENSITY FACTORS FOR SOME COMMON CRACK GEOMETRIES

This appendix summarizes some stress intensity factors for common machine element configurations and test specimen geometries. Most of this appendix is taken from Suresh (1998), although Kanninen and Popelar (1985) and Shigley et al. (2003) were also used in the compilation.

### C.1 INTERNALLY-CRACKED TENSION SPECIMEN

The stress intensity factor for a center crack of length $l_c$ in a plate of width $b$ is given by

$$K_I = \sigma \sqrt{\frac{\pi l_c}{2}} Y \tag{C.1}$$

The geometry correction factor $Y$ is shown in Figure C.1. If the tensile stress is remote ($h >> b$), then the geometry factor is given by

$$Y = \sec\left(\frac{\pi a}{W}\right)^{1/2} \tag{C.2}$$

If an internal crack of length $l_c$ exists a distance $d$ from the edge of the tension member of width $b$, the geometry correction factor is shown in Figure C.2.

### C.2 EDGE-CRACKED TENSION SPECIMEN

For a single edge-cracked tension specimen of crack width $l_c$ and width $b$, the stress intensity factor is given by

$$K_I = \sigma \sqrt{a} Y \tag{C.3}$$

The geometry factor $Y$ is shown in Figure C.3. If the tensile stress is remote ($h >> b$), then the geometry factor is given by

$$Y = 1.99 - 0.41 \frac{l_c}{b} + 18.7 \left(\frac{l_c}{b}\right)^2 - 38.48 \left(\frac{l_c}{b}\right)^3 + 53.83 \left(\frac{l_c}{b}\right)^4 \tag{C.4}$$

**Figure C.1**   Geometry factor for center-cracked tension specimen.

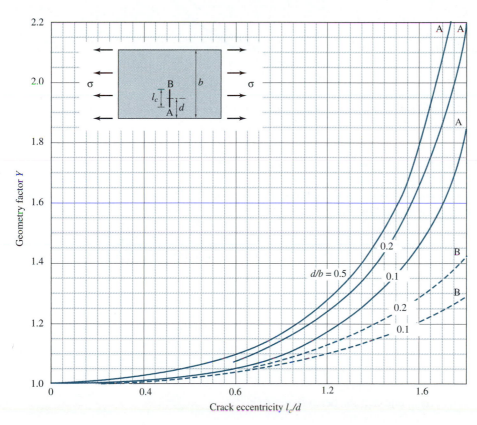

**Figure C.2**   Geometry factor for an asymmetrically-cracked tension specimen.

**Figure C.3**   Geometry factor for single edge-cracked tension specimen.

For a tensile specimen of width $b$ with two edge cracks, each of length $l_c$, the geometry factor for use in Equation (C.3) is given by

$$Y = 1.99 + 0.76\frac{l_c}{b} - 8.48\left(\frac{l_c}{b}\right)^2 + 27.36\left(\frac{l_c}{b}\right)^3 \tag{C.5}$$

## C.3   BENDING SPECIMEN

The stress intensity factor for an edge-cracked bending specimen is given by

$$K_I = \frac{2Pa}{w_t b^{3/2}} Y \tag{C.6}$$

where  $P$ = applied load, N (lb)
 $a$ = one-half the beam span, m (in)
 $w_t$ = beam width, m (in)
 $b$ = beam height, m (in)
 $Y$ = a geometry correction factor, given by

**Figure C.4**  Geometry factor for single edge-cracked bending specimen.

$$Y = \frac{3(l_c/b)^{1/2}}{2[1 + 2(l_c/b)][1 - (l_c/b)]^{3/2}} \left\{ 1.99 - \frac{l_c}{b}\left(1 - \frac{l_c}{b}\right)\left[2.15 - 3.93\frac{l_c}{b} + 2.7\left(\frac{l_c}{b}\right)^2\right] \right\}$$

(C.7)

The geometry correction factor $Y$ is shown in Figure C.4.

## C.4   CENTRAL HOLE WITH CRACK IN TENSION SPECIMEN

A common occurrence when a central hole is placed in a tension specimen is that a crack will initiate at the locations of maximum stress. The effect on the stress state of the tension member can be quite significant, and a very large apparent crack can develop. This is treated as a center crack [the stress intensity factor is given by Equation (C.1) but with a geometry correction factor as given in Figure C.5].

**Figure C.5**    Geometry factor for tension specimen with a central hole and cracks.

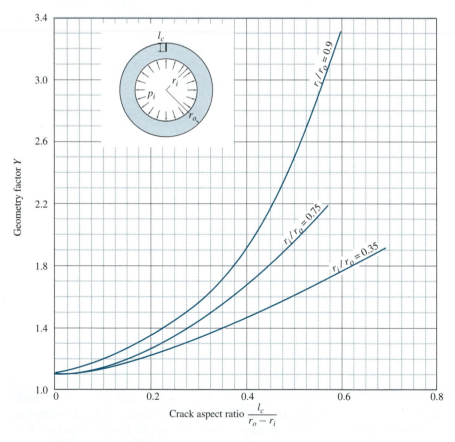

**Figure C.6**    Geometry factor for an internally pressurized cylinder.

## C.5     CRACK IN WALL OF PRESSURE VESSEL

In the circumstance where the internally applied pressure varies with time, a crack can develop and propagate within the wall of a pressure vessel. For such occurrences, the stress concentration factor is given by Equation (C.1) with a geometry correction factor as given in Figure C.6.

## REFERENCES

Kanninen, M. F., and Popelar, C. H. (1985) *Advanced Fracture Mechanics,* Oxford University Press, New York.

Shigley, J. E., Mischke, C. R., and Budynas, R. (2003) *Mechanical Engineering Design,* 6th ed., McGraw-Hill, New York.

Suresh, S. (1998) *Fatigue of Materials,* 2nd ed., Cambridge University Press, Cambridge.

# Index

Note: Major figures and tables are denoted by page numbers followed by *f* and *t*, respectively.

Centroid, area moment of inertia, and area for seven cross sections

| Cross section | Centroid | Area moment of inertia | Area |
|---|---|---|---|
| Circular area | $\bar{x} = 0$ <br> $\bar{y} = 0$ | $I_x = I_{\bar{x}} = \dfrac{\pi}{4} r^4$ <br> $I_y = I_{\bar{y}} = \dfrac{\pi}{4} r^4$ <br> $J = \dfrac{\pi}{2} r^4$ | $A = \pi r^2$ |
| Hollow circular area | $\bar{x} = 0$ <br> $\bar{y} = 0$ | $I_x = I_{\bar{x}} = \dfrac{\pi}{4} \left(r^4 - r_i^4\right)$ <br> $I_y = I_{\bar{y}} = \dfrac{\pi}{4} \left(r^4 - r_i^4\right)$ <br> $J = \dfrac{\pi}{2} \left(r^4 - r_i^4\right)$ | $A = \pi \left(r^2 - r_i^2\right)$ |
| Triangular area | $\bar{x} = \dfrac{a + b}{3}$ <br> $\bar{y} = \dfrac{h}{3}$ | $I_x = \dfrac{bh^3}{12}, I_{\bar{x}} = \dfrac{bh^3}{36}$ <br> $I_y = \dfrac{bh(b^2 + ab + a^2)}{12}$ <br> $I_{\bar{y}} = \dfrac{bh(b^2 - ab + a^2)}{36}$ <br> $\bar{J} = \dfrac{bh}{36}(b^2 + h^2)$ | $A = \dfrac{bh}{2}$ |
| Rectangular area | $\bar{x} = \dfrac{b}{2}$ <br> $\bar{y} = \dfrac{h}{2}$ | $I_x = \dfrac{bh^3}{3}, I_{\bar{x}} = \dfrac{bh^3}{12}$ <br> $I_y = \dfrac{bh^3}{3}, I_{\bar{y}} = \dfrac{bh^3}{12}$ <br> $\bar{J} = \dfrac{bh}{12}(b^2 + h^2)$ | $A = bh$ |
| Area of circular sector | $\bar{x} = \dfrac{2}{3} \dfrac{r \sin\alpha}{\alpha}$ | $I_x = \dfrac{r^4}{4} \left(\alpha - \dfrac{1}{2} \sin 2\alpha\right)$ <br> $I_y = \dfrac{r^4}{4} \left(\alpha + \dfrac{1}{2} \sin 2\alpha\right)$ <br> $J = \dfrac{1}{2} r^4 \alpha$ | $A = r^2 \alpha$ |
| Quarter-circular area | $\bar{x} = \bar{y} = \dfrac{4r}{3\pi}$ | $I_x = I_y = \dfrac{\pi r^4}{16}$ <br> $I_{\bar{x}} = I_{\bar{y}} = \left(\dfrac{x}{16} - \dfrac{4}{9\pi}\right) r^4$ <br> $J = \dfrac{\pi r^4}{8}$ | $A = \dfrac{\pi r^2}{4}$ |
| Area of elliptical quadrant | $\bar{x} = \dfrac{4a}{3\pi}$ <br> $\bar{y} = \dfrac{4b}{3\pi}$ | $I_x = \dfrac{\pi ab^3}{16}, I_{\bar{x}} = \left(\dfrac{\pi}{16} - \dfrac{4}{9\pi}\right) ab^3$ <br> $I_y = \dfrac{\pi a^3 b}{16}, I_{\bar{y}} = \left(\dfrac{\pi}{16} - \dfrac{4}{9\pi}\right) a^3 b$ <br> $J = \dfrac{\pi ab}{16}(a^2 + b^2)$ | $A = \dfrac{\pi ab}{4}$ |